standard catalog of®

V-8 ENGINES

1906-2002

John Gunnell

Published by

700 East State Street • Iola, WI 54990-0001
715-445-2214 • 888-457-2873
www.krause.com

Please, call or write us for our free catalog of publications.
Our toll-free number to place an order or obtain a free catalog is 800-258-0929
or please us our regular business telephone 715-445-2214.

ISBN: 0-87349-446-6
Library of Congress Number: 2002105754
Printed in the United States of America

Edited by: Karen O'Brien
Designed by: Brian Brogaard

Front Cover:
1955 Ford Y-Block V-8
1967 Chevrolet Camaro V-8
1951 Chrysler "Hemi" V-8

Back Cover:
1968 Hurst Oldsmobile V-8 (John Gunnell)

Title Page:
1965 Pontiac 2 + 2 V-8 (John Lee)

CONTENTS

BOILER PLATE

Changes in terminology

From time to time, publications listing automobile engine specifications changed their methods of presenting information. For instance, in 1919 the location of valves in the Oldsmobile V-8 was given as "in V." In 1920, the location of the valves was listed as "side." The engine was the same in both years, so it seems that the industry's terminology changed. There are many other instances of similar changes in terminology. It seems unwise for a modern researcher to attempt to standardize such terminology as this could, in some cases, lead to errors. Therefore, in this catalog we are presenting most V-8 engine specifications as they were presented in the year the car was new. This may lead to year-to-year variations in the amount of information listed or in the order in which certain specifications appear.

Trade name notes

Another consideration in creating a catalog like this one is the trade names used on parts such as a carburetor. In 1959, Oldsmobile called the Rochester 4GC carburetor a "Quadra-jet." In 1960, the same Rochester 4GC was described as a "Multi-jet" four-barrel carburetor. We have tried to follow the original descriptions, even if they vary from one year to the next.

Base versus optional

Automakers refer to some engines as the "base V-8." This does not mean the same thing as "standard equipment." For instance, a 231-cid V-6 was the standard engine in the 1984 Buick LeSabre. A 252-cid V-6 was an extra-cost option for the V-6 series. However, a 307-cid V-8 was considered "base" engine in the V-8 series. A 350-cid diesel V-8 was an extra-cost option.

Is the glass half full?

Through the years, certain equipment features gained greater acceptance and the methods of recording specifications were changed accordingly. For example, in 1961 the cooling system capacity of full-size Oldsmobiles was expressed as "19.3 qt., add 1 quart for heater." By 1962, the same specification was listed as "Cooling system capacity with heater: 20.3 qt." since almost all cars had heaters by that time.

General in nature

It goes without saying that both the engine specifications and the tune-up specifications presented here are general in nature for standard equipment on the typical car. If your car was ordered with extras such as a heavy-duty battery or heavy-duty cooling system, it might require a more powerful battery or have a larger cooling system capacity. In a similar manner, horsepower ratings and torque ratings can vary at different rpm levels and different sources sometimes recorded slightly different figures, especially in the 1970s. Likewise, tune up specs such as dwell angle are sometimes given as a range of values and sometimes as a single value in the middle of the range.

Horsepower and Torque (From 1979 Dodge data book and car selector)

For years, engine horsepower and torque ratings published by the automobile industry were the maximum ratings obtainable. They were gross horsepower and gross torque ratings.

In 1972, the automotive industry began using net horsepower and net torque in their publications. Net horsepower and net torque ratings are lower than gross ratings for the same engine. Here's the difference.

Gross horsepower and torque ratings are obtained at the flywheel with optimum ignition setting (manual instead of automatic advance) and without allowance for the power absorbed by such engine components as the fan, alternator, and car exhaust system.

Net horsepower and net torque ratings are measured at the flywheel of the engine—but with all of the accessories (as in a car) such as fan, alternator, automatic spark advance, and car exhaust system installed.

ACKNOWLEDGMENTS

MUSEUMS

Studebaker National Museum120 South St. Joseph St. South Bend, IN 46601 www.studebakermuseum.org

PEOPLE

Joe Abela (Ford-Mercury), Jim Benjaminson (Plymouth), Charles Blais (Lincoln), Richard Bowman (Chrysler-Mopar), Terry Boyce (Buick), Tony Buscemi (Chrysler-Mopar), Ron Crouch (Mercury), Rob Day (Mercury), Steve O. Day (Chevrolet), Bruce Dodds (Ford), Gary Esse (Chevrolet), Kit Foster (Society of Automotive Historians), Ray Funk (Ford), John Francis (Cadillac), Jim Griffin (Lincoln), Mike Grippo (Pontiac), Jon E. Hardgrove (Carburetion), Larry Hierholzer (Ford), Ron Hivalt (Plymouth Valiant), Beverly Rae Kimes (Prewar), Dave Lane (Batteries), Delores Lane (Batteries), Thomas J. Lester (Prewar), Jack Lewis (Plymouth Fury), Tom Lieb (Early Cadillac and Wills Ste. Clare), Chip Loree (Chrysler-Mopar), Josh B. Malks (Cord), Eric Marr (Chrysler-Mopar), Robert Mead (Lincoln), Larry G. Mitchell (AMC), Bob Palmer (Studebaker), Art Petridis (Chevrolet), Dick Power (Lincoln), George Riehl (Chrysler 300), Eleanor Riehl (Chrysler 300), Jerry Robbin (Mercury), John Sawruk (Pontiac), Doug Schellinger (Chrysler-Mopar), Stanley Sepiol (Chrysler-Mopar), Dennis Swenson (Ford), Greg Wallace (Cadillac Historical Collection) and Wayne Wooten (Dodge).

CLUBS

AMC

AMC World Clubs
7963 Depew St.
Arvada, CA 20003
www.amcwc.com

Buick

Buick Club of America
PO Box 401927
Hesperia, CA 92340-1927
www.buickclub.org

Chevrolet

Vintage Chevrolet Club
of America, Inc.
1751 E Route 66
Glendora CA 91740-3813
www.vcca.org

Chrysler-Mopar

Chrysler 300 Club
International, Inc.
4900 Jonesville Road
Jonesville, MI 49250-9439
www.chrysler300club.com

Chrysler Products Owners Club
100 Lynch St.
Rockville., MD 20850
(301) 340-7432

Daytona-Superbird Auto Club
13717 West Green Meadow Drive
New Berlin, WI 53151
E-Mail: dsac@execpc.com

National Chrysler Products Club
5844 West Eddy St.
Chicago, IL 60634
www.barthle.com/ncpc/

Dodge Charger National Registry
PO Box 184
Greenbay, VA 23942
www.dodge-charger.com/Charger-Registry-TDC.html

Wisconsin Mopar Club
1500 North Outgamie St.
Appleton, WI 54914

Walter P. Chrysler Club
PO Box 3504
Kalamazoo, MI 49003-3504
www.chryslerclub.org

Ford

Ford and Mercury Restorers Club
16712 Bell Creek Lane
Livonia, MI 48154
clubs.hemmings.com/clubsites/ford mercuryrestorers

Lincoln

Lincoln and Continental Owners Club
PO Box 570709
Dallas, TX 75357-0709
www.lcoc.org

Lincoln Owners Club
PO Box 1434
Minoqua, WI 54548
www.lincolnownersclub.com

Lincoln Zephyr Owners Club
PO Box 422
Hazel Green, AL 35750-0422
www.lzoc.org

Road Race Lincoln Register
726 E. Ind. Pk. Dr. #13
Manchester, NH 03109

Mercury

International Mercury Owners Association
6445 W. Grand Ave.
Chicago, IL 60707-3410
www.mercuryclub.com

Cougar Club of America
1637 Skyline Dr.
Norfolk, VA 23518-4327
www.cougarclub.org

Cyclone-Montego Registry
19 Glyn Dr.
Newark, DE 19713-4016
www.grantorinosport.atfreeweb.com/reg_day.htm

Plymouth

Plymouth Owners Club, Inc.
P.O. Box 416.
Cavalier, ND 58220
www.plymouthbulletin.com

The Golden Fin Society
13765 South 1300 W.
Riverton, UT 84065
www.forwardlook.net

Plymouth Valiant Club & Registry
2121 Seaport Circle #103.
Winter Park, FL 32792
Email: ValiantReg@aol.com

Pontiac

Pontiac Oakland Club International
P.O. Box 14057Bradenton, FL 34280-4057
www.poci.org

Studebaker

The Studebaker Drivers Club, Inc.
P.O. Box 1743
Maple Grove MN 55311
www.studebakerclubs.com

COMPANIESThe Antique Auto Battery Co.2320 Old Mill RdHudson, OH 44236
(330) 425-7580
(800) 426-7580
www.antiqueautobattery.com

The Carburetor Shop204 East 15th StreetEldon, MO 65026
(573) 392-7378
www.thecarburetorshop.com

The Eastwood Company
263 Shoemaker Road
Pottstown, PA 19464-6433
(610) 323-2200
www.eastwoodco.com

Tower Paint & Decorating
922 Oregon Street
PO Box 2345
Oshkosh, WI 54903
(920) 235-6520
www.towerpaint.com

FOREWORD

"Every car has an engine, so everyone restoring a collector car needs a Standard Catalog of engines!" That was the concept behind this book. We envisioned a long list of engines, specifications, and information that would help people identify and restore their engines; a book that would list "matching numbers," the bore and stroke, proper engine colors, and where engine restorers can get the parts and decals needed to detail an engine and make it look new.

We had no idea exactly how many different engines were made during a century or so of automobile manufacture. We also had no idea how little research had actually been done as to what older engines looked like when they left the factory. To make the project "do-able" we decided to start with V-8s only.

Our first step was to determine when the first V-8 was built. This was difficult. Some sources said Cadillac made the "first V-8" in 1914. Some references qualified this V-8 as "the first high-speed V-8." Other books credited Ford with the "first mass-produced V-8"—odd because we consider the 1914 Cadillac a mass-produced car, although it had a low production volume. Another source said the 1932 Ford engine was "the world's first production installed automobile single casting V-8," which seems like quite a mouthful.

Thanks to the Internet, we found that at least two V-8s were made in 1906. The first was in a motorcycle that Glenn Hammond Curtis used to set a land speed record. His V-8-powered bike went 137 mph at Ormond Beach, Florida on January 24, 1907. Another V-8 built in 1906 was used in a rare Rolls-Royce called the V8 Legalimit. Its 90-degree V-configuration engine had a 3-1/4-in. bore and stroke and displaced 3535 cc. The car had a top speed of 26 mph, which surpassed the legal speed limit in England at the time. Only three V8 Legalimits were built, but they seem to have been the first V-8-powered cars and they help to validate the title Standard Catalog of V-8s 1906-2002.

The book itself represents more than a year's worth of planning, compiling, and researching, but I will readily admit that the concept sounded simpler than the work proved to be. Various reference books listing V-8 engine specifications were filled with errors and inconsistencies. In some cases, engine specifications were modified or completely changed between the time that factory literature was published and the time that a car came out. Many automakers also built special racing engines that were never listed in sales literature.

After networking with many sources, car clubs, parts vendors, museums, and restorers, we compiled a list of American V-8s organized by model year and horsepower. The list provides basic technical specifications such as bore and stroke, displacement, compression ratio, horsepower, torque (in foot-pounds) and type of induction system for each V-8, along with available engine codes. The black-and-white illustrations of V-8s show what they looked like when new.

Engine detailing information is presented in articles in front of the book. The color information was provided by engine paint suppliers. We also felt that having the correct battery in the engine compartment is important to restorers, so we obtained applications data from companies that make reproduction batteries. Once this catalog is published, we're sure to get letters pointing out errors. We hope that such letters will take a constructive point of view and be aimed at improving this hard-to-compile database. We routinely transfer data between our "Standard Catalogs" with the aim of making them as accurate as possible and we are always happy to get information that can make them better books.

John Gunnell
June 2003

EASTWOOD AIDS ENGINE DETAILERS

Detailing an engine is the crowning touch in the restoration of a collectible vehicle. Most collectors of cars and trucks cut their automotive teeth on building plastic model kits, which always featured colorful engine blocks and heads, flat-black accessories and all kinds of chrome-plated add-ons. As the model makers progressed to real cars, they retained a strong appreciation for detailed engine bays.

Special heat-resistant paint for redoing engine the block and cylinder heads in real cars has been around for a long time. In the old days, most available colors were rather generic and the quality of the paint itself was not always the best. A few runs around the block would often — though not always — burn the new finish off of hot motor parts.

For years, old-car hobbyists have also tried their best to refinish other engine compartment items — like starters, generators, carburetors and radiators — as best as possible with regular paints. However, it is only in the past decade or so that finishes specifically made for engine detailing have been developed. One of the leading companies responsible for the push has been The Eastwood Company of Pottstown, Pennsylvania.

Curt Strohacker started Eastwood in the fall of 1978,. During high school, Curt had worked in a service station, bought and sold innumerable vehicles and learned some very valuable lessons about restoration — namely, the best ways to restore collector vehicles. When the auto restoration market took off in the 1970s, Curt realized that some specialized tools and techniques could help thousands of other restorers save time and money.

Curt's first catalog featured eight black-and-white pages of metal-finishing equipment. By the early 1980s, he gave his catalog mailing operation to a professional list house. By 1981, the list had 5,000 names and it kept growing. Today Eastwood mails over five million catalogs annually. Eastwood also sells its products directly through a retail outlet store, at select car shows, via space advertisements and television sponsorship and by utilizing the Internet.

Although engine-detailing products take up only a few pages of the Eastwood catalog, they represent a significant share of the firm's business because of their specialized nature. These products are generally not available in neighborhood auto parts stores or at local Wal Marts, but Eastwood can deliver them directly to a hobbyist's garage in as little as three days if the Expedited Shipping option is utilized.

As far as engine paint itself goes, Eastwood is not — and does not pretend to be — a supplier of all colors for all vintage cars. Eastwood carries 15 fast-dry engine paints, 12 of which are the "correct color" type for popular cars ranging from Model A Fords to 1983 Chrysler products. There are also "universal" colors like Silver, Gold and Clear. These paints are able to resist heat up to 300 degrees f. and come in 12-oz. aerosol cans.

Where the serious restorer really gets seriously interested in Eastwood products is when you get to the specialty products, which include radiator paint, manifold paint and finishes designed to duplicate the original look of a variety of engine parts made of different metals such as cast iron, aluminum, brass, copper and steel. There are also products designed to make detailing an engine easier.

For example, let's say that you are restoring an engine and your cast iron exhaust manifolds are rough textured and mostly rust colored. Eastwood offers a manifold smoothing kit that allows you to smooth out the rust pits with your die grinder. It includes abrasive cylinders, grit tapers and mandrels (a deluxe version adds a carbide bead burr) that can help smooth and shape the metal without biting through. After the manifold is as smooth as possible, you can order one of four different finishes to protect your manifold or header. These come in satin black, stainless steel gray, cast iron or silver colors and will resist temperatures as high as 1200-degrees f.

Another unique paint is Eastwood's No. 10040Z Radiator Black, which has a very thin consistency suitable for use on radiators, transmission coolers and oil coolers, where a thick paint could cause overheating problems. Although it's a thin paint, Radiator Black resists cooling system heat up to 250 degrees f. Two cans will cover most radiators and make them look just like they did when new. (We have found that this paint, because of its thinness, also looks great on plastic model cars!)

Other products in the catalog that come in handy when detailing an engine are Tricoat47 No. 34127Z (for the look of titanium), Detail Gray No. 10036Z (cast iron look), Aluma Blast No. 10109Z (aluminum look) and Spray Gray No. 10065ZP (cast iron look). Silver Exhaust Paint No. 10020Z is great for redoing exhaust pipes that can be seen in the engine compartment. Clear Zinc No. 10280Z and Zinc Phosphate No. 10281Z were developed to give diaphragms, hood latches and carb linkages a factory appearance.

Speaking of carburetors, Eastwood's No. 10179Z Bronze Carb Renew or No. 10187Z Silver Carb Renew restore an OEM look to rebuilt carburetors. Underhood Black, designed for use on pulleys, fans, fan hubs, brackets and other parts, comes in a quart can or an aerosol can and also gives such parts a factory appearance. Spray-on Silver Cad finish is also featured in the catalog. To replicate the wrinkly finish used on old-car heater boxes, simply spray on Eastwood's No. 10014Z Black Wrinkle Finish Paint.

Not all parts can be restored to a like-new appearance with spray paint. For example, to get the Gold Cad finish, you need a four-step No. 10031Z Gold Cad System kit that includes a base coat and three tints. You can order this with Diamond Clear if you want to protect your work. Eastwood also offers a Metal Blackening System that produces a true OEM-type black oxide finish at home. There's also a Tin-Zinc Electroplating System that restores the look of nuts and blots.

In addition to this cornucopia of paints and finishing systems, you can order painting prep solutions, metal cleaning solutions, mini sprayers and other restoration aids from the catalog. There is a kit for polishing aluminum intake manifolds, thread chaser sets, sanding and grinding tools, rust eating solutions, radiator fin pliers and rakes, at-home powdercoating kits and more. In fact, looking through the company's free catalog is always a fascinating experience, even if you don't have an engine to detail right now. Chances are it won't be long until your next old car somes along and then you'll want a copy.

Source

The Eastwood Company
263 Shoemaker Road
Pottstown, PA 19464
(800) 345-1178

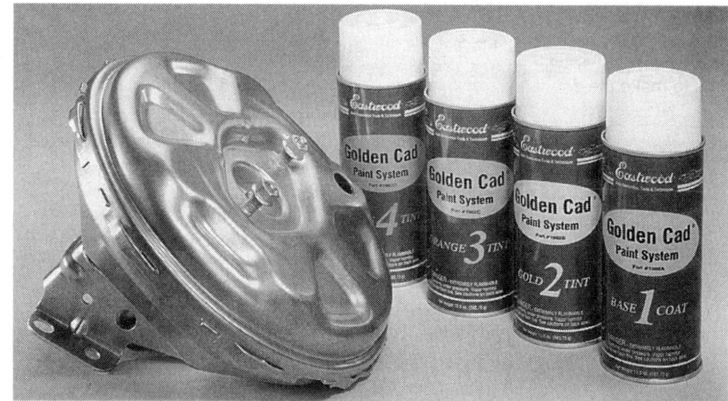

Eastwood's Golden Cad system let's you restore your cadmium-plated parts without expensive replating or replacement. The kit includes four cans of specially-formulated color sprays and how-to-do-it instructions.

Silver cadmium plating is used on auto engine parts for its rust protection and good looks. Eastwood has developed a spray-on coating that duplicates the look of silver cadmium plating.

TOWER ENGINE PAINT

Tower paint puts Custom Mixed Engine Paint in a Can

Tower Paint & Decorating of Oshkosh, Wis., has been in business since 1953. Although it sells everything from wallpaper to wooden blinds, one of the specialties of the firm is custom-matching paints for vintage auto restorations. The company sells paint for car and truck bodies, trim parts and engines.

Custom-filled spray cans containing paints and primer surfacers are a boon for enthusiasts who restore their own vehicles. Tower Paint & Decorating has the unique ability to put correct DuPont or PPG paint colors into 16-ounce spray cans (most store-bought cans hold 12 oz. of paint) and ships them daily. The company fits each of its custom-filled spray cans with a fan-pattern spray cap that reduces the chance of home-sprayed paint spotting or running.

"When you order our spray cans to paint your engine, you can be sure that you'll get a match of the topcoats that were originally applied to your vehicle's engine," says company president Greg Garl. By using a computerized database, Tower Paint & Decorating is able to call up original-equipment DuPont or PPG colors or save custom-made colors. "We have access to the mainframes at DuPont and PPG for all of their color histories and libraries," notes Garl. "DuPont's standard database goes back to the '30s and Tower Paint's own master database goes back even further; it includes colors taken from original parts that people sent us and we haven't lost a sent-in part since 1953."

Tower Paint & Decorating custom color library for older cars is maintained at company headquarters in Oshkosh. Two sets of binders, one covering domestic cars and one covering imported cars, have been saved since the company first started. They show the original body colors for most cars made since the 1930s and list the old formulas used to make those colors.

The color library includes information about the DuPont paint formulas used for some vintage V-8 engines. Tower Paint & Decorating has added to the factory information by working with parts sent in by restorers over the years. "When we get customer color samples, we custom color match their parts because we carry the complete Dupont and PPG mixing systems," says Garl. "We then save the new information about that engine color into our master database, in case someone else needs the same engine color later on."

As a DuPont and PPG refinish dealer, Tower Paint & Decorating can reference the paint companies' computerized databases to create modern matches for the obsolete color codes in the vintage color books, "We can also computer-adjust any formula to the size that the customer wants, from a half pint to 5 gallons," Garl advises. The custom-matched spray cans may be ordered by mail, by using a toll-free phone line or via the Internet.

Garl demonstrated how his computer searches the database for the paint formula. A special scale is then used to measure different colors to the nearest 10th of a gram. "Starting with an empty can, we combine the appropriate mixing tints to make the desired color," he explained. "Each spray can is individually filled and tested to assure a quality product. It is then packaged, sealed, labeled and shipped to the customer." None of the filled paint cans sit around in inventory. They are all shipped the same day they're made up so that the paint is very fresh when it reaches the customer.

As for accuracy, Garl claims perfection. "Our color matching is exact!" he says. "The average age of our employee is late 40s, so they have years of experience, plus we are used to working with history by working with the aircraft restorers at the Experimental Aircraft Assoc. Museum here in Oshkosh."

In addition to custom-filled spray cans, Tower Paint & Decorating handles a full line of automotive refinishing products that can help you properly prepare you vintage V-8 for a fresh coat of paint. In addition to selling direct to home restorers, the Oshkosh company does wholesale business with hobby suppliers, old car clubs and professional restoration shops. Company manager Kris Mork-Lee recalled a certain "funky gold" color used only on a particular Oldsmobile V-8. "We worked with Fusick's Automotive Products to get the color just right."

The tables below some of the information that Tower Paint & Decorating was able to supply for The Standard Catalog of V-8 Engines. In some cases the DuPont paint formulas for engine "motor enamels" of the '50s and '50s are given. Note that these apply to all engines, not V-8s only. In other cases, the tables show the modern suppliers of colors that have been determined to be correct for particular cars. The table indicates how to reach the suppliers. The last group of tables give some "generic" engine color information. Remember, if the engine you want to restore has traces of the original paint on some parts, you can send the parts to Tower Paint & Decorating to have them custom-matched.

The cost of a custom-matched and custom-filled spray can of enamel at the time this book was published was $14.95. UPS shipping for the first can was $8.95. Shipping on additional cans was 95 cents. Prices of course are subject to periodic revision. Tower Paint & Decorating is located at 922 Oregon St., PO Box 2345, Oshkosh, WI 54903-2345. You can call (920) 235-6520 or Email: info@towerpaint.com for additional information.

GENERAL NOTE: Table 1 shows information found on an original 1956 DuPont bulletin on Chevrolet engine enamels. It is edited for V-8s. Footnotes using + signs are information added later than 1956.

Table 1

DuPont Refinish Sales Bulletin

REFINISH JOBBERS REFINISH BRANCH MANAGERS AND REFINISH SALESMEN	BULLETIN No. 3-29 July 26, 1956

CHEVROLET - MOTOR ENAMELS

Listed below are the various 471-Line Motor Enamels used on 1956 Chevrolets:

(+) Note: Edited to give information on 8-cylinder passenger cars only

471-1810) 471-1823) 471-1816)	Red	8 cylinder passenger cars

The first above, 471-1810, is the production material on V-8 motors and is the same shade as 471-1823 and 471-1816, the latter being used at assembly plants for touch-up. Because field demand for these products is irregular and does not generally warrant a stocking program by RSWs or Refinish Jobbers, the possibility of offering 15-Line mixing machine formulas or suitable close matches to handle the situation has been investigated. Here are the results:

471-1810) 471-1823) 471-1816)	Red	Cannot be matched in 15-Line Offer 93-65883-H as a close match for which a M.M. formula is available

J.A. McVoy
Chicago Refinish
SalesSales Manager

(+) Note: This bulletin indicates that six-cylinder passenger cars with stick shift had gray engines (471-661); six-cylinder passenger cars with Powerglide had blue engines (471-1734); truck engines had yellow engines (471-1768) and heavy trucks had Power Red engines (655-1813)

GENERAL NOTE: Tabel 2 shows the information on an original 1958 DuPont bulletin. The information has been lightly edited for clarity. Footnotes using + signs are information added later than 1958.

Table 2

DuPont Refinish Sales Bulletin

REFINISH JOBBERS REFINISH BRANCH MANAGERS AND REFINISH SALESMEN	BULLETIN No. 3-16 April 11, 1958

MOTOR ENAMELS

The following information will assist you in handling inquiries for various Motor Enamels used on 1955-1958 model cars:

Buick - Hampton Blue – "Dulux" Mixing Formula 73108
Cadillac - Blue – close match *93-3380
Chevrolet - Gray (471-1661) Refinish match 471-7596
Red (471-1810/1823/1816) – Close match *93-65883-H
Blue (471-1734) "Dulux" Mixing Formula 73119
Green (471-1748) -"Dulux" Mixing Formula 73120
'58 Models: Green (471-1840) Close Match 93-90301 (File Stock #2424)
'58 Models: Red (+) (471-1834) - Close Match *93-6824 Orange also used in 1957

(+) Note: Tower Paint & Decorating has found the 1958 Engine Red to be Red/Orange

Chrysler Group Motors painted Aluminum and any of DuPont's regularly available "Dulux" or Varnish Enamels should prove satisfactory for reconditioning.

Ford - Red (M-1J-248) "Dulux" Mixing Formula 73100-H
Yellow (M-1J-477) "Dulux" Mixing Formula 73099
'58 Models: Green (M-1J-873) "Dulux" Mixing Formula 75917
'58 Models: Orange (M-1J-874) "Dulux" Mixing Formula 75918
'58 Models: Cream (M-1J-875) Close match *93-1222
'58 Models: Blue (M-1J-998) Close match *93-58740 (File Stock #2114)

Oldsmobile Dark Green "Dulux" Mixing Formula 73162 (**)

(++) Note: Tower Paint & Decorating suggests Fusick's Automotive products EP-60

Pontiac Dark Green "Dulux" Mixing Formula 73124

The special mixing formulas mentioned, with the exception of 75917 and 75918, should be in all jobbers' cabinets in the Motor Enamel Section (Orange cards). Advance copies of 75917 and 75918 are available at our RSWs and general distribution to jobbers will be made at a later date.

• DULUX Mixing Formula available.

J.A. McVoy
Chicago Refinish Sales
Sales Manager

GENERAL NOTE: Table 3 shows the information on an original 1961 DuPont bulletin. It has been edited lightly for clarity. Footnotes using + signs are information added later than 1961.

Table 3

DuPont Refinish Sales Bulletin

REFINISH JOBBERS
REFINISH BRANCH MANAGERS
AND REFINISH SALESMEN

BULLETIN No. 3-9
March 8,1961

MOTOR ENAMELS 1960-1961 MODEL CARS

The following information will assist you in handling inquiries for close matches to various Motor Enamels, other than Black (93-005) and Aluminum (354-789). Some of the colors listed are used on component parts such as air cleaners, valve covers, rocker arm covers, etc.

GM GROUP:

Buick	Hampton Blue	(93) -73108
Cadillac	Blue	(93) -3380
Chevrolet	Gray	471-7596 (stock item)
	Blue	(93) -73119-H
	Green	(93) -90301
	Red	*(93) -6824-H
Oldsmobile	Dark Green	*(93) -23943
	Gold Metallic	(181) –83647
	Red	93) –58785-H
Pontiac	Dark Green	(93) - 73124
	Blue (+)	(93) – 76981

(+) Note: A handwritten notation suggests B5020 was a later code for Pontiac Blue

GMC Truck	Alpine Green	(93) –73978 (Card in "Dulux" Sup. 3)

(+) Note: A handwritten notation suggests 73978D was a later code for Alpine Green

FORD GROUP: (All cars and trucks)

Cream (M2J-309)	*(93) -73209
Silver Met. (M2J-607)	(181)-56303-L
Gold Met. (M2J-842)	*(181)-58639
Red (M1J-870)	(93) –83954-H
Turquoise (M2J-951)	(93) –59225
Blue (M2J-998)	(93) –58740-H
Yellow (M2J-1293)	*(93) -24934

CHRYSLER GROUP:(All cars)

Red (C-7096-A)	*(93) –24314-H
Turquoise Met. (C-2247-C)	(181) -15657

* Formulas for these colors do not appear in the new simplified "Dulux" formula file, and are attached.

S. R. Miller
Regional Refinish
Sales Manager

GENERAL NOTE: Table 4 shows the information from an original 1966 DuPont bulletin. It has been edited lightly for clarity. Footnotes using + signs are information added later than 1966.

Table 4

DuPont Refinish Sales Bulletin

REFINISH JOBBERS
REFINISH BRANCH MANAGERS
AND REFINISH SALESMEN

BULLETIN No. LJ-3
February 21, 1966

1966 MOTOR ENAMELS

The current list of Motor Enamels and our suggested close matches are as follows:

GM GROUP:	COLOR	"DULUXE" FILE NO.
Buick	Hampton Blue	73108
Cadillac	Blue	3380
Chevrolet	Gray	471-7596 (factory package material only)
	Blue	73119
	Red	6824-H
Chev. Truck	Green	90301
Oldsmobile	Green	23943
	Gold Metallic	96910
	Red	58785-H
Pontiac	Blue/Gray Metallic	97219
GMC Truck	Alpine Green	73978

FORD GROUP:

All vehicles		
	Blue	76585A
	(Mixing Formula Below)	

(+) Note: A handwritten notation lists Ford Light Blue

	Light Blue	1111D (in Comp.)

CHRYSLER GROUP:

V-8	Red	061-H
V-6	Turquoise	51509
	Hi-Performance orange	70106-H

FORD BLUE (93) –76585 MIXING FORMULA
1 Qt.

"Dulux" Mixing Enamel	Scale Setting
VD-5450 Additive	35.5
#2 Black	70
#14-H Red	147
#3 Iron Blue	318
#23 Tinting White	569.5
#15 Blue	924.5

J. D. Ferm
Regional Refinish Manager

GENERAL NOTE: Table 5 shows specific colors available from Tower Paint & Decorating or from various hobby vendors. The addresses of the three suppliers are listed below and specific colors can be ordered directly from them. The other colors can be ordered from Tower Paint & Decorating.

Table 5

Make/Model	Model Years	Color	Source
Chevrolet	1957	Chevy Truck Yellow	Tower Paint & Decorating
Corvette	1953-1956	Corvette Engine Blue	Corvette Central
Ford	1941	Light Blue Engine Paint	Tower Paint & Decorating
Lincoln	1969	Lincoln Blue	Tower Paint & Decorating
Mopar	N.A.	Hemi Orange	Tower Paint & Decorating
Oldsmobile	1955-1958	Dark Green	Fusick's Automotive Products
Oldsmobile	1957-1958	Beige Metallic	Fusick's Automotive Products
Oldsmobile	1964-1972	Gold Metallic	Fusick's Automotive Products
Thunderbird	1967	T-Bird Red	Tower Paint & Decorating
Thunderbird	1964-1965	Gold (Valve Covers)	Bob's T-Bird Parts

Sources

Bob's Birdhouse
124 Watkin Avenue
Chadds Ford, PA 19317
(610) 358-3420

Fusick's Automotive Products
PO Box 655
22 Thompson Road
East Windsor, CT 06088
(860) 623-1589

Corvette Central
PO Box 16
13550 Three Oaks Road
Sawyer, MI 49125-9328
(800) 345-4122

Tower Paint & Decorating
922 Oregon St.
PO Box 2345
Oshkosh, WI 54903
(920) 235-6520

GENERAL NOTE: Table 6 shows some "generic" paint color information for AMC engines and later-model GM and Mopar engines. Original formulas for these colors were not available at our publication date. Tower Paint & Decorating has researched these colors from customer samples and has them in its master database. These colors can be ordered from Tower Paint & Decoratig.

Table 6

Make/Model	Model Years	Color	Source
AMC	1964	Silver	See below
	1964-1967	Cape Cod Blue Met.	See below
	1964-1966	Forum Blue	See below
	1964-1966	Lancelot Turquoise	See below
	1964-1966	Gray	See below
	1964-1967	Red	See below
	1964-1967	Rampert Red	See below
	1967	Gold	See below
	1967	Copper	See below
	1969-1974	Blue Metallic	See below
	1974-1978	Blue	See below
	1983-1988	Black	See below
Buick	1975-1977	Blue Metallic	See below
	1978	Blue	See below
Chevrolet	1975-1985	Blue Metallic	See below
	1978	Blue (1)	See below
Chrysler	1977-1978	Blue	See below
Plymouth	1973-1978	Blue (2)	See below
Pontiac	1975-1978	Blue Metallic (3)	See below

(1) Same as Buick
(2) Same as Chrysler
(3) Same as Buick

Source

Tower Paint & Decorating
922 Oregon St.
PO Box 2345
Oshkosh, WI 54903
(920) 235-6520

HIRSCH ENGINE PAINT

Hobbyists Have Used Hirsch Engine Paint for 32 Years

Bill Hirsch has been supplying engine paints to vintage-car restorers for 32 years. His Newark, New Jersey company provides the specific engine paint colors for a wide range of models of different years (see Table 1 below). The earliest specific match in Bill Hirsch's inventory is correct for a 1922 Buick; the latest specific match is for all 1983 Chrysler production engines.

In addition to such specific matches, Bill Hirsch supplies engine paint colors such as gray, black and aluminum that are used on many different car engines and engine parts. And the paints themselves will work on any engine in any car.

Hirsch's engine paints are rated for temperatures up to 600 degree F., while many others are only rated for 300 degrees F. "Ours are identical to the enamel found on military aircraft engines," says Bill. "They have the highest content of solids and are very high-gloss." His advertisements point out that these paints have a sprayed-on look, even when they are applied with a brush.

The engine paints that Hirsch sells are available in 14-ounce aerosol cans or in full quart cans. Prices at the time this book went to press were $12 for a spray can and $29 for a quart with shipping charges of $5 and $6, respectively. Bill Hirsch also sells an engine cleaner/degreaser and an engine primer in aerosol cans for $5.95 each. Shipping is slightly higher on foreign, COD and credit card orders and New Jersey residents must add the six percent states sales tax.

Prices of course are subject to periodic revision. Bill Hirsch is located at 396 Littleton Ave., Newark, NJ 07103. You can call (800) 828-2061 or (973) 642-2404 or Email: hirschauto@aol.com for additional information.

GENERAL NOTE: Table 1 shows the cars, model years and corresponding V-8 engine paint colors available from Bill Hirsch as of March 2003.

Table 1

CAR MAKE	YEARS	BILL HIRSCH COLOR NAME	Packaging
BUICK			
Buick V-8	1953-1965	Buick Green Late	Quart Only
Buick V-8	1966-1971	Buick Red Late	Aerosol & Quart
CADILLAC			
Cadillac V-8	1935-1948	Cadillac Green	Quart Only
Cadillac V-8	1949-1976	Cadillac Blue Dark	Aerosol & Quart
Cadillac V-8	1977-1981	Cadillac Blue Light	Aerosol & Quart
CHEVROLET - CORVETTE			
Chevrolet V-8	Most V-8s	Chevrolet Orange	Aerosol & Quart
Chevrolet V-8	1956-1957	Chevrolet Red	Aerosol & Quart
CHRYSLER – DODGE - PLYMOUTH			
Chrysler 273 cid V-8	1964-1969	Chrysler Red	Aerosol & Quart
Chrysler 340 cid V-8	1968-1969	Chrysler Red	Aerosol & Quart
Chrysler 318 cid V-8	To 1968	Chrysler Red	Aerosol & Quart
Chrysler (All engines)	1973-1983	Chrysler Blue	Aerosol & Quart
Chrysler 318-cid V-8	1969-1972	Chrysler Blue	Aerosol & Quart
Chrysler 340-cid V-8	1970-1971	Chrysler Hemi Orange	Aerosol & Quart
Chrysler 383-cid V-8	All	Chrysler Turquoise	Aerosol & Quart
Chrysler 383-cid Hi-Perf V-8	1969-1971	Chrysler Hemi Orange	Aerosol & Quart
Chrysler 426-cid V-8	1970-1971	Chrysler Hemi Orange	Aerosol & Quart
Chrysler 426 Street Hemi V-8	1966-1971	Chrysler Hemi Orange	Aerosol & Quart
Chrysler 440-cid Hi-Perf V-8	1969-1971	Chrysler Hemi Orange	Aerosol & Quart
Chrysler 440-cid V-8	All	Chrysler Turquoise	Aerosol & Quart
EDSEL			
Edsel V-8	1957-1959	Edsel Green	Quart Only
FORD			
Ford Flathead V-8	1941-1942	Ford Green	Quart Only
Ford Flathead V-8	1941-1948	Ford Blue	Quart Only
Thunderbird V-8	1955-1957	Ford Thunderbird Red	Quart Only
Ford Flathead V-8	1949-1951	Ford Bronze	Quart Only
Ford Flathead V-8	1952-53	Ford Bronze	Quart Only
Ford Flathead V-8	1952-53	Ford Orange	Quart Only
Ford V-8	1963-1967	Ford Medium Blue	Aerosol & Quart
Ford Mustang 302-cid V-8	1967-on	Ford Mustang Blue	Quart Only
Ford air cleaner & valve cover	1964-65 Black Block Mustang	Ford Gold	Aerosol & Quart
GENERAL MOTORS			
All Rear-Wheel-Drive V-8	1977-On	GM Blue	Aerosol & Quart
LINCOLN-MERCURY			
Lincoln L-Type V-8 (Engine and Firewall)	Up to 1929	Lincoln Gray	Quart Only
Lincoln V-8	1936-1948	Lincoln Green	Quart Only
Lincoln/Mercury V-8	1952-1956	Lincoln/Mercury Light Green	Aerosol & Quart
Lincoln/Mercury V-8 (Except Continental)	1954-1956	Lincoln/Mercury Light Green	Aerosol & Quart
Mercury Flathead V-8	1949-1951	Green	Quart Only
OLDSMOBILE			
Oldsmobile V-8	1949-1956	Olds. Green Late	Quart Only
Oldsmobile V-8	1957-1958	Olds. Gold	Quart Only
Oldsmobile 330-cid V-8	1964-1967	Olds. Gold Late	Aerosol & Quart
Oldsmobile 350-cid V-8	1968-1976	Olds. Gold Late	Aerosol & Quart
Oldsmobile 400-cid V-8	1965-1969	Olds. Bronze	Aerosol & Quart
Oldsmobile 455-cid V-8	1966-on	Olds. Metallic Blue	Aerosol & Quart
PACKARD			
Packard 320-cid V-8	1955	Packard Turquoise	Quart Only
Packard 352-cid V-8	1955	Packard Red	Quart Only
Packard 374-cid V-8	1956	Packard Bronze	Quart Only
Packard Big V-8	1955	Packard Ivory	Quart Only
PONTIAC			
Pontiac V-8 (except Bonneville)	1955-1957	Pontiac Blue/Green	Aerosol & Quart
Pontiac Bonneville V-8	1957	Pontiac Turquoise/Blue	Aerosol & Quart
Pontiac V-8	All 1958	Pontiac Turquoise/Blue	Aerosol & Quart
Pontiac V-8	1959-1965	Pontiac Light Blue	Aerosol & Quart
Pontiac V-8	1966-1972	Pontiac Metallic Blue	Aerosol & Quart

AMC/NASH RAMBLER
1955-1978

Nash and Hudson merged in 1954 to form American Motors Corp. The two nameplates continued to be separate brands until 1957. The first V-8 offered in AMC cars was actually a 320-cid Packard engine that was available in the 1955 Hudson Hornet Super and 1955 Nash Ambassador Series 80.

The first true AMC/Rambler V-8 appeared in 1956 as the 250-cid engine. This was a lightweight engine that tipped the scales at around 600 lbs. in most configurations. From the start, it was designed to be the stepping off point for a whole new family of V-8 engines originally known as Rambler V-8s and later thought of as AMC V-8s.

Large 4.75-in. bore spacing allowed for subsequent increases in displacement. The bore size varied from 3-1/2 inches in the original 250-cid version to four inches in a 327-cid version that bowed in 1957 and lasted through 1966 to 3-3/4 inches in the 287-cid version that bowed in 1963 and lasted through 1966. A carefully designed intake system promoted a high volumetric efficiency of 77 percent and uniform mixture distribution to all cylinders. Starting with the Special Edition 1966 $^1/_2$ Rambler American Rogue, a totally new "thinwall casting" engine family starting at 290 cid was introduced. This new engine series was lighter than the previous 250 through 327 cid engine family was. The old-style 287/327 cid V-8s were still used in the Classic and Ambassador until the end of the 1966 model run.

With the advent of AMC's entry into Trans-Am racing in 1968, and its 5.0-liter limit on maximum displacement, the AMC 290 was made available as an economical two-barrel regular fuel V-8 with 9.0:1 compression or a snorting four-barrel version with 10.0:1 compression and a thirst for premium fuel. In 1968, the racers bored this motor up to 302 cid to get the most out of it without breaking the Sports Car Club of America's (SCCA) Trans-Am rules.

Also in 1968 1/2, a new 390-cid V-8 was released. Like the other AMC "mills," it was a 90-degree overhead valve engine with high nickel-content Cast-iron head and block. It had a 4.17 x 3.57-in. bore and stroke and a 10.2:1 compression ratio and generated 315 hp. This would later climb to as high as 340 hp in cars like the Rebel Machine.

In 1970, the 290-cid engine was stroked to get 304 cid and the 343 was similarly modified to get 360 cid. AMC released its largest displacement ever 401-cid V-8 in Javelins, Matadors and Ambassadors in 1971 and made it available through 1974 except for approximately 84 401s installed in Matador fleet and police sedans. It is thought that 5 401s were installed in 1975 Matador coupes. The 401 was basically a stroked 390. AMC's Marketing Department wanted an over 400 cid motor to try and compete with the "Big Three's" big block V-8s. It should be noted that all the 290, 304, 343, 360, 401 V-8s were "medium-block"-sized engines and not small or big blocks in size and weight.

From 1955 forward, all optional four-barrel engines required premium leaded (ethyl) fuel due to the higher compression ratios.

The 270-hp V-8 engine used in the AMC Ambassador had a compression ratio of 9.7:1. It was equipped with a four-barrel carburetor that incorporated new refinements for better control of fuel under all conditions. The intake and exhaust valves were redesigned for quieter operation and longer service life.

1955 AMC

Ambassador Series 80
V-8. Overhead valves. Cast-iron block. Displacement: 320 cid. Bore & stroke: 3-13/16 x 3-1/2 in. Compression ratio: (Before engine no. P6001) 7.8:1; (After engine no. P6001) 8.25:1. Advertised hp: 208 at 4200 rpm. Advertised torque: 300 lbs.-ft. at 2300 rpm. Recommended idle speed: 450 rpm with Hydra-Matic; 500 rpm without Hydra-Matic. Five main bearings. Hydraulic valve lifters (non-adjustable). Carburetor: Carter WGD two-barrel model 2231S. Cooling system capacity: 20 qt. without heater; 21 qt. with heater. Crankcase capacity less filter: 5 qt. Packard V-8. Engine no. P-1001 and up.

1956 AMC

The 287-cid Rambler V-8 was related to the engine introduced in the Classic and Ambassador series in 1956, which had wide (4.75-in.) bore spacing that allowed for future displacement increases. Even the 3.75-in. bore of the 1965 version shown here left plenty of room for growth. This engine weighed about 600 pounds.

Ambassador Special Series 50; Midyear
V-8. Overhead valves. Cast-iron block. Displacement: 250 cid. Bore & stroke: 3-1/2 x 3-1/4 in. Compression ratio: 8.0:1. Advertised hp: 190 at 4900 rpm. Advertised torque: 240 lbs.-ft. at 2500 rpm. Recommended idle speed: 425 rpm with automatic transmission; 4509 rpm with automatic transmission and operating air conditioning. Five main bearings. Hydraulic valve lifters (non-adjustable). Carburetor: Carter WGD two-barrel model 2352SA. Cooling system capacity: 20 qt. without heater; 21 qt. with heater. Crankcase capacity less filter: 5 qt. Engine no. G-1001 and up.

Ambassador Series 80
V-8. Overhead valves. Cast-iron block. Displacement: 352 cid. Bore & stroke: 4 x 3-1/2 in. Compression ratio: 9.55:1. Advertised hp: 220 at 4800 rpm. Advertised torque: 320 lbs.-ft. at 2350 rpm. Recommended idle speed: 450 rpm. Five main bearings. Hydraulic valve lifters (non-adjustable). Carburetor: Carter WCFB four-barrel. Cooling system capacity: 27 qt. without heater; 28 qt. with heater. Crankcase capacity less filter: 5 qt. Packard V-8. Engine no. P-21001 and up.

1957 AMC

Ambassador Special Series 50
V-8. Overhead valves. Cast-iron block. Displacement: 250 cid. Bore & stroke: 3-1/2 x 3-1/4 in. Compression ratio: 8.0:1. Advertised hp: 190 at 4900 rpm. Advertised torque: 240 lbs.-ft. at 2500 rpm. Recommended idle speed: 425 rpm with automatic transmission; 4509 rpm with automatic transmission and operating air conditioning. Five main bearings. Hydraulic valve lifters (non-adjustable). Carburetor: Carter WGD two-barrel model 2352SA. Cooling system capacity: 20 qt. without heater; 21 qt. with heater. Crankcase capacity less filter: 5 qt. Engine no. G-7501 and up or engine no. GG-1001 and up.

Ambassador Special Series 80
V-8. Overhead valves. Cast-iron block. Displacement: 327 cid. Bore & stroke: 4 x 3-1/4 in. Compression ratio: 9.0:1. Advertised hp: 255 at 4700 rpm. Advertised torque: 345 lbs.-ft. at 2600 rpm. Recommended idle speed: 425 rpm with automatic transmission; 475 rpm with automatic transmission and operating air conditioning; 550 rpm with standard transmission. Five main bearings. Hydraulic valve lifters (non-adjustable). Carburetor: Carter WCFB four-barrel model 2593SA. Cooling system capacity: 19 qt. without heater; 20 qt. with heater. Crankcase capacity less filter: 5 qt. Equipped with dual exhaust system and 12-volt electrical system. Nash engine no. V-16501 and up; Ambassador engine no. N-1001 and up.

Rebel V-8
V-8. Overhead valves. Cast-iron block. Displacement: 327 cid. Bore & stroke: 4 x 3-1/4 in. Compression ratio: 9.0:1. Advertised hp: 255 at 4700 rpm. Advertised torque: 345 lbs.-ft. at 2600 rpm. Recommended idle speed: 425 rpm with automatic transmission; 475 rpm with automatic transmission and operating air conditioning; 550 rpm with standard transmission. Five main bearings. Hydraulic valve lifters (non-adjustable). Carburetor: Carter WCFB four-barrel model 2593SA. Cooling system capacity: 19 qt. without heater; 20 qt. with heater. Crankcase capacity less filter: 5 qt. Equipped with dual exhaust system and 12-volt electrical system. Engine no. CN-1001 and up.

1958 AMC

Rebel V-8
V-8. Overhead valves. Cast-iron block. Displacement: 250 cid. Bore & stroke: 3-1/2 x 3-1/4 in. Compression ratio: 8.7:1. Advertised hp: 215 at 4900 rpm. Advertised torque: 260 lbs.-ft. at 2500 rpm. Recommended idle speed: 550 rpm. Five main bearings. Hydraulic valve lifters (non-adjustable). Carburetor: Holley four-barrel model 4150C. Cooling system capacity: 20 qt. without heater; 21 qt. with heater. Crankcase capacity less filter: 4 qt. Equipped with dual exhaust system. Engine no. G-24001 and up.

Ambassador
V-8. Overhead valves. Cast-iron block. Displacement: 327 cid. Bore & stroke: 4 x 3-1/4 in. Compression ratio: 9.7:1. Advertised hp: 270 at 4700 rpm. Advertised torque: 360 lbs.-ft. at 2600 rpm. Recommended idle speed: 550 rpm. Five main bearings. Hydraulic valve lifters (non-adjustable). Carburetor: Holley four-barrel model 4150C. Cooling system capacity: 19 qt. without heater; 20 qt. with heater. Crankcase capacity less filter: 4 qt. Equipped with dual exhaust system. Engine no. N-17001 and up.

1959 AMC

Rebel V-8
V-8. Overhead valves. Cast-iron block. Displacement: 250 cid. Bore & stroke: 3-1/2 x 3-1/4 in. Compression ratio: 8.7:1. Advertised hp: 215 at 4900 rpm. Advertised torque: 260 lbs.-ft. at 2500 rpm. Recommended idle speed: 550 rpm. Five main bearings. Hydraulic valve lifters (non-adjustable). Carburetor: Holley four-barrel model 4150C on early production; 4160 on later production. Cooling system capacity: 20 qt. without heater; 21 qt. with heater. Crankcase capacity less filter: 4 qt. Single exhaust system standard, duals optional. Engine no. G-34501 and up.

Ambassador
V-8. Overhead valves. Cast-iron block. Displacement: 327 cid. Bore & stroke: 4 x 3-1/4 in. Compression ratio: 9.7:1. Advertised hp: 270 at 4700 rpm. Advertised torque: 360 lbs.-ft. at 2600 rpm. Recommended idle speed: 550 rpm. Five main bearings. Hydraulic valve lifters (non-adjustable). Carburetor: Holley four-barrel model 4150C on early

production; 4160 on later production. Cooling system capacity: 19 qt. without heater; 20 qt. with heater. Crankcase capacity less filter: 4 qt. Single exhaust system standard, duals optional. N-32501 and up.

1960 AMC

Rebel V-8
V-8. Overhead valves. Cast-iron block. Displacement: 250 cid. Bore & stroke: 3-1/2 x 3-1/4 in. Compression ratio: 8.7:1. Advertised hp: 200 at 4900 rpm. Advertised torque: 245 lbs.-ft. at 2500 rpm. Recommended idle speed: 550 rpm. Five main bearings. Hydraulic valve lifters (non-adjustable). Carburetor: Holley two-barrel model 2040. Cooling system capacity: 20 qt. without heater; 21 qt. with heater. Crankcase capacity less filter: 4 qt. Single exhaust system. Standard V-8. Engine no. A100001 and up.

Rebel V-8
V-8. Overhead valves. Cast-iron block. Displacement: 250 cid. Bore & stroke: 3-1/2 x 3-1/4 in. Compression ratio: 8.7:1. Advertised hp: 215 at 4900 rpm. Advertised torque: 260 lbs.-ft. at 2500 rpm. Recommended idle speed: 550 rpm. Five main bearings. Hydraulic valve lifters (non-adjustable). Carburetor: Holley four-barrel model 4150C on early production; 4160 on later production. Cooling system capacity: 20 qt. without heater; 21 qt. with heater. Crankcase capacity less filter: 4 qt. Single exhaust system, dual exhaust optional. Optional V-8. Engine no. A100001 and up.

Ambassador
V-8. Overhead valves. Cast-iron block. Displacement: 327 cid. Bore & stroke: 4 x 3-1/4 in. Compression ratio: 8.7:1. Advertised hp: 250 at 4700 rpm. Advertised torque: 340 lbs.-ft. at 2600 rpm. Recommended idle speed: 550 rpm. Five main bearings. Hydraulic valve lifters (non-adjustable). Carburetor: Carburetor: Holley two-barrel model 2040. Cooling system capacity: 19 qt. without heater; 20 qt. with heater. Crankcase capacity less filter: 4 qt. Single exhaust system. Engine code: E.

Ambassador
V-8. Overhead valves. Cast-iron block. Displacement: 327 cid. Bore & stroke: 4 x 3-1/4 in. Compression ratio: 9.7:1. Advertised hp: 270 at 4700 rpm. Advertised torque: 360 lbs.-ft. at 2600 rpm. Recommended idle speed: 550 rpm. Five main bearings. Hydraulic valve lifters (non-adjustable). Carburetor: Holley four-barrel model 4150C on early production; 4160 on later production. Cooling system capacity: 19 qt. without heater; 20 qt. with heater. Crankcase capacity less filter: 4 qt. Single exhaust system, duals optional. Optional V-8. Engine code: F.

1961 AMC

Classic V-8
V-8. Overhead valves. Cast-iron block. Displacement: 250 cid. Bore & stroke: 3-1/2 x 3-1/4 in. Compression ratio: 8.7:1. Advertised hp: 200 at 4900 rpm. Advertised torque: 245 lbs.-ft. at 2500 rpm. Recommended idle speed: 550 rpm. Five main bearings. Hydraulic valve lifters (non-adjustable). Carburetor: Holley two-barrel model 2040. Cooling system capacity: 19 qt. without heater; 20 qt. with heater. Crankcase capacity less filter: 4 qt. Single exhaust system. Standard V-8. Optional version with four-barrel Holley carb, 8.7:1 compression and dual exhaust- 215 hp at 4900 rpm. Torque: 260 lbs.-ft. at 2500 rpm.

Ambassador V-8
V-8. Overhead valves. Cast-iron block. Displacement: 327 cid. Bore & stroke: 4 x 3-1/4 in. Compression ratio: 8.7:1. Advertised hp: 250 at 4700 rpm. Advertised torque: 340 lbs.-ft. at 2600 rpm. Recommended idle speed: 550 rpm. Five main bearings. Hydraulic valve lifters (non-adjustable). Carburetor: Carburetor: Holley two-barrel model 2040. Cooling system capacity: 18 qt. without heater; 19 qt. with heater. Crankcase capacity less filter: 4 qt. Single exhaust system. Optional version with four-barrel Holley carb., 9.7:1 compression and dual exhaust- 270 hp at 4,700 rpm. Torque: 360 lbs.-ft. at 2600 rpm. Engine code: E.

Ambassador
V-8. Overhead valves. Cast-iron block. Displacement: 327 cid. Bore & stroke: 4 x 3-1/4 in. Compression ratio: 9.7:1. Advertised hp: 270 at 4700 rpm. Advertised torque: 360 lbs.-ft. at 2600 rpm. Recommended idle speed: 550 rpm. Five main bearings. Hydraulic valve lifters (non-adjustable). Carburetor: Holley four-barrel model 4150C on early production; 4160 on later production. Cooling system capacity: 19 qt. without heater; 20 qt. with heater. Crankcase capacity less filter: 4 qt. Single exhaust system, duals optional. Optional V-8. Engine code: F.

1962 AMC

Ambassador V-8
V-8. Overhead valves. Cast-iron block. Displacement: 327 cid. Bore & stroke: 4 x 3-1/4 in. Compression ratio: 8.7:1. Advertised hp: 250 at 4700 rpm. Advertised torque: 340 lbs.-ft. at 2600 rpm. Recommended idle speed: 550 rpm. Five main bearings. Hydraulic valve lifters (non-adjustable). Carburetor: Carburetor: Holley two-barrel model 2300-2442. Cooling system capacity: 18 qt. without heater; 19 qt. with heater. Crankcase capacity less filter: 4 qt. Single exhaust system. Standard V-8. Engine code: E.

Ambassador V-8
V-8. Overhead valves. Cast-iron block. Displacement: 327 cid. Bore & stroke: 4 x 3-1/4 in. Compression ratio: 9.7:1. Advertised hp: 270 at 4700 rpm. Advertised torque: 360 lbs.-ft. at 2600 rpm. Recommended idle speed: 550 rpm. Five main bearings. Hydraulic valve lifters (non-adjustable). Carburetor: Holley four-barrel model 4150-1957-1. Cooling system capacity: 18 qt. without heater; 19 qt. with heater. Crankcase capacity less filter: 4 qt. Dual exhaust system standard. Optional V-8. Engine code: F.

1963 AMC

Classic V-8
V-8. Overhead valves. Cast-iron block. Displacement: 287 cid. Bore & stroke: 3-3/4 x 31/4 in. Compression ratio: 8.7:1. Advertised hp: 198 at 4700 rpm. Advertised torque: 280 lbs.-ft. at 2600 rpm. Recommend idle speed: 475 rpm. Five main bearings. Hydraulic valve lifters (non-adjustable). Carburetor: Holley two-barrel model 2300-2442. Cooling system capacity: 18 qts. without heater; 19 qts. with heater. Crankcase capacity less filter: 4 qt. Single exhaust system. Engine code: G.

Ambassador V-8
V-8. Overhead valves. Cast-iron block. Displacement: 327 cid. Bore & stroke: 4 x 3-1/4 in. Compression ratio: 8.7:1. Advertised hp: 250 at 4700 rpm. Advertised torque: 340 lbs.-ft. at 2600 rpm. Recommended idle speed: 550 rpm. Five main bearings. Hydraulic valve lifters (non-adjustable). Carburetor: Carburetor: Holley two-barrel model 2300-2442. Cooling system capacity: 18 qt. without heater; 19 qt. with heater. Crankcase capacity less filter: 4 qt. Single exhaust system. Engine code: E.

Ambassador
V-8. Overhead valves. Cast-iron block. Displacement: 327 cid. Bore & stroke: 4 x 3-1/4 in. Compression ratio: 9.7:1. Advertised hp: 270 at 4700 rpm. Advertised torque: 360 lbs.-ft. at 2600 rpm. Recommended idle speed: 550 rpm. Five main bearings. Hydraulic valve lifters (non-adjustable). Carburetor: Holley four-barrel model 4150-1957-1. Cooling system capacity: 18 qt. without heater; 19 qt. with heater. Crankcase capacity less filter: 4 qt. Single exhaust system. Optional V-8. Engine code: F.

1964 AMC

Classic V-8
V-8. Overhead valves. Cast-iron block. Displacement: 287 cid. Bore & stroke: 3-3/4 x 31/4 in. Compression ratio: 8.7:1. Advertised hp: 198 at 4700 rpm. Advertised torque: 280 lbs.-ft. at 2600 rpm. Recommend idle speed: 475 rpm. Five main bearings. Hydraulic valve lifters (non-adjustable). Carburetor: Holley two-barrel model 2300-2442. Cooling system capacity: 18 qts without heater; 19 qts. with heater. Crankcase capacity less filter: 4 qt. Single exhaust system. Engine code: G.

Ambassador V-8
V-8. Overhead valves. Cast-iron block. Displacement: 327 cid. Bore & stroke: 4 x 3-1/4 in. Compression ratio: 8.7:1. Advertised hp: 250 at 4700 rpm. Advertised torque: 340 lbs.-ft. at 2600 rpm. Recommended idle speed: 475 rpm. Five main bearings. Hydraulic valve lifters (non-adjustable). Carburetor: Carburetor: Holley two-barrel model 2300-2442. Cooling system capacity: 18 qt. without heater; 19 qt. with heater. Crankcase capacity less filter: 4 qt. Single exhaust system. Engine code: E.

Ambassador V-8
V-8. Overhead valves. Cast-iron block. Displacement: 327 cid. Bore & stroke: 4 x 3-1/4 in. Compression ratio: 9.7:1. Advertised hp: 270 at 4700 rpm. Advertised torque: 360 lbs.-ft. at 2600 rpm. Recommended idle speed: 475 rpm. Five main bearings. Hydraulic valve lifters (non-adjustable). Carburetor: Holley four-barrel model 4150-1957-1 Cooling system capacity: 18 qt. without heater; 19 qt. with heater. Crankcase capacity less filter: 4 qt. Single exhaust system. Optional V-8. Engine code: F.

1965 AMC

The Rambler 327-cid V-8 was the 287-cid V-8 with a 4.00-in. bore. The 327 also had a four-barrel carburetor and a 9.7:1 compression ratio, which boosted output to 270 hp. Both AMC V-8s featured flat-top aluminum pistons with tin-plated wear surfaces and slipper-type skirts. In 1965, this engine required premium fuel.

Classic 550, 660, 770, 770-H, Ambassador, and Marlin V-8
V-8. Overhead valves. Cast-iron block. Displacement: 287 cid. Bore & stroke: 3-3/4 x 3-1/4 in. Compression ratio: 8.7:1. Advertised hp: 198 at 4700 rpm. Advertised torque: 280 lbs.-ft. at 2600 rpm. Recommended idle speed: 475 rpm. Five main bearings. Hydraulic valve lifters (non-adjustable). Carburetor: Carburetor: Holley two-barrel model 2290-2699. Cooling system capacity: 18 qt. without heater; 19 qt. with heater. Crankcase capacity less filter: 4 qt. Single exhaust. Standard V-8. Engine code: G.

Classic 550, 660, 770, 770-H, Ambassador, and Marlin V-8
V-8. Overhead valves. Cast-iron block. Displacement: 327 cid. Bore & stroke: 4 x 3-1/4 in. Compression ratio: 9.7:1. Advertised hp: 270 at 4700 rpm. Advertised torque: 360 lbs.-ft. at 2600 rpm. Recommended idle speed: 475 rpm. Five main bearings. Hydraulic valve lifters (non-adjustable). Carburetor: Holley four-barrel model 4150-3044 Cooling system capacity: 18 qt. without heater; 19 qt. with heater. Crankcase capacity less filter: 4 qt. Single exhaust system. Optional V-8. Engine code: F.

1966 AMC

For 1966, the AMC "Rambler 287" V-8 was continued and looked identical to the 1965 version of this 198-hp engine. The 287-cid V-8 developed 0.68 hp per cubic inch of displacement. It had an 8.7:1 compression ratio and weighed 600 lbs. giving a 3.03 weight-to-horsepower ratio.

Introduced as a mid-1966 option, the AMC "Typhoon 290" V-8 was a high-performance "thinwall" V-8 engine with smoother ports and redesigned combustion chambers. Compact in size, it also featured threaded rocker arm studs. It was used only in a limited number of 1966-1/2 Rambler Rogue hardtops, most of which were painted Sun Gold with black roof and trunk lid or Solar Yellow with black roof and trunk lid. The engine was a reaction to Ford's Mustang.

Classic 550, Classic 770, Rebel, Marlin, Ambassador 880, Ambassador 990, DPL V-8
V-8. Overhead valves. Cast-iron block. Displacement: 287 cid. Bore & stroke: 3-3/4 x 3-1/4 in. Compression ratio: 8.7:1. Advertised hp: 198 at 4700 rpm. Advertised torque: 280 lbs.-ft. at 2600 rpm. Recommended idle speed: 550 rpm. Five main bearings. Hydraulic valve lifters (non-adjustable). Carburetor: Carburetor: Holley two-barrel model 2209-3305. Cooling system capacity: 18 qt. without heater; 19 qt. with heater. Crankcase capacity less filter: 4 qt. Single exhaust. Engine code: G.

Classic 550, Classic 770, Rebel, Marlin, Ambassador 880, Ambassador 990, DPL V-8
V-8. Overhead valves. Cast-iron block. Displacement: 327 cid. Bore & stroke: 4 x 3-1/4 in. Compression ratio: 8.7:1. Advertised hp: 250 at 4700 rpm. Advertised torque: 340 lbs.-ft. at 2600 rpm. Recommended idle speed: 550 rpm. Five main bearings. Hydraulic valve lifters (non-adjustable). Carburetor: Carburetor: Holley two-barrel model 2209-3305. Cooling system capacity: 18 qt. without heater; 19 qt. with heater. Crankcase capacity less filter: 4 qt. Single exhaust system. Engine code: E.

Classic 550, Classic 770, Rebel, Marlin, Ambassador 880, Ambassador 990, DPL V-8
V-8. Overhead valves. Cast-iron block. Displacement: 327 cid. Bore & stroke: 4 x 3-1/4 in. Compression ratio: 9.7:1. Advertised hp: 270 at 4700 rpm. Advertised torque: 360 lbs.-ft. at 2600 rpm. Recommended idle speed: 475 rpm. Five main bearings. Hydraulic valve lifters (non-adjustable). Carburetor: Holley four-barrel model 4160-3201. Cooling system capacity: 18 qt. without heater; 19 qt. with heater. Crankcase capacity less filter: 4 qt. Single exhaust system. Optional V-8. Engine code: F.

1967 AMC

A sectional view of the new-for-1967 AMC 343-cid V-8 that was available in the Ambassador, Rebel, and Marlin when the model year kicked off and later made available in the American sedan, hardtop, and convertible with a four-barrel carburetor and four-speed manual transmission only. The four-barrel version cranked out 280 hp.

Marlin, Ambassador, DPL Except Convertible V-8
V-8. Overhead valves. Cast-iron block. Displacement: 290 cid. Bore & stroke: 3.75 x 3.28 in. Compression ratio: 9.0:1. Advertised hp: 200 at 4600 rpm. Advertised torque: 285 lbs.-ft. at 2800 rpm. Recommended idle speed: 600 rpm. Five main bearings. Hydraulic valve lifters (non-adjustable). Carburetor: Holley two-barrel model 2209-3308-1. Cooling system capacity: 12 qt. without heater; 13 qt. with heater. Crankcase capacity less filter: 4 qt. Single exhaust with crossover. Standard V-8. Engine code: H.

American 220, 440, Rogue V-8
V-8. Overhead valves. Cast-iron block. Displacement: 290 cid. Bore & stroke: 3.75 x 3.28 in. Compression ratio: 10.0:1. Advertised hp: 225 at 4700 rpm. Advertised torque: 300 lbs.-ft. at 3200 rpm. Recommended idle speed: 600 rpm. Five main bearings. Hydraulic valve lifters (non-adjustable). Carburetor: Carter AFB. Cooling system capacity: 12 qt. without heater; 13 qt. with heater. Crankcase capacity less filter: 4 qt. Exhaust: Single exhaust with crossover (duals not available). Optional V-8. Engine code: H.

Rebel 550-770, SST, Marlin, Ambassador 880-990, DPL V-8
V-8. Overhead valves. Cast-iron block. Displacement: 343 cid. Bore & stroke: 4.08 x 3.28 in. Compression ratio: 9.0:1. Advertised hp: 235 at 4400 rpm. Advertised torque: 345 lbs.-ft. at 2600 rpm. Recommended idle speed: 600 rpm. Five main bearings. Hydraulic valve lifters (non-adjustable). Carburetor: Holley two-barrel 2209. Cooling system capacity: 13 qts. Crankcase capacity less filter: 4 qts. Exhaust system: Single standard, duals optional. Engine code: J or V.

American 220-440, Rogue, Rebel 550-770, SST, Marlin, Ambassador 880-990, DPL V-8
V-8. Overhead valves. Cast-iron block. Displacement: 343 cid. Bore & stroke: 4.08 x 3.28 in. Compression ratio: 10.2:1. Advertised hp: 280 at 4800 rpm. Advertised torque: 365 lbs.-ft. at 3000 rpm. Recommended idle speed: 600 rpm. Five main bearings. Hydraulic valve lifters (non-adjustable). Carburetor: Carter four-barrel AFB. Cooling system capacity: 13 qts. Crankcase capacity less filter: 4 qts. Exhaust system: Single with crossover, standard, duals optional on all but American. Engine code: K or W.

1968 AMC

In 1968, AMC continued to offer the 343-cid V-8 in two-barrel (235-hp at 4400 rpm) and four-barrel (280-hp at 4800 rpm) versions. Compression ratios were upped to 9.0:1 with the two-barrel engine and 10.2:1 with the four-barrel version in order to keep up with America's galloping horsepower race.

Rebel SST, Ambassador SST V-8, American 220, 440 and Rogue, Javelin
V-8. Overhead valves. Cast-iron block. Displacement: 290 cid. Bore & stroke: 3.75 x 3.28 in. Compression ratio: 9.0:1. Advertised hp: 200 at 4600 rpm. Advertised torque: 285 lbs.-ft. at 2800 rpm. Recommended idle speed: 650 rpm. Five main bearings. Hydraulic valve lifters (non-adjustable). Carburetor: Motorcraft 8HM2. Cooling system capacity: 13 qts. without heater; 14 qts. with heater. Crankcase capacity less filter: 4 qts. Single exhaust with crossover, duals optional on all but American and all wagons. Standard V-8. Engine code: H.

AMX V-8
V-8. Overhead valves. Cast-iron block. Displacement: 290 cid. Bore & stroke: 3.75 x 3.28 in. Compression ratio: 10.0:1. Advertised hp: 225 at 4700 rpm. Advertised torque: 300 lbs.-ft. at 3200 rpm. Recommended idle speed: 650 rpm. Five main bearings. Hydraulic valve lifters (non-adjustable). Carburetor: Carter four-barrel AFB. Cooling system capacity: 13 qts. without heater; 14 qts. with heater. Crankcase capacity less filter: 4 qts. Exhaust system: Duals. Standard AMX V-8. Engine code: N.

Rebel 550-770, SST, Ambassador, DPL, SST V-8
V-8. Overhead valves. Cast-iron block. Displacement: 343 cid. Bore & stroke: 4.08 x 3.28 in. Compression ratio: 9.0:1. Advertised hp: 235 at 4400 rpm. Advertised torque: 345 lbs.-ft. at 2600 rpm. Recommended idle speed: 650 rpm. Five main bearings. Hydraulic valve lifters (non-adjustable). Carburetor: Motorcraft 6200. Cooling system capacity: 13 qts. Crankcase capacity less filter: 4 qts. Exhaust system: Single with crossover standard, duals options except wagons. Engine code: S.

Rebel 550-770, SST, AMX, Javelin SST, Ambassador, DPL, SST V-8
V-8. Overhead valves. Cast-iron block. Displacement: 343 cid. Bore & stroke: 4.08 x 3.28 in. Compression ratio: 10.2:1. Advertised hp: 280 at 4800 rpm. Advertised torque: 365 lbs.-ft. at 3000 rpm. Recommended idle speed: 650 rpm. Five main bearings. Hydraulic valve lifters (non-adjustable). Carburetor: Carter four-barrel. Cooling system capacity: 13 qts. Crankcase capacity less filter: 4 qts. Exhaust system: Single with crossover standard, duals optional except wagons. Engine code: Z.

Ambassador, DPL, SST V-8, AMX and Javelin
V-8. Overhead valves. Cast-iron block. Displacement: 390 cid. Bore & stroke: 4.17 x 3.57 in. Compression ratio: 10.2:1. Advertised hp: 315 at 4600 rpm. Advertised torque: 425 lbs.-ft. at 3200 rpm. Recommended idle speed: 650 rpm. Five main bearings. Hydraulic valve lifters (non-adjustable). Carburetor: Carter four-barrel AFB. Cooling system capacity: 13 qts. Crankcase capacity less filter: 4 qts. Exhaust system: Duals standard on all 390 V-8 engines. Engine code: W.

1969 AMC

The year 1969 brought a new 390-cid V-8 from AMC. This was again basically the same as the automaker's other V-8s, but with bigger bore and stroke dimensions. Compact (31 inches long, 24 inches wide, and 22 inches high) and high-performance-oriented, the Code W 390 V-8 produced .808 horsepower per cubic inch for a total of 315 hp.

Rambler 440 Rogue, Rebel, Rebel SST, Javelin, Javelin SST, Ambassador SST V-8
V-8. Overhead valves. Cast-iron block. Displacement: 290 cid. Bore & stroke: 3.75 x 3.28 in. Compression ratio: 9.0:1. Advertised hp: 200 at 4600 rpm. Advertised torque: 285 lbs.-ft. at 2800 rpm. Recommended idle speed: (Manual transmission) 650 rpm; (Automatic transmission) 550 rpm. Five main bearings. Hydraulic valve lifters. Carburetor: Motorcraft 6200.. Cooling system capacity: 13 qts. without heater; 14 qts. with heater. Crankcase capacity less filter: 4 qt. Exhaust system: single exhaust with crossover. Standard V-8. Engine code: M

AMX, Javelin, Javelin SST V-8
V-8. Overhead valves. Cast-iron block. Displacement: 290 cid. Bore & stroke: 3.75 x 3.28 in. Compression ratio: 10.0:1. Advertised hp: 225 at 4700 rpm. Advertised torque: 300 lbs.-ft. at 3200 rpm. Recommended idle speed: (Manual transmission) 650 rpm; (Automatic transmission) 550 rpm. Five main bearings. Hydraulic valve lifters. Carburetor: Carter four-barrel AFB. Cooling system capacity: 13 qts. without heater; 14 qts. with heater. Crankcase capacity less filter: 4 qts. Exhaust system: Single with crossover standard on Javelin, duals optional on Javelin, duals standard on AMX. Standard AMX V-8. Engine code: N.

Rebel, Rebel SST, Ambassador, Ambassador DPL, Ambassador SST V-8
V-8. Overhead valves. Cast-iron block. Displacement: 343 cid. Bore & stroke: 4.08 x 3.28 in. Compression ratio: 9.0:1. Advertised hp: 235 at 4400 rpm. Advertised torque: 345 lbs.-ft. at 2600 rpm. Recommended idle speed: (Manual transmission) 650 rpm; (Automatic transmission) 550 rpm. Five main bearings. Hydraulic valve lifters. Carburetor: Motocraft 6200. Cooling system capacity: 13 qts. Crankcase capacity less filter: 4 qts. Exhaust system: Single with crossover (no duals available with this engine). Engine code: S.

Rebel, Rebel SST, AMX, Javelin, Javelin SST, Ambassador, Ambassador DPL and SST V-8
V-8. Overhead valves. Cast-iron block. Displacement: 343 cid. Bore & stroke: 4.08 x 3.28 in. Compression ratio: 10.2:1. Advertised hp: 280 at 4800 rpm. Advertised torque: 365 lbs.-ft. at 3000 rpm. Recommended idle speed: (Manual transmission) 650 rpm; (Automatic transmission) 550 rpm. Five main bearings. Hydraulic valve lifters (non-adjustable). Carburetor: Carter AFB. Cooling system capacity: 13 qts. Crankcase capacity less filter: 4 qts. Exhaust system: Single with crossover standard except AMX (duals standard on AMX) optional all others. Engine code: T.

SC/Rambler, AMX, Javelin, Javelin SST, Ambassador, Ambassador DPL, Ambassador SST V-8
V-8. Overhead valves. Cast-iron block. Displacement: 390 cid. Bore & stroke: 4.17 x 3.57 in. Compression ratio: 10.2:1. Advertised hp: 315 at 4600 rpm. Advertised torque: 425 lbs.-ft. at 3200 rpm. Recommended idle speed: 650 rpm. Five main bearings. Hydraulic valve lifters (non-adjustable). Carburetor: Carter AFB. Cooling system capacity: 13 qts. Crankcase capacity less filter: 4 qts. Exhaust system: duals standard on all models with 390 engine. Engine code X.

1970 AMC

Hornet. Rebel, Rebel SST, Javelin, Javelin SST, Ambassador DPL, Ambassador SST V-8
V-8. Overhead valves. Cast-iron block. Displacement: 304 cid. Bore & stroke: 3.75 x 3.44 in. Compression ratio: 9.0:1. Advertised hp: 210 at 4400 rpm. Advertised torque: 305 lbs.-ft. at 2800 rpm. Recommended idle speed: (Manual transmission) 650 rpm; (Automatic transmission) 600 rpm. Five main bearings. Hydraulic valve lifters. Carburetor: Motorcraft 2100D two-barrel. Cooling system capacity: 13 qts. without heater; 14 qts. with heater. Crankcase capacity less filter: 4 qt. Exhaust system: Single with crossover. Engine code: H.

Javelin, Javelin SST, Rebel, Rebel SST, Ambassador, Ambassador SST V-8
V-8. Overhead valves. Cast-iron block. Displacement: 360 cid. Bore & stroke: 4.08 x 3.44 in. Compression ratio: 9.0:1. Advertised hp: 245 at 4500 rpm. Advertised torque: 365 lbs.-ft. at 2400 rpm. Recommended idle speed: (Manual transmission) 650 rpm; (Automatic transmission) 600 rpm. Five main bearings. Hydraulic valve lifters. Carburetor: AMC 2100 two-barrel. Cooling system capacity: 12 qts. without heater; 13 qts. with heater. Crankcase capacity less filter: 4 qt. Exhaust system: Single with crossover. Engine code: N.

[Javelin, Javelin SST, AMX, Rebel, Rebel SST, Ambassador and Ambassador SST]
V-8 Overhead valves. Cast-iron block. Displacement: 360 cid. Bore & stroke: 4.08 x 3.44 in. Compression ratio: 10.0:1. Advertised hp: 290 at 4800 rpm. Advertised torque: 395 lbs.-ft. at 3200 rpm. Recommended idle speed: (Manual transmission) 650 rpm; (Automatic transmission) 600 rpm. Five main bearings. Hydraulic valve lifters. Carburetor: Motorcraft 4300. Cooling system capacity: 12 qts without heater; 13 qts with heater. Crankcase capacity less filter: 4 qts. Exhaust system: Single standard, duals optional, duals standard on AMX. Engine code: P.

AMX, Javelin, Javelin SST, Rebel, Rebel SST, Machine and all Ambassador V-8
V-8. Overhead valves. Cast-iron block. Displacement: 390 cid. Bore & stroke: 4.17 x 3.57 in. Compression ratio: 10.0:1. Advertised hp: 325 at 5000 rpm. Advertised torque: 420 lbs.-ft. at 3200 rpm. Recommended idle speed: (Manual transmission) 650 rpm; (Automatic transmission) 600

As you can see, the hot 290-cid AMC V-8 looked identical to the 343-cid engine on the outside. It was available in a two-barrel version with 9.0:1 compression and 200 hp at 4600 rpm as well as a four-barrel version with 10.0:1 compression and 225 hp at 4700 rpm. The two-barrel churned up .069 hp per cube and the four-barrel .775 hp per cube.

rpm. Five main bearings. Hydraulic valve lifters. Carburetor: Motorcraft 4300. Cooling system capacity: 13 qts. Crankcase capacity less filter: 4 qts. Exhaust system: Duals standard on all models with 390 engine. Engine code: X.

Rebel Machine optional V-8

V-8. Overhead valves. Cast-iron block. Displacement: 390 cid. Bore & stroke: 4.17 x 3.57 in. Compression ratio: 10.0:1. Advertised hp: 340 at 5100 rpm. Advertised torque: 430 lbs.-ft. at 3600 rpm. Recommended idle speed: (Manual transmission) 650 rpm; (Automatic transmission) 600 rpm. Five main bearings. Hydraulic valve lifters. Carburetor: Four-barrel. Cooling system capacity: 13 qts. Crankcase capacity less filter: 4 qts. Exhaust system: Duals standard on all models with 390 engine. Engine code: Y.

1971 AMC

Hornet, Hornet SST, Javelin, Javelin SST, Matador, Ambassador DPL, Ambassador SST, Ambassador Brougham V-8

V-8. Overhead valves. Cast-iron block. Displacement: 304 cid. Bore & stroke: 3.75 x 3.44 in. Compression ratio: 8.4:1. Advertised hp: 210 at 4400 rpm. Advertised torque: 300 lbs.-ft. at 2600 rpm. Recommended idle speed: (Manual transmission) 750 rpm; (Automatic transmission) 650 rpm. Five main bearings. Hydraulic valve lifters. Carburetor: Motorcraft 2100 two-barrel. Cooling system capacity: 14 qts. Crankcase capacity less filter: 4 qts. Exhaust system: Single with crossover. Engine code: H.

Hornet, Hornet SST, SC/360, Javelin, Javelin SST, Javelin AMX, Matador, Ambassador DPL, Ambassador SST, Ambassador Brougham V-8

V-8. Overhead valves. Cast-iron block. Displacement: 360 cid. Bore & stroke: 4.08 x 3.44 in. Compression ratio: 8.5:1. Advertised hp: 245 at 4800 rpm. Advertised torque: 365 lbs.-ft. at 2600 rpm. Recommended idle speed: (Manual transmission) 750 rpm; (Automatic transmission) 650 rpm. Five main bearings. Hydraulic valve lifters. Carburetor: Motorcraft 2100 two-barrel. Cooling system capacity: 14 qts. Crankcase capacity less filter: 4 qts. Exhaust system: Single with crossover. Engine code: N.

For 1970, the AMX 390 V-8 was made more powerful and came in two versions. A formidable 325-hp (8.33 hp per cu. in.) was produced by the Code X "standard" version. The limited-edition Rebel "Machine" received a Code Y variation that produced 340 hp (.872 per cube). Intake and exhaust manifold modifications accounted for the 15 extra horses. Special high-po induction kits could be ordered from dealers. Options included an alloy high-riser with three- or four-barrel Holley or a cross-ram manifold that accommodated two Holley four-barrels.

Optional Hornet, Hornet SST, SC/360, Javelin, Javelin SST, Javelin AMX, Matador, Ambassador DPL, Ambassador SST, Ambassador Brougham V-8
V-8. Overhead valves. Cast-iron block. Displacement: 360 cid. Bore & stroke: 4.08 x 3.44 in. Compression ratio: 8.5:1. Advertised hp: 285 at 4800 rpm. Advertised torque: 330 lbs.-ft. at 5000 rpm. Recommended idle speed: (Manual transmission) 750 rpm; (Automatic transmission) 650 rpm. Five main bearings. Hydraulic valve lifters. Carburetor: Motorcraft: 4300. Cooling system capacity: 14 qts. Crankcase capacity less filter: 4 qts. Exhaust system: Single with crossover standard, duals optional (duals standard on SC/360 four-barrel. Engine code: P.

Optional Matador, Matador "Machine", Javelin, Javelin SST, Ambassador DPL, Ambassador SST and Ambassador Brougham V-8
V-8. Overhead valves. Cast-iron block. Displacement: 401 cid. Bore & stroke: 4.17 x 3.57 in. Compression ratio: 9.5:1. Advertised hp: 330 at 5000 rpm. Advertised torque: 430 lbs.-ft. at 3400 rpm. Recommended idle speed: (Manual transmission) 650 rpm; (Automatic transmission) 600 rpm. Five main bearings. Hydraulic valve lifters. Carburetor: Motorcraft 4300. Cooling system capacity: 14 qts. Crankcase capacity less filter: 4 qts. Exhaust system: Duals standard on all models with 401 engine. Engine code: Z.

1972 AMC

Gremlin, Hornet, Hornet SST, Javelin, Javelin SST, Javelin AMX, Matador, Ambassador SST and Ambassador Brougham V-8
V-8. Overhead valves. Cast-iron block. Displacement: 304 cid. Bore & stroke: 3.75 x 3.44 in. Compression ratio: 8.4:1. Net brake hp: 150 at 4200 rpm. Net torque: 245 lbs.-ft. at 2500 rpm. Recommended idle speed: (Manual transmission) 750 rpm; (Automatic transmission) 650 nationwide, 700 California. Five main bearings. Hydraulic valve lifters. Carburetor: Motorcraft 2100 two-barrel. Cooling system capacity: 14 qts. Crankcase capacity less filter: 4 qts. Exhaust system: Single with crossover. Engine code: H.

Optional Hornet, Hornet SST, Javelin SST, Javelin AMX, Matador, Ambassador SST and Ambassador Brougham V-8
V-8. Overhead valves. Cast-iron block. Displacement: 360 cid. Bore & stroke: 4.08 x 3.44 in. Compression ratio: 8.5:1. Net hp: 175 at 4000 rpm. Net torque: 285 lbs.-ft. at 2400 rpm. Recommended idle speed: (Manual and Automatic transmissions) 700 rpm. Five main bearings. Hydraulic valve lifters. Carburetor: Motorcraft 2100. Cooling system capacity: 14 qts. Crankcase capacity less filter: 4 qts. Exhaust system: Single with crossover. Engine code: N.

Optional Javelin SST, Matador, Ambassador SST and Ambassador Brougham V-8
V-8. Overhead valves. Cast-iron block. Displacement: 360 cid. Bore & stroke: 4.08 x 3.44 in. Compression ratio: 8.5:1. Net hp: 195 at 4400 rpm. Net torque: 295 lbs.-ft. at 2900 rpm. Recommended idle speed: (Manual transmission) 750 rpm; (Automatic transmission) 700 rpm. Five main bearings. Hydraulic valve lifters. Carburetor: Motorcraft 4300. Cooling system capacity: 13 qts. Crankcase capacity less filter: 4 qts. Exhaust system: Single with crossover. Engine code: P.

Optional Javelin SST, Matador, Ambassador SST and Ambassador Brougham V-8
V-8. Overhead valves. Cast-iron block. Displacement: 360 cid. Bore & stroke: 4.08 x 3.44 in. Compression ratio: 8.5:1. Net hp: 220 at 4400 rpm. Net torque: 315 lbs.-ft. at 3100 rpm. Recommended idle speed: (Manual transmission) 750 rpm; (Automatic transmission) 700 rpm. Five main bearings. Hydraulic valve lifters. Carburetor: Motorcraft 4300. Cooling system capacity: 13 qts. Crankcase capacity less filter: 4 qts. Exhaust system: Duals. Engine code: P.

Optional Javelin SST, Matador, Ambassador SST and Ambassador Brougham V-8
V-8. Overhead valves. Cast-iron block. Displacement: 401 cid. Bore & stroke: 4.17 x 3.68 in. Compression ratio: 8.5:1. Net hp: 255 at 4600 rpm. Net torque: 345 lbs.-ft. at 3300 rpm. Recommended idle speed: (Manual transmission) 750 rpm; (Automatic transmission) 650 rpm nationwide, 700 rpm California. Five main bearings. Hydraulic valve lifters. Carburetor: Motorcraft 4300. Cooling system capacity: 13 qts. Crankcase capacity less filter: 4 qts. Exhaust system: Duals standard on all models with 401. Engine code: Z.

1973 AMC

Hornet, Matador, Ambassador, Gremlin, Javelin base V-8
Overhead valves. Cast-iron block. Bore & stroke: 3.75 x 3.44 in. Displacement: 304 cid. Compression ratio: 8.4:1. Net brake hp: 150 at 4200 rpm. Net torque: 245 lbs.-ft. at 2500 rpm. Five main bearings. Hydraulic valve lifters. Crankcase capacity less filter: 4. Cooling system capacity (with heater): 14. Exhaust system: Single with crossover. Carburetor: AMC 2DA2-2100D two-barrel. Engine code: H.

Hornet, Javelin, Matador, Ambassador optional V-8
Overhead valves. Cast-iron block. Bore & stroke: 4.08 x 3.44 in. Displacement: 360 cid. Compression ratio: 8.5:1. Net hp: 175 at 4000 rpm. Net torque: 285 lbs.-ft. at 2400 rpm. Five main bearings. Hydraulic valve lifters. Crankcase capacity less filter: 4. Cooling system capacity (with heater): 13. Exhaust system: Single with crossover. Carburetor: AMC 2DA2-2100D two-barrel. Engine code: N.

Javelin, Matador, Ambassador optional V-8
Overhead valves. Cast-iron block. Bore & stroke: 4.08 x 3.44 in. Displacement: 360 cid. Compression ratio: 8.5:1. Net hp: 195 at 4400 rpm. Net torque: 295 lbs.-ft. at 2900 rpm. Five main bearings. Hydraulic valve lifters. Crankcase capacity less filter: 4. Cooling system capacity (with heater): 13. Exhaust system: Single with crossover. Carburetor: Four-barrel. Engine code: P.

Javelin, Matador, Ambassador optional V-8
Overhead valves. Cast-iron block. Bore & stroke: 4.08 x 3.44 in. Displacement: 360 cid. Compression ratio: 8.5:1. Net hp: 220 at 4400 rpm. Net torque: 315 lbs.-ft. at 3100 rpm. Five main bearings. Hydraulic valve lifters. Crankcase capacity less filter: 4. Cooling system capacity (with heater): 13. Exhaust system: Single with crossover. Carburetor: Four-barrel.

Javelin, Matador, Ambassador optional V-8
Overhead valves. Cast-iron block. Bore & stroke: 4.17 x 3.68 in. Displacement: 401 cid. Compression ratio: 8.5:1. Net hp: 255 at 4600 rpm. Net torque: 345 lbs.-ft. at 3300 rpm. Five main bearings. Hydraulic valve lifters. Crankcase capacity less filter: 4. Cooling system capacity (with heater): 13. Exhaust system: Single with crossover. Carburetor: Four-barrel. Engine code: Z.

1974 AMC

Hornet, Matador, Ambassador, Gremlin, Javelin base V-8
Overhead valves. Cast-iron block. Bore & stroke: 3.75 x 3.44 in. Displacement: 304 cid. Compression ratio: 8.4:1. Net brake hp: 150 at 4200 rpm. Net torque: 245 lbs.-ft. at 2500 rpm. Five main bearings. Hydraulic valve lifters. Crankcase capacity less filter: 4. Cooling system capacity with heater: (Hornet, Gremlin) 16 qt.; (Matador, Ambassador) 16.5 qt.; (Matador coupe) 18.5 qt.; (Javelin) 15.5 qt.. Exhaust system: Single with crossover. Carburetor: AMC 2DA2-2100D two-barrel. Engine code: H.

Hornet, Javelin, Matador, Ambassador optional V-8
Overhead valves. Cast-iron block. Bore & stroke: 4.08 x 3.44 in. Displacement: 360 cid. Compression ratio: 8.5:1. Net hp: 175 at 4000 rpm. Net torque: 285 lbs.-ft. at 2400 rpm. Five main bearings. Hydraulic valve lifters. Crankcase capacity less filter: 4. Cooling system capacity with heater: (Hornet, Javelin) 15 qt.; (Matador, Ambassador) 15.5 qt.; (Matador coupe) 17.54 qt. Exhaust system: Single with crossover. Carburetor: AMC 2DA2-2100D two-barrel. Engine code: N.

Javelin, Matador, Ambassador optional V-8
Overhead valves. Cast-iron block. Bore & stroke: 4.08 x 3.44 in. Displacement: 360 cid. Compression ratio: 8.5:1. Net hp: 195 at 4400 rpm. Net torque: 295 lbs.-ft. at 2900 rpm. Five main bearings. Hydraulic valve lifters. Crankcase capacity less filter: 4. Cooling system capacity with heater: (Hornet, Javelin) 15 qt.; (Matador, Ambassador) 15.5 qt.; (Matador coupe) 17.54 qt. Exhaust system: Single with crossover. Carburetor: Four-barrel. Engine code: P.

Javelin, Matador, Ambassador optional V-8
Overhead valves. Cast-iron block. Bore & stroke: 4.17 x 3.68 in. Displacement: 401 cid. Compression ratio: 8.5:1. Net hp: 235 at 4600 rpm. Net torque: 335 lbs.-ft. at 3200 rpm. Five main bearings. Hydraulic valve lifters. Crankcase capacity less filter: 4. Cooling system capacity

with heater: (Hornet, Javelin) 15 qt.; (Matador, Ambassador) 15.5 qt.; (Matador coupe) 17.54 qt. Exhaust system: Single with crossover. Carburetor: Four-barrel. Engine code: Z.

1975 AMC

Gremlin, Hornet, Matador base V-8
Overhead valves. Cast-iron block. Bore & stroke: 3.75 x 3.44 in. Displacement: 304 cid. Compression ratio: 8.4:1. Net brake hp: 150 at 4200 rpm. Net torque: 245 lbs.-ft. at 2500 rpm. Five main bearings. Hydraulic valve lifters. Crankcase capacity less filter: 4. Cooling system capacity with heater: (Hornet, Gremlin) 16 qt.; (Matador) 16.5 qt.; (Matador coupe) 18.5 qt. Exhaust system: Single with crossover. Carburetor: AMC 2DA2-2100D two-barrel. Engine code: H.

Matador optional V-8
Overhead valves. Cast-iron block. Bore & stroke: 4.08 x 3.44 in. Displacement: 360 cid. Compression ratio: 8.25:1. Net hp: 175 at 4000 rpm. Net torque: 285 lbs.-ft. at 2400 rpm. Five main bearings. Hydraulic valve lifters. Crankcase capacity less filter: 4. Cooling system capacity with heater: (Sedan, Wagon) 15.5 qt.; (Coupe) 17.5 qt. Exhaust system: Single with crossover. Carburetor: AMC 2DA2-2100D two-barrel. Engine code: N.

Matador optional V-8
Overhead valves. Cast-iron block. Bore & stroke: 4.08 x 3.44 in. Displacement: 360 cid. Compression ratio: 8.25:1. Net hp: 195 at 4400 rpm. Net torque: 295 lbs.-ft. at 2900 rpm. Five main bearings. Hydraulic valve lifters. Crankcase capacity less filter: 4. Cooling system capacity with heater: (Sedan, Wagon) 15.5 qt.; (Coupe) 17.5 qt. Exhaust system: Single with crossover. Carburetor: Four-barrel. Engine code: P.

Matador optional V-8
Overhead valves. Cast-iron block. Bore & stroke: 4.08 x 3.44 in. Displacement: 360 cid. Compression ratio: 8.25:1. Net hp: 220 at 4400 rpm. Net torque: 315 lbs.-ft. at 3100 rpm. Five main bearings. Hydraulic valve lifters. Crankcase capacity less filter: 4. Cooling system capacity with heater: (Sedan, Wagon) 15.5 qt.; (Coupe) 17.5 qt. Exhaust system: Dual. Carburetor: Four-barrel. Engine code: Z.

1976 AMC

Standard Matador Station Wagon V-8; Optional Gremlin, Hornet and other Matadors
Overhead valves. Cast-iron block. Bore & stroke: 3.75 x 3.44 in. Displacement: 304 cid (5.0 liters). Compression ratio: 8.25:1. Net brake hp: 120 at 3200 rpm. Net torque: 220 lbs.-ft. at 1600 rpm. Five main bearings. Hydraulic valve lifters. Crankcase capacity less filter: 4. Cooling system capacity with heater: (Gremlin) 21 qt.; (Hornet) 16 qt.; (Matador Sedan, Wagon) 16.5 qt.; (Matador coupe) 18.5 qt. Exhaust system: Single with crossover. Carburetor: Motorcraft 2100 two-barrel. Engine code: H.

Optional Matador V-8
Overhead valves. Cast-iron block. Bore & stroke: 4.08 x 3.44 in. Displacement: 360 cid (5.9 liter). Compression ratio: 8.25:1. Net hp: 140 at 3200 rpm. Net torque: 260 lbs.-ft. at 1600 rpm. Five main bearings. Hydraulic valve lifters. Crankcase capacity less filter: 4. Cooling system capacity with heater: (Matador Sedan, Wagon) 15.5 qt.; (Matador coupe) 17.5 qt. Exhaust system: Single with crossover. Carburetor: Motorcraft 2100 two-barrel. Engine code: N.

Optional Matador High-Performance V-8
Overhead valves. Cast-iron block. Bore & stroke: 4.08 x 3.44 in. Displacement: 360 cid (5.9 liter). Compression ratio: 8.25:1. Net hp: 180 at 3600 rpm. Net torque: 280 lbs.-ft. at 2800 rpm. Five main bearings. Hydraulic valve lifters. Crankcase capacity less filter: 4. Cooling system capacity with heater: (Matador Sedan, Wagon) 15.5 qt.; (Matador coupe) 17.5 qt. Exhaust system: Dual. Carburetor: Motorcraft 4350 four-barrel. Engine code: P.

Optional Matador Police Cars
Overhead valves. Cast-iron block. Bore & stroke: 4.17 x 3.68 in. Displacement: 401 cid. Compression ratio: 8.5:1. Net hp: 215 at 4200 rpm. Net torque: 320 lbs.-ft. at 2800. rpm. Five main bearings. Hydraulic valve lifters. Crankcase capacity less filter: 4. Cooling system capacity with heater: (Matador Sedan, Wagon) 15.5 qt.; (Matador coupe) 17.5 qt. Exhaust system: Dual. Carburetor: Motorcraft 4350 four-barrel. Engine code: Z.

1977 AMC

Matador base V-8 (station wagon)
Overhead valves. Cast-iron block. Bore & stroke: 3.75 x 3.44 in. Displacement: 304 cid (5.0 liters). Compression ratio: 8.4:1. Net brake hp: 121 at 3450 rpm. Net torque: 219 lbs.-ft. at 2000 rpm. Five main bearings. Hydraulic valve lifters. Crankcase capacity less filter: 4. Cooling system capacity with heater: (Hornet) 16 qt.; (Matador) 16.5 qt. Exhaust system: Single with crossover. Carburetor: Motorcraft 2100 two-barrel. Engine code: H.

Matador base V-8 (except station wagon)
Overhead valves. Cast-iron block. Bore & stroke: 3.75 x 3.44 in. Displacement: 304 cid (5.0 liters). Compression ratio: 8.4:1. Net brake hp: 126 at 3600 rpm. Net torque: 219 lbs.-ft. at 2000 rpm. Five main bearings. Hydraulic valve lifters. Crankcase capacity less filter: 4. Cooling system capacity with heater: (Matador Sedan) 16.5 qt.; (Matador coupe) 18.5 qt. Exhaust system: Single with crossover. Carburetor: Motorcraft 2100 two-barrel. Engine code: H.

Matador optional V-8
Overhead valves. Cast-iron block. Bore & stroke: 4.08 x 3.44 in. Displacement: 360 cid (5.9 liter). Compression ratio: 8.25:1. Net brake hp: 129 at 3700 rpm. Net torque: 245 lbs.-ft. at 1600 rpm. Five main bearings. Hydraulic valve lifters. Crankcase capacity less filter: 4. Cooling system capacity with heater: (Matador Sedan, Wagon) 16.5 qt.; (Matador coupe) 18.5 qt. Exhaust system: Single with crossover. Carburetor: Motorcraft 2100 two-barrel. Engine code: N.

1978 AMC

Gremlin, Pacer, Concord optional V-8
Overhead valves. Cast-iron block. Displacement: 304 cid (5.0 liters). Bore & stroke: 3.75 x 3.44 in. Compression ratio: 8.4:1. Net brake hp: 30 at 3200 rpm. Net torque: 238 lbs.-ft. at 2000 rpm. Five main bearings. Hydraulic valve lifters. Crankcase capacity less filter: 4. Cooling system capacity: Unknown. Exhaust system: Single with crossover. Carburetor: Motorcraft 2100 two-barrel. Engine code: H.

Matador station wagon base V-8; Matador sedan and coupe optional V-8
Overhead valves. Cast-iron block. Displacement: 360 cid (5.9 liter). Bore & stroke: 4.08 x 3.44 in. Compression ratio: 8.25:1. Net hp: 140 at 3350 rpm. Net torque: 278 lbs.-ft. at 2000 rpm. Five main bearings. Hydraulic valve lifters. Crankcase capacity less filter: 4. Cooling system capacity: Unknown. Exhaust system: Single with crossover. Carburetor: Motorcraft 2100 two-barrel. Engine code: N.

Selected AMC V-8 Tune-Up Specs

Year	Model	Spark Plugs Type	Spark Plugs Gap	Point Gap	Deg Cam Angle	Timing	Idle Speed	Compression Pressure	Battery (Delco)
Hudson									
1955	Hornet V-8	AL-AG5	.035	.017	Note 1	5B	550	140	Group 1
1956	Hornet V-8	AL-AL7	.035	.015-.017	Note 1	5B	550	140	11HS/2SM
1957	Hornet V-8	AL-AL7	.035	.015-.017	30	5B	550	140	11HS/2SM

Note 1: Distributor has two sets of breaker points. Total dwell 34 degrees, single set 27 degrees.
Note 2: Idle speed 550 rpm with synchromesh, 450 rpm with Ultramatic, 425 rpm with Hydra-Matic

Nash

Year	Model	Spark Plugs Type	Spark Plugs Gap	Point Gap	Deg Cam Angle	Timing	Idle Speed	Compression Pressure	Battery (Delco)
1955	Ambass. 80	AL-AG5	.035	.017	36	5B	550 rpm	140	Group 1
1956	Ambass. Spl. 50	AL-AL7	.035	.016	30	5B	550 rpm	150	Group 1
	Ambass. 80	CH-N18	.036	.016	31	5B	550 rpm	150	Group 1
1957	Ambass. 80	AL-AL7	.035	.016	30	5B	550 rpm	150	11HS/2SM
1958	5820	AL-AL7	.035	.016	28-32	5B	500 rpm	150	11HS-60
	5880	AL-AL7	.035	.016	28-32	5B	500 rpm	150	11HS-60

Idle speed 500 rpm with synchromesh; 475 rpm with automatic transmission in drive

Year	Model	Spark Plugs Type	Spark Plugs Gap	Point Gap	Deg Cam Angle	Timing	Idle Speed	Compression Pressure	Battery (Delco)
1959	5920	AL-AL7	.035	.016	30	5B	500 rpm	150	11HS-60
	5980	AL-AL7	.035	.016	30	5B	500 rpm	150	11HS-60

Idle speed 500 rpm with synchromesh; 475 rpm with automatic transmission in drive

Year	Model	Spark Plugs Type	Spark Plugs Gap	Point Gap	Deg Cam Angle	Timing	Idle Speed	Compression Pressure	Battery (Delco)
1960	6020	AL-AL7	.035	.017	30	TC	550 rpm	130	11HS-60
	6080	AL-AL7	.035	.017	30	TC	550 rpm	130	11HS-60

Idle speed 550 rpm with synchromesh and OD; 475 rpm with automatic transmission; 500 rpm with A/C

Timing 5B with four-barrel carb.

Year	Model	Spark Plugs Type	Spark Plugs Gap	Point Gap	Deg Cam Angle	Timing	Idle Speed	Compression Pressure	Battery (Delco)
1961	6120	AL-AL7J	.035	.017	30	TC	550 rpm	140	11HS-60
	6180	AL-AL7J	.035	.017	30	TC	550 rpm	140	11HS-60

Idle speed 550 rpm with synchromesh and OD; 475 rpm with automatic transmission; 500 rpm with A/C

Timing 5B with four-barrel carb.

Year	Model	Spark Plugs Type	Spark Plugs Gap	Point Gap	Deg Cam Angle	Timing	Idle Speed	Compression Pressure	Battery (Delco)
1962	6280	H-10	.035	.017	30	TC	550 rpm	145	MH-60

Idle speed 550 rpm with synchromesh and OD; 475 rpm with automatic transmission

Timing 5B with four-barrel carb.; 5B with two-barrel carb. and automatic trans.; 8B with premium fuel and two-barrel carb. And automatic trans; 3B with premium fuel and two-barrel carb. and OD

AMC V-8 Firing order

1-8-4-3-6-5-7-2

Cylinders numbered front to back. Right-hand (passenger side) bank numbered 2-4-6-8; left-hand (driver's side) bank numbered 1-3-5-7.

AMC's larger "Rambler 327" V-8 was available in a new two-barrel 250-hp version with a 8.7:1 compression ratio that was suitable for regular-fuel, as well as the previous 270-hp version. The two-barrel version developed .87 horsepower per cube. It weighed 605 pounds and had a 2.42 weight-to-horsepower ratio The four-barrel version was good for 0.82 horsepower per cubic inch. It tipped the scales at 610 pounds, giving a 2.26 weight-to-horsepower ratio.

BUICK
1953-1996

Famous for its "Fireball" valve-in-head straight eights, Buick got into the manufacture of V-8s starting in 1953. That original 322-cid V-8 became the predecessor of "big" Buick V-8s for years to come.

One of the design characteristics of the "nailhead" Buick V-8 was the pent-roof design of the pistons and "Fireball" combustion chambers with the spark plug at the top of the chamber and the valves at one side. In addition, each exhaust valve had its own individual port and branch, which created a free-flowing exhaust system.

The engine grew to 364 cid for 1957-1958 and reached the 401-cid level in top-of-the-line 1959 models. Beefing up the same basic block and overboring it for 1963 resulted in the 425-cid version, which was famous for its use in a sophisticated "hot rod" called the Riviera.

The big V-8 was re-engineered in 1967 to improve its breathing and reliability. This resulted in 400- and 430-cid power plants that made a great platform for the installation of factory-issued "Heavy-Duty Application" Stage I and Stage II performance kits. The hot ticket was the optional 430-cid dual-quad setup that produced 360 hp.

By 1970, Buick's 400-cid "Stage" V-8 was not keeping up with the muscle car competition and the 455-cid "Wedge" V-8 appeared on the (street) scene. The 455 engine oiled its rocker arms via the pushrods, as opposed to earlier designs in which the rockers were lubed by "weeping" oil through hollow rocker shaft passages. The 455 came in 350- and 360-hp versions and a few were fitted with Stage II kits including tunnel port heads.

The high-output 455s, which lasted through 1972, marked the peak of Buick V-8 development, at least in terms of muscle-car era performance. V-8 go-power went downhill from that point, although Buick would later bring the V-6 type engine to one of its highest forms. The big 455 lasted through 1976, but in a detuned version best-suited for moving big sedans, heavy ragtops, and kid-packed station wagons.

Smaller Buick V-8s were first offered in 1954, when the Special got a standard 264-cid engine, also offered in 1955. A later generation of small Buick V-8s evolved from the design of the innovative all-aluminum 215-cid V-8 introduced in 1961. Buick encountered production problems making the aluminum blocks, but the cast aluminum heads were retained and survived through 1963. By 1964, both the blocks and heads were Cast-iron and the displacement was up to 300 cid. A little stroking with no change in bore turned the 300-cid V-8 into the 340-cid V-8, with the two engines sharing many components.

In 1968, a 350-cid V-8 became Buick's latest small V-8. It lasted until 1981 and should not be confused with 350-cid Chevy V-8s used in later Buicks. Beginning in 1975—with a 260-cid Olds V-8—Buicks using engines sourced from other GM divisions were seen. These, of course, were not true Buick engines, nor was the Oldsmobile 350-cid diesel V-8 that bowed in 1980.

One of the interesting V-8 developments at Buick Div. in later years was the use of the "Corvette-type" LT1 Chevy V-8 in Roadmaster sedans and Estate Wagons, starting in 1994. The Buick version was similar, but not identical to, the Corvette motor. When it disappeared in 1996, so did the V-8-powered Buick. However, there are rumors of a return.

1953 BUICK

Series 50 V-8 (without Dynaflow)
Overhead valves. Cast-iron block and heads. Bore & stroke: 4.00 x 3.20 in. Displacement: 322 cid. Compression ratio: 8.0:1. Brake hp: 164 at 4000 rpm. Taxable hp: 51.2. Torque: 286 lbs.-ft. at 2200. Hydraulic valve lifters. Five main bearings. Crankcase capacity: 6 qt. (add 1 qt. if new filter installed). Cooling system capacity: 16 qt. (add 2 qt. for heater). Carburetor: Stromberg AAVB-26 two-barrel. Starting engine serial number: V-2415-5.

Series 50 V-8 (with Dynaflow)
Overhead valves. Cast-iron block and heads. Bore & stroke: 4.00 x 3.20 in. Displacement: 322 cid. Compression ratio: 8.5:1. Brake hp: 170 at 4000 rpm. Taxable hp: 51.2. Torque: 292 lbs.-ft. at 2200. Hydraulic valve lifters. Five main bearings. Crankcase capacity: 6 qt. (add 1 qt. if new filter installed). Cooling system capacity: 16 qt. (add 2 qt. for heater). Carburetor: Stromberg AAVB-26 two-barrel. Starting engine serial number: V-2415-5.

Series 70 V-8
Overhead valves. Cast-iron block and heads. Bore & stroke: 4.00 x 3.20 in. Displacement: 322 cid. Compression ratio: 8.5:1. Brake hp: 188 at 4000 rpm. Taxable hp: 51.2. Torque: 300 lbs.-ft. at 2400. Hydraulic valve lifters. Five main bearings. Crankcase capacity: 6 qt. (add 1 qt. if new filter installed). Cooling system capacity: 18 qt. (add 1 qt. for heater). Carburetor: Carter WCFB-996S four-barrel. Starting engine serial number: V-2001-7.

A F-263 Fireball Engine in a 1953 Buick Roadmaster.

1954 BUICK

Series 40 V-8 (without Dynaflow)
Overhead valves. Cast-iron block and heads. Bore & stroke: 3.625 x 3.20 in. Displacement: 264 cid. Compression ratio: 7.2:1. Brake hp: 143 at 4200 rpm. Taxable hp: 42.05. Torque: 228 lbs.-ft. at 2400. Hydraulic valve lifters. Five main bearings. Crankcase capacity: 6 qt. (add 1 qt. if new filter installed). Cooling system capacity: 16.5 qt. (add 1 qt. for heater). Carburetor: Stromberg AAVB-267 two-barrel. Starting engine number: 4.

Series 40 V-8 (with Dynaflow)
Overhead valves. Cast-iron block and heads. Bore & stroke: 3.625 x 3.20 in. Displacement: 264 cid. Compression ratio: 8.0:1. Brake hp: 150 at 4200 rpm. Taxable hp: 42.05. Torque: 240 lbs.-ft. at 2400. Hydraulic valve lifters. Five main bearings. Crankcase capacity: 6 qt. (add 1 qt. if new filter installed). Cooling system capacity: 18.5 qt. (add 1 qt. for heater). Carburetor: Stromberg AAVB-267 two-barrel. Starting engine number: 4.

Series 50 V-8 (without Dynaflow)
Overhead valves. Cast-iron block and heads. Bore & stroke: 4.00 x 3.20 in. Displacement: 322 cid. Compression ratio: 8.0:1. Brake hp: 177 at 4100 rpm. Taxable hp: 51.2. Torque: 295 lbs.-ft at 2000. Hydraulic valve lifters. Five main bearings. Crankcase capacity: 6 qt. (add 1 qt. if new filter installed). Cooling system capacity: 16.5 qt. (add 1 qt. for heater). Carburetor: Stromberg AAVB-267 two-barrel. Starting engine number: 5.

Series 50 V-8 (with Dynaflow)
Overhead valves. Cast-iron block and heads. Bore & stroke: 4.00 x 3.20 in. Displacement: 322 cid. Compression ratio: 8.5:1. Brake hp: 182 at 4100 rpm. Taxable hp: 51.2. Torque: 300 lbs.-ft. at 2000. Hydraulic valve lifters. Five main bearings. Crankcase capacity: 6 qt. (add 1 qt. if new filter installed). Cooling system capacity: 18.5 qt. (add 1 qt. for heater). Carburetor: Stromberg AAVB-267 two-barrel. Starting engine number: 5.

Series 60 V-8 (without Dynaflow)
Overhead valves. Cast-iron block and heads. Bore & stroke: 4.00 x 3.20 in. Displacement: 322 cid. Compression ratio: 8.0:1. Brake hp: 195 at 4100 rpm. Taxable hp: 51.2. Torque: 302 lbs.-ft. at 2400. Hydraulic valve lifters. Five main bearings. Crankcase capacity: 6 qt. (add 1 qt. if new filter installed). Cooling system capacity: 16.5 qt. (add 1 qt. for heater). Carburetor: Stromberg AUVB-267 four-barrel. Starting engine number: 6.

Series 60 V-8 (with Dynaflow)
Overhead valves. Cast-iron block and heads. Bore & stroke: 4.00 x 3.20 in. Displacement: 322 cid. Compression ratio: 8.5:1. Brake hp: 200 at 4100 rpm. Taxable hp: 51.2. Torque: 309 lbs.-ft. at 2400. Hydraulic valve lifters. Five main bearings. Crankcase capacity: 6 qt. (add 1 qt. if new filter installed). Cooling system capacity: 18.5 qt. (add 1 qt. for heater). Carburetor: Stromberg AUVB-267 four-barrel. Starting engine number: 6.

Series 70 V-8
Overhead valves. Cast-iron block and heads. Bore & stroke: 4.00 x 3.20 in. Displacement: 322 cid. Compression ratio: 8.5:1. Brake hp: 200 at 4100 rpm. Taxable hp: 51.2. Torque: 309 lbs.-ft. at 2400. Hydraulic valve lifters. Five main bearings. Crankcase capacity: 6 qt. (add 1 qt. if new filter installed). Cooling system capacity: 18.5 qt. (add 1 qt. for heater). Carburetor: Stromberg AUVB-267 four-barrel. Starting engine number: 7.

1955 BUICK

Series 40 V-8 (without Dynaflow)
Overhead valves. Cast-iron block and heads. Bore & stroke: 3.625 x 3.20 in. Displacement: 264 cid. Compression ratio: 7.5:1. Taxable hp: 42.05. Hydraulic valve lifters. Five main bearings. Crankcase capacity: 6 qt. (add 1 qt. if new filter installed). Cooling system capacity: 16.5 qt. (add 1 qt. for heater). Carburetor: Stromberg AAVB-267 two-barrel. Starting engine number: 4.
NOTE: Brake horsepower and torque ratings for several 1955-1960 Buick V-8s do not appear in many reference sources.

Series 40 V-8 (with Dynaflow)
Overhead valves. Cast-iron block and heads. Bore & stroke: 3.625 x 3.20 in. Displacement: 264 cid. Compression ratio: 8.4:1. Brake hp: 188 at 4800 rpm. Taxable hp: 42.05. Torque: 256 lbs.-ft. at 2400. Hydraulic valve lifters. Five main bearings. Crankcase capacity: 6 qt. (add 1 qt. if new filter installed). Cooling system capacity: 18.5 qt. (add 1 qt. for heater). Carburetor: Stromberg AAVB-267 two-barrel. Starting engine number: 4.

Series 50, 60 V-8 (without Dynaflow)
Overhead valves. Cast-iron block and heads. Bore & stroke: 4.00 x 3.20 in. Displacement: 322 cid. Compression ratio: 8.4:1. Taxable hp: 51.2. Hydraulic valve lifters. Five main bearings. Crankcase capacity: 6 qt. (add 1 qt. if new filter installed). Cooling system capacity: 16.5 qt. (add 1 qt. for heater). Carburetor: Rochester 4GC four-barrel. Starting engine number: (Series 50 Super) 5; (Series 60 Century) 6.
NOTE: Brake horsepower and torque ratings for several 1955-1960 Buick V-8s do not appear in many reference sources.

Series 50, 60 V-8 (with Dynaflow)
Overhead valves. Cast-iron block and heads. Bore & stroke: 4.00 x 3.20 in. Displacement: 322 cid. Compression ratio: 9.0:1. Brake hp: 236 at 4600 rpm. Taxable hp: 51.2. Torque: 330 lbs.-ft. at 3000. Hydraulic valve lifters. Five main bearings. Crankcase capacity: 6 qt. (add 1 qt. if new filter installed). Cooling system capacity: 18.5 qt. (add 1 qt. for heater). Carburetor: Rochester 4GC four-barrel. Starting engine number: (Series 50 Super) 5; (Series 60 Century) 6.

Series 70 V-8 (with Dynaflow)
Overhead valves. Cast-iron block and heads. Bore & stroke: 4.00 x 3.20 in. Displacement: 322 cid. Compression ratio: 9.0:1. Brake hp: 236 at 4600 rpm. Taxable hp: 51.2. Torque: 330 lbs.-ft. at 3000. Hydraulic valve lifters. Five main bearings. Crankcase capacity: 6 qt. (add 1 qt. if new filter installed). Cooling system capacity: 18.5 qt. (add 1 qt. for heater). Carburetor: Rochester 4GC four-barrel. Starting engine number: 7.

1956 BUICK

Series 40 V-8 (without Dynaflow)
Overhead valves. Cast-iron block and heads. Bore & stroke: 4.00 x 3.20 in. Displacement: 322 cid. Compression ratio: 7.6:1. Taxable hp: 51.2. Hydraulic valve lifters. Five main bearings. Crankcase capacity: 6 qt. (add 1 qt. if new filter installed). Cooling system capacity: 18 qt. (add 1 qt. for heater). Carburetor: Carter WGD two-barrel. Starting engine serial number, all series: 1,460,023 (followed by a suffix indicating the series).
NOTE: Brake horsepower and torque ratings for several 1955-1960 Buick V-8s do not appear in many reference sources.

Series 40 V-8 (with Dynaflow)
Overhead valves. Cast-iron block and heads. Bore & stroke: 4.00 x 3.20 in. Displacement: 322 cid. Compression ratio: 8.9:1. Brake hp: 220 at 4400 rpm. Taxable hp: 51.2. Torque: 319 lbs.-ft. at 2400. Hydraulic valve lifters. Five main bearings. Crankcase capacity: 6 qt. (add 1 qt. if new filter installed). Cooling system capacity: 18 qt. (add 1 qt. for heater). Carburetor: Carter WGD two-barrel. Starting engine serial number, all series: 1,460,023 (followed by a suffix indicating the series).

Series 50, 60, 70 V-8
Overhead valves. Cast-iron block and heads. Bore & stroke: 4.00 x 3.20 in. Displacement: 322 cid. Compression ratio: 9.5:1. Brake hp: 255 at 4400 rpm. Taxable hp: 51.2. Torque: 341 lbs.-ft. at 3200. Hydraulic valve lifters. Five main bearings. Crankcase capacity: 6 qt. (add 1 qt. if new filter installed). Cooling system capacity: 18 qt. (add 1 qt. for heater). Carburetor: Carter WCFB four-barrel. Starting engine serial number, all series: 1,460,023 (followed by a suffix indicating the series).

1957 BUICK

Series 40 V-8 (without Dynaflow)
Overhead valves. Cast-iron block and heads. Bore & stroke: 4.125 x 3.40 in. Displacement: 364 cid. Compression ratio: 8.0:1. Taxable hp: 54.4. Hydraulic valve lifters. Five main bearings. Crankcase capacity: 5 qt. (add 1 qt. if new filter installed). Cooling system capacity: 16.5 qt. (add 1 qt. for heater). Carburetor: Carter WGD two-barrel. The engine number is the same as the car serial number and starts with a 4.
NOTE: Brake horsepower and torque ratings for several 1955-1960 Buick V-8s do not appear in many reference sources.

Series 40 V-8 (with Dynaflow)
Overhead valves. Cast-iron block and heads. Bore & stroke: 4.125 x 3.40 in. Displacement: 364 cid. Compression ratio: 9.5:1. Brake hp: 250 at 4400 rpm. Taxable hp: 54.4. Torque: 380 lbs.-ft. at 2400. Hydraulic valve lifters. Five main bearings. Crankcase capacity: 5 qt. (add 1 qt. if new filter installed). Cooling system capacity: 16.5 qt. (add 1 qt. for heater). Carburetor: Carter WGD two-barrel. The engine number is the same as the car serial number and starts with a 4.

Under the hood of this 1957 Buick Century convertible at "Antiques on the Bay" at St. Ignace, Michigan, is the 364-cid 300-hp V-8 available in Series 50, 60, and 70 Buicks that year. With a 10.0:1 compression ratio and a Carter AFB four-barrel carburetor it produced 400 foot pounds of torque at 3200 rpm.

Series 50, 60, 70 V-8
Overhead valves. Cast-iron block and heads. Bore & stroke: 4.125 x 3.40 in. Displacement: 364 cid. Compression ratio: 10.0:1. Brake hp: 300 at 4600 rpm. Taxable hp: 54.4. Torque: 400 lbs.-ft. at 3200. Hydraulic valve lifters. Five main bearings. Crankcase capacity: 5 qt. (add 1 qt. if new filter installed). Cooling system capacity: 16.5 qt. (add 1 qt. for heater). Carburetor: Carter AFB-2507 four-barrel. The engine number is the same as the car serial number and starts with a 5, 6 or 7 depending on series.

1958 BUICK

Series 40 V-8 (without Dynaflow)
Overhead valves. Cast-iron block and heads. Bore & stroke: 4.125 x 3.40 in. Displacement: 364 cid. Compression ratio: 8.0:1. Taxable hp: 54.4. Hydraulic valve lifters. Five main bearings. Crankcase capacity: 5 qt. (add 1 qt. if new filter installed). Cooling system capacity: 16.5 qt. (add 1 qt. for heater). Carburetor: Carter WGD two-barrel. The engine number is the same as the car serial number and starts with a 4, 5, 6, 7 or 8 depending on series.
NOTE: Brake horsepower and torque ratings for several 1955-1960 Buick V-8s do not appear in many reference sources.

Series 40 V-8 (with Dynaflow)
Overhead valves. Cast-iron block and heads. Bore & stroke: 4.125 x 3.40 in. Displacement: 364 cid. Compression ratio: 9.5:1. Brake hp: 250 at 4400 rpm. Taxable hp: 54.4. Torque: 380 lbs.-ft. at 2400. Hydraulic valve lifters. Five main bearings. Crankcase capacity: 5 qt. (add 1 qt. if new filter installed). Cooling system capacity: 16.5 qt. (add 1 qt. for heater). Carburetor: Carter WGD two-barrel. The engine number is the same as the car serial number and starts with a 4, 5, 6, 7, or 8 depending on series.

Series 50, 60 V-8
Overhead valves. Cast-iron block and heads. Bore & stroke: 4.125 x 3.40 in. Displacement: 364 cid. Compression ratio: 10.0:1. Brake hp: 300 at 4600 rpm. Taxable hp: 54.4. Torque: 400 lbs.-ft. at 3200. Hydraulic valve lifters. Five main bearings. Crankcase capacity: 5 qt. (add 1 qt. if new filter installed). Cooling system capacity: 16.5 qt. (add 1 qt. for heater). Carburetor: Carter AFB-2507 or Rochester 4GC four-barrel. The engine number is the same as the car serial number and starts with a 4, 5, 6, 7, or 8 depending on series.

Series 70, Limited 700 (Serial number code 8) V-8
Overhead valves. Cast-iron block and heads. Bore & stroke: 4.125 x 3.40 in. Displacement: 364 cid. Compression ratio: 10.0:1. Brake hp: 300 at 4600 rpm. Taxable hp: 54.4. Torque: 400 lbs.-ft. at 3200. Hydraulic valve lifters. Five main bearings. Crankcase capacity: 5 qt. (add 1 qt. if new filter installed). Cooling system capacity: 16.5 qt. (add 1 qt. for heater). Carburetor: Carter AFB-2507 or Rochester 4GC four-barrel. The engine number is the same as the car serial number and starts with a 4, 5, 6, 7, or 8 depending on series.

1959 BUICK

LeSabre Series 4400 base V-8 (without Dynaflow)
Overhead valves. Cast-iron block and heads. Bore & stroke: 4.125 x 3.40 in. Displacement: 364 cid. Compression ratio: 8.5:1. Taxable hp: 54.4. Hydraulic valve lifters. Five main bearings. Crankcase capacity: 4 qt. (add 1 qt. if new filter installed). Cooling system capacity: 16.5 qt. (add 1 qt. for heater). Carburetor: Stromberg two-barrel. The engine number is the same as the car serial number and starts with a 4, 6, 7, or 8 depending on series.
NOTE: Brake horsepower and torque ratings for several 1955-1960 Buick V-8s do not appear in many reference sources.

LeSabre Series 4400 base V-8 (with Dynaflow)
Overhead valves. Cast-iron block and heads. Bore & stroke: 4.125 x 3.40 in. Displacement: 364 cid. Compression ratio: 10.5:1. Brake hp: 250 at 4400 rpm. Taxable hp: 54.4. Torque: 384 lbs.-ft. at 2400. Hydraulic valve lifters. Five main bearings. Crankcase capacity: 4 qt. (add 1 qt. if new filter installed). Cooling system capacity: 16.5 qt. (add 1 qt. for heater). Carburetor: Stromberg two-barrel. The engine number is the same as the car serial number and starts with a 4, 6, 7, or 8 depending on series.

Invicta Series 4600, Electra Series 4700, Electra 225 Series 4800 base V-8
Overhead valves. Cast-iron block and heads. Bore & stroke: 4.188 x 3.64 in. Displacement: 401 cid. Compression ratio: 10.50:1. Brake hp: 325 at 4400 rpm. Taxable hp: 56.1. Torque: 445 lbs.-ft. at 2800. Hydraulic valve lifters. Five main bearings. Crankcase capacity: 4 qt. (add 1 qt. if new filter installed). Cooling system capacity: 16.5 qt. (add 1 qt. for heater). Carburetor: Carter AFB or Rochester 4GC 7010070 four-barrel. The engine number is the same as the car serial number and starts with a 4, 6, 7, or 8 depending on series.

1960 BUICK

LeSabre Series 4400 optional V-8 (manual transmission)
Overhead valves. Cast-iron block and heads. Bore & stroke: 4.125 x 3.40 in. Displacement: 364 cid. Compression ratio: 8.50:1. Brake hp: 210 at 4000 rpm. Taxable hp: 54.4. Torque: 340 lbs.-ft. at 2400 rpm. Hydraulic valve lifters. Five main bearings. Crankcase capacity: 4 qt. (add 1 qt. if new filter installed). Cooling system capacity: 16.65 qt. (add 2.5 qt. for heater). Carburetor: Stromberg WW or Carter WGD two-barrel. Engine identification code: 3G.

LeSabre Series 4400 optional V-8 (regular fuel V-8 with Dynaflow only)
Overhead valves. Cast-iron block and heads. Bore & stroke: 4.125 x 3.40 in. Displacement: 364 cid. Compression ratio: 9.0:1. Brake hp: 235 at 4400 rpm. Taxable hp: 54.4. Torque: 375 at 2400. Hydraulic valve lifters. Five main bearings. Crankcase capacity: 4 qt. (add 1 qt. if new filter installed). Cooling system capacity: 16.5 qt. (add 2.5 qt. for heater). Carburetor: Stromberg two-barrel. Carburetor: Carter WGD or Stromberg WW two-barrel. Engine identification code: L3G.
NOTE: The above V-8 was a no-cost option.

LeSabre Series 4400 standard V-8
Overhead valves. Cast-iron block and heads. Bore & stroke: 4.125 x 3.40 in. Displacement: 364 cid. Compression ratio: 10.25:1. Brake hp: 250 at 4400 rpm. Torque: 384 at 2400. Hydraulic valve lifters. Five main bearings. Crankcase capacity: 4 qt. (add 1 qt. if new filter installed). Cooling system capacity: 16.5 qt. (add 2.5 qt. for heater). Carburetor: Stromberg two-barrel. Carburetor: Carter WGD or Stromberg WW two-barrel. Engine identification code: 3G.

LeSabre Series 4400 optional V-8
Overhead valves. Cast-iron block and heads. Bore & stroke: 4.125 x 3.40 in. Displacement: 364 cid. Compression ratio: 10.25:1. Brake hp: 300 at 4400 rpm. Torque: 405 at 2800. Hydraulic valve lifters. Five main bearings. Crankcase capacity: 4 qt. (add 1 qt. if new filter installed). Cooling system capacity: 16.5 qt. (add 2.5 qt. for heater). Carburetor: Stromberg two-barrel. Carburetor: Carter AFB or Rochester 4GC four-barrel. Dual exhausts. Engine identification code: 3G.

Invicta Series 4600, Electra Series 4700, Electra 225 Series 4800 base V-8
Overhead valves. Cast-iron block and heads. Bore & stroke: 4.188 x 3.64 in. Displacement: 401 cid. Compression ratio: 10.25:1. Brake hp: 325 at

4400 rpm. Taxable hp: 56.1. Torque: 445 lbs.-ft. at 2800. Hydraulic valve lifters. Five main bearings. Crankcase capacity: 4 qt. (add 1 qt. if new filter installed). Cooling system capacity: 16.5 qt. (add 2.5 qt. for heater). Carburetor: Carter AFB or Rochester 4GC four-barrel. Engine identification code: 4G.

1961 BUICK

Special base V-8
Overhead valves. Aluminum block and heads (*). Bore & stroke: 3.50 x 2.797 in. Displacement: 215 cid. Compression ratio: 8.80:1. Brake hp: 155 at 4600 rpm. Taxable hp: 39.2. Torque: 220 lbs.-ft. at 2400. Hydraulic valve lifters. Five main bearings. Crankcase capacity: 4 qt. (add 1 qt. if new filter installed). Cooling system capacity: 10.5 qt. (add 1.5 qt. for heater). Carburetor: Rochester 2GC two-barrel. Engine identification code: H.
NOTE: Due to production problems, the aluminum block was dropped almost immediately after its introduction, but the cast-aluminum heads were retained through 1963.

LeSabre base V-8 (manual transmission)
Overhead valves. Cast-iron block and heads. Bore & stroke: 4.125 x 3.40 in. Displacement: 364 cid. Compression ratio: 9.0:1. Brake hp: 235 at 4400 rpm. Taxable hp: 54.4. Torque: 375 lbs.-ft. at 2400. Hydraulic valve lifters. Five main bearings. Crankcase capacity: 4 qt. (add 1 qt. if new filter installed). Cooling system capacity: 17 qt. (add 1.5 qt. for heater). Carburetor: Stromberg two-barrel. Carburetor: Rochester 2GC or Stromberg WW two-barrel. Engine identification code: L3H.

LeSabre base V-8
Overhead valves. Cast-iron block and heads. Bore & stroke: 4.125 x 3.40 in. Displacement: 364 cid. Compression ratio: 10.25:1. Brake hp: 250 at 4400 rpm. Taxable hp: 54.4. Torque: 384 lbs.-ft. at 2400. Hydraulic valve lifters. Five main bearings. Crankcase capacity: 4 qt. (add 1 qt. if new filter installed). Cooling system capacity: 17 qt. (add 1.5 qt. for heater). Carburetor: Rochester 2GC or Stromberg WW two-barrel. Engine identification code: 3H.

Invicta, Electra, Electra 225 base V-8
Overhead valves. Cast-iron block and heads. Bore & stroke: 4.188 x 3.64 in. Displacement: 401 cid. Compression ratio: 10.25:1. Brake hp: 325 at 4400 rpm. Taxable hp: 56.1. Torque: 445 lbs.-ft. at 2800. Hydraulic valve lifters. Five main bearings. Crankcase capacity: 4 qt. (add 1 qt. if new filter installed). Cooling system capacity: 17 qt. (add 1.5 qt. for heater). Carburetor: Rochester 4GC or Carter AFB four-barrel. Engine identification code: 4H or L4H.

1962 BUICK

Special, Special Deluxe base V-8
Overhead valves. Cast-iron block and cast-aluminum heads. Bore & stroke: 3.50 x 2.797 in. Displacement: 215 cid. Compression ratio: 9.00:1. Brake hp: 155 at 4600 rpm. Taxable hp: 39.2. Torque: 220 lbs.-ft. at 2400. Hydraulic valve lifters. Five main bearings. Crankcase capacity: 4 qt. (add 1 qt. if new filter installed). Cooling system capacity: 11 qt. (add 1 qt. for heater). Carburetor: Rochester 2GC two-barrel. Engine identification code: I.

Special Skylark base V-8; Special, Special Deluxe optional V-8
Overhead valves. Cast-iron block and cast-aluminum heads. Bore & stroke: 3.50 x 2.797 in. Displacement: 215 cid. Compression ratio: 11.00:1. Brake hp: 190 at 4800 rpm. Taxable hp: 39.2. Torque: 235 lbs.-ft. at 3000. Hydraulic valve lifters. Five main bearings. Crankcase capacity: 4 qt. (add 1 qt. if new filter installed). Cooling system capacity: 12 qt. (add 1.5 qt. for heater). Carburetor: Rochester 4GC four-barrel. Engine identification code: HI.

LeSabre optional V-8
Overhead valves. Cast-iron block and heads. Bore & stroke: 4.188 x 3.64 in. Displacement: 401 cid. Compression ratio: 9.00:1. Brake hp: 265 at 4400 rpm. Taxable hp: 56.1. Torque: 412 lbs.-ft. at 2400. Hydraulic valve lifters. Five main bearings. Crankcase capacity: 4 qt. (add 1 qt. if new filter installed). Cooling system capacity: 17 qt. (add 1.5 qt. for heater). Carburetor: Rochester 2GC two-barrel. Engine identification code: L2I.

LeSabre base V-8
Overhead valves. Cast-iron block and heads. Bore & stroke: 4.188 x 3.64 in. Displacement: 401 cid. Compression ratio: 10.25:1. Brake hp: 280 at 4400 rpm. Taxable hp: 56.1. Torque: 424 lbs.-ft. at 2400. Hydraulic valve lifters. Five main bearings. Crankcase capacity: 4 qt. (add 1 qt. if new filter installed). Cooling system capacity: 17 qt. (add 1.5 qt. for heater). Carburetor: Rochester 2GC two-barrel. Engine identification code: 2I.

Invicta, Electra 225 base V-8; LeSabre optional V-8
Overhead valves. Cast-iron block and heads. Bore & stroke: 4.188 x 3.64 in. Displacement: 401 cid. Compression ratio: 10.25:1. Brake hp: 325 at 4400 rpm. Taxable hp: 56.1. Torque: 445 lbs.-ft. at 2800. Hydraulic valve lifters. Five main bearings. Crankcase capacity: 4 qt. (add 1 qt. if new filter installed). Cooling system capacity: 17 qt. (add 1.5 qt. for heater). Carburetor: Rochester 4GC four-barrel. Engine identification code: 4I.

1963 BUICK

Special, Special Deluxe base V-8
Overhead valves. Cast-iron block and cast-aluminum heads. Bore & stroke: 3.50 x 2.797 in. Displacement: 215 cid. Compression ratio: 9.00:1. Brake hp: 155 at 4600 rpm. Taxable hp: 39.2. Torque: 220 lbs.-ft. at 2400. Hydraulic valve lifters. Five main bearings. Crankcase capacity: 4 qt. (add 1 qt. if new filter installed). Cooling system capacity: 11 qt. (add 1 qt. for heater). Carburetor: Rochester 2GC two-barrel. Engine identification code: JM.

Special Skylark base V-8; Special, Special Deluxe optional V-8
Overhead valves. Cast-iron block and cast-aluminum heads. Bore & stroke: 3.50 x 2.797 in. Displacement: 215 cid. Compression ratio: 11.00:1. Brake hp: 200 at 5000 rpm. Taxable hp: 39.2. Torque: 240 lbs.-ft. at 3200. Hydraulic valve lifters. Five main bearings. Crankcase capacity: 4 qt. (add 1 qt. if new filter installed). Cooling system capacity: 12 qt. (add 1.5 qt. for heater). Carburetor: Rochester 4GC four-barrel. Engine identification code: JN.

LeSabre optional V-8
Overhead valves. Cast-iron block and heads. Bore & stroke: 4.188 x 3.64 in. Displacement: 401 cid. Compression ratio: 9.00:1. Brake hp: 265 at 4400 rpm. Taxable hp: 56.1. Torque: 412 lbs.-ft. at 2400. Hydraulic valve lifters. Five main bearings. Crankcase capacity: 4 qt. (add 1 qt. if new filter installed). Cooling system capacity: 17 qt. (add 1.5 qt. for heater). Carburetor: Rochester 2GC two-barrel. Engine identification code: JS.

LeSabre base V-8
Overhead valves. Cast-iron block and heads. Bore & stroke: 4.188 x 3.64 in. Displacement: 401 cid. Compression ratio: 10.25:1. Brake hp: 280 at 4400 rpm. Taxable hp: 56.1. Torque: 424 lbs.-ft. at 2400. Hydraulic valve lifters. Five main bearings. Crankcase capacity: 4 qt. (add 1 qt. if new filter installed). Cooling system capacity: 17 qt. (add 1.5 qt. for heater). Carburetor: Rochester 2GC two-barrel. Engine identification code: JR.

Invicta, Wildcat, Electra, Riviera base V-8; LeSabre optional V-8
Overhead valves. Cast-iron block and heads. Bore & stroke: 4.188 x 3.64 in. Displacement: 401 cid. Compression ratio: 10.25:1. Brake hp: 325 at 4400 rpm. Taxable hp: 56.1. Torque: 445 lbs.-ft. at 2800. Hydraulic valve lifters. Five main bearings. Crankcase capacity: 4 qt. (add 1 qt. if new filter installed). Cooling system capacity: 17 qt. (add 1.5 qt. for heater). Carburetor: Rochester 4GC four-barrel. Engine identification code: JT.

1964 BUICK

Special, Special Deluxe, LeSabre base V-8
Overhead valves. Cast-iron block and heads. Bore & stroke: 3.75 x 3.406 in. Displacement: 300 cid. Compression ratio: 9.00:1. Brake hp: 210 at 4600 rpm. Taxable hp: 45.0. Torque: 310 lbs.-ft. at 2400. Hydraulic valve lifters. Five main bearings. Crankcase capacity: 4 qt. (add 1 qt. if new filter installed). Cooling system capacity: 11 qt. (add 1 qt. for heater). Carburetor: Rochester 2GC two-barrel. Engine identification code: KL.

Special, Special Deluxe, Skylark, LeSabre optional V-8
Overhead valves. Cast-iron block and heads. Bore & stroke: 3.75 x 3.406 in. Displacement: 300 cid. Compression ratio: 11.00:1. Brake hp: 250 at 4800 rpm. Taxable hp: 45.0. Torque: 335 lbs.-ft. at 3000. Hydraulic valve lifters. Five main bearings. Crankcase capacity: 4 qt. (add 1 qt. if new filter installed). Cooling system capacity: 11 qt. (add 1 qt. for heater). Carburetor: Rochester 4GC four-barrel. Engine identification code: KP.

LeSabre station wagon, Wildcat, Electra 225 base V-8
Overhead valves. Cast-iron block and heads. Bore & stroke: 4.188 x 3.641 in. Displacement: 401 cid. Compression ratio: 10.25:1. Brake hp: 325 at 4400 rpm. Taxable hp: 56.1. Torque: 445 lbs.-ft. at 2800. Hydraulic valve lifters. Five main bearings. Crankcase capacity: 4 qt. (add 1 qt. if new filter installed). Cooling system capacity: 17 qt. (add 1.5 qt. for heater). Carburetor: Rochester 4GC or Carter AFB four-barrel. Engine identification code: KT.

Riviera base V-8; Wildcat, Electra 225 optional V-8
Overhead valves. Cast-iron block and heads. Bore & stroke: 4.312 x 3.641 in. Displacement: 425 cid. Compression ratio: 10.25:1. Brake hp: 340 at 4400 rpm. Taxable hp: 59.51. Torque: 465 lbs.-ft. at 2800. Hydraulic valve lifters. Five main bearings. Crankcase capacity: 4 qt. (add 1 qt. if new filter installed). Cooling system capacity: 17 qt. (add 1.5 qt. for heater). Carburetor: Carter AFB four-barrel. Engine identification code: KW.

Riviera, Wildcat, Electra 225 optional V-8
Overhead valves. Cast-iron block and heads. Bore & stroke: 4.312 x 3.641 in. Displacement: 425 cid. Compression ratio: 10.25:1. Brake hp: 360 at 4400 rpm. Taxable hp: 59.51. Torque: 465 lbs.-ft. at 2800. Hydraulic valve lifters. Five main bearings. Crankcase capacity: 4 qt. (add 1 qt. if new filter installed). Cooling system capacity: 17 qt. (add 1.5 qt. for heater). Carburetor: Two Carter AFB four-barrel. Engine identification code: KX.

The "Wildcat 465" V-8 featured overhead valves and cast-iron block and heads. It had a 4.312 x 3.641 inch bore and stroke. With a 10.25:1 compression ratio and a Carter AFB four-barrel carb, it cranked out 340 hp at 4400 rpm. (Karen O'Brien)

1965 BUICK

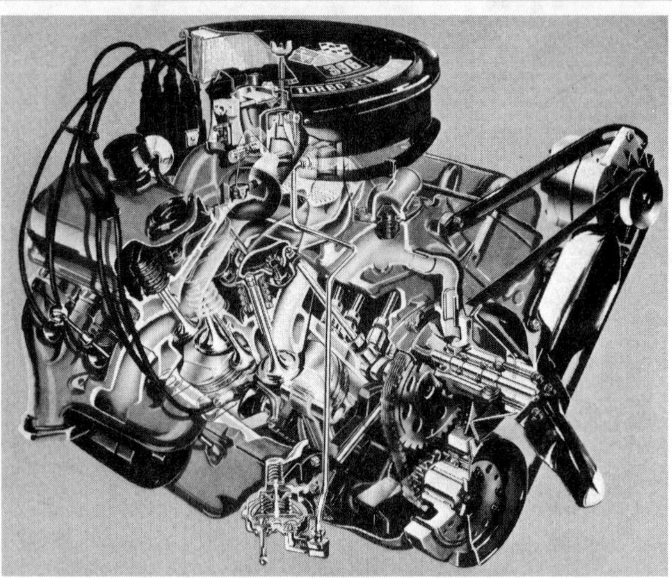

Sectional view of the big Buick V-8 of 1965. The big V-8 was available in 401- and 425-cid displacements, with both being essentially similar engines. The 425 had a larger bore diameter. The 401 came with a four-barrel carburetor, while the 425 was available with single or dual four-barrels. Weights of the three V-8s ranged from 640 to 660 pounds.

Special, Skylark, LeSabre base V-8
Overhead valves. Cast-iron block and heads. Bore & stroke: 3.75 x 3.406 in. Displacement: 300 cid. Compression ratio: 9.00:1. Brake hp: 210 at 4600 rpm. Taxable hp: 45.0. Torque: 310 lbs.-ft. at 2400. Hydraulic valve lifters. Five main bearings. Crankcase capacity: 4 qt. (add 1 qt. if new filter installed). Cooling system capacity: (Special) 11.75 qt.; (LeSabre) 11.5 qt. (add 1 qt. for heater). Carburetor: Rochester 2GC two-barrel. Engine identification code: IL.

Special, LeSabre optional V-8
Overhead valves. Cast-iron block and heads. Bore & stroke: 3.75 x 3.406 in. Displacement: 300 cid. Compression ratio: 11.00:1. Brake hp: 250 at 4800 rpm. Taxable hp: 45.0. Torque: 335 lbs.-ft. at 3000. Hydraulic valve lifters. Five main bearings. Crankcase capacity: 4 qt. (add 1 qt. if new filter installed). Cooling system capacity: 11.75 qt. (add 1 qt. for heater). Carburetor: Rochester 4GC four-barrel. Engine identification code: LP.

With higher cubic inches and horsepower ratings translating into added sales in 1964, Buick decided to use torque ratings to identify its V-8 engines and make them sound more powerful. This '64 Buick Electra 225 convertible has the "Wildcat 465" V-8, which was actually a 425-cid engine. It was standard in the Riviera and optional in the "Deuce-and-a-Quarter" ragtop. (Roger Huber)

The standard 401-cid V-8 in this 1965 Buick Wildcat four-door hardtop is being serviced by factory technicians using a period diagnostic machine. This V-8 had a 10.25:1 compression ratio. Brake horsepower was 325 at 4400 rpm.

Wildcat, Electra base V-8
Overhead valves. Cast-iron block and heads. Bore & stroke: 4.188 x 3.641 in. Displacement: 401 cid. Compression ratio: 10.25:1. Brake hp: 325 at 4400 rpm. Taxable hp: 56.1. Torque: 445 lbs.-ft. at 2800. Hydraulic valve lifters. Five main bearings. Crankcase capacity: 4 qt. (add 1 qt. if new filter installed). Cooling system capacity: 16.75 qt. (add 1.5 qt. for heater). Carburetor: Rochester 4GC or Carter AFB four-barrel. Engine identification code: LT. Called “400” V-8 in Skylark GS.

Riviera base V-8
Overhead valves. Cast-iron block and heads. Bore & stroke: 4.312 x 3.641 in. Displacement: 425 cid. Compression ratio: 10.25:1. Brake hp: 340 at 4400 rpm. Taxable hp: 59.51. Torque: 465 lbs.-ft. at 2800. Hydraulic valve lifters. Five main bearings. Crankcase capacity: 4 qt. (add 1 qt. if new filter installed). Cooling system capacity: 16.75 qt. (add 1.5 qt. for heater). Carburetor: Carter AFB four-barrel. Engine identification code: LW.

Riviera, Wildcat, Electra 225 optional V-8
Overhead valves. Cast-iron block and heads. Bore & stroke: 4.312 x 3.641 in. Displacement: 425 cid. Compression ratio: 10.25:1. Brake hp: 360 at 4400 rpm. Taxable hp: 59.51. Torque: 465 lbs.-ft. at 2800. Hydraulic valve lifters. Five main bearings. Crankcase capacity: 4 qt. (add 1 qt. if new filter installed). Cooling system capacity: 16.75 qt. (add 1.5 qt. for heater). Carburetor: Two Carter AFB four-barrel. Engine identification code: LX.

1966 BUICK

Special, Special Deluxe, Skylark base V-8
Overhead valves. Cast-iron block and heads. Bore & stroke: 3.75 x 3.406 in. Displacement: 300 cid. Compression ratio: 9.00:1. Brake hp: 210 at 4600 rpm. Taxable hp: 45.0. Torque: 310 lbs.-ft. at 2400. Hydraulic valve lifters. Five main bearings. Crankcase capacity: 4 qt. (add 1 qt. if new filter installed). Cooling system capacity: (Special) 11.75 qt.; (LeSabre) 11.5 qt. (add 1 qt. for heater). Carburetor: Rochester 2GC two-barrel. Engine identification code: ML.

Special Sport Wagon, LeSabre base V-8
Overhead valves. Cast-iron block and heads. Bore & stroke: 3.75 x 3.85 in. Displacement: 340 cid. Compression ratio: 9.00:1. Brake hp: 220 at 4400 rpm. Taxable hp: 45.0. Torque: 310 lbs.-ft. at 2400. Hydraulic valve lifters. Five main bearings. Crankcase capacity: 4 qt. (add 1 qt. if new filter installed). Cooling system capacity: (Special) 11.75 qt.; (LeSabre) 11.5 qt. (add 1 qt. for heater). Carburetor: Rochester 2GC two-barrel. Engine identification code: MA.

Special Deluxe, Skylark, Sport Wagon, LeSabre V-8
Overhead valves. Cast-iron block and heads. Bore & stroke: 3.75 x 3.85 in. Displacement: 340 cid. Compression ratio: 10.25:1. Brake hp: 260 at 4000 rpm. Taxable hp: 45.0. Torque: 365 lbs.-ft. at 2800. Hydraulic valve lifters. Five main bearings. Crankcase capacity: 4 qt. (add 1 qt. if new filter installed). Cooling system capacity: (Special) 11.75 qt.; (LeSabre) 11.5 qt. (add 1 qt. for heater). Carburetor: Carter AFB-4055 four-barrel. Engine identification code: MB.

Wildcat, Electra, Skylark Gran Sport base V-8
Overhead valves. Cast-iron block and heads. Bore & stroke: 4.188 x 3.641 in. Displacement: 401 cid. Compression ratio: 10.25:1. Brake hp: 325 at 4400 rpm. Taxable hp: 56.1. Torque: 445 lbs.-ft. at 2800. Hydraulic valve lifters. Five main bearings. Crankcase capacity: 4 qt. (add 1 qt. if new filter installed). Cooling system capacity: 16.75 qt. (add 1.5 qt. for heater). Carburetor: Carter AFB-4054 four-barrel. Engine identification code: MR or MT.

Wildcat, Electra, Riviera V-8
Overhead valves. Cast-iron block and heads. Bore & stroke: 4.312 x 3.641 in. Displacement: 425 cid. Compression ratio: 10.25:1. Brake hp 340 at 4400 rpm. Taxable hp: 59.51. Torque: 465 lbs.-ft. at 2800. Hydraulic valve lifters. Five main bearings. Crankcase capacity: 4 qt. (add 1 qt. if new filter installed). Cooling system capacity: 16.75 qt. (add 1.5 qt. for heater). Carburetor: Rochester 4GC four-barrel. Engine identification code: MW.

Wildcat, Electra, Riviera optional V-8
Overhead valves. Cast-iron block and heads. Bore & stroke: 4.312 x 3.641 in. Displacement: 425 cid. Compression ratio: 10.25:1. Brake hp 360 at 4400 rpm. Taxable hp: 59.51. Torque: 465 lbs.-ft. at 2800. Hydraulic valve lifters. Five main bearings. Crankcase capacity: 4 qt. (add 1 qt. if new filter installed). Cooling system capacity: 16.75 qt. (add 1.5 qt. for heater). Carburetor: Dual four-barrel.

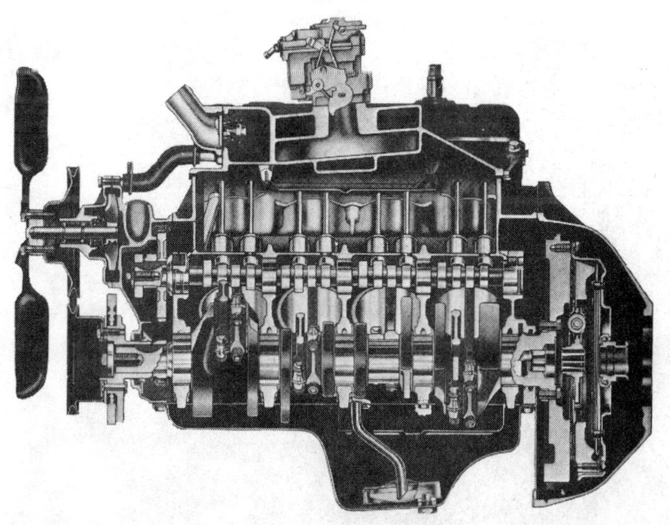

In 1966, the small-block Buick V-8 came in three options. Both the 300- and 340-cid engines shared a 3.75-in. bore but the stroke on the smaller block was 3.40 in., compared to 3.85 in. on the 340. A two-barrel carburetor was used on the smaller 210-hp V-8, while the 340-cid engine came in a 220-hp two-barrel version and a 260-hp four-barrel version. These engines tipped the scales at 470-485 pounds and were a lot lighter than the big V-8s.

NOTE: The above engine was a dealer option early in the model year and a factory option late in the model year.

1967 BUICK

The 400- and 430-cid big-block Buick V-8s were used in the GS 400, Wildcat, Electra, and Riviera models in 1967. Seen here without its air cleaner is the 400-cid four-barrel version, which produced 340 hp at 5000 rpm. This "nailhead" engine was in the same family tree as Buick's first V-8 issued in 1953. It weighed about 640 pounds.

Special, Special Deluxe, Skylark base V-8
Overhead valves. Cast-iron block and heads. Bore & stroke: 3.75 x 3.406 in. Displacement: 300 cid. Compression ratio: 9.00:1. Brake hp: 210 at 4600 rpm. Taxable hp: 45.0. Torque: 310 lbs.-ft. at 2400. Hydraulic valve lifters. Five main bearings. Crankcase capacity: 4 qt. (add 1 qt. if new filter installed). Cooling system capacity: 11.75 qt. (add 1 qt. for heater). Carburetor: Rochester 2GC two-barrel. Engine identification code: NL.

Skylark sedan, Sport Wagon, Skylark Custom, LeSabre base V-8
Overhead valves. Cast-iron block and heads. Bore & stroke: 3.75 x 3.85 in. Displacement: 340 cid. Compression ratio: 9.00:1. Brake hp: 220 at 4400 rpm. Taxable hp: 45.0. Torque: 340 lbs.-ft. at 2400. Hydraulic valve lifters. Five main bearings. Crankcase capacity: 4 qt. (add 1 qt. if new filter installed). Cooling system capacity: (Special) 11.75 qt.; (LeSabre) 12 qt. (add 1 qt. for heater). Carburetor: Rochester 2GC two-barrel. Engine identification code: NA.

This is a prototype version of the 1967 Buick 430-cid four-barrel V-8 with a single-snorkel style air cleaner. This more conventional air cleaner was similar or identical to the one used in production early in the model year.

Seen here with its exceptionally cool-looking late-in-the-year factory twin-scoop air cleaner is the 1967 Buick 430-cid V-8 with a four-barrel carburetor. This 650-pound V-8 produced 360 hp at 5000 rpm giving it a 1.91 weight-to-horsepower ratio. It generated 0.84 horsepower per cubic inch.

Special Deluxe, Skylark, Sport Wagon, Custom, LeSabre optional V-8
Overhead valves. Cast-iron block and heads. Bore & stroke: 3.75 x 3.85 in. Displacement: 340 cid. Compression ratio: 10.25:1. Brake hp: 260 at 4000 rpm. Taxable hp: 45.0. Torque: 365 lbs.-ft. at 2800. Hydraulic valve lifters. Five main bearings. Crankcase capacity: 4 qt. (add 1 qt. if new filter installed). Cooling system capacity: (Special) 11.75 qt.; (LeSabre) 12 qt. (add 1 qt. for heater). Carburetor: Carter AFB-4055 four-barrel. Engine identification code: NB.

GS 400 base V-8
Overhead valves. Cast-iron block and heads. Bore & stroke: 4.04 x 3.90 in. Displacement: 400 cid. Compression ratio: 10.25:1. Brake hp: 340 at 5000 rpm. Taxable hp: 51.9. Torque: 440 lbs.-ft. at 300. Hydraulic valve lifters. Five main bearings. Crankcase capacity: 4 qt. (add 1 qt. if new filter installed). Cooling system capacity: 17 qt. (add 1 qt. for heater). Carburetor: Rochester 4MV Quadra-Jet four-barrel. Engine identification code: NR.

Wildcat, Electra 225, Riviera GS base V-8
Overhead valves. Cast-iron block and heads. Bore & stroke: 4.188 x 3.90 in. Displacement: 430 cid. Compression ratio: 10.25:1. Brake hp: 360 at 5000 rpm. Taxable hp: 56.1. Torque: 475 lbs.-ft. at 3200. Hydraulic valve lifters. Five main bearings. Crankcase capacity: 4 qt. (add 1 qt. if new filter installed). Cooling system capacity: 17 qt. (add 1 qt. for heater). Carburetor: Rochester 4MV Quadra-Jet four-barrel. Engine identification codes: ND, MD.

1968 BUICK

NOTE: A heater was standard equipment on Buicks from 1968 on.

Special Deluxe, Skylark Custom, LeSabre base V-8
Overhead valves. Cast-iron block and heads. Bore & stroke: 3.80 x 3.85 in. Displacement: 350 cid. Compression ratio: 9.00:1. Brake hp: 230 at 4400 rpm. Taxable hp: 46.20. Torque: 350 lbs.-ft. at 2400. Hydraulic valve lifters. Five main bearings. Crankcase capacity: 4 qt. (add 1 qt. if new filter installed). Cooling system capacity with heater: (Skylark) 13.50 qt.; (LeSabre) 13.25 qt. Carburetor: Rochester 2GC two-barrel. Engine identification code: PO.

GS-350 base V-8; Special Deluxe, Skylark Custom, LeSabre optional V-8
Overhead valves. Cast-iron block and heads. Bore & stroke: 3.80 x 3.85 in. Displacement: 350 cid. Compression ratio: 10.25:1. Brake hp: 280 at 4600 rpm. Taxable hp: 46.20. Torque: 375 lbs.-ft. at 3200. Hydraulic valve lifters. Five main bearings. Crankcase capacity: 4 qt. (add 1 qt. if new filter installed). Cooling system capacity with heater: (Skylark) 13.50 qt. (LeSabre) 13.25 qt. Carburetor: Rochester 4MV Quadra-Jet four-barrel. Engine identification code: PP.

GS 400 base V-8; Sport Wagon optional V-8
Overhead valves. Cast-iron block and heads. Bore & stroke: 4.04 x 3.90 in. Displacement: 400 cid. Compression ratio: 10.25:1. Brake hp: 340 at 5000 rpm. Taxable hp: 52.23. Torque: 440 lbs.-ft. at 3200. Hydraulic valve lifters. Five main bearings. Crankcase capacity: 4 qt. (add 1 qt. if new filter installed). Cooling system capacity with heater: 16.75 qt. Carburetor: Rochester 4MV Quadra-Jet four-barrel. Engine identification code: PR.

Wildcat, Electra 225, Riviera GS base V-8
Overhead valves. Cast-iron block and heads. Bore & stroke: 4.188 x 3.90 in. Displacement: 430 cid. Compression ratio: 10.25:1. Brake hp: 360 at 5000 rpm. Taxable hp: 56.1. Torque: 475 lbs.-ft. at 3200. Hydraulic valve lifters. Five main bearings. Crankcase capacity: 4 qt. (add 1 qt. if new filter installed). Cooling system capacity with heater: 16.75 qt. Carburetor: Carter AFB-4054 four-barrel. Engine identification code: PD.

1969 BUICK

Special Deluxe, Skylark Custom, LeSabre base V-8
Overhead valves. Cast-iron block and heads. Bore & stroke: 3.80 x 3.85 in. Displacement: 350 cid. Compression ratio: 9.00:1. Brake hp: 230 at 4400 rpm. Taxable hp: 46.20. Torque: 350 lbs.-ft. at 2400. Hydraulic valve lifters. Five main bearings. Crankcase capacity: 4 qt. (add 1 qt. if new filter installed). Cooling system capacity with heater: (Skylark) 13.50 qt. (LeSabre) 13.25 qt. Carburetor: Rochester 2GC two-barrel. Engine identification code: RO.

GS-350 base V-8; Special Deluxe, Skylark Custom, LeSabre optional V-8
Overhead valves. Cast-iron block and heads. Bore & stroke: 3.80 x 3.85 in. Displacement: 350 cid. Compression ratio: 10.25:1. Brake hp: 280 at 4600 rpm. Taxable hp: 46.20. Torque: 375 lbs.-ft. at 3200. Hydraulic valve lifters. Five main bearings. Crankcase capacity: 4 qt. (add 1 qt. if new filter installed). Cooling system capacity with heater: (Skylark) 13.50 qt.; (LeSabre) 13.25 qt. Carburetor: Rochester 4MV Quadra-Jet four-barrel. Engine identification code: RP.

GS 400 base V-8; Sport Wagon optional V-8
Overhead valves. Cast-iron block and heads. Bore & stroke: 4.04 x 3.90 in. Displacement: 400 cid. Compression ratio: 10.25:1. Brake hp: 340 at 5000 rpm. Taxable hp: 52.23.Torque: 440 lbs.-ft. at 3200. Hydraulic valve lifters. Five main bearings. Crankcase capacity: 4 qt. (add 1 qt. if new filter installed). Cooling system capacity with heater: 16.75 qt. Carburetor: Rochester 4MV Quadra-Jet four-barrel. Engine identification code: RR.

The 340-cid small-block Buick V-8 was bored out to 350 cid in 1968 and was carried over into 1969 when it looked like this. This engine was actually in the same family tree as the 215-cid aluminum V-8 first offered in 1961. Production snafus caused Buick to quickly drop the aluminum block, but aluminum heads were retained through 1963.

Wildcat, Electra 225, Riviera GS base V-8
Overhead valves. Cast-iron block and heads. Bore & stroke: 4.188 x 3.90 in. Displacement: 430 cid. Compression ratio: 10.25:1. Brake hp: 360 at 5000 rpm. Taxable hp: 56.1. Torque: 475 lbs.-ft. at 3200. Hydraulic valve lifters. Five main bearings. Crankcase capacity: 4 qt. (add 1 qt. if new filter installed). Cooling system capacity with heater: 16.75 qt. Carburetor: Rochester 4MV Quadra-Jet four-barrel. Engine identification code: RD.

The 1969 Buick 430-cid four-barrel V-8 seen here was identical in appearance to the 400-cid engine. They varied only in bore size (4.040 in. to 4.1875 in.). Both engines were 33 in. long, 29 in. wide, and 24 in. tall.

1970 BUICK

Skylark 350, Skylark Custom, LeSabre, LeSabre Custom base V-8
Overhead valves. Cast-iron block and heads. Bore & stroke: 3.80 x 3.85 in. Displacement: 350 cid. Compression ratio: 9.00:1. Brake hp: 260 at 4600 rpm. Taxable hp: 46.20. Torque: 360 lbs.-ft. at 2600. Hydraulic valve lifters. Five main bearings. Crankcase capacity: 4 qt. (Add 1 qt. if new filter installed). Cooling system capacity with heater: (Skylark) 16.5 qt.; (LeSabre) 16.2 qt. Carburetor: Rochester 2GC two-barrel. Engine identification code: SO.

Skylark 350, Skylark 350 Custom, LeSabre, LeSabre Custom optional V-8
Overhead valves. Cast-iron block and heads. Bore & stroke: 3.80 x 3.85 in. Displacement: 350 cid. Compression ratio: 9.00:1. Brake hp: 285 at 4600 rpm. Taxable hp: 46.20. Torque: 375 lbs.-ft. at 3000. Hydraulic valve lifters. Five main bearings. Crankcase capacity: 4 qt. (Add 1 qt. if new filter installed). Cooling system capacity with heater: (Skylark) 16.5 qt.; (LeSabre) 16.2 qt. Carburetor: Rochester 4MV Quadra-Jet four-barrel. Engine identification code: SB.

GS base V-8
Overhead valves. Cast-iron block and heads. Bore & stroke: 3.80 x 3.85 in. Displacement: 350 cid. Compression ratio: 10.25:1. Brake hp: 315 at 4800 rpm. Taxable hp: 46.20. Torque: 410 lbs.-ft. at 3200. Hydraulic valve lifters. Five main bearings. Crankcase capacity: 4 qt. (Add 1 qt. if new filter installed). Cooling system capacity with heater: 16.5 qt. Carburetor: Rochester 4MV Quadra-Jet four-barrel. Engine identification code: SP.

GS optional Stage 1 V-8
Overhead valves. Cast-iron block and heads. Bore & stroke: 3.80 x 3.85 in. Displacement: 350 cid. Compression ratio: 10.25:1. Brake hp: 345 at 4800 rpm. Taxable hp: 46.20. Torque: 410 lbs.-ft. at 3200. Hydraulic valve lifters. Five main bearings. Crankcase capacity: 4 qt. (add 1 qt. if new filter installed). Cooling system capacity with heater: 19.2 qt. Carburetor: Rochester 4MV Quadra-Jet four-barrel. Dealer-installed kit including

high-lift camshaft, tubular push rods, heavy-duty valve springs and dampers, and a high-output oil pump. Engine identification code: SS.

GS 455, GSX base V-8
Overhead valves. Cast-iron block and heads. Bore & stroke: 4.31 x 3.90 in. Displacement: 455 cid. Compression ratio: 10.00:1. Brake hp: 350 at 4600 rpm. Taxable hp: 52.20. Torque: 510 lbs.-ft. at 2800. Hydraulic valve lifters. Five main bearings. Crankcase capacity: 4 qt. (add 1 qt. if new filter installed). Cooling system capacity with heater: 19.2 qt. Carburetor: Rochester 4MV Quadra-Jet four-barrel. Engine identification code: SR.

GS 455, GSX optional Stage 1 V-8
Overhead valves. Cast-iron block and heads. Bore & stroke: 4.31 x 3.90 in. Displacement: 455 cid. Compression ratio: 10.00:1. Brake hp: 360 at 4600 rpm. Taxable hp: 59.50. Torque: 510 lbs.-ft. at 2800. Hydraulic valve lifters. Five main bearings. Crankcase capacity: 4 qt. (add 1 qt. if new filter installed). Cooling system capacity with heater: 19.2 qt. Carburetor: Rochester 4MV Quadra-Jet four-barrel. Engine identification code: SS.

Buick's big "nailhead" V-8 had first appeared as a 322-cid motor in 1953. A half a decade later, it was bored out to 401 cid and it grew to 425 cid in 1963. The motor was re-engineered for model-year 1967 and then came in 400- and 430-cid versions. It finally grew to 455 cid in the 1970 configuration shown here. Versions with 350, 360, and 370 hp were offered.

Again in 1970, the 350-cid V-8 was Buick's small-block offering. However, this year's high-performance version of the engine gained 35 horsepower. It had a 10.25:1 compression ratio and a four-barrel carburetor good for 315 hp at 4600 rpm. The motor was only 28 in. long, 25 in. wide, and 21 in. high. The 315-hp edition (Code SP) produced .900 horsepower per cubic inch.

LeSabre 455, Wildcat Custom, Electa 225/225 Custom, Riviera, Estate Wagon base V-8
Overhead valves. Cast-iron block and heads. Bore & stroke: 4.31 x 3.90 in. Displacement: 455 cid. Compression ratio: 10.00:1. Brake hp: 370 at 4600 rpm. Taxable hp: 59.50. Torque: 510 lbs.-ft. at 2800. Hydraulic valve lifters. Five main bearings. Crankcase capacity: 4 qt. (add 1 qt. if new filter installed). Cooling system capacity with heater: 19.70 qt. Carburetor: Rochester 4MV Quadra-Jet four-barrel. Engine identification code: SF.

1971 BUICK

LeSabre base V-8
Overhead valves. Cast-iron block and heads. Bore & stroke: 3.80 x 3.85 in. Displacement: 350 cid. Compression ratio: 8.50:1. Brake hp: 230 at 4400 rpm. Taxable hp: 46.2. Torque: 155 lbs.-ft. at 2400. Hydraulic valve lifters. Five main bearings. Crankcase capacity: 4 qt. (add 1 qt. if new filter installed). Cooling system capacity with heater: 16.2 qt. Carburetor: Rochester 2GV two-barrel. Engine identification code: TO or TC.

Skylark, LeSabre optional V-8
Overhead valves. Cast-iron block and heads. Bore & stroke: 3.80 x 3.85 in. Displacement: 350 cid. Compression ratio: 8.50:1. Brake hp: 260 at 4600 rpm. Taxable hp: 46.2. Torque: 180 lbs.-ft. at 3800. Hydraulic valve lifters. Five main bearings. Crankcase capacity: 4 qt. (add 1 qt. if new filter installed). Cooling system capacity with heater: 16.2 qt. Carburetor: Rochester 2GV two-barrel. Engine identification code: TD or TB.

Skylark 350, Skylark 350 Custom, LeSabre, LeSabre Custom optional V-8
Overhead valves. Cast-iron block and heads. Bore & stroke: 3.80 x 3.85 in. Displacement: 350 cid. Compression ratio: 9.00:1. Brake hp: 285 at 4600 rpm. Taxable hp: 46.20. Torque: 375 lbs.-ft. at 3000. Hydraulic valve lifters. Five main bearings. Crankcase capacity: 4 qt. (add 1 qt. if new filter installed). Cooling system capacity with heater: (Skylark) 16.4 qt.; (LeSabre) 16.2 qt. Carburetor: Rochester 4MV four-barrel.

Centurion, Electra, Riviera base V-8
Overhead valves. Cast-iron block and heads. Bore & stroke: 4.31 x 3.90 in. Displacement: 455 cid. Compression ratio: 8.50:1. Brake hp: 315 at 4400 rpm. Taxable hp: 59.50. Torque: 230 lbs.-ft. at 4000. Hydraulic valve lifters. Five main bearings. Crankcase capacity: 4 qt. (add 1 qt. if new filter installed). Cooling system capacity with heater: 20.10 qt. Carburetor: Rochester 4MV four-barrel. Engine identification code: TR.

LeSabre 455, Wildcat Custom, Electa 225/225 Custom, Estate Wagon, Riviera base V-8
Overhead valves. Cast-iron block and heads. Bore & stroke: 4.31 x 3.90 in. Displacement: 455 cid. Compression ratio: 10.00:1. Brake hp: 370 at 4600 rpm. Taxable hp: 59.50. Torque: 510 lbs.-ft. at 2800. Hydraulic valve lifters. Five main bearings. Crankcase capacity: 4 qt. (Add 1 qt. if new filter installed). Cooling system capacity with heater: (LeSabre) 16.2 qt.; (Others) 19.75 qt. Carburetor: Rochester 4MV four-barrel. Engine identification code: TA or TS.

1972 BUICK

Skylark Custom base V-8
Overhead valves. Cast-iron block and heads. Bore & stroke: 3.80 x 3.85 in. Displacement: 350 cid. Compression ratio: 8.50:1. Net brake hp: 105 at 3800 rpm. Taxable hp: 46.20. Net torque: 265 lbs.-ft. at 2400. Hydraulic valve lifters. Five main bearings. Crankcase capacity: 4 qt. (add 1 qt. if new filter installed). Cooling system capacity with heater: 17.45 qt. Carburetor: Rochester 2GV two-barrel. Engine identification code: WC.

LeSabre Custom base V-8
Overhead valves. Cast-iron block and heads. Bore & stroke: 3.80 x 3.85 in. Displacement: 350 cid. Compression ratio: 8.50:1. Net brake hp: 155 at 3800 rpm. Taxable hp: 46.20. Torque: 270 lbs.-ft. at 2400. Hydraulic valve lifters. Five main bearings. Crankcase capacity: 4 qt. (add 1 qt. if new filter installed). Cooling system capacity with heater: 18.9 qt. Carburetor: Rochester 2GV two-barrel. Engine identification code: WC.

Skylark, Skylark Custom, LeSabre, LeSabre Custom base V-8
Overhead valves. Cast-iron block and heads. Bore & stroke: 3.80 x 3.85

in. Displacement: 350 cid. Compression ratio: 8.50:1. Net brake hp: 175 at 3800 rpm. Taxable hp: 46.20. Torque: 270 lbs.-ft. at 2400. Hydraulic valve lifters. Five main bearings. Crankcase capacity: 4 qt. (add 1 qt. if new filter installed). Cooling system capacity with heater: (midsize) 16.45 qt.; (LeSabre) 18.9 qt. Carburetor: Rochester 2GV two-barrel. Engine identification code: WC.

Skylark, Skylark Custom, LeSabre, LeSabre Custom base V-8
Overhead valves. Cast-iron block and heads. Bore & stroke: 3.80 x 3.85 in. Displacement: 350 cid. Compression ratio: 8.50:1. Net brake hp: 180 at 3800 rpm. Taxable hp: 46.20. Torque: 275 lbs.-ft. at 2400. Hydraulic valve lifters. Five main bearings. Crankcase capacity: 4 qt. (add 1 qt. if new filter installed). Cooling system capacity with heater: (midsize) 16.45 qt.; (LeSabre) 18.9 qt. Carburetor: Rochester 4MV four-barrel. Engine identification code: WB.

Skylark GS base V-8
Overhead valves. Cast-iron block and heads. Bore & stroke: 3.80 x 3.85 in. Displacement: 350 cid. Compression ratio: 8.50:1. Taxable hp: 46.20. Net brake hp: 190 at 4000 rpm. Net torque: 285 lbs.-ft. at 2800. Hydraulic valve lifters. Five main bearings. Crankcase capacity: 4 qt. (add 1 qt. if new filter installed). Cooling system capacity with heater: 16.45 qt. Carburetor: Rochester 4MV four-barrel. Engine identification code: WB.

Skylark GS optional V-8
Overhead valves. Cast-iron block and heads. Bore & stroke: 3.80 x 3.85 in. Displacement: 350 cid. Compression ratio: 8.50:1. Net brake hp: 195 at 4000 rpm. Taxable hp: 46.20. Net torque: 290 lbs.-ft. at 2800. Hydraulic valve lifters. Five main bearings. Crankcase capacity: 4 qt. (add 1 qt. if new filter installed). Cooling system capacity with heater: 16.45 qt. Carburetor: Rochester 4MV four-barrel. Engine identification code: WB.

Centurion, Electra 225, Electra Custom base V-8
Overhead valves. Cast-iron block and heads. Bore & stroke: 4.31 x 3.90 in. Displacement: 455 cid. Compression ratio: 8.50:1. Net brake hp: 225 at 4000 rpm. Taxable hp: 59.50. Net torque: 360 lbs.-ft. at 2600. Hydraulic valve lifters. Five main bearings. Crankcase capacity: 4 qt. (add 1 qt. if new filter installed). Cooling system capacity with heater: 18.7 qt. Carburetor: Rochester 4MV four-barrel. Engine identification code: WF.

Riviera base V-8
Overhead valves. Cast-iron block and heads. Bore & stroke: 4.31 x 3.90 in. Displacement: 455 cid. Compression ratio: 8.50:1. Net brake hp: 250 at 4000 rpm. Taxable hp: 59.50. Net torque: 375 lbs.-ft. at 2800. Hydraulic valve lifters. Five main bearings. Crankcase capacity: 4 qt. (add 1 qt. if new filter installed). Cooling system capacity with heater: 18.7 qt. Carburetor: Rochester 4MV four-barrel. Engine identification code: WS.

Skylark optional Stage 1 V-8
Overhead valves. Cast-iron block and heads. Bore & stroke: 4.31 x 3.90 in. Displacement: 455 cid. Compression ratio: 8.50:1. Net brake hp: 270 at 4400 rpm. Taxable hp: 59.50. Net torque: 390 lbs.-ft. at 3000. Hydraulic valve lifters. Five main bearings. Crankcase capacity: 4 qt. (add 1 qt. if new filter installed). Cooling system capacity with heater: 16.2 qt. Carburetor: Rochester 4MV four-barrel. Engine identification code: WA.

1973 BUICK

Apollo, Century, LeSabre base V-8
Overhead valves. Cast-iron block and heads. Bore & stroke: 3.80 x 3.85 in. Displacement: 350 cid. Compression ratio: 8.50:1. Net brake hp: 150 at 3800 rpm. Taxable hp: 46.20. Net torque: 265 lbs.-ft. at 2400. Hydraulic valve lifters. Five main bearings. Crankcase capacity: 4 qt. (add 1 qt. if new filter installed). Cooling system capacity with heater: (Apollo) 19.5 qt.; (Intermediates) 16.45 qt.; (LeSabre) 18.9 qt. Carburetor: Rochester 2GV two-barrel. Engine identification code: XC.

Centurion base V-8; Apollo, LeSabre optional V-8
Overhead valves. Cast-iron block and heads. Bore & stroke: 3.80 x 3.85 in. Displacement: 350 cid. Compression ratio: 8.50:1. Net brake hp: 175 at 3800 rpm. Taxable hp: 46.20. Net torque: 270 lbs.-ft. at 2400. Hydraulic valve lifters. Five main bearings. Crankcase capacity: 4 qt. (Add 1 qt. if new filter installed). Cooling system capacity with heater: (Apollo) 19.5 qt.; (Intermediates) 16.45 qt.; (LeSabre) 18.9 qt. Carburetor: Rochester 4MV four-barrel. Engine identification code: XB.

Century Gran Sport optional V-8
Overhead valves. Cast-iron block and heads. Bore & stroke: 3.80 x 3.85 in. Displacement: 350 cid. Compression ratio: 8.50:1. Net brake hp: 190 at 4000 rpm. Taxable hp: 46.20. Net torque: 285 lbs.-ft. at 2800. Hydraulic valve lifters. Five main bearings. Crankcase capacity: 4 qt. (Add 1 qt. if new filter installed). Cooling system capacity with heater: (Apollo) 19.5 qt.; (Intermediates) 16.45 qt.; (LeSabre) 18.9 qt. Carburetor: Rochester 4MV four-barrel. Engine identification code: XB.

Centurion wagon, Estate Wagon, Electra base V-8; Century, LeSabre, Centurion optional V-8
Overhead valves. Cast-iron block and heads. Bore & stroke: 4.31 x 3.90 in. Displacement: 455 cid. Compression ratio: 8.50:1. Net brake hp: 225 at 4000 rpm. Taxable hp: 59.50. Net torque: 360 lbs.-ft. at 2600. Hydraulic valve lifters. Five main bearings. Crankcase capacity: 4 qt. (Add 1 qt. if new filter installed). Cooling system capacity with heater: (GS) 16.2 qt.; (Full-size) 18.7 qt. Carburetor: Rochester 4MV four-barrel. Engine identification code: XF.

Riviera Series, base V-8; LeSabre, Centurion, Electra optional V-8
Overhead valves. Cast-iron block and heads. Bore & stroke: 4.31 x 3.90 in. Displacement: 455 cid. Compression ratio: 8.50:1. Net brake hp: 250 at 4000 rpm. Taxable hp: 59.50. Net torque: 375 lbs.-ft. at 2800. Hydraulic valve lifters. Five main bearings. Crankcase capacity: 4 qt. (add 1 qt. if new filter installed). Cooling system capacity with heater: (GS) 16.2 qt.; (Full-size) 18.7 qt. Carburetor: Rochester 4MV four-barrel. Engine identification code: XF.

Riviera Gran Sport optional V-8
Overhead valves. Cast-iron block and heads. Bore & stroke: 4.31 x 3.90 in. Displacement: 455 cid. Compression ratio: 8.50:1. Net brake hp: 260 at 4400 rpm. Taxable hp: 59.50. Net torque: 380 lbs.-ft. at 2800. Hydraulic valve lifters. Five main bearings. Crankcase capacity: 4 qt. (add 1 qt. if new filter installed). Cooling system capacity with heater: (GS) 16.2 qt.; (Full-size) 18.7 qt. Carburetor: Rochester 4MV four-barrel. Engine identification code: XA.

Century Gran Sport optional Stage 1 V-8
Overhead valves. Cast-iron block and heads. Bore & stroke: 4.31 x 3.90 in. Displacement: 455 cid. Compression ratio: 8.50:1. Net brake hp: 270 at 4400 rpm. Taxable hp: 59.50. Net torque: 390 lbs.-ft. at 3000. Hydraulic valve lifters. Five main bearings. Crankcase capacity: 4 qt. (add 1 qt. if new filter installed). Cooling system capacity with heater: (GS) 16.2 qt.; (Full-size) 18.7 qt. Carburetor: Rochester 4MV four-barrel. Engine identification code: XS.

1974 BUICK

Apollo, Century Luxus, LeSabre base V-8
Overhead valves. Cast-iron block and heads. Bore & stroke: 3.80 x 3.85 in. Displacement: 350 cid. Compression ratio: 8.50:1. Net brake hp: 150 at 3600 rpm. Taxable hp: 46.20. Net torque: 270 lbs.-ft. at 2000. Hydraulic valve lifters. Five main bearings. Crankcase capacity: 4 qt. (add 1 qt. if new filter installed). Cooling system capacity with heater: (Apollo) 16.5 qt.; (Intermediate) 17.3 qt.; (LeSabre) 17.3 qt. Carburetor: Rochester 2GV two-barrel. Engine identification code: XC.

Apollo, Century Luxus, LeSabre optional V-8
Overhead valves. Cast-iron block and heads. Bore & stroke: 3.80 x 3.85 in. Displacement: 350 cid. Compression ratio: 8.50:1. Net brake hp: 165 at 3800 rpm. Taxable hp: 46.20. Net torque: 285 lbs.-ft. at 2000. Hydraulic valve lifters. Five main bearings. Crankcase capacity: 4 qt. (add 1 qt. if new filter installed). Cooling system capacity with heater: (Apollo) 16.5 qt.; (Intermediate) 17.3 qt.; (LeSabre) 17.3 qt. Dual exhausts. Carburetor: Rochester 2GV two-barrel. Engine identification code: XC.

Apollo, Century Luxus, LeSabre optional V-8
Overhead valves. Cast-iron block and heads. Bore & stroke: 3.80 x 3.85 in. Displacement: 350 cid. Compression ratio: 8.50:1. Net brake hp: 175 at 3800 rpm. Taxable hp: 46.20. Net torque: 260 lbs.-ft. at 2000. Hydraulic valve lifters. Five main bearings. Crankcase capacity: 4 qt. (add 1 qt. if new filter installed). Cooling system capacity with heater: (Apollo) 16.5 qt.; (Intermediate) 17.3 qt.; (LeSabre) 17.3 qt. Carburetor: Rochester 4MV four-barrel. Engine identification code: XB.

Apollo, Century Luxus, LeSabre optional V-8
Overhead valves. Cast-iron block and heads. Bore & stroke: 3.80 x 3.85 in. Displacement: 350 cid. Compression ratio: 8.50:1. Net brake hp: 195 at 4000 rpm. Taxable hp: 46.20. Net torque: 280 lbs.-ft. at 2000. Hydraulic valve lifters. Five main bearings. Crankcase capacity: 4 qt. (add 1 qt. if new filter installed). Cooling system capacity with heater: (Apollo) 16.5 qt.; (Intermediate) 17.3 qt.; (LeSabre) 17.3 qt. Dual exhausts. Carburetor: Rochester 4MV four-barrel. Engine identification code: XB.

Estate Wagon base V-8; Century Luxus, LeSabre optional V-8
Overhead valves. Cast-iron block and heads. Bore & stroke: 4.31 x 3.90 in. Displacement: 455 cid. Compression ratio: 8.50:1. Net brake hp: 175 at 3400 rpm. Taxable hp: 59.50. Net torque: 355 lbs.-ft. at 2000. Hydraulic valve lifters. Five main bearings. Crankcase capacity: 4 qt. (add 1 qt. if new filter installed). Cooling system capacity with heater: (Intermediates)

19.4 qt.; (Full-size) 19.6 qt. Carburetor: Rochester 2GV two-barrel. Engine identification code: XF.

Century Luxus, LeSabre optional V-8
Overhead valves. Cast-iron block and heads. Bore & stroke: 4.31 x 3.90 in. Displacement: 455 cid. Compression ratio: 8.50:1. Net brake hp: 190 at 3600 rpm. Taxable hp: 59.50. Net torque: 370 lbs.-ft. at 2000. Hydraulic valve lifters. Five main bearings. Crankcase capacity: 4 qt. (add 1 qt. if new filter installed). Cooling system capacity with heater: (Intermediates) 19.4 qt.; (Full-size) 19.6 qt. Dual exhausts. Carburetor: Rochester 2GV two-barrel. Engine identification code: XF.

Riviera base V-8; Century Luxus, Electra optional V-8
Overhead valves. Cast-iron block and heads. Bore & stroke: 4.31 x 3.90 in. Displacement: 455 cid. Compression ratio: 8.50:1. Net brake hp: 210 at 3600 rpm. Taxable hp: 59.50. Net torque: 335 lbs.-ft. at 2200. Hydraulic valve lifters. Five main bearings. Crankcase capacity: 4 qt. (add 1 qt. if new filter installed). Cooling system capacity with heater: (Intermediates) 19.4 qt.; (Full-size) 19.6 qt. Carburetor: Rochester 4MV four-barrel. Engine identification code: XA.

Riviera base V-8; Century Luxus, Electra optional V-8
Overhead valves. Cast-iron block and heads. Bore & stroke: 4.31 x 3.90 in. Displacement: 455 cid. Compression ratio: 8.50:1. Net brake hp: 230 at 3800 rpm. Taxable hp: 59.50. Net torque: 355 lbs.-ft. at 2200. Hydraulic valve lifters. Five main bearings. Crankcase capacity: 4 qt. (add 1 qt. if new filter installed). Cooling system capacity with heater: (Intermediates) 19.4 qt.; (Full-size) 19.6 qt. Dual exhausts. Carburetor: Rochester 4MV four-barrel. Engine identification code: XA.

LeSabre, Electra, Riviera optional Stage 1 V-8
Overhead valves. Cast-iron block and heads. Bore & stroke: 4.31 x 3.90 in. Displacement: 455 cid. Compression ratio: 8.50:1. Net brake hp: 245 at 4000 rpm. Taxable hp: 59.50. Net torque: 360 lbs.-ft. at 2400. Hydraulic valve lifters. Five main bearings. Crankcase capacity: 4 qt. (add 1 qt. if new filter installed). Cooling system capacity with heater: (Intermediates) 19.4 qt.; (Full-size) 19.6 qt. Dual exhausts. Carburetor: Rochester 4MV four-barrel. Engine identification code: XS.

Century Luxus optional Stage 1 V-8
Overhead valves. Cast-iron block and heads. Bore & stroke: 4.31 x 3.90 in. Displacement: 455 cid. Compression ratio: 8.50:1. Net brake hp: 255 at 4400 rpm. Taxable hp: 59.50. Net torque: 370 lbs.-ft. at 2800. Hydraulic valve lifters. Five main bearings. Crankcase capacity: 4 qt. (add 1 qt. if new filter installed). Cooling system capacity with heater: (Intermediates) 19.4 qt.; (Full-size) 19.6 qt. Carburetor: Rochester 4MV four-barrel. Engine identification code: XS.

1975 BUICK

Apollo base V-8
Overhead valves. Cast-iron block and heads. Bore & stroke: 3.50 x 3.385 in. Displacement: 260 cid. Compression ratio: 8.00:1. Net brake hp: 110 at 3400 rpm. Taxable hp: 39.2. Net torque: 205 lbs.-ft. at 1600. Hydraulic valve lifters. Five main bearings. Crankcase capacity: 4 qt. (add 1 qt. if new filter installed). Cooling system capacity with heater: 22.4 qt. Carburetor: Rochester 2GV two-barrel. VIN Code: F.
NOTE: The above V-8 was built by Oldsmobile Div.

Century base V-8; Apollo optional V-8
Overhead valves. Cast-iron block and heads. Bore & stroke: 3.80 x 3.85 in. Displacement: 350 cid. Compression ratio: 8.00:1. Net brake hp: 145 at 3200 rpm. Taxable hp: 46.20. Net torque: 270 lbs.-ft. at 2000. Hydraulic valve lifters. Five main bearings. Crankcase capacity: 4 qt. (add 1 qt. if new filter installed). Cooling system capacity with heater: (Apollo and Skylark) 17.9 qt.; (Intermediates) 16.9 qt.; (Full-size) 16.9 qt. Carburetor: Rochester 2GV two-barrel. VIN Code: H.

Century wagon, LeSabre base V-8; Apollo, Skylark, Century optional V-8
Overhead valves. Cast-iron block and heads. Bore & stroke: 3.80 x 3.85 in. Displacement: 350 cid. Compression ratio: 8.00:1. Net brake hp: 165 at 3800 rpm. Taxable hp: 46.20. Net torque: 260 lbs.-ft. at 2200. Hydraulic valve lifters. Five main bearings. Crankcase capacity: 4 qt. (add 1 qt. if new filter installed). Cooling system capacity with heater: (Apollo and Skylark) 17.9 qt.; (Intermediates) 16.9 qt.; (Full-size) 16.9 qt. Carburetor: Rochester 4MV four-barrel. VIN Code: J.

LeSabre optional V-8
Overhead valves. Cast-iron block and heads. Bore & stroke: 4.12 x 3.75 in. Displacement: 400 cid. Compression ratio: 7.60:1. Net brake hp: 190 at 3400 rpm. Taxable hp: 54.34. Net torque: 350 lbs.-ft. at 2400. Hydraulic valve lifters. Five main bearings. Crankcase capacity: 4 qt. (add 1 qt. if new filter installed). Cooling system capacity with heater: 23.6 qt (25.6 qt. with heavy-duty cooling system). Carburetor: Rochester 4MV four-barrel.
NOTE: The above V-8 was built by Pontiac Motor Div.

LeSabre wagon, Estate Wagon, Electra, Riviera base V-8; LeSabre cars optional V-8
Overhead valves. Cast-iron block and heads. Bore & stroke: 4.31 x 3.90 in. Displacement: 455 cid. Compression ratio: 7.90:1. Net brake hp: 205 at 3800 rpm. Taxable hp: 59.50. Net torque: 345 lbs.-ft. at 2000. Hydraulic valve lifters. Five main bearings. Crankcase capacity: 4 qt. (add 1 qt. if new filter installed). Cooling system capacity with heater: 19.6 qt. Carburetor: Rochester 2GV two-barrel. VIN Code: T.

1976 BUICK

Skylark base V-8
Overhead valve. Cast-iron block and heads. Bore & stroke: 3.50 x 3.385 in. Displacement: 260 cid (4.3 liters). Compression ratio: 8.5:1. Net brake hp: 110 at 3400 rpm. Taxable hp: 39.2. Net torque: 210 at 1600 rpm. Five main bearings. Hydraulic valve lifters. Carburetor: Rochester two-barrel. VIN Code: F.
NOTE: The above V-8 was built by Oldsmobile Div.

Regal sedan, Century, Century Custom base V-8; Skylark, Regal Coupe optional V-8
Overhead valve. Cast-iron block and heads. Bore & stroke: 3.80 x 3.85 in. Displacement: 350 cid (5.7 liters). Compression ratio: 8.0:1. Net brake hp: 145 at 3200 rpm. Taxable hp: 46.20. Net torque: 270 lbs.-ft. at 2000 rpm. Five main bearings. Hydraulic valve lifters. Carburetor: Rochester 2GC two-barrel. VIN Code: H (except California).

Regal sedan, Century, Century Custom base V-8; Skylark, Regal Coupe optional V-8
Overhead valve. Cast-iron block and heads. Bore & stroke: 3.80 x 3.85 in. Displacement: 350 cid (5.7 liters). Compression ratio: 8.0:1. Net brake hp: 160 at 3800 rpm. Taxable hp: 46.20. Net torque: 260 lbs.-ft. at 2200 rpm. Five main bearings. Hydraulic valve lifters. Carburetor: Rochester M4MC four-barrel (California).

Century wagon, LeSabre Custom base V-8; Skylark, Century Custom, Regal optional V-8
Overhead valve. Cast-iron block and heads. Bore & stroke: 3.80 x 3.85 in. Displacement: 350 cid (5.7 liters). Compression ratio: 8.0:1. Net brake hp: 165 at 3800 rpm. Taxable hp: 46.20. Net torque: 260 lbs.-ft. at 2200 rpm. Five main bearings. Hydraulic valve lifters. Carburetor: Rochester M4MC four-barrel. VIN Code: J.

Electra, Estate Wagon, Riviera base V-8; Optional LeSabre Custom
Overhead valve. Cast-iron block and heads. Bore & stroke: 4.3125 x 3.9 in. Displacement: 455 cid. (7.5 liters). Compression ratio: 7.9:1. Net brake hp: 205 at 3800 rpm. Taxable hp: 59.50. Net torque: 345 lbs.-ft. at 2000 rpm. Five main bearings. Hydraulic valve lifters. Carburetor: Rochester M4MC four-barrel. VIN Code: T.

1977 BUICK

Skylark, LeSabre Sport Coupe base V-8; LeSabre (except Sport Coupe) optional V-8
Overhead valve. Cast-iron block and heads. Bore & stroke: 4.0 x 3.0 in. Displacement: 301 cid (5.0 liters). Compression ratio: 8.2:1. Net brake hp: 135 at 4000 rpm. Taxable hp: 51.2. Net torque: 245 lbs.-ft. at 2000 rpm. Five main bearings. Hydraulic valve lifters. Carburetor: Rochester M2MC two-barrel. RPO: L27. VIN Code: Y.
NOTE: The above V-8 was built by Pontiac Motor Div.

Skylark Century wagon alternate V-8
Overhead valve. Cast-iron block and heads. Bore & stroke: 3.736 x 3.48 in. Displacement: 305 cid (5.0 liters). Compression ratio: 8.5:1. Net brake hp: 145 at 3800 rpm. Taxable hp: 44.7. Net torque: 245 lbs.-ft. at 2400 rpm. Five main bearings. Hydraulic valve lifters. Carburetor: Rochester 2GC two-barrel. VIN Code: U.
NOTE: The above V-8 was built by Chevrolet Motor Div.

Regal sedan base V-8; Century, Century Custom coupe/sedan, Regal coupe optional V-8
Overhead valve. Cast-iron block and heads. Bore & stroke: 3.80 x 3.85

in. Displacement: 350 cid (5.7 liters). Compression ratio: 8.1:1. Net brake hp: 140 at 3200 rpm. Taxable hp: 46.20. Net torque: 280 lbs.-ft. at 1400 rpm. Five main bearings. Hydraulic valve lifters. Carburetor: Rochester 2GC two-barrel. RPO: L32. VIN Code: H.

Century wagon, Electra, Estate, Riviera base V-8; Skylark, Century, Regal coupe/sedan, LeSabre optional V-8
Overhead valve. Cast-iron block and heads. Bore & stroke: 3.80 x 3.85 in. Displacement: 350 cid (5.7 liters). Compression ratio: 8.0:1. Net brake hp: 155 at 3400 rpm. Taxable hp: 46.20. Net torque: 275 lbs.-ft. at 1800 rpm. Five main bearings. Hydraulic valve lifters. Carburetor: Rochester M4MC four-barrel. VIN Code: J.

Skylark, Century, Century Custom coupe/sedan, Regal, LeSabre alternate V-8
Overhead valve. Cast-iron block and heads. Bore & stroke: 4.057 x 3.385 in. Displacement: 350 cid (5.7 liters). Compression ratio: 7.9:1. Net brake hp: 170 at 3800 rpm. Taxable hp: 46.20. Net torque: 275 lbs.-ft. at 2000 rpm. Five main bearings. Hydraulic valve lifters. Carburetor: Rochester M4MC four-barrel. RPO: L34. VIN Code: R. (Not available in California.)
NOTE: The above V-8 was built by Oldsmobile Div.

Skylark, Century, Century Custom coupe/sedan, Regal, LeSabre alternate V-8
Overhead valve. Cast-iron block and heads. Bore & stroke: 4.00 x 3.48 in. Displacement: 350 cid (5.7 liters). Compression ratio: 8.5:1. Net brake hp: 170 at 3800 rpm. Taxable hp: 46.20. Net torque: 270 lbs.-ft. at 2400 rpm. Five main bearings. Hydraulic valve lifters. Carburetor: Rochester M4MC four-barrel. RPO: LM1. VIN Code: L.
NOTE: The above V-8 was built by Chevrolet Motor Div.

Century wagon, LeSabre, Electra, Estate Wagon, Riviera Optional V-8
Overhead valve. Cast-iron block and heads. Bore & stroke: 4.351 x 3.385 in. Displacement: 403 cid (6.6 liters). Compression ratio: 8.5:1. Net brake hp: 185 at 3600 rpm. Taxable hp: 60.57. Net torque: 320 lbs.-ft. at 2200 rpm. Five main bearings. Hydraulic valve lifters. Carburetor: Rochester M4MC four-barrel. RPO: L80. VIN Code: K.
NOTE: The above V-8 was built by Oldsmobile Div.

1978 BUICK

LeSabre base V-8
Overhead valve. Cast-iron block and heads. Bore & stroke: 4.0 x 3.0 in. Displacement: 301 cid (5.0 liters). Compression ratio: 8.2:1. Net brake hp: 165. Taxable hp: 51.2. Five main bearings. Hydraulic valve lifters. Carburetor: Rochester M2MC two-barrel. RPO: L27. VIN Code: Y.
NOTE: The above V-8 was built by Pontiac Motor Div.

Skylark, Century, Regal base V-8; Regal, LeSabre optional V-8
Overhead valve. Cast-iron block and heads. Bore & stroke: 3.736 x 3.48 in. Displacement: 305 cid (5.0 liters). Compression ratio: 8.5:1. Net brake hp: 145 at 3800 rpm. Taxable hp: 44.7. Net torque: 245 lbs.-ft. at 2400 rpm. Five main bearings. Hydraulic valve lifters. Carburetor: Rochester 2GC two-barrel. RPO: LG3. VIN Code: U.
NOTE: The above V-8 was built by Chevrolet Motor Div.

Regal alternate base V-8
Overhead valve. Cast-iron block and heads. Bore & stroke: 3.736 x 3.48 in. Displacement: 305 cid (5.0 liters). Compression ratio: 8.5:1. Net brake hp: 160 at 4000 rpm. Taxable hp: 44.7. Net torque: 285 lbs.-ft. at 2400 rpm. Five main bearings. Hydraulic valve lifters. Carburetor: Rochester 4GC four-barrel. RPO: LG4. VIN Code: H.
NOTE: The above V-8 was built by Chevrolet Motor Div.

Riviera base V-8
Overhead valve. Cast-iron block and heads. Bore & stroke: 3.80 x 3.85 in. Displacement: 350 cid (5.7 liters). Compression ratio: 8.0:1. Net brake hp: 155 at 3400 rpm. Taxable hp: 46.20. Net torque: 280 lbs.-ft. at 1800 rpm. Five main bearings. Hydraulic valve lifters. Carburetor: Rochester M4MC four-barrel. RPO: L77. VIN Code: X.

LeSabre, Electra, Estate, Riviera alternate V-8; Skylark optional V-8
Overhead valve. Cast-iron block and heads. Bore & stroke: 4.057 x 3.385 in. Displacement: 350 cid (5.7 liters). Compression ratio: 8.0:1. Net brake hp: 170 at 3800 rpm. Taxable hp: 46.20. Net torque: 275 lbs.-ft. at 2000 rpm. Five main bearings. Hydraulic valve lifters. Carburetor: Rochester M4MC four-barrel. RPO: L34. VIN Code: R.

Skylark, Century optional V-8
Overhead valve. Cast-iron block and heads. Bore & stroke: 4.00 x 3.48 in. Displacement: 350 cid (5.7 liters). Compression ratio: 8.2:1. Net brake hp: 170 at 3800 rpm. Taxable hp: 46.20. Net torque: 275 lbs.-ft. at 2000 rpm. Five main bearings. Hydraulic valve lifters. Carburetor: Rochester M4MC four-barrel. RPO: LM1. VIN Code: L.
NOTE: The above V-8 was built by Chevrolet Motor Div.

LeSabre wagon, Electra base V-8; LeSabre, Estate, Riviera optional V-8
Overhead valve. Cast-iron block and heads. Bore & stroke: 4.351 x 3.385 in. Displacement: 403 cid (6.6 liters). Compression ratio: 8.0:1. Net brake hp: 185 at 3600 rpm. Taxable hp: 60.57. Net torque: 320 lbs.-ft. at 2000 rpm. Five main bearings. Hydraulic valve lifters. Carburetor: four-barrel Rochester M4MC. RPO: L80. VIN Code: K.
NOTE: The above V-8 was built by Oldsmobile Div.

Century, Regal, LeSabre base V-8
Overhead valve. Cast-iron block and heads. Bore & stroke: 4.0 x 3.0 in. Displacement: 301 cid (4.9 liters). Compression ratio: 8.1:1. Net brake hp: 140 at 3600 rpm. Taxable hp: 51.2. Net torque: 235 lbs.-ft. at 2000 rpm. Five main bearings. Hydraulic valve lifters. Carburetor: Rochester M2MC two-barrel. RPO: L27. VIN Code: Y.
NOTE: The above V-8 was built by Pontiac Motor Div.

1979 BUICK

Century, Regal, LeSabre optional V-8
Overhead valve. Cast-iron block and heads. Bore & stroke: 4.0 x 3.0 in. Displacement: 301 cid (4.9 liters). Compression ratio: 8.1:1. Net brake hp: 150 at 4000 rpm. Taxable hp: 51.2. Net torque: 240 lbs.-ft. at 2000 rpm. Five main bearings. Hydraulic valve lifters. Carburetor: Rochester M4MC four-barrel. RPO: L37. VIN Code: W.
NOTE: The above V-8 was built by Pontiac Motor Div.

Skylark base V-8
Overhead valve. Cast-iron block and heads. Bore & stroke: 3.736 x 3.48 in. Displacement: 305 cid (5.0 liters). Compression ratio: 8.4:1. Net brake hp: 130 at 3200 rpm. Taxable hp: 44.7. Net torque: 245 lbs.-ft. at 2000 rpm. Five main bearings. Hydraulic valve lifters. Carburetor: Rochester M2MC two-barrel. RPO: LG3. VIN Code: U.
NOTE: The above V-8 was built by Chevrolet Motor Div.

Century, Regal optional V-8 in California and high-altitude counties
Overhead valve. Cast-iron block and heads. Bore & stroke: 3.736 x 3.48 in. Displacement: 305 cid (5.0 liters). Compression ratio: 8.4:1. Net brake hp: 115 at 4000 rpm. Taxable hp: 44.7. Net torque: 225 lbs.-ft. at 2400 rpm. Five main bearings. Hydraulic valve lifters. Carburetor: Rochester M4MC four-barrel. RPO: LG4. VIN Code: H.
NOTE: The above V-8 was built by Chevrolet Motor Div.

Electra, Estate, Riviera base V-8; LeSabre optional V-8
Overhead valve. Cast-iron block and heads. Bore & stroke: 3.80 x 3.85 in. Displacement: 350 cid (5.7 liters). Compression ratio: 8.0:1. Net brake hp: 155 at 3400 rpm. Taxable hp: 46.20. Net torque: 280 lbs.-ft. at 1800 rpm. Five main bearings. Hydraulic valve lifters. Carburetor: Rochester M4MC four-barrel. RPO: L77. VIN Code: X.

California and high-altitude Estate, Riviera base V-8; LeSabre, Electra, Riviera S base V-8
Overhead valve. Cast-iron block and heads. Bore & stroke: 4.057 x 3.385 in. Displacement: 350 cid (5.7 liters). Compression ratio: 8.0:1. Net brake hp: 160 at 3800 rpm. Taxable hp: 46.20. Net torque: 270 lbs.-ft. at 2000 rpm. Five main bearings. Hydraulic valve lifters. Carburetor: Rochester M4MC four-barrel. RPO: L34. VIN Code: R.
NOTE: The above V-8 was built by Oldsmobile Div.

Century wagon base V-8; Electra, Riviera alternate base V-8; Skylark optional V-8
Overhead valve. Cast-iron block and heads. Bore & stroke: 4.00 x 3.48 in. Displacement: 350 cid (5.7 liters). Compression ratio: 8.2:1. Net brake hp: 170 at 3800 rpm. Taxable hp: 46.20. Net torque: 260 lbs.-ft. at 2400 rpm. Five main bearings. Hydraulic valve lifters. Carburetor: Rochester M4MC four-barrel. RPO: LM1. VIN Code: L.
NOTE: The above V-8 was built by Chevrolet Motor Div.

Electra optional V-8
Overhead valve. Cast-iron block and heads. Bore & stroke: 4.351 x 3.385 in. Displacement: 403 cid (6.6 liters). Compression ratio: 8.0:1. Net brake hp: 175 at 3600 rpm. Taxable hp: 60.57. Net torque: 310 lbs.-ft. at 2000 rpm. Five main bearings. Hydraulic valve lifters. Carburetor: Rochester, M4MC four-barrel. RPO: L80. VIN Code: K.
NOTE: The above V-8 was built by Oldsmobile Div.

1980 BUICK

Century, Regal optional V-8
Overhead valve. Cast-iron block and heads. Bore & stroke: 3.75 x 3.00 in. Displacement: 265 cid (4.3 liters). Compression ratio: 8.0:1. Net brake hp: 120 at 3600 rpm. Taxable hp: 45.00. Net torque: 210 lbs.-ft. at 1800 rpm. Four main bearings. Hydraulic valve lifters. Carburetor: Rochester M2ME two-barrel. VIN Code: S.
NOTE: The above V-8 was built by Chevrolet Motor Div.

Estate Wagon base V-8; Century, Regal, LeSabre optional V-8
Overhead valve. Cast-iron block and heads. Bore & stroke: 4.0 x 3.0 in. Displacement: 301 cid (4.9 liters). Taxable hp: 51.20. Compression ratio: 8.2:1. Net brake hp: 140 at 4000 rpm. Net torque: 240 lbs.-ft. at 1800 rpm. Five main bearings. Hydraulic valve lifters. Carburetor: Rochester M4ME four-barrel. RPO: L37. VIN Code: W.
NOTE: The above V-8 was built by Pontiac Motor Div.

Century, Regal optional V-8
Overhead valve. Cast-iron block and heads. Bore & stroke: 3.736 x 3.48 in. Displacement: 305 cid (5.0 liters). Compression ratio: 8.6:1. Net brake hp: 155 at 4000 rpm. Taxable hp: 44.70. Net torque: 240 lbs.-ft. at 1600 rpm. Five main bearings. Hydraulic valve lifters. Carburetor: Rochester M4MC four-barrel. RPO: LG4. VIN Code: H.
NOTE: The above V-8 was built by Chevrolet Motor Div.

Riviera base V-8; LeSabre, Estate, Electra, Riviera S optional V-8
Overhead valve. Cast-iron block and heads. Bore & stroke: 3.80 x 3.85 in. Displacement: 350 cid (5.7 liters). Compression ratio: 8.0:1. Net brake hp: 155 at 3400 rpm. Taxable hp: 52.70. Net torque: 280 lbs.-ft. at 1600 rpm. Five main bearings. Hydraulic valve lifters. Carburetor: Rochester M4MC four-barrel. RPO: L77. VIN Code: X.

Riviera alternate base V-8; LeSabre, Estate, Electra, Riviera S alternate optional V-8
Overhead valve. Cast-iron block and heads. Bore & stroke: 4.057 x 3.385 in. Displacement: 350 cid (5.7 liters). Compression ratio: 8.3:1. Net brake hp: 160 at 3600 rpm. Taxable hp: 52.70. Net torque: 270 lbs.-ft. at 1600 rpm. Five main bearings. Hydraulic valve lifters. Carburetor: Rochester M4MC four-barrel. RPO: L34. VIN Code: R.
NOTE: The above V-8 was built by Oldsmobile Div.

Electra, Estate optional diesel V-8
Overhead valve. Cast-iron block and heads. Bore & stroke: 4.057 x 3.385 in. Displacement: 350 cid (5.7 liters). Compression ratio: 22.5:1. Net brake hp: 105 at 3200 rpm. Taxable hp: 52.70. Net torque: 205 lbs.-ft. at 1600 rpm. Five main bearings. Hydraulic valve lifters. Fuel injection. RPO: LF9. VIN Code: N.
NOTE: The above V-8 was built by Oldsmobile Div.

1981 BUICK

Century, Regal base V-8
Overhead valve. Cast-iron block and heads. Bore & stroke: 3.75 x 3.00 in. Displacement: 265 cid (4.3 liters). Compression ratio: 8.0:1. Net brake hp: 119 at 4000 rpm. Taxable hp: 45.00. Net torque: 203 lbs.-ft. at 2000 rpm. Five main bearings. Hydraulic valve lifters. Carburetor: Rochester M2ME two-barrel. RPO: LS5. VIN Code: S.
NOTE: The above V-8 was built by Pontiac Motor Div.

LeSabre, Estate Wagon, Electra base V-8; Riviera optional V-8
Overhead valve. Cast-iron block and heads. Bore & stroke: 3.80 x 3.385 in. Displacement: 307 cid (5.0 liters). Compression ratio: 8.0:1. Net brake hp: 140 at 3600 rpm. Taxable hp: 46.20. Net torque: 240 lbs.-ft. at 1600 rpm. Five main bearings. Hydraulic valve lifters. Carburetor: Rochester M4MC four-barrel. RPO: LV2. VIN Code: Y.
NOTE: The above V-8 was built by Oldsmobile Div.

Century, LeSabre, Electra, Estate, Riviera Z57 optional diesel V-8
Overhead valve. Cast-iron block and heads. Bore & stroke: 4.057 x 3.385 in. Displacement: 350 cid (5.7 liters). Compression ratio: 22.5:1. Net brake hp: 105 at 3200 rpm. Taxable hp: 52.70. Net torque: 200 lbs.-ft. at 1600 rpm. Five main bearings. Hydraulic valve lifters. Fuel injection. RPO: LF9. VIN Code: N.
NOTE: The above V-8 was built by Oldsmobile Div.

1982 BUICK

Estate Wagon, LeSabre, Electra, Riviera base V-8
Overhead valve. Cast-iron block and heads. Bore & stroke: 3.80 x 3.385 in. Displacement: 307 cid (5.0 liters). Compression ratio: 8.0:1. Net brake hp: 140 at 3600 rpm. Taxable hp: 46.20. Net torque: 240 lbs.-ft. at 1600 rpm. Five main bearings. Hydraulic valve lifters. Carburetor: Rochester M4MC four-barrel. RPO: LV2. VIN Code: Y.
NOTE: The above V-8 was built by Oldsmobile Div.

Regal, LeSabre, Electra, Estate, Riviera optional diesel V-8
Overhead valve. Cast-iron block and heads. Bore & stroke: 4.057 x 3.385 in. Displacement: 350 cid (5.7 liters). Compression ratio: 21.6:1. Net brake hp: 105 at 3200 rpm. Taxable hp: 52.70. Net torque: 200 lbs.-ft. at 1600 rpm. Five main bearings. Hydraulic valve lifters. Fuel injection. RPO: LF9. VIN Code: N.
NOTE: The above V-8 was built by Oldsmobile Div.

1983 BUICK

Estate Wagon, LeSabre, Electra, Riviera base V-8
Overhead valve. Cast-iron block and heads. Bore & stroke: 3.80 x 3.385 in. Displacement: 307 cid (5.0 liters). Compression ratio: 8.0:1. Net brake hp: 140 at 3600 rpm. Taxable hp: 46.20. Net torque: 240 lbs.-ft. at 1600 rpm. Five main bearings. Hydraulic valve lifters. Carburetor: Rochester four-barrel. VIN Code: Y.
NOTE: The above V-8 was built by Oldsmobile Div.

Regal, LeSabre, Electra, Estate wagon, Riviera optional diesel V-8
Overhead valve. Cast-iron block and heads. Bore & stroke: 4.057 x 3.385 in. Displacement: 350 cid (5.7 liters). Compression ratio: 21.6:1. Net brake hp: 105 at 3200 rpm. Taxable hp: 52.70. Net torque: 200 lbs.-ft. at 1600 rpm. Five main bearings. Hydraulic valve lifters. Fuel injection. VIN Code: N.
NOTE: The above V-8 was built by Oldsmobile Div.

1984 BUICK

Base V-8 Estate Wagon, LeSabre, Electra, Riviera
Overhead valve. Cast-iron block and heads. Bore & stroke: 3.80 x 3.385 in. Displacement: 307 cid (5.0-liters). Compression ratio: 8.0:1. Net brake hp: 140 at 3600 rpm. Taxable hp: 46.20. Net torque: 240 lbs.-ft. at 1600 rpm. Five main bearings. Hydraulic valve lifters. Carburetor: Rochester M4ME four-barrel. VIN Code: Y.
NOTE: The above V-8 was built by Oldsmobile Div.

Optional diesel V-8 LeSabre, Electra, Estate Wagon, Riviera
Overhead valve. Cast-iron block and heads. Bore & stroke: 4.057 x 3.385 in. Displacement: 350 cid 5.7-liters). Compression ratio: 21.6:1. Net brake hp: 105 at 3200 rpm. Taxable hp: 52.70. Net torque: 200 lbs.-ft. at 1600 rpm. Five main bearings. Hydraulic valve lifters. Fuel injection. VIN Code: N.
NOTE: The above V-8 was built by Oldsmobile Div.

1985 BUICK

LeSabre, Estate, Riviera base V-8
Overhead valve. Cast-iron block and heads. Bore & stroke: 3.80 x 3.385 in. Displacement: 307 cid 5.0-liters). Compression ratio: 8.0:1. Net brake hp: 140 at 3200 rpm; LeSabre, 140 at 3600 rpm. Taxable hp: 46.20. Net torque: 255 lbs.-ft. at 2000 rpm; LeSabre, 240 lbs.-ft. at 1600 rpm. Five main bearings. Hydraulic valve lifters. Carburetor: Four-barrel. VIN Code: Y or H.
NOTE: The above V-8 was built by Oldsmobile Div.

LeSabre, Estate, Riviera optional diesel V-8
Overhead valve. Cast-iron block and heads. Bore & stroke: 4.057 x 3.385 in. Displacement: 350 cid 5.7-liters). Compression ratio: 21.6:1. Net brake hp: 105 at 3200 rpm. Taxable hp: 52.70. Net torque: 200 lbs.-ft. at 1600 rpm. Five main bearings. Hydraulic valve lifters. Fuel injection. VIN Code:
NOTE: The above V-8 was built by Oldsmobile Div.

1986 BUICK

Estate wagon, Regal base V-8
Overhead valve. Cast-iron block and heads. Bore & stroke: 3.80 x 3.385 in. Displacement: 307 cid 5.0 liters). Compression ratio: 8.0:1. Net brake hp: 140 at 3200 rpm. Taxable hp: 46.20. Net torque: 255 lbs.-ft. at 2000 rpm. Five main bearings. Hydraulic valve lifters. Carburetor: Four-barrel. VIN Code: Y or H.
NOTE: The above V-8 was built by Oldsmobile Div.

1987 BUICK

Regal, LeSabre Estate Wagon, Electra Estate Wagon base V-8
Overhead valve. Cast-iron block and heads. Bore & stroke: 3.8 x 3.39 in. Displacement: 307 cid (5.0 liters). Compression ratio: 8.0:1. Net brake hp: 140 at 3200 rpm. Taxable hp: 46.20. Net torque: 255 lbs.-ft. at 2000 rpm. Five main bearings. Hydraulic valve lifters. Carburetor: Four-barrel. VIN Code: Y.
NOTE: The above V-8 was built by Oldsmobile Div.

1988 BUICK

LeSabre Estate Wagon, Electra Estate Wagon base V-8
Overhead valve. Cast-iron block and heads. Bore & stroke: 3.8 x 3.39 in. Displacement: 307 cid (5.0 liters). Compression ratio: 8.0:1. Net brake hp: 140 at 3200 rpm. Taxable hp: 46.20. Net torque: 255 lbs.-ft. at 2000 rpm. Five main bearings. Hydraulic valve lifters. Carburetor: Four-barrel. VIN Code: Y.
NOTE: The above V-8 was built by Oldsmobile Div.

1989 BUICK

LeSabre Estate Wagon, Electra Estate Wagon base V-8
Overhead valve. Cast-iron block and heads. Bore & stroke: 3.8 x 3.39 in. Displacement: 307 cid (5.0 liters). Compression ratio: 8.0:1. Net brake hp: 140 at 3200 rpm. Taxable hp: 46.20. Net torque: 255 lbs.-ft. at 2000 rpm. Five main bearings. Hydraulic valve lifters. Carburetor: Four-barrel. VIN Code: Y.
NOTE: The above V-8 was built by Oldsmobile Div.

1990 BUICK

Estate Wagon base V-8
Overhead valve. Cast-iron block and heads. Bore & stroke: 3.8 x 3.39 in. Displacement: 307 cid (5.0 liters). Compression ratio: 8.0:1. Net brake hp: 140 at 3200 rpm. Taxable hp: 46.20. Net torque: 255 lbs.-ft. at 2000 rpm. Five main bearings. Hydraulic valve lifters. Carburetor: Four-barrel. VIN Code: Y.
NOTE: The above V-8 was built by Oldsmobile Div.

1991 BUICK

Roadmaster Estate Wagon base V-8
Overhead valve. Cast-iron block and heads. Bore & stroke: 3.74 x 3.48 in. Displacement: 305 cid (5.0 liters). Compression ratio: 9.3:1. Net brake hp: 170 at 4200 rpm. Taxable hp: 44.80. Net torque: 255 lbs.-ft. at 2400 rpm. Throttle-body fuel injection. VIN Code: E.
NOTE: The above V-8 was built by Chevrolet Motor Div.

1992 BUICK

Roadmaster base V-8
Overhead valve. Cast-iron block and heads. Bore & stroke: 4.00 x 3.48 in. Displacement: 350 cid (5.7 liters). Compression ratio: 9.8:1. Net brake hp: 180 at 4000 rpm. Taxable hp: 51.20. Net torque: 300 lbs.-ft. at 2400 rpm. Throttle-body fuel injection. VIN Code: 7.
NOTE: The above V-8 was built by Chevrolet Motor Div.

1993 BUICK

Roadmaster base V-8
Overhead valve. Cast-iron block and heads. Bore & stroke: 4.00 x 3.48 in. Displacement: 350 cid (5.7 liters). Compression ratio: 9.8:1. Net brake hp: 180 at 4000 rpm. Taxable hp: 51.20. Net torque: 300 lbs.-ft. at 2400 rpm. Throttle-body fuel injection. VIN Code: 7.
NOTE: The above V-8 was built by Chevrolet Motor Div.

1994 BUICK

Roadmaster sedan/Estate Wagon base V-8
Overhead valve. Cast-iron block and aluminum heads. Bore & stroke: 4.00 x 3.48 in. Displacement: 350 cid (5.7 liters). Compression ratio: 10.5:1. Taxable hp: 51.20. Net brake hp: 260 at 5000 rpm. Net torque: 335 lbs.-ft. at 3200 rpm. Sequential fuel injection. RPO: LT1. VIN Code: P.
NOTE: The above V-8 was built by Chevrolet Motor Div. It is very similar to, but not identical to the Corvette LT-1 V-8.

1995 BUICK

Roadmaster sedan/Estate Wagon base V-8
Overhead valve. Cast-iron block and aluminum heads. Bore & stroke: 4.00 x 3.48 in. Displacement: 350 cid (5.7 liters). Compression ratio: 10.5:1. Net brake hp: 260 at 5000 rpm. Taxable hp: 51.20. Net torque: 335 lbs.-ft. at 3200 rpm. Sequential fuel injection. RPO: LT1. VIN Code: P.
NOTE: The above V-8 was built by Chevrolet Motor Div. It is very similar to, but not identical to the Corvette LT-1 V-8.

1996 BUICK

Roadmaster sedan/Estate Wagon base V-8
Overhead valve. Cast-iron block and aluminum heads. Bore & stroke: 4.00 in. x 3.48 in. Displacement: 350 cid (5.7 liters). Compression ratio: 10.5:1. Net brake hp: 260 at 5000 rpm. Taxable hp: 51.20. Net torque: 330 lbs.-ft. at 2400 rpm. Sequential fuel injection. RPO: LT1. VIN Code: P. **NOTE:** The above V-8 was built by Chevrolet Motor Div. It is very similar to, but not identical to the Corvette LT-1 V-8. Last year for V-8 engine in a Buick.

Buick V-8 Firing Order (1949-1976)

1953-1960

All V-8 1-2-7-8-4-5-6-3
Note: *Cylinders are numbered front to back left 2-4-6-8; right 1-3-5-7*

1960-1963

215-cid V-8 1-8-4-3-6-5-7-2

All other V-8s 1-2-7-8-4-5-6-3
Note: *Cylinders are numbered front to back left 2-4-6-8; right 1-3-5-7*

1964-1966

300, 340 V-8 1-8-4-3-6-5-7-2

401, 425 V-8 1-2-7-8-4-5-6-3
Note: *Cylinders are numbered front to back left 2-4-6-8; right 1-3-5-7*

1967-1968

All V-8 1-8-4-3-6-5-7-2
Note: Cylinders are numbered front to back left 2-4-6-8; right 1-3-5-7

Selected Buick V-8 Tune-Up Specs 1953-1965

Year	Model	Spark Plugs Type	Spark Plugs Gap	Point Gap	Deg Cam Angle	Timing	Idle Speed	Compression Pressure	Battery (Delco)
1953	50, 70	AC-44-5	.032	.015	29.5	5B	450 rpm.	150	2EE
1954	40	AC-44-5	.032	.015	29.5	5B	450 rpm.	130	2EE
	50, 60, 70	AC-44-5	.032	.015	29.5	5B	450 rpm.	150	2EE
1955	40	AC-44-5	.032	.015	29.5	5B	450 rpm.	140	3EA/3KM
	50, 60	AC-44-5	.032	.015	29.5	5B	450 rpm.	155	3EA/3KM
	70	AC-44-5	.032	.015	29.5	5B	450 rpm.	155	3EA/3KM
1956	40	AC-44	.032	.015	29.5	5B	450 rpm.	155	3EA/3KM
	50, 60, 70	AC-44	.032	.015	29.5	5B	450 rpm.	155	3EA/3KM
1957	40 w/o Dynaflow	AC-44	.032	.015	30	5B	485 rpm.	140	3EA/3KM
	40 w Dynaflow	AC-44	.032	.015	30	5B	485 rpm.	170	3EA/3KM 50,
60, 70	AC-44	.032	.015	30	5B	485 rpm.	185	3EA/3KM	
1958	40	AC-45S	.030-.035	.016	28-32	5B/12B	475/550	NA	Delco 560
	50, 60, 70	AC-45S	.030-.035	.016	28-32	5B/12B	475/550	NA	Delco 560
Note: *Set 1958 ignition timing 5 degrees before top dead center 5B) for early production and 12B for late production.*									
Note: *Recommended idle speed is for synchromesh above slash/for automatic below slash.*									
1959	4400	AC-44S	.030-.035	.016	28-32	5B/12B	475/550	NA	Delco 1980562
	600, 4700, 4800	AC-44S	.030-.035	.016	28-32	5B/12B	475/550	NA	Delco 1980562
Note: *Set 1959 ignition timing 5B for cars with synchromesh transmission and 12B for cars with automatic transmission.*									
Note: *Recommended idle speed is for synchromesh above slash/for automatic below slash.*									
1960	4400		AC-44S	.032	.012-.017	28-32	5B/12B	475/550	NA Delco 562
4	600, 4700, 4800	AC-44S	.032	.012-.017	28-32	5B/12B	475/550	NA	Delco 562
Note: *Set 1960 ignition timing 5B for cars with synchromesh transmission and 12B for cars with automatic transmission.*									
Note: *Recommended idle speed is for synchromesh above slash/for automatic below slash.*									
1961	Special	AC45FFS	.035	.016	28-32	5B	525/575	160/180	Delco 455
	Skylark	AC44FFS	.035	.016	28-32	5B	525/575	160/180	Delco 562
	All others	AC-44S	.032	.016	28-32	12B	525/575	160/180	Delco 562
Note: *Recommended idle speed is for cars without air conditioning above slash/for cars with air conditioning below slash.*									
1962	Special	AC45FFS .035	.016	28-32	5B	525/575	160	Delco 455	
	Skylark	AC44FFS.035	.016	28-32	5B	525/575	160	Delco 562	
	All others	AC-44S.032	.016	28-32	12B	525/575	160	Delco 562	
Note: *Recommended idle speed is for cars without air conditioning above slash/for cars with air conditioning below slash.*									
1963	Special	AC45FFS	035	.016	28-32	5B	525/575	160	Delco 455
	Skylark	AC44FFS	.035	.016	28-32	5B	525/575	160	Delco 562
	All others manual)	AC-44S	.032	.016	28-32	5B	525/575	160	Delco 562
	All others auto.)	AC-44S	.032	.016	28-32	12B	525/575	160	Delco 562
Note: *Recommended idle speed is for cars without air conditioning above slash/for cars with air conditioning below slash.*									
1964	300-cid Skylark	AC44FFS .035	.016	30	5B	550	160	35M	
	300-cid 2Bbl. V-8	AC44FFS	.032	.016	30	2.5B	550	160	35M
	300-cid 4Bbl. V-8	AC44S	.032	.016	30	2.5B	500	170	35M
	401-cid 4Bbl. V-8	AC44S	.032	.016	30	2.5B	500	170	35M
	425-cid 4Bbl. V-8	AC44S	.032	.016	30	2.5B	525	170	35M
	425-cid 2x4 Bbl V-8	AC44S	.032	.016	30	12B	525	170	35M
1965	300-cid Skylark	AC44FFS	.035	.016	30	2.5B	550	160	35M
	400-cid Skylark	AC44S	.035	.016	30	2.5B	500	160	35M
	300-cid 2Bbl. V-8	AC44FFS	.032	.016	30	2.5B	550	160	35M
	300-cid 4Bbl. V-8	AC44S	.032	.016	30	2.5B	500	170	35M
	401-cid 4Bbl. V-8	AC44S	.032	.016	30	2.5B	500	170	35M
	425-cid 4Bbl. V-8	AC44S	.032	.016	30	2.5B	525	170	35M
	425-cid 2x4 Man.	AC44S	.032	.016	30	2.5B	525	170	35M
	425-cid 2x4 Auto.	AC44S	.032	.016	30	12B	525	170	35M

CADILLAC
1915-2002

The Henry Leland-designed Cadillac V-8 was introduced as an innovation for the 1915 line in September 1914. It was the first mass-produced V-8-powered automobile. The 314-cid 90-degree V-8 had cast-iron cylinder in two blocks of four located exactly opposite each other on an aluminum-copper alloy crankcase. The water jackets and combustion chambers were integral. Water circulation and temperature control was provided by an impeller type pump with a thermostat for each block of cylinders.

There were three bearings on the crankshaft with the throws all in one plane and fork-and-blade connecting rods. Three-ring, solid wall pistons were standard. The single camshaft had eight cams with the shaft, as well as the generator, driven by a silent chain. An updraft carburetor sat on the water-heated intake manifold and the log-type dual exhaust manifolds were inside the vee. The valves (flat-type exhaust and tulip-type intake) were activated by rocker arms with rollers riding on the cam. The V-8 featured a re-circulating lubrication system, which was pressure fed from a gear-type oil pump. The oil flowed through drilled holes in the crankshaft. Other than the addition of detachable cylinder heads in 1917, the 314-cid V-8 lasted without major change until 1925, when improvements began with the last 2,000 V-63 engines.

While retaining the same displacement, the "new 90-degree V-8" of 1926 added a crankcase ventilation system, an oil filter, a screw-type oil filler cap, full-pressure lubrication to the camshaft bearings, the elimination of rocker arms, and other advances. The following year, practically everything on the outside of the motor was relocated and, in 1928, the bore and stroke dimensions changed. The new 341-cid Cadillac V-8 was practically the same as the V-8 used in the company's new LaSalle. It featured offset cylinder blocks, side-by-side connecting rods, single exhaust and numerous accessory changes. The horsepower rose to 90 from 80. In 1930 the bore size was increased to reach 353 cid.

Big changes were made to the Cadillac V-8 in 1932 and resulted in a 21 percent increase in power, most of which came from revised manifolding and a new carburetor design. While the displacement remained at 353 cid, brake horsepower climbed to 115 at 3000 rpm. Lynite aluminum pistons, a higher compression ratio, intake improvements, and a new carburetor produced 120 hp in the 1934 V-8, which was upped to 130 hp in 1935.

There were big changes in 1936, when a new 322-cid, 125-hp V-8 powered Series 60 Cadillacs. Fleetwood bodied cars used an engine of the same design with a 3/8-in. larger bore, 346 cid and 135 hp. The "346" became the only V-8 offered in 1937 and got a new carburetor with a fully automatic electric choke and an oil bath air cleaner. The 346 lasted until after World War II, by which time it was good for 150 hp.

History was made with the all-new, overhead-valve, hydraulic-lifter V-8 that powered all 1949 Cadillacs. Along with the Olds "Rocket 88," it represented the first mass-produced, high-speed, high-compression OHV engine used in an American automobile. The engine served Cadillac extremely well, with only minor changes, for many years. By 1962 it had grown to 390 cid and 325 hp. Although neither engine displacement nor horsepower output changed, the 1963 Cadillac V-8 was completely redesigned. Quieter, smoother and more efficient, the new engine was one inch lower, four inches narrower, and 1-1/4 inches shorter than the 1962 V-8. It also weighed about 82 less pounds, due to the use of aluminum accessory drives.

By the mid-'60s, the second-generation 390-cid Cadillac V-8 grew to 429 cid. This engine produced just 340 hp, but had a high power-to-weight ratio because of its lightweight construction, which featured a cored-out pearlitic malleable iron crankshaft and die-cast aluminum chain cover and accessory drive. To protect its "Standard of the World" tradition, Cadillac built the engines in air-conditioned areas to insure precision and an extremely high level of quality control. The engine grew to 472 cid in 1968 and horsepower hit 375. Even more impressive was a torque rating of 525 lbs.-ft. at 3000 rpm. Ultimately, a version of this V-8 used in the Eldorado was stroked to 500 cid in 1970.

This big V-8 remained in the Eldorado and eventually found its way into other Cadillacs, but history was made in 1975 when a 350-cid Olds V-8 was employed to power the new Seville. Cadillac's move to such a "corporate" engine was a sign of what was to come. However, Cadillac was by no means through with making its own engines, such as the 425-cid V-8 (a de-stroked 472) that bowed in 1977, the 350-cid Cadillac V-8 first used in the 1978 Seville and the 368-cid V-8 offered in 1980 models such as the DeVille, Brougham, and Eldorado.

Innovative, even if not infallible, Cadillac tarnished its image a bit with its "variable-displacement" V-8 of 1981. Developed by Eaton Corp, the engine, which could run on four, six or eight cylinders, was prompted by fuel conservation efforts. A micro-processor was programmed to determine which cylinders weren't necessary under different driving conditions. A solenoid-activated blocker plate then inactivated certain valves in accordance with performance demands. Also new-for-1981 was an Olds-built diesel V-8. Cadillac's next trick, in 1982, was a new 249-cid V-8 for the Seville. By 1985, this would be the only Cadillac-built V-8, but it would not be the last. A 4.5-liter (273-cid) Cadillac-manufactured V-8 was base engine in the DeVille, Fleetwood, Seville, and Eldorado in 1988.

The 1990s brought two additional V-8s that were true "Cadillac" engines. The first was the renowned 4.6-liter (279-cid) Northstar V-8 that bowed in the 1993 Eldorado. It was promoted as a motor that could give 100,000 miles of use without a tune up. And in 1995, a 4.9-liter (300-cid) V-8 became standard equipment in the Sedan DeVille.

1915 CADILLAC

Cadillac Model 51 base V-8
L-head. Heads not detachable. Cast-iron blocks of four on aluminum crankcase. Bore & stroke: 3-1/8 x 5-1/8 in. Displacement: 314 cid. Compression ratio: 4.25:1. Brake hp: 70 at 2400 rpm. Taxable hp: 31.25. Torque or Compression Pressure: 180 lbs.-ft. at 1800-2200 rpm. Main bearings: three. Rockers with roller cam follower. Crankcase capacity: 1.5 gal. Cooling system capacity: 5.25 gal. Carburetor: float feed, auxiliary air control; manufactured by Cadillac under C.F. Johnson patents.

1916 CADILLAC

Cadillac's 314-cid V-8 of 1916 had cast iron cylinders in two blocks of four located exactly opposite each other on an aluminum-copper alloy crankcase. The water jackets and combustion chambers were integral. Water circulation and temperature control were provided by an impeller type pump with a thermostat for each block of cylinders. An updraft carburetor sat on the water-heated intake manifold and the log-type dual exhaust manifolds were inside the vee. The valves (flat-type exhaust and tulip-type intake) were activated by rocker arms with rollers riding on the cam. The V-8 featured a re-circulating lubrication system. Other than the addition of detachable cylinder heads in 1917, the 314-cid V-8 lasted without major change until 1925. (Tom Lieb)

Cadillac Model 53 base V-8
L-head. Heads not detachable. Cast-iron blocks of four on aluminum crankcase. Bore & stroke: 3-1/8 x 5-1/8 in. Displacement: 314 cid. Compression ratio: 4.25:1. Brake hp: 77 at 2600 rpm. Taxable hp: 31.25. Torque or Compression Pressure: 180 lbs.-ft. at 1800-2200 rpm. Main bearings: three. Rockers with roller cam follower. Crankcase capacity: 1.5 gal. Cooling system capacity: 5.25 gal. Carburetor: float feed, auxiliary air control; manufactured by Cadillac under C.F. Johnson patents.

1917 CADILLAC

Cadillac Model 55 base V-8
L-head. Heads not detachable. Cast-iron blocks of four on aluminum crankcase. Bore & stroke: 3-1/8 x 5-1/8 in. Displacement: 314 cid. Compression ratio: 4.25:1. Brake hp: 77 at 2600 rpm. Taxable hp: 31.25. Torque or Compression Pressure: 180 lbs.-ft. at 1800-2200 rpm. Main bearings: three. Rockers with roller cam follower. Crankcase capacity: 1.5 gal. Cooling system capacity: 5.25 gal. Carburetor: float feed, auxiliary air control; manufactured by Cadillac under C.F. Johnson patents.

1918 CADILLAC

Cadillac Model 57 base V-8
L-head. Heads detachable. Cast-iron blocks of four on aluminum crankcase. Bore & stroke: 3-1/8 x 5-1/8 in. Displacement: 314 cid. Compression ratio: 4.25:1. Brake hp: 77 at 2600 rpm. Taxable hp: 31.25. Torque or Compression Pressure: 180 lbs.-ft. at 1800-2200 rpm. Main bearings: three. Rockers with roller cam follower. Crankcase capacity: 1.5 gal. Cooling system capacity: 5.25 gal. Carburetor: float feed, auxiliary air control; manufactured by Cadillac under C.F. Johnson patents.

1919 CADILLAC

Cadillac Model 57 base V-8
L-head. Heads detachable. Cast-iron blocks of four on aluminum crankcase. Bore & stroke: 3-1/8 x 5-1/8 in. Displacement: 314 cid. Compression ratio: 4.25:1. Brake hp: 77 at 2600 rpm. Taxable hp: 31.25. Torque or Compression Pressure: 180 lbs.-ft. at 1800-2200 rpm. Main bearings: three. Rockers with roller cam follower. Crankcase capacity: 1.5 gal. Cooling system capacity: 5.25 gal. Carburetor: float feed, auxiliary air control; manufactured by Cadillac under C.F. Johnson patents.

1920 CADILLAC

Cadillac Model 59 base V-8
L-head. Cast-iron blocks of four on aluminum crankcase. Bore & stroke: 3-1/8 x 5-1/8 in. Displacement: 314 cid. Horsepower: 79 at 2700 rpm. Taxable hp: 31.25. Main bearings: three. Rockers with roller cam follower. Cooling system capacity: 6 gal. Carburetor: float feed, auxiliary air control; manufactured by Cadillac under C.F. Johnson patents.

1921 CADILLAC

Cadillac Model 59 base V-8
L-head. Cast-iron blocks of four on aluminum crankcase. Bore & stroke: 3-1/8 x 5-1/8 in. Displacement: 314 cid. Horsepower: 79 at 2700 rpm. Taxable hp: 31.25. Main bearings: three. Rockers with roller cam follower. Cooling system capacity: 6 gal. Carburetor: float feed, auxiliary Air control; manufactured by Cadillac under C.F. Johnson patents.

1922 CADILLAC

Cadillac Model 61 base V-8
L-head. Cast-iron blocks of four on aluminum crankcase. Bore & stroke: 3-1/8 x 5-1/8 in. Displacement: 314 cid. Brake hp: 79 at 2760 rpm. Taxable hp: 31.25. Compression Pressure: 180 lbs.-ft. at 2,000 rpm. Main bearings: three. Rockers with roller cam follower. Torque compression 180 lbs.-ft. at 2,000 rpm. Crankcase capacity: 6 qt. Cooling system capacity: 6 gal. Carburetor: float feed, auxiliary air control; manufactured by Cadillac under C.F. Johnson patents.

1923 CADILLAC

Cadillac Model 61 base V-8
L-head. Cast-iron blocks of four on aluminum crankcase. Bore & stroke: 3-1/8 x 5-1/8 in. Displacement: 314 cid. Brake hp: 79 at 2760 rpm. Taxable hp: 31.25. Compression Pressure: 180 lbs.-ft. at 2000 rpm. Main bearings: three. Valve lifters: Rockers with roller cam follower. Crankcase capacity: 7 qt. Cooling system capacity: 6 gal. Carburetor: float feed, auxiliary air control; manufactured by Cadillac under C.F. Johnson patents.

1924 CADILLAC

Cadillac Model V-63 base V-8
L-head. Heads detachable. Cast-iron blocks of four on aluminum crankcase. Bore & stroke: 3-1/8 x 5-1/8 in. Displacement: 314 cid. Compression ratio: 4.25:1. Brake hp: 80 at 3000 rpm. Taxable hp: 31.25. Main bearings: three. Rockers with roller cam follower. Crankcase capacity: 7 qt. Cooling system capacity: 6 gal. Carburetor: float feed, auxiliary air control; manufactured by Cadillac under C.F. Johnson patents. Two-plane crankshaft 90-degree counter-balanced.

1925 CADILLAC

Cadillac Model V-63 base V-8
L-head. Heads detachable. Cast-iron blocks of four on aluminum crankcase. Bore & stroke: 3-1/8 x 5-1/8 in. Displacement: 314 cid. Compression ratio: 4.25:1. Brake hp: 80 at 3000 rpm. Taxable hp: 31.25. Main bearings: three. Rockers with roller cam follower. Crankcase capacity: 7 qt. Cooling system capacity: 6 gal. Carburetor: float feed, auxiliary air control; manufactured by Cadillac under C.F. Johnson patents. Two-plane crankshaft 90-degree counter-balanced.
Note: Crankcase ventilation system to eliminate dilution and condensation used on the last 2,000 units of the Cadillac V-63 models built in late 1925 and continued on the 1926 and later Model 314 "New Ninety Degree Cadillac."

1926 CADILLAC

Cadillac Series 314 base V-8
L-head. Cast-iron blocks of four on aluminum crankcase. Bore & stroke: 3-1/8 x 5-1/8 in. Displacement: 314.4 cid. Compression ratio: 5.0:1. Brake hp: 87 at 3000 rpm. Taxable hp: 31.25. Main bearings: three. Rockers with roller cam follower. Crankcase capacity: 8 qt. Cooling system capacity: 5.5 gal. Carburetor: float feed, auxiliary air control; manufactured by Cadillac under C.F. Johnson patents.
Note: A crankcase ventilation system to eliminate dilution and condensation continued on the 1926 and later Cadillacs using this engine family. Oil filter added. Oil level indicator on right side of crankcase instead of inside "vee." Oil filler cap screw type instead of hinged. Camshaft bearing fed full oil pressure, not overflow from regulator. Rocker arms eliminated. Valves and tappets placed at angle to the cylinder bores to line up with cams. One water pump at left. Detachable elbow on cylinder heads. Oil pump at right front corner of crankcase, in place of second water pump. Oil and water pumps driven directly from cross shaft. Starter and generator separate units for first time on Cadillac. Starter vertical at top of flywheel housing, driving through teeth on rear face of flywheel. Generator in front, in vee. Generator/fan driven by belt, eliminating one chain. Tension on single chain maintained by idler sprocket. Front cover of engine made of steel instead of aluminum. Intake manifold now a separate piece.

1927 CADILLAC

Cadillac Series 314-A base V-8
L-head. Cast-iron blocks of four on aluminum crankcase. Bore & stroke: 3-1/8 x 5-1/8 in. Displacement: 314 cid. Compression ratio: 4.25:1. Brake hp: 87 ay 3000 rpm. Taxable hp: 31.25. Main bearings: three. Rockers with roller cam follower. Crankcase capacity: 8 qt. Cooling system capacity: 4.5 gal. Carburetor: float feed, auxiliary air control; manufactured by Cadillac under C.F. Johnson patents.

1928 CADILLAC

Cadillac Series 341 base V-8
L-head. Cast-iron block on copper/aluminum crankcase. Bore & stroke: 3-5/16 x 4-15/16 in. Displacement: 341 cid. Compression ratio: 4.8:1. Brake hp: 90 at 3000 rpm. Taxable hp: 35.1. Torque: 208 lbs.-ft. Compression 90-92 psi at 1000 rpm. Main bearings: three. Valve lifters: Mechanical with rollers riding on cams. Crankcase capacity: 8 qt. Cooling system capacity: 6 gal. Carburetor: manufactured by Cadillac under C.F. Johnson patents.

Cadillac Series 341 optional V-8
L-head. High compression cylinder head. Cast-iron block on copper/aluminum crankcase. Bore & stroke: 3-5/16 x 4-15/16 in. Displacement: 341 cid. Compression ratio: 5.3:1. Brake hp: 90 at 3000 rpm. Taxable hp: 35.1. Torque: 208 lbs.-ft. Compression pressure: 105-107 psi at 1000 rpm. Main bearings: three. Valve lifters: Mechanical with rollers riding on cams. Crankcase capacity: 8 qt. Cooling system capacity: 6 gal. Carburetor: manufactured by Cadillac under C.F. Johnson patents.

1929 CADILLAC

Cadillac Series 341 B base V-8
L-head. Cast-iron block on copper/aluminum crankcase. Bore & stroke: 3-5/16 x 4-15/16 in. Displacement: 341 cid. Compression ratio: 4.8:1. Brake hp: 90 at 3000 rpm. Taxable hp: 35.1. Compression 90-92 psi at 1000 rpm. Main bearings: three. Valve lifters: Mechanical with rollers riding on cams. Crankcase capacity: 8 qt. Cooling system capacity: 6 gal. Carburetor: manufactured by Cadillac under C.F. Johnson patents.

Cadillac Series 341 B optional V-8
L-head. Cast-iron block on copper/aluminum crankcase. Bore & stroke: 3-5/16 x 4-15/16 in. Displacement: 341 cid. Compression ratio: 5.3:1. Brake hp: 90 at 3000 rpm. Taxable hp: 35.1. Compression pressure 105-107 psi at 1000 rpm. Main bearings: three. Valve lifters: Mechanical with rollers riding on cams. Crankcase capacity: 8 qt. Cooling system capacity: 6 gal. Carburetor: manufactured by Cadillac under C.F. Johnson patents.

1930 CADILLAC

Cadillac Series 353 base V-8
L-head. Cast-iron block on silicon/aluminum crankcase. Bore & stroke: 3-3/8 x 4-15/16 in. Displacement: 353 cid. Compression ratio: 5.15:1. Brake hp: 96+ at 3000 rpm. Taxable hp: 36.45. Main bearings: three. Valve lifters: Mechanical with rollers riding on cams. Crankcase capacity: 8 qt. Cooling system capacity: 6 gal. Carburetor: manufactured by Cadillac under C.F. Johnson patents.

Cadillac Series 353 optional V-8
L-head. Cast-iron block on silicon/aluminum crankcase. Bore & stroke: 3-3/8 x 4-15/16 in. Displacement: 353 cid. Compression ratio: 5.03:1. Brake hp: 96+ at 3000 rpm. Taxable hp: 36.45. Main bearings: three. Valve lifters: Mechanical with rollers riding on cams. Crankcase capacity: 8 qt.

Cooling system capacity: 6 gal. Carburetor: manufactured by Cadillac under C.F. Johnson patents.

1931 CADILLAC

Cadillac Series 355 base V-8
L-head. Cast-iron block on aluminum crankcase. Bore & stroke: 3-3/8 x 4-15/16 in. Displacement: 353 cid. Compression ratio: 5.26:1. Brake hp: 95+ at 3000 rpm. Taxable hp: 36.45. Compression pressure: 104 at 1000 rpm. Main bearings: three. Valve lifters: Mechanical. Crankcase capacity: 8 qt. Cooling system capacity: 6 gal. Carburetor: Cadillac/Johnson with intake silencer.

Cadillac Series 355 optional V-8
L-head. Cast-iron block on aluminum crankcase. Bore & stroke: 3-3/8 x 4-15/16 in. Displacement: 353 cid. Compression ratio: 5.35:1. Brake hp: 95+ at 3000 rpm. Taxable hp: 36.45. Compression pressure: 104 at 1000 rpm. Main bearings: three. Valve lifters: Mechanical. Crankcase capacity: 8 qt. Cooling system capacity: 6 gal. Carburetor: Cadillac/Johnson with intake silencer.

1932 CADILLAC

Cadillac Series 355B base V-8
L-head. Cast-iron block on aluminum crankcase. Bore & stroke: 3-3/8 x 4-15/16 in. Displacement: 353 cid. Compression ratio: 5.38:1. Brake hp: 115 at 3000 rpm. Taxable hp: 38.1. Compression pressure: 112 at 1000 rpm. Main bearings: three. Valve lifters: Mechanical. Crankcase capacity: 8 qt. Cooling system capacity: 6.5 gal. Carburetor: Cadillac/Johnson.

Cadillac Series 355B optional V-8
L-head. Cast-iron block on aluminum crankcase. Bore & stroke: 3-3/8 x 4-15/16 in. Displacement: 353 cid. Compression ratio: 5.70:1. Brake hp: 115+ at 3000 rpm. Taxable hp: 38.1. Compression pressure: 112 at 1000 rpm. Main bearings: three. Valve lifters: Mechanical. Crankcase capacity: 8 qt. Cooling system capacity: 6.5 gal. Carburetor: Cadillac/Johnson.

Cadillac Series 355B optional V-8
L-head. Cast-iron block on aluminum crankcase. Bore & stroke: 3-3/8 x 4-15/16 in. Displacement: 353 cid. Compression ratio: 5.20:1. Brake hp: 115 at 3000 rpm. Taxable hp: 38.1. Compression pressure: 112 at 1000 rpm. Main bearings: three. Valve lifters: Mechanical. Crankcase capacity: 8 qt. Cooling system capacity: 6.5 gal. Carburetor: Cadillac/Johnson.

1933 CADILLAC

Cadillac Series 355C base V-8
L-head. Cast-iron block on aluminum crankcase. Bore & stroke: 3-3/8 x 4-15/16 in. Displacement: 353 cid. Compression ratio: 5.4:1. Brake hp: 115 at 3000 rpm. Taxable hp: 38.1. Compression pressure: 104 at 1000 rpm. Main bearings: three. Valve lifters: Mechanical. Crankcase capacity: 8 qt. Cooling system capacity: 6.5 gal. Carburetor: Cadillac/Johnson.

Cadillac Series 355C optional V-8
L-head. Cast-iron block on aluminum crankcase. Bore & stroke: 3-3/8 x 4-15/16 in. Displacement: 353 cid. Compression ratio: 5.70:1. Brake hp: 115+ at 3000 rpm. Taxable hp: 38.1. Compression pressure: 104 at 1000 rpm. Main bearings: three. Valve lifters: Mechanical. Crankcase capacity: 8 qt. Cooling system capacity: 6.5 gal. Carburetor: Cadillac/Johnson.

1934 CADILLAC

Cadillac Series 355D base V-8
L-head. Cast-iron block on aluminum crankcase. Bore & stroke: 3-3/8 x 4-15/16 in. Displacement: 353 cid. Compression ratio: 6.25:1. Brake hp: 120 at 3000 rpm. Taxable hp: 36.45. Compression pressure: 148 at 1000 rpm. Main bearings: three. Valve lifters: Mechanical. Crankcase capacity: 8 qt. Cooling system capacity: 5 gal. Carburetor: Detroit Lubricator Model 51.

Cadillac Series 355D optional V-8
L-head. Cast-iron block on aluminum crankcase. Bore & stroke: 3-3/8 x 4-15/16 in. Displacement: 353 cid. Compression ratio: 5.75:1. Brake hp: 120 at 3000 rpm. Taxable hp: 36.45. Compression pressure: 148 at 1000 rpm. Main bearings: three. Valve lifters: Mechanical. Crankcase capacity: 8 qt. Cooling system capacity: 5 qt. Carburetor: Detroit Lubricator Model 51.

Big changes were made to the Cadillac V-8 in 1932 and resulted in a 21 percent increase in power, most of which came from revised manifolding and a new carburetor design. While the displacement remained at 353 cid, brake horsepower climbed to 115 at 3000 rpm. The radiator badge reveals this to be a 1932 V-8-powered phaeton.

1935 CADILLAC

Cadillac Series 355E base V-8
L-head. Cast-iron block on aluminum crankcase. Bore & stroke: 3-3/8 x 4-15/16 in. Displacement: 353 cid. Compression ratio: 6.25:1. Brake hp: 130 at 3400 rpm. Taxable hp: 36.45. Compression pressure: 148 at 1000 rpm. Main bearings: three. Valve lifters: Mechanical. Crankcase capacity: 8 qt. Cooling system capacity: 5 qt. Carburetor: Detroit Lubricator Model 51.

Cadillac Series 355E optional V-8
L-head. Cast-iron block on aluminum crankcase. Bore & stroke: 3-3/8 x 4-15/16 in. Displacement: 353 cid. Compression ratio: 5.75:1. Brake hp: 130 at 3400 rpm. Taxable hp: 36.45. Compression pressure: 148 at 1000 rpm. Main bearings: three. Valve lifters: Mechanical. Crankcase capacity: 8 qt. Cooling system capacity: 5 qt. Carburetor: Detroit Lubricator Model 51.

1936 CADILLAC

Cadillac Series 36-60 base V-8
L-head. Cast-iron block on blocks cast enbloc with crankcase. Bore & stroke: 3-3/8 x 4-1/2 in. Displacement: 322 cid. Compression ratio: 6.25:1. Brake hp: 125 at 3400 rpm. Taxable hp: 36.45. Compression pressure: 155 at 1000 rpm. Main bearings: three. Valve lifters: Hydraulic. Crankcase capacity: 7 qt. Cooling system capacity: 30 qt. Carburetor: Stromberg EE-25 dual downdraft.

Cadillac Series 36-70 and 36-75 base V-8
L-head. Cast-iron block on blocks cast enbloc with crankcase. Bore & stroke: 3-1/2 x 4-1/2 in. Displacement: 346 cid. Compression ratio: 6.25:1. Brake hp: 135 at 3400 rpm. Taxable hp: 39.20. Compression pressure: 170 at 1000 rpm. Main bearings: three. Valve lifters: Hydraulic. Crankcase capacity: 7 qt. Cooling system capacity: 29 qt. Carburetor: Stromberg EE-25 dual downdraft.

1937 CADILLAC

Cadillac base V-8
L-head. Cast-iron block on blocks cast enbloc with crankcase. Bore & stroke: 3-1/2 x 4-1/2 in. Displacement: 346 cid. Compression ratio: 6.25:1. Brake hp: 135 at 3400 rpm. Taxable hp: 39.20. Compression pressure: 155 at 1000 rpm. Main bearings: three. Valve lifters: Hydraulic. Crankcase capacity: 7 qt. Cooling system capacity: 25 qt. Carburetor: Stromberg AA-25 dual downdraft.

Cadillac optional V-8
L-head. Cast-iron block on blocks cast enbloc with crankcase. Bore & stroke: 3-1/2 x 4-1/2 in. Displacement: 346 cid. Compression ratio: 5.75:1. Brake hp: 135 at 3400 rpm. Taxable hp: 39.20. Main bearings: three. Valve lifters: Hydraulic. Crankcase capacity: 7 qt. Cooling system capacity: 25 qt. Carburetor: Stromberg AA-25 dual downdraft.

1938 CADILLAC

Cadillac Series 60, 60S, 65 base V-8
L-head. Cast-iron block on blocks cast enbloc with crankcase. Bore & stroke: 3-1/2 x 4-1/2 in. Displacement: 346 cid. Compression ratio: 6.25:1. Brake hp: 135 at 3400 rpm. Taxable hp: 39.20. Compression pressure: 155 at 1000. Main bearings: three. Valve lifters: Hydraulic. Crankcase capacity: 7 qt. Cooling system capacity: (Series 60) 24 qt.; (Series 65) 25 qt. Carburetor: Stromberg AAV-25 dual downdraft.

Cadillac Series 75 base V-8
L-head. Cast-iron block on blocks cast enbloc with crankcase. Bore & stroke: 3-1/2 x 4-1/2 in. Displacement: 346 cid. Compression ratio: 6.70:1. Brake hp: 140 at 3400 rpm. Taxable hp: 39.20. Compression pressure: 170 at 1000 rpm. Main bearings: three. Valve lifters: Hydraulic. Crankcase capacity: 7 qt. Cooling system capacity: 25 qt. Carburetor: Stromberg AAV-25 dual downdraft.

1939 CADILLAC

Cadillac Series 60S, 61 base V-8
L-head. Cast-iron block on blocks cast enbloc with crankcase. Bore & stroke: 3-1/2 x 4-1/2 in. Displacement: 346 cid. Compression ratio: 6.25:1. Brake hp: 135 at 3400 rpm. Taxable hp: 39.20. Torque: 250 lbs.-ft. at 1700 rpm. Main bearings: three. Valve lifters: Hydraulic. Crankcase capacity: 7 qt. Cooling system capacity: 24.5 qt. Carburetor: Stromberg AAV-26 dual downdraft.

Cadillac Series 75 base V-8
L-head. Cast-iron block on blocks cast enbloc with crankcase. Bore & stroke: 3-1/2 x 4-1/2 in. Displacement: 346 cid. Compression ratio: 6.70:1. Brake hp: 140 at 3400 rpm. Taxable hp: 39.20. Main bearings: three. Valve lifters: Hydraulic. Crankcase capacity: 7 qt. Cooling system capacity: 24.5 qt. Carburetor: Stromberg AAV-26 dual downdraft.

1940 CADILLAC

Cadillac Series 60S, 62 base V-8
L-head. Cast-iron block on blocks cast enbloc with crankcase. Bore & stroke: 3-1/2 x 4-1/2 in. Displacement: 346 cid. Compression ratio: 6.25:1. Brake hp: 135 at 3400 rpm. Taxable hp: 39.20. Torque: 250 lbs.-ft. at 1700 rpm. Main bearings: three. Valve lifters: Hydraulic. Crankcase capacity: 7 qt. Cooling system capacity: 24.5 qt. Carburetor: Stromberg AAV-26 dual downdraft.

Cadillac Series 72, 75 base V-8
L-head. Cast-iron block on blocks cast enbloc with crankcase. Bore & stroke: 3-1/2 x 4-1/2 in. Displacement: 346 cid. Compression ratio: 6.70:1. Brake hp: 140 at 3400 rpm. Taxable hp: 39.20. Torque: 270 lbs.-ft. at 1700 rpm. Main bearings: three. Valve lifters: Hydraulic. Crankcase capacity: 7 qt. Cooling system capacity: 24.5 qt. Carburetor: Stromberg AAV-26 dual downdraft.

1941 CADILLAC

Cadillac base V-8
L-head. Cast-iron block on blocks cast enbloc with crankcase. Bore & stroke: 3-1/2 x 4-1/2 in. Displacement: 346 cid. Compression ratio: 7.25:1. Brake hp: 150 at 3400 rpm. Taxable hp: 39.20. Torque: 283 lbs.-ft. at 1700. Main bearings: three. Valve lifters: Hydraulic. Crankcase capacity: 7 qt. Cooling system capacity: 24.5 qt. Carburetor: Stromberg AAV-26 dual downdraft or Carter WDO 506-S.

1942 CADILLAC

Cadillac base V-8
L-head. Cast-iron block on blocks cast enbloc with crankcase. Bore & stroke: 3-1/2 x 4-1/2 in. Displacement: 346 cid. Compression ratio: 7.25:1. Brake hp: 150 at 3400 rpm. Taxable hp: 39.20. Torque: 283 lbs.-ft. at 1700. Main bearings: three. Valve lifters: Hydraulic. Crankcase capacity: 7 qt. Cooling system capacity: 24.5 qt. Carburetor: Stromberg AAV-26 dual downdraft or Carter WDO 486-S.

1946 CADILLAC

Cadillac base V-8
L-head. Cast-iron block. Displacement: 346 cid. Bore & stroke: 3-1/2 x 4-1/2 in. Compression ratio: 7.25:1. Brake hp: 150 at 3600 rpm. Taxable hp: 39.20. Torque: 274 lbs.-ft. at 1600 rpm. Three main bearings. Hydraulic valve lifters. Crankcase capacity: 7 qt. (add 1 qt. if new filter installed). Cooling system capacity: 25 qt. (add 1 qt. for heater). Carburetor: Carter WCD two-barrel Models 595-S or 595-SA or Stromberg AAV 26 two-barrel Models 380154 or 380871.

1947 CADILLAC

Cadillac base V-8
L-head. Cast-iron block. Displacement: 346 cid. Bore & stroke: 3-1/2 x 4-1/2 in. Compression ratio: 7.25:1. Brake hp: 150 at 3600 rpm. Taxable hp: 39.20. Torque: 274 lbs.-ft. at 1600 rpm. Three main bearings. Hydraulic valve lifters. Crankcase capacity: 7 qt. (add 1 qt. if new filter installed). Cooling system capacity: 25 qt. (add 1 qt. for heater). Carburetor: Carter WCD two-barrel Models 595-S or 595-SA or Stromberg AAV 26 two-barrel Models 380154 or 380871.

1948 CADILLAC

Cadillac base V-8
L-head. Cast-iron block. Displacement: 346 cid. Bore & stroke: 3-1/2 x 4-1/2 in. Compression ratio: 7.25:1. Brake hp: 150 at 3400 rpm. Taxable hp: 39.20. Torque: 283 lbs.-ft. at 1600 rpm. Three main bearings. Hydraulic valve lifters. Crankcase capacity: 7 qt. (add 1 qt. if new filter installed). Cooling system capacity: 25 qt. (add 1 qt. for heater). Carburetor: Carter WCD two-barrel Models 595-S or 595-SA or Stromberg AAV 26 two-barrel Models 380154 or 380871.

1949 CADILLAC

Cadillac's 300,000th overhead-valve V-8 built is now on display in the Cadillac Historical Collection. Cadillac sold 92,554 cars with this engine in 1949, 103,857 in 1950, and 110,340 in 1951. The overhead-valve V-8 weighed 188 pounds less than the previous flathead V-8.

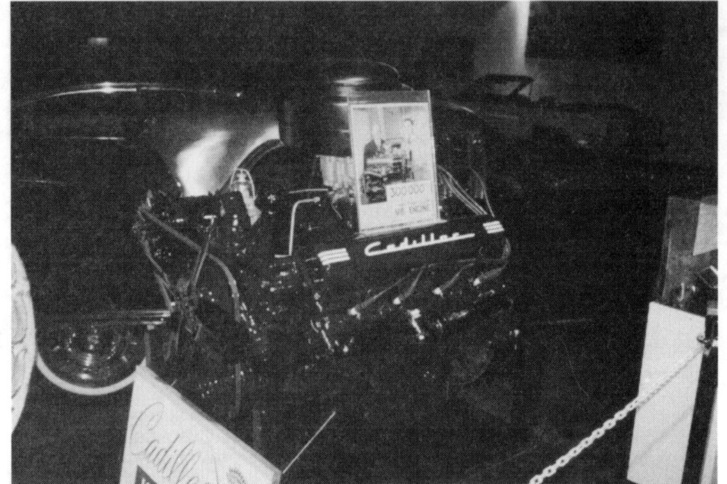
History was made with the all-new, overhead-valve, hydraulic-lifter V-8 that powered all 1949 Cadillacs. Along with the Olds "Rocket 88," it represented the first mass-produced, high-speed, high-compression OHV engine used in an American automobile. The engine served Cadillac extremely well, with only minor changes, for many years. The 331-cid engine produced 160 hp.

Cadillac base V-8
Overhead valves. Cast-iron block. Displacement: 331 cid. Bore & stroke: 3-13/16 x 3-5/8 in. Compression ratio: 7.5:1. Brake hp: 160 at 3800 rpm. Taxable hp: 46.5. Torque: 312 lbs.-ft. at 1800 rpm. Five main bearings. Hydraulic valve lifters. Crankcase capacity: 5 qt. (add 1 qt. if new filter installed). Cooling system capacity: 18 qt. (add 1 qt. for heater). Carburetor: Carter WCD two-barrel Model 742-S.

1950 CADILLAC

Cadillac base V-8
Overhead valves. Cast-iron block. Displacement: 331 cid. Bore & stroke: 3-13/16 x 3-5/8 in. Compression ratio: 7.5:1. Brake hp: 160 at 3600 rpm. Taxable hp: 46.5. Torque: 312 lbs.-ft. at 1800 rpm. Five main bearings. Hydraulic valve lifters. Crankcase capacity: 5 qt. (add 1 qt. if new filter installed). Cooling system capacity: 17.25 qt. (add 1 qt. for heater). Carburetor: Carter WCD two-barrel Models 682-S or 722-S.

1951 CADILLAC

Cadillac base V-8
Overhead valves. Cast-iron block. Displacement: 331 cid. Bore & stroke: 3-13/16 x 3-5/8 in. Compression ratio: 7.5:1. Brake hp: 160 at 3800 rpm. Taxable hp: 46.5. Torque: 312 lbs.-ft. at 1800 rpm. Five main bearings. Hydraulic valve lifters. Crankcase capacity: 5 qt. (add 1 qt. if new filter installed). Cooling system capacity: 18 qt. (add 1 qt. for heater). Carburetor: Carter WCD two-barrel Model 845-S or Rochester BB two-barrel Model 7004200.

1952 CADILLAC

Cadillac base V-8
Overhead valves. Cast-iron block. Displacement: 331 cid. Bore & stroke: 3-13/16 x 3-5/8 in. Compression ratio: 7.5:1. Brake hp: 190 at 4000 rpm. Taxable hp: 46.5. Torque: 322 lbs.-ft. at 2400 rpm. Five main bearings. Hydraulic valve lifters. Crankcase capacity: 5 qt. (add 1 qt. if new filter installed). Cooling system capacity: 18 qt. (add 1 qt. for heater). Carburetor: Carter WFCB four-barrel Model 896-S or Rochester 4GC four-barrel Model 7004500.

1953 CADILLAC

Cadillac base V-8
Overhead valves. Cast-iron block. Displacement: 331 cid. Bore & stroke: 3-13/16 x 3-5/8 in. Compression ratio: 8.25:1. Brake hp: 210 at 4150 rpm. Taxable hp: 46.5. Torque: 330 lbs.-ft. at 2700 rpm. Five main bearings. Hydraulic valve lifters. Crankcase capacity: 5 qt. (add 1 qt. if new filter installed). Cooling system capacity: 19.75 qt. (add 1 qt. for heater). Carburetor: (Hydra-Matic) Carter WFCB four-barrel Model 2005-S or Rochester 4GC four-barrel Model 7005100; (Dynaflow) Carter WFCB four-barrel Model 2088-S or Model 2119-S or Rochester 4GC four-barrel Model 7006215.
Note: Buick Dynaflow transmissions were used for a short while in Cadillacs built after a fire in General Motors' Hyrdra-Matic transmission factory in Livonia, Michigan shut down production of Hydra-Matic transmissions late in 1953.

1954 CADILLAC

Cadillac base V-8
Overhead valves. Cast-iron block. Displacement: 331 cid. Bore & stroke: 3-13/16 x 3-5/8 in. Compression ratio: 8.25:1. Brake hp: 230 at 4400 rpm. Taxable hp: 46.5. Torque: 330 lbs.-ft. at 2700 rpm. Five main bearings. Hydraulic valve lifters. Crankcase capacity: 5 qt. (add 1 qt. if new filter installed). Cooling system capacity: 19.75 qt. (add 1 qt. for heater). Carburetor: Carter WFCB four-barrel Models 2143-S, 2109-S or 2110-S or Rochester 4GC four-barrel Model 7006963 (with air conditioning) or Model 7006962 (without air conditioning).

1955 CADILLAC

Cadillac (except Eldorado) base V-8
Overhead valves. Cast-iron block. Displacement: 331 cid. Bore & stroke: 3-13/16 x 3-5/8 in. Compression ratio: 9.0:1. Brake hp: 250 at 4600 rpm. Taxable hp: 46.5. Torque: 345 lbs.-ft. at 2800 rpm. Five main bearings. Hydraulic valve lifters. Crankcase capacity: 5 qt. (add 1 qt. if new filter installed). Cooling system capacity: 18 qt. (add 2 qt. for heater). Carburetor: Carter WFCB four-barrel Models 2355-S, 2354-S, 2185-S, 2186-S, 2266-S, 2267-S and 2255 or Rochester 4GC four-barrel Model 7007971 (with air conditioning) or Model 7007970 (without air conditioning).

Eldorado base V-8
Overhead valves. Cast-iron block. Displacement: 331 cid. Bore & stroke: 3-13/16 x 3-5/8 in. Compression ratio: 9.0:1. Brake hp: 270 at 4800 rpm. Taxable hp: 46.5. Torque: 345 lbs.-ft. at 3200 rpm. Five main bearings. Hydraulic valve lifters. Crankcase capacity: 5 qt. (add 1 qt. if new filter installed). Cooling system capacity: 18 qt. (add 2 qt. for heater). Carburetor: Two Rochester 4GC four-barrels (front, all) Model 7007240; (rear without air conditioning) Model 7007240; (rear with air conditioning) Model 7007241.
Note: The 270-hp Eldorado engine was available as an optional "Power Package" in other 1955 Cadillac models for $161 extra.

1956 CADILLAC

Cadillac (except Eldorado) base V-8
Overhead valves. Cast-iron block. Displacement: 365 cid. Bore & stroke: 4.00 x 3.63 in. Compression ratio: 9.75:1. Brake hp: 285 at 4600 rpm. Taxable hp: 51.2. Torque: 400 lbs.-ft. at 2800 rpm. Five main bearings. Hydraulic valve lifters. Crankcase capacity: 5 qt. (add 1 qt. if new filter installed). Cooling system capacity: 17.5 qt. (add 2 qt. for heater). Carburetor: Carter WFCB four-barrel Models 2370S.

Eldorado base V-8
Overhead valves. Cast-iron block. Displacement: 365 cid. Bore & stroke: 4.00 x 3.63 in. Compression ratio: 9.75:1. Brake hp: 305 at 4700 rpm. Taxable hp: 51.2. Torque: 400 lbs.-ft. at 3200 rpm. Five main bearings. Hydraulic valve lifters. Crankcase capacity: 5 qt. (add 1 qt. if new filter installed). Cooling system capacity: 17.5 qt. (add 2 qt. for heater). Carburetor: Two Carter WFCB four-barrel Model 2371. Carburetor: Carter WFCB four-barrel Model 2370S.
Note: The 305-hp Eldorado engine was available as an optional "Power Package" in other 1956 Cadillac models at extra cost.

Cadillac promoted its 1956 V-8 as a "new" engine. The 365-cid power plant had a larger 4.00-inch bore size. With a 9.75:1 compression ratio it produced 285 hp. The Eldorado version, which featured dual four-barrel carburetors, produced 305 hp. "The increased horsepower, high displacement power plant provides a major torque increase in city speed ranges where safety often depends on split-second acceleration," said Cadillac. Note the heavy-duty type AC air cleaner.

1957 CADILLAC

This is the 1957 Cadillac power plant featuring the most powerful high-performance engine ever offered by Cadillac. With a 10.0:1 compression ratio, four-inch bore, and a displacement of 365 cid, this engine produced 300 hp. The transmission it's attached to was an improved "contolled coupling" Hydra-Matic. Cadillac's Q-code engine, available on 1957 Eldorados, used dual four-barrel carburetors and was rated at 325 hp.

Cadillac (except Eldorado) base V-8
Overhead valves. Cast-iron block. Displacement: 365 cid. Bore & stroke: 4.00 x 3.625 in. Compression ratio: 10.0:1. Brake hp: 300 at 4800 rpm. Taxable hp: 51.2. Torque: 400 lbs.-ft. at 3200 rpm. Five main bearings. Hydraulic valve lifters. Crankcase capacity: 5 qt. (add 1 qt. if new filter installed). Cooling system capacity: 19.5 qt. (add 2 qt. for heater). Dual exhaust. Carburetor: Rochester four-barrel Model 7015701 or Carter WFCB four-barrel Model 2370-S.

Eldorado base V-8
Overhead valves. Cast-iron block. Displacement: 365 cid. Bore & stroke: 4.00 x 3.625 in. Compression ratio: 10.0:1. Brake hp: 325 at 4800 rpm. Taxable hp: 51.2. Torque: 400 lbs.-ft. at 3200 rpm. Five main bearings. Hydraulic valve lifters. Crankcase capacity: 5 qt. (add 1 qt. if new filter installed). Cooling system capacity: 19.5 qt. (add 2 qt. for heater). Dual

exhaust. Carburetor: Two Carter four barrels (front) Model 2584S; (rear) Model 2583S. Carburetor: Carter WFCB four-barrel Models 2370-S.
Note: The 325-hp Eldorado Brougham engine was available as an option on the Eldorado Seville and Biarritz only. In normal installations on these models the front carburetor was changed to a Carter Model 258-S. When air conditioning was also installed the front carburetor was the same model used on the Brougham, which came with air conditioning as standard equipment.

1958 CADILLAC

Note the "batwing" air cleaner on this 1958 Cadillac Eldorado V-8 with three two-barrel carburetors. Batwing air cleaners were used with "Eldo" V-8s from 1955, but this particular style was a one-year-only item. Twin air cleaners are mounted at the outer ends of the wings and the carbs run up the middle. The unit is hollow and the wings duct air to the outboard air cleaners. This 335-hp V-8 was standard on the Eldorado and optional in all other Cadillacs. It was again a 365-cid engine with a 4.00 x 3.625-in. bore and stroke, but it gained 10 horsepower for 1958.

Cadillac (except Eldorado) base V-8
Overhead valve. Cast-iron block. Displacement: 365 cid. Bore & stroke: 4.00 x 3.625 in. Compression ratio: 10.25:1. Brake hp: 310 at 4800 rpm. Taxable hp: 51.2. Torque: 405 lbs.-ft. at 3100 rpm. Five main bearings. Hydraulic valve lifters. Crankcase capacity: 5 qt. (Add 1 qt. if new filter installed). Cooling system capacity: (Series 75) 22 qt.; (Series 60 and Series 62) 19.5 qt. (add 1 qt. for heater). Dual exhaust. Carburetor: Carter AFB four-barrel (without air conditioning) Model 2696S; (with air conditioning) Model 2697S-2802S. Alternate carburetor: Rochester 4GC (early production without air conditioning) Model 7012010; (late production without air conditioning) Model 7012910; (early production with air conditioning) Model 7012011; (late production with air conditioning) Model 7012811.

Eldorado base V-8
Overhead valve. Cast-iron block. Displacement: 365 cid. Bore & stroke: 4.00 x 3.625 in. Compression ratio: 10.25:1. Brake hp: 335 at 4800 rpm. Taxable hp: 51.2. Torque: 405 lbs.-ft. at 3400 rpm. Five main bearings. Hydraulic valve lifters. Crankcase capacity: 5 qt. (add 1 qt. if new filter installed). Cooling system capacity: (Series 75) 22 qt.; (Series 60 and Series 62) 19.5 qt. (add 1 qt. for heater). Dual exhaust. Carburetion: Rochester 2GC Triple Power Pack (front carburetor with or without air conditioning) Model 7012201 used in early production and Model 7012901 used in late production; (center carburetor without air conditioning) Model 7012202 used in early production and Model 7012902 used in late production; (center carburetor with air conditioning) Model 7012205 used in early production and Model 7012905 used in late production; (rear carburetor with or without air conditioning) Model 7012203 used in early production and Model 7012903 used in late production.
Note: The 335-hp Eldorado engine was used on all Eldorados as standard equipment. This engine was also optional for all other Cadillacs.

1959 CADILLAC

Cadillac (except Eldorado) base V-8
Overhead valve. Cast-iron block. Displacement: 390 cid. Bore & stroke: 4.00 x 3.875 in. Compression ratio: 10.5:1. Brake hp: 325 at 4800 rpm. Taxable hp: 51.2. Torque: 430 lbs.-ft. at 3100 rpm. Five main bearings. Hydraulic valve lifters. Crankcase capacity: 5 qt. (add 1 qt. if new filter installed). Cooling system capacity: (Series 60 and Series 62) 18.5 qt. without heater; 19.5 qt. with heater; (Series 67) 18.5 qt. without heater; 20.75 qt. with heater; (Series 69) 19.5 qt. with heater only. Dual exhaust. Carburetor: Carter AFB four-barrel (without air conditioning) Model 2814S; (with air conditioning) Model 2815S. Alternate carburetor: Rochester 4GC (without air conditioning) Model 7013030; (with air conditioning) Model 7013031.

Eldorado base V-8
Overhead valve. Cast-iron block. Displacement: 390 cid. Bore & stroke: 4.00 x 3.875 in. Compression ratio: 10.5:1. Brake hp: 345 at 4800 rpm. Taxable hp: 51.2. Torque: 435 lbs.-ft. at 34000. Five main bearings. Hydraulic valve lifters. Crankcase capacity: 5 qt. (add 1 qt. if new filter installed). Cooling system capacity: (Series 63) 18.5 qt. without heater; 19.5 qt. with heater. Dual exhaust. Carburetion: Rochester 2GC Triple Power Pack (front carburetor with or without air conditioning) Model 7013033; (center carburetor without air conditioning) Model 7013034; (center carburetor with air conditioning) Model 7013037 used in early production; (rear carburetor with or without air conditioning) Model 7013035.
Note: The 345-hp Eldorado engine was optional for all other Cadillacs at $134.30 extra.

1960 CADILLAC

Cadillac (except Eldorado) base V-8
Overhead valve. Cast-iron block. Displacement: 390 cid. Bore & stroke: 4.00 x 3.875 in. Compression ratio: 10.5:1. Brake hp: 325 at 4800 rpm. Taxable hp: 51.2. Torque: 430 lbs.-ft. at 3100 rpm. Five main bearings. Hydraulic valve lifters. Crankcase capacity: 5 qt. (add 1 qt. if new filter installed). Cooling system capacity: (without heater) 18.5 qt.; (with heater) 19.25 qt.; (with air conditioning) 19.75 qt. Dual exhaust. Carburetor: Carter four-barrel (without air conditioning) Model 2951S; (with air conditioning) Model 2952S. Alternate carburetor: Rochester 4GC (without air conditioning) Model 7013030; (with air conditioning) Model 7013031.

Eldorado/Eldorado Brougham base V-8
Overhead valve. Cast-iron block. Displacement: 390 cid. Bore & stroke: 4.00 x 3.875 in. Compression ratio: 10.5:1. Brake hp: 345 at 4800 rpm. Taxable hp: 51.2. Torque: 435 lbs.-ft. at 34000. Five main bearings. Hydraulic valve lifters. Crankcase capacity: 5 qt. (add 1 qt. if new filter installed). Cooling system capacity: (without heater) 18.5 qt.; (with heater) 19.25 qt.; (with air conditioning) 19.75 qt. Dual exhaust. Rochester 2GC Triple Power Pack Carburetion: (front carburetor with or without air conditioning) Model 7013033; (center carburetor without air conditioning) Model 7013034; (center carburetor with air conditioning) Model 7013037 used in early production; (rear carburetor with or without air conditioning) Model 7013035.
Note: The 345-hp Eldorado engine was optional for all other Cadillacs at $134.30 extra.

1961 CADILLAC

Cadillac base V-8
Overhead valve. Cast-iron block. Displacement: 390 cid. Bore & stroke: 4.00 x 3.875 in. Compression ratio: 10.5:1. Brake hp: 325 at 4800 rpm. Taxable hp: 51.2. Torque: 430 lbs.-ft. at 3100 rpm. Five main bearings. Hydraulic valve lifters. Crankcase capacity: 5 qt. (add 1 qt. if new filter installed). Cooling system capacity: (all without heater) 18.25 qt.; (Series 60 and Series 62 with heater) 19.25 qt.; (Series 75 with heater only) 20.75 qt. Single exhaust with crossover. Carburetor: Carter four-barrel (without air conditioning) Model 3177S; (with air conditioning) Model 3178S. Alternate carburetor: Rochester four-barrel (without air conditioning) Model 7019030; (with air conditioning) Model 7019031.

1962 CADILLAC

Cadillac base V-8
Overhead valve. Cast-iron block. Displacement: 390 cid. Bore & stroke: 4.00 x 3.875 in. Compression ratio: 10.5:1. Brake hp: 325 at 4800 rpm. Taxable hp: 51.2. Torque: 430 lbs.-ft. at 3100 rpm. Five main bearings. Hydraulic valve lifters. Crankcase capacity: 4 qt. (add 1 qt. if new filter installed). Cooling system capacity: (Series 60, Series 62 and Series 63 with heater only) 19.33 qt.; (Series 67 with heater only) 20.89 qt. Single exhaust with crossover. Carburetor: Carter four-barrel (without air conditioning) Model 3351S; (with air conditioning) Model 3352S. Alternate carburetor: Rochester four-barrel (without air conditioning) Model 7019030; (with air conditioning) Model 7019031.

1963 CADILLAC

Cadillac base V-8
Overhead valve. Cast-iron block. Displacement: 390 cid. Bore & stroke: 4.00 x 3.875 in. Compression ratio: 10.5:1. Brake hp: 325 at 4800 rpm. Taxable hp: 51.2. Torque: 430 lbs.-ft. at 3100 rpm. Five main bearings. Hydraulic valve lifters. Crankcase capacity: 4 qt. (add 1 qt. if new filter installed). Cooling system capacity: 17.25 qt. Single exhaust with crossover. Carburetor: Rochester 4GC four-barrel (without air conditioning) Model 7019030; (with air conditioning) Model 7019031.
Note: Although neither engine displacement or horsepower output changed, the 1963 Cadillac V-8 was completely redesigned. Quieter, smoother and more efficient, the new engine was one inch lower, four inches narrower and 1-1/4 inches shorter than the 1962 V-8. It also weighed about 82 less pounds, due to the use of aluminum accessory drives.

1964 CADILLAC

Cadillac base V-8
Overhead valve. Cast-iron block. Displacement: 429 cid. Bore & stroke: 4.13 x 4.00 in. Compression ratio: 10.5:1. Brake hp: 340 at 4600 rpm. Taxable hp: 54.6. Torque: 480 lbs.-ft. at 3000 rpm. Five main bearings. Hydraulic valve lifters. Crankcase capacity: 4 qt. (add 1 qt. if new filter installed). Cooling system capacity: 17.25 qt. Single exhaust with crossover. Carburetor: Carter AFB four-barrel Model 3655S. Alternate carburetor: Rochester 4GC four-barrel.

1965 CADILLAC

Cadillac base V-8
Overhead valve. Cast-iron block. Displacement: 429 cid. Bore & stroke: 4.13 x 4.00 in. Compression ratio: 10.5:1. Brake hp: 340 at 4600 rpm. Taxable hp: 54.6. Torque: 480 lbs.-ft. at 3000 rpm. Five main bearings. Hydraulic valve lifters. Crankcase capacity: 4 qt. (add 1 qt. if new filter installed). Cooling system capacity: 17.25 qt. Single exhaust with crossover. Carburetor: Carter AFB four-barrel Model 3903S. Alternate carburetor: Rochester 4GC four-barrel.

1966 CADILLAC

Cadillac base V-8
Overhead valve. Cast-iron block. Displacement: 429 cid. Bore & stroke: 4.13 x 4.00 in. Compression ratio: 10.5:1. Brake hp: 340 at 4600 rpm. Taxable hp: 54.6. Torque: 480 lbs.-ft. at 3000 rpm. Five main bearings. Hydraulic valve lifters. Crankcase capacity: 4 qt. (add 1 qt. if new filter installed). Cooling system capacity: 18 qt. Single exhaust with crossover. Carburetor: Carter AFB four-barrel: (Fleetwood, Calais and DeVille) Model 4168S; (Fleetwood 75 limousine) 4169S; (Fleetwood Eldorado) 4171S. Alternate carburetor: Rochester 4GC four-barrel (Fleetwood, Calais and DeVille) Model 7026030.

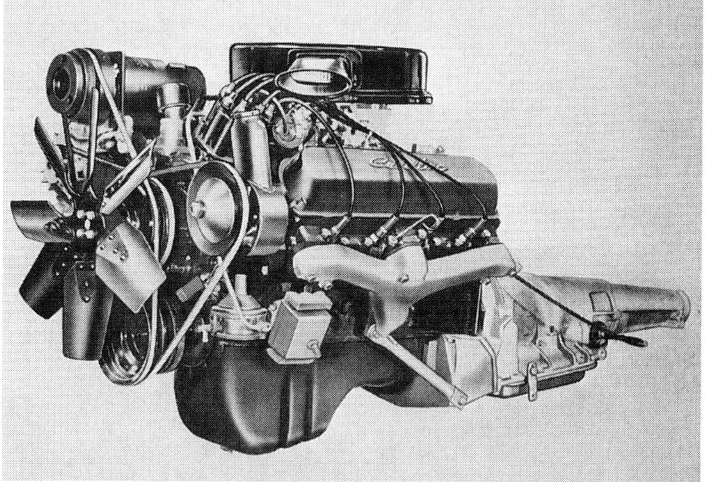

Despite the large size of the 429-cid Cadillac V-8 of 1966, it had a very high weight-to-horsepower ratio of 1.77. This engine generated 340 hp at 4600 rpm and tipped the scales at 600 pounds. It produced 0.79 horsepower per cubic inch. The Cadillac name was carried on the valve covers in script. This power plant was related to the famous 1948 engine, although there had been so many refinements over the years that the two looked very little alike.

1967 CADILLAC

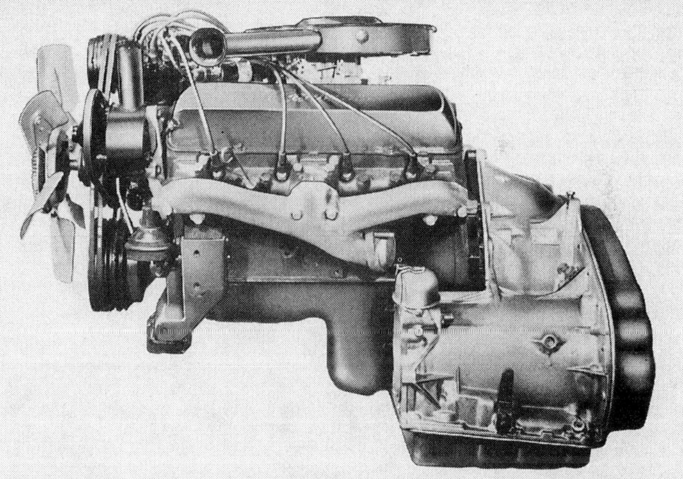

Here is a view of the 429-cid Cadillac V-8 equipped with the front-wheel-drive transaxle setup used in the 1967 Eldorado. The engine was identical to the 340-hp version used in conventional rear-wheel-drive Cadillacs, but parts such as the oil pan, exhaust manifold, and motor mounts were specifically designed for the new FWD car. The engine alone weighed about 600 pounds.

Cadillac base V-8
Overhead valve. Cast-iron block. Displacement: 429 cid. Bore & stroke: 4.13 x 4.00 in. Compression ratio: 10.5:1. Brake hp: 340 at 4600 rpm. Taxable hp: 54.6. Torque: 480 lbs.-ft. at 3000 rpm. Five main bearings. Hydraulic valve lifters. Crankcase capacity: 4 qt. (add 1 qt. if new filter installed). Cooling system capacity: (Eldorado) 17 qt.; (Other models) 18 qt. Dual exhaust on Eldorado, single exhaust system with crossover on other models. Carburetor: Rochester 4MV Quadrajet four-barrel: (Calais, deVille and Fleetwood) Model 7028230; (Fleetwood 75 limousine) Model 7027233; (Fleetwood Eldorado) Model 7027234.
Note: An air-injection-reactor emissions system was $34.13 extra for all 1967 Cadillacs and was required on cars built for California sale. Closed positive crankcase ventilators were a $4 extra on all Cadillacs and required on all cars built for California sale. California cars also had a

different Rochester 4MV carburetor and recommended idle speed setting was 550 rpm.

1968 CADILLAC

This sectional-view drawing shows the inner workings of 1968's new 472-cid Cadillac V-8. This engine had a new 4.30 x 4.06-in. bore and stroke and a 10.5:1 compression ratio. With a single four-barrel carburetor it produced 375 hp at 4400 rpm and 525 foot-pounds of torque at 3000 rpm. It delivered .795 horsepower per cubic inch of displacement.

Cadillac base V-8
Overhead valve. Cast-iron block. Displacement: 472 cid. Bore & stroke: 4.30 x 4.06 in. Compression ratio: 10.5:1. Brake hp: 375 at 4400 rpm. Taxable hp: 59.2. Torque: 525 lbs.-ft. at 3000 rpm. Five main bearings. Hydraulic valve lifters. Crankcase capacity: (Eldorado) 5 qt.; (other models) 4 qt. (add 1 qt. if new filter installed). Cooling system capacity: (Series 75) 24.8 qt.; (Eldorado) 20 qt.); (All Cadillacs other series) 21.3 qt. Dual exhaust on Eldorado, single exhaust system with crossover on other models. Carburetor: Rochester 4MV Quadrajet four-barrel: (Calais, DeVille and Fleetwood) Model 7028230; (Eldorado) Model 7028234.
Note: The new engine featured a metal temperature monitoring device; an air-injection emissions control system; an integral air conditioning system (still optional); a cast crankshaft and cast connecting rods; an integral water crossover pipe with thermostatic passages; engine bearings with more surface area and a 15-plate battery.

1969 CADILLAC

Cadillac base V-8
Overhead valve. Cast-iron block. Displacement: 472 cid. Bore & stroke: 4.30 x 4.06 in. Compression ratio: 10.5:1. Brake hp: 375 at 4400 rpm. Taxable hp: 59.2. Torque: 525 lbs.-ft. at 3000 rpm. Five main bearings. Hydraulic valve lifters. Crankcase capacity: (Eldorado) 5 qt.; (other models) 4 qt. (add 1 qt. if new filter installed). Cooling system capacity: (Series 75) 24.8 qt.; (Eldorado) 21.8 qt. with visual capacity tank; (All Cadillacs other series) 21.3 qt. Dual exhaust on Eldorado, single exhaust system with crossover on other models. Carburetor: Rochester 4MV Quadrajet four-barrel Model (Calais, DeVille and Fleetwood) Model 7028230; (Eldorado) Model 7028234.
Note: On 1969-1974 Cadillacs, the air conditioning must be turned off before making hot idle speed adjustments. In addition, disconnect the hose at the vacuum release cylinder and close the hot idle compensator by pressing finger or eraser end of a pencil on the compensator.

The 472-cid Cadillac V-8 was also used in the front-wheel-drive Eldorado. In this case, the transmission was mounted alongside the engine on the driver side of the car. Shown here is the 1969 version. The 472-cid V-8 was a big square engine measuring 28 in. long, 28 in. wide and 23 in. high. Engine code Number 9 identified this motor.

1970 CADILLAC

In 1970, the new front-wheel-drive Cadillac Eldorado came with a huge 500-cid V-8 which was the 472-cid V-8 with a large 4.304-in. piston stroke. This engine had the same exterior dimensions and appearance as the 472, but it produced .800 horsepower per cubic inch compared to the 472's .794. It was rated for 400 hp at 4400 rpm and 550 foot-pounds of torque at 3000 rpm.

Cadillac (except Eldorado) base V-8
Overhead valve. Cast-iron block. Displacement: 472 cid. Bore & stroke: 4.30 x 4.06 in. Compression ratio: 10.0:1. Brake hp: 375 at 4400 rpm. Taxable hp: 59.2. Torque: 525 lbs.-ft. at 3000 rpm. Five main bearings. Hydraulic valve lifters. Crankcase capacity: 4 qt. (add 1 qt. if new filter installed). Cooling system capacity: (Series 75) 24.8 qt.; (All Cadillacs other series except Eldorado) 21.3 qt. Single exhaust system with crossover. Carburetor: Rochester 4MV Quadrajet four-barrel: Model 7028230.

Eldorado base V-8
Overhead valve. Cast-iron block. Displacement: 500 cid. Bore & stroke: 4.30 x 4.30 in. Compression ratio: 10.0:1. Brake hp: 400 at 4400 rpm. Taxable hp: 59.2. Torque: 550 lbs.-ft. at 3000 rpm. Five main bearings. Hydraulic valve lifters. Crankcase capacity: 5 qt. (add 1 qt. if filter installed). Cooling system capacity: 21.8 qt. with visual capacity tank. Dual exhaust. Carburetor: Rochester 4MV Quadrajet four-barrel Model 7028230.
Note: On 1969-1974 Cadillacs, the air conditioning must be turned off before making hot idle speed adjustments. In addition, disconnect the hose at the vacuum release cylinder and close the hot idle compensator by pressing finger or eraser end of a pencil on the compensator.

1971 CADILLAC

Cadillac (except Eldorado) base V-8
Overhead valve. Cast-iron block. Displacement: 472 cid. Bore & stroke: 4.30 x 4.06 in. Compression ratio: 8.8:1. Brake hp: 345 at 4400 rpm. Net brake hp: 220 at 4000 rpm. Taxable hp: 59.2. Torque: 500 lbs.-ft. at 2800 rpm. Net torque: 380 lbs.-ft. at 2400 rpm. Five main bearings. Hydraulic valve lifters. Crankcase capacity: 4 qt. (add 1 qt. if new filter installed). Cooling system capacity: (Series 75) 24.8 qt.; (All Cadillacs other series except Eldorado) 21.3 qt. Single exhaust system with crossover. Carburetor: Rochester 4MV Quadrajet four-barrel Model 7028230.

Eldorado base V-8
Overhead valve. Cast-iron block. Displacement: 500 cid. Bore & stroke: 4.30 x 4.30 in. Compression ratio: 9.0:1. Brake hp: 365 at 4400 rpm. Taxable hp: 59.2. Net brake hp: 235 at 3800 rpm. Torque: 535 lbs.-ft. at 2800 rpm. Net torque: 410 lbs.-ft. at 2400 rpm. Five main bearings. Hydraulic valve lifters. Crankcase capacity: 5 qt. (add 1 qt. if new filter installed). Cooling system capacity: 21.8 qt. with visual capacity tank. Single exhaust system with crossover. Carburetor: Rochester 4MV Quadrajet four-barrel Model 7028230.
Note: On 1969-1974 Cadillacs, the air conditioning must be turned off before making hot idle speed adjustments. In addition, disconnect the hose at the vacuum release cylinder and close the hot idle compensator by pressing finger or eraser end of a pencil on the compensator.

1972 CADILLAC

Cadillac (except Eldorado) base V-8
Overhead valve. Cast-iron block. Displacement: 472 cid. Bore & stroke: 4.30 x 4.06 in. Compression ratio: 8.5:1. Gross hp: 345 at 4400 rpm. Net brake hp: 220 at 4000 rpm. Taxable hp: 59.2. Torque: 365 lbs.-ft. at 2400 rpm. Five main bearings. Hydraulic valve lifters. Crankcase capacity: 4 qt. (add 1 qt. if new filter installed). Cooling system capacity: (Series 75) 24.8 qt.; (All Cadillacs other series except Eldorado) 21.3 qt. with heater only; 21.8 qt. with air conditioning. Single exhaust system with crossover. Carburetor: Rochester Quadrajet four-barrel.

Eldorado base V-8
Overhead valve. Cast-iron block. Displacement: 500 cid. Bore & stroke: 4.30 x 4.30 in. Compression ratio: 8.5:1. Brake hp: 365 at 4400 rpm. Taxable hp: 59.2. Net brake hp: 235 at 3800 rpm. Torque: 385 lbs.-ft. at 2400. Five main bearings. Hydraulic valve lifters. Crankcase capacity: 5 qt. (add 1 qt. if new filter installed). Cooling system capacity: 21.3 qt. with heater only; 21.8 qt. with air conditioning. Single exhaust system with crossover. Carburetor: Rochester Quadrajet four-barrel (Eldorado Special).
Note: On 1969-1974 Cadillacs, the air conditioning must be turned off before making hot idle speed adjustments. In addition, disconnect the hose at the vacuum release cylinder and close the hot idle compensator by pressing finger or eraser end of a pencil on the compensator.

1973 CADILLAC

Cadillac (except Eldorado) base V-8
Overhead valve. Cast-iron block. Displacement: 472 cid. Bore & stroke: 4.30 x 4.06 in. Compression ratio: 8.5:1. Gross hp: 345 at 4400 rpm. Net brake hp: 220 at 4000 rpm. Taxable hp: 59.2. Net torque: 385 lbs.-ft. at 2400 rpm. Five main bearings. Hydraulic valve lifters. Crankcase capacity: 4 qt. (add 1 qt. if new filter installed). Cooling system capacity: (Series 75 with air conditioning) 26.8 qt.; (All Cadillacs other series) 21.3 qt. with heater only; 23.8 qt. with air conditioning. Single exhaust system with crossover. Carburetor: Rochester Quadrajet four-barrel.

Eldorado base V-8
Overhead valve. Cast-iron block. Displacement: 500 cid. Bore & stroke: 4.30 x 4.30 in. Compression ratio: 8.5:1. Net brake hp: 365 at 4400 rpm. Taxable hp: 59.2. Net brake hp: 235 at 3800 rpm. Net torque: 385 lbs.-ft. at 2400 rpm. Five main bearings. Hydraulic valve lifters. Crankcase capacity: 5 qt. (add 1 qt. if new filter installed). Cooling system capacity: 21.3 qt. with heater only; 23.8 qt. with air conditioning. Single exhaust system with crossover. Carburetor: Rochester Quadrajet four-barrel (Eldorado Special).
Note: On 1969-1974 Cadillacs, the air conditioning must be turned off before making hot idle speed adjustments. In addition, disconnect the hose at the vacuum release cylinder and close the hot idle compensator by pressing finger or eraser end of a pencil on the compensator.

1974 CADILLAC

Cadillac (except Eldorado) base V-8
Overhead valve. Cast-iron block. Displacement: 472 cid. Bore & stroke: 4.30 x 4.06 in. Compression ratio: 8.25:1. Net brake hp: 205 at 3600 rpm. Taxable hp: 59.2. Net torque: 365 lbs.-ft. at 2000 rpm. Five main bearings. Hydraulic valve lifters. Crankcase capacity: 4 qt. (add 1 qt. if new filter installed). Cooling system capacity: (Series 75 with air conditioning) 26.8 qt.; (All Cadillacs other series) 21.3 qt. with heater only; 23.8 qt. with air conditioning. Single exhaust system with crossover. Carburetor: Four-barrel. VIN Code: R. Built by Cadillac

Eldorado base V-8
Overhead valve. Cast-iron block. Displacement: 500 cid. Bore & stroke: 4.30 x 4.30 in. Compression ratio: 8.25:1. Net brake hp: 210 at 3600 rpm. Taxable hp: 59.2. Net torque: 380 lbs.-ft. at 2000 rpm. Five main bearings. Hydraulic valve lifters. Crankcase capacity: 5 qt. (add 1 qt. if new filter installed). Cooling system capacity: 21.3 qt. with heater only; 23.8 qt. with air conditioning. Single exhaust system with crossover. Carburetor: Four-barrel. VIN Code: S. Built by Cadillac.
Note: On 1969-1974 Cadillacs, the air conditioning must be turned off before making hot idle speed adjustments. In addition, disconnect the hose at the vacuum release cylinder and close the hot idle compensator by pressing finger or eraser end of a pencil on the compensator.

1975 CADILLAC

Seville base V-8
Overhead valve. Cast-iron block. Displacement: 350 cid. Bore & stroke: 4.057 x 3.385 in. Compression ratio: 8.0:1. Net brake hp: 180 at 4400 rpm. Net torque: 360 lbs.-ft. at 2000. Hydraulic valve lifters. Crankcase capacity: 4 qt. Cooling system capacity: (with heater and air conditioner) 18.9 qt. Induction: Fuel-injection. VIN Code: R. Built by Oldsmobile exclusively for Cadillac. Due to midyear release in May 1975 this engine does not show up in many references until 1976.

Cadillac (except Seville) base V-8
Overhead valve. Cast-iron block. Displacement: 500 cid. Bore & stroke: 4.30 x 4.30 in. Compression ratio: 8.5:1. Net brake hp: 190 at 3800 rpm. Taxable hp: 59.2. Net torque: 360 lbs.-ft. at 2000 rpm. Five main bearings. Hydraulic valve lifters. Crankcase capacity: 5 qt. (add 1 qt. if new filter installed). Cooling system capacity: (Series 75 with heater and air conditioner) 25.8 qt.; (Others except SeVille with heater and air conditioner) 23 qt. Carburetor: Four-barrel. VIN Code: S. Built by Cadillac.

1976 CADILLAC

Seville base V-8
Overhead valve. Cast-iron block. Displacement: 350 cid (5.7 liters). Bore & stroke: 4.057 x 3.385 in. Compression ratio: 8.0:1. Net brake hp: 180 at 4400 rpm. Taxable hp: 52.6. Net torque: 275 lbs.-ft. at 2000 rpm. Five main bearings. Hydraulic valve lifters. Crankcase capacity: 4 qt. Cooling system capacity: (with heater and air conditioner) 18.9 qt. Induction: Electronic fuel injection (speed-density; port injected). VIN Code: R. Built by Oldsmobile exclusively for Cadillac.

Cadillac (except Seville) base V-8
Overhead valve. Cast-iron block. Displacement: 500 cid (8.2 liters). Bore & stroke: 4.300 x 4.304 in. Compression ratio: 8.5:1. Net brake hp: 190 at 3600 rpm. Taxable hp: 59.2. Net torque: 360 lbs.-ft. at 2000 rpm. Five main bearings. Hydraulic valve lifters. Crankcase capacity: 5 qt. (add 1 qt. if new filter installed). Cooling system capacity: (Series 75 with heater and air conditioner) 25.8 qt.; (Others except SeVille with heater and air

conditioner) 23 qt. Carburetor: Rochester four barrel Model M4ME. VIN Code: S. Built by Cadillac.

Cadillac (except Seville) optional V-8
Overhead valve. Cast-iron block. Displacement: 500 cid (8.2 liters). Bore & stroke: 4.300 x 4.304 in. Compression ratio: 8.5:1. Net brake hp: 215 at 3600 rpm. Taxable hp: 59.2. Net torque: 400 lbs.-ft. at 2000 rpm. Five main bearings. Hydraulic valve lifters. Crankcase capacity: 5 qt. (add 1 qt. if new filter installed). Cooling system capacity: (Series 75 with heater and air conditioner) 25.8 qt.; (Others except SeVille with heater and air conditioner) 23 qt. Induction: Fuel-injection. VIN Code: S. Built by Cadillac.

1977 CADILLAC

Seville base V-8
Overhead valve. Cast-iron block. Displacement: 350 cid (5.7 liters). Bore & stroke: 4.057 x 3.385 in. Compression ratio: 8.0:1. Net brake hp: 180 at 4400 rpm. Taxable hp: 52.6. Net torque: 275 lbs.-ft. at 2000 rpm. Five main bearings. Hydraulic valve lifters. Induction: Electronic fuel injection (speed-density; port injected). VIN Code: R. Built by Oldsmobile exclusively for Cadillac.

Cadillac (except Seville) base V-8
Overhead valve. Cast-iron block. Displacement: 425 cid (7.0 liters). Bore & stroke: 4.082 x 4.06 in. Compression ratio: 8.2:1. Net brake hp: 180 at 4000 rpm. Taxable hp: 53.31. Net torque: 320 lbs.-ft. at 2000 rpm. Five main bearings. Hydraulic valve lifters. Carburetor: Four barrel. VIN Code: S. Built by Cadillac

DeVille, Brougham, Eldorado optional V-8
Overhead valve. Cast-iron block. Displacement: 425 cid (7.0 liters). Bore & stroke: 4.082 x 4.06 in. Compression ratio: 8.2:1. Net brake hp: 195 at 3800 rpm. Taxable hp: 53.31. Net torque: 320 lbs.-ft. at 2400 rpm. Five main bearings. Hydraulic valve lifters. Induction: Electronic fuel-injection. VIN Code: T. Built by Cadillac

1978 CADILLAC

Seville base V-8
Overhead valve. Cast-iron block. Displacement: 350 cid (5.7 liters). Bore & stroke: 4.057 x 3.385 in. Compression ratio: 8.0:1. Net brake hp: 170 at 4400 rpm. Taxable hp: 52.6. Net torque: 270 lbs.-ft. at 2000 rpm. Five main bearings. Hydraulic valve lifters. Induction: Electronic fuel injection. VIN Code: B. Built by Cadillac.

Cadillac (except Seville) base V-8
Overhead valve. Cast-iron block. Displacement: 425 cid (7.0 liters). Bore & stroke: 4.082 x 4.06 in. Compression ratio: 8.2:1. Net brake hp: 180 at 4000 rpm. Taxable hp: 53.31. Net torque: 320 lbs.-ft. at 2000 rpm. Five main bearings. Hydraulic valve lifters. Carburetor: Four barrel. VIN Code: S. Built by Cadillac.

DeVille, Brougham, Eldorado optional V-8
Overhead valve. Cast-iron block. Displacement: 425 cid (7.0 liters). Bore & stroke: 4.082 x 4.06 in. Compression ratio: 8.2:1. Net brake hp: 195 at 3800 rpm. Taxable hp: 53.31. Net torque: 320 lbs.-ft. at 2400 rpm. Five main bearings. Hydraulic valve lifters. Induction: Electronic fuel injection. VIN Code: T. Built by Cadillac.

DeVille, Brougham, Eldorado optional diesel V-8
Overhead valve. Cast-iron block. Displacement: 350 cid (5.7 liters). Bore & stroke: 4.057 x 3.385 in. Compression ratio: 22.5:1. Net brake hp: 120 at 3600 rpm. Taxable hp: 52.6. Net torque: 220 lbs.-ft. at 1600 rpm. Five main bearings. Hydraulic valve lifters. Induction: Electronic fuel injection. VIN Code: N. Built by Oldsmobile.

1979 CADILLAC

Seville, Eldorado base V-8
Overhead valve. Cast-iron block. Displacement: 350 cid (5.7 liters). Bore & stroke: 4.057 x 3.385 in. Compression ratio: 8.0:1. Net brake hp: 170 at 4400 rpm. Taxable hp: 52.6. Net torque: 270 lbs.-ft. at 2000 rpm. Five main bearings. Hydraulic valve lifters. Induction: Electronic fuel injection. VIN Code: B. Built by Cadillac.

DeVille, Brougham, Fleetwood base V-8
Overhead valve. Cast-iron block. Displacement: 425 cid (7.0 liters). Bore & stroke: 4.082 x 4.06 in. Compression ratio: 8.2:1. Net brake hp: 180 at 4000 rpm. Taxable hp: 53.31. Net torque: 320 lbs.-ft. at 2000 rpm. Five main bearings. Hydraulic valve lifters. Carburetor: Four barrel. VIN Code: S. Built by Cadillac.

DeVille, Brougham optional V-8
Overhead valve. Cast-iron block. Displacement: 425 cid (7.0 liters). Bore & stroke: 4.082 x 4.06 in. Compression ratio: 8.2:1. Net brake hp: 195 at 3800 rpm. Taxable hp: 53.31. Net torque: 320 lbs.-ft. at 2400 rpm. Five main bearings. Hydraulic valve lifters. Induction: Electronic fuel-injection. VIN Code: T. Built by Cadillac.

DeVille, Brougham, Eldorado optional diesel V-8
Overhead valve. Cast-iron block. Displacement: 350 cid (5.7 liters). Bore & stroke: 4.057 x 3.385 in. Compression ratio: 22.5:1. Net brake hp: 125 at 3600 rpm. Taxable hp: 52.6. Net torque: 225 lbs.-ft. at 1600 rpm. Five main bearings. Hydraulic valve lifters. Induction: Electronic fuel injection. VIN Code: N. Built by Oldsmobile.

1980 CADILLAC

Seville base diesel V-8; DeVille, Brougham, Eldorado optional diesel V-8
Overhead valve. Cast-iron block. Displacement: 350 cid (5.7 liters). Bore & stroke: 4.057 x 3.385 in. Compression ratio: 22.5:1. Net brake hp: 105 at 3200 rpm. Taxable hp: 52.6. Net torque: 205 lbs.-ft. at 1600 rpm. Five main bearings. Hydraulic valve lifters. Induction: Electronic fuel injection. VIN Code: N. Built by Oldsmobile.

DeVille, Brougham, Fleetwood base V-8
Overhead valve. Cast-iron block. Displacement: 368 cid (6.0 liters). Bore & stroke: 3.80 x 4.06 in. Compression ratio: 8.2:1. Net brake hp: 150 at 3800 rpm. Taxable hp: 46.20. Net torque: 265 lbs.-ft. at 1600 rpm. Five main bearings. Hydraulic valve lifters. Carburetion: Four-barrel. VIN Code: 6. Built by Cadillac.

Eldorado base V-8; Seville optional V-8
Overhead valve. Cast-iron block. Displacement: 368 cid (6.0 liters). Bore & stroke: 3.80 x 4.06 in. Compression ratio: 8.2:1. Net brake hp: 145 at 3600 rpm. Taxable hp: 46.20. Net torque: 270 lbs.-ft. at 2000 rpm. Five main bearings. Hydraulic valve lifters. Induction: Digital fuel injection. VIN Code: 9. Built by Cadillac.

Eldorado base V-8 (California)
Overhead valve. Cast-iron block. Displacement: 350 cid (5.7 liters). Bore & stroke: 4.057 x 3.385 in. Compression ratio: 8.0:1. Net brake hp: 160 at 4400 rpm. Taxable hp: 52.6. Net torque: 265 lbs.-ft. at 1600 rpm. Five main bearings. Hydraulic valve lifters. Induction: Electronic fuel injection. VIN Code: 8. Built by Cadillac.
Note: For all models except limousines, an optional new 4.1-liter Buick built V-6 was introduced late in the year. It was the first non-V-8 engine used in a Cadillac in 66 years. This engine does not lie within the scope of this book.

1981 CADILLAC

Seville base diesel V-8; DeVille, Brougham, Eldorado optional diesel V-8
Overhead valve. Cast-iron block. Displacement: 350 cid (5.7 liters). Bore & stroke: 4.057 x 3.385 in. Compression ratio: 22.5:1. Net brake hp: 105

at 3200 rpm. Taxable hp: 52.6. Net torque: 200 lbs.-ft. at 1600 rpm. Five main bearings. Hydraulic valve lifters. Induction: Electronic fuel injection. VIN Code: N. Built by Oldsmobile.

DeVille, Brougham, Eldorado base Variable Displacement V-8-6-4
Seville optional Variable Displacement V-8-6-4
Overhead valve. Cast-iron block. Displacement: 368 cid (6.0 liters). Bore & stroke: 3.80 x 4.06 in. Compression ratio: 8.2:1. Net brake hp: 140 at 3800 rpm. Taxable hp: 46.20. Net torque: 265 lbs.-ft. at 1400 rpm. Five main bearings. Hydraulic valve lifters. Induction: Digital fuel injection. VIN Code: 9. Built by Cadillac.

Cadillac commercial chassis optional V-8
Overhead valve. Cast-iron block. Displacement: 368 cid (6.0 liters). Bore & stroke: 3.80 x 4.06 in. Compression ratio: 8.2:1. Net brake hp: 150 at 3800 rpm. Taxable hp: 46.20. Net torque: 265 lbs.-ft. at 1600 rpm. Five main bearings. Hydraulic valve lifters. Carburetion: Four-barrel. VIN Code: 6. Built by Cadillac.
Note: A 1.8-liter four-cylinder engine was introduced in the "Cimarron by Cadillac" in May 1981. This engine is not within the scope of this book.

1982 CADILLAC

Seville, DeVille, Brougham, Eldorado base HT4100 V-8
Overhead valve. Aluminum block with cast-iron liners. Displacement: 249 cid (4.1 liters). Bore & stroke: 3.465 x 3.307 in. Compression ratio: 8.5:1. Net brake hp: 125 at 4200 rpm. Taxable hp: 39.2. Net torque: 190 lbs.-ft. at 2000 rpm. Five main bearings. Hydraulic valve lifters. Induction: Digital fuel injection. VIN Code: 8. Built by Cadillac.

Seville, DeVille, Brougham, Eldorado optional diesel V-8
Overhead valve. Cast-iron block. Displacement: 350 cid (5.7 liters). Bore & stroke: 4.057 x 3.385 in. Compression ratio: 21.6:1. Net brake hp: 105 at 3200 rpm. Taxable hp: 52.6. Net torque: 200 lbs.-ft. at 1600 rpm. Five main bearings. Hydraulic valve lifters. Induction: Electronic fuel injection. VIN Code: N. Built by Oldsmobile.

Cadillac limousine base V-8-6-4 Variable Displacement engine
Overhead valve. Cast-iron block. Displacement: 368 cid (6.0 liters). Bore & stroke: 3.80 x 4.06 in. Compression ratio: 8.2:1. Net brake hp: 140 at 3800 rpm. Taxable hp: 46.20. Net torque: 200 lbs.-ft. at 1400 rpm. Five main bearings. Hydraulic valve lifters. Induction: Digital fuel injection. VIN Code: 9. Built by Cadillac.

Cadillac commercial chassis optional V-8
Overhead valve. Cast-iron block. Displacement: 368 cid (6.0 liters). Bore & stroke: 3.80 x 4.06 in. Compression ratio: 8.2:1. Net brake hp: 150 at 3800 rpm. Taxable hp: 46.20. Net torque: 265 lbs.-ft. at 1600 rpm. Five main bearings. Hydraulic valve lifters. Carburetion: Four-barrel. VIN Code: 6. Built by Cadillac.

1983 CADILLAC

Seville, DeVille, Brougham, Eldorado base V-8
Overhead valve. Aluminum block with cast-iron liners. Displacement: 249 cid (4.1 liters). Bore & stroke: 3.47 x 3.31 in. Compression ratio: 8.5:1. Net brake hp: 135 at 4400 rpm. Taxable hp: 39.2. Net torque: 200 lbs.-ft. at 2000 rpm. Five main bearings. Hydraulic valve lifters. Induction: Digital fuel injection. VIN Code: 8. Built by Cadillac.

Seville, DeVille, Brougham, Eldorado optional diesel V-8
Overhead valve. Cast-iron block. Displacement: 350 cid (5.7 liters). Bore & stroke: 4.057 x 3.385 in. Compression ratio: 22.5:1. Net brake hp: 105 at 3200 rpm. Taxable hp: 52.6. Net torque: 200 lbs.-ft. at 1600 rpm. Five main bearings. Hydraulic valve lifters. Induction: Electronic fuel injection. VIN Code: N. Built by Oldsmobile.

Cadillac limousine base V-8-6-4 Variable Displacement engine
Overhead valve. Cast-iron block. Displacement: 368 cid (6.0 liters). Bore & stroke: 3.80 x 4.06 in. Compression ratio: 8.2:1. Net brake hp: 140 at 3800 rpm. Taxable hp: 46.20. Net torque: 200 lbs.-ft. at 1400 rpm. Five main bearings. Hydraulic valve lifters. Induction: Digital fuel injection. VIN Code: 9. Built by Cadillac.

Cadillac commercial chassis optional V-8
Overhead valve. Cast-iron block. Displacement: 368 cid (6.0 liters). Bore & stroke: 3.80 x 4.06 in. Compression ratio: 8.2:1. Net brake hp: 150 at 3800 rpm. Taxable hp: 46.20. Net torque: 265 lbs.-ft. at 1600 rpm. Five main bearings. Hydraulic valve lifters. Carburetion: Four-barrel. VIN Code: 6. Built by Cadillac.

1984 CADILLAC

Seville, DeVille, Brougham, Eldorado base V-8
Overhead valve. Aluminum block with cast-iron liners. Displacement: 249 cid (4.1 liters). Bore & stroke: 3.47 x 3.31 in. Compression ratio: 8.5:1. Net brake hp: 135 at 4400 rpm. Taxable hp: 39.2. Net torque: 200 lbs.-ft. at 2000 rpm. Five main bearings. Hydraulic valve lifters. Induction: Digital fuel injection. VIN Code: 8. Built by Cadillac.

Seville, DeVille, Brougham, Eldorado optional diesel V-8
Overhead valve. Cast-iron block. Displacement: 350 cid (5.7 liters). Bore & stroke: 4.057 x 3.385 in. Compression ratio: 22.5:1. Net brake hp: 105 at 3200 rpm. Taxable hp: 52.6. Net torque: 200 lbs.-ft. at 1600 rpm. Five main bearings. Hydraulic valve lifters. Induction: Electronic fuel injection. VIN Code: N. Built by Oldsmobile.

Cadillac limousine base V-8-6-4 Variable Displacement engine
Overhead valve. Cast-iron block. Displacement: 368 cid (6.0 liters). Bore & stroke: 3.80 x 4.06 in. Compression ratio: 8.2:1. Net brake hp: 140 at 3800 rpm. Taxable hp: 46.20. Net torque: 200 lbs.-ft. at 1400 rpm. Five main bearings. Hydraulic valve lifters. Induction: Digital fuel injection. VIN Code: 9. Built by Cadillac.

1985 CADILLAC

Here's a bird's-eye view of the 1985 Cadillac HT4100 digital fuel-injected V-8. It was the only tranverse-mounted V-8 offered in any front-wheel-drive production car at that time. Cadillac said that it "combines the power of a 4.1-liter V-8 engine and the quick response of Digital Fuel Injection and the smoothness of a refined automatic overdrive transmission."

Seville, Brougham, Eldorado base V-8
Overhead valve. Aluminum block with cast-iron liners. Displacement: 249 cid (4.1 liters). Bore & stroke: 3.47 x 3.31 in. Compression ratio: 8.5:1. Net brake hp: 135 at 4400 rpm. Taxable hp: 39.2. Net torque: 200 lbs.-ft. at 2200 rpm. Five main bearings. Hydraulic valve lifters. Induction: Digital fuel injection. VIN Code: 8. Built by Cadillac.

DeVille, Fleetwood base V-8
Overhead valve. Aluminum block with cast-iron liners. Displacement: 249 cid (4.1 liters). Bore & stroke: 3.47 x 3.31 in. Compression ratio: 8.5:1. Net brake hp: 125 at 4200 rpm. Taxable hp: 39.2. Net torque: 190 lbs.-ft. at 2200 rpm. Five main bearings. Hydraulic valve lifters. Induction: Digital fuel injection. VIN Code: 8. Built by Cadillac.

Seville, DeVille, Brougham, Eldorado optional diesel V-8
Overhead valve. Cast-iron block. Displacement: 350 cid (5.7 liters). Bore & stroke: 4.057 x 3.385 in. Compression ratio: 22.7:1. Net brake hp: 105 at 3200 rpm. Taxable hp: 52.6. Net torque: 200 lbs.-ft. at 1600 rpm. Five main bearings. Hydraulic valve lifters. Induction: Electronic fuel injection. VIN Code: N. Built by Oldsmobile.

1986 CADILLAC

Seville, DeVille, Fleetwood, Limousine, Eldorado base V-8
Overhead valve. Aluminum block with cast-iron liners. Displacement: 249 cid (4.1 liters). Bore & stroke: 3.47 x 3.31 in. Compression ratio: 8.5:1. Net brake hp: 130 at 4200 rpm. Taxable hp: 39.2. Net torque: 200 lbs.-ft. at 2200 rpm. Five main bearings. Hydraulic valve lifters. Induction: Digital fuel injection. VIN Code: 8. Built by Cadillac.

Brougham base V-8
Overhead valve. Cast-iron block. Displacement: 307 cid (5.0 liters). Bore & stroke: 3.80 x 3.39 in. Compression ratio: 8.0:1. Net brake hp: 140 at 3200 rpm. Taxable hp: 46.20. Net torque: 346 lbs.-ft. at 2000 rpm. Five main bearings. Hydraulic valve lifters. Carburetion: Four-barrel. VIN Code: Y.

1987 CADILLAC

DeVille, Fleetwood, Seville, Eldorado base V-8
Overhead valve. Aluminum block with cast-iron liners. Displacement: 249 cid (4.1 liters). Bore & stroke: 3.47 x 3.31 in. Compression ratio: 8.5:1. Net brake hp: 130 at 4200 rpm. Taxable hp: 39.2. Net torque: 200 lbs.-ft. at 2200 rpm. Five main bearings. Hydraulic valve lifters. Induction: Digital fuel injection. VIN Code: 8. Built by Cadillac.

Allante base V-8
Overhead valve. Aluminum block with cast-iron liners. Displacement: 249 cid (4.1 liters). Bore & stroke: 3.47 x 3.31 in. Compression ratio: 8.5:1. Net brake hp: 170 at 4300 rpm. Taxable hp: 39.2. Net torque: 235 lbs.-ft. at 3200 rpm. Five main bearings. Roller hydraulic valve lifters. Induction: Sequential multi-port fuel injection. VIN Code: 7. Built by Cadillac.

Brougham base V-8
Overhead valve. Cast-iron block. Displacement: 307 cid (5.0 liters). Bore & stroke: 3.80 x 3.39 in. Compression ratio: 8.0:1. Net brake hp: 140 at 3200 rpm. Taxable hp: 46.20. Net torque: 346 lbs.-ft. at 2000 rpm. Five main bearings. Hydraulic valve lifters. Carburetion: Four-barrel. VIN Code: Y. Built by Oldsmobile.

Cadillac optional V-8
Overhead valve. Cast-iron block. Displacement: 307 cid (5.0 liters). Taxable hp: 46.20. Bore & stroke: 3.80 x 3.39 in. Five main bearings. Hydraulic valve lifters. Carburetion: Four-barrel. VIN Code: 9. Built by Oldsmobile.

1988 CADILLAC

DeVille, Fleetwood, Seville, Eldorado base V-8
Overhead valve. Aluminum block with Cast-iron liners. Displacement: 273 cid (4.5 liters). Bore & stroke: 3.62 x 3.31 in. Compression ratio: 9.0:1. Net brake hp: 155 at 4000 rpm. Taxable hp: 42.00. Net torque: 240 lbs.-ft. at 2800 rpm. Five main bearings. Roller hydraulic valve lifters. Induction: Digital fuel injection. VIN Code: 5. Built by Cadillac.

Allante base V-8
Overhead valve. Aluminum block with Cast-iron liners. Displacement: 249 cid (4.1 liters). Bore & stroke: 3.47 x 3.31 in. Compression ratio: 8.5:1. Net brake hp: 170 at 4300 rpm. Taxable hp: 39.2. Net torque: 235 lbs.-ft. at 3200 rpm. Five main bearings. Roller hydraulic valve lifters. Induction: Sequential multi-port fuel injection. VIN Code: 7. Built by Cadillac.

Brougham base V-8
Overhead valve. Cast-iron block. Displacement: 307 cid (5.0 liters). Bore & stroke: 3.80 x 3.39 in. Compression ratio: 8.0:1. Net brake hp: 140 at 3200 rpm. Taxable hp: 46.20. Net torque: 346 lbs.-ft. at 2000 rpm. Five main bearings. Hydraulic valve lifters. Carburetion: Four-barrel. VIN Code: Y. Built by Oldsmobile.

Cadillac optional V-8
Overhead valve. Cast-iron block. Displacement: 307 cid (5.0 liters). Bore & stroke: 3.80 x 3.39 in. Taxable hp: 46.20. Five main bearings. Hydraulic valve lifters. Carburetion: Four-barrel. VIN Code: 9. Built by Oldsmobile.

1989 CADILLAC

DeVille, Fleetwood, Seville, Eldorado base V-8
Overhead valve. Aluminum block with cast-iron liners. Displacement: 273 cid (4.5 liters). Bore & stroke: 3.62 x 3.31 in. Compression ratio: 9.0:1. Net brake hp: 155 at 4000 rpm. Taxable hp: 42.00. Net torque: 240 lbs.-ft. at 2800 rpm. Five main bearings. Roller hydraulic valve lifters. Induction: Digital fuel injection. VIN Code: 5. Built by Cadillac.

Allante base V-8
Overhead valve. Aluminum block with cast-iron liners. Displacement: 273 cid (4.5 liters). Bore & stroke: 3.62 x 3.31 in. Compression ratio: 9.0:1. Net brake hp: 200 at 4300 rpm. Taxable hp: 42.00. Net torque: 270 lbs.-ft. at 3200 rpm. Five main bearings. Roller hydraulic valve lifters. Induction: Sequential multiport fuel injection. VIN Code: 8. Built by Cadillac.

Brougham base V-8
Overhead valve. Cast-iron block. Displacement: 307 cid (5.0 liters). Bore & stroke: 3.80 x 3.39 in. Compression ratio: 8.0:1. Net brake hp: 140 at 3200 rpm. Taxable hp: 46.20. Net torque: 346 lbs.-ft. at 2000 rpm. Five main bearings. Hydraulic valve lifters. Carburetion: Four-barrel. VIN Code: Y. Built by Oldsmobile.

Cadillac optional V-8
Overhead valve. Cast-iron block. Displacement: 307 cid (5.0 liters). Bore & stroke: 3.80 x 3.39 in. Taxable hp: 46.20. Five main bearings. Hydraulic valve lifters. Carburetion: Four-barrel. VIN Code: 9. Built by Oldsmobile. **Note:** Cadillac returned to using only V-8 engines this year.

1990 CADILLAC

DeVille, Fleetwood, Seville, Eldorado base V-8
Overhead valve. Aluminum block with cast-iron liners. Displacement: 273 cid (4.5 liters). Bore & stroke: 3.62 x 3.31 in. Compression ratio: 9.5:1. Net brake hp: 180 at 4300 rpm. Taxable hp: 42.00. Net torque: 240 lbs.-ft. at 3000 rpm. Five main bearings. Roller hydraulic valve lifters. Induction: Sequential port fuel injection. VIN Code: 6. Built by Cadillac.

Allante base V-8
Overhead valve. Aluminum block with cast-iron liners. Displacement: 273 cid (4.5 liters). Bore & stroke: 3.62 x 3.31 in. Compression ratio: 9.0:1. Net brake hp: 200 at 4400 rpm. Taxable hp: 42.00. Net torque: 270 lbs.-ft. at 3200 rpm. Five main bearings. Roller hydraulic valve lifters. Induction: Sequential multiport fuel injection. VIN Code: 8. Built by Cadillac.

Base V-8 Brougham
Overhead valve. Cast-iron block. Displacement: 307 cid (5.0 liters). Bore & stroke: 3.80 x 3.39 in. Compression ratio: 8.0:1. Net brake hp: 140 at 3200 rpm. Taxable hp: 46.20. Net torque: 346 lbs.-ft. at 2000 rpm. Five main bearings. Hydraulic valve lifters. Carburetion: Four-barrel. VIN Code: Y. Built by Oldsmobile.

Cadillac optional V-8
Overhead valve. Cast-iron block. Displacement: 307 cid (5.0 liters). Bore

& stroke: 3.80 x 3.39 in. Taxable hp: 46.20. Five main bearings. Hydraulic valve lifters. Carburetion: Four-barrel. VIN Code: 9. Built by Oldsmobile.

Available V-8—coachbuilder/towing
Overhead valve. Cast-iron block. Displacement: 350 cid (5.7 liters). Bore & stroke: 4.00 x 3.48 in. Compression ratio: 9.3:1. Net brake hp: 175 at 4200 rpm. Taxable hp: 51.20. Net torque: 400 lbs.-ft. at 2000 rpm. Five main bearings. Hydraulic valve lifters. Induction: Fuel injection. VIN Code: 7. Built by Chevrolet.

1991 CADILLAC

DeVille, Fleetwood, Seville, Eldorado base V-8
Overhead valve. Aluminum block with cast-iron liners. Displacement: 300 cid (4.9 liters). Bore & stroke: 3.62 x 3.62 in. Compression ratio: 9.5:1. Net brake hp: 200 at 4100 rpm. Taxable hp: 41.93. Net torque: 275 lbs.-ft. at 3000 rpm. Five main bearings. Roller hydraulic valve lifters. Induction: Multiport fuel injection. VIN Code: B; Option Code L26. Built by Cadillac.

Allante base V-8
Overhead valve. Aluminum block with cast-iron liners. Displacement: 273 cid (4.5 liters). Bore & stroke: 3.62 x 3.31 in. Compression ratio: 9.0:1. Net brake hp: 200 at 4400 rpm. Taxable hp: 42.00.Net torque: 270 lbs.-ft. at 3200 rpm. Five main bearings. Roller hydraulic valve lifters. Induction: Multiport fuel injection. VIN Code: 8; Option Code L06. Built by Cadillac.

Brougham base V-8
Overhead valve. Cast-iron block. Displacement: 305 cid (5.0 liters). Bore & stroke: 3.74 x 3.48 in. Compression ratio: 9.3:1. Net brake hp: 170 at 4200 rpm. Taxable hp: 44.8. Net torque: 255 lbs.-ft. at 2400 rpm. Five main bearings. Roller hydraulic valve lifters. Induction: Throttle Body Injection. VIN Code: E; Option Code L03.

Brougham coachbuilder/towing packages base V-8; Cadillac optional V-8
Overhead valve. Cast-iron block. Displacement: 350 cid (5.7 liters). Bore & stroke: 4.00 x 3.48 in. Compression ratio: 9.8:1. Net brake hp: 185 at 4200 rpm. Taxable hp: 51.20. Net torque: 300 lbs.-ft. at 2400 rpm. Five main bearings. Roller hydraulic valve lifters. Induction: Throttle Body Injection. VIN Code: 7; Option Code LLO. Built by Chevrolet.

1992 CADILLAC

The high-tech look of the 1992 Cadillac 4.6-liter Northstar V-8 reflects the character of the technology embodied in this engine. The Northstar power train was standard equipment on the Allante, Seville Touring Sedan, Eldorado Touring coupe, and Eldorado Sport Coupe.

Base V-8 DeVille, Fleetwood, Seville, Eldorado
Overhead valve. Aluminum block with cast-iron liners. Displacement: 300 cid (4.9 liters). Bore & stroke: 3.62 x 3.62 in. Compression ratio: 9.5:1. Net brake hp: 200 at 4100 rpm. Taxable hp: 41.93. Net torque: 275 lbs.-ft. at 3000 rpm. Five main bearings. Roller hydraulic valve lifters. Induction: Multiport fuel injection. VIN Code: B; Option Code L26. Built by Cadillac.

Allante base V-8
Overhead valve. Aluminum block with cast-iron liners. Displacement: 273 cid (4.5 liters). Bore & stroke: 3.62 x 3.31 in. Compression ratio: 9.0:1. Net brake hp: 200 at 4400 rpm. Taxable hp: 42.00.Net torque: 270 lbs.-ft. at 3200 rpm. Five main bearings. Roller hydraulic valve lifters. Induction: Multiport fuel injection. VIN Code: 8; Option Code L06. Built by Cadillac.

Brougham base V-8
Overhead valve. Cast-iron block. Displacement: 305 cid (5.0 liters). Bore & stroke: 3.74 x 3.48 in. Compression ratio: 9.3:1. Net brake hp: 170 at 4200 rpm. Taxable hp: 44.8. Net torque: 255 lbs.-ft. at 2400 rpm. Five main bearings. Roller hydraulic valve lifters. Induction: Throttle Body Injection. VIN Code: E; Option Code L03.

Brougham coachbuilder/towing packages base V-8; Cadillac optional V-8
Overhead valve. Cast-iron block. Displacement: 350 cid (5.7 liters). Bore & stroke: 4.00 x 3.48 in. Compression ratio: 9.8:1. Net brake hp: 185 at 4200 rpm. Taxable hp: 51.20. Net torque: 300 lbs.-ft. at 2400 rpm. Five main bearings. Roller hydraulic valve lifters. Induction: Throttle Body Injection. VIN Code: 7; Option Code L05. Built by Chevrolet.

1993 CADILLAC

Eldorado Sport Coupe with Sport Performance Package base V-8
Northstar Overhead valve. Dual overhead cam. Aluminum block with cast-iron liners. Aluminum cylinder heads. Displacement: 279 cid (4.6 liters). Taxable hp: 42.90. Bore & stroke: 3.66 x 3.31 in. Compression ratio: 10.3:1. Net brake hp: 270 at 5600 rpm. Net torque: 300 lbs.-ft. at 4000 rpm. Direct acting hydraulic tappets. Induction: Tuned port injection. VIN Code: 9; Option Code L37. Built by Cadillac.

Eldorado Touring Coupe, Seville Touring Sedan, Allante base V-8
Northstar Overhead valve. Dual overhead cam. Aluminum block with cast-iron liners. Aluminum cylinder heads. Displacement: 279 cid (4.6 liters). Taxable hp: 42.90. Bore & stroke: 3.66 x 3.31 in. Compression ratio: 10.3:1. Net brake hp: 295 at 6000 rpm. Net torque: 290 lbs.-ft. at 4400 rpm. Direct acting hydraulic tappets. Induction: Tuned port injection. VIN Code: Y; Option Code LD8. Built by Cadillac.

DeVille, Sixty Special, Seville, Eldorado base V-8
Overhead valve. Aluminum block with cast-iron liners. Displacement: 300 cid (4.9 liters). Bore & stroke: 3.62 x 3.62 in. Compression ratio: 9.5:1. Net brake hp: 200 at 4100 rpm. Taxable hp: 41.93. Net torque: 275 lbs.-ft. at 3000 rpm. Five main bearings. Roller hydraulic valve lifters. Induction: Multiport fuel injection. VIN Code: B; Option Code L26. Built by Cadillac.

Fleetwood base V-8
Overhead valve. Cast-iron block. Displacement: 350 cid (5.7 liters). Bore & stroke: 4.00 x 3.48 in. Compression ratio: 9.8:1. Net brake hp: 185 at 4200 rpm. Taxable hp: 51.20. Net torque: 300 lbs.-ft. at 2400 rpm. Five main bearings. Roller hydraulic valve lifters. Induction: Throttle Body Injection. VIN Code: 7; Option Code L07. Built by Chevrolet.

1994 CADILLAC

DeVille Concours, Eldorado, Seville Luxury Sedan base V-8
Northstar Overhead valve. Dual overhead cam. Aluminum block with cast-iron liners. Aluminum cylinder heads. Displacement: 279 cid (4.6 liters). Bore & stroke: 3.66 x 3.31 in. Compression ratio: 10.3:1. Net brake hp: 270 at 5600 rpm. Taxable hp: 42.90. Net torque: 300 lbs.-ft. at 4000 rpm. Direct acting hydraulic tappets. Induction: Tuned port injection. VIN Code: 9; Option Code L37. Built by Cadillac.

Eldorado Touring Coupe, Seville Touring Sedan base V-8
Northstar Overhead valve. Dual overhead cam. Aluminum block with cast-iron liners. Aluminum cylinder heads. Displacement: 279 cid (4.6 liters). Bore & stroke: 3.66 x 3.31 in. Compression ratio: 10.3:1. Net brake hp: 295 at 6000 rpm. Taxable hp: 42.90. Net torque: 290 lbs.-ft. at 4400 rpm. Direct acting hydraulic tappets. Induction: Tuned port injection. VIN Code: Y; Option Code LD8. Built by Cadillac.

Sedan DeVille base V-8
Overhead valve. Aluminum block with cast-iron liners. Displacement: 300 cid (4.9 liters). Bore & stroke: 3.62 x 3.62 in. Compression ratio: 9.5:1. Net brake hp: 200 at 4100 rpm. Taxable hp: 41.93. Net torque: 275 lbs.-ft. at 3000 rpm. Five main bearings. Roller hydraulic valve lifters. Induction: Multiport port injection. VIN Code: B; Option Code L26. Built by Cadillac.

Fleetwood base V-8
Overhead valve. Cast-iron block. Displacement: 350 cid (5.7 liters). Bore & stroke: 4.00 x 3.48. Compression ratio: 9.7:1. Net brake hp: 260 at 5000 rpm. Taxable hp: 51.20. Net torque: 335 lbs.-ft. at 2400 rpm. Roller hydraulic valve lifters. Induction: Multiport port Injection. VIN Code: P; Option Code LT1. Built by Chevrolet.

1995 CADILLAC

DeVille Concours, Eldorado, Seville Luxury Sedan base V-8
Northstar Overhead valve. Dual overhead cam. Aluminum block with cast-iron liners. Aluminum cylinder heads. Displacement: 279 cid (4.6 liters). Bore & stroke: 3.66 x 3.31 in. Compression ratio: 10.3:1. Net brake hp: 275 at 5600 rpm. Taxable hp: 42.90. Net torque: 300 lbs.-ft. at 4000 rpm. Direct acting hydraulic tappets. Induction: Tuned port injection. VIN Code: 9; Option Code L37. Built by Cadillac.

Eldorado Touring Coupe, Seville Touring Sedan base V-8
Northstar Overhead valve. Dual overhead cam. Aluminum block with cast-iron liners. Aluminum cylinder heads. Displacement: 279 cid (4.6 liters). Bore & stroke: 3.66 x 3.31 in. Compression ratio: 10.3:1. Net brake hp: 300 at 6000 rpm. Taxable hp: 42.90. Net torque: 295 lbs.-ft. at 4400 rpm. Direct acting hydraulic tappets. Induction: Tuned port injection. VIN Code: Y; Option Code LD8. Built by Cadillac.

Sedan DeVille base V-8
Overhead valve. Aluminum block with cast-iron liners. Displacement: 300 cid (4.9 liters). Bore & stroke: 3.62 x 3.62 in. Compression ratio: 9.5:1. Net brake hp: 200 at 4100 rpm. Taxable hp: 41.93. Net torque: 275 lbs.-ft. at 3000 rpm. Five main bearings. Roller hydraulic valve lifters. Induction: Multiport port injection. VIN Code: B; Option Code L26. Built by Cadillac.

Fleetwood base V-8
Overhead valve. Cast-iron block. Displacement: 350 cid (5.7 liters). Bore & stroke: 4.00 x 3.48 in. Compression ratio: 10.5:1. Net brake hp: 260 at 5000 rpm. Taxable hp: 51.20. Net torque: 335 lbs.-ft. at 2400 rpm. Roller hydraulic valve lifters. Induction: Multiport port Injection. VIN Code: P; Option Code LT1. Built by Chevrolet.

1996 CADILLAC

Sedan DeVille, Eldorado, Seville Luxury Sedan base V-8
Northstar Overhead valve. Dual overhead cam. Aluminum block with cast-iron liners. Aluminum cylinder heads. Displacement: 279 cid (4.6 liters). Bore & stroke: 3.66 x 3.31 in. Compression ratio: 10.3:1. Net brake hp: 275 at 5600 rpm. Taxable hp: 42.90. Net torque: 300 lbs.-ft. at 4000 rpm. Direct acting hydraulic tappets. Induction: Tuned port injection. VIN Code: 9; Option Code L37. Built by Cadillac.

DeVille Concours, Eldorado Touring Coupe, Seville Touring Sedan base V-8
Northstar Overhead valve. Dual overhead cam. Aluminum block with cast-iron liners. Aluminum cylinder heads. Displacement: 279 cid (4.6 liters). Bore & stroke: 3.66 x 3.31 in. Compression ratio: 10.3:1. Net brake hp: 300 at 6000 rpm. Taxable hp: 42.90. Net torque: 295 lbs.-ft. at 4400 rpm. Direct acting hydraulic tappets. Induction: Tuned port injection. VIN Code: Y; Option Code LD8. Built by Cadillac.

Fleetwood base V-8
Overhead valve. Cast-iron block. Displacement: 350 cid (5.7 liters). Bore & stroke: 4.00 x 3.48 in. Compression ratio: 10.0:1. Net brake hp: 260 at 5000 rpm. Taxable hp: 51.20. Net torque: 330 lbs.-ft. at 2400 rpm. Roller hydraulic valve lifters. Induction: Multiport port Injection. VIN Code: P; Option Code LT1. Built by Chevrolet.

1997 CADILLAC

DeVille d'Elegance, Eldorado, Seville Luxury Sedan base V-8
Northstar Overhead valve. Dual overhead cam. Aluminum block with cast-iron liners. Aluminum cylinder heads. Displacement: 279 cid (4.6 liters). Bore & stroke: 3.66 x 3.31 in. Compression ratio: 10.3:1. Net brake hp: 275 at 5600 rpm. Taxable hp: 42.90. Net torque: 300 lbs.-ft. at 4000 rpm. Direct acting hydraulic tappets. Induction: Tuned port injection. VIN Code: 9; Option Code L37. Built by Cadillac.

DeVille Concours, Eldorado Touring Coupe, Seville Touring Sedan base V-8
Northstar Overhead valve. Dual overhead cam. Aluminum block with cast-iron liners. Aluminum cylinder heads. Displacement: 279 cid (4.6 liters). Bore & stroke: 3.66 x 3.31 in. Compression ratio: 10.3:1. Net brake hp: 300 at 6000 rpm. Taxable hp: 42.90. Net torque: 295 lbs.-ft. at 4400 rpm. Direct acting hydraulic tappets. Induction: Tuned port injection. VIN Code: Y; Option Code LD8. Built by Cadillac.
Note: The Cadillac Catera was introduced this year, but its 3.0-liter 200-hp V-6 is not within the scope of this book.

1998 CADILLAC

DeVille d'Elegance, Eldorado, Seville Luxury Sedan base V-8
Northstar Overhead valve. Dual overhead cam. Aluminum block with cast-iron liners. Aluminum cylinder heads. Displacement: 279 cid (4.6 liters). Bore & stroke: 3.66 x 3.31 in. Compression ratio: 10.3:1. Net brake hp: 275 at 5600 rpm. Taxable hp: 42.90. Net torque: 300 lbs.-ft. at 4000 rpm. Direct acting hydraulic tappets. Induction: Tuned port injection. VIN Code: 9; Option Code L37. Built by Cadillac.

DeVille Concours, Eldorado Touring Coupe, Seville Touring Sedan base V-8
Northstar Overhead valve. Dual overhead cam. Aluminum block with cast-iron liners. Aluminum cylinder heads. Displacement: 279 cid (4.6 liters). Bore & stroke: 3.66 x 3.31 in. Compression ratio: 10.3:1. Net brake hp: 300 at 6000 rpm. Taxable hp: 42.90. Net torque: 295 lbs.-ft. at 4400 rpm. Direct acting hydraulic tappets. Induction: Tuned port injection. VIN Code: Y; Option Code LD8. Built by Cadillac.

1999 CADILLAC

DeVille d'Elegance, Eldorado, Seville Luxury Sedan base V-8
Northstar Overhead valve. Dual overhead cam. Aluminum block with cast-iron liners. Aluminum cylinder heads. Displacement: 279 cid (4.6 liters). Bore & stroke: 3.66 x 3.31 in. Compression ratio: 10.3:1. Net brake hp: 275 at 5600 rpm. Taxable hp: 42.90. Net torque: 300 lbs.-ft. at 4000 rpm. Direct acting hydraulic tappets. Induction: Tuned port injection. VIN Code: 9; Option Code L37. Built by Cadillac.

DeVille Concours, Eldorado Touring Coupe, Seville Touring Sedan base V-8
Northstar Overhead valve. Dual overhead cam. Aluminum block with cast-iron liners. Aluminum cylinder heads. Displacement: 279 cid (4.6 liters). Bore & stroke: 3.66 x 3.31 in. Compression ratio: 10.3:1. Net brake hp: 300 at 6000 rpm. Taxable hp: 42.90. Net torque: 295 lbs.-ft. at 4400 rpm. Direct acting hydraulic tappets. Induction: Tuned port injection. VIN Code: Y; Option Code LD8. Built by Cadillac.

2000 CADILLAC

DeVille d'Elegance, Eldorado, Seville Luxury Sedan base V-8
Northstar Overhead valve. Dual overhead cam. Aluminum block with cast-iron liners. Aluminum cylinder heads. Displacement: 279 cid (4.6 liters). Bore & stroke: 3.66 x 3.31 in. Compression ratio: 10.3:1. Net

brake hp: 275 at 5600 rpm. Taxable hp: 42.90. Net torque: 300 lbs.-ft. at 4000 rpm. Direct acting hydraulic tappets. Induction: Tuned port injection. VIN Code: 9; Option Code L37. Built by Cadillac.

DeVille Concours, Eldorado Touring Coupe, Seville Touring Sedan base V-8
Northstar Overhead valve. Dual overhead cam. Aluminum block with cast-iron liners. Aluminum cylinder heads. Displacement: 279 cid (4.6 liters). Bore & stroke: 3.66 x 3.31 in. Compression ratio: 10.3:1. Net brake hp: 300 at 6000 rpm. Taxable hp: 42.90. Net torque: 295 lbs.-ft. at 4400 rpm. Direct acting hydraulic tappets. Induction: Tuned port injection. VIN Code: Y; Option Code LD8. Built by Cadillac.

2001 CADILLAC

DeVille d'Elegance, Eldorado, Seville Luxury Sedan base V-8
Northstar Overhead valve. Dual overhead cam. Aluminum block with cast-iron liners. Aluminum cylinder heads. Displacement: 279 cid (4.6 liters). Bore & stroke: 3.66 x 3.31 in. Compression ratio: 10.3:1. Net brake hp: 275 at 5600 rpm. Taxable hp: 42.90. Net torque: 300 lbs.-ft. at 4000 rpm. Direct acting hydraulic tappets. Induction: Tuned port injection. VIN Code: 9; Option Code L37. Built by Cadillac.

DeVille Concours, Eldorado Touring Coupe, Seville Touring Sedan base V-8
Northstar Overhead valve. Dual overhead cam. Aluminum block with cast-iron liners. Aluminum cylinder heads. Displacement: 279 cid (4.6 liters). Bore & stroke: 3.66 x 3.31 in. Compression ratio: 10.3:1. Net brake hp: 300 at 6000 rpm. Taxable hp: 42.90. Net torque: 295 lbs.-ft. at 4400 rpm. Direct acting hydraulic tappets. Induction: Tuned port injection. VIN Code: Y; Option Code LD8. Built by Cadillac.

2002 CADILLAC

DeVille, DHS, Seville SLS, Eldorado base V-8
Northstar Overhead valve. Dual overhead cam. Aluminum block with cast-iron liners. Aluminum cylinder heads. Displacement: 279 cid (4.6 liters). Bore & stroke: 3.66 x 3.31 in. Compression ratio: 10.0:1. Net brake hp: 275 at 5600 rpm. Taxable hp: 42.90. Net torque: 300 lbs.-ft. at 4000 rpm. Direct acting hydraulic tappets. Induction: Sequential fuel injection. VIN Code: 9; Option Code L37. Built by Cadillac.

DeVille DTS, Seville STS, Eldorado Touring Coupe base V-8
Northstar Overhead valve. Dual overhead cam. Aluminum block with cast-iron liners. Aluminum cylinder heads. Displacement: 279 cid (4.6 liters). Bore & stroke: 3.66 x 3.31 in. Compression ratio: 10.0:1. Net brake hp: 300 at 6000 rpm. Taxable hp: 42.90. Net torque: 295 lbs.-ft. at 4400 rpm. Direct acting hydraulic tappets. Induction: Sequential fuel injection. VIN Code: Y; Option Code LD8. Built by Cadillac.

Selected Cadillac V-8 Tune-Up Specs 1927-1965

Year	Model	Spark Plugs Type	Spark Plugs Gap	Point Gap	Deg Cam Angle	Timing	Idle Speed	Compression Pressure	Battery (Delco)	
1927	341	7/8-18	.026	.022-.027	NA	IG/A2/6'	NA	NA	3-LXRV-15-2G	
1928	29	341-B	AC-Y	.025-.028	.022-.027	NA	IG/A2/6'	NA	90-92 @ 1000	3-LXRV-15-2G
1930	353	AC-G-10	.025-.028	.018-.022	NA	IG/A2/6'	NA	NA	3-XLV-15-2G	
1931	355	AC-G-10	.025-.028	.018-.022	NA	IG/A2/6'	NA	104 @ 1000	15-CW	
1932	355-C	AC-D-8	.025-.028	.015	31 deg.	9deg. 12'	NA	104 @ 1000	17BW	
1933	355-C	AC-D-8	.025-.028	.015	31 deg.	9deg. 12'	NA	104 @ 1000	17CF	
1934	355-D (10, 20, 30)	AC-G-7	.025-.027	.015	31 deg.	4B	NA	148 @ 1000	17D	
1935	355-D (10, 20, 30)	AC-G-6	.025-.027	.015	31 deg.	4B	NA	148 @ 1000	17D	
1936	60	AC-K7	.028	.015	31 deg.	4B	6 mph	155 @ 1000	17K	
	70, 75	AC-K7	.028	.015	31 deg.	4B	6 mph	170 @ 1000	17D	
1937	60, 65	AC-45	.028	.0125	31 deg.	5B	6 mph	155 @ 1000	17K	
	70, 75	AC-45	.028	.0125	31 deg.	5B	6 mph	170 @ 1000	17D	
1938	60, 60S, 65	AC-45	.026	.0125	31 deg.	5B	7-8 mph	155 @ 1000	17K	
	75	AC-45	.026	.0125	31 deg.	5B	7-8 mph	170 @ 1000	17D	
1939	60S, 61	AC-104	.030	.015	31 deg.	5B	7-8 mph	155 @ 1000	17K1	
	75	AC-104	.030	.015	31 deg.	5B	7-8 mph	170 @ 1000	17K1	
1940	60S, 62	AC-104	.030	.015	31 deg.	5B	375 rpm	155 @ 1000	Group 2	
	72, 75	AC-104	.030	.015	31 deg.	5B	375 rpm	170 @ 1000	Group 2	
1941	60S, 61	AC-104	.030	.015	31 deg.	5B	375 rpm	182 @ 1000	Group 2	
	62, 63	AC-104	.030	.015	31 deg.	5B	375 rpm	182 @ 1000	Group 2	
	67	AC-104	.030	.015	31 deg.	5B	375 rpm	182 @ 1000	Group 2	
	75	AC-104	.030	.015	31 deg.	5B	375 rpm	182 @ 1000	Group 2	
1942	60S	AC-104	.030	.015	31 deg.	5B	375 rpm	180 @ 1000	Group 2	
	62, 63	AC-104	.030	.015	31 deg.	5B	375 rpm	180 @ 1000	Group 2	
	67	AC-104	.030	.015	31 deg.	5B	375 rpm	180 @ 1000	Group 2	
	75	AC-104	.030	.015	31 deg.	5B	375 rpm	180 @ 1000	Group 2	
1946	60S, 61, 62	AC-104	.030	.015	31 deg.	5B	375 rpm	182 @ 1000	Group 2	
	75	AC-104	.030	.015	31 deg.	5B	375 rpm	180 @ 1000	Group 2	
1947	60S, 61, 62	AC-104	.030	.015	31 deg.	5B	375 rpm	180 @ 1000	Group 2	
	75	AC-104	.030	.015	31 deg.	5B	375 rpm	180 @ 1000	Group 2	
1948	60S, 61, 62	AC-104	.030	.015	31=1-1/2	5B	375 rpm	180 @ 1000	Group 2	
	75	AC-104	.030	.015	31=1-1/2	5B	375 rpm	180 @ 1000	Group 2	
1949	60S, 61, 62	AC-48X	.035	.015	31=1-1/2	5B	375 rpm	194 @ 1000	Group 2	
	75	AC-48X	.035	.015	31=1-1/2	5B	375 rpm	194 @ 1000	Group 2	

Selected Cadillac V-8 Tune-Up Specs 1927-1965 (continueed)

Year	Model	Spark Plugs Type	Spark Plugs Gap	Point Gap	Deg Cam Angle	Timing	Idle Speed	Compression Pressure	Battery (Delco)
1950	60S, 61, 62	AC-48X	.035	.015	31=1-1/2	5B	375 rpm	194 @ 1000	Group 2
	75	AC-48X	.035	.015	31=1-1/2	5B	375 rpm	194 @ 1000	Group 2
1951	60S, 61, 62	AC-48X	.035	.015	31=1-1/2	5B	375 rpm	194 @ 1000	Group 2
	75	AC-48X	.035	.015	31=1-1/2	5B	375 rpm	194 @ 1000	Group 2
1952	60, 62, 75	AC-48X	.035	.015	31=1-1/2	5B	375 rpm	140	Group 2
1953	60, 62, 75	AC-46-5	.035	.015	31=1-1/2	5B	375 rpm	150	2EE/3EM
1954	60, 62, 75	AC-46-5	.035	.015	31=1-1/2	5B	375 rpm	150	2EE/3EM
1955	60, 62, 75	AC-46-5	.035	.015	31=1-1/2	5B	375 rpm	175	2EE/3EM
1956	60, 62, 75	AC-46-5	.035	.015	31=1-1/2	5B	375 rpm	175	2EE/3EM
1957	60, 62, 75	AC-44	.035	.016	28-32	5B	500 rpm	180	2EE/3EM

Adjust idle with transmission in D1 or D2 to 500 rpm with two Carter four-barrel carbs.; to 420 rpm with one Rochester or Carter four-barrel carb. If car has A/C it should be on

Year	Model	Spark Plugs Type	Spark Plugs Gap	Point Gap	Deg Cam Angle	Timing	Idle Speed	Compression Pressure	Battery (Delco)
1958	60, 62, 75	AC-44	.033-.038	.016	28-32	5B	450 rpm	NA	562

Timing 10B with three two-barrel carburetors

Year	Model	Spark Plugs Type	Spark Plugs Gap	Point Gap	Deg Cam Angle	Timing	Idle Speed	Compression Pressure	Battery (Delco)
1959	60, 62, 63, 67, 69	AC-44	.033-.038	.016	28-32	5B	450 rpm	NA	562

Timing 7.5B with three two-barrel carburetors

Year	Model	Spark Plugs Type	Spark Plugs Gap	Point Gap	Deg Cam Angle	Timing	Idle Speed	Compression Pressure	Battery (Delco)
1960	60, 62, 63, 67, 68, 69	AC-44	.035	.016	28-32	5B	450 rpm	NA	562
1961	60, 62, 75	AC-44	.035	.016	28-32	5B	480 rpm	165-185	562
1962	60, 62, 75	AC-44	.035	.016	28-32	5B	480 rpm	165-185	562
1963	60, 62, 75	AC-44	.035	.016	28-32	5B	480 rpm	165-185	NA
1964	60, 62, 75	AC-44	.035	.016	30	5B	480 rpm	180	3KMB
1965	60, 62, 75	AC-44	.035	.016	30	5B	480 rpm	180	3KMB

Cadillac V-8 Firing Order

1927-1931
1L-4R-4L-2L-3R-3L-2R-1R

1932-1935
1L-4R-4L-2L-3R-3L-2R-1R or 1-2-7-8-4-5-6-3 (*)

1936-1948
1-8-7-3-6-5-4-2

1949-1962
1-8-4-3-6-5-7-2

1963-1967
1-8-7-2-6-5-4-3

1968-1973
1-5-6-3-4-2-7-8

1974
(with high energy ignition)
1-5-6-3-4-2-7-8 (**)

(without high energy ignition)
1-5-6-3-4-2-7-8 (**)

1975-1976
(350 cid)
1-8-4-3-6-5-7-2

(500 cid)
1-5-6-3-4-2-7-8

(*) Prior to 1932 the firing order was numbered using numbers with "L" and "R" suffixes to indicate order on the left (L) and right ® banks of the engine. For example, 1L indicated the first spark plug on the left-hand (driver's) side of the engine. In 1932, this was changed to numbers only, with even-number cylinders 2-4-6-8 on the left-hand (driver's) side of the engine and odd-number cylinders (1-3-5-7) on the right-hand (passenger's) side of the engine. The new firing order was the same as before, but it was expressed differently.

(**) Firing order is the same but position of distributor towers relative to front center of engine differs.

CHEVROLET
1917-2002

Chevrolet's first V-8 powered the 1917-1919 Model D. A grand total of 2,781 of these cars were built over three years, with 2,502 made in America and 279 assembled in Canada. Only 15 cars were listed on the 1917-1919 Model D register as of May 27, 1998. The specifications of the engine were as follows:

A Chevrolet V-8 wasn't available again until the introduction of the legendary small-block in 1955. That new-from-the-ground-up 265-cid engine, with up to 180 hp in "power pack" format, was an engineering milestone known for its performance, smoothness, durability and compact size. It offered 211 hp in 1956 and was bored out to 283 cid the following year. A fuel-injection option with 283 advertised hp gave one horsepower per cubic inch. The hot little 283 went under the hood of full-size Chevys and Corvettes and enthusiasts were soon dropping "small blocks" into all types of machines.

The 348-cid W-block V-8 bowed in 1958 and could be teamed up with a Turboglide automatic transmission. Horsepower ratings ranged to 315 with three two-barrel carburetors, but the high-revving small-block was still very competitive against the new "big-block" V-8, which was often scorned as a "truck engine."

The '59s were hotter than ever and a "Special Super Turbo Thrust" version of the 348 with 11.25:1 compression and "three deuces" cranked up 350 hp at 6,000 rpm. The Impala SS (Super Sport) bowed in 1961 with a big-block V-8 that was punched out to 409 cid. One SS blew the doors off the other cars at the 1961 Winternationals Drag Racing Championships in Pomona, Calif. For 1962, the SS came with an extra-cost 409-cid 409-hp option running dual four-barrel carburetors.

Small-block Chevy V-8s were not overshadowed by the big engines during the early 1960s "Super Stock" era. With up to 360 hp on tap in "fuelie" Corvettes, Chevy power ruled. Carbureted 327- and 350-cid V-8s found their way into the small Camaro and Nova SS models and turned them into formidable drag cars, whether at the strip or the "Stoplight Grand Prix." The 327 shared the 283's overbore, but had a longer stroke as well. The 350 also had the same bore as the 327, but a longer stroke.

Corvettes and mid-sized Chevelles of the mid-1960s offered big-block 427s and 396s pumping out up to 425 hp. The big-block ultimately became the 454-cid and, by 1970, came in an LS6 version with 450 hp. Chevrolet adapted "net" horsepower ratings in 1971, as the "green" movement swept away muscle car mania and changed the high-performance world forever. Geo-political rumblings erupted in the Arab oil embargo of 1973, causing gas prices to zoom upwards. Fuel economy gained greater importance with car buyers than elapsed times and this led to demand for smaller, more efficient cars. The 1975 Monza, derived from Chevy's first sub-compact, the Vega, brought sportiness and performance to the econo-car niche with a 262-cid engine—the smallest V-8 in history.

The 1990 Corvette ZR1's LT5 V-8 sported four cams and 32 valves. It generated a pavement-pounding 375 hp, which grew to 405 hp by 1993. However, a second-generation small-block base V-8 that arrived in 1992 turned out to be a sleeper for Chevrolet. This LT1 engine was so good that it reduced demand for the much more expensive ZR1. It was a 90-degree overhead valve engine with an aluminum cylinder head and multi-port fuel injection. It produced 300 hp at 5000 rpm and 340 foot-pounds of torque at 3600 rpm. The 1993 LT1-powered Corvette did 0-to-60 mph in 5.4 seconds. The engine also found its way into Camaros, Caprices and the hot Impala SS.

Introduced in 1997 Corvettes and used in all V-8-equipped Chevys made after 1998 was a powerful new 346-cid 16-valve Chevrolet LS1 V-8 with an aluminum block and aluminum cylinder heads. This high-tech V-8 has GM's engine-control system with port fuel injection. It generates 320 hp in the stock Camaro and 345 hp in the Corvette.

This 1965 Chevy 283-cid Turbo-Fire V-8 was designed to put a lot of power into a compact package. The 90-degree cylinder block measured only 21.78 inches long and nine inches from the crankcase porting line. Cylinder bore spacing was 4.4 inches, giving Chevy plenty of room to bore and stroke it to 327 cid.

Chevrolet V-8s
Includes Chevelle, Chevy II, and Camaro arranged according to horsepower.

1917 CHEVROLET

1917 Series D
Overhead valve. Cast-iron block and head. Bore & stroke: 3-3/8 x 4 in. Displacement: 288 cid. Taxable hp: 36. Three main bearings. Hydraulic valve lifters. Carburetor: Zenith vertical double jet.

1918 CHEVROLET

1918 Series D
Overhead valve. Cast-iron block and head. Bore & stroke: 3-3/8 x 4 in. Displacement: 288 cid. Taxable hp: 36. Three main bearings. Hydraulic valve lifters. Carburetor: Zenith vertical double jet.

1919 CHEVROLET

1919 Series D
Overhead valve. Cast-iron block and head. Bore & stroke: 3-3/8 x 4 in. Displacement: 288 cid. Taxable hp: 36. Three main bearings. Hydraulic valve lifters. Carburetor: Zenith vertical double jet.

This 265-cid 162-hp V-8 was Chevrolet's first overhead-valve V-8, but the automaker had offered an L-head V-8 much earlier. Chevrolet built nearly 3,000 cars fitted with the earlier V-8 engine in the 1917-1919 period. The first of the famous small-block Chevy V-8s seen here was introduced in 1955 and also came in a Power Pack version with 180 hp. A third version used in the Corvette offered 195 hp.

1955 CHEVROLET

The 1955 Chevy 265-cid V-8 was painted orange by the factory. The air cleaner and other accessories were black. The optional canister type oil filter, seen just behind the radiator hose, had a blue body and bracket, with the top painted orange. The new-from-the-ground-up 265-cid engine was an engineering milestone known for its performance, smoothness, durability, and compact size.

Returning to its 265-cid V-8 roots, Chevrolet concocted this special 265-cid 330-hp smog-legal prototype V-8 engine for display at the Specialty Equipment Market assoc. show in 1995. It was used in a customized 1955 Chevrolet constructed as a Chevrolet Engineering Test and Development vehicle. Note the use of authentic finned Corvette valve covers that were popular on custom cars in the '50s and '60s.

Chevrolet "Turbo-Fire" base V-8
Overhead valve. Cast-iron block and head. Bore & stroke: 3.75 x 3.00 in. Displacement: 265 cid. Compression ratio: 8.0:1. Brake hp: 162 at 4400 rpm. Taxable hp: 45.0. Torque: 257 lbs.-ft. at 2200. Five main bearings. Hydraulic valve lifters. Crankcase capacity: 4 qt. (add 1 qt. for new filter). Cooling system capacity: 16 qt. (add 1 qt. for heater). Carburetor: Rochester 2 BC two-barrel. Engine code: G, GC, GF, GG, GJ, GK, F, or FC.

Chevrolet "Turbo-Fire" optional V-8
Overhead valve. Cast-iron block and head. Bore & stroke: 3.75 x 3.00 in. Displacement: 265 cid. Compression ratio: 8.0:1. Brake hp: 180 at 4600 rpm. Taxable hp: 45.0. Torque: 260 lbs.-ft. at 2800. Five main bearings. Hydraulic valve lifters. Crankcase capacity: 4 qt. (add 1 qt. for new filter). Cooling system capacity: 16 qt. (add 1 qt. for heater). Carburetor: Rochester four-barrel. Engine code: GD, GE, GL, or GM.

1955 CORVETTE

Chevrolet "Corvette" V-8
Overhead valve. Cast-iron block and head. Special camshaft. Bore & stroke: 3.75 x 3.00 in. Displacement: 265 cid. Compression ratio: 8.0:1. Brake hp: 195 at 5000 rpm. Taxable hp: 45.0. Torque: 260 lbs.-ft. at 3000 rpm. Five main bearings. Solid valve lifters. Crankcase capacity: 4 qt. (add 1 qt. for new filter). Cooling system capacity: 16 qt. (add 1 qt. for heater). Carburetor: Carter WCFB four-barrel. Dual exhausts. Engine code: FG, GR.

1956 CHEVROLET

Chevrolet "Turbo-Fire" base V-8 (manual transmission)
Overhead valve. Cast-iron block and head. Bore & stroke: 3.75 x 3.00 in. Displacement: 265 cid. Compression ratio: 8.0:1. Brake hp: 162 at 4400 rpm. Taxable hp: 45.0. Torque: 257 lbs.-ft. at 2200 rpm. Five main bearings. Hydraulic valve lifters. Crankcase capacity: 4 qt. (add 1 qt. for new filter). Cooling system capacity: 16 qt. (add 1 qt. for heater). Carburetor: Rochester 2BC two-barrel. Engine code: G, GC, GQ, GF, GJ, and GK.

Chevrolet "Turbo-Fire" base V-8 (Powerglide transmission)
Overhead valve. Cast-iron block and head. Bore & stroke: 3.75 x 3.00 in. Displacement: 265 cid. Compression ratio: 8.0:1. Brake hp: 170 at 4400 rpm. Taxable hp: 45.0. Torque: 257 lbs.-ft. at 2400 rpm. Five main bearings. Hydraulic valve lifters. Crankcase capacity: 4 qt. (add 1 qt. for new filter). Cooling system capacity: 16 qt. (add 1 qt. for heater). Carburetor: Rochester 2BC or Carter 2286-S two-barrel. Engine code: GL, F, FC, or FD.

Chevrolet "Turbo-Fire 205" optional V-8
Overhead valve. Cast-iron block and head. Bore & stroke: 3.75 x 3.00 in. Displacement: 265 cid. Compression ratio: 9.25:1. Brake hp: 205 at 4600 rpm. Taxable hp: 45.0. Torque: 268 lbs.-ft. at 3000 rpm. Five main bearings. Hydraulic valve lifters. Crankcase capacity: 4 qt. (add 1 qt. for new filter). Cooling system capacity: 16 qt. (add 1 qt. for heater). Carburetor: Rochester 4 BC four-barrel. Dual exhausts. Engine code: GE, GM, GN, or FB.

Chevrolet "Corvette" optional V-8; "Turbo-Fire 225" optional V-8
Overhead valve. Cast-iron block and head. Bore & stroke: 3.75 x 3.00 in. Displacement: 265 cid. Compression ratio: 9.25:1. Brake hp: 225 at 5200 rpm. Taxable hp: 45.0. Torque: 270 lbs.-ft. at 3600 rpm. Five main bearings. Solid valve lifters. Crankcase capacity: 4 qt. (add 1 qt. for new filter). Cooling system capacity: 16 qt. (add 1 qt. for heater). Carburetors: Dual Carter four-barrel. Dual exhausts. Engine: GR and FG.

1956 CORVETTE

Chevrolet "Corvette" V-8
Overhead valve. Cast-iron block and head. Bore & stroke: 3.75 x 3.00 in. Displacement: 265 cid. Compression ratio: 9.25:1. Brake hp: 210 at 5200 rpm. Taxable hp: 45.0. Torque: 270 lbs.-ft. at 3200 rpm. Five main bearings. Hydraulic valve lifters. Crankcase capacity: 4 qt. (add 1 qt. for new filter). Cooling system capacity: 16 qt. (add 1 qt. for heater). Carburetor: Carter WCFB four-barrel. Engine code: FK or GV.

Chevrolet "Corvette" optional V-8; "Turbo-Fire 225" optional V-8
Overhead valve. Cast-iron block and head. Bore & stroke: 3.75 x 3.00 in. Displacement: 265 cid. Compression ratio: 9.25:1. Brake hp: 225 at 5200 rpm. Taxable hp: 45.0. Torque: 270 lbs.-ft. at 3600 rpm. Five main bearings. Solid valve lifters. Crankcase capacity: 4 qt. (add 1 qt. for new filter). Cooling system capacity: 16 qt. (add 1 qt. for heater). Carburetors: Dual Carter four-barrel. Dual exhausts. Engine: GR and FG.

Chevrolet "Corvette" optional V-8; "Turbo-Fire 240" optional V-8
Overhead valve. Cast-iron block and head. High-lift camshaft. Bore & stroke: 3.75 x 3.00 in. Displacement: 265 cid. Compression ratio: 9.25:1. Brake hp: 240. Taxable hp: 45.0. Five main bearings. Solid valve lifters. Crankcase capacity: 4 qt. (add 1 qt. for new filter). Cooling system capacity: 16 qt. (add 1 qt. for heater). Carburetors: Dual Carter four-barrel. Dual exhausts. Engine: GR and FG.

1957 CHEVROLET

Two fuel-injected 283-cid small-block V-8s were offered in 1957 Chevys and Corvettes. The "Ram-Jet" version had 9.5:1 compression and developed 250 hp at 5000 rpm. The "Super Ram-Jet" version had a 10.5:1 compression ratio and generated one horsepower per cubic inch, a fact that Chevy promoted. This fuel-injection system was developed by General Motors Rochester Products Divsion. (Courtesy Blackhawk Auto Collection)

Chevrolet base "Turbo-Fire" V-8 (manual transmission or "Touchdown" overdrive)
Overhead valve. Cast-iron block and head. Bore & stroke: 3.75 x 3.00 in. Displacement: 265 cid. Compression ratio: 8.0:1. Brake hp: 162 at 4400 rpm. Taxable hp: 45.0. Torque: 257 lbs.-ft. at 2200 rpm. Five main bearings. Hydraulic valve lifters. Crankcase capacity: 4 qt. (add 1 qt. for new filter). Cooling system capacity: 16 qt. (add 1 qt. for heater). Carburetor: Rochester 2 BC two-barrel. Engine code: C, CD, or CE.

Chevrolet base "Turbo-Fire 283" V-8 (Powerglide transmission)
Overhead valve. Cast-iron block and head. Bore & stroke: 3.875 x 3.00 in. Displacement: 283 cid. Compression ratio: 8.50:1. Brake hp: 185 at 4600 rpm. Taxable hp: 48.0. Torque: 275 lbs.-ft. at 2400 rpm. Five main bearings. Hydraulic valve lifters. Crankcase capacity: 4 qt. (add 1 qt. for new filter). Cooling system capacity: 16 qt. (add 1 qt. for heater). Carburetor: Carter 2286-S or Rochester 2 BC two-barrel. Engine code: C, CD, CB, F, D, FA, G, or E.

Chevrolet Turbo-Fire optional V-8
Overhead valve. Cast-iron block and head. Bore & stroke: 3.875 x 3.00 in. Displacement: 283 cid. Compression ratio: 9.50:1. Brake hp: 220 at 4800 rpm. Taxable hp: 48.0. Torque: 300 lbs.-ft. at 3000 rpm. Five main bearings. Hydraulic valve lifters. Crankcase capacity: 4 qt. (add 1 qt. for new filter). Cooling system capacity: 16 qt. (add 1 qt. for heater). Dual exhaust. Carburetor: Carter WCFB or Rochester 4 BC four-barrel.

Chevrolet "Turbo-Fire" optional V-8
Overhead valve. Cast-iron block and head. Bore & stroke: 3.875 x 3.00 in. Displacement: 283 cid. Compression ratio: 9.50:1. Brake hp: 245 at 5000 rpm. Taxable hp: 48.0. Torque: 300 lbs.-ft. at 3800 rpm. Five main bearings. Hydraulic valve lifters. Crankcase capacity: 4 qt. (add 1 qt. for new filter). Cooling system capacity: 16 qt. (add 1 qt. for heater). Dual exhaust. Carburetors: Dual Carter WCFB four-barrels. Engine code: EA, FD, GD, EH, CT, FG, or DG.

Chevrolet "Ram-Jet Fuel-Injected" optional V-8
Overhead valve. Cast-iron block and head. Bore & stroke: 3.875 x 3.00 in. Displacement: 283 cid. Compression ratio: 9.50:1. Brake hp: 250 at 5000 rpm. Taxable hp: 48.0. Torque: 305 lbs.-ft. at 3800 rpm. Five main bearings. Hydraulic valve lifters. Crankcase capacity: 4 qt. (add 1 qt. for new filter). Cooling system capacity: 16 qt. (add 1 qt. for heater). Dual exhaust. Induction system: Rochester fuel injection. Engine code: EJ, CH, FJ, DP, GF, EC, CY, EM, CR, FK, or DH.

Chevrolet "Super Turbo-Fire" optional V-8
Overhead valve. Cast-iron block and head. Special camshaft. Bore & stroke: 3.875 x 3.00 in. Displacement: 283 cid. Compression ratio: 9.50:1. Brake hp: 270 at 6000 rpm. Taxable hp: 48.0. Torque: 285 lbs.-ft. at 4200 rpm. Five main bearings. Hydraulic valve lifters. Crankcase capacity: 4 qt. (add 1 qt. for new filter). Cooling system capacity: 16 qt. (add 1 qt. for heater). Dual exhaust. Carburetors: Dual Carter WCFB four-barrels. Engine code: RB or EG.

Chevrolet "Super Ram-Jet Fuel-Injected" optional V-8
Overhead valve. Cast-iron block and head. Bore & stroke: 3.875 x 3.00 in. Displacement: 283 cid. Compression ratio: 10.50:1. Brake hp: 283 at 6200 rpm. Taxable hp: 48.0. Torque: 290 lbs.-ft. at 4400 rpm. Five main bearings. Hydraulic valve lifters. Crankcase capacity: 4 qt. (add 1 qt. for new filter). Cooling system capacity: 16 qt. (add 1 qt. for heater). Dual exhaust. Induction system: Rochester fuel injection. Engine code: EK, CJ, CZ, or EL.
NOTE: This famous V-8 was one of the first to offer one horsepower per cubic inch.

1957 CORVETTE

Corvette "Turbo-Fire 283" V-8 (Powerglide transmission)
Overhead valve. Cast-iron block and head. Bore & stroke: 3.875 x 3.00 in. Displacement: 283 cid. Compression ratio: 8.50:1. Brake hp: 185 at 4600 rpm. Taxable hp: 48.0. Torque: 275 lbs.-ft. at 2400 rpm. Five main bearings. Hydraulic valve lifters. Crankcase capacity: 4 qt. (add 1 qt. for new filter). Cooling system capacity: 16 qt. (add 1 qt. for heater). Carburetor: Carter 2286-S or Rochester 2 BC two-barrel. Engine code: FH.

Corvette V-8
Overhead valve. Cast-iron block and head. Bore & stroke: 3.875 x 3.00 in. Displacement: 283 cid. Compression ratio: 9.50:1. Brake hp: 220 at 4800 rpm. Taxable hp: 48.0. Torque: 300 lbs.-ft. at 3000 rpm. Five main bearings. Hydraulic valve lifters. Crankcase capacity: 4 qt. (add 1 qt. for new filter). Cooling system capacity: 16 qt. (add 1 qt. for heater). Dual exhaust. Carburetor: Carter WCFB or Rochester 4 BC four-barrel.

Corvette "Turbo-Fire" optional V-8
Overhead valve. Cast-iron block and head. Bore & stroke: 3.875 x 3.00 in. Displacement: 283 cid. Compression ratio: 9.50:1. Brake hp: 245 at 5000 rpm. Taxable hp: 48.0. Torque: 300 lbs.-ft. at 3800 rpm. Five main bearings. Hydraulic valve lifters. Crankcase capacity: 4 qt. (add 1 qt. for new filter). Cooling system capacity: 16 qt. (add 1 qt. for heater). Dual exhaust. Carburetors: Dual Carter WCFB four-barrels. Engine code: EA, FD, GD, EH, CT, FG, or DG.

Corvette "Ram-Jet Fuel-Injected" optional V-8
Overhead valve. Cast-iron block and head. Bore & stroke: 3.875 x 3.00 in. Displacement: 283 cid. Compression ratio: 9.50:1. Brake hp: 250 at 5000 rpm. Taxable hp: 48.0. Torque: 305 lbs.-ft. at 3800 rpm. Five main bearings. Hydraulic valve lifters. Crankcase capacity: 4 qt. (add 1 qt. for new filter). Cooling system capacity: 16 qt. (add 1 qt. for heater). Dual exhaust. Induction system: Rochester fuel injection. Engine code: EJ, CH, FJ, DP, GF, EC, CY, EM, CR, FK, or DH.

Corvette "Super Turbo-Fire" optional V-8
Overhead valve. Cast-iron block and head. Special camshaft. Bore & stroke: 3.875 x 3.00 in. Displacement: 283 cid. Compression ratio: 9.50:1. Brake hp: 270 at 6000 rpm. Taxable hp: 48.0. Torque: 285 lbs.-ft. at 4200 rpm. Five main bearings. Solid valve lifters. Crankcase capacity: 4 qt. (add 1 qt. for new filter). Cooling system capacity: 16 qt. (add 1 qt. for heater). Dual exhaust. Carburetors: Dual Carter WCFB four-barrels. Engine code: RB or EG.

Corvette "Super Ram-Jet Fuel-Injected" optional V-8
Overhead valve. Cast-iron block and head. Bore & stroke: 3.875 x 3.00 in. Displacement: 283 cid. Compression ratio: 10.50:1. Brake hp: 283 at 6200 rpm. Taxable hp: 48.0. Torque: 290 lbs.-ft. at 4400 rpm. Five main bearings. Solid valve lifters. Crankcase capacity: 4 qt. (add 1 qt. for new filter). Cooling system capacity: 16 qt. (add 1 qt. for heater). Dual exhaust. Induction system: Rochester fuel injection. Engine code: EK, CJ, CZ, or EL.
NOTE: This famous V-8 was one of the first to offer one horsepower per cubic inch.

1958 CHEVROLET

Chevrolet "Turbo-Fire 283" base V-8
Overhead valve. Cast-iron block and head. Bore & stroke: 3.875 x 3.00 in. Displacement: 283 cid. Compression ratio: 8.50:1. Brake hp: 185 at 4600 rpm. Taxable hp: 48.0. Torque: 275 lbs.-ft. at 2400 rpm. Five main bearings. Hydraulic valve lifters. Crankcase capacity: 4 qt. (add 1 qt. for new filter). Cooling system capacity: 16 qt. (add 1 qt. for heater). Carburetor: Rochester 2 BC two-barrel. Engine code: C, CD, CB, F, D, FA, G, or E.

Chevrolet "Super Turbo-Fire 283" optional V-8
Overhead valve. Cast-iron block and head. Bore & stroke: 3.875 x 3.00 in. Displacement: 283 cid. Compression ratio: 9.50:1. Brake hp: 230 at 4800 rpm. Taxable hp: 48.0. Torque: 300 lbs.-ft. at 3000 rpm. Five main bearings. Hydraulic valve lifters. Crankcase capacity: 4 qt. (add 1 qt. for new filter). Cooling system capacity: 16 qt. (add 1 qt. for heater). Dual exhaust. Carburetor: Carter WCFB or Rochester 4GC four-barrel. Engine code: CM, DB, EB, CQ, CG, DM, EJ, and DG.

Chevrolet "Turbo-Thrust 348" optional V-8
Overhead valve. Cast-iron block and head. Bore & stroke: 4.125 x 3.25 in. Displacement: 348 cid. Compression ratio: 9.50:1. Brake hp: 250 at 4400 rpm. Taxable hp: 54.5. Torque: 355 lbs.-ft. at 2800 rpm. Five main bearings. Hydraulic valve lifters. Crankcase capacity: 4 qt. (add 1 qt. for new filter). Cooling system capacity: 22 qt. (add 1 qt. for heater). Dual exhaust. Carburetor: Carter WCFB four-barrel. Engine code: G or H.

Chevrolet "283 Fuel-Injected" optional V-8
Overhead valve. Cast-iron block and head. Bore & stroke: 3.875 x 3.00 in. Displacement: 283 cid. Compression ratio: 9.50:1. Brake hp: 250 at 5000 rpm. Taxable hp: 48.0. Torque: 305 lbs.-ft. at 3800 rpm. Five main bearings. Hydraulic valve lifters. Crankcase capacity: 4 qt. (add 1 qt. for new filter). Cooling system capacity: 16 qt. (add 1 qt. for heater). Dual exhaust. Induction system: Rochester fuel injection. Engine code: CR.

Chevrolet "Ram-Jet Fuel-Injected" optional V-8
Overhead valve. Cast-iron block and head. Bore & stroke: 3.875 x 3.00 in. Displacement: 283 cid. Compression ratio: 9.50:1. Brake hp: 250 at 5000 rpm. Taxable hp: 48.0. Torque: 305 lbs.-ft. at 3800 rpm. Five main bearings. Hydraulic valve lifters. Crankcase capacity: 4 qt. (add 1 qt. for new filter). Cooling system capacity: 16 qt. (add 1 qt. for heater). Induction: Rochester fuel-injection. Engine code: EJ, CH, FJ, DP, GF, EC, CY, EM, CR, FK, or DH.

Chevrolet Super Turbo-Thrust optional V-8
Overhead valve. Cast-iron block and head. Bore & stroke: 4.125 x 3.25 in. Displacement: 348 cid. Compression ratio: 9.50:1. Brake hp: 280 at 4800 rpm. Taxable hp: 54.5. Torque: 355 lbs.-ft. at 3200 rpm. Five main bearings. Hydraulic valve lifters. Crankcase capacity: 4 qt. (add 1 qt. for new filter). Cooling system capacity: 22 qt. (add 1 qt. for heater). Dual exhaust. Carburetor: Three two-barrel. Engine code: FA, GB, or HA.

Chevrolet "Special Solid Lifter Ram-Jet Fuel-Injected" optional V-8 (manual transmission)
Overhead valve. Cast-iron block and head. Bore & stroke: 3.875 x 3.00 in. Displacement: 283 cid. Compression ratio: 11.50:1. Brake hp: 290 at 6200 rpm. Taxable hp: 48.0. Torque: 290 lbs.-ft. at 4400 rpm. Five main bearings. Solid valve lifters. Crankcase capacity: 4 qt. (add 1 qt. for new filter). Cooling system capacity: 16 qt. (add 1 qt. for heater). Induction: Fuel-injection. Engine code: EK, CJ, CZ, or CS.

Chevrolet Super Turbo-Thrust optional V-8
Overhead valve. Cast-iron block and head. Bore & stroke: 4.125 x 3.25 in. Displacement: 348 cid. Compression ratio: 11.00:1. Brake hp: 300 at 5600 rpm. Taxable hp: 54.5. Torque: 350 lbs.-ft. at 3600 rpm. Five main bearings. Solid valve lifters. Crankcase capacity: 4 qt. (add 1 qt. for new filter). Cooling system capacity: 22 qt. (add 1 qt. for heater). Dual exhaust. Carburetor: Four-barrel. Engine code: FD.

Chevrolet Super Turbo-Thrust optional V-8
Overhead valve. Cast-iron block and head. Bore & stroke: 4.125 x 3.25 in. Displacement: 348 cid. Compression ratio: 11.00:1. Brake hp: 315 at 5600 rpm. Taxable hp: 54.5. Torque: 356 lbs.-ft. at 3600 rpm. Five main bearings. Solid valve lifters. Crankcase capacity: 4 qt. (add 1 qt. for new filter). Cooling system capacity: 22 qt. (add 1 qt. for heater). Dual exhaust. Carburetor: Three two-barrel. Engine code: FB.

1958 CORVETTE

Corvette "Super Turbo-Fire 283" optional V-8
Overhead valve. Cast-iron block and head. Bore & stroke: 3.875 x 3.00 in. Displacement: 283 cid. Compression ratio: 9.50:1. Brake hp: 230 at 4800 rpm. Taxable hp: 48.0. Torque: 300 lbs.-ft. at 3000 rpm. Five main bearings. Hydraulic valve lifters. Crankcase capacity: 4 qt. (add 1 qt. for

new filter). Cooling system capacity: 16 qt. (add 1 qt. for heater). Dual exhaust. Carburetor: Carter WCFB or Rochester 4GC four-barrel. Engine code: CM, DB, EB, CQ, CG, DM, EJ, and DG.

Corvette "Dual Four-Barrel 283" optional V-8
Overhead valve. Cast-iron block and head. Bore & stroke: 3.875 x 3.00 in. Displacement: 283 cid. Compression ratio: 9.50:1. Brake hp: 245 at 5000 rpm. Taxable hp: 48.0. Torque: 300 lbs.-ft. at 3800 rpm. Five main bearings. Hydraulic valve lifters. Crankcase capacity: 4 qt. (add 1 qt. for new filter). Cooling system capacity: 16 qt. (add 1 qt. for heater). Dual exhaust. Carburetor: Dual Carter WCFB four-barrels. Engine code: EH, CT, FG, or DJ.

Chevrolet "Corvette 283" optional V-8
Overhead valve. Cast-iron block and head. Bore & stroke: 3.875 x 3.00 in. Displacement: 283 cid. Compression ratio: 9.50:1. Brake hp: 250 at 5000 rpm. Taxable hp: 48.0. Torque: 305 lbs.-ft. at 3800 rpm. Five main bearings. Hydraulic valve lifters. Crankcase capacity: 4 qt. (add 1 qt. for new filter). Cooling system capacity: 16 qt. (add 1 qt. for heater). Dual exhaust. Induction system: Rochester fuel injection. Engine code: CR.

Corvette "Ram-Jet Fuel-Injected" optional V-8
Overhead valve. Cast-iron block and head. Bore & stroke: 3.875 x 3.00 in. Displacement: 283 cid. Compression ratio: 9.50:1. Brake hp: 250 at 5000 rpm. Taxable hp: 48.0. Torque: 305 lbs.-ft. at 3800 rpm. Five main bearings. Hydraulic valve lifters. Crankcase capacity: 4 qt. (add 1 qt. for new filter). Cooling system capacity: 16 qt. (add 1 qt. for heater). Induction: Rochester fuel-injection. Engine code: EJ, CH, FJ, DP, GF, EC, CY, EM, CR, FK, or DH.

Corvette "Dual Four-Barrel" optional V-8
Overhead valve. Cast-iron block and head. Bore & stroke: 3.875 x 3.00 in. Displacement: 283 cid. Compression ratio: 9.50:1. Brake hp: 270 at 6000 rpm. Taxable hp: 48.0. Torque: 285 lbs.-ft. at 4200 rpm. Five main bearings. Solid valve lifters. Crankcase capacity: 4 qt. (add 1 qt. for new filter). Cooling system capacity: 16 qt. (add 1 qt. for heater). Dual exhaust. Induction system: Rochester fuel injection. Engine code: CU.

Corvette "Special Solid Lifter Ram-Jet Fuel-Injected" optional V-8 (manual transmission)
Overhead valve. Cast-iron block and head. Bore & stroke: 3.875 x 3.00 in. Displacement: 283 cid. Compression ratio: 11.50:1. Brake hp: 290 at 6200 rpm. Taxable hp: 48.0. Torque: 290 lbs.-ft. at 4400 rpm. Five main bearings. Solid valve lifters. Crankcase capacity: 4 qt. (add 1 qt. for new filter). Cooling system capacity: 16 qt. (add 1 qt. for heater). Induction: Fuel-injection. Engine code: EK, CJ, CZ, or CS.

1959 CHEVROLET

Chevrolet base V-8
Overhead valve. Cast-iron block and head. Bore & stroke: 3.875 x 3.00 in. Displacement: 283 cid. Compression ratio: 8.50:1. Brake hp: 185 at 4600 rpm. Taxable hp: 48.0. Torque: 275 lbs.-ft. at 2400. Five main bearings. Hydraulic valve lifters. Crankcase capacity: 4 qt. (add 1 qt. for new filter). Cooling system capacity: 18.5 qt. (add 1 qt. for heater). Carburetor: Rochester 2 BC two-barrel. Engine code: C, CD, CQ, or D.

Chevrolet Super Turbo-Fire optional V-8
Overhead valve. Cast-iron block and head. Bore & stroke: 3.875 x 3.00 in. Displacement: 283 cid. Compression ratio: 9.50:1. Brake hp: 230 at 4800 rpm. Taxable hp: 48.0. Torque: 300 lbs.-ft. at 3000 rpm. Five main bearings. Hydraulic valve lifters. Crankcase capacity: 4 qt. (add 1 qt. for new filter). Cooling system capacity: 18.5 qt. (add 1 qt. for heater). Dual exhaust. Carburetor: Dual Carter WCFB four-barrels. Engine code: CT, CU, or DJ.

Chevrolet Turbo-Thrust optional V-8
Overhead valve. Cast-iron block and head. Bore & stroke: 4.125 x 3.25 in. Displacement: 348 cid. Compression ratio: 9.50:1. Brake hp: 250 at 4400 rpm. Taxable hp: 54.5. Torque: 355 lbs.-ft. at 2800 rpm. Five main bearings. Hydraulic valve lifters. Crankcase capacity: 4 qt. (add 1 qt. for new filter). Cooling system capacity: 22 qt. (add 1 qt. for heater). Dual exhaust. Carburetor: Carter WCFB four-barrel. Engine code: F, G, or H.

Chevrolet "Corvette Ram-Jet Fuel-Injected" optional V-8
Overhead valve. Cast-iron block and head. Bore & stroke: 3.875 x 3.00 in. Displacement: 283 cid. Compression ratio: 9.50:1. Brake hp: 250 at 5000 rpm. Taxable hp: 48.0. Torque: 305 lbs.-ft. at 3800 rpm. Five main bearings. Hydraulic valve lifters. Crankcase capacity: 4 qt. (add 1 qt. for new filter). Cooling system capacity: 18.5 qt. (add 1 qt. for heater). Dual exhaust. Induction system: Rochester fuel injection. Engine code: CR or DH.

Chevrolet Super Turbo-Thrust optional V-8
Overhead valve. Cast-iron block and head. Bore & stroke: 4.125 x 3.25 in. Displacement: 348 cid. Compression ratio: 9.50:1. Brake hp: 280 at 4800 rpm. Taxable hp: 54.5. Torque: 355 lbs.-ft. at 3200 rpm. Five main bearings. Hydraulic valve lifters. Crankcase capacity: 4 qt. (add 1 qt. for new filter). Cooling system capacity: 22 qt. (add 1 qt. for heater). Dual exhaust. Carburetor: Three Rochester two-barrels. Engine code: FA, GB, or HA.

Chevrolet "Special Corvette Ram-Jet Fuel-Injected" optional V-8
Overhead valve. Cast-iron block and head. Bore & stroke: 3.875 x 3.00 in. Displacement: 283 cid. Compression ratio: 10.50:1. Brake hp: 290 at 6200 rpm. Taxable hp: 48.0. Torque: unknown at unknown. Five main bearings. Hydraulic valve lifters. Crankcase capacity: 4 qt. (add 1 qt. for new filter). Cooling system capacity: 18.5 qt. (add 1 qt. for heater). Dual exhaust. Induction system: Rochester fuel injection. Engine code: CS

Chevrolet Special Turbo-Thrust optional V-8
Overhead valve. Cast-iron block and head. Bore & stroke: 4.125 x 3.25 in. Displacement: 348 cid. Compression ratio: 11.00:1. Brake hp: 300 at 5600 rpm. Taxable hp: 54.5. Torque: 350 lbs.-ft. at 3200 rpm. Five main bearings. Solid valve lifters. Crankcase capacity: 4 qt. (add 1 qt. for new filter). Cooling system capacity: 22 qt. (add 1 qt. for heater). Dual exhaust. Carburetors: Carter AFB four-barrel. Engine code: FG.

Chevrolet Special Turbo-Thrust optional V-8
Overhead valve. Cast-iron block and head. Bore & stroke: 4.125 x 3.25 in. Displacement: 348 cid. Compression ratio: 11.00:1. Brake hp: 305 at 5600 rpm. Taxable hp: 54.5. Torque: 350 lbs.-ft. at 3600. Five main bearings. Solid valve lifters. Crankcase capacity: 4 qt. (add 1 qt. for new filter). Cooling system capacity: 21 qt. (add 1 qt. for heater). Dual exhaust. Carburetors: Carter AFB four-barrel. Engine code: GD.

Chevrolet Special Super Turbo-Thrust optional V-8
Overhead valve. Cast-iron block and head. Bore & stroke: 4.125 x 3.25 in. Displacement: 348 cid. Compression ratio: 11.00:1. Brake hp: 315 at 5600 rpm. Taxable hp: 54.5. Torque: unknown at unknown. Five main bearings. Solid valve lifters. Crankcase capacity: 4 qt. (add 1 qt. for new filter). Cooling system capacity: 22 qt. (add 1 qt. for heater). Dual exhaust. Carburetor: Three Rochester two-barrels. Engine code: FA.

Chevrolet Special Turbo-Thrust optional V-8
Overhead valve. Cast-iron block and head. Bore & stroke: 4.125 x 3.25 in. Displacement: 348 cid. Compression ratio: 11.25:1. Brake hp: 320 at 5600 rpm. Taxable hp: 54.5. Torque: 358 lbs.-ft. at 3600 rpm. Five main bearings. Solid valve lifters. Crankcase capacity: 4 qt. (add 1 qt. for new filter). Cooling system capacity: 22 qt. (add 1 qt. for heater). Dual exhaust. Carburetors: Carter AFB four-barrel. Engine code: FD.

Chevrolet Special Super Turbo-Thrust optional V-8
Overhead valve. Cast-iron block and head. Bore & stroke: 4.125 x 3.25 in. Displacement: 348 cid. Compression ratio: 11.25:1. Brake hp: 335 at 5800 rpm. Taxable hp: 54.5. Torque: 362 lbs.-ft. at 3600 rpm. Five main bearings. Solid valve lifters. Crankcase capacity: 4 qt. (add 1 qt. for new filter). Cooling system capacity: 22 qt. (add 1 qt. for heater). Dual exhaust. Carburetor: Three Rochester two-barrels: (front) 7013973; (center) 7013974; (rear) 7013975. Engine code: FB.

Chevrolet Special Super Turbo-Thrust optional V-8
Overhead valve. Cast-iron block and head. Bore & stroke: 4.125 x 3.25 in. Displacement: 348 cid. Compression ratio: 11.25:1. Brake hp: 350 at 6000 rpm. Taxable hp: 54.5. Torque: 364 lbs.-ft. at 3600 rpm. Five main bearings. Solid valve lifters. Crankcase capacity: 4 qt. (add 1 qt. for new filter). Cooling system capacity: 22 qt. (add 1 qt. for heater). Dual exhaust. Carburetor: Three Rochester two-barrels. Engine code: Unknown.

1959 CORVETTE

Chevrolet Super Turbo-Fire optional V-8
Overhead valve. Cast-iron block and head. Bore & stroke: 3.875 x 3.00 in. Displacement: 283 cid. Compression ratio: 9.50:1. Brake hp: 230 at 4800 rpm. Taxable hp: 48.0. Torque: 300 lbs.-ft. at 3000 rpm. Five main bearings. Hydraulic valve lifters. Crankcase capacity: 4 qt. (add 1 qt. for new filter). Cooling system capacity: 18.5 qt. (add 1 qt. for heater). Dual exhaust. Carburetor: Dual Carter WCFB four-barrels. Engine code: CQ or DG.

Chevrolet Dual Four-Barrel optional V-8
Overhead valve. Cast-iron block and head. Bore & stroke: 3.875 x 3.00 in. Displacement: 283 cid. Compression ratio: 9.50:1. Brake hp: 245 at 5000 rpm. Taxable hp: 48.0. Torque: 300 lbs.-ft. at 3800 rpm. Five main bearings. Hydraulic valve lifters. Crankcase capacity: 4 qt. (add 1 qt. for

new filter). Cooling system capacity: 18.5 qt. (add 1 qt. for heater). Dual exhaust. Carburetor: Dual Carter WCFB four-barrels. Engine code: EH, CT, FJ, or DJ.

Chevrolet "Corvette Ram-Jet Fuel-Injected" optional V-8
Overhead valve. Cast-iron block and head. Bore & stroke: 3.875 x 3.00 in. Displacement: 283 cid. Compression ratio: 9.50:1. Brake hp: 250 at 5000 rpm. Taxable hp: 48.0. Torque: 305 lbs.-ft. at 3800 rpm. Five main bearings. Hydraulic valve lifters. Crankcase capacity: 4 qt. (add 1 qt. for new filter). Cooling system capacity: 18.5 qt. (add 1 qt. for heater). Dual exhaust. Induction system: Rochester fuel injection. Engine code: CR.

Corvette "Dual Four-Barrel" optional V-8
Overhead valve. Cast-iron block and head. Bore & stroke: 3.875 x 3.00 in. Displacement: 283 cid. Compression ratio: 9.50:1. Brake hp: 270 at 6000 rpm. Taxable hp: 48.0. Torque: 285 lbs.-ft. at 4200 rpm. Five main bearings. Solid valve lifters. Crankcase capacity: 4 qt. (add 1 qt. for new filter). Cooling system capacity: 16 qt. (add 1 qt. for heater). Dual exhaust. Induction system: Rochester fuel injection. Engine code: CU.

Chevrolet "Special Corvette Ram-Jet Fuel-Injected" optional V-8
Overhead valve. Cast-iron block and head. Bore & stroke: 3.875 x 3.00 in. Displacement: 283 cid. Compression ratio: 10.50:1. Brake hp: 290 at 6200 rpm. Taxable hp: 48.0. Torque: unknown at unknown. Five main bearings. Hydraulic valve lifters. Crankcase capacity: 4 qt. (add 1 qt. for new filter). Cooling system capacity: 18.5 qt. (add 1 qt. for heater). Dual exhaust. Induction system: Rochester fuel injection. Engine code: CS.

1960 CHEVROLET

Chevrolet base V-8 (manual transmission)
Overhead valve. Cast-iron block and head. Bore & stroke: 3.875 x 3.00 in. Displacement: 283 cid. Compression ratio: 8.5:1. Brake hp: 170 at 4200 rpm. Taxable hp: 48.0. Torque: 275 lbs.-ft. at 2200 rpm. Five main bearings. Hydraulic valve lifters. Crankcase capacity: 4 qt. (add 1 qt. for new filter). Cooling system capacity: 17.5 qt. (add 1 qt. for heater). Carburetor: Rochester 2GC two-barrel. Engine code: C, CD, CL, D, DK, E, and EG.

Chevrolet Super Turbo-Fire optional V-8
Overhead valve. Cast-iron block and head. Bore & stroke: 3.875 x 3.00 in. Displacement: 283 cid. Compression ratio: 9.50:1. Brake hp: 230 at 4800 rpm. Taxable hp: 48.0. Torque: 300 lbs.-ft. at 3000 rpm. Five main bearings. Hydraulic valve lifters. Crankcase capacity: 4 qt. (add 1 qt. for new filter). Cooling system capacity: 17.5 qt. (add 1 qt. for heater). Dual exhaust. Carburetor: Rochester 4GC or Carter WCFB four-barrel. Engine code: CF, CG, CM, DB, DM, EB, and EJ.

Chevrolet Turbo-Thrust optional V-8
Overhead valve. Cast-iron block and head. Bore & stroke: 4.125 x 3.25 in. Displacement: 348 cid. Compression ratio: 9.50:1. Brake hp: 250 at 4400 rpm. Taxable hp: 54.5. Torque: 355 lbs.-ft. at 2800 rpm. Five main bearings. Hydraulic valve lifters. Crankcase capacity: 4 qt. (add 1 qt. for new filter). Cooling system capacity: 21 qt. (add 1 qt. for heater). Dual exhaust. Carburetor: Carter WCFB four-barrel. Engine code: F, G, and H.

Chevrolet Super Turbo-Thrust optional V-8
Overhead valve. Cast-iron block and head. Bore & stroke: 4.125 x 3.25 in. Displacement: 348 cid. Compression ratio: 9.50:1. Brake hp: 280 at 4800 rpm. Taxable hp: 54.5. Torque: 355 lbs.-ft. at 3200 rpm. Five main bearings. Hydraulic valve lifters. Crankcase capacity: 4 qt. (add 1 qt. for new filter). Cooling system capacity: 21 qt. (add 1 qt. for heater). Dual exhaust. Carburetor: Three Rochester two-barrel: (front) 7013015; (center) 7013020; (rear) 7013017. Engine code: FA, GB, and HA.

Chevrolet Special Turbo-Thrust optional V-8
Overhead valve. Cast-iron block and head. Bore & stroke: 4.125 x 3.25 in. Displacement: 348 cid. Compression ratio: 11.00:1. Brake hp: 305 at 5600 rpm. Taxable hp: 54.5. Torque: 350 lbs.-ft. at 3600 rpm. Five main bearings. Solid valve lifters. Crankcase capacity: 4 qt. (add 1 qt. for new filter). Cooling system capacity: 21 qt. (add 1 qt. for heater). Dual exhaust. Carburetors: Carter AFB four-barrel. Engine code: GD.

Chevrolet Special Turbo-Thrust optional V-8
Overhead valve. Cast-iron block and head. Bore & stroke: 4.125 x 3.25 in. Displacement: 348 cid. Compression ratio: 11.00:1. Brake hp: 320 at 5600 rpm. Taxable hp: 54.5. Torque: 358 lbs.-ft. at 3600 rpm. Five main bearings. Solid valve lifters. Crankcase capacity: 4 qt. (add 1 qt. for new filter). Cooling system capacity: 21 qt. (add 1 qt. for heater). Dual exhaust. Carburetors: Carter AFB four-barrel. Engine code: FG.

Chevrolet Special Super Turbo-Thrust optional V-8
Overhead valve. Cast-iron block and head. Bore & stroke: 4.125 x 3.25 in. Displacement: 348 cid. Compression ratio: 11.25:1. Brake hp: 335 at 5800 rpm. Taxable hp: 54.5. Torque: 362 lbs.-ft. at 3600 rpm. Five main bearings. Solid valve lifters. Crankcase capacity: 4 qt. (add 1 qt. for new filter). Cooling system capacity: 21 qt. (add 1 qt. for heater). Dual exhaust. Carburetor: Three Rochester two-barrels; (front) 7013973; (center) 7013974; (rear) 7013975. Engine code: FE.

Chevrolet Special Super Turbo-Thrust optional V-8
Overhead valve. Cast-iron block and head. Bore & stroke: 4.125 x 3.25 in. Displacement: 348 cid. Compression ratio: 11.25:1. Brake hp: 340 at 5800 rpm. Taxable hp: 54.5. Torque: 362 lbs.-ft. at 3600 rpm. Five main bearings. Solid valve lifters. Crankcase capacity: 4 qt. (add 1 qt. for new filter). Cooling system capacity: 21 qt. (add 1 qt. for heater). Dual exhaust. Carburetor: Carter AFB four-barrel. Engine code: FJ.

Chevrolet Special Super Turbo-Thrust optional V-8
Overhead valve. Cast-iron block and head. Bore & stroke: 4.125 x 3.25 in. Displacement: 348 cid. Compression ratio: 11.25:1. Brake hp: 350 at 6000 rpm. Taxable hp: 54.5. Torque: 364 lbs.-ft. at 3600 rpm. Five main bearings. Solid valve lifters. Crankcase capacity: 4 qt. (add 1 qt. for new filter). Cooling system capacity: 21 qt. (add 1 qt. for heater). Dual exhaust. Carburetor: Three Rochester two-barrels. Engine code: FH.

1960 CORVETTE

Chevrolet Super Turbo-Fire optional V-8
Overhead valve. Cast-iron block and head. Bore & stroke: 3.875 x 3.00 in. Displacement: 283 cid. Compression ratio: 9.50:1. Brake hp: 230 at 4800 rpm. Taxable hp: 48.0. Torque: 300 lbs.-ft. at 3000 rpm. Five main bearings. Hydraulic valve lifters. Crankcase capacity: 4 qt. (add 1 qt. for new filter). Cooling system capacity: 17.5 qt. (add 1 qt. for heater). Dual exhaust. Carburetor: Rochester 4GC or Carter WCFB four-barrel. Engine code: CQ and DG.

Corvette optional V-8
Overhead valve. Cast-iron block. Bore & stroke: 3.87 x 3.00 in. Displacement: 283 cid. Compression ratio: 9.50:1. Brake hp: 245 at 5000 rpm. Taxable hp: 48.00. Torque: 300 lbs.-ft. at 3800 rpm. Five main bearings. Hydraulic valve lifters. Crankcase capacity: 5 qt. (add 1 qt. for filter). Cooling system capacity: 15.5 qt. (add 1 qt. for heater). Dual exhaust. Carburetor: Dual Carter WCFB. Engine code: CT and DJ.

Chevrolet "Corvette Ram-Jet Fuel-Injected" optional V-8
Overhead valve. Cast-iron block and head. Bore & stroke: 3.875 x 3.00 in. Displacement: 283 cid. Compression ratio: 9.50:1. Brake hp: 250 at 5000 rpm. Taxable hp: 48.0. Torque: 305 lbs.-ft. at 3800 rpm. Five main bearings. Hydraulic valve lifters. Crankcase capacity: 4 qt. (add 1 qt. for new filter). Cooling system capacity: 18.5 qt. (add 1 qt. for heater). Dual exhaust. Induction system: Rochester fuel injection. Engine code: CR.

Corvette optional V-8
Overhead valve. Cast-iron block. Bore & stroke: 3.87 x 3.00 in. Displacement: 283 cid. Compression ratio: 9.50:1. Brake hp: 270 at 6000 rpm. Taxable hp: 48.00. Torque: 285 lbs.-ft. at 4200 rpm. Five main bearings. Solid valve lifters. Special high-lift camshaft. Crankcase capacity: 5 qt. (add 1 qt. for filter). Cooling system capacity: 15.5 qt. (add 1 qt. for heater). Dual exhaust. Carburetor: Dual Carter WCFB. Engine code: CU.

Corvette optional V-8
Overhead valve. Cast-iron block. Bore & stroke: 3.87 x 3.00 in. Displacement: 283 cid. Compression ratio: 11.00:1. Brake hp: 275 at 5200 rpm. Taxable hp: 48.00. Torque: 305 lbs.-ft. at 4400 rpm. Five main bearings. Solid valve lifters. Hydraulic valve lifters. Crankcase capacity: 5 qt. (add 1 qt. for filter). Cooling system capacity: 15.5 qt. (add 1 qt. for heater). Dual exhaust. Induction: Rochester fuel injection 7017310. Engine code: CR or CY.

Chevrolet "Special Corvette Ram-Jet Fuel-Injected" optional V-8
Overhead valve. Cast-iron block and head. Bore & stroke: 3.875 x 3.00 in. Displacement: 283 cid. Compression ratio: 10.50:1. Brake hp: 290 at 6200 rpm. Taxable hp: 48.0. Torque: unknown at unknown. Five main bearings. Hydraulic valve lifters. Crankcase capacity: 4 qt. (add 1 qt. for new filter). Cooling system capacity: 18.5 qt. (add 1 qt. for heater). Dual exhaust. Induction system: Rochester fuel injection. Engine code: CS.

Corvette optional V-8
Overhead valve. Cast-iron block. Bore & stroke: 3.87 x 3.00 in. Displacement: 283 cid. Compression ratio: 11.00:1. Brake hp: 315 at 6200 rpm. Taxable hp: 48.00. Torque: 295 lbs.-ft. at 4900 rpm. Five main bearings. Solid valve lifters. Crankcase capacity: 5 qt. (add 1 qt. for filter). Cooling system capacity: 15.5 qt. (add 1 qt. for heater). Dual exhaust. Induction: Rochester fuel injection 7017320. Engine code: CS or CZ.

1961 CHEVROLET

The 409-cid Chevy "big-block" V-8 was originally developed in 1961 as a racing engine for use in NASCAR stock cars. This basic engine bowed as the 348-cid V-8 of 1958 and was actually derived from the Chevrolet truck engine family. It was used in the rare 1961 Impala Super Sport model and came in four versions that year. After GM pulled out of racing in 1963, de-tuned versions of the 409 racing engine were released for full-size Chevy street-performance cars.

The 1961 Turbo-Thrust V-8 had a bore and stroke of 4.125 x 3.25 in. With a 9.5:1 compression ratio and single Carter WCFB four-barrel carburetor it produced 250 hp at 4400 rpm. The 348-cid V-8 featured five main bearings and hydraulic valve lifters. Dual exhaust were fitted. Codes F, GE, and H identified this muscular mill.

Chevrolet base V-8
Overhead valve. Cast-iron block and head. Bore & stroke: 3.875 x 3.00 in. Displacement: 283 cid. Compression ratio: 8.5:1. Brake hp: 170 at 4200 rpm. Taxable hp: 48.0. Torque: 275 lbs.-ft. at 2200 rpm. Five main bearings. Hydraulic valve lifters. Crankcase capacity: 4 qt. (add 1 qt. for new filter). Cooling system capacity: 17.5 qt. (add 1 qt. for heater). Carburetor: Rochester 2GC two-barrel. Engine code: C, CD, CL, D, DK, E, and EG.

Chevrolet optional V-8
Overhead valve. Cast-iron block and head. Bore & stroke: 3.875 x 3.00 in. Displacement: 283 cid. Compression ratio: 9.50:1. Brake hp: 230 at 4800 rpm. Taxable hp: 48.0. Torque: 300 lbs.-ft. at 3000 rpm. Five main bearings. Hydraulic valve lifters. Crankcase capacity: 4 qt. (add 1 qt. for new filter). Cooling system capacity: 17.5 qt. (add 1 qt. for heater). Dual exhaust. Carburetor: Rochester 4GC or Carter WCFB four-barrel. Engine code: CF, CG, CM, DB, DM, EB, and EJ.

Optional 1961 Chevrolet V-8s included the 348-cid 280-hp job seen under the hood of this Impala with 20,000 original miles. This was called the "Turbo-Thrust" V-8. A Chevrolet bow-tie emblem is stamped into the valve covers and a single-snorkel air cleaner is used. Note that Chevrolet had not yet switched from a generator to an alternator, as some of its competitors had.

Base V-8 for 1961 Chevrolets was this 283-cid Turbo-Fire small-block. The economy version came with a two-barrel carburetor and 170 hp. The Super Turbo-Fire version, with a single four-barrel carburetor, cranked up 230 hp. This engine was the descendant of the classic 265-cid V-8. The 265 was last used in 1957 models with stick shift or overdrive.

Chevrolet optional V-8
Overhead valve. Cast-iron block and head. Bore & stroke: 4.125 x 3.25 in. Displacement: 348 cid. Compression ratio: 9.50:1. Brake hp: 250 at 4400 rpm. Taxable hp: 54.5. Torque: 355 lbs.-ft. at 2800 rpm. Five main bearings. Hydraulic valve lifters. Crankcase capacity: 4 qt. (add 1 qt. for new filter). Cooling system capacity: 21 qt. (add 1 qt. for heater). Dual exhaust. Carburetor: Carter WCFB four-barrel. Engine code: F, GE, and H.

Chevrolet optional V-8
Overhead valve. Cast-iron block and head. Bore & stroke: 4.125 x 3.25 in. Displacement: 348 cid. Compression ratio: 9.50:1. Brake hp: 280 at 4800 rpm. Taxable hp: 54.5. Torque: 355 lbs.-ft. at 3200 rpm. Five main bearings. Hydraulic valve lifters. Crankcase capacity: 4 qt. (add 1 qt. for new filter). Cooling system capacity: 21 qt. (add 1 qt. for heater). Dual exhaust. Carburetor: Three Rochester two-barrel. Engine code: FA and HA.

Chevrolet optional V-8
Overhead valve. Cast-iron block and head. Bore & stroke: 4.125 x 3.25 in. Displacement: 348 cid. Compression ratio: 9.50:1. Brake hp: 305 at 5200 rpm. Taxable hp: 54.5. Torque: 355 lbs.-ft. at 3400 rpm. Five main bearings. Hydraulic valve lifters. Crankcase capacity: 4 qt. (add 1 qt. for new filter). Cooling system capacity: 21 qt. (add 1 qt. for heater). Dual exhaust. Carburetors: Carter AFB four-barrel. Sales code: RPO 572. Engine code: FL and GD.

Chevrolet optional V-8
Overhead valve. Cast-iron block and head. Bore & stroke: 4.125 x 3.25 in. Displacement: 348 cid. Compression ratio: 11.25:1. Brake hp: 340 at 5800 rpm. Taxable hp: 54.5. Torque: 362 lbs.-ft. at 3800 rpm. Five main bearings. Solid valve lifters. Crankcase capacity: 4 qt. (add 1 qt. for new filter). Cooling system capacity: 21 qt. (add 1 qt. for heater). Dual exhaust. Carburetors: Carter AFB four-barrel. Sales code: RPO 590. Engine code: FJ.

Chevrolet optional V-8
Overhead valve. Cast-iron block and head. Bore & stroke: 4.125 x 3.25 in. Displacement: 348 cid. Compression ratio: 11.25:1. Brake hp: 350 at 6000 rpm. Taxable hp: 54.5. Torque: 364 lbs.-ft. at 3600 rpm. Five main bearings. Solid valve lifters. Crankcase capacity: 4 qt. (add 1 qt. for new filter). Cooling system capacity: 21 qt. (add 1 qt. for heater). Dual exhaust. Carburetor: Three Rochester two-barrel. Sales code: RPO 573B. Engine code: FH.

Chevrolet optional V-8 (Impala SS only)
Overhead valve. Cast-iron block and head. Bore & stroke: 4.312 x 3.50 in. Displacement: 409 cid. Compression ratio: 11.25:1. Brake hp: 360 at 5800 rpm. Taxable hp: 59.50. Torque: 409 lbs.-ft. at 3600 rpm. Five main bearings. Solid valve lifters. Crankcase capacity: 4 qt. (add 1 qt. for new filter). Cooling system capacity: 21 qt. (add 1 qt. for heater). Dual exhaust. Carburetor: Carter AFB four-barrel. Sales code: RPO 580. Engine code: Q and QA.

1961 CORVETTE

Corvette base V-8
Overhead valve. Cast-iron block. Bore & stroke: 3.87 x 3.00 in. Displacement: 283 cid. Compression ratio: 9.50:1. Brake hp: 230 at 4800 rpm. Taxable hp: 48.00. 300 at 3000 rpm. Five main bearings. Hydraulic valve lifters. Crankcase capacity: 5 qt. (add 1 qt. for filter). Cooling system capacity: 15.5 qt. Carburetor: Carter WCFB four-barrel. Engine code: CQ or DG.

Corvette optional V-8
Overhead valve. Cast-iron block. Bore & stroke: 3.87 x 3.00 in. Displacement: 283 cid. Compression ratio: 9.50:1. Brake hp: 245 at 5000 rpm. Taxable hp: 48.00. Torque: 300 lbs.-ft. at 3800 rpm. Five main bearings. Hydraulic valve lifters. Crankcase capacity: 5 qt. (add 1 qt. for filter). Cooling system capacity: 15.5 qt. (add 1 qt. for heater). Dual exhaust. Carburetor: Two Carter WCFB four-barrel carburetors. Engine code: CT or DJ.

Corvette optional V-8
Overhead valve. Cast-iron block. Bore & stroke: 3.87 x 3.00 in. Displacement: 283 cid. Compression ratio: 9.50:1. Brake hp: 270 at 6000 rpm. Taxable hp: 48.00. Torque: 285 lbs.-ft. at 4200 rpm. Five main bearings. Solid valve lifters. Special high-lift camshaft. Crankcase capacity: 5 qt. (add 1 qt. for filter). Cooling system capacity: 15.5 qt. (add 1 qt. for heater). Dual exhaust. Carburetor: Two Carter WCFB four-barrel carburetors. Engine code: CU.

Corvette optional V-8
Overhead valve. Cast-iron block. Bore & stroke: 3.87 x 3.00 in. Displacement: 283 cid. Compression ratio: 11.00:1. Brake hp: 275 at 5200 rpm. Taxable hp: 48.00. Torque: 305 lbs.-ft. at 4400 rpm. Five main bearings. Solid valve lifters. Hydraulic valve lifters. Crankcase capacity: 5 qt. (add 1 qt. for filter). Cooling system capacity: 15.5 qt. (add 1 qt. for heater). Dual exhaust. Induction: Rochester fuel injection. Engine code: CR or CY.

Corvette optional V-8
Overhead valve. Cast-iron block. Bore & stroke: 3.87 x 3.00 in. Displacement: 283 cid. Compression ratio: 11.00:1. Brake hp: 315 at 6200 rpm. Taxable hp: 48.00. Torque: 295 lbs.-ft. at 4900 rpm. Five main bearings. Solid valve lifters. Crankcase capacity: 5 qt. (add 1 qt. for filter). Cooling system capacity: 15.5 qt. (add 1 qt. for heater). Dual exhaust. Induction: Rochester fuel injection. Engine code: CS or CZ.

1962 CHEVROLET

Chevy II, Chevrolet base V-8
Overhead valve. Cast-iron block and head. Bore & stroke: 3.875 x 3.00 in. Displacement: 283 cid. Compression ratio: 8.5:1. Brake hp: 170 at 4200 rpm. Taxable hp: 48.0. Torque: 275 lbs.-ft. at 2200 rpm. Five main bearings. Hydraulic valve lifters. Crankcase capacity: 4 qt. (add 1 qt. for new filter). Cooling system capacity: 17.5 qt. (add 1 qt. for heater). Carburetor: Rochester 7020007 two-barrel. Engine code: C, CB, CL, CD, D, or DK.

Chevy II, Chevrolet optional V-8
Overhead valve. Cast-iron block and head. Bore & stroke: 4.00 x 3.25 in. Displacement: 327 cid. Compression ratio: 10.50:1. Brake hp: 250 at 4400 rpm. Taxable hp: 51.20. Torque: 350 lbs.-ft. at 2800 rpm. Five main bearings. Hydraulic valve lifters. Crankcase capacity: 4 qt. (add 1 qt. for new filter). Cooling system capacity: 17.5 qt. (add 1 qt. for heater). Carburetor: Carter 3788246 four-barrel. Engine code: R, RA, S, or SA.

Chevy II, Chevrolet optional V-8
Overhead valve. Cast-iron block and head. Bore & stroke: 4.00 x 3.25 in. Displacement: 327 cid. Compression ratio: 10.50:1. Brake hp: 300 at 5000 rpm. Taxable hp: 51.20. Torque: 360 lbs.-ft. at 3200 rpm. Five main bearings. Hydraulic valve lifters. Crankcase capacity: 4 qt. (add 1 qt. for new filter). Cooling system capacity: 17.5 qt. (add 1 qt. for heater). Dual exhaust. Carburetor: Carter 3797699 four-barrel. Engine code: RB or SB.

Chevy II optional "Corvette" V-8
Overhead valve. Cast-iron block and head. Bore & stroke: 4.00 x 3.25 in. Displacement: 327 cid. Compression ratio: 11.25:1. Brake hp: 340 at 6000 rpm. Taxable hp: 51.20. Torque: 344 lbs.-ft. at 4000 rpm. Five main bearings. Hydraulic valve lifters. Crankcase capacity: 4 qt. (add 1 qt. for new filter). Cooling system capacity: 17.5 qt. (add 1 qt. for heater). Dual exhaust. Carburetor: Carter 3797699 four-barrel. Engine code: RE.

The 1962 Chevrolet 348-cid V-8 had the company's bow-tie emblem stamped prominently on its large valve covers. Above the emblem was a "Turbo-Fire" sticker and below it was a "409" sticker. Two versions of this engine, both with a high 11.0:1 compression ratio, were offered. Seen here is the single four-barrel 380-hp option (engine code QA) in Gordon Van Vechten's 1962 Impala SS.

This is another view of the 380-hp 409-cid V-8 in Gordon Van Vechten's 1962 Impala SS. The second (code QB) version of the 409 featured dual four-barrel carburetors and was good for 409 hp at 6000 rpm and 420 foot-pounds of torque at 4000 rpm. Carter AFB carburetors of different models were used on each of these big-block V-8s.

Chevy II optional "Corvette" V-8
Overhead valve. Cast-iron block and head. Bore & stroke: 4.00 x 3.25 in. Displacement: 327 cid. Compression ratio: 11.25:1. Brake hp: 360 at 6000 rpm. Taxable hp: 51.20. Torque: 352 lbs.-ft. at 4000 rpm. Five main bearings. Hydraulic valve lifters. Crankcase capacity: 4 qt. (add 1 qt. for new filter). Cooling system capacity: 17.5 qt. (add 1 qt. for heater). Dual exhaust. Induction: Rochester 7017360 fuel injection. Engine code: RF.

Chevrolet optional V-8 (All full-size Chevrolet)
Overhead valve. Cast-iron block and head. Bore & stroke: 4.312 x 3.50 in. Displacement: 409 cid. Compression ratio: 11.00:1. Brake hp: 380 at 5800 rpm. Taxable hp: 59.50. Torque: 420 lbs.-ft. at 3200 rpm. Five main bearings. Solid valve lifters. Crankcase capacity: 4 qt. (add 1 qt. for new filter). Cooling system capacity: 21 qt. (add 1 qt. for heater). Dual exhaust. Carburetor: Carter AFB 3345-SA four-barrel. Sales code: RPO 580. Engine code: QA.

Chevrolet optional V-8 (All full-size Chevrolet)
Overhead valve. Cast-iron block and head. Bore & stroke: 4.312 x 3.50 in. Displacement: 409 cid. Compression ratio: 11.00:1. Brake hp: 409 at 6000 rpm. Taxable hp: 59.50. Torque: 420 lbs.-ft. at 4000 rpm. Five main bearings. Solid valve lifters. Crankcase capacity: 4 qt. (add 1 qt. for new filter). Cooling system capacity: 21 qt. (add 1 qt. for heater). Dual exhaust. Carburetors: Dual Carter AFB four-barrel.; (front) 3361-S; (rear) 3362-S. Sales code: RPO 587. Engine code: QB.

Gordon Van Vechten has his 409 V-8 restored to the same degree of quality reflected on the outside of his 1962 Impala SS. The dual-snorkel air cleaner was finished in black and carries a white decal from GM's AC carburetor division.

1962 CORVETTE

Corvette base V-8
Overhead valve. Cast-iron block. Bore & stroke: 4.00 x 3.25 in. Displacement: 327 cid. Compression ratio: 10.5:1. Brake hp: 250 at 4400 rpm. Taxable hp: 51.20. Torque: 350 lbs.-ft. at 2800 rpm. Five main bearings. Hydraulic valve lifters. Crankcase capacity: 5 qt. (add 1 qt. for filter). Cooling system capacity: 17.5 qt. (add 1 qt. for heater). Carburetor: Carter AFB 3788246 four-barrel. Engine code: RC or SC.

Corvette optional V-8
Overhead valve. Cast-iron block. Bore & stroke: 4.00 x 3.25 in. Displacement: 327 cid. Compression ratio: 10.50:1. Brake hp: 300 at 5000 rpm. Taxable hp: 51.20. Torque: 360 lbs.-ft. at 3200 rpm. Five main bearings. Hydraulic valve lifters. Crankcase capacity: 5 qt. (add 1 qt. for filter). Cooling system capacity: 17.5 qt. (add 1 qt. for heater). Carburetor: Carter AFB 3797699 four-barrel. Engine code: RD or SD.

Corvette optional V-8
Overhead valve. Cast-iron block. Bore & stroke: 4.00 x 3.25 in. Displacement: 327 cid. Compression ratio: 11.25:1. Brake hp: 340 at 6000 rpm. Taxable hp: 51.20. Torque: 344 lbs.-ft. at 4000 rpm. Five main bearings. Hydraulic valve lifters. Crankcase capacity: 5 qt. (add 1 qt. for filter). Cooling system capacity: 17.5 qt. (add 1 qt. for heater). Carburetor: Carter AFB 3797699 four-barrel. Engine code RE.

Corvette optional fuel-injected V-8
Overhead valve. Cast-iron block. Bore & stroke: 4.00 x 3.25 in. Displacement: 327 cid. Compression ratio: 11.25:1. Brake hp: 360 at 6000 rpm. Taxable hp: 51.20. Torque: 352 lbs.-ft. at 4000 rpm. Five main bearings. Hydraulic valve lifters. Crankcase capacity: 5 qt. (add 1 qt. for filter). Cooling system capacity: 17.5 qt. (add 1 qt. for heater). Induction: Rochester fuel injection 7017360. Engine code: RF.

1963 CHEVROLET

The 327-cid V-8 introduced in 1963 was carried over in 1964 and in 1965. This is the latter version. Note the large "327" call-out on the valve cover sticker. This engine was another member of the 265-283 small-block — or "Turbo-Fire" — family. It was available with various cams, compression ratios, and induction systems, including the four-barrel carburetor seen here.

Chevrolet base V-8
Overhead valve. Cast-iron block and head. Bore & stroke: 3.875 x 3.00 in. Displacement: 283 cid. Compression ratio: 9.25:1. Brake hp: 195 at 4800 rpm. Taxable hp: 48.0. Torque: 285 lbs.-ft. at 2400 rpm. Five main bearings. Hydraulic valve lifters. Crankcase capacity: 4 qt. (add 1 qt. for new filter). Cooling system capacity: 17.5 qt. (add 1 qt. for heater). Carburetor: Rochester 7023007 two-barrel. Engine code: C, CB, CD, CL, D, or DK.

Chevrolet optional V-8
Overhead valve. Cast-iron block and head. Bore & stroke: 4.00 x 3.25 in. Displacement: 327 cid. Compression ratio: 10.50:1. Brake hp: 250 at 4400 rpm. Taxable hp: 51.20. Torque: 350 lbs.-ft. at 2800 rpm. Five main bearings. Hydraulic valve lifters. Crankcase capacity: 4 qt. (add 1 qt. for new filter). Cooling system capacity: 17.5 qt. (add 1 qt. for heater). Carburetor: Carter 3826003 four-barrel. Engine code: R or RA.

Chevrolet optional V-8
Overhead valve. Cast-iron block and head. Bore & stroke: 4.00 x 3.25 in. Displacement: 327 cid. Compression ratio: 10.50:1. Brake hp: 300 at 5000 rpm. Taxable hp: 51.20. Torque: 360 lbs.-ft. at 3200 rpm. Five main bearings. Hydraulic valve lifters. Crankcase capacity: 4 qt. (add 1 qt. for new filter). Cooling system capacity: 17.5 qt. (add 1 qt. for heater). Dual exhaust. Carburetor: Carter 3826004 four-barrel. Engine code: RB, RK, SB, or SG.

Chevrolet optional V-8 (All full-size Chevrolet)
Overhead valve. Cast-iron block and head. Bore & stroke: 4.312 x 3.50 in. Displacement: 409 cid. Compression ratio: 10.00:1. Brake hp: 340 at 5000 rpm. Taxable hp: 59.50. Torque: 420 lbs.-ft. at 3200 rpm. Five main bearings. Hydraulic valve lifters. Crankcase capacity: 4 qt. (add 1 qt. for new filter). Cooling system capacity: 21 qt. (add 1 qt. for heater). Dual exhaust. Carburetor: Carter AFB 3345-SA four-barrel. Sales code: RPO L33. Engine code: QC or QG.

Chevy II optional "Corvette" V-8
Overhead valve. Cast-iron block and head. Bore & stroke: 4.00 x 3.25 in. Displacement: 327 cid. Compression ratio: 11.25:1. Brake hp: 360 at 6000 rpm. Taxable hp: 51.20. Torque: 352 lbs.-ft. at 4000 rpm. Five main bearings. Hydraulic valve lifters. Crankcase capacity: 4 qt. (add 1 qt. for new filter). Cooling system capacity: 17.5 qt. (add 1 qt. for heater). Dual exhaust. Induction: Rochester 7017360 fuel injection. Engine code: RF.

Chevrolet optional V-8
Overhead valve. Cast-iron block and head. Bore & stroke: 4.312 x 3.50 in. Displacement: 409 cid. Compression ratio: 11.00:1. Brake hp: 400 at 5800 rpm. Taxable hp: 59.50. Torque: 425 lbs.-ft. at 3600 rpm. Five main bearings. Solid valve lifters. Crankcase capacity: 4 qt. (add 1 qt. for new filter). Cooling system capacity: 21 qt. (add 1 qt. for heater). Dual exhaust. Carburetor: Carter 3820580 four-barrel. Sales code: RPO L31. Engine code: QA.

Chevrolet optional V-8
Overhead valve. Cast-iron block and head. Bore & stroke: 4.312 x 3.50 in. Displacement: 409 cid. Compression ratio: 11.00:1. Brake hp: 425 at 6000 rpm. Taxable hp: 59.50. Torque: 425 lbs.-ft. at 4200 rpm. Five main bearings. Solid valve lifters. Crankcase capacity: 4 qt. (add 1 qt. for new filter). Cooling system capacity: 21 qt. (add 1 qt. for heater). Dual exhaust. Carburetors: Dual Carter AFB four-barrels: (front) 3815403; (rear) 3815404. Sales code: RPO L80. Engine code: QB.

Chevrolet optional V-8
Overhead valve. Cast-iron block and head. Bore & stroke: 4.312 x 3.650 in. Displacement: 427 cid. Compression ratio: 12.50-13.50:1. Brake hp: 430 at 6000 rpm. Taxable hp: 57.80. Torque: Unknown at Unknown. Five main bearings. Solid valve lifters. Crankcase capacity: 4 qt. (add 1 qt. for new filter). Cooling system capacity: 21 qt. (add 1 qt. for heater). Dual exhaust. Carburetors: Dual Carter AFB four-barrel. Sales code: RPO Z11. Engine code: QM.
NOTE: The Z11 package introduced August 1, 1962 included aluminum sheet metal and other special equipment. Only 59 cars with this option were made for drag racing only.

The fuel-injected 1963 Corvette V-8 looked like this one in a Split Window Coupe. Note the finned cover with the Corvette crossed-racing-flags emblem and the "360" horsepower call-out on the finned valve cover, below the "Corvette" name stamped on the center. This engine was as special in appearance as it was in performance. (Blackhawk Auto Collection)

1963 CORVETTE

Corvette base V-8
Overhead valve. Cast-iron block. Bore & stroke: 4.00 x 3.25 in. Displacement: 327 cid. Compression ratio: 10.5:1. Brake hp: 250 at 4400 rpm. Taxable hp: 51.20. Torque: 350 lbs.-ft. at 2800 rpm. Five main bearings. Hydraulic valve lifters. Crankcase capacity: 4 qt. (add 1 qt. for filter). Cooling system capacity: 17.5 qt. (add 1 qt. for heater). Carburetor: Carter AFB 3826003 four-barrel. Engine code: RC or SC.

Corvette optional V-8
Overhead valve. Cast-iron block. Bore & stroke: 4.00 x 3.25 in. Displacement: 327 cid. Compression ratio: 10.50:1. Brake hp: 300 at 5000 rpm. Taxable hp: 51.20. Torque: 360 lbs.-ft. at 3200 rpm. Five main bearings. Hydraulic valve lifters. Crankcase capacity: 4 qt. (add 1 qt. for filter). Cooling system capacity: 17.5 qt. (add 1 qt. for heater). Carburetor: Carter AFB 3826004 four-barrel. Sales code: L75. Engine code: RD or SD.

Corvette optional V-8
Overhead valve. Cast-iron block. Bore & stroke: 4.00 x 3.25 in. Displacement: 327 cid. Compression ratio: 11.25:1. Brake hp: 340 at 6000 rpm. Taxable hp: 51.20. Torque: 344 lbs.-ft. at 4000 rpm. Five main bearings. Hydraulic valve lifters. Crankcase capacity: 4 qt. (add 1 qt. for filter). Cooling system capacity: 17.5 qt. (add 1 qt. for heater). Carburetor: Carter AFB 3826004 four-barrel. Sales code: L76. Engine code: RE.

Three versions of the 327-cid Chevy V-8 were available in this 1963 Corvette Split-Window Coupe. The first two had 10.5:1 compression ratios and single four-barrel carburetors. Horsepower ratings were 250 and 300 respectively. The third option was a single four-barrel engine with 11.25:1 compression and 340 hp. A 360-hp fuel-injected version of the hotter engine was also available. (Blackhawk Auto Collection)

A view of the 1963 fuel-injected 327-cid V-8 from the driver's side of a Split Window Coupe. Note that the engine block and cylinder heads were finished in Chevrolet orange and the huge air cleaner is glossy black. This RF code engine produced 352 foot-pounds of torque at 4000 rpm and the fuel-injection system was again built by Rochester Products Division. (Blackhawk Auto Collection)

Corvette optional fuel-injection V-8
Overhead valve. Cast-iron block. Bore & stroke: 4.00 x 3.25 in. Displacement: 327 cid. Compression ratio: 11.25:1. Brake hp: 360 at 6000 rpm. Taxable hp: 51.20. Torque: 352 at 4000 rpm. Five main bearings. Hydraulic valve lifters. Crankcase capacity: 4 qt. (add 1 qt. for filter). Cooling system capacity: 17.5 qt. (add 1 qt. for heater). Induction: Rochester fuel injection 7017360. Sales code: L84. Engine code: RF.

1964 CHEVROLET

Chevy II 100, Nova, Chevelle 300, Malibu, Malibu SS base V-8
Biscayne, Bel Air, Impala, Impala SS base V-8
Overhead valve. Cast-iron block and head. Bore & stroke: 3.875 x 3.00 in. Displacement: 283 cid. Compression ratio: 9.25:1. Brake hp: 195 at 4800 rpm. Taxable hp: 48.0. Torque: 285 lbs.-ft. at 2400 rpm. Five main bearings. Hydraulic valve lifters. Crankcase capacity: 4 qt. (add 1 qt. for new filter). Cooling system capacity: 16 qt. (add 1 qt. for heater). Carburetor: Rochester 7024101 two-barrel. Engine code: (Full-size Chevrolet) C, CB, or D; (Chevelle) J, JA, or JD; (Chevy II) CH, CJ, CF, CG, DE, or DF.

Chevelle 300, Malibu, Malibu SS optional V-8
Overhead valve. Cast-iron block and head. Bore & stroke: 3.875 x 3.00 in. Displacement: 283 cid. Compression ratio: 9.25:1. Brake hp: 220 at 4800 rpm. Taxable hp: 48.0. Torque: 295 lbs.-ft. at 3200 rpm. Five main bearings. Hydraulic valve lifters. Crankcase capacity: 4 qt. (add 1 qt. for new filter). Cooling system capacity: 16 qt. (add 1 qt. for heater). Carburetor: Rochester 7024024 four-barrel. Engine code: JG or JH.

Chevelle 300, Malibu, Malibu SS, Biscayne, Bel Air, Impala, Impala SS optional V-8
Overhead valve. Cast-iron block and head. Bore & stroke: 4.00 x 3.25 in. Displacement: 327 cid. Compression ratio: 10.50:1. Brake hp: 250 at 4400 rpm. Taxable hp: 51.20. Torque: 350 lbs.-ft. at 2800 rpm. Five main bearings. Hydraulic valve lifters. Crankcase capacity: 4 qt. (add 1 qt. for new filter). Cooling system capacity: 15 qt. (add 1 qt. for heater). Carburetor: (Chevelle) Rochester 7024125 four-barrel.; (Chevrolet full-size) Carter 3846247 four-barrel.. Engine code: (Full-size Chevrolet) R or S; (Chevelle) JQ, JT, or SR.

Chevelle 300, Malibu, Malibu SS, Biscayne, Bel Air, Impala, Impala SS optional V-8
Overhead valve. Cast-iron block and head. Bore & stroke: 4.00 x 3.25 in. Displacement: 327 cid. Compression ratio: 10.50:1. Brake hp: 300 at 5000 rpm. Taxable hp: 51.20. Torque: 360 lbs.-ft. at 3200 rpm. Five main bearings. Hydraulic valve lifters. Crankcase capacity: 4 qt. (add 1 qt. for new filter). Cooling system capacity: 15 qt. (add 1 qt. for heater). Dual exhaust. Carburetor: Carter 3826004 four-barrel. Engine code: (Full-size Chevrolet) RB or SB; (Chevelle) JR or SS.

Biscayne, Bel Air, Impala, Impala SS optional V-8
Overhead valve. Cast-iron block and head. Bore & stroke: 4.312 x 3.50 in. Displacement: 409 cid. Compression ratio: 10.00:1. Brake hp: 340 at 5000 rpm. Taxable hp: 59.50. Torque: 420 lbs.-ft. at 3200 rpm. Five main bearings. Hydraulic valve lifters. Crankcase capacity: 5 qt. (add 1 qt. for new filter). Cooling system capacity: 21 qt. (add 1 qt. for heater). Dual exhaust. Carburetor: Rochester 7024164 four-barrel. Sales code: RPO L33. Engine code: QC, QG, QQ, or QR.

Four-bolt main bearings are among the high-performance hardware shown in this photograph of the component parts of the 396-cid Chevrolet "Turbo-Jet" V-8. The crankshaft is made of forged steel, the camshaft is of cast-alloy construction, the intake valves are made of alloy steel and the exhaust valves are made of a special, high alloy steel that resists heat and wear. The pistons have domed heads with valve-clearance cutouts and slipper-type skirts.

Chevelle Malibu SS optional V-8
Overhead valve. Cast-iron block and head. Bore & stroke: 4.00 x 3.25 in. Displacement: 327 cid. Compression ratio: 11.00:1. Brake hp: 365 at 6200 rpm. Taxable hp: 51.20. Torque: 350 lbs.-ft. at 4000 rpm. Five main bearings. Hydraulic valve lifters. Crankcase capacity: 4 qt. (add 1 qt. for new filter). Cooling system capacity: 15 qt. (add 1 qt. for heater). Dual exhaust. Carburetor: Holley 3858399 four-barrel. Engine code: RR, RT, or RU.

Biscayne, Bel Air, Impala, Impala SS optional V-8
Overhead valve. Cast-iron block and head. Bore & stroke: 4.312 x 3.50 in. Displacement: 409 cid. Compression ratio: 11.00:1. Brake hp: 400 at 5800 rpm. Taxable hp: 59.50. Torque: 425 lbs.-ft. at 3600 rpm. Five main bearings. Solid valve lifters. Crankcase capacity: 5 qt. (add 1 qt. for new filter). Cooling system capacity: 21 qt. (add 1 qt. for heater). Dual exhaust. Carburetor: Carter 3827324 four-barrel. Sales code: RPO L31. Engine code: QA or QN.

Biscayne, Bel Air, Impala, Impala SS optional V-8
Overhead valve. Cast-iron block and head. Bore & stroke: 4.312 x 3.50 in. Displacement: 409 cid. Compression ratio: 11.00:1. Brake hp: 425 at 6000 rpm. Taxable hp: 59.50. Torque: 425 lbs.-ft. at 4200 rpm. Five main bearings. Solid valve lifters. Crankcase capacity: 5 qt. (add 1 qt. for new filter). Cooling system capacity: 21 qt. (add 1 qt. for heater). Dual exhaust. Carburetors: Dual Carter AFB four-barrels: (front) 3815403; (rear) 3815404. Sales code: RPO L80. Engine code: QB or QP.

1964 CORVETTE

Corvette base V-8
Overhead valve. Cast-iron block. Bore & stroke: 4.00 x 3.25 in. Displacement: 327 cid. Compression ratio: 10.5:1. Brake hp: 250 at 4400 rpm. Taxable hp: 51.20. Torque: 350 lbs.-ft. at 2800 rpm. Five main bearings. Hydraulic valve lifters. Crankcase capacity: 4 qt. (add 1 qt. for filter). Cooling system capacity: 18 qt. (add 1 qt. for heater). Carburetor: Carter WCFB 3846247 four-barrel. Engine code: RC, SC, or SK.

Corvette optional V-8
Overhead valve. Cast-iron block. Bore & stroke: 4.00 x 3.25 in. Displacement: 327 cid. Compression ratio: 10.50:1. Brake hp: 300 at 5000 rpm. Taxable hp: 51.20. Torque: 360 lbs.-ft. at 3200 rpm. Five main bearings. Hydraulic valve lifters. Crankcase capacity: 4 qt. (add 1 qt. for filter). Cooling system capacity: 18 qt. (add 1 qt. for heater). Carburetor: Carter AFB 3826004 four-barrel. Sales code: L75. Engine code: RD, RQ, SD, or SL.

Corvette optional V-8
Overhead valve. Cast-iron block. Bore & stroke: 4.00 x 3.25 in. Displacement: 327 cid. Compression ratio: 11.00:1. Brake hp: 365 at 6200 rpm. Taxable hp: 51.20. Torque: 350 lbs.-ft. at 4000 rpm. Five main bearings. Mechanical valve lifters. High-lift camshaft. Crankcase capacity: 5 qt. (add 1 qt. for filter). Cooling system capacity: 18 qt. (add 1 qt. for heater). Carburetor: Holley 3849804 four-barrel. Sales code: L76. Engine code: RE, RR, RT, or RU.

Corvette optional fuel-injection V-8
Overhead valve. Cast-iron block. Bore & stroke: 4.00 x 3.25 in. Displacement: 327 cid. Compression ratio: 11.00:1. Brake hp: 375 at 6200 rpm. Taxable hp: 51.20. Torque: 350 lbs.-ft. at 4400 rpm. Five main bearings. Mechanical valve lifters. High-lift camshaft. Crankcase capacity: 5 qt. (add 1 qt. for filter). Cooling system capacity: 18 qt. (add 1 qt. for heater). Induction: Rochester Ram-Jet fuel injection 7017380. Sales code: L84. Engine code: RF or RX.

1965 CHEVROLET

Chevy II 100, Nova, Nova SS, Chevelle 300, 300 Deluxe, Malibu, Malibu SS base V-8
Biscayne, Bel Air, Impala, Impala SS base V-8
Overhead valve. Cast-iron block and head. Bore & stroke: 3.875 x 3.00 in. Displacement: 283 cid. Compression ratio: 9.25:1. Brake hp: 195 at 4800 rpm. Taxable hp: 48.0. Torque: 285 lbs.-ft. at 2400 rpm. Five main bearings. Hydraulic valve lifters. Crankcase capacity: 4 qt. (add 1 qt. for new filter). Cooling system capacity: 16 qt. (add 1 qt. for heater). Carburetor: Rochester 7024101 two-barrel. Engine code: (Chevy II) PD,

PF, PL, PM, PN, or PP; (Chevelle) DA, DB, or DE; (Chevrolet) GA, GC, or GF;

Chevy II, Chevelle 300, Malibu, Malibu SS, Chevrolet optional V-8
Overhead valve. Cast-iron block and head. Bore & stroke: 3.875 x 3.00 in. Displacement: 283 cid. Compression ratio: 9.25:1. Brake hp: 220 at 4800 rpm. Taxable hp: 48.0. Torque: 295 lbs.-ft. at 3200 rpm. Five main bearings. Hydraulic valve lifters. Crankcase capacity: 4 qt. (add 1 qt. for new filter). Cooling system capacity: 16 qt. (add 1 qt. for heater). Carburetor: Rochester 7024024 four-barrel. Engine code: (Chevy II) PE, PG, PO, PQ, PK, or PB; (Chevelle) DG or DH; (Chevrolet) GK or GL.

Chevy II 100, Nova, Nova SS, Chevelle 300, 300 Deluxe, Malibu, Malibu SS optional V-8
Biscayne, Bel Air, Impala, Impala SS optional V-8
Overhead valve. Cast-iron block and head. Bore & stroke: 4.00 x 3.25 in. Displacement: 327 cid. Compression ratio: 10.50:1. Brake hp: 250 at 4400 rpm. Taxable hp: 51.20. Torque: 350 lbs.-ft. at 2800 rpm. Five main bearings. Hydraulic valve lifters. Crankcase capacity: 4 qt. (add 1 qt. for new filter). Cooling system capacity: (Chevy II) 16 qt.; (others) 15 qt. (add 1 qt. for heater). Carburetor: (Chevy II) Rochester 7024225; (Chevelle) Rochester 7024127 four-barrel.; (Chevrolet full-size) Carter 3846247 four-barrel. Engine code: (Chevy II) ZA, ZE, ZK, or ZM; (Chevelle) EA or EE; (Chevrolet) HA or HC.

Chevy II 100, Nova, Nova SS, Chevelle, Malibu, Malibu SS optional V-8
Biscayne, Bel Air, Impala, Impala SS optional V-8
Overhead valve. Cast-iron block and head. Bore & stroke: 4.00 x 3.25 in. Displacement: 327 cid. Compression ratio: 10.50:1. Brake hp: 300 at 5000 rpm. Taxable hp: 51.20. Torque: 360 lbs.-ft. at 3200 rpm. Five main bearings. Hydraulic valve lifters. Crankcase capacity: 4 qt. (add 1 qt. for new filter). Cooling system capacity: 15 qt. (add 1 qt. for heater). Dual exhaust. Carburetor: Carter 3851761 four-barrel. Engine code: (Chevy II) ZB, ZF, ZL, or ZN; (Chevelle) EB or EF; (Chevrolet) HB or HD.

Biscayne, Bel Air, Impala, Impala SS optional V-8
Overhead valve. Cast-iron block and head. Bore & stroke: 4.09 x 3.76 in. Displacement: 396 cid. Compression ratio: 10.25:1. Brake hp: 325 at 4800 rpm. Taxable hp: 53.60. Torque: 410 lbs.-ft. at 3200 rpm. Five main bearings. Hydraulic valve lifters. Crankcase capacity: 4 qt. (add 1 qt. for new filter). Cooling system capacity: 21 qt. (add 1 qt. for heater). Carburetor: Holley 3868864 or Rochester 7025200 four-barrel. Engine code: IA, LF, IC, IG, II, LB, IV, IC, IW, or IB.

Biscayne, Bel Air, Impala, Impala SS optional V-8
Overhead valve. Cast-iron block and head. Bore & stroke: 4.312 x 3.50 in. Displacement: 409 cid. Compression ratio: 10.00:1. Brake hp: 340 at 5000 rpm. Taxable hp: 59.50. Torque: 420 lbs.-ft. at 3200 rpm. Five main bearings. Hydraulic valve lifters. Crankcase capacity: 5 qt. (add 1 qt. for new filter). Cooling system capacity: 21 qt. (add 1 qt. for heater). Dual exhaust. Carburetor: Rochester 7025123 four-barrel. Sales code: RPO L33. Engine code: JB, JC, JE, or JF.

Chevelle Malibu SS optional V-8
Overhead valve. Cast-iron block and head. Bore & stroke: 4.00 x 3.25 in. Displacement: 327 cid. Compression ratio: 11.00:1. Brake hp: 350 at 6000 rpm. Taxable hp: 51.20. Torque: 360 lbs.-ft. at 3200 rpm. Five main bearings. Hydraulic valve lifters. Crankcase capacity: 5 qt. (add 1 qt. for new filter). Cooling system capacity: 15 qt. (add 1 qt. for heater). Dual exhaust. Carburetor: Holley 3863150 four-barrel. Engine code: EC or ED.

Chevelle SS-396 (Z11) optional V-8
Overhead valve. Cast-iron block and head. Bore & stroke: 4.09 x 3.76 in. Displacement: 396 cid. Compression ratio: 11.0:1. Brake hp: 375 at 5600 rpm. Taxable hp: 53.60. Torque: 415 lbs.-ft. at 3600 rpm. Five main bearings. Solid valve lifters. Crankcase capacity: 4 qt. (add 1 qt. for new filter). Cooling system capacity: 22 qt. (add 1 qt. for heater). Carburetor: Holley 3893229 four-barrel. Sales code: L37. Engine code: IX.
NOTE: Midyear option; only 201 Chevelle Z11 coupes with the L37 V-8 were built in late 1965.

Biscayne, Bel Air, Impala, Impala SS optional V-8
Overhead valve. Cast-iron block and head. Bore & stroke: 4.312 x 3.50 in. Displacement: 409 cid. Compression ratio: 11.00:1. Brake hp: 400 at 5800 rpm. Taxable hp: 59.50. Torque: 425 lbs.-ft. at 3600 rpm. Five main bearings. Solid valve lifters. Crankcase capacity: 5 qt. (add 1 qt. for new filter). Cooling system capacity: 21 qt. (add 1 qt. for heater). Dual exhaust. Carburetor: Carter 3855581 four-barrel. Sales code: RPO L31. Engine code: JA or JD.

1965 CORVETTE

Corvette base V-8
Overhead valve. Cast-iron block. Bore & stroke: 4.00 x 3.25 in. Displacement: 327 cid. Compression ratio: 10.5:1. Brake hp: 250 at 4400 rpm. Taxable hp: 51.20. Torque: 350 lbs.-ft. at 2800 rpm. Five main bearings. Hydraulic valve lifters. Crankcase capacity: 4 qt. (add 1 qt. for filter). Cooling system capacity: 15.5 qt. (add 1 qt. for heater). Carburetor: Carter WCFB 3846247 four-barrel. Engine code: HE, HI, HO, or HQ.

Corvette optional V-8
Overhead valve. Cast-iron block. Bore & stroke: 4.00 x 3.25 in. Displacement: 327 cid. Compression ratio: 10.50:1. Brake hp: 300 at 5000 rpm. Taxable hp: 51.20. Torque: 360 lbs.-ft. at 3200 rpm. Five main bearings. Hydraulic valve lifters. Crankcase capacity: 4 qt. (add 1 qt. for filter). Cooling system capacity: 15.5 qt. (add 1 qt. for heater). Carburetor: Carter AFB 3851761 four-barrel. Sales code: L75. Engine code: HF, HJ, HP, or HR.

Corvette optional V-8
Overhead valve. Cast-iron block. Bore & stroke: 4.00 x 3.25 in. Displacement: 327 cid. Compression ratio: 11.00:1. Brake hp: 350 at 5800 rpm. Taxable hp: 51.20. Torque: 360 lbs.-ft. at 3600 rpm. Five main bearings. Hydraulic valve lifters. High-lift camshaft. Crankcase capacity: 5 qt. (add 1 qt. for filter). Cooling system capacity: 15.5 qt. (add 1 qt. for heater). Dual exhausts. Carburetor: Carter 3849804 four-barrel. Sales code: L79. Engine code: HT, HU, HV, or HW.

Corvette optional V-8
Overhead valve. Cast-iron block. Bore & stroke: 4.00 x 3.25 in. Displacement: 327 cid. Compression ratio: 11.00:1. Brake hp: 365 at 6200 rpm. Taxable hp: 51.20. Torque: 350 lbs.-ft. at 4000 rpm. Five main bearings. Mechanical valve lifters. High-lift camshaft. Crankcase capacity: 5 qt. (add 1 qt. for filter). Cooling system capacity: 15.5 qt. (add 1 qt. for heater). Carburetor: Holley 3849804 four-barrel. Sales code: L76. Engine code: HK, HL, or HM.

Corvette optional V-8
Overhead valve. Cast-iron block. Bore & stroke: 4.00 x 3.25 in. Displacement: 327 cid. Compression ratio: 11.00:1. Brake hp: 375 at 6200 rpm. Taxable hp: 51.20. Torque: 350 lbs.-ft. at 4400 rpm. Five main bearings. Mechanical valve lifters. High-lift camshaft. Crankcase capacity: 5 qt. (add 1 qt. for filter). Cooling system capacity: 15.5 qt. (add 1 qt. for heater). Induction: Rochester Ram-Jet fuel injection 7017380. Sales code: L84. Engine code: HG or HN.

A flat, round chrome air cleaner, an aluminum-finished intake, finned "Corvette" valve covers with a "350" horsepower call-out and orange paint on the block and heads characterize the physical appearance of this 327-cid V-8 in a 1965 Corvette. This engine featured a high-lift camshaft and a single four-barrel carburetor. (Blackhawk Auto Collection)

1965 1/2 CORVETTE

Corvette optional V-8
Overhead valve. Cast-iron block. Bore & stroke: 4.094 x 3.76 in. Displacement: 396 cid. Compression ratio: 11.00:1. Brake hp: 425 at 6400 rpm. Taxable hp: 53.60. Torque: 415 lbs.-ft. at 4000 rpm. Five main bearings. Mechanical valve lifters. High-lift camshaft. Crankcase capacity: 5 qt. (add 1 qt. for filter). Cooling system capacity: 21 qt. (add 1 qt. for heater). Carburetor: Holley 4150 four-barrel. Sales code: L78. Engine code: IF.

1966 CHEVROLET

Chevy II 100, Nova, Nova SS, Chevelle 300, 300 Deluxe, Malibu base V-8
Biscayne, Bel Air, Impala, Impala SS, Caprice base V-8
Overhead valve. Cast-iron block and head. Bore & stroke: 3.875 x 3.00 in. Displacement: 283 cid. Compression ratio: 9.25:1. Brake hp: 195 at 4800 rpm. Taxable hp: 48.0. Torque: 285 lbs.-ft. at 2400 rpm. Five main bearings. Hydraulic valve lifters. Crankcase capacity: 4 qt. (add 1 qt. for new filter). Cooling system capacity: 16 qt. (add 1 qt. for heater). Carburetor: Rochester 7024101 two-barrel. Engine code: (Chevy II) PO,

The large cylinder heads on this 1966 Chevrolet 409-cid V-8 were noted for their depth. This contributed to a large flow of coolant and large, free-breathing ports. The heads did not contain a recessed combustion chamber. There was a slight recess around the valves, but otherwise, the flat surface of the cylinder head formed the roof of the combustion chamber.

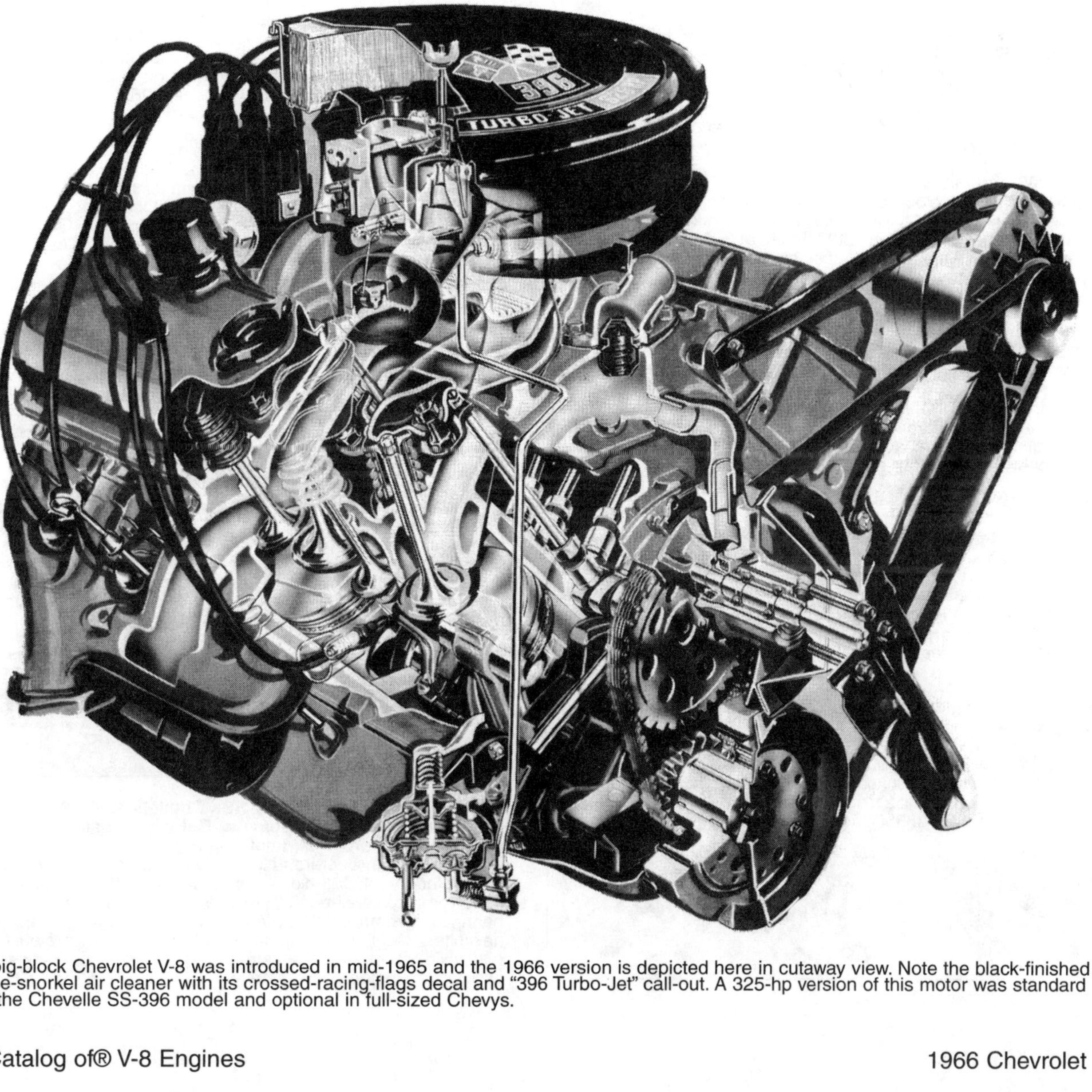

This 396-cid big-block Chevrolet V-8 was introduced in mid-1965 and the 1966 version is depicted here in cutaway view. Note the black-finished standard single-snorkel air cleaner with its crossed-racing-flags decal and “396 Turbo-Jet” call-out. A 325-hp version of this motor was standard equipment in the Chevelle SS-396 model and optional in full-sized Chevys.

PF, PL, PM, PE, PG, PQ, PS, PN, PP, PV, and PO; (Chevelle) DA, DB, DI, DJ, DK, DF, and DE; (Chevrolet) GA, GC, GK, GS, GF, and GT.

Chevy II, Chevelle 300, 300 Deluxe, Malibu, Chevrolet optional V-8
Overhead valve. Cast-iron block and head. Bore & stroke: 3.875 x 3.00 in. Displacement: 283 cid. Compression ratio: 9.25:1. Brake hp: 220 at 4800 rpm. Taxable hp: 48.0. Torque: 295 lbs.-ft. at 3200 rpm. Five main bearings. Hydraulic valve lifters. Crankcase capacity: 4 qt. (add 1 qt. for new filter). Cooling system capacity: (Chevy II, Chevelle) 15 qt.; (Chevrolet) 16 qt. (add 1 qt. for heater). Carburetor: Rochester 7025127 four-barrel. Engine code: (Chevy II) QA, QB, QC, QF, PK, PP, QD, and QE; (Chevelle) DG, DL, DM, DH; (Chevrolet) GW, GX, GL, and GZ.

Chevy II 100, Nova, Nova SS, Chevelle 300, 300 Deluxe, Malibu optional V-8
Biscayne, Bel Air, Impala, Impala SS, Caprice base V-8
Overhead valve. Cast-iron block and head. Bore & stroke: 4.00 x 3.25 in. Displacement: 327 cid. Compression ratio: 10.25:1. Brake hp: 275 at 4800 rpm. Taxable hp: 51.20. Torque: 355 lbs.-ft. at 3200 rpm. Five main bearings. Hydraulic valve lifters. Crankcase capacity: 4 qt. (add 1 qt. for new filter). Cooling system capacity: 14 qt. (add 1 qt. for heater). Carburetor: Holley 3876747 or Carter 3876749 four-barrel. Sales code: L30. Engine code: (Chevy II) ZA, ZE, ZB, ZC, ZK, ZM, ZD, and ZF; (Chevelle) Unknown; (Chevrolet) HA, HB, HC, and HF.

Chevelle SS-396 base V-8; Biscayne, Bel Air, Impala, Impala SS optional V-8
Overhead valve. Cast-iron block and head. Bore & stroke: 4.09 x 3.76 in. Displacement: 396 cid. Compression ratio: 10.25:1. Brake hp: 325 at 4800 rpm. Taxable hp: 53.60. Torque: 410 lbs.-ft. at 3200 rpm. Five main bearings. Hydraulic valve lifters. Crankcase capacity: 4 qt. (add 1 qt. for new filter). Cooling system capacity: 22 qt. (add 1 qt. for heater). Carburetor: Holley 3874898 or Rochester 7026201 four-barrel. Sales code: L35. Engine code: (Chevelle) ED, EH, EK, and EM; (Chevrolet) IA, IB, IG, IV, IC, and IN.

Nova SS optional V-8
Overhead valve. Cast-iron block and head. Bore & stroke: 4.00 x 3.25 in. Displacement: 327 cid. Compression ratio: 11.00:1. Brake hp: 350 at 6000 rpm. Taxable hp: 51.20. Torque: 360 lbs.-ft. at 3200 rpm. Five main bearings. Hydraulic valve lifters. Crankcase capacity: 4 qt. (add 1 qt. for new filter). Cooling system capacity: 14 qt. (add 1 qt. for heater). Dual exhaust. Carburetor: Holley 3877413 four-barrel. Sales code: L79. Engine code: ZI, ZG, ZJ, and ZH.

Chevelle SS-396 optional V-8
Overhead valve. Cast-iron block and head. Bore & stroke: 4.09 x 3.76 in. Displacement: 396 cid. Compression ratio: 10.25:1. Brake hp: 360 at 5200 rpm. Taxable hp: 53.60. Torque: 420 lbs.-ft. at 3600 rpm. Five main bearings. Hydraulic valve lifters. Crankcase capacity: 4 qt. (add 1 qt. for new filter). Cooling system capacity: 22 qt. (add 1 qt. for heater). Carburetor: Holley 3886087 or Rochester 7026201 four-barrel. Sales code: L34. Engine code: EF, EJ, EL, and EN.

Chevelle SS-396 optional V-8
Overhead valve. Cast-iron block and head. Bore & stroke: 4.09 x 3.76 in. Displacement: 396 cid. Compression ratio: 11.0:1. Brake hp: 375 at 5600 rpm. Taxable hp: 53.60. Torque: 415 lbs.-ft. at 3600 rpm. Five main bearings. Hydraulic valve lifters. Crankcase capacity: 4 qt. (add 1 qt. for new filter). Cooling system capacity: 22 qt. (add 1 qt. for heater). Carburetor: Holley 3893229 four-barrel. Sales code: L78. Engine code: EG.
NOTE: The L78 V-8 option was released at midyear.

Biscayne, Bel Air, Impala, Impala SS, Caprice optional V-8
Overhead valve. Cast-iron block and head. Bore & stroke: 4.25 x 3.76 in. Displacement: 427 cid. Compression ratio: 11.0:1. Brake hp: 390 at 5200 rpm. Taxable hp: 57.80. Torque: 460 lbs.-ft. at 3600 rpm. Five main bearings. Hydraulic valve lifters. Crankcase capacity: 4 qt. (add 1 qt. for new filter). Cooling system capacity: 21 qt. (add 1 qt. for heater). Dual exhaust. Carburetor: Rochester 7025123 four-barrel. Engine code: IH, IJ, and II.

Biscayne, Bel Air, Impala, Impala SS, Caprice optional V-8
Overhead valve. Cast-iron block and head. Bore & stroke: 4.25 x 3.76 in. Displacement: 427 cid. Compression ratio: 10.25:1. Brake hp: 425 at 5600 rpm. Taxable hp: 57.80. Torque: 460 lbs.-ft. at 4000 rpm. Five main bearings. Solid valve lifters. Crankcase capacity: 4 qt. (add 1 qt. for new filter). Cooling system capacity: 21 qt. (add 1 qt. for heater). Dual exhaust. Carburetor: Holley 3885067 four-barrel. Engine code: ID and IO.

1966 CORVETTE

This is the 327-cid 300-hp Turbo-Fire Chevy V-8 in a 1966 Corvette. With the horsepower race in full gear, this was the Corvette's base engine and looks a bit plainer than other versions with orange painted valve covers. The block and water pump are also orange. This motor carried a single Holley four-barrel carburetor. (Blackhawk Auto Collection)

Corvette base V-8
Overhead valve. Cast-iron block. Bore & stroke: 4.00 x 3.25 in. Displacement: 327 cid. Compression ratio: 10.50:1. Brake hp: 300 at 5000 rpm. Taxable hp: 51.20. Torque: 360 lbs.-ft. at 3400 rpm. Five main bearings. Hydraulic valve lifters. Crankcase capacity: 5 qt. (add 1 qt. for filter). Cooling system capacity: 18 qt. (add 1 qt. for heater). Carburetor: Holley 3884505 four-barrel. Sales code: L75. engine code: HE, HH, HR, HO, or HP.

Corvette optional V-8
Overhead valve. Cast-iron block. Bore & stroke: 4.00 x 3.25 in. Displacement: 327 cid. Compression ratio: 11.00:1. Brake hp: 350 at 5800 rpm. Taxable hp: 51.20. Torque: 360 lbs.-ft. at 3000 rpm. Five main bearings. Hydraulic valve lifters. High-lift camshaft. Crankcase capacity: 5 qt. (add 1 qt. for filter). Cooling system capacity: 18 qt. (add 1 qt. for heater). Carburetor: Holley 3884505 four-barrel. Sales code: L79. Engine code: HD, HT, or HK.

Corvette optional V-8
Overhead valve. Cast-iron block. Displacement: 427 cid. Bore & stroke: 4.251 x 3.76 in. Compression ratio: 10:25:1. Brake hp: 390 at 5200 rpm. Taxable hp: 57.80. Torque: 460 lbs.-ft. at 3600 rpm. Five main bearings. Hydraulic valve lifters. High-performance camshaft. Crankcase capacity: 5 qt. (add 1 qt. for filter). Cooling system capacity: 21 qt. (add 1 qt. for heater). Carburetor: Holley 3882835 four-barrel. Sales code: L30. Engine code: IL, IM, IQ, or IR.

Corvette optional V-8
Overhead valve. Cast-iron block. Displacement: 427 cid. Bore & stroke: 4.251 x 3.76 in. Compression ratio: 11.00:1. Brake hp: 425 at 5600 rpm. Taxable hp: 57.80. Torque: 460 lbs.-ft. at 4000 rpm. Five main bearings. Mechanical valve lifters. Special-performance camshaft. Crankcase capacity: 5 qt. (add 1 qt. for filter). Cooling system capacity: 21 qt. (add 1 qt. for heater). Carburetor: Holley 3886101 four-barrel. Sales code: L72. Engine code: IK or IP.

1967 CHEVROLET

Chevy II 100, Nova, Nova SS, Chevelle 300, 300 Deluxe, Malibu, Concours base V-8; Biscayne, Bel Air, Impala, Impala SS, Caprice base V-8; Camaro V-8 (delete option)
Overhead valve. Cast-iron block and head. Bore & stroke: 3.875 x 3.00 in. Displacement: 283 cid. Compression ratio: 9.25:1. Brake hp: 195 at 4800 rpm. Taxable hp: 48.0. Torque: 285 lbs.-ft. at 2400 rpm. Five main bearings. Hydraulic valve lifters. Crankcase capacity: 4 qt. (add 1 qt. for new filter). Cooling system capacity: (Chevy II, Chevrolet) 15 qt.; (Chevelle) 16 qt. (add 1 qt. for heater). Carburetor: Rochester 7027101 two-barrel. Engine code: (Chevy II) PD, PF, PM, PE, PQ, PL, PN, PP,

This is the cylinder head used on the special 302-cid V-8 that was standard equipment in the 1967 Camaro Z/28. This engine had a 4.00 x 3.00-in. bore and stroke and an 11.0:1 compression ratio. Brake horsepower was 290 at 5800 rpm and torque was 290 foot-pounds at 4200 rpm. It had five four-bolt main bearings.

The components of the 302-cid Chevy V-8 offered in the 1967 Camaro Z/28. It featured five four-bolt main bearings with nodular iron main caps, solid valve lifters, an 800-cfm Holley no. 3923289 (R4055A) four-barrel carburetor, a special solid lifter camshaft, a cast-aluminum "high-rise" intake manifold, a high-capacity oil pump, and special baffling in oil pan. The 302 was standard in the Camaro Z/28, but was not available in other Camaros.

The 396-cid Chevy V-8 had "porcupine" valves, which means they were "splayed" away from each other in both axis to improve the flow of intake and exhaust gases. The cylinder heads were of semi-wedge design with spark plugs located nearly in the center. In addition to air cleaner decals, with crossed flags and the "396 Turbo-Jet" description, there were metallic stickers on the valve covers that carried the red Chevrolet bow-tie emblem.

PU; (Chevelle) DA, DB, DI, DJ, DK, DN, and DE; (Camaro) MD and MJ; (Chevrolet) GA, GC, GK, GS, GU, GF, GO, and GT.

Camaro base V-8
Overhead valve. Cast-iron block and head. Bore & stroke: 4.00 x 3.25 in. Displacement: 327 cid. Compression ratio: 8.75:1. Brake hp: 210 at 4600 rpm. Taxable hp: 51.20. Torque: 320 lbs.-ft. at 2400 rpm. Five main bearings. Hydraulic valve lifters. Crankcase capacity: 4 qt. (add 1 qt. for new filter). Cooling system capacity: 16 qt. (add 1 qt. for heater). Carburetor: Rochester 7027101 two-barrel. Engine code: MA, MB, ME, and MF.

Biscayne, Bel Air, Impala, Impala SS, Caprice optional V-8
Overhead valve. Cast-iron block and head. Bore & stroke: 3.875 x 3.00 in. Displacement: 283 cid. Compression ratio: 9.25:1. Brake hp: 220 at 4800 rpm. Taxable hp: 48.0. Torque: 295 lbs.-ft. at 3200 rpm. Five main bearings. Hydraulic valve lifters. Crankcase capacity: 4 qt. (add 1 qt. for new filter). Cooling system capacity: 15 qt. (add 1 qt. for heater). Carburetor: Four-barrel. Engine code: Unknown.

Additional special components of this 302-cid Chevy V-8 offered in the 1967 Camaro Z/28 included a dual-belt-drive cooling system with an external bypass water pump and a five-blade, thermostatically-controlled fan with an operating range of 2300-2600 rpm. Also included were a chrome-plated air cleaner, chrome rocker covers, a chrome oil filler tube, and a chrome oil filler cap. A 2-1/4-in. dual exhaust system with dual deep-tone mufflers was fitted.

Tucked below the hood of Dave and Dawn Barta's 1967 Chevelle SS-396 is the optional 350-hp version of the 396 Turbo-Jet V-8. This engine had a 10.25:1 compression ratio and carried a single four-barrel carburetor. It developed .885 horsepower per cubic inch of displacement. The intake manifold, block, and water pump are painted orange, while the valve covers and air cleaner are chrome plated.

Chevy II 100, Nova, SS, Camaro, Chevelle 300, 300 Deluxe, Malibu, Concours optional V-8
Biscayne, Bel Air, Impala, Impala SS, Caprice optional V-8
Overhead valve. Cast-iron block and head. Bore & stroke: 4.00 x 3.25 in. Displacement: 327 cid. Compression ratio: 10.25:1. Brake hp: 275 at 4800 rpm. Taxable hp: 51.20. Torque: 355 lbs.-ft. at 3200 rpm. Five main bearings. Hydraulic valve lifters. Crankcase capacity: 4 qt. (add 1 qt. for new filter). Cooling system capacity: 16 qt. (add 1 qt. for heater). Carburetor: Four-barrel. Sales code: L30. Engine code: (Chevy II) ZA, ZB, ZD, ZE, ZK, and ZM; (Camaro) MK, ML, MM, and MN; (Chevelle) EA, EB, EQ, EE, and EC; (Chevrolet) HA, HB, HC, HF, KE, K, L, and KM.

Camaro Z/28 base V-8
Overhead valve. Cast-iron block and head. Bore & stroke: 4.00 x 3.00 in. Displacement: 302 cid. Compression ratio: 11.00:1. Brake hp: 290 at 5800 rpm. Taxable hp: 51.20. Torque: 290 lbs.-ft. at 4200 rpm. Five main bearings. Hydraulic valve lifters. Crankcase capacity: 4 qt. (add 1 qt. for new filter). Cooling system capacity: 16 qt. (add 1 qt. for heater). Five four-bolt main bearings with nodular iron main caps. Solid valve lifters. Induction system: 800-cfm Holley no. 3923289 (R4055A) four-barrel carburetor. Four-bolt main bearings. Special solid lifter camshaft. Cast-aluminum "high-rise" intake manifold. High capacity oil pump. Special baffling in oil pan. Cooling system includes dual-belt fan drive, external bypass water pump and five-blade thermostatically controlled fan with operating range of 2300-2600 rpm. Also includes chrome-plated air cleaner, rocker covers, filler tube and cap. Dual exhaust system (2-1/4-in. diameter) with dual deep-tone mufflers. A 10.34-in. diameter bent-finger diaphragm spring clutch with a 2300-2600-lb. rating. Standard in Camaro

Z/28; not available in other Camaros. Engine code: MO and MP.

Camaro base V-8
Overhead valve. Cast-iron block and head. Bore & stroke: 4.00 x 3.48 in. Displacement: 350 cid. Compression ratio: 10.25:1. Brake hp: 295 at 4800 rpm. Taxable hp: 51.20. Torque: 380 lbs.-ft. at 3200 rpm. Five main bearings. Hydraulic valve lifters. Crankcase capacity: 4 qt. (add 1 qt. for new filter). Cooling system capacity: 14 qt. (add 1 qt. for heater). Carburetor: Four-barrel. Engine code: MS, MT, MU, and MV.

Chevelle 300, 300 Deluxe, Malibu, Concours optional V-8
Overhead valve. Cast-iron block and head. Bore & stroke: 4.00 x 3.25 in. Displacement: 327 cid. Compression ratio: 11.00:1. Brake hp: 325 at 5600 rpm. Taxable hp: 51.20. Torque: 355 lbs.-ft. at 3600 rpm. Five main bearings. Hydraulic valve lifters. Crankcase capacity: 4 qt. (add 1 qt. for new filter). Cooling system capacity: 14 qt. (add 1 qt. for heater). Carburetor: Four-barrel. Sales code: L30.

Chevelle SS-396 base V-8; Camaro, Biscayne, Bel Air, Impala, Impala SS optional V-8
Overhead valve. Cast-iron block and head. Bore & stroke: 4.09 x 3.76 in. Displacement: 396 cid. Compression ratio: 10.25:1. Brake hp: 325 at 4800 rpm. Taxable hp: 53.60. Torque: 410 lbs.-ft. at 3200 rpm. Five main bearings. Hydraulic valve lifters. Crankcase capacity: 4 qt. (add 1 qt. for new filter). Cooling system capacity: 22 qt. (add 1 qt. for heater). Carburetor: Rochester 7027201 four-barrel. Sales code: L35. Engine code: (Camaro) MW, MX, MY, and MZ; (Chevelle) ER, ES, EP, ED, EF, EH, EM, ET, and EV; (Chevrolet) IA, IB, IG, IV, IC, and IN.

Nova SS optional V-8
Overhead valve. Cast-iron block and head. Bore & stroke: 4.00 x 3.25 in. Displacement: 327 cid. Compression ratio: 11.00:1. Brake hp: 350 at 6000 rpm. Taxable hp: 51.20. Torque: 360 lbs.-ft. at 3200 rpm. Five main bearings. Hydraulic valve lifters. Crankcase capacity: 4 qt. (add 1 qt. for new filter). Cooling system capacity: 14 qt. (add 1 qt. for heater). Dual exhaust. Carburetor: Holley 3877413 four-barrel. Sales code: L79. Engine code: ZI, ZG, and ZJ.

Camaro, Chevelle SS-396 optional V-8
Overhead valve. Cast-iron block and head. Bore & stroke: 4.09 x 3.76 in. Displacement: 396 cid. Compression ratio: 10.25:1. Brake hp: 350 at 5200 rpm. Taxable hp: 53.60. Torque: 415 lbs.-ft. at 3400 rpm. Five main bearings. Hydraulic valve lifters. Crankcase capacity: 4 qt. (add 1 qt. for new filter). Cooling system capacity: 22 qt. (add 1 qt. for heater). Carburetor: Four-barrel. Sales code: L34. Engine code: (Camaro) EI, EQ, and EY; (Chevelle) EK, EL, EN, EU, and EW.

Camaro, Chevelle SS-396 optional V-8
Overhead valve. Cast-iron block and head. Bore & stroke: 4.09 x 3.76 in. Displacement: 396 cid. Compression ratio: 11.00:1. Brake hp: 375 at 5600 rpm. Taxable hp: 53.60. Torque: 415 lbs.-ft. at 3600 rpm. Five main bearings. Solid valve lifters. Crankcase capacity: 4 qt. (add 1 qt. for new filter). Cooling system capacity: 23 qt. (add 1 qt. for heater). Carburetor: Four-barrel. Engine code: (Camaro) MQ or MR; (Chevelle) EG or EX. **NOTE:** This engine was not listed in sales literature but had limited installations.

Impala SS-427 base V-8; Biscayne, Bel Air, Impala, Impala SS, Caprice optional V-8
Overhead valve. Cast-iron block and head. Bore & stroke: 4.25 x 3.76 in. Displacement: 427 cid. Compression ratio: 11.0:1. Brake hp: 385 at 5200 rpm. Taxable hp: 57.80. Torque: 460 lbs.-ft. at 3400 rpm. Five main bearings. Hydraulic valve lifters. Crankcase capacity: 4 qt. (add 1 qt. for new filter). Cooling system capacity: 21 qt. (add 1 qt. for heater). Dual exhaust. Carburetor: Four-barrel. Sales code: L36. Engine code: IE, IH, IJ, IS, II, IX, IF, and IO.

Impala SS-427 optional V-8
Overhead valve. Cast-iron block and head. Bore & stroke: 4.25 x 3.76 in. Displacement: 427 cid. Compression ratio: 10.25:1. Brake hp: 425 at 5600 rpm. Taxable hp: 57.80. Torque: 460 lbs.-ft. at 4000 rpm. Five main bearings. Solid valve lifters. Crankcase capacity: 4 qt. (add 1 qt. for new filter). Cooling system capacity: 21 qt. (add 1 qt. for heater). Dual exhaust. Carburetor: Holley 3885067 four-barrel. Sales code L72. Engine code: ID and IK.

1967 CORVETTE

Corvette base V-8
Overhead valve. Cast-iron block. Bore & stroke: 4.00 x 3.25 in. Displacement: 327 cid. Compression ratio: 10.00:1. Brake hp: 300 at 5000 rpm. Taxable hp: 51.20. Torque: 360 lbs.-ft. at 3400 rpm. Five main bearings. Hydraulic valve lifters. Crankcase capacity: 5 qt. (add 1 qt. for filter). Cooling system capacity: 18 qt. (add 1 qt. for heater). Carburetor: Rochester 3096631 four-barrel. Sales code: L75. Engine code: HE, HH, HR, and HO.

Corvette optional V-8
Overhead valve. Cast-iron block. Bore & stroke: 4.00 x 3.25 in. Displacement: 327 cid. Compression ratio: 11.00:1. Brake hp: 350 at 5800 rpm. Taxable hp: 51.20. Torque: 360 lbs.-ft. at 3600 rpm. Five main bearings. Hydraulic valve lifters. High-lift camshaft. Crankcase capacity: 5 qt. (add 1 qt. for filter). Cooling system capacity: 18 qt. (add 1 qt. for heater). Carburetor: Holley four-barrel. Sales code: L79. Engine code: HD, HT, HP, and KH.

Corvette optional V-8
Overhead valve. Cast-iron block. Displacement: 427 cid. Bore & stroke: 4.251 x 3.76 in. Compression ratio: 10:25:1. Brake hp: 390 at 5400 rpm. Taxable hp: 57.80. Torque: 460 lbs.-ft. at 3600 rpm. Five main bearings. Hydraulic valve lifters. High-performance camshaft. Crankcase capacity: 5 qt. (add 1 qt. for filter). Cooling system capacity: 21 qt. (add 1 qt. for heater). Carburetor: Holley four-barrel. Sales code: L36. Engine code: IL, IM, IQ, and IR.

Corvette optional V-8
Overhead valve. Cast-iron block. Displacement: 427 cid. Bore & stroke: 4.251 x 3.76 in. Compression ratio: 10.25:1. Brake hp: 400 at 5400 rpm. Taxable hp: 57.80. Torque: 460 lbs.-ft. at 3600 rpm. Five main bearings. Hydraulic valve lifters. Special-performance camshaft. Crankcase capacity: 5 qt. (add 1 qt. for filter). Cooling system capacity: 21 qt. (add 1 qt. for heater). Carburetor: Three Holley two-barrel. Sales code: L68. Engine code: JC, JF, JD, and JG.

Corvette optional V-8
Overhead valve. Cast-iron block. Displacement: 427 cid. Bore & stroke: 4.251 x 3.76 in. Compression ratio: 11.0:1. Brake hp: 425 at 5600 rpm. Taxable hp: 57.80. Torque: 460 lbs.-ft. at 3800 rpm. Five main bearings. Mechanical valve lifters. Special-performance camshaft. Crankcase capacity: 5 qt. (add 1 qt. for filter). Cooling system capacity: 21 qt. (add 1 qt. for heater). Carburetor: Four-barrel. Sales code: L72. Engine code: IT.

Corvette optional V-8
Overhead valve. Cast-iron block. Displacement: 427 cid. Bore & stroke: 4.251 x 3.76 in. Compression ratio: 11.0:1. Brake hp: 435 at 5800 rpm. Taxable hp: 57.80. Torque: 460 lbs.-ft. at 4000 rpm. Five main bearings. Mechanical valve lifters. Special-performance camshaft. Crankcase capacity: 5 qt. (add 1 qt. for filter). Cooling system capacity: 21 qt. (add 1 qt. for heater). Carburetor: Three Holley two-barrels. Sales code: L71. Engine code: JE, IU, JH, and JA.

Corvette optional V-8
Overhead valve. Cast-iron block. Displacement: 427 cid. Bore & stroke: 4.251 x 3.76 in. Compression ratio: 11.0:1. Brake hp: 435 at 5800 rpm. Taxable hp: 57.80. Torque: 460 lbs.-ft. at 4000 rpm. Five main bearings. Mechanical valve lifters. Aluminum cylinder heads. Extra-large exhaust valves. Special-performance camshaft. Crankcase capacity: 5 qt. (add 1 qt. for filter). Cooling system capacity: 21 qt. (add 1 qt. for heater). Carburetor: Three Holley two-barrels. Sales code: L89. Engine code: IU. **NOTE:** Only 16 Corvettes with this engine were built in 1967.

Corvette 1967-1/2 optional V-8
Overhead valve. Displacement: 427 cid. Bore & stroke: 4.251 x 3.76 in. Compression ratio: 12.50:1. Brake hp: 560 at 6400 rpm. Taxable hp: 57.80. Torque: unknown at unknown rpm. Five main bearings. Mechanical valve lifters. Special-ultra-high-performance camshaft with .5365-in. intakes. Crankcase capacity: 5 qt. (add 1 qt. for filter). Cooling system capacity: 21 qt. (add 1 qt. for heater). Carburetor: Single Holley 850CFM four-barrel. Sales code: L88.
NOTE: Only 20 Corvettes with this engine were built in 1967.

1968 CHEVROLET

Chevy II 100, Nova, Chevelle 300, 300 Deluxe, Malibu, Nomad, Concours base V-8; Biscayne, Bel Air, Impala, Caprice base V-8
Overhead valve. Cast-iron block and head. Bore & stroke: 3.88 x 3.25 in. Displacement: 307 cid. Compression ratio: 9.00:1. Brake hp: 200 at 4600 rpm. Taxable hp: 48.0. Torque: 300 lbs.-ft. at 2400. Five main bearings. Hydraulic valve lifters. Crankcase capacity: 4 qt. (add 1 qt. for new filter). Cooling system capacity: 16 qt. (add 1 qt. for heater). Carburetor: Rochester 7028101 two-barrel. Engine code: (Chevy II) DA, DB, DE, MB, MC, and MD; (Chevelle) DA, DB, DE, DF, or DN; (Chevrolet) DH, DK, DO, DP, DS, DQ, and DR.

Camaro base V-8
Overhead valve. Cast-iron block and head. Bore & stroke: 4.00 x 3.25 in.

The V-8 used in the 1968 Camaro Z/28 was not as much of a performance brute as a 396 or a 427. Instead, it was a compact 302-cid V-8 that was as high-tech as engines came at the time. It was designed to get maximum performance out of a particular small-block V-8 and to conform to Trans-Am racing regulations. It was capable of pulling .960 horsepower from every cubic inch.

Displacement: 327 cid. Compression ratio: 8.75:1. Brake hp: 210 at 4600 rpm. Taxable hp: 51.20. Torque: 320 lbs.-ft. at 2400 rpm. Five main bearings. Hydraulic valve lifters. Crankcase capacity: 4 qt. (add 1 qt. for new filter). Cooling system capacity: 16 qt. (add 1 qt. for heater). Carburetor: Rochester 7028101 two-barrel. Engine code: MA or ME.

Chevelle 300, 300 Deluxe, Malibu, Nomad, Concours optional V-8; Biscayne, Bel Air, Impala, Caprice optional V-8
Overhead valve. Cast-iron block and head. Bore & stroke: 4.00 x 3.25 in. Displacement: 327 cid. Compression ratio: 8.75:1. Brake hp: 250 at 4800 rpm. Taxable hp: 51.20. Torque: 335 lbs.-ft. at 3200 rpm. Five main bearings. Hydraulic valve lifters. Crankcase capacity: 4 qt. (add 1 qt. for new filter). Cooling system capacity: (full-size) 14 qt.; (others) 16 qt. (add 1 qt. for heater). Carburetor: Four-barrel. Engine code: (Chevelle) EA, EC, EE, or EO; (Chevrolet) HA, HB, HC, HF, HG, HH, HK, HL, HM, and HN.

Chevy II, Malibu, Nomad, Concours, Camaro optional V-8 Biscayne, Bel Air, Impala, Caprice optional V-8
Overhead valve. Cast-iron block and head. Bore & stroke: 4.00 x 3.25 in. Displacement: 327 cid. Compression ratio: 10.25:1. Brake hp: 275 at 4800 rpm. Taxable hp: 51.20. Torque: 355 lbs.-ft. at 3200 rpm. Five main bearings. Hydraulic valve lifters. Crankcase capacity: 4 qt. (add 1 qt. for new filter). Cooling system capacity: (full-size) 14 qt.; (others) 16 qt. (add 1 qt. for heater). Carburetor: Four-barrel. Sales code: L30. Engine code: (Chevy II) EA, MK, EE, MM, and EC; (Chevelle) EH, EI, and EJ; (Camaro) EE, EA, and EC; (Chevrolet) HI or HJ.

Camaro Z/28 base V-8
Overhead valve. Cast-iron block and head. Bore & stroke: 4.00 x 3.00 in. Displacement: 302 cid. Compression ratio: 11.00:1. Brake hp: 290 at 5800 rpm. Taxable hp: 51.20. Torque: 290 lbs.-ft. at 4200 rpm. Five main bearings. Hydraulic valve lifters. Crankcase capacity: 4 qt. (add 1 qt. for new filter). Cooling system capacity: 16 qt. (add 1 qt. for heater). Five four-bolt main bearings with nodular iron main caps. Solid valve lifters. Induction system: 800-cfm Holley no. 3923289 (R4055A) four-barrel carburetor. Four-bolt main bearings. Special solid lifter camshaft. Cast-aluminum "high-rise" intake manifold. High capacity oil pump. Special baffling in oil pan. Cooling system includes dual-belt fan drive, external bypass water pump and five-blade thermostatically controlled fan with operating range of 2300-2600 rpm. Also includes chrome-plated air cleaner, rocker covers, filler tube, and cap. Dual exhaust system (2-1/4-in. diameter) with dual deep-tone mufflers. A 10.34-in. diameter bent-finger diaphragm spring clutch with a 2300-2600-lb. rating. Standard in Camaro Z/28; not available in other Camaros. Engine code: MO.

Chevy II Nova, Camaro optional V-8
Overhead valve. Cast-iron block and head. Bore & stroke: 4.00 x 3.48 in. Displacement: 350 cid. Compression ratio: 10.25:1. Brake hp: 295 at 4800 rpm. Taxable hp: 51.20. Torque: 380 lbs.-ft. at 3200 rpm. Five main bearings. Hydraulic valve lifters. Crankcase capacity: 4 qt. (add 1 qt. for new filter). Cooling system capacity: 15 qt. (add 1 qt. for heater). Carburetor: Four-barrel. Engine code: (Chevy II) MS or MU; (Camaro) MS or MU.

Chevy II, Chevelle 300, 300 Deluxe, Malibu, Nomad, Concours optional V-8
Overhead valve. Cast-iron block and head. Bore & stroke: 4.00 x 3.25 in. Displacement: 327 cid. Compression ratio: 11.00:1. Brake hp: 325 at 5600 rpm. Taxable hp: 51.20. Torque: 355 lbs.-ft. at 3600 rpm. Five main bearings. Hydraulic valve lifters. Crankcase capacity: 4 qt. (add 1 qt. for new filter). Cooling system capacity: 14 qt. (add 1 qt. for heater). Carburetor: Four-barrel. Sales code: L30. Engine code: (Chevy II) ML or EP; (Chevelle) EP or ES.

Chevelle SS-396 base V-8; Camaro, Biscayne, Bel Air, Impala, Caprice optional V-8
Overhead valve. Cast-iron block and head. Bore & stroke: 4.09 x 3.76 in. Displacement: 396 cid. Compression ratio: 10.25:1. Brake hp: 325 at 4800 rpm. Taxable hp: 53.60. Torque: 410 lbs.-ft. at 3200 rpm. Five main bearings. Hydraulic valve lifters. Crankcase capacity: 4 qt. (add 1 qt. for new filter). Cooling system capacity: 23 qt. (add 1 qt. for heater). Carburetor: Rochester 7028211 four-barrel. Sales code: L35. Engine code: (Chevelle) ET, EK, and EO; (Camaro) MY; (Chevrolet) IA, IG, IV, IK, IN, and IF.

Chevy II, Chevelle SS-396, Camaro optional V-8
Overhead valve. Cast-iron block and head. Bore & stroke: 4.09 x 3.76 in. Displacement: 396 cid. Compression ratio: 10.25:1. Brake hp: 350 at 5200 rpm. Taxable hp: 53.60. Torque: 415 lbs.-ft. at 3400 rpm. Five main bearings. Hydraulic valve lifters. Crankcase capacity: 4 qt. (add 1 qt. for new filter). Cooling system capacity: 23 qt. (add 1 qt. for heater). Carburetor: Four-barrel. Sales code: L34. Engine code: (Chevy II) MX, (Chevelle) EF, EL, and EU; (Camaro) MR or MX.

Chevy II, Camaro, Chevelle SS-396 optional V-8
Overhead valve. Cast-iron block and head. Bore & stroke: 4.09 x 3.76 in. Displacement: 396 cid. Compression ratio: 11.00:1. Brake hp: 375 at 5600 rpm. Taxable hp: 53.60. Torque: 415 lbs.-ft. at 3600 rpm. Five main bearings. Solid valve lifters. Crankcase capacity: 4 qt. (add 1 qt. for new filter). Cooling system capacity: 23 qt. (add 1 qt. for heater). Carburetor: Four-barrel. Engine code: (Chevy II) MQ or MR; (Chevelle) EG; (Camaro) MQ, MT, and MW.

Biscayne, Bel Air, Impala, Impala SS, Caprice optional V-8
Overhead valve. Cast-iron block and head. Bore & stroke: 4.25 x 3.76 in. Displacement: 427 cid. Compression ratio: 11.0:1. Brake hp: 385 at 5200 rpm. Taxable hp: 57.80. Torque: 460 lbs.-ft. at 3400 rpm. Five main bearings. Hydraulic valve lifters. Crankcase capacity: 4 qt. (add 1 qt. for new filter). Cooling system capacity: 21 qt. (add 1 qt. for heater). Dual exhaust. Carburetor: Four-barrel. Sales code: L36. Engine code: IB, IE, IH, IJ, IS, and IC.

Impala SS-427 optional V-8
Overhead valve. Cast-iron block and head. Bore & stroke: 4.25 x 3.76 in. Displacement: 427 cid. Compression ratio: 10.25:1. Brake hp: 425 at 5600 rpm. Taxable hp: 57.80. Torque: 460 lbs.-ft. at 4000 rpm. Five main bearings. Solid valve lifters. Crankcase capacity: 4 qt. (add 1 qt. for new filter). Cooling system capacity: 21 qt. (add 1 qt. for heater). Dual exhaust. Carburetor: Holley 3885067 four-barrel. Sales code L72. Engine code: ID.

1968 CORVETTE

Corvette base V-8
Overhead valve. Cast-iron block. Bore & stroke: 4.00 x 3.25 in. Displacement: 327 cid. Compression ratio: 10.00:1. Brake hp: 300 at 5000 rpm. Taxable hp: 51.20. Torque: 360 lbs.-ft. at 3400 rpm. Five main bearings. Hydraulic valve lifters. Crankcase capacity: 5 qt. (add 1 qt. for filter). Cooling system capacity: 13 qt. (add 1 qt. for heater). Carburetor: Rochester Type 4MV four-barrel Model 7028207. Sales code: L75. Engine code: HE, HO, and HP.

Corvette optional V-8
Overhead valve. Cast-iron block. Bore & stroke: 4.00 x 3.25 in. Displacement: 327 cid. Compression ratio: 11.00:1. Brake hp: 350 at 5800 rpm. Taxable hp: 51.20. Torque: 360 lbs.-ft. at 3600 rpm. Five main bearings. Hydraulic valve lifters. High-lift camshaft. Crankcase capacity: 5 qt. (add 1 qt. for filter). Cooling system capacity: 13 qt. (add 1 qt. for heater). Carburetor: Holley four-barrel. Sales code: L79. Engine code: HT.

Corvette optional V-8
Overhead valve. Cast-iron block. Bore & stroke: 4.251 x 3.76 in. Displacement: 427 cid. Compression ratio: 10:25:1. Brake hp: 390 at 5400 rpm. Taxable hp: 57.80. Torque: 460 lbs.-ft. at 3600 rpm. Five main bearings. Hydraulic valve lifters. High-performance camshaft. Crankcase

For all-out performance in your totally-redesigned 1968 Corvette you just couldn't beat the 427-cid Chevy big-block V-8. The 435-hp tri-power version used this unusual triangular air cleaner, which was painted blue. The decal on the front point showed a red flag and a black-and-white checkered flag crossing each other with a red-and-silver decal below stating "TURBO-JET 427 3 x 2 435 HORSEPOWER." The standard valve covers seen here carried the same orange paint as the block, water pump, and timing chain cover, while the intake was aluminum finished.

capacity: 5 qt. (add 1 qt. for filter). Cooling system capacity: 20 qt. (add 1 qt. for heater). Carburetor: Holley four-barrel. Sales code: L36. Engine code: IL or IQ.

Corvette optional V-8
Overhead valve. Cast-iron block. Bore & stroke: 4.251 x 3.76 in. Displacement: 427 cid. Compression ratio: 10.25:1. Brake hp: 400 at 5400 rpm. Taxable hp: 57.80. Torque: 460 lbs.-ft. at 3600 rpm. Five main bearings. Hydraulic valve lifters. Special-performance camshaft. Crankcase capacity: 5 qt. (add 1 qt. for filter). Cooling system capacity: 20 qt. (add 1 qt. for heater). Carburetor: Three Holley two-barrel. Sales code: L68. Engine code: IM or IO.

Corvette optional V-8
Overhead valve. Cast-iron block. Bore & stroke: 4.251 x 3.76 in. Displacement: 427 cid. Compression ratio: 11.0:1. Brake hp: 435 at 5800 rpm. Taxable hp: 57.80. Torque: 460 lbs.-ft. at 4000 rpm. Five main bearings. Mechanical valve lifters. Special-performance camshaft. Crankcase capacity: 5 qt. (add 1 qt. for filter). Cooling system capacity: 20 qt. (add 1 qt. for heater). Carburetor: Three Holley two-barrels. Sales code: L71. Engine code IR.

Corvette optional V-8
Overhead valve. Cast-iron block. Bore & stroke: 4.251 x 3.76 in. Displacement: 427 cid. Compression ratio: 11.0:1. Brake hp: 435 at 5800 rpm. Taxable hp: 57.80. Torque: 460 lbs.-ft. at 4000 rpm. Five main bearings. Mechanical valve lifters. Aluminum cylinder heads. Extra-large exhaust valves. Special-performance camshaft. Crankcase capacity: 5 qt. (add 1 qt. for filter). Cooling system capacity: 21 qt. (add 1 qt. for heater). Carburetor: Three Holley two-barrels. Sales code: L71/L89. Engine code IU.

Corvette optional V-8
Overhead valve. Bore & stroke: 4.251 x 3.76 in. Displacement: 427 cid. Compression ratio: 12.50:1. Brake hp: (Advertised) 430 hp; (Actual) 560 at 6400 rpm. Taxable hp: 57.80. Torque: unknown at unknown rpm. Five main bearings. Mechanical valve lifters. Special-ultra-high-performance camshaft with .5365-in. intakes. Crankcase capacity: 5 qt. (add 1 qt. for filter). Cooling system capacity: 20 qt. (add 1 qt. for heater). Carburetor: Single Holley 850CFM four-barrel. Sales code: L88. Engine code: IT.

1969 CHEVROLET

Nova, Chevelle base V-8; Camaro V-8 (delete option)
Overhead valve. Cast-iron block and head. Bore & stroke: 3.88 x 3.25 in. Displacement: 307 cid. Compression ratio: 9.00:1. Brake hp: 200 at 4600 rpm. Taxable hp: 48.0. Torque: 300 lbs.-ft. at 2400 rpm. Five main bearings. Hydraulic valve lifters. Crankcase capacity: 4 qt. (add 1 qt. for new filter). Cooling system capacity: 16 qt (add 1 qt. for heater). Carburetor: Two-barrel. Engine code: (Chevy II) DA, DC, DD, and DE; (Chevelle) DA, DC, DD, and DE; (Camaro) DA, DE, DC, and DD.

Here's a rare Chevrolet V-8! It's a 427-cid 425-hp "big-block" in a 1969 Yenko "SYC" Nova. The initials SYC stood for Super Yenko Chevrolet. Don Yenko was a Chevrolet dealer from Canonsburg, Pennsylvania, who specialized in muscle cars and race cars. He had enough clout to order batches of SPO (special production order) cars from the factory. Note the "Yenko Tuned" sticker on the chrome valve cover. (Jerry Heasley photo)

In 1969, Chevrolet introduced the 350-cid Turbo-Fire small-block V-8 to join the venerable 327. This was essentially a 327 with its stroke increased from 3.25 in. to 3.48 in. Externally, this engine was the same size as the 307- and 327-cid V-8s, which meant that it was 28 in. long, 25 in. wide and 22 in. tall. It was offered in 255-, 300-, and 350-hp versions.

Camaro base V-8
Overhead valve. Cast-iron block and head. Bore & stroke: 4.00 x 3.25 in. Displacement: 327 cid. Compression ratio: 9.00:1. Brake hp: 210 at 4600 rpm. Taxable hp: 51.20. Torque: 320 lbs.-ft. at 2400 rpm. Five main bearings. Hydraulic valve lifters. Crankcase capacity: 4 qt. (add 1 qt. for new filter). Cooling system capacity: 16 qt. (add 1 qt. for heater) Carburetor: Two-barrel. Sales code: FJ, FK, and FL.

Chevy II, Chevelle, Camaro, Chevrolet optional V-8
Overhead valve. Cast-iron block and head. Bore & stroke: 4.00 x 3.48 in. Displacement: 350 cid. Compression ratio: 9.00:1. Brake hp: 250 at 4800 rpm. Taxable hp: 51.20. Torque: 345 lbs.-ft. at 2800 rpm. Five main bearings. Hydraulic valve lifters. Crankcase capacity: 4 qt. (add 1 qt. for new filter). Cooling system capacity: (full-size) 14 qt.; (others) 15 qt. (add 1 qt. for heater). Carburetor: Two-barrel. Engine code: (Chevy II) HC, HD, and HF; (Camaro) HC, HD, and HF; (Chevelle) HC, HD, and HF; (Chevrolet) HI, HL, HF, HD, HJ, HM, IL, IM, and IN.

Nova, Camaro, Chevelle, Malibu, Biscayne, Bel Air, Impala, Caprice optional V-8
Overhead valve. Cast-iron block and head. Bore & stroke: 4.00 x 3.48 in.

Chevrolet's huge 427-cid "big-block" V-8 was basically a 396 with the bore increased from 4.094 in. to 4.25 in. Both engines had a 3.76-in. stroke. With a good selection of factory and aftermarket speed equipment, the 427 could easily be setup to deliver 600 hp! This is a 1969 version.

This 1969 version of the Chevy 396-cid V-8 was offered in a 265-hp option with a two-barrel carburetor, as well as 325- and 350-hp four-barrel versions. This engine measured 30 inches in length, 28 in wide, and 22 in. tall. The least powerful version developed .668 horsepower per cubic inch, while the 350-hp job was good for .885 cubes per pound.

Displacement: 350 cid. Compression ratio: 9.00:1. Brake hp: 255 at 4800 rpm. Taxable hp: 51.20. Torque: 365 lbs.-ft. at 3200 rpm. Five main bearings. Hydraulic valve lifters. Crankcase capacity: 4 qt. (add 1 qt. for new filter). Cooling system capacity: (full-size) 14 qt.; (others) 15 qt. (add 1 qt. for heater). Carburetor: Four-barrel. Engine code: (Chevy II) HQ, HR, and HS; (Chevelle) HQ, HR, and HS; (Camaro) HQ, HR, and HS; (Chevrolet) HT, HU, IA, HV, IW, IX, IY, and IZ.

Biscayne, Bel Air, Impala, Caprice optional V-8
Overhead valve. Cast-iron block and head. Bore & stroke: 4.09 x 3.76 in. Displacement: 396 cid. Compression ratio: 9.00:1. Brake hp: 265 at 4800 rpm. Taxable hp: 53.60. Torque: 400 lbs.-ft. at 2800 rpm. Five main bearings. Hydraulic valve lifters. Crankcase capacity: 4 qt. (add 1 qt. for new filter). Cooling system capacity: 23 qt. (add 1 qt. for heater). Carburetor: Two-barrel. Sales code: L35. Engine code: JN, JG, JP, JO, JR, and JT.

Camaro Z/28 base V-8
Overhead valve. Cast-iron block and head. Bore & stroke: 4.00 x 3.00 in. Displacement: 302 cid. Compression ratio: 11.00:1. Brake hp: 290 at 5800 rpm. Taxable hp: 51.20. Torque: 290 lbs.-ft. at 4200 rpm. Five main bearings. Hydraulic valve lifters. Crankcase capacity: 4 qt. (add 1 qt. for new filter). Cooling system capacity: 16 qt. (add 1 qt. for heater). Five four-bolt main bearings with nodular iron main caps. Solid valve lifters. Induction system: 800-cfm Holley no. 3923289 (R4055A) four-barrel carburetor. Four-bolt main bearings. Special solid lifter camshaft. Cast-aluminum "high-rise" intake manifold. High capacity oil pump. Special baffling in oil pan. Cooling system includes dual-belt fan drive, external bypass water pump and five-blade thermostatically controlled fan with operating range of 2300-2600 rpm. Also includes chrome-plated air cleaner, rocker covers, filler tube and cap. Dual exhaust system (2-1/4-in. diameter) with dual deep-tone mufflers. A 10.34-in. diameter bent-finger diaphragm spring clutch with a 2300-2600-lb. rating. Standard in Camaro Z/28; not available in other Camaros. Engine code: DZ.

Nova, Camaro, Chevelle, Malibu, Biscayne, Bel Air, Impala, Caprice optional V-8
Overhead valve. Cast-iron block and head. Bore & stroke: 4.00 x 3.48 in. Displacement: 350 cid. Compression ratio: 9.00:1. Brake hp: 300 at 4800 rpm. Taxable hp: 51.20. Torque: 380 lbs.-ft. at 3200 rpm. Five main bearings. Hydraulic valve lifters. Crankcase capacity: 4 qt. (add 1 qt. for new filter). Cooling system capacity: (full-size) 14 qt.; (others) 15 qt. (add 1 qt. for heater). Carburetor: Four-barrel. Engine code: (Chevy II) HA, HB, and HE; (Chevelle) HA, HB, and HE; (Camaro) HA, HB, HE, and HP; (Chevrolet) HG, HK, HH, HN, IR, IS, IT, IQ, HO, and IV.

Camaro SS-396, Chevelle SS-396 optional V-8
Overhead valve. Cast-iron block and head. Bore & stroke: 4.09 x 3.76 in. Displacement: 396 cid (actually 402 cid). Compression ratio: 10.25:1. Brake hp: 325 at 4800 rpm. Taxable hp: 53.60. Torque: 410 lbs.-ft. at 3200 rpm. Five main bearings. Hydraulic valve lifters. Crankcase capacity: 4 qt. (add 1 qt. for new filter). Cooling system capacity: 23 qt. (add 1 qt. for heater). Carburetor: Four-barrel. Engine code: (Chevelle) JK, KG, KH, JV, and KI; (Camaro) JB, JG, and JU.

Impala SS optional V-8 (As part of SS-427 package only)
Overhead valve. Cast-iron block and head. Bore & stroke: 4.25 x 3.76 in. Displacement: 427 cid. Compression ratio: 10.25:1. Brake hp: 335 at 4800 rpm. Taxable hp: 57.80. Torque: 460 lbs.-ft. at 3600 rpm. Five main bearings. Hydraulic valve lifters. Crankcase capacity: 4 qt. (add 1 qt. for new filter). Cooling system capacity: 21 qt. (add 1 qt. for heater). Dual exhaust. Carburetor: Holley 3885067 four-barrel. Sales code: L72. Engine code: LB, LE, LK, LJ, LY, and MA.

Chevy II, Camaro SS-396, Chevelle SS-396 optional V-8
Overhead valve. Cast-iron block and head. Bore & stroke: 4.09 x 3.76 in. Displacement: 396 cid (actually 402 cid). Compression ratio: 10.25:1. Brake hp: 350 at 5200 rpm. Taxable hp: 53.60. Torque: 415 lbs.-ft. at 3400 rpm. Five main bearings. Hydraulic valve lifters. Crankcase capacity: 4 qt. (add 1 qt. for new filter). Cooling system capacity: 23 qt. (add 1 qt. for heater). Carburetor: Four-barrel. Engine code: (Chevy II) JF, JI, JM, JU, KA, and KE; (Camaro) JF, JI, and KA; (Chevelle) JC, JE, and KB.

Chevy II, Camaro SS-396, Chevelle SS-396 optional V-8
Overhead valve. Cast-iron block and head. Bore & stroke: 4.09 x 3.76 in. Displacement: 396 cid (actually 402 cid). Compression ratio: 11.00:1. Brake hp: 375 at 5600 rpm. Taxable hp: 53.60. Torque: 415 lbs.-ft. at 3600 rpm. Five main bearings. Hydraulic valve lifters. Crankcase capacity: 4 qt. (add 1 qt. for new filter). Cooling system capacity: 23 qt. (add 1 qt. for heater). Carburetor: Four-barrel. Sales code: L78. Engine code: (Chevy II) JH, JL, and KC; (Camaro) JH, KL, JL, KC, JJ, JM, and KE; (Chevelle) JD, KF, and KD.

Camaro SS-396, Chevelle SS-396 optional V-8
Overhead valve. Cast-iron block. Aluminum head with large valves. Bore & stroke: 4.09 x 3.76 in. Displacement: 396 cid (actually 402 cid). Compression ratio: 11.00:1. Brake hp: 375 at 5600 rpm. Taxable hp: 53.60. Torque: 415 lbs.-ft. at 3600 rpm. Five main bearings. Solid valve lifters. Crankcase capacity: 4 qt. (add 1 qt. for new filter). Cooling system capacity: 23 qt. (add 1 qt. for heater). Carburetor: Holley four-barrel carburetor on high-rise aluminum intake manifold. Sales code: L78/L89.

Impala SS optional V-8 (As part of SS-427 package only)
Overhead valve. Cast-iron block and head. Bore & stroke: 4.25 x 3.76 in. Displacement: 427 cid. Compression ratio: 10.25:1. Brake hp: 390 at 4800 rpm. Taxable hp: 57.80. Torque: 460 lbs.-ft. at 3600 rpm. Five main bearings. Hydraulic valve lifters. Crankcase capacity: 4 qt. (add 1 qt. for new filter). Cooling system capacity: 21 qt. (add 1 qt. for heater). Dual exhaust. Carburetor: Holley 3885067 four-barrel. Sales code: L72. Engine code: LA, LC, LG, LF, LZ, MB, LH, MC, and LI.

Impala SS optional V-8 (As part of SS-427 package only)
Overhead valve. Cast-iron block and head. Bore & stroke: 4.25 x 3.76 in. Displacement: 427 cid. Compression ratio: 11.00:1. Brake hp: 425 at 5600 rpm. Taxable hp: 57.80. Torque: 460 lbs.-ft. at 4000 rpm. Five main bearings. Hydraulic valve lifters. Crankcase capacity: 4 qt. (add 1 qt. for new filter). Cooling system capacity: 21 qt. (add 1 qt. for heater). Dual exhaust. Carburetor: Holley 800-cfm four-barrel. Sales code: Special COPO only version. Engine code: LD, LS, and MD.

Camaro optional V-8 (COPO)
Overhead valve V-8. Aluminum block and "open-chamber" aluminum heads. Bore & stroke: 4.251 x 3.76 in. Displacement: 427 cid. Compression ratio: 12.5:1. Brake horsepower: 430 at 5800 rpm. Torque: 460 lbs.-ft. at 4000 rpm. Five main bearings. Solid valve lifters. Induction

system: special 850-cfm Holley four-barrel carburetor on top of a special high-rise aluminum manifold. This very special engine package was originally released for Fred Gibb Chevrolet under Central Office Production Order (COPO) 9560. A total of only 69 Camaros received this special engine package. Engine code: ME and MG.

1969 CORVETTE

The bottom end of the ZL1 version of the 427-cid V-8 had four-bolt main bearings with a forged steel crank and rods with 7/16-inch bolts, Spiralock washers, and full floating pins. The pistons were even higher domed than the L88's, yielding a titanic compression ratio of 12.5:1. The cylinder heads were also aluminum and featured open combustion chambers, round exhaust ports, and 2.19-inch/1.88-inch valves (a configuration adopted by the L88 in mid-1969). The aluminum dual plane intake was topped by an 850-cfm Holley "double-pumper" four-barrel carburetor featuring mechanical secondaries. The ZL1's solid-lifter camshaft was so radical that the engine could live in the upper revs.

A 427-cid Chevrolet "Rat" engine in a limited-edition Corvette ZL1. This was perhaps the wildest, most exotic, high-performance muscle engine ever offered to the public. It was an all-aluminum 427. Just 69 engines were installed in 1969 Camaros and two in 1969 Corvettes. The L88 engine itself was a race option, but in ZL1 metal, it featured thicker walls and main webbing, along with dry-sump lubrication provisions.

Corvette base V-8
Overhead valve. Cast-iron block. Bore & stroke: 4.00 x 3.48 in. Displacement: 350 cid. Compression ratio: 10.25:1. Brake hp: 300 at 4800 rpm. Taxable hp: 51.20. Torque: 380 lbs.-ft. at 3200 rpm. Five main bearings. Hydraulic valve lifters. Crankcase capacity: 4 qt. (add 1 qt. for filter). Cooling system capacity: 14 qt. (add 1 qt. for heater). Carburetor: Rochester 7029203 four-barrel. Sales code: ZQ3. Engine code: HY or HZ.

Corvette optional V-8
Overhead valve. Cast-iron block. Bore & stroke: 4.00 x 3.48 in. Displacement: 350 cid. Compression ratio: 11.0:1. Brake hp: 350 at 5600 rpm. Taxable hp: 51.20. Torque: 380 lbs.-ft. at 3600 rpm. Five main bearings. Hydraulic valve lifters. Crankcase capacity: 4 qt. (add 1 qt. for filter). Cooling system capacity: 14 qt. (add 1 qt. for heater). Carburetor: Four-barrel. Sales code: L46. Engine code: HW or HX.

Corvette optional V-8
Overhead valve. Cast-iron block. Bore & stroke: 4.00 x 3.48 in. Displacement: 350 cid. Compression ratio: 11.0:1. Brake hp: 370 at 6000 rpm. Taxable hp: 51.20. Torque: 380 lbs.-ft. at 4000 rpm. Five main bearings. Solid valve lifters. High-performance camshaft. Full transistor ignition. Carburetor: Holley four-barrel on aluminum intake manifold. Sales code: LT1. Engine code: Unknown.
NOTE: Listed in early 1969 Corvette specifications, but cancelled as a factory option. Available only as a dealer-installed parts kit with L71, L88 and L89 engines only.

Corvette optional V-8
Overhead valve. Cast-iron block. Bore & stroke: 4.251 x 3.76 in. Displacement: 427 cid. Compression ratio: 10:25:1. Brake hp: 390 at 5400 rpm. Taxable hp: 57.80. Torque: 460 lbs.-ft. at 3600 rpm. Five main bearings. Hydraulic valve lifters. High-performance camshaft. Crankcase capacity: 5 qt. (add 1 qt. for filter). Cooling system capacity: 20 qt. (add 1 qt. for heater). Carburetor: Holley four-barrel. Sales code: L36. Engine code: LM, LL, MH or MI.

Corvette optional V-8
Overhead valve. Cast-iron block. Bore & stroke: 4.251 x 3.76 in. Displacement: 427 cid. Compression ratio: 10.25:1. Brake hp: 400 at 5400 rpm. Taxable hp: 57.80. Torque: 460 lbs.-ft. at 3600 rpm. Five main bearings. Hydraulic valve lifters. Special-performance camshaft. Crankcase capacity: 5 qt. (add 1 qt. for filter). Cooling system capacity: 20 qt. (add 1 qt. for heater). Carburetor: Three Holley two-barrel. Sales code: L68. Engine code: Engine code: LQ, LN, MJ or MK.

Corvette optional V-8
Overhead valve. Cast-iron block. Bore & stroke: 4.251 x 3.76 in. Displacement: 427 cid. Compression ratio: 11.0:1. Brake hp: 435 at 5800 rpm. Taxable hp: 57.80. Torque: 460 lbs.-ft. at 4000 rpm. Five main bearings. Mechanical valve lifters. Special-performance camshaft. Crankcase capacity: 5 qt. (add 1 qt. for filter). Cooling system capacity: 20 qt. (add 1 qt. for heater). Carburetor: Three Holley two-barrels. Sales code: L71. Engine code: LR, LX, LP, LW, LT and LU.

Corvette optional V-8
Overhead valve. Cast-iron block. Bore & stroke: 4.251 x 3.76 in. Displacement: 427 cid. Compression ratio: 11.0:1. Brake hp: 435 at 5800 rpm. Taxable hp: 57.80. Torque: 460 lbs.-ft. at 4000 rpm. Five main bearings. Mechanical valve lifters. Aluminum cylinder heads. Extra-large exhaust valves. Special-performance camshaft. Crankcase capacity: 5 qt. (add 1 qt. for filter). Cooling system capacity: 21 qt. (add 1 qt. for heater). Carburetor: Three Holley two-barrels. Sales code: L71/L89.

Corvette optional V-8
Overhead valve. Bore & stroke: 4.251 x 3.76 in. Displacement: 427 cid. Compression ratio: 12.50:1. Brake hp: (Advertised) 430 hp; (Actual) 560 at 6400 rpm. Taxable hp: 57.80. Torque: unknown at unknown rpm. Five main bearings. Mechanical valve lifters. Special-ultra-high-performance camshaft with .5365-in. intakes. Crankcase capacity: 5 qt. (add 1 qt. for filter). Cooling system capacity: 20 qt. (add 1 qt. for heater). Carburetor: Single Holley 850CFM four-barrel. Sales code: L88. Engine code: ME, MG, and MR.

1970 CHEVROLET

Nova, Camaro, Malibu base V-8
Overhead valve. Cast-iron block and head. Bore & stroke: 3.88 x 3.25 in. Displacement: 307 cid. Compression ratio: 9.00:1. Brake hp: 200 at 4600 rpm. Taxable hp: 48.0. Torque: 300 lbs.-ft. at 2400 rpm. Five main bearings. Hydraulic valve lifters. Crankcase capacity: 4 qt. (add 1 qt. for new filter). Cooling system capacity: 14 qt (add 1 qt. for heater). Carburetor: Two-barrel. Engine code: (Nova) CNC, CNG, CND, CNH, CNE, and CNF; (Chevelle) CNC, CND, CNE, and CNF; (Camaro) CNC, CND, CNG, CNE, CNH, and CNF.

Monte Carlo, Chevelle, Caprice base V-8
Overhead valve. Cast-iron block and head. Bore & stroke: 4.00 x 3.48 in. Displacement: 350 cid. Compression ratio: 9.00:1. Brake hp: 250 at 4800 rpm. Taxable hp: 51.20. Torque: 345 lbs.-ft. at 2800 rpm. Five main bearings. Hydraulic valve lifters. Crankcase capacity: 4 qt. (add 1 qt. for new filter). Cooling system capacity: (full-size) 14 qt.; (others) 15 qt. (add 1 qt. for heater). Carburetor: Two-barrel. Engine code: (Nova) CNI, CNM, and CNN; (Camaro) CNI, CNM, and CNN; (Monte Carlo) CNI, CNM, and CNN; (Chevelle) CNI, CNM, and CNN; (Chevrolet) CND, CNP, CNU, CNV, CNW, CNX, and CNO.

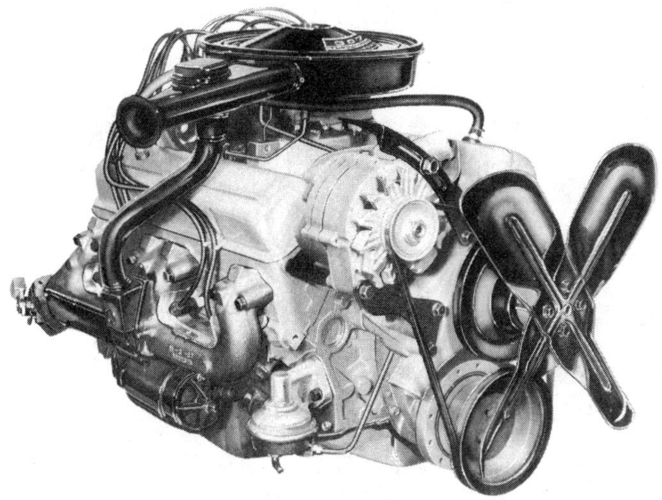

For 1970, the 307-cid Turbo-Fire V-8 was Chevrolet's smallest V-8 engine. It was a direct descendent of the 1955 Chevrolet 265-cid V-8. It came only with a two-barrel carburetor and delivered 200 hp at 4600 rpm and 300 foot-pounds of torque at 2400 rpm.

This is the Turbo-Fire 350 version of the 1970 Chevy small-block V-8. With a four-barrel carburetor and 300 hp, it was the company's "workhorse" V-8. It was available in Novas, Camaros, Monte Carlos, Chevelles, and full-sized Chevrolets.

The 400-cid 265-hp was another V-8 option for 1970 Chevrolet models. It was designed to give good service and to provide strong torque ratings while using regular gasoline.

This unrestored 350-cid V-8 resides in Tim Steblow's 1970 Chevrolet Impala. Although this was a simple engine, you can see the various hoses and contraptions that were starting to crowd up the engine compartment as new emissions regulations took effect.

Chevelle, Biscayne, Bel Air, Impala, Caprice, Monte Carlo optional V-8

Overhead valve. Cast-iron block and head. Bore & stroke: 4.13 x 3.75 in. Displacement: 400 cid. Compression ratio: 9.00:1. Brake hp: 265 at 4400 rpm. Taxable hp: 54.40. Torque: 400 lbs.-ft. at 2400 rpm. Five main bearings. Hydraulic valve lifters. Crankcase capacity: 4 qt. (add 1 qt. for new filter). Cooling system capacity: 23 qt. (add 1 qt. for heater). Carburetor: Two-barrel. Engine code: (Chevelle) CZX and CRH; (Monte Carlo) CZX and CRH; (Chevrolet) CGR or CNX.

Nova, Camaro, Chevelle, Malibu, Biscayne, Bel Air, Impala, Caprice optional V-8

Overhead valve. Cast-iron block and head. Bore & stroke: 4.00 x 3.48 in. Displacement: 350 cid. Compression ratio: 9.00:1. Brake hp: 300 at 4800 rpm. Taxable hp: 51.20. Torque: 380 lbs.-ft. at 3200 rpm. Five main bearings. Hydraulic valve lifters. Crankcase capacity: 4 qt. (add 1 qt. for new filter). Cooling system capacity: (full-size) 14 qt.; (others) 15 qt. (add 1 qt. for heater). Carburetor: Four-barrel. Engine code: (Nova) CNJ, CNK, and CRE; (Camaro) CNJ, CNK, and CRE; (Chevelle) CNJ, CNK, and CRE; (Monte Carlo) CNJ, CNK, and CRE; (Chevrolet) CNQ, CNR, CNS, and CNT.

Malibu, Monte Carlo optional V-8

Overhead valve. Cast-iron block and head. Bore & stroke: 4.13 x 3.76 in. Displacement: 402 cid (400 cid). Compression ratio: 10.25:1. Brake hp: 330 at 4800 rpm. Taxable hp: 54.50. Torque: 410 lbs.-ft. at 3200 rpm. Five main bearings. Hydraulic valve lifters. Crankcase capacity: 4 qt. (add 1 qt. for new filter). Cooling system capacity: 22 qt. (add 1 qt. for heater). Carburetor: Four-barrel. Engine code: (Chevelle) CKN, CKR, and CKS; (Monte Carlo) CKN, CKR, and CKS.

Biscayne, Bel Air, Impala, Caprice optional V-8

Overhead valve. Cast-iron block and head. Bore & stroke: 4.25 x 4.00 in. Displacement: 454 cid. Compression ratio: 10.25:1. Brake hp: 345 at 4400 rpm. Taxable hp: 57.80. Torque: 500 lbs.-ft. at 3000 rpm. Five main bearings. Hydraulic valve lifters. Crankcase capacity: 4 qt. (add 1 qt. for new filter). Cooling system capacity: 21 qt. (add 1 qt. for heater). Dual exhaust. Carburetor: Four-barrel. Engine code: CGV or CGS.

Nova, Malibu optional V-8

Overhead valve. Cast-iron block and head. Bore & stroke: 4.13 x 3.76 in. Displacement: 402 cid (396 cid). Compression ratio: 10.25:1. Brake hp: 350 at 5200 rpm. Taxable hp: 54.50. Torque: 415 lbs.-ft. at 3400 rpm. Five main bearings. Hydraulic valve lifters. Crankcase capacity: 4 qt. (add 1 qt. for new filter). Cooling system capacity: 22 qt. (add 1 qt. for heater). Carburetor: Four-barrel. Engine code: (Nova) CTW, CTX, or CTZ; (Malibu) CTW, CTX, or CTZ; (Camaro) CJF, CJI, CTW, and CTX.

Camaro Z/28 base V-8

Overhead valve. Cast-iron block and head. Bore & stroke: 4.00 x 3.48 in. Displacement: 350 cid. Compression ratio: 11.00:1. Brake hp: 360 at 6000 rpm. Taxable hp: 51.20. Torque: 380 lbs.-ft. at 4000 rpm. Five main bearings. Solid valve lifters. Crankcase capacity: 4 qt. (add 1 qt. for new filter). Cooling system capacity: (full-size) 14 qt.; (others) 15 qt. (add 1 qt. for heater). Induction system: Holley four-barrel carburetor. Four-bolt main bearings. Big valve heads. Special cam. Aluminum high-rise

For 1970, the 427-cid Chevy big-block V-8 was dropped as a passenger car engine. Replacing it was a new 454-cid engine. This could be considered a 427 V-8 with an increase in stroke measurement. There were five horsepower options on the 454-cid block: 345-, 360-, 390-, 450-, and 460-hp.

manifold. Standard in Camaro Z/28; not available in other Camaros. Engine code: CTB or CTC.

Malibu, Monte Carlo optional V-8
Overhead valve. Cast-iron block and head. Bore & stroke: 4.25 x 4.00 in. Displacement: 454 cid. Compression ratio: 10.25:1. Brake hp: 360 at 4400 rpm. Taxable hp: 57.80. Torque: 500 lbs.-ft. at 3200 rpm. Five main bearings. Hydraulic valve lifters. Crankcase capacity: 4 qt. (add 1 qt. for new filter). Cooling system capacity: 22 qt. (add 1 qt. for heater). Dual exhaust. Carburetor: Four-barrel. Engine code: (Chevelle) CRN, CGT, CRQ, CRM, CRT, CGU, and CRU; (Monte Carlo) CRU, CRN, CRT, CRQ, CGW, and CGT.

Biscayne, Bel Air, Impala, Caprice optional V-8
Overhead valve. Cast-iron block and head. Bore & stroke: 4.25 x 4.00 in. Displacement: 454 cid. Compression ratio: 10.25:1. Brake hp: 390 at 4800 rpm. Taxable hp: 57.80. Torque: 500 lbs.-ft. at 3400 rpm. Five main bearings. Hydraulic valve lifters. Crankcase capacity: 4 qt. (add 1 qt. for new filter). Cooling system capacity: 21 qt. (add 1 qt. for heater). Dual exhaust. Carburetor: Four-barrel. Engine code: CGT or CGU.

Nova, Camaro SS-396, Chevelle SS-396 optional V-8
Overhead valve. Cast-iron block and head. Bore & stroke: 4.09 x 3.76 in. Displacement: 396 cid (actually 402 cid). Compression ratio: 11.00:1. Brake hp: 375 at 5600 rpm. Taxable hp: 53.60. Torque: 415 lbs.-ft. at 3600 rpm. Five main bearings. Hydraulic valve lifters. Crankcase capacity: 4 qt. (add 1 qt. for new filter). Cooling system capacity: 23 qt. (add 1 qt. for heater). Carburetor: Four-barrel. Sales code: L78. Engine code: (Chevy II) CTY, CKO, CKP, CTY, CKQ, CKT, and CKU; (Camaro) CJL, CJH, CTY, and CKO; (Chevelle) CTY, CKN, CKO, CKD, CKP, CKQ, CKT, and CKU.

Chevelle SS-454 optional V-8
Overhead valve. Cast-iron block. Bore & stroke: 4.251 x 4.00 in. Displacement: 454 cid. Compression ratio: 11:25:1. Brake hp: 450 at 5600 rpm. Taxable hp: 57.80. Torque: 500 lbs.-ft. at 3600 rpm. Five main bearings. Solid valve lifters. High-performance camshaft. Crankcase capacity: 5 qt. (add 1 qt. for filter). Cooling system capacity: 21 qt. (add 1 qt. for heater). Carburetor: Holley four-barrel. Sales code: LS6. Engine code: CRR, CRY, CRV, CRW, and CRX.

1970 CORVETTE

NOTE: The awesome LS6 V-8 was added to the Chevelle options list at midyear.

Corvette base V-8
Overhead valve. Cast-iron block. Bore & stroke: 4.00 x 3.48 in. Displacement: 350 cid. Compression ratio: 10.25:1. Brake hp: 300 at 4800 rpm. Taxable hp: 51.20. Torque: 380 lbs.-ft. at 3200 rpm. Five main bearings. Hydraulic valve lifters. Crankcase capacity: 4 qt. (add 1 qt. for

This is the hot Corvette LT-1 version of the 350-cid V-8. With 370 hp, it was the most powerful small-block available in 1970 models. It became available in April 1970 as part of the Corvette's ZR1 package and included an aluminum intake manifold.

filter). Cooling system capacity: 14 qt. (add 1 qt. for heater). Carburetor: Rochester 7029203 four-barrel. Sales code: ZQ3. Engine code: CTG, CTD, CTL, and CTM.

Corvette optional V-8
Overhead valve. Cast-iron block. Bore & stroke: 4.00 x 3.48 in. Displacement: 350 cid. Compression ratio: 11.0:1. Brake hp: 350 at 5600 rpm. Taxable hp: 51.20. Torque: 380 lbs.-ft. at 3600 rpm. Five main bearings. Hydraulic valve lifters. Crankcase capacity: 4 qt. (add 1 qt. for filter). Cooling system capacity: 14 qt. (add 1 qt. for heater). Carburetor: Four-barrel. Sales code: L46. Engine code: CTN, CTO, CTH, CTJ, CTP, and CTQ.

Corvette optional V-8
Overhead valve. Cast-iron block. Bore & stroke: 4.00 x 3.48 in. Displacement: 350 cid. Compression ratio: 11.0:1. Brake hp: 370 at 6000 rpm. Taxable hp: 51.20. Torque: 380 lbs.-ft. at 4000 rpm. Five main bearings. Solid valve lifters. High-performance camshaft. Full transistor ignition. Crankcase capacity: 4 qt. (add 1 qt. for filter). Cooling system capacity: 17 qt. (add 1 qt. for heater). Carburetor: Holley four-barrel on aluminum intake manifold. Sales code: LT1. Engine code: CTR, CTU, CTK, and CTV.
NOTE: Available as a 1970 production option beginning in April 1970 as part of ZR1 option package.

Corvette optional V-8
Overhead valve. Cast-iron block. Bore & stroke: 4.251 x 4.00 in. Displacement: 454 cid. Compression ratio: 10:25:1. Brake hp: 390 at 4800 rpm. Taxable hp: 57.80. Torque: 500 lbs.-ft. at 3400 rpm. Five main bearings. Hydraulic valve lifters. High-performance camshaft. Crankcase capacity: 5 qt. (add 1 qt. for filter). Cooling system capacity: 21 qt. (add 1 qt. for heater). Carburetor: Rochester 750CFM Quadra-Jet four-barrel. Sales code: LS5. Engine code: CGW, CZU, CRJ, and CRI.

Proposed Corvette optional V-8
Overhead valve. Cast-iron block. Bore & stroke: 4.251 x 4.00 in. Displacement: 454 cid. Compression ratio: 11:25:1. Brake hp: 465 at 5200 rpm. Taxable hp: 57.80. Torque: 490 lbs.-ft. at 3400 rpm. Five main bearings. Solid valve lifters. High-performance camshaft. Crankcase capacity: 5 qt. (add 1 qt. for filter). Cooling system capacity: 21 qt. (add 1 qt. for heater). Carburetor: Holley 800CFM four-barrel. Sales code: LS7. Engine code CZL or CZN.
NOTE: Only one car with the LS7 engine was built. Sports Car Graphic editor Paul Van Valkenburgh drove it 2,500 miles from a press conference at Riverside, Calif., to Detroit and raved about it. The car did the quarter mile in 13.8 seconds at 108 mph. However, GM's policies against ultra-high-performance cars at this time led to the option being stillborn. The LS7 option is listed in some early 1970 sales literature but none were ever sold.

1971 CHEVROLET

The optional Turbo-Jet 454 V-8 in this 1971 Chevrolet produced 365 hp, as indicated by a large air cleaner decal. The chrome goodies of the past were not as much in evidence on 1970s high-performance V-8s.

Nova, Camaro, Malibu optional V-8
Overhead valve. Cast-iron block and head. Bore & stroke: 3.88 x 3.25 in. Displacement: 307 cid. Compression ratio: 8.50:1. Gross brake horsepower: 200 at 4600 rpm. Net brake hp: 140 at 4400 rpm. Taxable hp: 48.0. Gross torque: 300 lbs.-ft. at 2400. Net torque. 235 lbs.-ft. at 2400 rpm. Five main bearings. Hydraulic valve lifters. Crankcase capacity: 4 qt. (add 1 qt. for new filter). Cooling system capacity: (Camaro) 14 qt.; (Chevelle, Monte Carlo) 15 qt.; (add 1 qt. for heater). Carburetor: Two-barrel. Engine code (all) CCA or CCC.

Monte Carlo base V-8; Chevrolet optional V-8
Overhead valve. Cast-iron block and head. Bore & stroke: 4.00 x 3.48 in. Displacement: 350 cid. Compression ratio: 8.50:1. Gross brake horsepower: 245 at 4800 rpm. Net brake hp: 165 at 4000 rpm. Taxable hp: 51.20. Gross torque: 350 lbs.-ft. at 2800. Net torque: 280 lbs.-ft. at 2400 rpm. Five main bearings. Hydraulic valve lifters. Crankcase capacity: 4 qt. (add 1 qt. for new filter). Cooling system capacity: (full-size) 14 qt.; (others) 15 qt. (add 1 qt. for heater). Carburetor: Two-barrel. Engine code: (Nova) CGA, CGB, CGC; (Camaro) CGB, CGA, CGC; (Chevelle) CGA, CGB, CGC; (Monte Carlo) CGA and CGB; (Chevrolet) CGA, CGB, CGC, CGJ, CJB, and CLT.

Nova, Camaro, Chevelle, Monte Carlo, Chevrolet optional V-8
Overhead valve. Cast-iron block and head. Bore & stroke: 4.00 x 3.48 in. Displacement: 350 cid. Compression ratio: 8.50:1. Gross brake horsepower: 270 at 4800 rpm. Taxable hp: 51.20. Gross torque: 360 lbs.-ft. at 3200 rpm. Five main bearings. Hydraulic valve lifters. Crankcase capacity: 4 qt. (add 1 qt. for new filter). Cooling system capacity: (full-size) 14 qt.; (others) 15 qt. (add 1 qt. for heater). Carburetor: Four-barrel. Engine code: (Nova) CGK, CGL, CJD, and CJG; (Camaro) CGK, CGL, CJD, and CJG; (Chevelle) CGK, CGL, CJD, and CJJ; (Monte Carlo) CGK, CGL, and CJG; (Chevrolet) CJH and CJD.

Camaro, Chevelle, Malibu, Monte Carlo, Chevrolet optional V-8
Overhead valve. Cast-iron block and head. Bore & stroke: 4.13 x 3.76 in. Displacement. 402 cid (396 cid). Compression ratio. 8.50:1. Gross brake horsepower: 300 at 4800 rpm. Net brake hp: 260 at 4400 rpm. Taxable hp: 54.50. Gross torque: 400 lbs.-ft. at 3200 rpm. Net torque. 345 lbs.-ft. at 3200 rpm. Five main bearings. Hydraulic valve lifters. Crankcase capacity: 4 qt. (add 1 qt. for new filter). Cooling system capacity: 22 qt. (add 1 qt. for heater). Carburetor: Four-barrel. Engine code: (Camaro) CLD, CLC, CLA, and CLB; (Chevelle) CLP, CLB, CLL, CLR, and CLS; (Monte Carlo) CLA, CLB, CLJ, CLK, CLL, and CLP; (Chevrolet) CLP, CLK, CLR, and CLB,

Camaro Z28 base V-8
Overhead valve V-8. Cast-iron block and head. Bore & stroke: 4.00 x 3.48 in. Displacement: 350 cid (5.7 liters). Compression ratio: 9.0:1. Gross horsepower: 330 at 5600 rpm. Net horsepower: 275 at 5600 rpm. Gross torque: 360 lbs.-ft. at 4000 rpm. Net torque: 300 lbs.-ft. at 4000 rpm. Five main bearings. Hydraulic valve lifters. Induction system: Rochester no. 7042202 four-barrel carburetor. Four-bolt main bearings. Big valve heads. Special cam. Aluminum high-rise manifold. Standard in Camaro Z/28; not available in other Camaros. Engine code: CGR or CGP.

Chevelle, Monte Carlo, Chevrolet optional V-8
Overhead valve. Cast-iron block and head. Bore & stroke: 4.25 x 4.00 in. Displacement: 454 cid. Compression ratio: 8.50:1. Gross brake horsepower: 365 at 4800 rpm. Net brake hp: 285 at 4000 rpm. Taxable hp: 57.80. Gross torque: 465 lbs.-ft. at 3200 rpm. Net torque. 390 lbs.-ft. at 3200 rpm. Five main bearings. Hydraulic valve lifters. Crankcase capacity: 4 qt. (add 1 qt. for new filter). Cooling system capacity: 21 qt. (add 1 qt. for heater). Dual exhaust. Carburetor: Four-barrel. Engine code: (Chevelle) CPA, CPG, and CPD; (Monte Carlo) CPA, CPG, CPP, CPR, and CPD; (Chevrolet) CPG or CPD.

Chevelle, Monte Carlo optional V-8
Overhead valve. Cast-iron block and head. Bore & stroke: 4.25 x 4.00 in. Displacement: 454 cid. Compression ratio: 9.00:1. Gross brake horsepower: 425 at 5600 rpm. Net brake hp: 325 at 5600 rpm. Taxable hp: 57.80. Gross torque: 475 lbs.-ft. at 4000 rpm. Net torque. 390 lbs.-ft. at 3600 rpm. Five main bearings. Hydraulic valve lifters. Crankcase capacity: 4 qt. (add 1 qt. for new filter). Cooling system capacity: 21 qt. (add 1 qt. for heater). Dual exhaust. Carburetor: Four-barrel. Engine code: (Chevelle) CPP, CPR, and CPZ.

1971 CORVETTE

Corvette base V-8
Overhead valve. Cast-iron block and head. Bore & stroke: 4.00 x 3.48 in. Displacement: 350 cid. Compression ratio: 8.50:1. Gross brake horsepower: 270 at 4800 rpm. Taxable hp: 51.20. Gross torque: 360 lbs.-ft. at 3200 rpm. Five main bearings. Hydraulic valve lifters. Crankcase capacity: 4 qt. (add 1 qt. for new filter). Cooling system capacity: (full-size) 14 qt.; (others) 15 qt. (add 1 qt. for heater). Carburetor: Four-barrel. Engine code: CJL, CGT, and CJK.

Corvette optional V-8
Overhead valve. Cast-iron block. Bore & stroke: 4.00 x 3.48 in. Displacement: 350 cid. Compression ratio: 9.00:1. Brake hp: 330 at 5600 rpm. Net brake hp: 275 at 5600 rpm. Taxable hp: 51.20. Torque: 360 lbs.-ft. at 4000 rpm. Taxable hp: 51.20. Net torque: 300 at 4000 rpm. Five main bearings. Solid valve lifters. High-performance camshaft. Crankcase capacity: 4 qt. (add 1 qt. for filter). Cooling system capacity: 14 qt. (add 1 qt. for heater). Carburetor: Rochester Rochester Type Quadra-Jet four-barrel Model 4MV. Sales code: LT1. Engine code: CGZ, CGY, and CJK.

Corvette optional V-8
Overhead valve. Cast-iron block. Bore & stroke: 4.00 x 3.48 in. Displacement: 350 cid. Compression ratio: 9.00:1. Brake hp: 330 at 5600 rpm. Taxable hp: 51.20. Torque: 360 lbs.-ft. at 4000 rpm. Five main bearings. Hydraulic valve lifters. Crankcase capacity: 4 qt. (add 1 qt. for filter). Cooling system capacity: 14 qt. (add 1 qt. for heater). Carburetor: Rochester Type Quadra-Jet four-barrel Model 4MV. Sales code: LT1/ZR1. **NOTE:** Special competition engine; only 8 assembled.

Corvette optional V-8
Overhead valve. Cast-iron block. Bore & stroke: 4.251 x 4.00 in. Displacement: 454 cid. Compression ratio: 8.50:1. Brake hp: 365 at 4800 rpm. Net brake hp: 285 at 4000 rpm. Taxable hp: 57.80. Torque: 465 lbs.-ft. at 3200 rpm. Net torque: 390 at 3200 rpm. Five main bearings. Hydraulic valve lifters. High-performance camshaft. Crankcase capacity: 5 qt. (add 1 qt. for filter). Cooling system capacity: 21 qt. (add 1 qt. for heater). Carburetor: Rochester 750CFM Quadra-Jet four-barrel. Sales code: LS5. Engine code: CPH or CPJ.

Corvette optional V-8
Overhead valve. Cast-iron block. Bore & stroke: 4.251 x 4.00 in. Displacement: 454 cid. Compression ratio: 8.5:1. Brake hp: 425 at 5600 rpm. Net brake hp: 325 at 6500 rpm. Taxable hp: 57.80. Torque: 475 lbs.-ft. at 4000 rpm. Net torque: 390 at 3600 rpm. Five main bearings. Hydraulic valve lifters. High-performance camshaft. Crankcase capacity: 5 qt. (add 1 qt. for filter). Cooling system capacity: 21 qt. (add 1 qt. for heater). Carburetor: Holley 880CFM four-barrel. Sales code: LS6. Engine code: CPW or CPX.

1972 CHEVROLET

Nova, Camaro, Chevelle base V-8
Overhead valve. Cast-iron block and head. Bore & stroke: 3.88 x 3.25 in. Displacement: 307 cid. Compression ratio: 8.50:1. Net brake hp: 130 at 4000 rpm. Taxable hp: 48.0. Net torque. 230 lbs.-ft. at 2400 rpm. Five main bearings. Hydraulic valve lifters. Crankcase capacity: 4 qt. (add 1 qt. for new filter). Cooling system capacity: 14 qt.; (add 1 qt. for heater). Carburetor: Two-barrel. Engine code: (Nova) CKG, CKH, CTK, CAY, CAZ and CMA; (Camaro) CKG, CKH, CTK, CAY, CAZ, and CMA; (Chevelle) CKG, CKH, CAY, and CTK. VIN code: F.

Monte Carlo, Chevrolet base V-8; Nova, Camaro, Chevelle optional V-8
Overhead valve. Cast-iron block and head. Bore & stroke: 4.00 x 3.48 in.

Displacement: 350 cid. Compression ratio: 8.50:1. Net brake hp: 165 at 4000 rpm. Taxable hp: 51.20. Net torque: 280 lbs.-ft. at 2400 rpm. Five main bearings. Hydraulic valve lifters. Crankcase capacity: 4 qt. (add 1 qt. for new filter). Cooling system capacity: 15 qt. (add 1 qt. for heater). Carburetor: Two-barrel. Engine code: (Nova) CKA, CTL, CDA, and CMD; (Camaro) CKA, CTL, CRG, CRD, CMH, and CMB; (Chevelle) CKA, CTL, CSH, CDA, CMD, CAR, and CDB; (Monte Carlo) CKA, CTL, CDA, CMD, CKB, and CDB; (Chevrolet)CBL, CSJ, CKB, CAR, CSH, and CDB. VIN code: H.

Chevrolet optional V-8
Overhead valve. Cast-iron block and head. Bore & stroke: 4.13 x 3.75 in. Displacement: 400 cid. Compression ratio: 8.50:1. Net brake hp: 170 at 3400 rpm. Taxable hp: 54.40. Net torque. 325 lbs.-ft. at 2000 rpm. Five main bearings. Hydraulic valve lifters. Crankcase capacity: 4 qt. (add 1 qt. for new filter). Cooling system capacity: 15 qt. (add 1 qt. for heater). Carburetor: Two-barrel. RPO: LF6. Engine code: CKP, CAT, CDL, and CDM. VIN code: R.

Chevelle, Monte Carlo optional V-8
Overhead valve. Cast-iron block and head. Bore & stroke: 4.00 x 3.48 in. Displacement: 350 cid. Compression ratio: 8.50:1. Net brake hp: 175 at 4000 rpm. Taxable hp: 51.20. Net torque: 280 lbs.-ft. at 2400 rpm. Five main bearings. Hydraulic valve lifters. Crankcase capacity: 4 qt. (add 1 qt. for new filter). Cooling system capacity: 15 qt. (add 1 qt. for heater). Carburetor: Four-barrel. Engine code: (Chevelle) CKK, CKD, CDG, CDD, and CKB; (Monte Carlo) CKD or CDD. VIN code: J.

Nova, Camaro optional V-8
Overhead valve. Cast-iron block and head. Bore & stroke: 4.00 x 3.48 in. Displacement: 350 cid. Compression ratio: 8.50:1. Net brake hp: 200 at 4400 rpm. Taxable hp: 51.20. Net torque: 300 lbs.-ft. at 2800 rpm. Five main bearings. Hydraulic valve lifters. Crankcase capacity: 4 qt. (add 1 qt. for new filter). Cooling system capacity: 15 qt. (add 1 qt. for heater). Carburetor: Four-barrel. Engine code: (Nova) CKK, CKD, CRL, CRK, CDG, and CDD; (Camaro) CKK, CKD, CDG, and CDD. VIN code: J.

Chevrolet optional V-8
Overhead valve. Cast-iron block and head. Bore & stroke: 4.13 x 3.76 in. Displacement: 402 cid (396 cid). Compression ratio: 8.50:1. Net brake hp: 210 at 4400 rpm. Taxable hp: 54.50. Net torque. 320 lbs.-ft. at 2400 rpm. Five main bearings. Hydraulic valve lifters. Crankcase capacity: 4 qt. (add 1 qt. for new filter). Cooling system capacity: 22 qt. (add 1 qt. for heater). Carburetor: Four-barrel. Engine code: (Chevrolet) CLB, CLR, CTB, and CTJ. VIN code: S.

Chevelle, Monte Carlo, Chevrolet optional V-8
Overhead valve. Cast-iron block and head. Bore & stroke: 4.25 x 4.00 in. Displacement: 454 cid. Compression ratio: 8.50:1. Net brake hp: 230 at 4000 rpm. Taxable hp: 57.80. Net torque. 360 lbs.-ft. at 3200 rpm. Five main bearings. Hydraulic valve lifters. Crankcase capacity: (Corvette) 5 qt.; (Others) 4 qt. (add 1 qt. for new filter). Cooling system capacity: (Chevelle, Monte Carlo) 22 qt.; (Chevrolet) 21 qt. (add 1 qt. for heater). Carburetor: Four-barrel. RPO: LS5. VIN code: V.

Camaro, Chevelle, Monte Carlo optional V-8
Overhead valve. Cast-iron block and head. Bore & stroke: 4.13 x 3.76 in. Displacement: 402 cid (400 cid). Compression ratio: 8.50:1. Net brake hp: 240 at 4400 rpm. Taxable hp: 54.50. Net torque. 345 lbs.-ft. at 3200 rpm. Five main bearings. Hydraulic valve lifters. Crankcase capacity: 4 qt. (add 1 qt. for new filter). Cooling system capacity: (Camaro) 23 qt.; (Chevele, Monte Carlo, Chevrolet) 22 qt. (add 1 qt. for heater). Carburetor: Four-barrel. Engine code: (Camaro) CLA, CLB, CTA, and CTB; (Chevelle) CLA, CLB, CTA, CTB, CLS, CTJ, and CTH; (Monte Carlo) CLB, CTJ, and CTB. VIN code: U.

For 1972, the Turbo-Fire 400 V-8 came only in 170- and 210--hp versions. It was the base V-8 for the 1972 Chevrolet Caprice. The engine in Tim Steblow's Caprice coupe is shown here. It is the higher-output version.

Camaro optional V-8
Overhead valve. Cast-iron block and head. Bore & stroke: 4.00 x 3.48 in. Displacement: 350 cid. Compression ratio: 9.00:1. Net brake hp: 255 at 5600 rpm. Taxable hp: 51.20. Net torque: 280 lbs.-ft. at 4000 rpm. Five main bearings. Hydraulic valve lifters. Crankcase capacity: 4 qt. (add 1 qt. for new filter). Cooling system capacity: (Camaro) 15 qt.; (Corvette) 16 qt. (add 1 qt. for heater). Carburetor: Four-barrel. Engine code: CKT or CLA. VIN code: L.

Chevelle, Monte Carlo, Chevrolet optional V-8
Overhead valve. Cast-iron block and head. Bore & stroke: 4.25 x 4.00 in. Displacement: 454 cid. Compression ratio: 8.50:1. Net brake hp: 270 at 4000 rpm. Taxable hp: 57.80. Net torque. 390 lbs.-ft. at 3200 rpm. Five main bearings. Hydraulic valve lifters. Crankcase capacity: (Corvette) 5 qt.; (Others) 4 qt. (add 1 qt. for new filter). Cooling system capacity: (Chevelle, Monte Carlo) 22 qt.; (Chevrolet) 21 qt. (add 1 qt. for heater). Dual exhaust. Carburetor: Four-barrel. (Chevelle) CPA, CPD, CRX, and CRW; (Monte Carlo) CPD, CRX and CRW; (Chevrolet) CPD, CPG, CRW, and CRY. RPO: LS5. VIN code: W.

1972 CORVETTE

Corvette base V-8
Overhead valve. Cast-iron block. Bore & stroke: 4.00 x 3.48 in. Displacement: 350 cid. Compression ratio: 8.5:1. Net brake hp: 200 at 4400 rpm. Taxable hp: 51.20. Torque: 300 lbs.-ft. at 2800 rpm. Five main bearings. Hydraulic valve lifters. Crankcase capacity: 5 qt. (add 1 qt. for filter). Cooling system capacity: 16 qt. (add 1 qt. for heater). Carburetor: Rochester Type Quadra-Jet four-barrel Model 4MV. Sales code: ZQ3. Engine code: CRS, CRW, CDH, CKX, and CDJ. VIN code: K.

Corvette base V-8
Overhead valve. Cast-iron block. Bore & stroke: 4.00 x 3.48 in. Displacement: 350 cid. Compression ratio: 9.0:1. Net brake hp: 255 at 5600 rpm. Taxable hp: 51.20. Torque: 280 lbs.-ft. at 4000 rpm. Five main bearings. Forged steel crankshaft. Solid valve lifters. High-performance camshaft. Crankcase capacity: 5 qt. (add 1 qt. for filter). Cooling system capacity: 16 qt. (add 1 qt. for heater). Carburetor: Holley four-barrel on aluminum intake manifold. 2.50-inch diameter dual exhaust system. Sales code LT1. Engine code: CKY, CRT, and CKZ.
NOTE: The LT1 V-8 was available with the 255-hp ZR1 off-road package intended for competition.

Corvette optional V-8
Overhead valve. Cast-iron block. Bore & stroke: 4.251 x 4.00 in. Displacement: 454 cid. Compression ratio: 8.5:1. Net brake hp: 270 at 4000 rpm. Taxable hp: 51.20. Torque: 390 lbs.-ft. at 3200 rpm. Five main bearings. Hydraulic valve lifters. High-performance camshaft. Crankcase capacity: 5 qt. (add 1 qt. for filter). Cooling system capacity: 22 qt. (add 1 qt. for heater). Carburetor: Rochester 750CFM Quadra-Jet four-barrel. Sales code: LS6. Engine code: CPH, CPJ, CSR, and CSS. VIN code: W.

1973 CHEVROLET

Nova, Chevelle, Camaro base V-8
Overhead valve. Cast-iron block and head. Bore & stroke: 3.88 x 3.25 in. Displacement: 307 cid. Compression ratio: 8.50:1. Net brake hp: 115 at 4000 rpm. Taxable hp: 48.0. Net torque. 205 lbs.-ft. at 2000 rpm. Five main bearings. Hydraulic valve lifters. Crankcase capacity: 4 qt. (add 1 qt. for new filter). Cooling system capacity: 15 qt. (add 1 qt. for heater). Carburetor: Two-barrel. Engine code: (Nova) CHB, CHH, CHC, and CHD; (Camaro) CHB, CHH, CHJ, and CHK; (Chevelle) CHD, CHA, CHB, CHC, and CMA. Vin code: F.

Monte Carlo, Chevrolet base V-8; Nova, Chevelle, Camaro optional V-8
Overhead valve. Cast-iron block and head. Bore & stroke: 4.00 x 3.48 in. Displacement: 350 cid. Compression ratio: 8.50:1. Net brake hp: 145 at 4000 rpm. Taxable hp: 51.20. Net torque: 255 lbs.-ft. at 2400 rpm. Five main bearings. Hydraulic valve lifters. Crankcase capacity: 4 qt. (add 1 qt. for new filter). Cooling system capacity: (Nova) 17 qt.; (Chevrolet) 16 qt.; (others) 15 qt. (add 1 qt. for heater). Carburetor: Two-barrel. Engine code: (Nova) CKA, CKW, CKC, and CKK; (Camaro) CKA, CKW, CKX, and CKY; (Chevelle) CKA, CKL, CKB, CKC, and CKK; (Monte Carlo)

CKA, CKL, CKR, CKB, CKC, and CKK; (Chevrolet) CKL, CKS, CKK, CLT, CLU, CLW, and CLX. VIN code: H.

Chevrolet optional V-8
Overhead valve. Cast-iron block and head. Bore & stroke: 4.13 x 3.75 in. Displacement: 400 cid. Compression ratio: 8.50:1. Net brake hp: 150 at 3200 rpm. Taxable hp: 54.40. Net torque. 295 lbs.-ft. at 2000 rpm. Five main bearings. Hydraulic valve lifters. Crankcase capacity: 4 qt. (add 1 qt. for new filter). Cooling system capacity: 16.5 qt. (add 1 qt. for heater). Carburetor: Two-barrel. (Chevrolet) CSJ, CSA, CSB, CSC, CSD, CSK, CSL, and CSM. VIN code: R.

Caprice wagon base V-8; Nova, Chevelle, Camaro, Monte Carlo, Chevrolet optional V-8
Overhead valve. Cast-iron block and head. Bore & stroke: 4.00 x 3.48 in. Displacement: 350 cid. Compression ratio: 8.50:1. Net brake hp: 175 at 4000 rpm. Taxable hp: 51.20. Net torque: 270 lbs.-ft. at 2400 rpm. Five main bearings. Hydraulic valve lifters. Crankcase capacity: 4 qt. (add 1 qt. for new filter). Cooling system capacity: (Nova) 17 qt.; (Chevrolet) 16 qt.; (others) 15 qt. (add 1 qt. for heater). Carburetor: Four-barrel. Engine code: (Nova) CKU, CKB, CKD, CKH, and CKV; (Camaro) CKU, CKB, CKD, and CKH; (Chevelle) CKM, CKR, CKH, CKD, and CKJ; (Monte Carlo) CKH, CKD, and CKJ; (Chevrolet) CKR, CKD, CKJ, and CKM. VIN code: K.

Chevrolet optional V-8
Overhead valve. Cast-iron block and head. Bore & stroke: 4.25 x 4.00 in. Displacement: 454 cid. Compression ratio: 8.50:1. Net brake hp: 215 at 4000 rpm. Taxable hp: 57.80. Net torque. 345 lbs.-ft. at 2400 rpm. Five main bearings. Hydraulic valve lifters. Crankcase capacity: 4 qt. (add 1 qt. for new filter). Cooling system capacity: 23 qt. (add 1 qt. for heater). Single exhaust. Carburetor: Four-barrel. RPO: L54. VIN code: X.

Camaro optional V-8
Overhead valve. Cast-iron block and head. Bore & stroke: 4.00 x 3.48 in. Displacement: 350 cid. Compression ratio: 9.00:1. Net brake hp: 245 at 5200 rpm. Taxable hp: 51.20. Net torque: 280 lbs.-ft. at 4000 rpm. Five main bearings. Hydraulic valve lifters. Crankcase capacity: 4 qt. (add 1 qt. for new filter). Cooling system capacity: 15 qt. (add 1 qt. for heater). Carburetor: Four-barrel. RPO: L82. Engine code: (Camaro) CLJ, CLK, CLL, and CLM. VIN code: T.

Chevrolet optional V-8
Overhead valve. Cast-iron block and head. Bore & stroke: 4.25 x 4.00 in. Displacement: 454 cid. Compression ratio: 8.50:1. Net brake hp: 245 at 4000 rpm. Taxable hp: 57.80. Net torque. 375 lbs.-ft. at 2800 rpm. Five main bearings. Hydraulic valve lifters. Crankcase capacity: 4 qt. (add 1 qt. for new filter). Cooling system capacity: (Chevelle, Monte Carlo) 22 qt.; (Chevrolet) 23 qt. (add 1 qt. for heater). Dual exhaust. Carburetor: Four-barrel. Engine code: (Chevelle) CWC, CWD, CWR, CWA, and CWB; (Monte Carlo) CWC, CWD, CWA, and CWB; (Chevrolet) CWD, CWJ, CWL, and CWK. RPO: L54. VIN code: Y.

1973 CORVETTE

Corvette base V-8
Overhead valve. Cast-iron block. Bore & stroke: 4.00 x 3.48 in. Displacement: 350 cid. Compression ratio: 8.5:1. Net brake hp: 190 at 4400 rpm. Taxable hp: 51.20. Torque: 270 lbs.-ft. at 2800 rpm. Five main bearings. Hydraulic valve lifters. Crankcase capacity: 5 qt. (add 1 qt. for filter). Cooling system capacity: 17 qt. (add 1 qt. for heater). Carburetor: Rochester Type Quadra-Jet four-barrel Model 4MV. Sales code: ZQ3. Engine code: CKZ, CLB, CLA, and CLC.

Corvette optional V-8
Overhead valve V-8. Cast-iron block and head. Bore & stroke: 4.00 x 3.48 in. Displacement: 350 cid. Compression ratio: 9.00:1. Net brake hp: 250 at 5200 rpm. Taxable hp: 51.20. Net torque: 285 lbs.-ft. at 4000 rpm. Five main bearings. Hydraulic valve lifters. Crankcase capacity: 4 qt. (add 1 qt. for new filter). Cooling system capacity: 17 qt. (add 1 qt. for heater). Carburetor: Four-barrel. Sales code: L82. Engine code: CLR, CLS, CLD, and CLH.

Corvette optional V-8
Overhead valve V-8. Cast-iron block and head. Bore & stroke: 4.25 x 4.00 in. Displacement: 454 cid. Compression ratio: 8.50:1. Net brake hp: 275 at 4000 rpm. Taxable hp: 57.80. Net torque. 395 lbs.-ft. at 2800 rpm. Five main bearings. Hydraulic valve lifters. Crankcase capacity: 5 qt. (add 1 qt. for new filter). Cooling system capacity: 23 qt. (add 1 qt. for heater). Dual exhaust. Carburetor: Four-barrel. Sales code: LS4. Engine code: CWS, CWT, CWM, and CWR.

1974 CHEVROLET

Nova, Chevelle, Camaro, Monte Carlo, Chevrolet base V-8
Overhead valve. Cast-iron block and head. Bore & stroke: 4.00 x 3.48 in. Displacement: 350 cid. Compression ratio: 8.50:1. Net brake hp: 145 at 3600 rpm. Taxable hp: 51.20. Net torque: 250 lbs.-ft. at 2200 rpm. Five main bearings. Hydraulic valve lifters. Crankcase capacity: 4 qt. (add 1 qt. for new filter). Cooling system capacity with heater: 18 qt. Carburetor: Two-barrel. Engine code: (Nova) CMC and CMA; (Chevelle) CTC, CSU, and CSX; (Camaro) CMA and CMC; (Monte Carlo) CMC, CMA, and CMR; (Chevrolet) CMA and CMD. VIN code: F.

Chevelle, Monte Carlo, Chevrolet optional V-8
Overhead valve. Cast-iron block and head. Bore & stroke: 4.13 x 3.75 in. Displacement: 400 cid. Compression ratio: 8.50:1. Net brake hp: 150 at 3200 rpm. Taxable hp: 54.40. Net torque. 295 lbs.-ft. at 2000 rpm. Five main bearings. Hydraulic valve lifters. Crankcase capacity: 4 qt. (add 1 qt. for new filter). Cooling system capacity with heater: (Chevrolet) 16 qt.; (others) 18 qt. Carburetor: Two-barrel. VIN code: R. Engine code: (Chevelle) CTA; (Monte Carlo) CTA; (Chevrolet) CTA, CTB, CSU, and CSW.

Nova, Chevelle, Camaro, Monte Carlo, Chevrolet optional V-8
Overhead valve. Cast-iron block and head. Bore & stroke: 4.00 x 3.48 in. Displacement: 350 cid. Compression ratio: 8.50:1. Net brake hp: 160 at 3800 rpm. Taxable hp: 51.20. Net torque: 250 lbs.-ft. at 2400 rpm. Five main bearings. Hydraulic valve lifters. Crankcase capacity: 4 qt. (add 1 qt. for new filter). Cooling system capacity with heater: (Chevrolet) 16 qt.; (others) 18 qt. Carburetor: Four-barrel. VIN code: K. Engine code: (Nova) CKD; (Chevelle) CKH and CKD; (Camaro) CKH and CKD; (Monte Carlo) CKH, CKD; (Chevrolet) CMK, CMH, CMJ, and CKD.

Caprice Classic and station wagon base V-8; Chevelle, Monte Carlo optional V-8
Overhead valve. Cast-iron block and head. Bore & stroke: 4.13 x 3.75 in. Displacement: 400 cid. Compression ratio: 8.50:1. Net brake hp: 180 at 3800 rpm. Taxable hp: 54.40. Net torque. 290 lbs.-ft. at 2400 rpm. Five main bearings. Hydraulic valve lifters. Crankcase capacity: 4 qt. (add 1 qt. for new filter). Cooling system capacity with heater: 16.5 qt. Carburetor: Four-barrel. VIN code: R. Engine code: (Chevelle) CTC, CSU, and CSX; (Monte Carlo) CTC, CSU, and CSX; (Chevrolet) CTC, CTD, CTK, and CTJ.

Camaro optional V-8
Overhead valve. Cast-iron block and head. Bore & stroke: 4.00 x 3.48 in. Displacement: 350 cid. Compression ratio: 8.50:1. Net brake hp: 185 at 4000 rpm. Taxable hp: 51.20. Net torque: 270 lbs.-ft. at 2600 rpm. Five main bearings. Hydraulic valve lifters. Crankcase capacity: 4 qt. (add 1 qt. for new filter). Cooling system capacity with heater: 18 qt. Carburetor: Four-barrel. VIN code: T. Engine code: (Nova) CKB, CKH, and CKU; (Camaro) CKU, CKH, CKD, and CLK.

Chevrolet optional V-8
Overhead valve. Cast-iron block and head. Bore & stroke: 4.25 x 4.00 in. Displacement: 454 cid. Compression ratio: 8.50:1. Net brake hp: 235 at 4000 rpm. Taxable hp: 57.80. Net torque. 360 lbs.-ft. at 2800 rpm. Five main bearings. Hydraulic valve lifters. Crankcase capacity: 5 qt. (add 1 qt. for new filter). Cooling system capacity with heater: 23 qt. Dual exhaust. Carburetor: Four-barrel. VIN code: X. Engine code: (Chevelle) CWA, CWX, CWD, CXM, CXR, and CXS; (Monte Carlo) CWA, CWX, and CWD; (Chevrolet) CWU, CWW, CWY, CXA, and CXB.

Camaro Z28 base V-8; Camaro optional V-8
Overhead valve V-8. Cast-iron block and head. Bore & stroke: 4.00 x 3.48 in. Displacement: 350 cid (5.7 liters). Compression ratio: 9.0:1. Brake horsepower: 245 at 5200 rpm. Torque: 285 lbs.-ft. at 4000 rpm. Five main bearings. Hydraulic valve lifters. Induction system: Rochester four-barrel carburetor. Standard in Camaro Z/28; not available in other Camaros. VIN Code: T. Engine code: CMT or CMS.

1974 CORVETTE

Corvette base V-8
Overhead valve. Cast-iron block. Bore & stroke: 4.00 x 3.48 in. Displacement: 350 cid. Compression ratio: 8.5:1. Net brake hp: 195 at 4400 rpm. Taxable hp: 51.20. Torque: 275 lbs.-ft. at 2800 rpm. Five main bearings. Hydraulic valve lifters. Crankcase capacity: 5 qt. (add 1 qt. for

filter). Cooling system capacity: 17 qt. (add 1 qt. for heater). Carburetor: Rochester Type Quadra-Jet four-barrel Model 4MV. Sales code: ZQ3. Engine code: CKZ, CLB, CLA, and CLC.

Corvette optional V-8
Overhead valve V-8. Cast-iron block and head. Bore & stroke: 4.00 x 3.48 in. Displacement: 350 cid. Compression ratio: 9.00:1. Net brake hp: 250 at 5200 rpm. Taxable hp: 51.20. Net torque: 285 lbs.-ft. at 4000 rpm. Five main bearings. Hydraulic valve lifters. Crankcase capacity: 4 qt. (add 1 qt. for new filter). Cooling system capacity: 17 qt. (add 1 qt. for heater). Carburetor: Four-barrel. Sales code: L82. Engine code: CLR and CLD.

Corvette optional V-8
Overhead valve V-8. Cast-iron block and head. Bore & stroke: 4.25 x 4.00 in. Displacement: 454 cid. Compression ratio: 8.50:1. Net brake hp: 270 at 4400 rpm. Taxable hp: 57.80. Net torque. 380 lbs.-ft. at 2800 rpm. Five main bearings. Hydraulic valve lifters. Crankcase capacity: 4 qt. (add 1 qt. for new filter). Cooling system capacity: 24 qt. (add 1 qt. for heater). Dual exhaust. Carburetor: Four-barrel. Sales code: LS4. Engine code: CWM, CWR, and CWS.

1975 CHEVROLET

Monza base V-8; Nova optional V-8
Overhead valve. Cast-iron block and head. Bore & stroke: 3.67 x 3.10 in. Displacement: 262 cid. Compression ratio: 8.50:1. Net brake hp: 110 at 3600 rpm. Taxable hp: 43.1. Net torque: 200 lbs.-ft. at 2200 rpm. Five main bearings. Hydraulic valve lifters. Crankcase capacity: 4 qt. (add 1 qt. for new filter). Cooling system capacity with heater: 18 qt. Carburetor: Two-barrel. VIN code: G.

Monza optional V-8
Overhead valve. Cast-iron block and head. Bore & stroke: 4.00 x 3.48 in. Displacement: 350 cid. Compression ratio: 8.50:1. Net brake hp: 125 at 3600 rpm. Taxable hp: 51.20. Net torque: 235 lbs.-ft. at 2000 rpm. Five main bearings. Hydraulic valve lifters. Crankcase capacity: 4 qt. (add 1 qt. for new filter). Cooling system capacity with heater: 18 qt. Carburetor: Two-barrel. VIN code: H.

Nova, Chevelle, Camaro, Monte Carlo, Chevrolet (car) base V-8
Overhead valve. Cast-iron block and head. Bore & stroke: 4.00 x 3.48 in. Displacement: 350 cid. Compression ratio: 8.50:1. Net brake hp: 145 at 3800 rpm. Taxable hp: 51.20. Net torque: 250 lbs.-ft. at 2200 rpm. Five main bearings. Hydraulic valve lifters. Crankcase capacity: 4 qt. (add 1 qt. for new filter). Cooling system capacity with heater: (Nova, Camaro) 15.5 qt.; (Chevelle, Monte Carlo, Chevrolet) 16 qt. Carburetor: Two-barrel. Sales code: L65. VIN code: H.

Nova, Chevelle, Camaro, Monte Carlo, Chevrolet optional V-8
Overhead valve. Cast-iron block and head. Bore & stroke: 4.00 x 3.48 in. Displacement: 350 cid. Compression ratio: 8.50:1. Net brake hp: 155 at 3800 rpm. Taxable hp: 51.20. Net torque: 250 lbs.-ft. at 2400 rpm. Five main bearings. Hydraulic valve lifters. Crankcase capacity: 4 qt. (add 1 qt. for new filter). Cooling system capacity with heater: (Nova, Camaro) 15.5 qt.; (Chevelle, Monte Carlo, Chevrolet) 16 qt. Carburetor: Four-barrel. Sales code: LM1. VIN code: K.

Chevrolet station wagon base V-8; Chevelle, Monte Carlo, Chevrolet (car) optional V-8
Overhead valve. Cast-iron block and head. Bore & stroke: 4.13 x 3.75 in. Displacement: 400 cid. Compression ratio: 8.50:1. Net brake hp: 175 at 3600 rpm. Taxable hp: 54.40. Net torque. 305 lbs.-ft. at 2000 rpm. Five main bearings. Hydraulic valve lifters. Crankcase capacity: 4 qt. (add 1 qt. for new filter). Cooling system capacity with heater: (Chevelle, Monte Carlo) 16 qt.; Chevrolet: 16.5 qt. Carburetor: Four-barrel. VIN code: R.

Chevelle, Monte Carlo, Chevrolet optional V-8
Overhead valve. Cast-iron block and head. Bore & stroke: 4.25 x 4.00 in. Displacement: 454 cid. Compression ratio: 8.15:1. Net brake hp: 215 at 4000 rpm. Taxable hp: 57.80. Net torque. 350 lbs.-ft. at 2400 rpm. Five main bearings. Hydraulic valve lifters. Crankcase capacity: 5 qt. (add 1 qt. for new filter). Cooling system capacity with heater: (Chevelle, Monte Carlo) 24 qt.; (Chevrolet) 23 qt. Dual exhaust. Carburetor: Four-barrel. VIN code: X.

1975 CORVETTE

Corvette base V-8
Overhead valve. Cast-iron block. Bore & stroke: 4.00 x 3.48 in. Displacement: 350 cid. Compression ratio: 8.5:1. Net brake hp: 165 at 3800 rpm. Taxable hp: 51.20. Net torque: 255 lbs.-ft. at 2400 rpm. Five main bearings. Hydraulic valve lifters. Crankcase capacity: 4 qt. (add 1 qt. for new filter). Cooling system capacity with heater: 18 qt. Carburetor: Rochester Type Quadra-Jet four-barrel Model 4MV. Sales code: L48. VIN code: T.

Corvette optional V-8
Overhead valve. Cast-iron block. Bore & stroke: 4.00 x 3.48 in. Displacement: 350 cid. Compression ratio: 9.0:1. Net brake hp: 205 at 4800 rpm. Taxable hp: 51.20. Net torque: 255 lbs.-ft. at 3600 rpm. Five main bearings. Forged steel crankshaft. Hydraulic valve lifters. High-performance camshaft. Crankcase capacity: 4 qt. (add 1 qt. for new filter). Cooling system capacity with heater: 18 qt. Carburetor: Four-barrel. Sales code: L82. VIN code: Y.

1976 CHEVROLET

Monza base V-8
Overhead valve. Cast-iron block and head. Bore & stroke: 3.67 x 3.10 in. Displacement: 262 cid. Compression ratio: 8.50:1. Net brake hp: 110 at 3600 rpm. Taxable hp: 43.1. Net torque: 195 lbs.-ft. at 2000 rpm. Five main bearings. Hydraulic valve lifters. Crankcase capacity: 4 qt. (add 1 qt. for new filter). Cooling system capacity with heater: 18 qt. Carburetor: Two-barrel. VIN code: G.

Nova, Chevelle (cars), Camaro, Monte Carlo base V-8; Monza optional V-8
Overhead valve. Cast-iron block and head. Bore & stroke: 4.00 x 3.48 in. Displacement: 305 cid. Compression ratio: 8.50:1. Net brake hp: 140 at 3800 rpm. Taxable hp: 44.66. Net torque: 245 lbs.-ft. at 2000 rpm. Five main bearings. Hydraulic valve lifters. Crankcase capacity: 4 qt. (add 1 qt. for new filter). Cooling system capacity with heater: (Nova, Camaro) 15.5 qt.; (Chevelle, Monte Carlo, Chevrolet) 16 qt. Carburetor: Two-barrel. Sales code: LG3. VIN code: Q.

Chevelle station wagon, Chevrolet (car) base V-8; Monte Carlo optional V-8
Overhead valve. Cast-iron block and head. Bore & stroke: 4.00 x 3.48 in. Displacement: 350 cid. Compression ratio: 8.50:1. Net brake hp: 145 at 3800 rpm. Taxable hp: 51.20. Net torque: 250 lbs.-ft. at 2200 rpm. Five main bearings. Hydraulic valve lifters. Crankcase capacity: 4 qt. (add 1 qt. for new filter). Cooling system capacity with heater: (Nova, Camaro) 15.5 qt.; (Chevelle, Monte Carlo, Chevrolet) 16 qt. Carburetor: Two-barrel. VIN code: H.

Nova, Chevelle (all), Camaro, Monte Carlo, Chevrolet optional V-8
Overhead valve. Cast-iron block and head. Bore & stroke: 4.00 x 3.48 in. Displacement: 350 cid. Compression ratio: 8.50:1. Net brake hp: 165 at 3800 rpm. Taxable hp: 51.20. Net torque: 260 lbs.-ft. at 2400 rpm. Five main bearings. Hydraulic valve lifters. Crankcase capacity: 4 qt. (add 1 qt. for new filter). Cooling system capacity with heater: (Nova, Camaro) 15.5 qt.; (Chevelle, Monte Carlo, Chevrolet) 16 qt. Carburetor: Four-barrel. Sales code: LM1. VIN code: L.

Chevrolet station wagon base V-8; Chevelle, Monte Carlo optional V-8
Overhead valve. Cast-iron block and head. Bore & stroke: 4.13 x 3.75 in. Displacement: 400 cid. Compression ratio: 8.50:1. Net brake hp: 175 at 3600 rpm. Taxable hp: 54.40. Net torque. 305 lbs.-ft. at 2000 rpm. Five main bearings. Hydraulic valve lifters. Crankcase capacity: 4 qt. (add 1 qt. for new filter). Cooling system capacity with heater: (Chevelle, Monte Carlo) 16 qt.; Chevrolet: 16.5 qt. Carburetor: Four-barrel. VIN code: U.

Chevrolet optional V-8
Overhead valve. Cast-iron block and head. Bore & stroke: 4.25 x 4.00 in. Displacement: 454 cid. Compression ratio: 8.50:1. Net brake hp: 235 at 4000 rpm. Taxable hp: 57.80. Net torque. 360 lbs.-ft. at 2800 rpm. Five main bearings. Hydraulic valve lifters. Crankcase capacity: 5 qt. (add 1 qt. for new filter). Cooling system capacity with heater: (Chevelle, Monte Carlo) 24 qt.; (Chevrolet) 23 qt. Dual exhaust. Carburetor: Four-barrel. VIN code: S.

1976 CORVETTE

Corvette base V-8
Overhead valve. Cast-iron block and head. Bore & stroke: 4.00 x 3.48 in. Displacement: 350 cid (5.7 liters). Compression ratio: 8.5:1. Net brake hp: 180 at 4000 rpm. Taxable hp: 51.20. Net torque: 270 lbs.-ft. at 2400 rpm. Five main bearings. Hydraulic valve lifters. Crankcase capacity: 4 qt. (add 1 qt. for new filter). Cooling system capacity with heater: 21 qt. Carburetor: Rochester M4MC. Sales code: L48. VIN code: L.

Corvette optional V-8
Overhead valve. Cast-iron block and head. Bore & stroke: 4.00 x 3.48 in. Displacement: 350 cid (5.7 liters). Compression ratio: 9.0:1. Net brake hp: 210 at 5200 rpm. Taxable hp: 51.20. Net torque: 255 lbs.-ft. at 3600 rpm. Five main bearings. Hydraulic valve lifters. Crankcase capacity: 4 qt. (add 1 qt. for new filter). Cooling system capacity with heater: 21 qt. Carburetor: Rochester M4MC. Sales code: L82. VIN code: X

1977 CHEVROLET

Monte Carlo, Monza, Nova, Chevelle, Camaro, Impala, Caprice base V-8
Overhead valve. Cast-iron block and head. Bore & stroke: 3.74 x 3.48 in. Displacement: 305 cid (5.0 liters). Compression ratio: 8.5:1. Net brake hp: 145 at 3800 rpm. Taxable hp: 44.66. Net torque: 245 lbs.-ft. at 2400 rpm. Five main bearings. Hydraulic valve lifters. Carburetor: Rochester 2GC two-barrel. VIN Code: U.
NOTE: There was no extra cost for the 5.0-liter V-8 in the Malibu, Impala or Caprice station wagon.

Malibu Classic wagon base V-8; Nova, Camaro, Chevelle, Monte Carlo, Impala, Caprice optional V-8
Overhead valve. Cast-iron block and head. Bore & stroke: 4.00 x 3.48 in. Displacement: 350 cid (5.7 liters). Compression ratio: 8.5:1. Net brake hp: 170 at 3800 rpm. Taxable hp: 51.20. Net torque: 270 lbs.-ft. at 2400 rpm. Five main bearings. Hydraulic valve lifters. Carburetor: Rochester M4MC four-barrel. VIN Code: L.

1977 CORVETTE

Corvette base V-8
Overhead valve. Cast-iron block and head. Bore & stroke: 4.00 x 3.48 in. Displacement: 350 cid (5.7 liters). Compression ratio: 8.5:1. Net brake hp: 180 at 4000 rpm. Taxable hp: 51.20. Net torque: 270 lbs.-ft. at 2400 rpm. Five main bearings. Hydraulic valve lifters. Carburetor: Rochester M4MC. Sales code: L48.

Corvette optional V-8
Overhead valve. Cast-iron block and head. Bore & stroke: 4.00 x 3.48 in. Displacement: 350 cid (5.7 liters). Compression ratio: 9.0:1. Net brake hp: 210 at 5200 rpm. Taxable hp: 51.20. Net torque: 255 lbs.-ft. at 3600 rpm. Five main bearings. Hydraulic valve lifters. Carburetor: Rochester M4MC. Sales code: L82.

1978 CHEVROLET

Monza, Nova, Camaro, Malibu, Monte Carlo, Impala, Caprice base V-8
Overhead valve. Cast-iron block and head. Bore & stroke: 3.74 x 3.48 in. Displacement: 305 cid (5.0 liters). Compression ratio: 8.4:1. Net brake hp: 145 at 3800 rpm. Taxable hp: 44.66. Net torque: 245 lbs.-ft. at 2400 rpm. Five main bearings. Hydraulic valve lifters. Carburetor: Rochester 2GC two-barrel. VIN Code: U.
NOTE: There was no extra cost for the 5.0-liter V-8 in the Impala or Caprice station wagons.

Camaro Z28 base V-8; Nova, Camaro, Malibu wagon, Impala, Caprice optional V-8
Overhead valve. Cast-iron block and head. Bore & stroke: 4.00 x 3.48 in. Displacement: 350 cid (5.7 liters). Compression ratio: 8.2:1. Net brake hp: 170 at 3800 rpm. Taxable hp: 51.20. Net torque: 270 lbs.-ft. at 2400 rpm. Five main bearings. Hydraulic valve lifters. Carburetor: Rochester M4MC four-barrel. VIN Code: L.

Camaro Z28 optional V-8
Overhead valve. Cast-iron block and head. Bore & stroke: 4.00 x 3.48 in. Displacement: 350 cid (5.7 liters). Compression ratio: 8.2:1. Net brake hp: 185 at 4000 rpm. Taxable hp: 51.20. Net torque: 280 lbs.-ft. at 2400 rpm. Five main bearings. Hydraulic valve lifters. Carburetor: Rochester M4MC four-barrel. VIN Code: L.

1978 CORVETTE

Corvette base V-8 (Federal)
Overhead valve. Cast-iron block and head. Bore & stroke: 4.00 x 3.48 in. Displacement: 350 cid (5.7 liters). Compression ratio: 8.2:1. Net brake hp: 185 at 4000 rpm. Taxable hp: 51.20. Net torque: 280 lbs.-ft. at 2400 rpm. Five main bearings. Hydraulic valve lifters. Carburetor: Rochester M4MC. Sales code: L48.

Corvette base V-8 (California)
V-8. Overhead valve. Cast-iron block and head. Bore & stroke: 4.00 x 3.48 in. Displacement: 350 cid (5.7 liters). Compression ratio: 8.2:1. Net brake hp: 175 at 4000 rpm. Taxable hp: 51.20. Net torque: 270 lbs.-ft. at 2400 rpm. Five main bearings. Hydraulic valve lifters. Carburetor: Rochester M4MC. California emissions and high-altitude emissions. Sales code: L48.

Corvette optional V-8 (Federal)
Overhead valve. Cast-iron block and head. Bore & stroke: 4.00 x 3.48 in. Displacement: 350 cid (5.7 liters). Compression ratio: 8.9:1. Net brake hp: 220 at 5200 rpm. Taxable hp: 51.20. Net torque: 260 lbs.-ft. at 3600 rpm. Five main bearings. Hydraulic valve lifters. Carburetor: Rochester M4MC. Sales code: L82.

1979 CHEVROLET

Malibu, Monte Carlo base V-8
Overhead valve. Cast-iron block and head. Bore & stroke: 3.50 x 3.48 in. Displacement: 267 cid (4.4 liters). Compression ratio: 8.2:1. Net brake hp: 125 at 3800 rpm. Taxable hp: 39.20. Net torque: 215 lbs.-ft. at 2400 rpm. Five main bearings. Hydraulic valve lifters. Carburetor: Rochester M2MC two-barrel. VIN code: J.

Monza, Nova, Camaro, Impala, Caprice base V-8
Overhead valve. Cast-iron block and head. Bore & stroke: 3.74 x 3.48 in. Displacement: 305 cid (5.0 liters). Compression ratio: 8.4:1. Net brake hp: 130 at 3200 rpm. Taxable hp: 44.66. Net torque: 245 lbs.-ft. at 2000 rpm. Five main bearings. Hydraulic valve lifters. Carburetor: Rochester M2MC two-barrel. VIN code: G.
NOTE: There was no extra cost for the 5.0-liter V-8 in the Impala or Caprice station wagons.

Malibu, Monte Carlo optional V-8
Overhead valve. Cast-iron block and head. Bore & stroke: 3.74 x 3.48 in. Displacement: 305 cid (5.0 liters). Compression ratio: 8.4:1. Net brake hp: 160 at 4000 rpm. Taxable hp: 44.66. Net torque: 235 lbs.-ft. at 2400 rpm. Five main bearings. Hydraulic valve lifters. Carburetor: Rochester M4MC four-barrel. VIN code: H.

Nova, Malibu wagon, Impala, Caprice optional V-8
Overhead valve. Cast-iron block and head. Bore & stroke: 4.00 x 3.48 in. Displacement: 350 cid (5.7 liters). Compression ratio: 8.2:1. Net brake hp: 165-170 at 3800 rpm. Taxable hp: 51.20. Net torque: 260-270 lbs.-ft. at 2400 rpm. Five main bearings. Hydraulic valve lifters. Carburetor: Rochester M4MC four-barrel. VIN code: L.

Camaro Z28 base V-8; Base Camaro optional V-8
Overhead valve. Cast-iron block and head. Bore & stroke: 4.00 x 3.48 in. Displacement: 350 cid (5.7 liters). Compression ratio: 8.2:1. Net brake hp: 175 at 4000 rpm. Taxable hp: 51.20. Net torque: 270 lbs.-ft. at 2400 rpm. Five main bearings. Hydraulic valve lifters. Carburetor: Rochester M4MC four-barrel. VIN code: L.

1979 CORVETTE

Corvette base V-8
Overhead valve. Cast-iron block and head. Bore & stroke: 4.00 x 3.48 in. Displacement: 350 cid (5.7 liters). Compression ratio: 8.2:1. Net brake hp: 195 at 4000 rpm. Taxable hp: 51.20. Net torque: 285 lbs.-ft. at 3200 rpm. Five main bearings. Hydraulic valve lifters. Carburetor: Rochester M4MC. Sales code: L48.

Corvette optional V-8
Overhead valve. Cast-iron block and head. Bore & stroke: 4.00 x 3.48 in. Displacement: 350 cid (5.7 liters). Compression ratio: 8.9:1. Net brake hp: 225 at 5200 rpm. Taxable hp: 51.20. Net torque: 270 lbs.-ft. at 3600 rpm. Five main bearings. Hydraulic valve lifters. Carburetor: Rochester M4MC. Sales code: L82.

1980 CHEVROLET

Impala/Caprice wagon optional diesel V-8
Overhead valve. Cast-iron block and head. Bore & stroke: 4.06 x 3.39 in. Displacement: 350 cid (5.7 liters). Compression ratio: 22.5:1. Net brake hp: 105 at 3200 rpm. Taxable hp: 51.20. Net torque: 205 lbs.-ft. at 1600 rpm. Five main bearings. Hydraulic valve lifters. Fuel injection. VIN code: N.

Camaro, Malibu, Monte Carlo, Impala, Caprice base V-8
Overhead valve. Cast-iron block and head. Bore & stroke: 3.50 x 3.48 in. Displacement: 267 cid (4.4 liters). Compression ratio: 8.3:1. Net brake hp: 120 at 3600 rpm. Taxable hp: 39.20. Net torque: 215 lbs.-ft. at 2000 rpm. Five main bearings. Hydraulic valve lifters. Carburetor: Rochester M2ME two-barrel. VIN code: J.

Camaro, Malibu, Monte Carlo, Impala, Caprice optional V-8
Overhead valve. Cast-iron block and head. Bore & stroke: 3.74 x 3.48 in. Displacement: 305 cid (5.0 liters). Compression ratio: 8.6:1. Net brake hp: 155 at 4000 rpm. Taxable hp: 44.66. Net torque: 240 lbs.-ft. at 1600 rpm. Five main bearings. Hydraulic valve lifters. Carburetor: Rochester M4ME four-barrel. VIN code: H.

Camaro Z28 base V-8
Overhead valve. Cast-iron block and head. Bore & stroke: 4.00 x 3.48 in. Displacement: 350 cid (5.7 liters). Compression ratio: 8.2:1. Net brake hp: 190 at 4200 rpm. Taxable hp: 51.20. Net torque: 280 lbs.-ft. at 2400 rpm. Five main bearings. Hydraulic valve lifters. Carburetor: Rochester M4ME four-barrel. VIN code: L.

1980 CORVETTE

Corvette base V-8 (California)
Overhead valve. Cast-iron block and head. Bore & stroke: 3.74 x 3.48 in. Displacement: 305 cid (5.0 liters). Compression ratio: 8.5:1. Net brake hp: 180 at 4200 rpm. Taxable hp: 44.64. Net torque: 255 lbs.-ft. at 2000 rpm. Five main bearings. Hydraulic valve lifters. Carburetor: Rochester Quadrajet. Sales code: LG4.

Corvette base V-8 (Federal)
Overhead valve. Cast-iron block and head. Bore & stroke: 4.00 x 3.48 in.

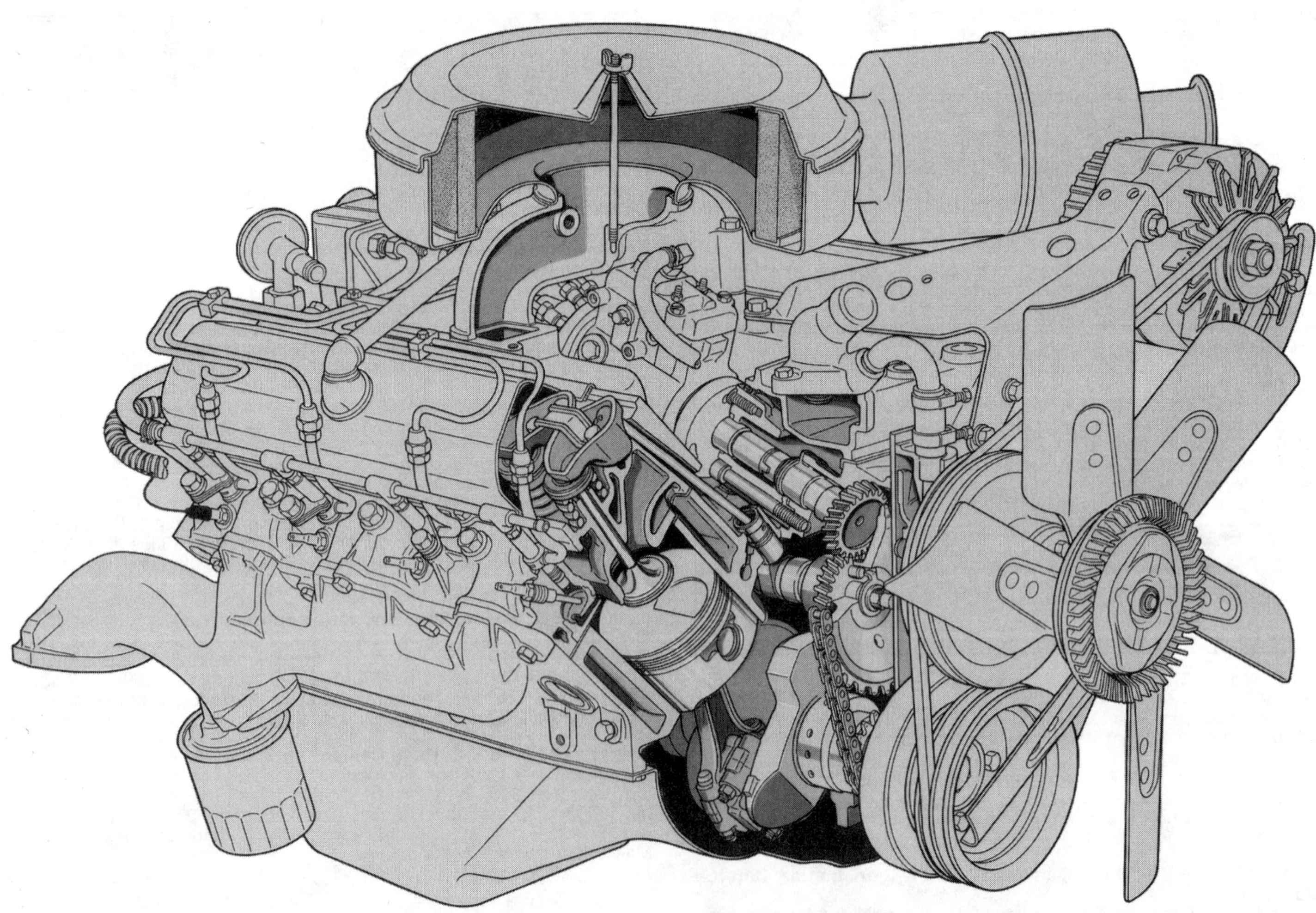

Chevrolet Motor Division's public relations department released this cutaway drawing of the 350-cid diesel V-8 on September 15, 1977. Chevrolet manufactured this new version of the 350, although it was officially promoted as the "GM-built 5.7-liter (350 cid.) diesel V-8 engine. It was used in Oldsmobiles starting in 1978 and added to some Cadillacs in 1979. Full-size Chevrolets didn't get a diesel until 1980.

Displacement: 350 cid (5.7 liters). Compression ratio: 8.2:1. Net brake hp: 190 at 4200 rpm. Taxable hp: 51.20. Net torque: 280 lbs.-ft. at 2400 rpm. Five main bearings. Hydraulic valve lifters. Carburetor: Rochester Quadrajet. Sales code: L48.

Corvette optional V-8 (Federal; automatic transmission mandatory)
Overhead valve. Cast-iron block and head. Bore & stroke: 4.00 x 3.48 in. Displacement: 350 cid (5.7 liters). Compression ratio: 9.0:1. Net brake hp: 230 at 5200 rpm. Taxable hp: 51.20. Net torque: 275 lbs.-ft. at 3600 rpm. Five main bearings. Hydraulic valve lifters. Carburetor: Rochester Quadrajet. Sales code: L82.

1981 CHEVROLET

Impala/Caprice optional diesel V-8
Overhead valve. Cast-iron block and head. Bore & stroke: 4.057 x 3.385 in. Displacement: 350 cid (5.7 liters). Compression ratio: 22.5:1. Net brake hp: 105 at 3200 rpm. Taxable hp: 51.20. Net torque: 200 lbs.-ft. at 1600 rpm. Five main bearings. Hydraulic valve lifters. Induction system: fuel injection. Olds-built. VIN code: N.

Camaro, Malibu, Monte Carlo, Impala, Caprice base V-8
Overhead valve. Cast-iron block and head. Bore & stroke: 3.50 x 3.48 in. Displacement: 267 cid (4.4 liters). Compression ratio: 8.3:1. Net brake hp: 115 at 4000 rpm. Taxable hp: 39.20. Net brake hp: 120 at 3600 rpm. Net torque: 200 lbs.-ft. at 2400 rpm. Five main bearings. Hydraulic valve lifters. Carburetor: Rochester 2ME two-barrel. VIN code: J.
NOTE: There was no extra cost for the 4.4-liter V-8 in the Impala or Caprice station wagons.

Camaro, Malibu, Monte Carlo, Impala, Caprice optional V-8
Overhead valve. Cast-iron block and head. Bore & stroke: 3.74 x 3.48 in. Displacement: 305 cid (5.0 liters). Compression ratio: 8.6:1. Net brake hp: 150 at 3800 rpm. Taxable hp: 44.66. Net torque: 240 lbs.-ft. at 2400 rpm. Five main bearings. Hydraulic valve lifters. Carburetor: Rochester 4ME four-barrel. VIN code: H.

Camaro Z28 base V-8
Overhead valve. Cast-iron block and head. Bore & stroke: 3.74 x 3.48 in. Displacement: 305 cid (5.0 liters). Compression ratio: 8.6:1. Net brake hp: 165 at 4000 rpm. Taxable hp: 44.66. Net torque: 245 lbs.-ft. at 2400 rpm. Five main bearings. Hydraulic valve lifters. Carburetor: Rochester 4ME four-barrel. VIN code: H.

Camaro Z28 optional V-8
Overhead valve. Cast-iron block and head. Bore & stroke: 4.00 x 3.48 in. Displacement: 350 cid (5.7 liters). Compression ratio: 8.2:1. Net brake hp: 175 at 4000 rpm. Taxable hp: 51.20. Net torque: 275 lbs.-ft. at 2400 rpm. Five main bearings. Hydraulic valve lifters. Carburetor: Rochester 4ME four-barrel. VIN code: L.

1981 CORVETTE

Corvette base V-8
Overhead valve. Cast-iron block and head. Bore & stroke: 4.00 x 3.48 in. Displacement: 350 cid (5.7 liters). Compression ratio: 8.2:1. Net brake hp: 190 at 4200 rpm. Taxable hp: 51.20. Net torque: 280 lbs.-ft. at 1600 rpm. Five main bearings. Hydraulic valve lifters. Carburetor: Rochester Quadrajet. Sales code: L81.

1982 CHEVROLET

Impala/Caprice optional diesel V-8
Overhead valve. Cast-iron block and head. Bore & stroke: 4.057 x 3.385 in. Displacement: 350 cid (5.7 liters). Compression ratio: 22.5:1. Net brake hp: 105 at 3200 rpm. Taxable hp: 51.20. Net torque: 200 lbs.-ft. at 1600 rpm. Five main bearings. Hydraulic valve lifters. Induction system: fuel injection. Olds-built. VIN Code: N.

Malibu, Monte Carlo, Impala, Caprice base V-8
Overhead valve. Cast-iron block and head. Bore & stroke: 3.50 x 3.48 in. Displacement: 267 cid (4.4 liters). Compression ratio: 8.3:1. Net brake hp: 115 at 4000 rpm. Taxable hp: 39.20. Net brake hp: 120 at 3600 rpm. Net torque: 205 lbs.-ft. at 2400 rpm. Five main bearings. Hydraulic valve lifters. Carburetor: Rochester E2ME two-barrel. VIN Code: J.
NOTE: There was no extra cost for the 4.4-liter V-8 in the Impala or Caprice station wagons.

Camaro Z28 base V-8; Camaro, Malibu, Monte Carlo, Impala, Caprice optional V-8
Overhead valve. Cast-iron block and head. Bore & stroke: 3.74 x 3.48 in. Displacement: 305 cid (5.0 liters). Compression ratio: 8.6:1. Net brake hp: 145 at 4000 rpm. Taxable hp: 44.66. Net torque: 240 lbs.-ft. at 2000 rpm. Five main bearings. Hydraulic valve lifters. Carburetor: Rochester E4ME four-barrel. VIN code: H.

Camaro Z28 optional V-8
Overhead valve. Cast-iron block and head. Bore & stroke: 3.74 x 3.48 in. Displacement: 305 cid (5.0 liters). Compression ratio: 8.6:1. Net brake hp: 165 at 4200 rpm. Taxable hp: 44.66. Net torque: 240 lbs.-ft. at 2400 rpm. Five main bearings. Hydraulic valve lifters. Carburetor: Rochester E4ME four-barrel. VIN code: 7.

1982 CORVETTE

Corvette base V-8
Overhead valve. Cast-iron block and head. Bore & stroke: 4.00 x 3.48 in. Displacement: 350 cid (5.7 liters). Compression ratio: 9.0:1. Net brake hp: 200 at 4200 rpm. Taxable hp: 51.20. Net torque: 285 lbs.-ft. at 2800 rpm. Five main bearings. Hydraulic valve lifters. Induction: Rochester 750 cfm dual throttle body fuel injection. Sales code: L83.

1983 CHEVROLET

Impala/Caprice optional diesel V-8
Overhead valve. Cast-iron block and head. Bore & stroke: 4.057 x 3.385 in. Displacement: 350 cid (5.7 liters). Compression ratio: 22.5:1. Net brake hp: 105 at 3200 rpm. Taxable hp: 51.20. Net torque: 200 lbs.-ft. at 1600 rpm. Five main bearings. Hydraulic valve lifters. Induction system: fuel injection. Olds-built. VIN code: N.

Camaro, Malibu, Monte Carlo, Impala, Caprice base V-8
Overhead valve. Cast-iron block and head. Bore & stroke: 3.74 x 3.48 in. Displacement: 305 cid (5.0 liters). Compression ratio: 8.6:1. Net brake hp: 150 at 4000 rpm. Taxable hp: 44.66. Net torque: 240 lbs.-ft. at 2400 rpm. Five main bearings. Hydraulic valve lifters. Carburetor: Rochester E4ME four-barrel. VIN code: H.
NOTE: There was no extra cost for this 5.0-liter V-8 in the Impala or Caprice station wagons.

Monte Carlo SS, Camaro optional H.O. V-8
Overhead valve. Cast-iron block and head. Bore & stroke: 3.74 x 3.48 in. Displacement: 305 cid (5.0 liters). Compression ratio: 8.6:1. Net brake hp: 175 at 4800 rpm. Taxable hp: 44.66. Net torque: 235 lbs.-ft. at 3200 rpm. Five main bearings. Hydraulic valve lifters. Carburetor: Rochester E4ME four-barrel. VIN code: H.

Camaro Z28 base V-8
Overhead valve. Cast-iron block and head. Displacement: 305 cid (5.0 liters). Bore & stroke: 3.74 x 3.48 in. Compression ratio: 8.6:1. Net brake hp: 175 at 4200 rpm. Taxable hp: 44.66. Net torque: 250 lbs.-ft. at 2800 rpm. Five main bearings. Hydraulic valve lifters. Induction: Cross Fire fuel injection. VIN code: S.

1984 CHEVROLET

Impala/Caprice optional diesel V-8
Overhead valve. Cast-iron block and head. Bore & stroke: 4.057 x 3.385 in. Displacement: 350 cid (5.7 liters). Compression ratio: 22.5:1. Net brake hp: 105 at 3200 rpm. Taxable hp: 51.20. Net torque: 200 lbs.-ft. at 1600 rpm. Five main bearings. Hydraulic valve lifters. Induction system: fuel injection. Olds-built. VIN code: N.

Camaro, Monte Carlo, Impala, Caprice base V-8
Overhead valve. Cast-iron block and head. Bore & stroke: 3.74 x 3.48 in. Displacement: 305 cid (5.0 liters). Compression ratio: 8.6:1. Net brake hp: 150 at 4000 rpm. Taxable hp: 44.66. Net torque: 240 lbs.-ft. at 2400 rpm. Five main bearings. Hydraulic valve lifters. Carburetor: four-barrel. VIN code: H.
NOTE: There was no extra cost for this 5.0-liter V-8 in the Camaro Z28 or Caprice station wagon.

Monte Carlo SS optional H.O. V-8
Overhead valve. Cast-iron block and head. Bore & stroke: 3.74 x 3.48 in. Displacement: 305 cid (5.0 liters). Compression ratio: 8.6:1. Net brake hp: 180 at 4800 rpm. Taxable hp: 44.66. Net torque: 235 lbs.-ft. at 3200 rpm. Five main bearings. Hydraulic valve lifters. Carburetor: Rochester E4ME four-barrel. VIN code: G.

Camaro Z28 optional H.O. V-8
Overhead valve. Cast-iron block and head. Bore & stroke: 3.74 x 3.48 in. Displacement: 305 cid (5.0 liters). Compression ratio: 9.5:1. Net brake hp: 190 at 4800 rpm. Taxable hp: 44.66. Net torque: 240 lbs.-ft. at 3200 rpm. Five main bearings. Hydraulic valve lifters. Carburetor: Rochester E4ME four-barrel. VIN code: G.

1984 CORVETTE

Corvette base V-8
Overhead valve. Cast-iron block and head. Bore & stroke: 4.00 x 3.48 in. Displacement: 350 cid (5.7 liters). Compression ratio: 9.0:1. Net brake hp: 205 at 4200 rpm. Taxable hp: 51.20. Net torque: 290 lbs.-ft. at 2800 rpm. Five main bearings. Hydraulic valve lifters. Induction: Cross-fire fuel injection (Rochester 750 cfm dual throttle body fuel injection). Sales code: L83.

1985 CHEVROLET

Impala/Caprice optional diesel V-8
Overhead valve. Cast-iron block and head. Bore & stroke: 4.057 x 3.385 in. Displacement: 350 cid (5.7 liters). Compression ratio: 22.5:1. Net brake hp: 105 at 3200 rpm. Taxable hp: 51.20.Net torque: 200 lbs.-ft. at 1600 rpm. Five main bearings. Hydraulic valve lifters. Induction system: fuel injection. Olds-built. VIN code: N.

Monte Carlo base V-8
Overhead valve. Cast-iron block and head. Bore & stroke: 3.74 x 3.48 in. Displacement: 305 cid (5.0 liters). Compression ratio: 9.5:1. Net brake hp: 150 at 4000 rpm. Taxable hp: 44.66. Net torque: 240 lbs.-ft. at 2000 rpm. Five main bearings. Hydraulic valve lifters. Carburetor: four-barrel. VIN code: H.

Camaro base V-8
Overhead valve. Cast-iron block and head. Bore & stroke: 3.74 x 3.48 in. Displacement: 305 cid (5.0 liters). Compression ratio: 9.5:1. Net brake hp: 155 at 4200 rpm. Taxable hp: 44.66. Net torque: 245 lbs.-ft. at 2000 rpm. Five main bearings. Hydraulic valve lifters. Carburetor: four-barrel. VIN code: H.
NOTE: There was no extra cost for this 5.0-liter V-8 in the Camaro Z28.

Impala, Caprice base V-8
Overhead valve. Cast-iron block and head. Bore & stroke: 3.74 x 3.48 in. Displacement: 305 cid (5.0 liters). Compression ratio: 9.5:1. Net brake hp: 165 at 4200 rpm. Taxable hp: 44.66. Net torque: 245 lbs.-ft. at 2400 rpm. Five main bearings. Hydraulic valve lifters. Carburetor: four-barrel. VIN code: H.
NOTE: There was no extra cost for this 5.0-liter V-8 in the Camaro Z28 or Caprice station wagon.

Monte Carlo SS optional H.O. V-8
Overhead valve. Cast-iron block and head. Bore & stroke: 3.74 x 3.48 in. Displacement: 305 cid (5.0 liters). Compression ratio: 8.6:1. Net brake hp: 180 at 4800 rpm. Taxable hp: 44.66. Net torque: 235 lbs.-ft. at 3200 rpm. Five main bearings. Hydraulic valve lifters. Carburetor: Rochester E4ME four-barrel. VIN code: G.

Camaro Z28 optional H.O. V-8
Overhead valve. Cast-iron block and head. Bore & stroke: 3.74 x 3.48 in. Displacement: 305 cid (5.0 liters). Compression ratio: 9.5:1. Net brake hp: 190 at 4800 rpm. Taxable hp: 44.66. Net torque: 240 lbs.-ft. at 3200 rpm. Five main bearings. Hydraulic valve lifters. Carburetor: Rochester E4ME four-barrel. VIN code: G.

Camaro IROC Z28 optional TPI V-8
Overhead valve. Cast-iron block and head. Bore & stroke: 3.74 x 3.48 in. Displacement: 305 cid (5.0 liters). Compression ratio: 9.5:1. Net brake hp: 215 at 4400 rpm. Taxable hp: 44.66. Net torque: 275 lbs.-ft. at 3200 rpm. Five main bearings. Hydraulic valve lifters. Carburetor: Rochester E4ME four-barrel. VIN code: F.

1985 CORVETTE

Corvette base V-8
Overhead valve. Cast-iron block and head. Bore & stroke: 4.00 x 3.48 in. Displacement: 350 cid (5.7 liters). Compression ratio: 9.0:1. Net brake hp: 230 at 4000 rpm. Taxable hp: 51.20. Net torque: 330 lbs.-ft. at 3200 rpm. Five main bearings. Hydraulic valve lifters. Induction: Tuned-port-induction (TPI) system. Sales code: L98.

1986 CHEVROLET

Monte Carlo base V-8
Overhead valve. Cast-iron block and head. Bore & stroke: 3.74 x 3.48 in. Displacement: 305 cid (5.0 liters). Compression ratio: 9.5:1. Net brake hp: 150 at 4000 rpm. Taxable hp: 44.66. Net torque: 240 lbs.-ft. at 2000 rpm. Five main bearings. Hydraulic valve lifters. Carburetor: four-barrel. VIN code: H.

Caprice station wagon base V-8
Overhead valve. Cast-iron block and head. Bore & stroke: 3.74 x 3.48 in. Displacement: 305 cid (5.0 liters). Compression ratio: 9.5:1. Net brake hp: 155 at 4200 rpm. Taxable hp: 44.66. Net torque: 245 lbs.-ft. at 2000 rpm. Five main bearings. Hydraulic valve lifters. Carburetor: four-barrel. VIN code: Y.

Camaro base V-8
Overhead valve. Cast-iron block and head. Bore & stroke: 3.74 x 3.48 in. Displacement: 305 cid (5.0 liters). Compression ratio: 9.5:1. Net brake hp: 165 at 4400 rpm. Taxable hp: 44.66. Net torque: 250 lbs.-ft. at 2000 rpm. Five main bearings. Hydraulic valve lifters. Carburetor: four-barrel. VIN code: H.

Caprice base V-8
Overhead valve. Cast-iron block and head. Displacement: 305 cid (5.0 liters). Bore & stroke: 3.74 x 3.48 in. Compression ratio: 9.5:1. Net brake hp: 165 at 4200 rpm. Taxable hp: 44.66. Net torque: 245 lbs.-ft. at 2400 rpm. Five main bearings. Hydraulic valve lifters. Carburetor: four-barrel. VIN code: Y.
NOTE: There was no extra cost for this 5.0-liter V-8 in the Camaro Z28 or Caprice station wagon.

Monte Carlo SS optional H.O. V-8
Overhead valve. Cast-iron block and head. Bore & stroke: 3.74 x 3.48 in. Displacement: 305 cid (5.0 liters). Compression ratio: 8.6:1. Net brake hp: 180 at 4800 rpm. Taxable hp: 44.66. Net torque: 225 lbs.-ft. at 3200 rpm. Five main bearings. Hydraulic valve lifters. Carburetor: Rochester E4ME four-barrel. VIN code: G.

Camaro Z28 optional H.O. V-8
Overhead valve. Cast-iron block and head. Bore & stroke: 3.74 x 3.48 in. Displacement: 305 cid (5.0 liters). Compression ratio: 9.5:1. Net brake hp: 190 at 4800 rpm. Taxable hp: 44.66. Net torque: 240 lbs.-ft. at 3200 rpm. Five main bearings. Hydraulic valve lifters. Carburetor: Rochester E4ME four-barrel. VIN code: G.

Camaro IROC Z28 optional TPI V-8
Overhead valve. Cast-iron block and head. Bore & stroke: 3.74 x 3.48 in. Displacement: 305 cid (5.0 liters). Compression ratio: 9.5:1. Net brake hp: 190 at 4000 rpm. Taxable hp: 44.66. Net torque: 285 lbs.-ft. at 3200 rpm. Five main bearings. Hydraulic valve lifters. Carburetor: Rochester E4ME four-barrel. VIN code: F.

1986 CORVETTE

Corvette base V-8
Overhead valve. Cast-iron block and head. Bore & stroke: 4.00 x 3.48 in. Displacement: 350 cid (5.7 liters). Compression ratio: 9.0:1. Net brake hp: 230 at 4000 rpm. Taxable hp: 51.20. Net torque: 330 at 3200 rpm. Five main bearings. Hydraulic valve lifters. Induction: Tuned-port-induction (TPI) system. Sales code: L98.

1987 CHEVROLET

Caprice wagon base V-8
Overhead valve. Cast-iron block and head. Bore & stroke: 3.80 x 3.38 in. Displacement: 307 cid (5.0 liters). Compression ratio: 8.0:1 Net brake hp: 140 at 3200 rpm. Taxable hp: 48.0. Net torque: 225 lbs.-ft. at 2000 rpm. Five main bearings. Hydraulic roller valve lifters. Carburetor: four-barrel.

Monte Carlo base V-8
Overhead valve. Cast-iron block and head. Bore & stroke: 3.74 x 3.48 in. Displacement: 305 cid (5.0 liters). Compression ratio: 9.5:1. Net brake hp: 150 at 4000 rpm. Taxable hp: 44.66. Net torque: 240 lbs.-ft. at 2000 rpm. Five main bearings. Hydraulic valve lifters. Carburetor: four-barrel.

Camaro base V-8
Overhead valve. Cast-iron block and head. Bore & stroke: 3.74 x 3.48 in. Displacement: 305 cid (5.0 liters). Compression ratio: 9.3:1. Net brake hp: 165 at 4400 rpm. Taxable hp: 44.66. Net torque: 245/250 lbs.-ft. at 2800 rpm. Five main bearings. Hydraulic valve lifters. Carburetor: four-barrel.
NOTE: There was no extra cost for this 5.0-liter V-8 in the Camaro Z28.

Caprice base V-8
Overhead valve. Cast-iron block and head. Bore & stroke: 3.74 x 3.48 in. Displacement: 305 cid (5.0 liters). Compression ratio: 9.5:1. Net brake hp: 170 at 4400 rpm. Taxable hp: 44.66. Net torque: 250 lbs.-ft. at 2800 rpm. Five main bearings. Hydraulic valve lifters. Carburetor: four-barrel.

Monte Carlo optional H.O. V-8
Overhead valve. Cast-iron block and head. Bore & stroke: 3.74 x 3.48 in. Displacement: 305 cid (5.0 liters). Compression ratio: 9.5:1. Net brake hp: 180 at 4800 rpm. Taxable hp: 44.66. Net torque: 225 lbs.-ft. at 3200 rpm. Five main bearings. Hydraulic valve lifters. Carburetor: four-barrel.

Camaro Z28 optional H.O. V-8 (automatic transmission)
Overhead valve. Cast-iron block and head. Bore & stroke: 3.74 x 3.48 in. Displacement: 305 cid (5.0 liters). Compression ratio: 9.5:1. Net brake hp: 190 at 4000 rpm. Taxable hp: 44.66. Net torque: 295 lbs.-ft. at 2800 rpm. Five main bearings. Hydraulic valve lifters. Induction: Tuned Port Injection.

Camaro Z28 optional H.O. V-8 (manual transmission)
Overhead valve. Cast-iron block and head. Bore & stroke: 3.74 x 3.48 in. Displacement: 305 cid (5.0 liters). Compression ratio: 9.5:1. Net brake hp: 215 at 4400 rpm. Taxable hp: 44.66. Net torque: Monte 250 lbs.-ft. at 3200 rpm. Five main bearings. Hydraulic valve lifters. Induction: Tuned Port Injection.

Camaro IROC-Z optional 5.7L V-8
Overhead valve. Cast-iron block and head. Bore & stroke: 4.00 x 3.48 in. Displacement: 350 cid (5.7 liters). Compression ratio: 9.0:1 Net brake hp: 225 at 4400 rpm. Taxable hp: 51.20. Net torque: 330 lbs.-ft. at 2800 rpm. Five main bearings. Hydraulic valve lifters. Induction: Induction: Tuned Port Injection.

1987 CORVETTE

Corvette base V-8
Overhead valve. Cast-iron block and head. Bore & stroke: 4.00 x 3.48 in. Displacement: 350 cid (5.7 liters). Compression ratio: 9.5:1. Net brake hp: 240 at 4000 rpm. Taxable hp: 51.20. Net torque: 345 lbs.-ft. at 3200 rpm. Five main bearings. Aluminum cylinder head. Hydraulic valve lifters. Induction: Tuned-port-induction (TPI) system. Sales code: L98.

1988 CHEVROLET

Caprice wagon base V-8
Overhead valve. Cast-iron block and head. Bore & stroke: 3.80 x 3.38 in. Displacement: 307 cid (5.0 liters). Compression ratio: 8.0:1 Net brake hp: 140 at 3200 rpm. Taxable hp: 44.8. Net torque: 225 lbs.-ft. at 2000 rpm. Five main bearings. Hydraulic roller valve lifters. Carburetor: four-barrel.

Monte Carlo base V-8
Overhead valve. Cast-iron block and head. Bore & stroke: 3.74 x 3.48 in. Displacement: 305 cid (5.0 liters). Compression ratio: 9.3:1. Net brake hp: 150 at 4000 rpm. Taxable hp: 44.8. Taxable hp: 44.66. Net torque: 240 lbs.-ft. at 2000 rpm. Five main bearings. Hydraulic valve lifters. Carburetor: four-barrel.

Caprice base V-8
Overhead valve. Cast-iron block and head. Bore & stroke: 3.74 x 3.48 in. Displacement: 305 cid (5.0 liters). Compression ratio: 9.3:1. Net brake hp: 170 at 4400 rpm. Taxable hp: 44.8. Net torque: 250 lbs.-ft. at 2800 rpm. Five main bearings. Hydraulic valve lifters. Carburetor: four-barrel.

Camaro convertible and IROC-Z base V-8; base Camaro optional V-8
Overhead valve. Cast-iron block and head. Bore & stroke: 3.74 x 3.48 in. Displacement: 305 cid (5.0 liters). Compression ratio: 9.3:1. Net brake hp: 170 at 4000 rpm. Taxable hp: 44.8. Net torque: 225 lbs.-ft. at 2400 rpm. Five main bearings. Hydraulic valve lifters. Induction: Electronic fuel injection.

Camaro IROC-Z optional H.O. V-8 (automatic transmission)
Overhead valve. Cast-iron block and head. Bore & stroke: 3.74 x 3.48 in. Displacement: 305 cid (5.0 liters). Compression ratio: 9.5:1. Net brake hp: 195 at 4000 rpm. Taxable hp: 44.8. Net torque: 295 lbs.-ft. at 2800 rpm. Five main bearings. Hydraulic valve lifters. Induction: Tuned Port Injection.

Camaro IROC-Z optional H.O. V-8 (manual transmission)
Overhead valve. Cast-iron block and head. Bore & stroke: 3.74 x 3.48 in. Displacement: 305 cid (5.0 liters). Compression ratio: 9.5:1. Net brake hp: 220 at 4400 rpm. Taxable hp: 44.8. Net torque: Monte 290 lbs.-ft. at 3200 rpm. Five main bearings. Hydraulic valve lifters. Induction: Tuned Port Injection.

Monte Carlo SS optional H.O. V-8
Overhead valve. Cast-iron block and head. Bore & stroke: 3.74 x 3.48 in. Displacement: 305 cid (5.0 liters). Compression ratio: 9.3:1. Net brake hp: 180 at 4800 rpm. Taxable hp: 44.8. Net torque: 225 lbs.-ft. at 3200 rpm. Five main bearings. Hydraulic valve lifters. Induction: Electronic fuel injection.

Camaro IROC-Z optional 5.7L V-8
Overhead valve. Cast-iron block and head. Bore & stroke: 4.00 x 3.48 in. Displacement: 350 cid (5.7 liters). Compression ratio: 9.3:1. Net brake hp: 230 at 4400 rpm. Taxable hp: 51.20. Net torque: 330 lbs.-ft. at 3200 rpm. Five main bearings. Hydraulic valve lifters. Induction: Tuned Port Injection.

1988 CORVETTE

Corvette base V-8
Overhead valve. Cast-iron block and head. Bore & stroke: 4.00 x 3.48 in. Displacement: 350 cid (5.7 liters). Compression ratio: 9.5:1. Net brake hp: 240/245 (*) at 4000 rpm. Taxable hp: 51.20. Net torque: 345 lbs.-ft. at 3200 rpm. Five main bearings. Aluminum cylinder head. Hydraulic valve lifters. Induction: Tuned-Port-Injection system. (*) Engine rating is 245-hp for hatchbacks using 3.07:1 rear axle due to use of low-restriction mufflers with this option combination only. Sales code: L98.

1989 CHEVROLET

Caprice wagon base V-8
Overhead valve. Cast-iron block and head. Bore & stroke: 3.80 x 3.38 in. Displacement: 307 cid (5.0 liters). Compression ratio: 8.0:1 Net brake hp: 140 at 3200 rpm. Taxable hp: 48.0. Net torque: 255 lbs.-ft. at 2000 rpm. Five main bearings. Hydraulic roller valve lifters. Carburetor: four-barrel.

Caprice base V-8
Overhead valve. Cast-iron block and head. Bore & stroke: 3.74 x 3.48 in. Displacement: 305 cid (5.0 liters). Compression ratio: 9.3:1. Net brake hp: 170 at 4000 rpm. Taxable hp: 44.8. Net torque: 250 lbs.-ft. at 2400 rpm. Five main bearings. Hydraulic valve lifters. Induction: Throttle-body fuel injection.

Camaro RS convertible and IROC-Z base V-8; Camaro RS coupe optional V-8
Overhead valve. Cast-iron block and head. Bore & stroke: 3.74 x 3.48 in. Displacement: 305 cid (5.0 liters). Compression ratio: 9.3:1. Net brake hp: 170 at 4000 rpm. Taxable hp: 44.8. Net torque: 255 lbs.-ft. at 2400 rpm. Five main bearings. Hydraulic valve lifters. Induction: Throttle-body fuel injection.

Camaro IROC-Z optional H.O. V-8 (automatic transmission)
Overhead valve. Cast-iron block and head. Bore & stroke: 3.74 x 3.48 in. Displacement: 305 cid (5.0 liters). Compression ratio: 9.5:1. Net brake hp: 195 at 4000 rpm. Taxable hp: 44.8. Net torque: 295 lbs.-ft. at 2800 rpm. Five main bearings. Hydraulic valve lifters. Induction: Tuned Port Injection.

Camaro IROC-Z optional H.O. V-8 (manual transmission)
Overhead valve. Cast-iron block and head. Bore & stroke: 3.74 x 3.48 in. Displacement: 305 cid (5.0 liters). Compression ratio: 9.5:1. Net brake hp: 220 at 4400 rpm. Taxable hp: 44.8. Net torque: Monte 290 lbs.-ft. at 3200 rpm. Five main bearings. Hydraulic valve lifters. Induction: Tuned Port Injection.

Camaro IROC-Z optional 5.7L V-8
Overhead valve. Cast-iron block and head. Bore & stroke: 4.00 x 3.48 in. Displacement: 350 cid (5.7 liters). Compression ratio: 9.3:1. Net brake hp: 230 at 4400 rpm. Taxable hp: 51.20. Net torque: 330 lbs.-ft. at 3200 rpm. Five main bearings. Hydraulic valve lifters. Induction: Tuned Port Injection.

1989 CORVETTE

Corvette base V-8
Overhead valve. Cast-iron block and head. Bore & stroke: 4.00 x 3.48 in. Displacement: 350 cid (5.7 liters). Compression ratio: 9.5:1. Net brake hp: 240/245 (*) at 4300 rpm. Taxable hp: 51.20. Net torque: 340 lbs.-ft. at 3200 rpm. Five main bearings. Aluminum cylinder head. Hydraulic valve lifters. Induction: Tuned-Port-Injection system. (*) Engine rating is 245-hp

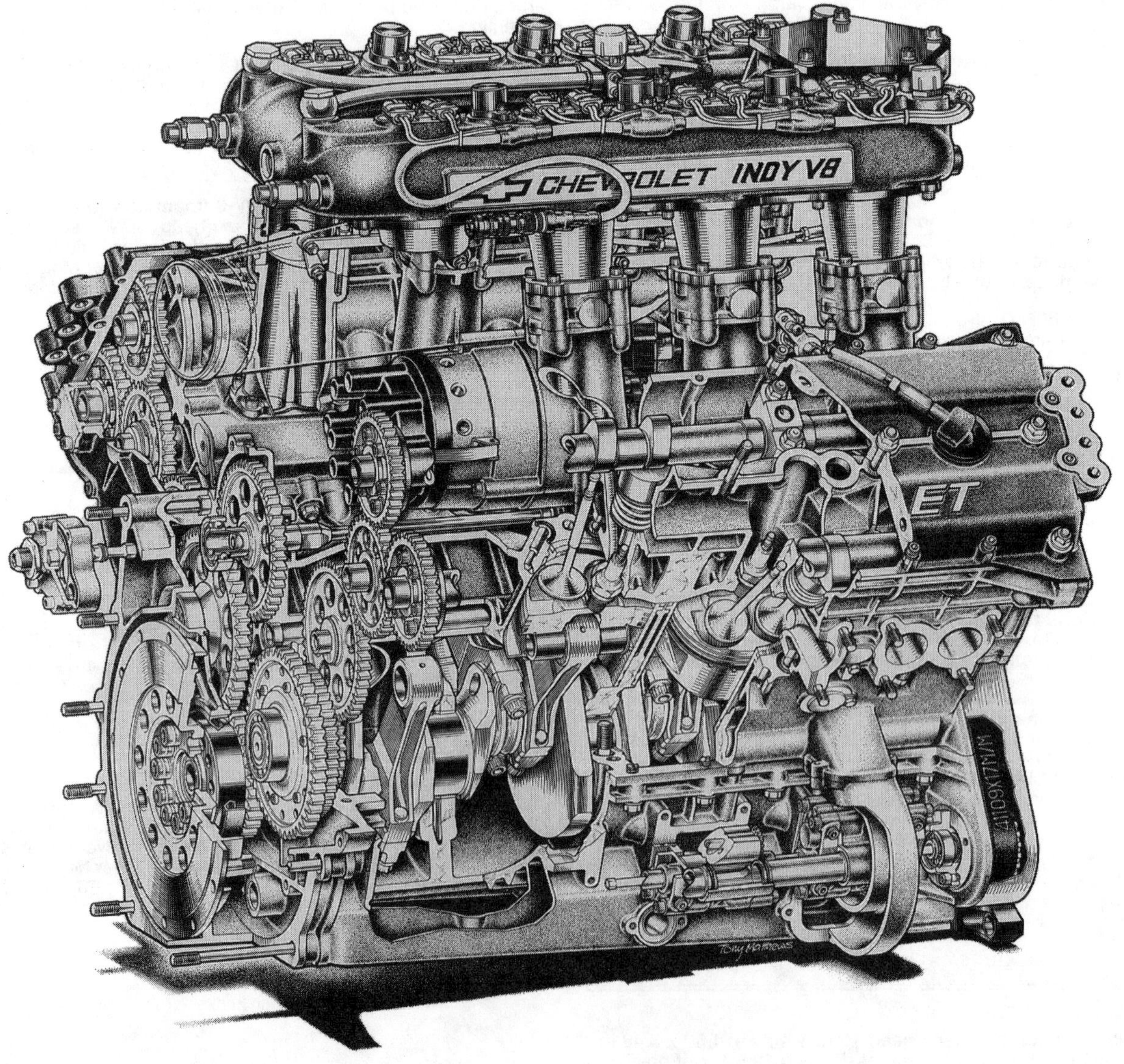

The 1990 Chevrolet Indy V-8 was a special racing engine. It was based on the engine used in the 1986 Corvette Indy concept car. That engine had a 5735 cc displacement and produced 380 hp at 6000 rpm and 370 foot-pounds of torque at 3800 rpm.

for hatchbacks using 3.07:1 rear axle due to use of low-restriction mufflers with this option combination only. Sales code: L98.

1990 CHEVROLET

Caprice wagon base V-8
Overhead valve. Cast-iron block and head. Bore & stroke: 3.80 x 3.38 in. Displacement: 307 cid (5.0 liters). Compression ratio: 8.0:1. Net brake hp: 140 at 3200 rpm. Taxable hp: 48.0. Net torque: 255 lbs.-ft. at 2000 rpm. Five main bearings. Hydraulic roller valve lifters. Carburetor: four-barrel.

Caprice sedan base V-8
Overhead-valve V-8. Cast-iron block and head. Bore & stroke: 3.74 x 3.48 in. Displacement: 305 cid (5.0 liters). Compression ratio: 9.3:1. Net brake hp: 170 at 4000 rpm. Taxable hp: 44.8. Net torque: 255 lbs.-ft. at 2400 rpm. Five main bearings. Hydraulic roller valve lifters. Induction: Throttle-body fuel injection.

Camaro RS convertible and IROC-Z base V-8; Camaro RS coupe optional V-8
Overhead valve. Cast-iron block and head. Bore & stroke: 3.74 x 3.48 in. Displacement: 305 cid (5.0 liters). Compression ratio: 9.3:1. Net brake hp: 170 at 4000 rpm. Taxable hp: 44.8. Net torque: 255 lbs.-ft. at 2400 rpm. Five main bearings. Hydraulic valve lifters. Induction: Throttle-body fuel injection.

Camaro IROC-Z optional H.O. V-8 (manual transmission)
Overhead valve. Cast-iron block and head. Bore & stroke: 3.74 x 3.48 in. Displacement: 305 cid (5.0 liters). Compression ratio: 9.5:1. Net brake hp: 220 at 4400 rpm. Taxable hp: 44.8. Net torque: Monte 290 lbs.-ft. at 3200 rpm. Five main bearings. Hydraulic valve lifters. Induction: Tuned Port Injection.

Camaro IROC-Z optional 5.7L V-8
Overhead valve. Cast-iron block and head. Bore & stroke: 4.00 x 3.48 in. Displacement: 350 cid (5.7 liters). Compression ratio: 9.3:1. Net brake hp: 230 at 4400 rpm. Taxable hp: 51.20. Net torque: 330 lbs.-ft. at 3200 rpm. Five main bearings. Hydraulic valve lifters. Induction: Tuned Port Injection.

1990 CORVETTE

Corvette base V-8
Overhead valve. Cast-iron block and head. Bore & stroke: 4.00 x 3.48 in. Displacement: 350 cid (5.7 liters). Compression ratio: 9.5:1. Net brake hp: 240/245 (*) at 4300 rpm. Taxable hp: 51.20. Net torque: 340 lbs.-ft. at 3200 rpm. Five main bearings. Aluminum cylinder head. Hydraulic valve lifters. Induction: Tuned-Port-Injection system. (*) Engine rating is 245-hp for hatchbacks using 3.07:1 rear axle due to use of low-restriction mufflers with this option combination only. Sales code: L98.

Corvette ZR-1 base V-8
Overhead valve with four valves per cylinder. Four overhead camshafts. Cast-iron block and head. Bore & stroke: 3.90 x 3.66 in. Displacement: 350 cid (5.7 liters). Compression ratio: 11.0:1. Net brake hp: 375 at 5800 rpm. Taxable hp: 48.67. Net torque: 370 lbs.-ft. at 5600 rpm. Five main bearings. Aluminum cylinder head. Hydraulic valve lifters. Induction: Tuned-Port-Injection system. LT5 engines were manufactured and assembled by Mercury Marine Corporation in Stillwater, Okla., and shipped to the Corvette factory for installation in ZR-1 Corvettes. Sales code: LT5.

1991 CHEVROLET

Camaro RS convertible and Z28 base V-8; Camaro RS coupe optional V-8
Overhead valve. Cast-iron block and head. Bore & stroke: 3.74 x 3.48 in. Displacement: 305 cid (5.0 liters). Compression ratio: 9.3:1. Net brake hp: 170 at 4000. Taxable hp: 44.8. Net torque: 255 lbs.-ft. at 2400 rpm. Five main bearings. Hydraulic valve lifters. Induction: Throttle body fuel injection. VIN code: H.

Camaro RS convertible, Z28, Caprice, Caprice Classic, Caprice Classic LTZ base V-8
Overhead-valve V-8. Cast-iron block and head. Bore & stroke: 3.74 x 3.48 in. Displacement: 305 cid (5.0 liters). Compression ratio: 9.3:1. Net brake hp: 230 at 4400 rpm (205 hp with automatic transmission). Taxable hp: 44.8. Net torque: 255 lbs.-ft. at 2400 rpm. Hydraulic valve lifters. Induction: (Camaro) Tuned port fuel injection; (Caprice) Throttle body fuel injection). VIN code: F.

Camaro Z28 coupe, Caprice Classic LTZ optional V-8
Overhead-valve V-8. Cast-iron block and head. Displacement: 350 cid (5.7 liters). Bore & stroke: 4.00 x 3.48 in. Compression ratio: 9.3:1. Net brake hp: 245 at 4400 rpm. Taxable hp: 51.20. Net torque: 345 lbs.-ft. at 3200 rpm. Hydraulic valve lifters. Induction: Tuned port fuel injection. VIN code: 8.

1991 CORVETTE

Corvette base V-8
Overhead valve. Cast-iron block and head. Bore & stroke: 4.00 x 3.48 in. Displacement: 350 cid (5.7 liters). Compression ratio: 10.0:1. Net brake hp: 245 at 4000 rpm. Taxable hp: 51.20. Net torque: 340 lbs.-ft. at 3200 rpm. Five main bearings. Aluminum cylinder head. Hydraulic valve lifters. Induction: Tuned-Port-Injection system. (*) Engine rating is 250-hp for hatchbacks using 3.07:1 rear axle due to use of low-restriction mufflers with this option combination only. Sales code: L98.

Corvette ZR-1 base V-8
Overhead valve with four valves per cylinder. Four overhead camshafts. Cast-iron block and head. Displacement: 350 cid (5.7 liters). Bore & stroke: 3.90 x 3.66 in. Compression ratio: 11.0:1. Net brake hp: 375 at 5800 rpm. Taxable hp: 48.67. Net torque: 370 lbs.-ft. at 5600 rpm. Five main bearings. Aluminum cylinder head. Hydraulic valve lifters. Induction: Tuned-Port-Injection system. Sales code: LT5.

1992 CHEVROLET

Caprice, Caprice Classic, Caprice Classic LTZ base V-8
Overhead-valve V-8. Cast-iron block and head. Bore & stroke: 3.74 x 3.48 in. Displacement: 305 cid (5.0 liters). Compression ratio: 9.3:1. Net brake hp: 170 at 4200 rpm. Taxable hp: 44.8. Net torque: 255 lbs.-ft. at 2400 rpm for Caprice. Hydraulic valve lifters. Induction: Throttle-body fuel injection. Sales code: LO3. VIN code: H.

Camaro RS optional V-8
Overhead-valve V-8. Cast-iron block and head. Bore & stroke: 3.74 x 3.48 in. Displacement. 305 cid (5.0 liters). Compression ratio. 9.3.1. Net brake hp: 170 at 4000 rpm. Taxable hp: 44.8. Net torque: 285 lbs.-ft. at 2400 rpm. Hydraulic valve lifters. Induction: Tuned port fuel injection. Sales code: LO3. VIN code: H.

Caprice wagon optional V-8
Overhead-valve V-8. Cast-iron block and head. Bore & stroke: 4.00 x 3.48 in. Displacement: 350 cid (5.7 liters). Compression ratio: 9.3:1. Net brake hp: 180 at 4000 rpm. Taxable hp: 51.20. Net torque: 300 lbs.-ft. at 2400 rpm. Hydraulic valve lifters. Induction: Throttle-body fuel injection.

Camaro Z28 base V-8
Overhead-valve V-8. Cast-iron block and head. Bore & stroke: 3.74 x 3.48 in. Displacement: 305 cid (5.0 liters). Compression ratio: 9.3:1. Net brake hp: 230 at 4400 rpm. Taxable hp: 44.8. Net torque: 300 lbs.-ft. at 2400 rpm. Hydraulic valve lifters. Induction: Tuned port fuel injection. Sales code: LO3. VIN code: H.

Camaro Z28 coupe optional V-8
Overhead-valve V-8. Cast-iron block and head. Bore & stroke: 4.00 x 3.48 in. Displacement: 350 cid (5.7 liters). Compression ratio: 9.8:1 for Camaro. Net brake hp: 245 at 4400 rpm for Camaro. Taxable hp: 51.20. Net torque: 345 lbs.-ft. at 3200 rpm for Camaro. Hydraulic valve lifters. Induction: Tuned port fuel injection.

1992 CORVETTE

Corvette base V-8
Overhead valve. Cast-iron block and head. Bore & stroke: 4.00 x 3.48 in. Displacement: 350 cid (5.7 liters). Compression ratio: 10.3:1. Net brake hp: 300 at 5000 rpm. Taxable hp: 51.20. Net torque: 330 lbs.-ft. at 4000 rpm. Five main bearings. Aluminum cylinder head. Hydraulic valve lifters. Induction: Multiport-Fuel-Injection system. Sales code: LT1.

Corvette ZR-1 base V-8
Overhead valve with four valves per cylinder. Four overhead camshafts. Cast-iron block and head. Bore & stroke: 3.90 x 3.66 in. Displacement: 350 cid (5.7 liters). Compression ratio: 11.0:1. Net brake hp: 375 at 5800 rpm. Taxable hp: 48.67. Net torque: 370 lbs.-ft. at 5600 rpm. Five main bearings. Aluminum cylinder head. Hydraulic valve lifters. Induction: Multiport-Fuel-Injection system. Sales code: LT5.

1993 CHEVROLET

The base V-8 in the 1993 Camaro Z28 was this LT1 engine. This code P 350-cid motor had a 4.00 x 3.48-in. bore and stroke and a 10.5:1 compression ratio. It developed 275 net brake horsepower at 5000 rpm and 325 net foot-pounds of torque at 2400 rpm. A TPI (Tuned Port Injection) sytem was used on the new LT1.

Caprice Classic base V-8
Overhead-valve V-8. Cast-iron block and head. Displacement: 305 cid (5.0 liters). Bore & stroke: 3.74 x 3.48 in. Compression ratio: 9.3:1. Net brake hp: 170 at 4200 rpm. Taxable hp: 44.8. Net torque: 255 lbs.-ft. at 2400 rpm. Hydraulic valve lifters. Induction: Throttle-body fuel injection. Sales code: L03. VIN code: E.

Caprice Classic LTZ base V-8
Overhead-valve V-8. Cast-iron block and head. Bore & stroke: 4.00 x 3.48 in. Displacement: 350 cid (5.7 liters). Compression ratio: 9.3:1. Net brake hp: 180 at 4000 rpm. Taxable hp: 51.20. Net torque: 300 lbs.-ft. at 2400 rpm. Hydraulic valve lifters. Induction: Throttle-body fuel injection. Sales code: L05. VIN code: 7

Camaro Z28 base V-8
Overhead-valve V-8. Cast-iron block and head. Bore & stroke: 4.00 x 3.48 in. Displacement: 350 cid (5.7 liters). Compression ratio: 10.5:1. Net brake hp: 275 at 5000 rpm. Taxable hp: 51.20. Net torque: 325 lbs.-ft. at 2400 rpm. Hydraulic valve lifters. Induction: Tuned port fuel injection. Sales code: LT1. VIN code: P.

1993 CORVETTE

Corvette base V-8
Overhead valve. Cast-iron block and head. Bore & stroke: 4.00 x 3.48 in. Displacement: 350 cid (5.7 liters). Compression ratio: 10.5:1. Net brake hp: 300 at 5000 rpm. Taxable hp: 51.20. Net torque: 340 lbs.-ft. at 3600 rpm. Five main bearings. Aluminum cylinder head. Hydraulic valve lifters. Induction: Multiport fuel-injection system. Sales code: LT1.

Corvette ZR-1 base V-8
Overhead valve with four valves per cylinder. Four overhead camshafts. Cast-iron block and head. Bore & stroke: 3.90 x 3.66 in. Displacement: 350 cid (5.7 liters). Compression ratio: 11.0:1. Net brake hp: 405 at 5800 rpm. Taxable hp: 48.67. Net torque: 385 lbs.-ft. at 5200 rpm. Five main bearings. Aluminum cylinder head. Hydraulic valve lifters. Induction: Multiport fuel-injection system. Sales code: LT5.

1994 CHEVROLET

Caprice Classic, Caprice Classic LS base V-8
Overhead-valve V-8. Cast-iron block and head. Bore & stroke: 3.75 x 3.48 in. Displacement: 265 cid (4.3 liters). Compression ratio: 9.9:1. Net brake hp: 200 at 5200 rpm. Taxable hp: 44.8. Net torque: 245 lbs.-ft. at 2400 rpm. Hydraulic valve lifters. Induction: Sequential fuel injection. Sales code: L99. VIN code: W.

Impala SS base V-8; Caprice Classic, Caprice Classic LS optional V-8
Overhead-valve V-8. Cast-iron block and head. Displacement: 350 cid (5.7 liters). Bore & stroke: 4.00 x 3.48 in. Compression ratio: 10.5:1. Net brake hp: 260 at 5000 rpm. Taxable hp: 51.20. Net torque: 330 lbs.-ft. at 3200 rpm. Hydraulic valve lifters. Induction: Sequential fuel injection. Sales code: LT1. VIN code: P.

Camaro Z28 base V-8
Overhead-valve V-8. Cast-iron block and head. Displacement: 350 cid (5.7 liters). Bore & stroke: 4.00 x 3.48 in. Compression ratio: 10.5:1. Net brake hp: 275 at 5000 rpm. Taxable hp: 51.20. Net torque: 325 lbs.-ft. at 2000 rpm. Hydraulic valve lifters. Sequential fuel injection. Sales code: LT1. VIN code: P.

1994 CORVETTE

Corvette base V-8
Overhead valve. Cast-iron block and head. Bore & stroke: 4.00 x 3.48 in. Displacement: 350 cid (5.7 liters). Compression ratio: 10.5:1. Net brake hp: 300 at 5000 rpm. Taxable hp: 51.20. Net torque: 340 lbs.-ft. at 3600 rpm. Five main bearings. Aluminum cylinder head. Hydraulic valve lifters. Induction: Sequential multiport fuel injection. Sales code: LT1.

Corvette ZR-1 base V-8
Overhead valve with four valves per cylinder. Four overhead camshafts. Cast-iron block and head. Bore & stroke: 3.90 x 3.66 in. Displacement: 350 cid (5.7 liters). Compression ratio: 11.0:1. Net brake hp: 405 at 5800 rpm. Taxable hp: 48.67. Net torque: 385 lbs.-ft. at 5200 rpm. Five main bearings. Aluminum cylinder head. Hydraulic valve lifters. Induction: Sequential multiport-fuel-injection system. Sales code: LT5.

1995 CHEVROLET

Caprice Classic base V-8
Overhead-valve V-8. Cast-iron block and head. Bore & stroke: 3.75 x 3.48 in. Displacement: 265 cid (4.3 liters). Compression ratio: 9.9:1. Net brake hp: 200 at 5200 rpm. Taxable hp: 44.8. Net torque: 245 lbs.-ft. at 2400 rpm. Hydraulic valve lifters. Induction: Sequential fuel injection.

Sales code: L99. VIN code: W.

Impala SS base V-8; Caprice Classic, Caprice Classic LS optional V-8
Overhead-valve V-8. Cast-iron block and head. Displacement: 350 cid (5.7 liters). Bore & stroke: 4.00 x 3.48 in. Compression ratio: 10.0:1. Net brake hp: 260 at 5000 rpm. Taxable hp: 51.20. Net torque: 330 lbs.-ft. at 3200 rpm. Hydraulic valve lifters. Induction: Sequential fuel injection. Sales code: LT1. VIN code: P.

Camaro Z28 base V-8
Overhead-valve V-8. Cast-iron block and head. Displacement: 350 cid (5.7 liters). Bore & stroke: 4.00 x 3.48 in. Compression ratio: 10.5:1. Net brake hp: 275 at 5000 rpm. Taxable hp: 51.20. Net torque: 325 lbs.-ft. at 2000 rpm. Hydraulic valve lifters. Induction: Sequential fuel injection. Sales code: LT1. VIN code: P.

1995 CORVETTE

Corvette base V-8
Overhead valve. Cast-iron block and head. Bore & stroke: 4.00 x 3.48 in. Displacement: 350 cid (5.7 liters). Compression ratio: 10.5:1. Net brake hp: 300 at 5000 rpm. Taxable hp: 51.20. Net torque: 340 lbs.-ft. at 3600 rpm. Five main bearings. Aluminum cylinder head. Hydraulic valve lifters. Induction: Sequential multiport fuel injection. Sales code: LT1.

Corvette ZR-1 base V-8
Overhead valve with four valves per cylinder. Four overhead camshafts. Cast-iron block and head. Bore & stroke: 3.90 x 3.66 in. Displacement: 350 cid (5.7 liters). Compression ratio: 11.0:1. Net brake hp: 405 at 5800 rpm. Taxable hp: 48.67. Net torque: 385 lbs.-ft. at 5200 rpm. Five main bearings. Aluminum cylinder head. Hydraulic valve lifters. Induction: Sequential multiport fuel-injection system. Sales code: LT5.

1996 CHEVROLET

Caprice Classic base V-8
Overhead-valve V-8. Cast-iron block and head. Bore & stroke: 3.75 x 3.48 in. Displacement: 265 cid (4.3 liters). Compression ratio: 9.9:1. Net brake hp: 200 at 5200 rpm. Taxable hp: 44.8. Net torque: 245 lbs.-ft. at 2400 rpm. Hydraulic valve lifters. Induction: Sequential fuel injection. Sales code: L99. VIN code: W.

Impala SS, Caprice wagon base V-8; Caprice Classic, Caprice Classic LS optional V-8
Overhead-valve V-8. Cast-iron block and head. Displacement: 350 cid (5.7 liters). Bore & stroke: 4.00 x 3.48 in. Compression ratio: 10.0:1. Net brake hp: 260 at 5000 rpm. Taxable hp: 51.20. Net torque: 330 lbs.-ft. at 3200 rpm. Hydraulic valve lifters. Induction: Sequential fuel injection. Sales code: LT1. VIN code: P.

Camaro Z28 base V-8
Overhead-valve V-8. Cast-iron block and head. Displacement: 350 cid (5.7 liters). Bore & stroke: 4.00 x 3.48 in. Compression ratio: 10.5:1. Net brake hp: 275 at 5000 rpm. Taxable hp: 51.20. Net torque: 325 lbs.-ft. at 2400 rpm. Hydraulic valve lifters. Induction: Sequential fuel injection. Sales code: LT1. VIN code: P.

Camaro SS base V-8
Overhead-valve V-8. Cast-iron block and head. Displacement: 350 cid (5.7 liters). Bore & stroke: 4.00 x 3.48 in. Compression ratio: 10.5:1. Net brake hp: 305 at 5500 rpm. Taxable hp: 51.20. Net torque: 325 lbs.-ft. at 2400 rpm. Hydraulic valve lifters. Induction: Sequential fuel injection. Sales code: LT1. VIN code: P.

Camaro SS optional V-8
Overhead-valve V-8. Cast-iron block and head. Displacement: 350 cid (5.7 liters). Bore & stroke: 4.00 x 3.48 in. Compression ratio: 10.5:1. Net brake hp: 310 at 5500 rpm. Taxable hp: 51.20. Net torque: 325 lbs.-ft. at 2400 rpm. Hydraulic valve lifters. Induction: Sequential fuel injection. Performance exhaust system. Sales code: LT1. VIN code: P.

1996 CORVETTE

Corvette base V-8
Overhead valve. Cast-iron block and head. Bore & stroke: 4.00 x 3.48 in. Displacement: 350 cid (5.7 liters). Compression ratio: 10.4:1. Net brake hp: 300 at 5000 rpm. Taxable hp: 51.20. Net torque: 335 lbs.-ft. at 4000 rpm. Five main bearings. Aluminum cylinder head. Hydraulic valve lifters. Induction: Sequential multiport fuel injection. Sales code: LT1.

Corvette optional V-8
Overhead valve with four valves per cylinder. Four overhead camshafts. Cast-iron block and head. Bore & stroke: 4.00 x 3.48 in. Displacement: 350 cid (5.7 liters). Compression ratio: 10.8:1. Net brake hp: 330 at 5800 rpm. Taxable hp: 51.20. Net torque: 340 lbs.-ft. at 4500 rpm. Five main bearings. Aluminum cylinder head. Hydraulic valve lifters. Induction: Sequential multiport fuel-injection system. Sales code: LT4.

1997 CHEVROLET

Camaro Z28 base V-8
Overhead-valve V-8. Cast-iron block and head. Displacement: 350 cid (5.7 liters). Bore & stroke: 4.00 x 3.48 in. Compression ratio: 10.4:1. Net brake hp: 285 at 5200 rpm. Taxable hp: 51.20. Net torque: 325 lbs.-ft. at 2400 rpm. Hydraulic valve lifters. Induction: Sequential fuel injection. Sales code: LT1. VIN code: P.

Camaro SS base V-8
Overhead-valve V-8. Cast-iron block and head. Displacement: 350 cid (5.7 liters). Bore & stroke: 4.00 x 3.48 in. Compression ratio: 10.5:1. Net brake hp: 305 at 5500 rpm. Taxable hp: 51.20. Net torque: 325 lbs.-ft. at 2400 rpm. Hydraulic valve lifters. Induction: Sequential fuel injection. Sales code: LT1. VIN code: P.

Camaro SS optional V-8
Overhead-valve V-8. Cast-iron block and head. Displacement: 350 cid (5.7 liters). Bore & stroke: 4.00 x 3.48 in. Compression ratio: 10.5:1. Net brake hp: 310 at 5500 rpm. Taxable hp: 51.20. Net torque: 325 lbs.-ft. at 2400 rpm. Hydraulic valve lifters. Induction: Sequential fuel injection. Performance exhaust system. Sales code: LT1. VIN code: P.

The LS1 V-8 bowed in the 1997 Corvette and was first added to the limited-edition Camaro SS one year later. This 5.7-liter (350-cid) small-block featured sequential fuel injecton. Seen here is the 1999 LS1 that was available in three versions. The least powerful was a 305-nhp engine that came in the regular Camaro Z28.

1997 CORVETTE

Corvette base V-8
Overhead valve. Cast aluminum block and head. Displacement: 346 cid (5.7 liters). Bore & stroke: 3.90 x 3.62 in. Compression ratio: 10.1:1. Net brake hp: 345 at 5600 rpm. Taxable hp: 49.00. Net torque: 350 lbs.-ft. at 4400 rpm. Hydraulic valve lifters. Crankcase capacity with filter: 6.5 qt. Cooling system capacity with heater: (automatic) 11.5 qt.; (manual) 11.8 qt. Induction: Sequential fuel injection. Sales code: LS1.

1998 CHEVROLET

Camaro Z28 base V-8
Overhead-valve V-8. Cast-iron block and head. Bore & stroke: 3.898 x 3.622 in. Displacement: 346 cid (5.7 liters). Compression ratio: 10.1:1. Net brake hp: 305 at 5200 rpm. Taxable hp: 49.00. Net torque: 335 lbs.-ft. at 4000 rpm. Hydraulic valve lifters. Induction: Sequential fuel injection. Standard in Z28. Sales code: LS1. VIN code: G.

Camaro SS base V-8
Overhead-valve V-8. Cast-iron block and head. Bore & stroke: 3.898 x 3.622 in. Displacement: 346 cid (5.7 liters). Net brake hp: 315. Taxable hp: 49.00. Hydraulic valve lifters. Induction: Sequential fuel injection. Sales code: LS1. VIN code: G.

Camaro SS optional V-8
Overhead-valve V-8. Cast-iron block and head. Bore & stroke: 3.898 x 3.622 in. Displacement: 346 cid (5.7 liters). Net brake hp: 320. Taxable hp: 49.00. Hydraulic valve lifters. Induction: Sequential fuel injection. Performance exhaust system. Sales code: LS1. VIN code: G.

1998 CORVETTE

Corvette base V-8
Overhead valve. Cast aluminum block and head. Displacement: 346 cid (5.7 liters). Bore & stroke: 3.90 x 3.62 in. Compression ratio: 10.1:1. Net brake hp: 345 at 5600 rpm. Taxable hp: 49.00. Net torque: 350 lbs.-ft. at 4400 rpm. Hydraulic valve lifters. Crankcase capacity with filter: 6.5 qt. Cooling system capacity with heater: (automatic) 11.5 qt.; (manual) 11.8 qt. Induction: Sequential fuel injection. Sales code: LS1.

1999 CHEVROLET

Camaro Z28 base V-8
Overhead-valve V-8. Cast-iron block and head. Bore & stroke: 3.898 x 3.622 in. Displacement: 346 cid (5.7 liters). Compression ratio: 10.1:1. Net brake hp: 305 at 5200 rpm. Taxable hp: 49.00. Net torque: 335 lbs.-ft. at 4000 rpm. Hydraulic valve lifters. Induction: Sequential fuel injection. Standard in Z28. Sales code: LS1. VIN code: G.

Camaro SS optional V-8
Overhead-valve V-8. Cast-iron block and head. Bore & stroke: 3.898 x 3.622 in. Displacement: 346 cid (5.7 liters). Net brake hp: 320. Taxable hp: 49.00. Hydraulic valve lifters. Induction: Sequential fuel injection. Performance exhaust system. Sales code: LS1. VIN code: G.

This is another view of the 305-hp 1999 Camaro Z28's engine. The Camaro SS, built by SLP Engineering and sold by Chevrolet dealers, was available with a 320-hp version that featured a performance exhaust system and other upgrades. Also available was the Corvette LS1, which carried a 345-nhp rating.

A cutaway illustration of the special 1999 Chevrolet Kimble Vortec V-8.

1999 CORVETTE

Corvette base V-8
Overhead valve. Cast aluminum block and head. Displacement: 346 cid (5.7 liters). Bore & stroke: 3.90 x 3.62 in. Compression ratio: 10.1:1. Net brake hp: 345 at 5600 rpm. Taxable hp: 49.00. Net torque: 350 lbs.-ft. at 4400 rpm. Hydraulic valve lifters. Crankcase capacity with filter: 6.5 qt. Cooling system capacity with heater: (automatic) 11.5 qt.; (manual) 11.8 qt. Induction: Sequential fuel injection. Sales code: LS1.

2000 CHEVROLET

Camaro Z28 base V-8
Overhead-valve V-8. Cast-iron block and head. Bore & stroke: 3.898 x 3.622 in. Displacement: 346 cid (5.7 liters). Compression ratio: 10.1:1. Net brake hp: 305 at 5200 rpm. Taxable hp: 49.00. Net torque: 335 lbs.-ft. at 4000 rpm. Hydraulic valve lifters. Induction: Sequential fuel injection. Standard in Z28. Sales code: LS1. VIN code: G.

Camaro SS optional V-8
Overhead-valve V-8. Cast-iron block and head. Bore & stroke: 3.898 x 3.622 in. Displacement: 346 cid (5.7 liters). Net brake hp: 320. Taxable hp: 49.00. Hydraulic valve lifters. Induction: Sequential fuel injection. Performance exhaust system. Sales code: LS1. VIN code: G.

2000 CORVETTE

Corvette base V-8
Overhead valve. Cast aluminum block and head. Displacement: 346 cid (5.7 liters). Bore & stroke: 3.90 x 3.62 in. Compression ratio: 10.1:1. Net brake hp: 345 at 5600 rpm. Taxable hp: 49.00. Net torque: 350 lbs.-ft. at 4400 rpm. Hydraulic valve lifters. Crankcase capacity with filter: 6.5 qt. Cooling system capacity with heater: (automatic) 11.5 qt.; (manual) 11.8 qt. Induction: Sequential fuel injection. Sales code: LS1.

2001 CHEVROLET

Camaro Z28 base V-8
Overhead-valve V-8. Cast-iron block and head. Bore & stroke: 3.898 x 3.622 in. Displacement: 346 cid (5.7 liters). Compression ratio: 10.1:1. Net brake hp: 310 at 5200 rpm. Taxable hp: 49.00. Net torque: 340 lbs.-ft. at 4000 rpm. Hydraulic valve lifters. Induction: Sequential fuel injection. Standard in Z28. Sales code: LS1. VIN code: G.

Camaro SS optional V-8
Overhead-valve V-8. Cast-iron block and head. Bore & stroke: 3.898 x 3.622 in. Displacement: 346 cid (5.7 liters). Compression ratio: 10.5:1 Net brake hp: 325 at 5200 rpm. Taxable hp: 49.00. Net torque: 350 lbs.-ft. at 4000 rpm. Hydraulic valve lifters. Induction: Sequential fuel injection. Performance exhaust system. Sales code: LS1. VIN code: G.

2001 CORVETTE

Corvette base V-8
Overhead valve. Cast aluminum block and head. Displacement: 346 cid (5.7 liters). Bore & stroke: 3.90 x 3.62 in. Compression ratio: 10.1:1. Net brake hp: 350 at 5600 rpm. Taxable hp: 49.00. Net torque: (automatic) 360 lbs.-ft. at 4000 rpm; (manual) 375 lbs.-ft. at 4400 rpm. Hydraulic valve lifters. Crankcase capacity with filter: 6.5 qt. Cooling system capacity with heater: (automatic) 11.5 qt.; (manual) 11.8 qt. Induction: Sequential multiport fuel injection. Sales code: LS1.

Chevrolet V-8 Firing Order (1955-1976)

<u>1955-1976</u>

(All V-8)
1-8-4-3-6-5-7-2

NOTE: Cylinders are numbered front to back (left) 1-3-5-7; (right) 2-4-6-8

Corvette Z06 base V-8
Overhead valve. Cast aluminum block and head. Displacement: 346 cu. in. (5.7 liters). Bore & stroke: 3.90 x 3.62 in. Compression ratio: 10.5:1. Net brake hp: 385 at 6000 rpm. Taxable hp: 49.00. Net torque: 385 lbs.-ft. at 4800 rpm. Hydraulic valve lifters. Induction: Sequential multiport fuel injection. Sales code: LS6.

2002 CHEVROLET

Camaro Z28 base V-8
Overhead-valve V-8. Cast-iron block and head. Bore & stroke: 3.898 x 3.622 in. Displacement: 346 cid (5.7 liters). Compression ratio: 10.1:1. Net brake hp: 310 at 5200 rpm. Taxable hp: 49.00. Net torque: 340 lbs.-ft. at 4000 rpm. Hydraulic valve lifters. Induction: Sequential fuel injection. Standard in Z28. Sales code: LS1. VIN code: G.

Camaro SS optional V-8
Overhead-valve V-8. Cast-iron block and head. Bore & stroke: 3.898 x 3.622 in. Displacement: 346 cid (5.7 liters). Compression ratio: 10.5:1 Net brake hp: 325 at 5200 rpm. Taxable hp: 49.00. Net torque: 350 lbs.-ft. at 4000 rpm. Hydraulic valve lifters. Induction: Sequential fuel injection. Performance exhaust system. Sales code: LS1. VIN code: G.

2002 CORVETTE

Corvette base V-8
Overhead valve. Cast aluminum block and head. Displacement: 346 cid (5.7 liters). Bore & stroke: 3.90 x 3.62 in. Compression ratio: 10.1:1. Net brake hp: 350 at 5600 rpm. Taxable hp: 49.00. Net torque: (automatic) 360 lbs.-ft. at 4000 rpm; (manual) 375 lbs.-ft. at 4400 rpm. Hydraulic valve lifters. Crankcase capacity with filter: 6.5 qt. Cooling system capacity with heater: (automatic) 11.5 qt.; (manual) 11.8 qt. Induction: Sequential multiport fuel injection. Sales code: LS1.

Corvette Z06 base V-8
Overhead valve. Cast aluminum block and head. Displacement: 346 cid (5.7 liters). Bore & stroke: 3.90 x 3.62 in. Compression ratio: 10.5:1. Net brake hp: 405 at 6000 rpm. Taxable hp: 49.00. Net torque: 400 lbs.-ft. at 4800 rpm. Hydraulic valve lifters. Crankcase capacity with filter: 6.5 qt. Cooling system capacity with heater: (automatic) 11.5 qt.; (manual) 11.8 qt. Induction: Sequential multiport fuel injection. Sales code: LS6.

Chevrolet offered a two-barrel version of the 350-cid V-8 and used it as the base V-8 in Monte Carlos, Chevelles, and Caprices. This engine had a 9.0:1 compression ratio and single exhaust system. It produced 250 hp at 4800 rpm and 345 foot-pounds of torque at 2800 rpm.

Selected Chevrolet V-8 Tune-Up Specs 1955-1965

Year	Model	Spark Plugs		Point	Deg Cam		Idle	Compression	Battery
		Type	Gap	Gap	Angle	Timing	Speed	Pressure	(Delco)
1955 (All 265 cid)									
	V-8 (manual trans)	AC44-5	.035	.019	30	4B	475 rpm	150	2SM
	V-8 (Powerglide)	AC44-5	.035	.019	30	4B	500 rpm	150	2SM
1956 (All 265 cid)									
	V-8 (manual trans)	AC44	.035	.019	32	4B	475 rpm	150	2SM
	V-8 (Powerglide)	AC44	.035	.019	32	4B	500 rpm	150	2SM
	V-8 (2 four barrel)	AC44	.035	.019	32	12B	500 rpm	160	2SM
1957									
	265 V-8 (manual)	AC44	.035	.019	30	4B	475 rpm	150	2SM
	283 V-8	AC44	.035	.019	30	4B	475 rpm	155	2SM
	V-8 (2 four barrel)	AC44	.035	.019	30	12B	500 rpm	160	2SM
NOTE 1: Point gap setting is for new points. For old points use .016 in.									
1958 (All V-8s)									
	V-8	AC44	.033-.038	.019	30	4B	475 rpm	NA	2SMR-53
NOTE 2: Point gap setting is for new points. For old points use .016 in.									
1959									
	283 V-8	AC44	.035	.019	30	4B	450 rpm	NA	2SMR-53
	283 V-8 (spl cam)	AC44	.035	.019	30	4B	450 rpm	NA	2SMR-53
	348 V-8	AC44	.035	.019	30	8B	450 rpm	NA	2SMR-53
NOTE 3: Point gap setting is for new points. For old points use .016 in.									
1960									
	283/170-hp V-8	AC44	.033-.038	.016-.019	30	4B	475 rpm	140	Type 458
	283/230-hp V-8	AC44	.033-.038	.016-.019	30	4B	475 rpm	140	Type 458
	283 spec. cam V-8	AC44	.033-.038	.016-.019	30	4B	475 rpm	150	Type 458
	348/250-hp V-8	AC44N	.033-.038	.016-.019	30	4B	475 rpm	150	Type 458
	348/280-hp V-8	AC44N	.033-.038	.016-.019	NOTE 4	4B	475 rpm	150	Type 458
	348/305-hp V-8	AC44N	.033-.038	.016-.019	30	8B	475 rpm	150	Type 668
	348/320-hp V-8	AC44N	.033-.038	.016-.019	30	8B	475 rpm	150	Type 668
	348/335-hp V-8	AC43N	.033-.038	.016-.019	NOTE 4	12B	475 rpm	150	Type 668
NOTE 4: Set Chevrolet distributor with dual breaker points to 28-30 per set of points; total dwell both sets of points is 33-35 degrees.									
1961									
	283/170-hp V-8	AC44	.033-.038	.016-.019	30	4B	475 rpm	140	1980458
	283/230-hp V-8	AC44	.033-.038	.016-.019	30	4B	475 rpm	140	1980458
	283 spec. cam V-8	AC44	.033-.038	.016-.019	30	4B	475 rpm	150	1980458
	348/250-hp V-8	AC44N	.033-.038	.016-.019	30	4B	475 rpm	150	1980558
	348/280-hp V-8	AC44N	.033-.038	.016-.019	NOTE 5	4B	475 rpm	150	Type 458
	348/305-hp V-8	AC44N	.033-.038	.016-.019	30	8B	475 rpm	150	1980558
	348/320-hp V-8	AC44N	.033-.038	.016-.019	30	8B	475 rpm	150	1980558
	348/335-hp V-8	AC43N	.033-.038	.016-.019	NOTE 5	12B	475 rpm	150	1980558
	348/340-hp V-8	AC44N	.033-.038	.016-.019	30	12B	475 rpm	150	1980558
	348/350-hp V-8	AC44N	.033-.038	.016-.019	NOTE 5	8B	475 rpm	150	1980558
	348/360-hp V-8	AC43N	.033-.038	.016-.019	30	12B	475 rpm	150	1980558
NOTE 5: Set Chevrolet distributor with dual breaker points to 28-30 per set of points; total dwell both sets of points is 33-35 degrees.									
1962									
	283 V-8	AC46	.35	.019-.023	30	4B	475 rpm	140	1980458
	327 V-8	AC44N	.35	.019-.023	30	4B	475 rpm	160	1980558
	327 spec. cam V-8	AC44N	.35	.019-.023	30	8B	475 rpm	160	1980558
	409 V-8	AC43N	.35	.019-.023	NOTE 6	12B	650 rpm	150	1980458
NOTE 6: Set Chevrolet 409-cid distributor with dual breaker points to 28-30 per set of points; total dwell both sets of points is 33-35 degrees.									
1963									
	283 V-8	AC46	.35	.016-.019	30	4B	475 rpm	140	1980458
	327 V-8 (250 hp)	AC44N	.35	.016-.019	30	4B	475 rpm	160	1980558
	327 V-8 (300 hp)	AC44N	.35	.016-.019	30	8B	475 rpm	160	1980558
	409 V-8 (340 hp)	AC43N	.35	.016-.019	30	6B	650 rpm	150	1980458
	409 V-8 (400 hp)	AC43N	.35	.016-.019	30	12B	650 rpm	150	1980458
	409 V-8 (425 hp)	AC43N	.35	.016-.019	30	12B	650 rpm	150	1980458
1964									
	283 V-8	AC45	.35	.019	30	4B	500 rpm	150	2SMR
	327 V-8 (std)	AC44	.35	.019	30	4B	500 rpm	160	2SMD
	327 V-8 (hi-perf)	AC44	.35	.019	30	8B	500 rpm	160	2SMD
	409 V-8 (std)	AC43N	.35	.019	30	6B	700 rpm	150	2SMD
	409 V-8 (hi-perf)	AC43N	.35	.019	30	12B	700 rpm	150	2SMD
1965									
	283 V-8	AC45	.35	.019	30	4B	500 rpm	150	2SMR
	327 V-8 (250 hp)	AC44	.35	.019	30	4B	550 rpm	160	2SMD
	327 V-8 (300 hp)	AC44	.35	.019	30	8B	550 rpm	160	2SMD
	327 V-8 (350 hp)	AC44	.35	.019	30	12B	800 rpm	160	2SMD
	409 V-8 (std)	AC43N	.35	.019	30	6B	500 rpm	150	2SMD
	409 V-8 (hi-perf)	AC43N	.35	.019	30	12B	700 rpm	150	2SMD

CHRYSLER
1951-1989

When Chrysler decided to offer a V-8 in 1951, it got off on the right foot with the famous "Firepower" Hemi. This modern (for the time) short-stroke, overhead-valve design was unique because of its hemispherical combustion chambers. While it had the same cubic-inch displacement as the Cadillac V-8 it produced quite a bit more horsepower. Automotive writer Tom McCahill highlighted its potential with a 10.9-second 0-to-60 mph run in a '51 New Yorker.

The first major change in the Hemi came in 1954 when, without any added displacement, the New Yorker offered 235 hp. The engine was enlarged to 392-cid in 1957, when the top power option was 325 hp. That climbed to 345 hp in the 1958 New Yorker.

Early drag racers flocked to the Chrysler Firepower V-8 and set many speed records with it, as did NASCAR drivers and Mexican Road Race participants. However, by the late 1950s, Chrysler realized that the first-generation Hemi was too expensive to make and too heavy to use in contemporary passenger cars. In 1955, the company introduced its new "A" series "Spitfire" V-8 with polyspherical combustion chambers for smaller Chryslers and other corporate brands. This change ultimately sired the famous 273- and 318-cid Chrysler small-block V-8s, but it did not really solve the heavy Hemi problem. Mopar engineers then designed a wedge-head V-8 that offered many of the Hemi's high-performance characteristics. These "B" Series intermediate V-8s—like the 350, 361, 383, and 413—bridged the performance gap. They found their way to America's drag strips in the early-to-mid 1960s.

The Hemi was revised in 1964 as a racing-only option and as a factory-installed option in some 1966 Chrysler Corp. cars. However, it was not offered in Chrysler brand cars, which instead got a 440-cid wedge-head big-block V-8 starting in 1966. The 440 was a direct descendant of the 413 with a longer stroke and other component and casting upgrades. It generated up to 375 hp.

By 1971, with high insurance rates and new anti-pollution rules weighing on automakers, Chrysler began to retreat to a 360-cid V-8 that was a stroked version of the 318-cid small-block. The following year's new V-8 was a 400-cid job. When the Cordoba came out in 1975, its engine was the 318-cid V-8 and it produced 135 net horsepower. Something about those numbers sounded a little un-Chrysler like. However, the 360-, 400- and 440-cid V-8s were retained through 1978. After that, Chrysler buyers had a choice of the 318 or 360 until 1980. The 318 was the only V-8 offered from 1981 until the end of the Chrysler V-8 in 1989.

In 1951, advertising copywriters described the 180-hp Chrysler Firepower V-8 as a "sensational" engine. The heart of this power plant was the hemispherical combustion chamber, which Chrysler engineers developed to produce one of the most powerful, most efficient, best-constructed V-8s ever seen. The engine was designed to run on regular gasoline and provide full combustion, full compression, and full energy in an almost carbon-free operating environment.

NOTE: Chrysler V-8s are listed in order of horsepower, not cubic-inch displacement. Imperial V-8s are listed following Chrysler V-8s, even in years when the Imperial used the New Yorker V-8.

1951 CHRYSLER

Saratoga, New Yorker base "Firepower" V-8
Overhead valve. Hemispherical combustion chambers. Bore & stroke: 3.81 x 3.63 in. Displacement: 331.1 cid. Compression ratio: 7.50:1. Brake hp: 180 at 4000 rpm. Taxable hp: 46.5. Torque: 312 lbs.-ft. at 2000 rpm. Five main bearings. Hydraulic valve lifters. Crankcase capacity: 5 qt. (add 1 qt. for new filter). Cooling system capacity: 25 qt. (add 1 qt. for heater). Carburetors: Early cars used Carter WCD 830S, 830SA, 830SB or 931 SC. Later cars used Carter WCD 931 S, 931 SA, 931 SB, or 931 SC two-barrel. Engine code: (New Yorker) C52; (Saratoga) C55.

Imperial base "Firepower" V-8
Overhead valve. Hemispherical combustion chambers. Bore & stroke: 3.81 x 3.63 in. Displacement: 331.1 cid. Compression ratio: 7.50:1. Brake hp: 180 at 4000 rpm. Taxable hp: 46.5. Torque: 312 lbs.-ft. at 2000 rpm. Five main bearings. Hydraulic valve lifters. Crankcase capacity: 5 qt. (add 1 qt. for new filter). Cooling system capacity: 25 qt. (add 1 qt. for heater). Carburetor: Carter two-barrel Model WCD-830S two-barrel. Engine code: (Imperial) C54; (Crown Imperial) C53.

1952 CHRYSLER

Saratoga, New Yorker base "Firepower" V-8
Overhead valve. Hemispherical combustion chambers. Bore & stroke: 3.81 x 3.63 in. Displacement: 331.1 cid. Compression ratio: 7.50:1. Brake hp: 180 at 4000 rpm. Taxable hp: 46.5. Torque: 312 lbs.-ft. at 2000 rpm.

Five main bearings. Hydraulic valve lifters. Crankcase capacity: 5 qt. (add 1 qt. for new filter). Cooling system capacity: 25 qt. (add 1 qt. for heater). Carburetors: Carter WCD 830-S, 931 S, 931 SA, 931 SB, or 931 SC two-barrel. Engine code: (New Yorker) C52; (Saratoga) C55. Engine code: (New Yorker) C52; (Saratoga) C55.

Imperial base "Firepower" V-8
Overhead valve. Hemispherical combustion chambers. Bore & stroke: 3.81 x 3.63 in. Displacement: 331.1 cid. Compression ratio: 7.50:1. Brake hp: 180 at 4000 rpm. Taxable hp: 46.5. Torque: 312 lbs.-ft. at 2000 rpm. Five main bearings. Hydraulic valve lifters. Crankcase capacity: 5 qt. (add 1 qt. for new filter). Cooling system capacity: 25 qt. (add 1 qt. for heater). Carburetor: Carter two-barrel Model WCD-884S, 830-S two-barrel. Engine code: (Imperial) C54; (Crown Imperial) C53.

1953 CHRYSLER

New Yorker base "Firepower" V-8
Overhead valve. Hemispherical combustion chambers. Bore & stroke: 3.81 x 3.63 in. Displacement: 331.1 cid. Compression ratio: 7.5:1. Brake hp: 180 at 4000 rpm. Taxable hp: 46.5. Torque: 312 lbs.-ft. at 2000 rpm. Five main bearings. Hydraulic valve lifters. Crankcase capacity: 5 qt. (add 1 qt. for new filter). Cooling system capacity: 25 qt. (add 1 qt. for heater). Carburetors: Carter WCD 935-S, 935-SA two-barrel. Engine code: (New Yorker) C56-1; (New Yorker Deluxe) C56-2.

Imperial base "Firepower" V-8
Overhead valve. Hemispherical combustion chambers. Bore & stroke: 3.81 x 3.63 in. Displacement: 331.1 cid. Compression ratio: 7.5:1. Brake hp: 180 at 4000 rpm. Taxable hp: 46.5. Torque: 312 lbs.-ft. at 2000 rpm. Five main bearings. Hydraulic valve lifters. Crankcase capacity: 5 qt. (add 1 qt. for new filter). Cooling system capacity: 25 qt. (add 1 qt. for heater). Carburetor: Carter WCD-992-S two-barrel. Engine code: (Crown Imperial) C59; (Custom Imperial) C58.

1954 CHRYSLER

New Yorker base "Firepower" V-8
Overhead valve. Hemispherical combustion chambers. Bore & stroke: 3.81 x 3.63 in. Displacement: 331.1 cid. Compression ratio: 7.5:1. Brake hp: 195 at 4400 rpm. Taxable hp: 46.5. Torque: 320 lbs.-ft. at 2000 rpm. Five main bearings. Hydraulic valve lifters. Crankcase capacity: 5 qt. (add 1 qt. for new filter). Cooling system capacity: 25 qt. (add 1 qt. for heater). Carburetor: Carter BBD-919S two-barrel. Engine code: C63-1.

New Yorker Deluxe base "Firepower" V-8
Overhead valve. Hemispherical combustion chambers. Bore & stroke: 3.81 x 3.63 in. Displacement: 331.1 cid. Compression ratio: 7.5:1. Brake hp: 235 at 4400 rpm. Taxable hp: 46.5. Torque: 330 lbs.-ft. at 2600 rpm. Five main bearings. Hydraulic valve lifters. Crankcase capacity: 5 qt. (add 1 qt. for new filter). Cooling system capacity: 25 qt. (add 1 qt. for heater). Dual exhaust system. Carburetor: WCFB Carter 2041S four-barrel. Engine code: C63-2.

Imperial base "Firepower" V-8
Overhead valve. Hemispherical combustion chambers. Bore & stroke: 3.81 x 3.63 in. Displacement: 331.1 cid. Compression ratio: 7.5:1. Brake hp: 235 at 4400 rpm. Taxable hp: 46.5. Torque: 330 lbs.-ft. at 2600 rpm. Five main bearings. Hydraulic valve lifters. Crankcase capacity: 5 qt. (add 1 qt. for new filter). Cooling system capacity: 25 qt. (add 1 qt. for heater). Dual exhaust system. Carburetor: WCFB Carter 2041S four-barrel. Engine code: (Custom Imperial) C64; (Crown Imperial) C66.

1955 CHRYSLER

Windsor Deluxe base "Spitfire" V-8
Overhead valve. Cast-iron block. Bore & stroke: 3.625 x 3.625 in. Displacement: 301 cid. Compression ratio: 8.0:1. Brake hp: 188 at 4400 rpm. Taxable hp: 42.2. Torque: 275 lbs.-ft. at 2400 rpm. Five main bearings. Hydraulic valve lifters. Crankcase capacity: 5 qt. (add 1 qt. for new filter). Cooling system capacity: 25 qt. (add 1 qt. for heater). Carburetors: (manual transmission) Carter BBD 2180S, 2180SA, 2180SB two-barrel. (PowerFlite automatic transmission) BBD 2162S, 2162SA, 2162SB two-barrel. Engine code: C67.

New Yorker base "Firepower" V-8
Overhead valve. Hemispherical combustion chambers. Bore & stroke: 3.81 x 3.63 in. Displacement: 331.1 cid. Compression ratio: 8.5:1. Brake hp: 250 at 4600 rpm. Taxable hp: 46.5. Torque: 340 lbs.-ft. at 2800 rpm. Five main bearings. Hydraulic valve lifters. Crankcase capacity: 5 qt. (add 1 qt. for new filter). Cooling system capacity: 25 qt. (add 1 qt. for heater). Dual exhaust system. Carburetor: Carter WCFB 2126S four-barrel. Engine code: C68.

Chrysler C-300 base "Firepower" V-8
Overhead valve. Hemispherical combustion chambers. Bore & stroke: 3.81 x 3.63 in. Displacement: 331.1 cid. Compression ratio: 8.5:1. Brake hp: 300 at 5200 rpm. Taxable hp: 49.7. Torque: 385 lbs.-ft. at 3400 rpm. Five main bearings. Hydraulic valve lifters. Crankcase capacity: 5 qt. (add 1 qt. for new filter). Cooling system capacity: 25 qt. (add 1 qt. for heater). Dual exhaust system. Solid lifters with full-race camshaft. Two four-barrel carburetors. Engine code: C68-300.

Imperial base "Firepower" V-8
Overhead valve. Hemispherical combustion chambers. Bore & stroke: 3.81 x 3.63 in. Displacement: 331.1 cid. Compression ratio: 8.5:1. Brake hp: 250 at 4600 rpm. Taxable hp: 46.5. Torque: 340 lbs.-ft. at 2800 rpm. Five main bearings. Hydraulic valve lifters. Crankcase capacity: 5 qt. (add 1 qt. for new filter). Cooling system capacity: 25 qt. (add 1 qt. for heater). Dual exhaust system. Carburetor: Carter four-barrel Model WCFB-2126S four-barrel. Engine code: (Custom Imperial) C69; (Crown Imperial) C70.

1956 CHRYSLER

Windsor base "Spitfire" V-8
Overhead valve. Cast-iron block. Bore & stroke: 3.81 x 3.63 in. Displacement: 331.1 cid. Compression ratio: 8.5:1. Brake hp: 225 at 4400 rpm. Taxable hp: 46.5. Torque: 310 lbs.-ft. at 2400. Five main bearings. Hydraulic valve lifters. Crankcase capacity: 5 qt. (add 1 qt. for new filter). Cooling system capacity: 25 qt. (add 1 qt. for heater). Carburetors: (manual transmission) Carter BBD 2312S two-barrel; (PowerFlite automatic transmission) Carter 2313S two-barrel. Engine code: C71.

Windsor power-pack "Spitfire" V-8
Overhead valve. Cast-iron block. Bore & stroke: 3.81 x 3.63 in. Displacement: 331.1 cid. Compression ratio: 8.5:1. Brake hp: 250 at 4600 rpm. Taxable hp: 46.5. Torque: 340 lbs.-ft. at 2800. Five main bearings. Hydraulic valve lifters. Crankcase capacity: 5 qt. (add 1 qt. for new filter). Cooling system capacity: 25 qt. (add 1 qt. for heater). Carburetors: Carter WCFB 2367S or 2367SA four-barrel. Engine code: C71.

New Yorker base "Firepower" V-8
Overhead valve. Hemispherical combustion chambers. Cast-iron block. Bore & stroke: 3.94 x 3.63 in. Displacement: 354 cid. Compression ratio: 9.0:1. Brake hp: 280 at 4600 rpm. Taxable hp: 49.7. Torque: 380 lbs.-ft. at 2800 rpm. Five main bearings. Hydraulic valve lifters. Crankcase capacity: 5 qt. (add 1 qt. for new filter). Cooling system capacity: 25 qt. (add 1 qt. for heater). Carburetor: Carter WCFB 2314S or 2314SA four-barrel. Engine code: C72.

Chrysler 300-B base "Firepower" V-8
Overhead valve. Hemispherical combustion chambers. Cast-iron block. Bore & stroke: 3.94 x 3.63 in. Displacement: 354 cid. Compression ratio: 9.0:1. Brake hp: 340 at 5200 rpm. Taxable hp: 49.7. Torque: 385 lbs.-ft. at 3400 rpm. Five main bearings. Solid lifters with full-race camshaft. Crankcase capacity: 5 qt. (add 1 qt. for new filter). Cooling system capacity: 25 qt. (add 1 qt. for heater). Carburetor: Two four-barrel carburetors. Engine code: C72-300.

Chrysler 300-B optional "Firepower" V-8
Overhead valve. Hemispherical combustion chambers. High-lift camshaft. Extra stiff valve springs. Cast-iron block. Bore & stroke: 3.94 x 3.63 in. Displacement: 354 cid. Compression ratio: 10.0:1. Taxable hp: 49.7. Brake hp: 355 at 5200 rpm. Torque: unknown. Five main bearings. Solid lifters with full-race camshaft. Crankcase capacity: 5 qt. (add 1 qt. for new filter). Cooling system capacity: 25 qt. (add 1 qt. for heater). Carburetor: Two Carter four-barrel carburetors. Engine code: C72-300.

Imperial base "Firepower" V-8
Overhead valve. Hemispherical combustion chambers. Cast-iron block. Bore & stroke: 3.94 x 3.63 in. Displacement: 354 cid. Compression ratio: 9.0:1. Brake hp: 280 at 4600 rpm. Taxable hp: 49.7. Torque: 380 lbs.-ft. at 2800 rpm. Five main bearings. Hydraulic valve lifters. Crankcase capacity: 5 qt. (add 1 qt. for new filter). Cooling system capacity: 25 qt.

(add 1 qt. for heater). Carburetor: Carter four-barrel Model WCFB-2314S four-barrel. Engine code: (Imperial) C73; (Crown Imperial) C70.

1957 CHRYSLER

Windsor base "Spitfire" V-8
Overhead valve. Cast-iron block. Bore & stroke: 3.94 x 3.63 in. Displacement: 354 cid. Compression ratio: 9.25:1. Brake hp: 285 at 4600 rpm. Taxable hp: 49.7. Torque: 365 lbs.-ft. at 2400 rpm. Five main bearings. Hydraulic valve lifters. Crankcase capacity: 5 qt. (add 1 qt. for new filter). Cooling system capacity: 21 qt. (add 1 qt. for heater). Carburetor: Carter BBD Model 2527S two-barrel. Engine code: C-75-1.

Saratoga base "Spitfire" V-8
Overhead valve. Cast-iron block. Bore & stroke: 3.94 x 3.63 in. Displacement: 354 cid. Compression ratio: 9.25:1. Brake hp: 295 at 4600 rpm. Taxable hp: 49.7. Torque: 390 lbs.-ft. at 2800 rpm. Five main bearings. Hydraulic valve lifters. Crankcase capacity: 5 qt. (add 1 qt. for new filter). Cooling system capacity: 21 qt. (add 1 qt. for heater). Dual exhaust. Carburetor: Carter WCFB Model 2589-S four-barrel. Engine code: C-75-2.

New Yorker base "Firepower" V-8
Overhead valve. Hemispherical combustion chambers. Cast-iron block. Bore & stroke: 4.00 x 3.90 in. Displacement: 392 cid. Compression ratio: 9.25:1. Brake hp: 325 at 4600 rpm. Taxable hp: 51.2. Torque: 430 lbs.-ft. at 2800 rpm. Five main bearings. Hydraulic valve lifters. Crankcase capacity: 5 qt. (add 1 qt. for new filter). Cooling system capacity: 24 qt. (add 1 qt. for heater). Dual exhaust. Carburetor: Carter WCFB Model 2590-S four-barrel. Engine code: C-76.

Chrysler 300-C base "Firepower" V-8
Overhead valve. Hemispherical combustion chambers. High-lift camshaft. Extra stiff valve springs. Cast-iron block. Bore & stroke: 4.00 x 3.90 in. Displacement: 392 cid. Compression ratio: 9.25:1. Brake hp: 375 at 5200 rpm. Taxable hp: 51.2.Torque: 420 lbs.-ft. at 4000 rpm. Five main bearings. Hydraulic valve lifters. Crankcase capacity: 5 qt. (add 1 qt. for new filter). Cooling system capacity: 24 qt. (add 1 qt. for heater). Dual exhaust. Carburetor: Two Carter Model WCFB 2334-S four-barrel (1119168 with air conditioning). Engine code: C-76-300.

Chrysler 300-C special "Firepower" V-8
Overhead valve. Hemispherical combustion chambers. High-lift camshaft. Extra stiff valve springs. Cast-iron block. Bore & stroke: 4.00 x 3.90 in. Displacement: 392 cid. Compression ratio: 10.0:1. Brake hp: 390 at 5400 rpm. Taxable hp: 51.2. Torque: unknown. Five main bearings. Hydraulic valve lifters. Crankcase capacity: 5 qt. (add 1 qt. for new filter). Cooling system capacity: 24 qt. (add 1 qt. for heater). Carburetor: Two Carter Model WCFB 2334S four-barrel. This extra-cost solid lifter competition engine was intended mainly for acceleration trials and stock racing cars. It was available only with manual transmission (adapted from a Dodge column-mounted three-speed) and no power options. It had four-bolt Cast-iron exhaust headers and a 2-1/2-inch low back-pressure dual exhaust system. Engine code: C-76-300 *.

Crown Imperial, Imperial LeBaron base "Firepower" V-8
Overhead valve. Hemispherical combustion chambers. Cast-iron block. Bore & stroke: 4.00 x 3.90 in. Displacement: 392 cid. Compression ratio: 9.25:1. Brake hp: 325 at 4600 rpm. Taxable hp: 51.2. Torque: 430 lbs.-ft. at 2800 rpm. Five main bearings. Hydraulic valve lifters. Crankcase capacity: 5 qt. (add 1 qt. for new filter). Cooling system capacity: 24 qt. (add 1 qt. for heater). Carburetor: Carter Model WCFB-2590-S four-barrel. Engine code: (Imperial) IM-1; (Crown Imperial) IM-2 and (Imperial LeBaron) IM-4.

1958 CHRYSLER

Windsor base "Spitfire" V-8
Overhead valve. Cast-iron block. Bore & stroke: 3.94 x 3.63 in. Displacement: 354 cid. Compression ratio: 10.0:1. Brake hp: 290 at 4600 rpm. Taxable hp: 49.7. Torque: 385 lbs.-ft. at 2000 rpm. Five main bearings. Hydraulic valve lifters. Crankcase capacity: 4 qt. (add 1 qt. for new filter). Cooling system capacity: 21 qt. (add 1 qt. for heater). Carburetor: Carter BBD Model 2733-S two-barrel. Engine code: LC1-L.

Saratoga base "Spitfire" V-8
Overhead valve. Cast-iron block. Bore & stroke: 3.94 x 3.63 in. Displacement: 354 cid. Compression ratio: 10.0:1. Brake hp: 310 at 4600 rpm. Taxable hp: 49.7. Torque: 405 lbs.-ft. at 3200 rpm. Five main bearings. Hydraulic valve lifters. Crankcase capacity: 4 qt. (add 1 qt. for new filter). Cooling system capacity: 21 qt. (add 1 qt. for heater). Carburetor: Carter AFB Model 2650-S four-barrel. Engine code: LC2-M.

New Yorker base "Firepower" V-8
Overhead valve. Hemispherical combustion chambers. Cast-iron block. Bore & stroke: 4.00 x 3.90 in. Displacement: 392 cid. Compression ratio: 10.0:1. Brake hp: 345 at 4600 rpm. Taxable hp: 51.2. Torque: 450 lbs.-ft. at 2800 rpm. Five main bearings. Hydraulic valve lifters. Crankcase capacity: 5 qt. (add 1 qt. for new filter). Cooling system capacity: 24 qt. (add 1 qt. for heater). Dual exhaust. Carburetor: Carter Type AFB Model 2651-S four-barrel. Engine code: LC3-H.

Chrysler 300-D base "Firepower" V-8
Overhead valve. Hemispherical combustion chambers. High-lift camshaft. Extra stiff valve springs. Cast-iron block. Bore & stroke: 4.00 x 3.90 in. Displacement: 392 cid. Compression ratio: 10.0:1. Brake hp: 380 at 5200 rpm. Taxable hp: 51.2. Torque: 435 lbs.-ft. at 3600 rpm. Five main bearings. Hydraulic valve lifters. Crankcase capacity: 5 qt. (add 1 qt. for new filter). Cooling system capacity: 24 qt. (add 1 qt. for heater). Dual exhaust. Carburetor: Two Carter Model WCFB (front) Model 2741-S four-barrel; (rear) Model 2742-S four-barrel. Engine code: LC3-S.

Chrysler 300-D fuel-injected "Firepower" V-8
Overhead valve. Hemispherical combustion chambers. High-lift camshaft. Extra stiff valve springs. Cast-iron block. Bore & stroke: 4.00 x 3.90 in. Displacement: 392 cid. Compression ratio: 10.0:1. Brake hp: 390 at 5200 rpm. Taxable hp: 51.2. Torque: unknown. Five main bearings. Hydraulic valve lifters. Crankcase capacity: 5 qt. (add 1 qt. for new filter). Cooling system capacity: 24 qt. (add 1 qt. for heater). Dual exhaust. Induction system: Bendix fuel injection. (NOTE: Available only for a short time and recalled. Recalled cars were retrofitted with two four-barrel Carter carburetors). Engine code: LC3-S.

Crown Imperial, Imperial LeBaron base "Firepower" V-8
Overhead valve. Hemispherical combustion chambers. Cast-iron block. Bore & stroke: 4.00 x 3.90 in. Displacement: 392 cid. Compression ratio: 10.0:1. Brake hp: 345 at 4600 rpm. Taxable hp: 51.2. Torque: 450 lbs.-ft. at 2800 rpm. Five main bearings. Hydraulic valve lifters. Crankcase capacity: 5 qt. (add 1 qt. for new filter). Cooling system capacity: 24 qt. (add 1 qt. for heater). Dual exhaust. Carburetor: Carter Type AFB Model 2651-S four-barrel. Engine code: (Imperial) LY1-L; (Crown Imperial) LY1-M and (Imperial LeBaron) LY1-H.

1959 CHRYSLER

Windsor base V-8
Overhead valve. Cast-iron block. Bore & stroke: 4.03 x 3.75 in. Displacement: 383 cid. Compression ratio: 10.10:1. Brake hp: 305 at 4600 rpm. Taxable hp: 52.0. Torque: 410 lbs.-ft. at 2400 rpm. Five main bearings. Hydraulic valve lifters. Crankcase capacity: 5 qt. (add 1 qt. for new filter). Cooling system capacity: 17 qt. (add 1 qt. for heater). Carburetor: Carter BBD Model 2072-S two-barrel. Engine code: MC1-L.

Saratoga base V-8
Overhead valve. Cast-iron block. Bore & stroke: 4.03 x 3.75 in. Displacement: 383 cid. Compression ratio: 10.10:1. Brake hp: 325 at 4600 rpm. Taxable hp: 52.0. Torque: 425 lbs.-ft. at 2800 rpm. Five main bearings. Hydraulic valve lifters. Crankcase capacity: 5 qt. (add 1 qt. for new filter). Cooling system capacity: 17 qt. (add 1 qt. for heater). Carburetor: Carter Type AFB Model 2797-S four-barrel. Engine code: MC2-M.

New Yorker base V-8
Overhead valve. Cast-iron block. Bore & stroke: 4.00 x 3.90 in. Displacement: 392 cid. Compression ratio: 10.10:1. Brake hp: 350 at 4600 rpm. Taxable hp: 55.90. Torque: 470 lbs.-ft. at 2800 rpm. Five main bearings. Hydraulic valve lifters. Crankcase capacity: 5 qt. (add 1 qt. for new filter). Cooling system capacity: 16 qt. (add 1 qt. for heater). Dual exhaust. Carburetor: Carter Type AFB Model 2651-S four-barrel. Engine code: MC3-H.

Chrysler 300-E base V-8
Overhead valve. Cast-iron block. Bore & stroke: 4.00 x 3.90 in. Displacement: 392 cid. Compression ratio: 10.10:1. Brake hp: 380 at 5000 rpm. Taxable hp: 55.90. Torque: 450 lbs.-ft. at 3600 rpm. Five main bearings. Hydraulic valve lifters. Crankcase capacity: 5 qt. (add 1 qt. for new filter). Cooling system capacity: 16 qt. (add 1 qt. for heater). Dual exhaust. Carburetor: Two Carter Type AFB (front) Model 2798-S four-barrel; (rear) Model 2799-S four-barrel. Engine code: MC3-S.

Imperial Custom, Crown Imperial, Imperial LeBaron base V-8
Overhead valve. Cast-iron block. Bore & stroke: 4.00 x 3.90 in. Displacement: 392 cid. Compression ratio: 10.10:1. Brake hp: 350 at 4600 rpm. Taxable hp: 55.90. Torque: 470 lbs.-ft. at 2800 rpm. Five main bearings. Hydraulic valve lifters. Crankcase capacity: 5 qt. (add 1 qt. for new filter). Cooling system capacity: 16 qt. (add 1 qt. for heater). Dual exhaust. Carburetor: Carter four-barrel Model AFB-2797-S. Engine code: (Imperial) MY1-L; (Crown Imperial) MY1-M; and (Imperial LeBaron) MY1-H.

1960 CHRYSLER

Windsor base V-8
Overhead valve. Cast-iron block. Bore & stroke: 4.03 x 3.75 in. Displacement: 383 cid. Compression ratio: 10.10:1. Brake hp: 305 at 4600 rpm. Taxable hp: 52.0. Torque: 410 lbs.-ft. at 2400 rpm. Five main bearings. Hydraulic valve lifters. Crankcase capacity: 5 qt. (add 1 qt. for new filter). Cooling system capacity: 16 qt. (add 1 qt. for heater). Carburetor: Carter BBD Model 2924-S two-barrel. Engine code: PR-38.

Saratoga base V-8
Overhead valve. Cast-iron block. Bore & stroke: 4.03 x 3.75 in. Displacement: 383 cid. Compression ratio: 10.10:1. Brake hp: 325 at 4600 rpm. Taxable hp: 52.0. Torque: 425 at 2800 rpm. Five main bearings. Hydraulic valve lifters. Crankcase capacity: 5 qt. (add 1 qt. for new filter). Cooling system capacity: 16 qt. (add 1 qt. for heater). Carburetor: Carter Type AFB Model 2927-S four-barrel. PR-38.

New Yorker base V-8; Saratoga optional V-8
Overhead valve. Cast-iron block. Bore & stroke: 4.18 x 3.75 in. Displacement: 413 cid. Compression ratio: 10.10:1. Brake hp: 350 at 4600 rpm. Taxable hp: 55.90. Torque: 470 lbs.-ft. at 2800 rpm. Five main bearings. Hydraulic valve lifters. Crankcase capacity: 5 qt. (add 1 qt. for new filter). Cooling system capacity: 16 qt. (add 1 qt. for heater). Dual exhaust. Carburetor: Carter AFB-2903-S four-barrel. Engine code: P-41.

Chrysler 300-F base "Long Ram" V-8
Overhead valve. Cast-iron block. Bore & stroke: 4.18 x 3.75 in. Displacement: 413 cid. Compression ratio: 10.10:1. Brake hp: 375 at 5000 rpm. Taxable hp: 55.90. Torque: 495 lbs.-ft. at 2800 rpm. Five main bearings. Hydraulic valve lifters. Crankcase capacity: 5 qt. (add 1 qt. for new filter). Cooling system capacity: 16 qt. (add 1 qt. for heater). Dual exhaust. Carburetor: "Long" Ram Induction intake manifold with two Carter AFB-2903-S four-barrel carburetors. Engine code: P-41.

Chrysler 300-F optional "Short Ram" V-8
Overhead valve. Cast-iron block. Bore & stroke: 4.18 x 3.75 in. Displacement: 413 cid. Compression ratio: 10.10:1. Brake hp: 400 at 5200 rpm. Taxable hp: 55.90. Torque: 465 lbs.-ft. at 3600 rpm. Five main bearings. Solid valve lifters. Crankcase capacity: 5 qt. (add 1 qt. for new filter). Cooling system capacity: 16 qt. (add 1 qt. for heater). Dual exhaust. Carburetor: "Short" Ram Induction intake manifold with two Carter AFB-2903-S four-barrel carburetors. Engine code: P-41.

NOTE: Limited production V-8. The external length of the "long" and "short" rams was the same, but the interior length varied.

Imperial Custom, Crown Imperial base V-8
Overhead valve. Cast-iron block. Bore & stroke: 4.18 x 3.75 in. Displacement: 413 cid. Compression ratio: 10.10:1. Brake hp: 350 at 4600 rpm. Taxable hp: 55.90. Torque: 470 lbs.-ft. at 2800 rpm. Five main bearings. Hydraulic valve lifters. Crankcase capacity: 5 qt. (add 1 qt. for new filter). Cooling system capacity: 16 qt. (add 1 qt. for heater). Dual exhaust. Carburetor: Carter AFB-2927-S four-barrel. Engine code: P-41.

1961 CHRYSLER

Newport base V-8
Overhead valve. Cast-iron block. Bore & stroke: 4.125 x 3.375 in. Displacement: 361 cid. Compression ratio: 9.0:1. Brake hp: 265 at 4400 rpm. Taxable hp: 54.3. Torque: 380 lbs.-ft. at 2400 rpm. Five main bearings. Hydraulic valve lifters. Crankcase capacity: 5 qt. (add 1 qt. for new filter). Cooling system capacity: 16 qt. (add 1 qt. for heater). Carburetor: Stromberg WWC-3188 two-barrel. Engine code: R-36.

Windsor base V-8
Overhead valve. Cast-iron block. Bore & stroke: 4.25 x 3.375 in. Displacement: 383 cid. Compression ratio: 10.00:1. Brake hp: 305 at 4800 rpm. Taxable hp: 57.80. Torque: 395 lbs.-ft. at 3000 rpm. Five main bearings. Hydraulic valve lifters. Crankcase capacity: 5 qt. (add 1 qt. for new filter). Cooling system capacity: 16 qt. (add 1 qt. for heater). Carburetor: Carter BBD 2923SA two-barrel. Engine code: R-38.

New Yorker base V-8; Newport optional V-8
Overhead valve. Cast-iron block. Bore & stroke: 4.18 x 3.75 in. Displacement: 413 cid. Compression ratio: 10.10:1. Brake hp: 350 at 4600 rpm. Taxable hp: 55.90. Torque: 470 lbs.-ft. at 2800 rpm. Five main bearings. Hydraulic valve lifters. Crankcase capacity: 5 qt. (add 1 qt. for new filter). Cooling system capacity: 16 qt. (add 1 qt. for heater). Dual exhaust. Carburetor: Carter AFB-3108-S four-barrel. Engine code: R-41.

New Yorker optional V-8
Overhead valve. Cast-iron block. Bore & stroke: 4.19 x 3.75 in. Displacement: 413 cid. Compression ratio: 11.00:1. Brake hp: 365 at 4600 rpm. Taxable hp: 55.90. Torque: 460 lbs.-ft. at 2800 rpm. Five main bearings. Hydraulic valve lifters. Crankcase capacity: 4 qt. (add 1 qt. if new filter installed). Cooling system capacity: 16 qt. (add 1 qt. for heater). Carburetor: Carter AFB 3251-S four-barrel. Engine code: S-41.

Chrysler 300-G base V-8 (Long Ram)
Overhead valve. Cast-iron block. Bore & stroke: 4.18 x 3.75 in. Displacement: 413 cid. Compression ratio: 10.10:1. Brake hp: 375 at 5000 rpm. Taxable hp: 55.90. Torque: 495 lbs.-ft. at 2800 rpm. Five main bearings. Hydraulic valve lifters. Crankcase capacity: 5 qt. (add 1 qt. for new filter). Cooling system capacity: 16 qt. (add 1 qt. for heater). Dual exhaust. Carburetors: Two Carter Type AFB Model 2903S four-barrel. Engine code: R-41.

Chrysler 300-G base V-8 (Short Ram)
Overhead valve. Cast-iron block. Bore & stroke: 4.18 x 3.75 in. Displacement: 413 cid. Compression ratio: 10.10:1. Brake hp: 400 at 5000 rpm. Taxable hp: 55.90. Torque: 465 lbs.-ft. at 3600 rpm. Five main bearings. Solid valve lifters. Crankcase capacity: 5 qt. (add 1 qt. for new filter). Cooling system capacity: 16 qt. (add 1 qt. for heater). Dual exhaust. Carburetors: Two Carter Type AFB Model 2903S four-barrel. Engine code: R-41.

Imperial, Crown Imperial, Imperial LeBaron base V-8
Overhead valve. Cast-iron block. Bore & stroke: 4.18 x 3.75 in. Displacement: 413 cid. Compression ratio: 10.10:1. Brake hp: 350 at 4600 rpm. Taxable hp: 55.90. Torque: 470 lbs.-ft. at 2800 rpm. Five main bearings. Hydraulic valve lifters. Crankcase capacity: 5 qt. (add 1 qt. for new filter). Cooling system capacity: 16 qt. (add 1 qt. for heater). Dual exhaust. Carburetor: Carter AFB-3108-S four-barrel. Engine code: R-41.

1962 CHRYSLER

Newport base V-8
Overhead valve. Cast-iron block. Bore & stroke: 4.125 x 3.375 in. Displacement: 361 cid. Compression ratio: 9.0:1. Brake hp: 265 at 4400 rpm. Taxable hp: 54.3. Torque: 380 lbs.-ft. at 2400 rpm. Five main bearings. Hydraulic valve lifters. Crankcase capacity: 5 qt. (add 1 qt. for new filter). Cooling system capacity: 16 qt. (add 1 qt. for heater). Carburetor: Stromberg Type WWC3 Model 201-A two-barrel. Engine code: S-36.

Chrysler 300 Sport base V-8
Overhead valve. Cast-iron block. Bore & stroke: 4.25 x 3.375 in. Displacement: 383 cid. Compression ratio: 10.0:1. Brake hp: 305 at 4600 rpm. Taxable hp: 57.80. Torque: 410 lbs.-ft. at 2400 rpm. Five main bearings. Hydraulic valve lifters. Crankcase capacity: 5 qt. (add 1 qt. for new filter). Cooling system capacity: 16 qt. (add 1 qt. for heater). Carburetor: Carter Type BBD Model 3244-S two-barrel. Engine code: S-38.

New Yorker base V-8; Sport 300, Imperial optional V-8
Overhead valve. Cast-iron block. Bore & stroke: 4.18 x 3.75 in. Displacement: 413 cid. Compression ratio: 10.10:1. Brake hp: 340 at 4600 rpm. Taxable hp: 55.90. Torque: 470 lbs.-ft. at 2800 rpm. Five main bearings. Hydraulic valve lifters. Crankcase capacity: 5 qt. (add 1 qt. for new filter). Cooling system capacity: 16 qt. (add 1 qt. for heater). Dual exhaust. Carburetor: Carter AFB-3251-S four-barrel. Engine code: S-41.

Chrysler 300-H base V-8; Sport 300 optional V-8
Overhead valve. Cast-iron block. Bore & stroke: 4.18 x 3.75 in. Displacement: 413 cid. Compression ratio: 10.10:1. Brake hp: 380 at 5200 rpm. Taxable hp: 55.90. Torque: 450 lbs.-ft. at 3600 rpm. Five main bearings. Solid valve lifters. Crankcase capacity: 5 qt. (add 1 qt. for new filter). Cooling system capacity: 16 qt. (add 1 qt. for heater). Dual exhaust. Carburetor: Two Carter carburetors, (front) AFB Model 3258-S

Standard V-8 in the 1962 Chrysler Newport (and Dodge 880) was the 361-cid 265-hp Chrysler "B" series engine. This power plant was first used in certain Dodge, Plymouth, and DeSoto models in 1958, when it was offered both in 350- and 361-cid versions. It had a very short stroke, short rods, integral valve guides, and stamped rocker arms. It was cheaper to produce than MoPar's earlier "dome head" V-8s.

four-barrel; (rear) AFB Model 3259-S four-barrel. Engine code: S-41.

Imperial, Crown Imperial, Imperial LeBaron base V-8
Overhead valve. Cast-iron block. Bore & stroke: 4.18 x 3.75 in. Displacement: 413 cid. Compression ratio: 10.10:1. Brake hp: 380 at 5200 rpm. Taxable hp: 55.90. Torque: 450 lbs.-ft. at 3600 rpm. Five main bearings. Solid valve lifters. Crankcase capacity: 5 qt. (add 1 qt. for new filter). Cooling system capacity: 16 qt. (add 1 qt. for heater). Dual exhaust. Carburetor: Two Carter carburetors, (front) AFB Model 3258-S four-barrel; (rear) AFB Model 3259-S four-barrel. Engine code: S-41.

Chrysler 300, 300-H optional V-8 (Short Ram)
Overhead valve. Cast-iron block. Bore & stroke: 4.18 x 3.75 in. Displacement: 413 cid. Compression ratio: 10.10:1. Brake hp: 405. Taxable hp: 55.90. Five main bearings. Solid valve lifters. Crankcase capacity: 5 qt. (add 1 qt. for new filter). Cooling system capacity: 16 qt. (add 1 qt. for heater). Dual exhaust. Carburetors: Two Carter Type AFB Model 2903S four-barrel. Engine code: R-41.

NOTE: A very interesting item was mentioned in Motor Trend's "Spotlight on Detroit" column in the magazine's June 1962 issue. It read, "You can now order 426 cubic inches on any Chrysler 300 or 300-H. Factory 413-cubic-inch blocks bored .060 are available on special order, with

A 1960-1962 Chrysler wedge head V-8 with the wild-looking ram-induction intake manifolding. These came with "long" and "short" rams, which measured the same on the outside, but the inner "effective length" varied. With the log rams, the torque curve rose quickly at 2000 rpm, but dropped off rapidly beyond 2800 rpm. Take special note of the unique exhaust headers used on these motors.

component combinations including forged pistons of 12-to-1 compression, 292-degree-duration cam with solid lifters, big exhaust valves, streamlined exhaust headers, and the original dual-four-barrel ram manifolds with passages shortened to "tune" above 4000 rpm. This top engine is rated 421 hp at 5400 rpm. Chrysler Division isn't going in for any dragging like Dodge and Plymouth, but they will supply the hot stuff if you want it." The 426-cid V-8 came in 373-, 385-, 413-, and 421-hp versions with the top options featuring dual four-barrel carburetors. The 426 engine carried the code S-42.

1963 CHRYSLER

Newport base V-8
Overhead valve. Cast-iron block. Bore & stroke: 4.125 x 3.375 in. Displacement: 361 cid. Compression ratio: 9.0:1. Brake hp: 265 at 4400 rpm. Taxable hp: 54.3. Torque: 380 lbs.-ft. at 2400 rpm. Five main bearings. Hydraulic valve lifters. Crankcase capacity: 5 qt. (add 1 qt. for new filter). Cooling system capacity: 16 qt. (add 1 qt. for heater). Carburetor: Stromberg Type WWC3 Model 221-A two-barrel. Engine code: T-36.

Chrysler 300 Sport base V-8
Overhead valve. Cast-iron block. Bore & stroke: 4.25 x 3.375 in. Displacement: 383 cid. Compression ratio: 10.0:1. Brake hp: 305 at 4600 rpm. Taxable hp: 57.80. Torque: 410 lbs.-ft. at 2400 rpm. Five main bearings. Hydraulic valve lifters. Crankcase capacity: 5 qt. (add 1 qt. for new filter). Cooling system capacity: 16 qt. (add 1 qt. for heater). Carburetor: Carter Type BBD Model 3476-S two-barrel. Engine code: T-38.

New Yorker base V-8
Overhead valve. Cast-iron block. Bore & stroke: 4.18 x 3.75 in. Displacement: 413 cid. Compression ratio: 10.10:1. Brake hp: 340 at 4600 rpm. Taxable hp: 56.20. Torque: 470 lbs.-ft. at 2800 rpm. Five main bearings. Hydraulic valve lifters. Crankcase capacity: 5 qt. (add 1 qt. for new filter). Cooling system capacity: 16 qt. (add 1 qt. for heater). Dual exhaust. Carburetor: Carter AFB-3256-S four-barrel. Engine code: T-41.

Chrysler 300 Sport optional V-8
Overhead valve. Cast-iron block. Bore & stroke: 4.18 x 3.75 in. Displacement: 413 cid. Compression ratio: 10.10:1. Brake hp: 360 at 4800 rpm. Taxable hp: 56.20. Torque: 470 lbs.-ft. at 3200 rpm. Five main bearings. Hydraulic valve lifters. Crankcase capacity: 5 qt. (add 1 qt. for new filter). Cooling system capacity: 16 qt. (add 1 qt. for heater). Dual exhaust. Carburetor: Carter AFB-3478-S four-barrel. Engine code: T-41.

Chrysler 300-J base V-8
Overhead valve. Cast-iron block. Bore & stroke: 4.18 x 3.75 in. Displacement: 413 cid. Compression ratio: 9.60:1. Brake hp: 390 at 4800 rpm. Taxable hp: 56.20. Torque: 485 lbs.-ft. at 3600 rpm. Five main bearings. Solid valve lifters. Crankcase capacity: 5 qt. (add 1 qt. for new filter). Cooling system capacity: 16 qt. (add 1 qt. for heater). Dual exhaust. Carburetor: Two Carter Type AFB four-barrel Model 3505-S. Engine code: T-41.

Imperial Custom, Crown Imperial, Imperial LeBaron base V-8
Overhead valve. Cast-iron block. Bore & stroke: 4.18 x 3.75 in. Displacement: 413 cid. Compression ratio: 10.10:1. Brake hp: 340 at 4600 rpm. Taxable hp: 56.20. Torque: 470 lbs.-ft. at 2800 rpm. Five main bearings. Hydraulic valve lifters. Crankcase capacity: 5 qt. (add 1 qt. for new filter). Cooling system capacity: 16 qt. (add 1 qt. for heater). Dual exhaust. Carburetor: Carter four-barrel Model AFB-3251-S. Engine code: T-41

NOTE: See note under 1962. Some sources indicate that the 426-cid V-8 was offered for 1963 Chryslers with 373-, 415-, and 425-hp ratings. The 426 engine code was T-42.

1964 CHRYSLER

Newport base V-8
Overhead valve. Cast-iron block. Bore & stroke: 4.125 x 3.375 in. Displacement: 361 cid. Compression ratio: 9.0:1. Brake hp: 265 at 4400 rpm. Taxable hp: 54.3. Torque: 380 lbs.-ft. at 2400 rpm. Five main bearings. Hydraulic valve lifters. Crankcase capacity: 5 qt. (add 1 qt. for new filter). Cooling system capacity: 16 qt. (add 1 qt. for heater). Carburetor: Stromberg Type WWC3 Model 244 two-barrel.

Engine code: V-36.

Chrysler 300 Sport base V-8
Overhead valve. Cast-iron block. Bore & stroke: 4.25 x 3.375 in. Displacement: 383 cid. Compression ratio: 10.0:1. Brake hp: 305 at 4600 rpm. Taxable hp: 57.80. Torque: 410 lbs.-ft. at 2400 rpm. Five main bearings. Hydraulic valve lifters. Crankcase capacity: 5 qt. (add 1 qt. for new filter). Cooling system capacity: 16 qt. (add 1 qt. for heater). Carburetor: Carter Type BBD Model 3685-S two-barrel. Engine code: V-38.

New Yorker, New Yorker Salon base V-8
Overhead valve. Cast-iron block. Bore & stroke: 4.18 x 3.75 in. Displacement: 413 cid. Compression ratio: 10.10:1. Brake hp: 340 at 4600 rpm. Taxable hp: 56.20. Torque: 470 lbs.-ft. at 2800 rpm. Five main bearings. Hydraulic valve lifters. Crankcase capacity: 5 qt. (add 1 qt. for new filter). Cooling system capacity: 16 qt. (add 1 qt. for heater). Dual exhaust. Carburetor: Carter AFB-3615-S four-barrel. Engine code: V-41.

Chrysler 300-K base V-8; Sport 300 optional V-8
Overhead valve. Cast-iron block. Bore & stroke: 4.18 x 3.75 in. Displacement: 413 cid. Compression ratio: 10.10:1. Brake hp: 360 at 4800 rpm. Taxable hp: 56.20. Torque: 470 lbs.-ft. at 3200 rpm. Five main bearings. Hydraulic valve lifters. Crankcase capacity: 5 qt. (add 1 qt. for new filter). Cooling system capacity: 16 qt. (add 1 qt. for heater). Dual exhaust. Carburetor: Carter AFB-3614-S four-barrel. Engine code: V-41.

Chrysler 300-K optional V-8
Overhead valve. Cast-iron block. Bore & stroke: 4.18 x 3.75 in. Displacement: 413 cid. Compression ratio: 9.60:1. Brake hp: 390 at 4800 rpm. Taxable hp: 56.20. Torque: 485 lbs.-ft. at 3600 rpm. Five main bearings. Solid valve lifters. Crankcase capacity: 5 qt. (add 1 qt. for new filter). Cooling system capacity: 16 qt. (add 1 qt. for heater). Dual exhaust. Carburetor: Two Carter Type AFB four-barrel Model 3505-SA. Engine code: V-41.

Crown Imperial, Imperial LeBaron base V-8
Overhead valve. Cast-iron block. Bore & stroke: 4.18 x 3.75 in. Displacement: 413 cid. Compression ratio: 10.10:1. Brake hp: 340 at 4600 rpm. Taxable hp: 56.20. Torque: 470 lbs.-ft. at 2800 rpm. Five main bearings. Hydraulic valve lifters. Crankcase capacity: 5 qt. (add 1 qt. for new filter). Cooling system capacity: 16 qt. (add 1 qt. for heater). Dual exhaust. Carburetor: Carter four-barrel Model AFB-3644-S. Engine code: V-41.

1965 CHRYSLER

Newport base V-8
Overhead valve. Cast-iron block. Bore & stroke: 4.25 x 3.375 in. Displacement: 383 cid. Compression ratio: 9.20:1. Brake hp: 270 at 4400 rpm. Taxable hp: 57.80. Torque: 390 lbs.-ft. at 2800 rpm. Five main bearings. Hydraulic valve lifters. Crankcase capacity: 4 qt. (add 1 qt. for new filter). Cooling system capacity: 16 qt. (add 1 qt. for heater). Carburetor: Carter Type BBD Model 3849-S two-barrel. Engine code: A-383.

Chrysler Sport 300 base V-8; Newport optional V-8
Overhead valve. Cast-iron block. Bore & stroke: 4.25 x 3.375 in. Displacement: 383 cid. Compression ratio: 10.0:1. Brake hp: 315 at 4400 rpm. Taxable hp: 57.80. Torque: 420 lbs.-ft. at 2800 rpm. Five main bearings. Hydraulic valve lifters. Crankcase capacity: 4 qt. (add 1 qt. for new filter). Cooling system capacity: 16 qt. (add 1 qt. for heater). Carburetor: Carter Type BBD Model 3855-S two-barrel. Engine code: A-413.

New Yorker base V-8
Overhead valve. Cast-iron block. Bore & stroke: 4.18 x 3.75 in. Displacement: 413 cid. Compression ratio: 10.10:1. Brake hp: 340 at 4600 rpm. Taxable hp: 56.20. Torque: 470 lbs.-ft. at 2800 rpm. Five main bearings. Hydraulic valve lifters. Crankcase capacity: 4 qt. (add 1 qt. for new filter). Cooling system capacity: 16 qt. (add 1 qt. for heater). Dual exhaust. Carburetor: Carter AFB-3858-S four-barrel. Engine code: A-413.

Chrysler 300-L base V-8; Chrysler Sport 300, New Yorker optional V-8
Overhead valve. Cast-iron block. Bore & stroke: 4.18 x 3.75 in. Displacement: 413 cid. Compression ratio: 10.10:1. Brake hp: 360 at 4800 rpm. Taxable hp: 56.20. Torque: 470 lbs.-ft. at 3200 rpm. Five main bearings. Hydraulic valve lifters. Crankcase capacity: 4 qt. (add 1 qt. for new filter). Cooling system capacity: 16 qt. (add 1 qt. for heater). Dual exhaust. Carburetor: (Sport 300) Carter AFB-3859-S four-barrel; (New Yorker, 300-L) Carter AFB-3860-S four-barrel. Engine code: A-413.

Crown Imperial, Imperial LeBaron base V-8
Overhead valve. Cast-iron block. Bore & stroke: 4.18 x 3.75 in. Displacement: 413 cid. Compression ratio: 10.10:1. Brake hp: 340 at 4600 rpm. Taxable hp: 56.20. Torque: 470 lbs.-ft. at 2800 rpm. Five main bearings. Hydraulic valve lifters. Crankcase capacity: 4 qt. (add 1 qt. for new filter). Cooling system capacity: 16 qt. (add 1 qt. for heater). Dual exhaust. Carburetor: Carter AFB-3871-S four-barrel. Engine code: A-413.

1966 CHRYSLER

Newport base V-8
Overhead valve. Cast-iron block. Bore & stroke: 4.25 x 3.375 in. Displacement: 383 cid. Compression ratio: 9.20:1. Brake hp: 270 at 4400 rpm. Taxable hp: 57.80. Torque: 390 lbs.-ft. at 2800 rpm. Five main bearings. Hydraulic valve lifters. Crankcase capacity: 4 qt. (add 1 qt. for new filter). Cooling system capacity: 17 qt. (with heater standard). Carburetor: Carter Type BBD Model 4125-S two-barrel. Engine code: B-383. VIN code: G.

Chrysler Sport 300 base V-8; Newport optional V-8
Overhead valve. Cast-iron block. Bore & stroke: 4.25 x 3.375 in. Displacement: 383 cid. Compression ratio: 10.0:1. Brake hp: 325 at 4800 rpm. Taxable hp: 57.80. Torque: 425 lbs.-ft. at 2800 rpm. Five main bearings. Hydraulic valve lifters. Crankcase capacity: 4 qt. (add 1 qt. for new filter). Cooling system capacity: 17 qt. (with heater standard). Carburetor: Carter Type AFB four-barrel Model 4130-S. Engine code: B-383. VIN code: G.

New Yorker base V-8
Overhead valve. Cast-iron block. Bore & stroke: 4.32 x 3.75 in. Displacement: 440 cid. Compression ratio: 10.10:1. Brake hp: 350 at 4400 rpm. Taxable hp: 59.70. Torque: 480 lbs.-ft. at 2800 rpm. Five main bearings. Hydraulic valve lifters. Crankcase capacity: 4 qt. (add 1 qt. for new filter). Cooling system capacity: 17 qt. (with heater standard). Carburetor: Carter AFB-4131-S four-barrel. Engine code: B-413. VIN code: J.

Newport, Sport 300, New Yorker optional V-8
Overhead valve. Cast-iron block. Bore & stroke: 4.32 x 3.75 in. Displacement: 440 cid. Compression ratio: 10.10:1. Brake hp: 365 at 4600 rpm. Taxable hp: 59.70. Torque: 480 lbs.-ft. at 3200 rpm. Five main bearings. Hydraulic valve lifters. Crankcase capacity: 4 qt. (add 1 qt. for new filter). Cooling system capacity: 17 qt. (with heater standard). Carburetor: Carter AFB-4136-S four-barrel. Engine code: B-413. VIN code: J.

Crown Imperial, Imperial LeBaron base V-8
Overhead valve. Cast-iron block. Bore & stroke: 4.32 x 3.75 in. Displacement: 440 cid. Compression ratio: 10.10:1. Brake hp: 350 at 4400 rpm. Taxable hp: 59.70. Torque: 480 lbs.-ft. at 2800 rpm. Five main bearings. Hydraulic valve lifters. Crankcase capacity: 4 qt. (add 1 qt. for new filter). Cooling system capacity: 17 qt. (with heater standard). Carburetor: Carter AFB-4131-S four-barrel. Engine code: B-413. VIN code: J.

1967 CHRYSLER

Newport, Newport Custom base V-8
Overhead valve. Cast-iron block. Bore & stroke: 4.25 x 3.375 in. Displacement: 383 cid. Compression ratio: 9.20:1. Brake hp: 270 at 4400 rpm. Taxable hp: 57.80. Torque: 390 lbs.-ft. at 2800 rpm. Five main bearings. Hydraulic valve lifters. Crankcase capacity: 4 qt. (add 1 qt. for new filter). Cooling system capacity: 17 qt. (with heater standard). Carburetor: Carter Type BBD Model 4296-S two-barrel. VIN code: G.

Newport, Newport Custom, Town & Country optional V-8
Overhead valve. Cast-iron block. Bore & stroke: 4.25 x 3.375 in. Displacement: 383 cid. Compression ratio: 10.0:1. Brake hp: 325 at 4800 rpm. Taxable hp: 57.80. Torque: 425 lbs.-ft. at 2800 rpm. Five main bearings. Hydraulic valve lifters. Crankcase capacity: 4 qt. (add 1 qt. for new filter). Cooling system capacity: 17 qt. (with heater standard). Carburetor: Carter AFB four-barrel. VIN code: H.

Chrysler 300, New Yorker base V-8; Town & Country optional V-8
Overhead valve. Cast-iron block. Bore & stroke: 4.32 x 3.75 in. Displacement: 440 cid. Compression ratio: 10.10:1. Brake hp: 350 at 4400 rpm. Taxable hp: 59.70. Torque: 480 lbs.-ft. at 2800 rpm. Five main

The 1967 Chrysler 383-cid "B" V-8 was a big-block and was offered in two horsepower ratings that year. The 270-hp version was standard in Chryslers and the 325-hp version was optional in Chryslers, Plymouths, and Dodges. These massive V-8s weighed 650-660 lbs.

bearings. Hydraulic valve lifters. Crankcase capacity: 4 qt. (add 1 qt. for new filter). Cooling system capacity: 17 qt. (with heater standard). Carburetor: Holley Model R-3667A four-barrel. VIN code: K.

Newport, Newport Custom, Chrysler 300, New Yorker optional V-8
Overhead valve. Cast-iron block. Bore & stroke: 4.32 x 3.75 in. Displacement: 440 cid. Compression ratio: 10.10:1. Brake hp: 375 at 4600 rpm. Taxable hp: 59.70. Torque: 480 lbs.-ft. at 3200 rpm. Five main bearings. Hydraulic valve lifters. Crankcase capacity: 4 qt. (add 1 qt. for new filter). Cooling system capacity: 17 qt. (with heater standard). Carburetor: Carter AFB four-barrel. VIN code: L.

Imperial, Crown Imperial, Imperial LeBaron base V-8
Overhead valve. Cast-iron block. Bore & stroke: 4.32 x 3.75 in. Displacement: 440 cid. Compression ratio: 10.10:1. Brake hp: 350 at 4400 rpm. Taxable hp: 59.70. Torque: 480 lbs.-ft. at 2800 rpm. Five main bearings. Hydraulic valve lifters. Crankcase capacity: 4 qt. (add 1 qt. for new filter). Cooling system capacity: 18 qt. (with heater standard). Carburetor: Holley Model R-3667A four-barrel. VIN code: K.

1968 CHRYSLER

Newport, Newport Custom base V-8
Overhead valve. Cast-iron block. Bore & stroke: 4.25 x 3.375 in. Displacement: 383 cid. Compression ratio: 9.20:1. Brake hp: 290 at 4400 rpm. Taxable hp: 57.80. Torque: 380 lbs.-ft. at 2400 rpm. Five main bearings. Hydraulic valve lifters. Crankcase capacity: 4 qt. (add 1 qt. for new filter). Cooling system capacity: 17 qt. (with heater standard). Carburetor: Carter Type BBD Model 4422-S two-barrel. VIN code: G.

Newport, Newport Custom, Town & Country optional V-8
Overhead valve. Cast-iron block. Bore & stroke: 4.25 x 3.375 in. Displacement: 383 cid. Compression ratio: 10.0:1. Brake hp: 330 at 5000 rpm. Taxable hp: 57.80. Torque: 425 lbs.-ft. at 3200 rpm. Five main bearings. Hydraulic valve lifters. Crankcase capacity: 4 qt. (add 1 qt. for new filter). Cooling system capacity: 17 qt. (with heater standard). Carburetor: Carter AFB four-barrel. VIN code: H.

Chrysler 300, New Yorker base V-8; Town & Country optional V-8
Overhead valve. Cast-iron block. Bore & stroke: 4.32 x 3.75 in. Displacement: 440 cid. Compression ratio: 10.10:1. Brake hp: 350 at 4400 rpm. Taxable hp: 59.70. Torque: 480 lbs.-ft. at 2800 rpm. Five main bearings. Hydraulic valve lifters. Crankcase capacity: 4 qt. (add 1 qt. for new filter). Cooling system capacity: 18 qt. (with heater standard). Carburetor: Holley Model R-3918A four-barrel. VIN code: K.

Newport, Newport Custom, Chrysler 300, New Yorker optional V-8
Overhead valve. Cast-iron block. Bore & stroke: 4.32 x 3.75 in. Displacement: 440 cid. Compression ratio: 10.10:1. Brake hp: 375 at 4600 rpm. Taxable hp: 59.70. Torque: 480 lbs.-ft. at 3200 rpm. Five main bearings. Hydraulic valve lifters. Crankcase capacity: 4 qt. (add 1 qt. for new filter). Cooling system capacity: 17 qt. (with heater standard). Carburetor: Carter AFB four-barrel. VIN code: L.

Crown Imperial, Imperial LeBaron base V-8
Overhead valve. Cast-iron block. Bore & stroke: 4.32 x 3.75 in. Displacement: 440 cid. Compression ratio: 10.10:1. Brake hp: 350 at 4400 rpm. Taxable hp: 59.70. Torque: 480 lbs.-ft. at 2800 rpm. Five main bearings. Hydraulic valve lifters. Crankcase capacity: 4 qt. (add 1 qt. for new filter). Cooling system capacity: 17 qt. (with heater standard). Carburetor: Holley Model R-3918A four-barrel. VIN code: K.

1969 CHRYSLER

Newport, Newport Custom, Town & Country base V-8
Overhead valve. Cast-iron block. Bore & stroke: 4.25 x 3.375 in. Displacement: 383 cid. Compression ratio: 9.20:1. Brake hp: 290 at 4400 rpm. Taxable hp: 57.80. Torque: 380 lbs.-ft. at 2400 rpm. Five main bearings. Hydraulic valve lifters. Crankcase capacity: 4 qt. (add 1 qt. for new filter). Cooling system capacity: 16 qt. (with heater standard). Carburetor: Carter Type BBD Model 4613-S two-barrel. VIN code: G.

Firepower 440 TNT V-8

In 1969, the Firepower 440 TNT V-8 that was optional in Chrysler brand cars featured a chrome plated oil filler cap, a black crackle-finished twin-snorkel air cleaner with a TNT nameplate, a high-performance camshaft, special low-restriction exhaust manifolds, larger-diameter intake and exhaust valves, a dual exhaust system with oversize 2-1/2-in. diameter exhaust pipes and 2-in. diameter tailpipes, an oil pan windage tray, and a high-performance camshaft.

Firepower 383 V-8 *Firepower 383 XP V-8*

For 1969, Chrysler's engine lineup started big. The Firepower 383 big-block on the left was standard equipment in Newport and Newport Custom models, as well as in Town & Country station wagons. This engine provided terrific big-car performance with quick acceleration and adequate passing speeds and had the advantage of being suitable to operate on regular gas. The premium-fuel Firepower 383 XP V-8 on the right had a higher 10.0:1 compression ratio and dual exhaust system that added 40 hp. Note the dual-snorkel air cleaner.

1969 Chrysler Engines and Transmissions

ENGINE	HORSEPOWER	AVAILABILITY	TRANSMISSIONS*
383-cu.-in. V-8 2-barrel, Single Exhaust	290	Standard on Newport, Newport Custom and Town & Country Wagon	3-speed manual, 3-speed TorqueFlite automatic
383-cu.-in. V-8 4-barrel, Dual Exhaust	330	Optional on Newport, Newport Custom and Town & Country Wagon	3-speed TorqueFlite automatic
440-cu.-in. V-8 4-barrel, Single Exhaust	350	Standard on 300 and New Yorker	3-speed TorqueFlite automatic
440-cu.-in. V-8 4-barrel, Dual Exhaust	350	Optional on Town & Country Wagon	3-speed TorqueFlite automatic
440-cu.-in. V-8 4-barrel TNT, Dual Exhaust	375	Optional on Newport, Newport Custom, 300, and New Yorker	3-speed TorqueFlite automatic

C-21 *See standard and optional list for transmission availability.

In 1969, Chrysler offered two versions of the 383-cid V-8 and three versions of the 440-cid V-8 for Chrysler brand cars and wagons. The most powerful was the 440 TNT option, with 375 hp, that was optional on the Newport, Newport Custom, 300, and New Yorker.

Chrysler's 1969 Firepower 440 V-8 was a superior performer at high speeds because of its excellent breathing ability. Features included a five-inch carburetor throat, extra-large intake manifold passages, and big intake and exhaust valve ports that let the fuel-air mixture flow in and the exhaust gases flow out with the least possible resistance. It utilized an air cleaner with one snorkel that pointed to the left-hand side of the engine bay. The 440 was the largest V-8 engine ever offered by Chrysler Corporation, although not the most powerful.

Newport, Newport Custom optional V-8
Overhead valve. Cast-iron block. Bore & stroke: 4.25 x 3.375 in. Displacement: 383 cid. Compression ratio: 10.0:1. Brake hp: 330 at 5000 rpm. Taxable hp: 57.80. Torque: 425 lbs.-ft. at 3200 rpm. Five main bearings. Hydraulic valve lifters. Crankcase capacity: 4 qt. (add 1 qt. for new filter). Cooling system capacity: 16 qt. (with heater standard). Carburetor: Four-barrel. VIN code: H.

Chrysler 300, New Yorker base V-8; Town & Country optional V-8
Overhead valve. Cast-iron block. Bore & stroke: 4.32 x 3.75 in. Displacement: 440 cid. Compression ratio: 10.10:1. Brake hp: 350 at 4400 rpm. Taxable hp: 59.70. Torque: 480 lbs.-ft. at 2800 rpm. Five main bearings. Hydraulic valve lifters. Crankcase capacity: 4 qt. (add 1 qt. for new filter). Cooling system capacity: 17 qt. (with heater standard). Carburetor: Holley Model R-4166A four-barrel. VIN code: K.

Newport, Newport Custom, Chrysler 300, New Yorker optional V-8
Overhead valve. Cast-iron block. Bore & stroke: 4.32 x 3.75 in. Displacement: 440 cid. Compression ratio: 10.10:1. Brake hp: 375 at 4600 rpm. Taxable hp: 59.70. Torque: 480 lbs.-ft. at 3200 rpm. Five main bearings. Hydraulic valve lifters. Crankcase capacity: 4 qt. (add 1 qt. for new filter). Cooling system capacity: 17 qt. (with heater standard). Carburetor: Four-barrel. VIN code: L.

Crown Imperial, Imperial LeBaron base V-8
Overhead valve. Cast-iron block. Bore & stroke: 4.32 x 3.75 in. Displacement: 440 cid. Compression ratio: 10.10:1. Brake hp: 350 at 4400 rpm. Taxable hp: 59.70. Torque: 480 lbs.-ft. at 2800 rpm. Five main bearings. Hydraulic valve lifters. Crankcase capacity: 4 qt. (add 1 qt. for new filter). Cooling system capacity: 18 qt. (with heater standard). Carburetor: Holley Model R-4166A four-barrel. VIN code: K.

1970 CHRYSLER

ChrylserNewport, Newport Custom, Town & Country base V-8
Overhead valve. Cast-iron block. Bore & stroke: 4.25 x 3.375 in. Displacement: 383 cid. Compression ratio: 8.70:1. Brake hp: 290 at 4400 rpm. Taxable hp: 57.80. Torque: 390 lbs.-ft. at 2800 rpm. Five main bearings. Hydraulic valve lifters. Crankcase capacity: 4 qt. (add 1 qt. for new filter). Cooling system capacity: 14.5 qt. (with heater standard). Carburetor: Carter Type BBD Model 4894-S two-barrel. VIN code: L.

Newport, Newport Custom optional V-8
Overhead valve. Cast-iron block. Bore & stroke: 4.25 x 3.375 in. Displacement: 383 cid. Compression ratio: 9.50:1. Brake hp: 330 at 5000 rpm. Taxable hp: 57.80. Torque: 425 lbs.-ft. at 3200 rpm. Five main bearings. Hydraulic valve lifters. Crankcase capacity: 4 qt. (add 1 qt. for new filter). Cooling system capacity: 14.5 qt. (with heater standard). Carburetor: Four-barrel. VIN code: N.

Chrysler 300, New Yorker base V-8; Town & Country optional V-8
Overhead valve. Cast-iron block. Bore & stroke: 4.32 x 3.75 in. Displacement: 440 cid. Compression ratio: 9.70:1. Brake hp: 350 at 4400 rpm. Taxable hp: 59.70. Torque: 480 lbs.-ft. at 2800 rpm. Five main bearings. Hydraulic valve lifters. Crankcase capacity: 4 qt. (add 1 qt. for new filter). Cooling system capacity: 15.5 qt. (with heater standard). Carburetor: Holley Model R-4166A four-barrel. VIN code: T.

Newport, Newport Custom, Chrysler 300, New Yorker optional V-8
Overhead valve. Cast-iron block. Bore & stroke: 4.32 x 3.75 in.

Displacement: 440 cid. Compression ratio: 9.70:1. Brake hp: 375 at 4600 rpm. Taxable hp: 59.70. Torque: 480 lbs.-ft. at 3200 rpm. Five main bearings. Hydraulic valve lifters. Crankcase capacity: 4 qt. (add 1 qt. for new filter). Cooling system capacity: (Newport) 14.5 qt.; (New Yorker, 300) 15.5 qt. (with heater standard). Carburetor: Four-barrel. VIN code: U.

Crown Imperial, Imperial LeBaron base V-8
Overhead valve. Cast-iron block. Bore & stroke: 4.32 x 3.75 in. Displacement: 440 cid. Compression ratio: 9.70:1. Brake hp: 350 at 4400 rpm. Taxable hp: 59.70. Torque: 480 lbs.-ft. at 2800 rpm. Five main bearings. Hydraulic valve lifters. Crankcase capacity: 4 qt. (add 1 qt. for new filter). Cooling system capacity: 15.5 qt. (with heater standard). Carburetor: Holley Model R-4166A four-barrel. VIN code: T.

1971 CHRYSLER

Newport Royal base V-8
Overhead valve. Cast-iron block. Bore & stroke: 3.91 x 3.58 in. Displacement: 360 cid. Compression ratio: 8.70:1. Brake hp: 255. Taxable hp: 51.20. Five main bearings. Hydraulic valve lifters. Crankcase capacity: 4 qt. (add 1 qt. for new filter). Cooling system capacity: 14.5 qt. (with heater standard). Carburetor: Carburetor: Carter BBD Model 4961-S two-barrel. VIN code: K.

Newport, Newport Custom, Town & Country base V-8
Overhead valve. Cast-iron block. Bore & stroke: 4.25 x 3.375 in. Displacement: 383 cid. Compression ratio: 8.50:1. Brake hp: 275 at 4400 rpm. Taxable hp: 57.80. Torque: 375 lbs.-ft. at 2800 rpm. Five main bearings. Hydraulic valve lifters. Crankcase capacity: 4 qt. (add 1 qt. for new filter). Cooling system capacity: 14.5 qt. (with heater standard). Carburetor: Carter BBD Model 4961-S two-barrel. VIN code: L.

Newport optional V-8
Overhead valve. Cast-iron block. Bore & stroke: 4.25 x 3.375 in. Displacement: 383 cid. Compression ratio: 8.50:1. Brake hp: 300 at 4800 rpm. Taxable hp: 57.80. Torque: 410 lbs.-ft. at 3400 rpm. Five main bearings. Hydraulic valve lifters. Crankcase capacity: 4 qt. (add 1 qt. for new filter). Cooling system capacity: 14.5 qt. (with heater standard). Carburetor: Carter BBD Model 4961-S two-barrel. VIN code: N.

Chrysler 300, New Yorker base V-8; Town & Country base V-8
Overhead valve. Cast-iron block. Bore & stroke: 4.32 x 3.75 in. Displacement: 440 cid. Compression ratio: 8.80:1. Brake hp: 335 at 4400 rpm. Taxable hp: 59.70. Torque: 460 lbs.-ft. at 3200 rpm. Five main bearings. Hydraulic valve lifters. Crankcase capacity: 4 qt. (add 1 qt. for new filter). Cooling system capacity: 15.5 qt. (with heater standard). Carburetor: Carter AVS Model 4967-S four-barrel. VIN code: T.

Chrysler 300, New Yorker optional V-8
Overhead valve. Cast-iron block. Bore & stroke: 4.32 x 3.75 in. Displacement: 440 cid. Compression ratio: 9.50:1. Brake hp: 370 at 4600 rpm. Taxable hp: 59.70. Torque: 480 lbs.-ft. at 3200 rpm. Five main bearings. Hydraulic valve lifters. Crankcase capacity: 4 qt. (add 1 qt. for new filter). Cooling system capacity: 15.5 qt. (with heater standard). Carburetor: Carter AVS Model 4967-S four-barrel. VIN code: U.

Imperial LeBaron base V-8
Overhead valve. Cast-iron block. Bore & stroke: 4.32 x 3.75 in. Displacement: 440 cid. Compression ratio: 8.80:1. Brake hp: 335 at 4400 rpm. Taxable hp: 59.70. Torque: 460 lbs.-ft. at 3200 rpm. Five main bearings. Hydraulic valve lifters. Crankcase capacity: 4 qt. (add 1 qt. for new filter). Cooling system capacity: 15.5 qt. (with heater standard). Carburetor: Carter AVS Model 4966-S four-barrel. VIN code: T.

1972 CHRYSLER

Newport Royal base V-8
Overhead valve. Cast-iron block. Bore & stroke: 4.00 x 3.58 in. Displacement: 360 cid. Compression ratio: 8.80:1. Net brake hp: 175 at 4000 rpm. Taxable hp: 51.20. Net torque: 285 lbs.-ft. at 2400 rpm. Five main bearings. Hydraulic valve lifters. Crankcase capacity: 4 qt. (add 1 qt. for new filter). Cooling system capacity: 14.5 qt. (with heater standard). Carburetor: Carburetor: Carter two-barrel. VIN code: K.

Newport Custom, New Yorker Brougham, Town & Country base V-8
Overhead valve. Cast-iron block. Bore & stroke: 4.34 x 3.38 in. Displacement: 400 cid. Compression ratio: 8.20:1. Net brake hp: 190 at 4400 rpm. Taxable hp: 60.30. Net torque: 310 lbs.-ft. at 2400 rpm. Five main bearings. Hydraulic valve lifters. Crankcase capacity: 4 qt. (add 1 qt. for new filter). Cooling system capacity: 14.5 qt. (with heater standard). Carburetor: Carburetor: Carter two-barrel. VIN code: M.

New Yorker base V-8
Overhead valve. Cast-iron block. Bore & stroke: 4.32 x 3.75 in. Displacement: 440 cid. Compression ratio: 8.20:1. Brake hp: 225 at 4400 rpm. Taxable hp: 59.70. Torque: 340 lbs.-ft. at 3200 rpm. Five main bearings. Hydraulic valve lifters. Crankcase capacity: 4 qt. (add 1 qt. for new filter). Cooling system capacity: 15.5 qt. (with heater standard). Carburetor: Four-barrel. VIN code: T.

Imperial LeBaron base V-8
Overhead valve. Cast-iron block. Bore & stroke: 4.32 x 3.75 in. Displacement: 440 cid. Compression ratio: 8.20:1. Brake hp: 225 at 4400 rpm. Taxable hp: 59.70. Torque: 340 lbs.-ft. at 3200 rpm. Five main bearings. Hydraulic valve lifters. Crankcase capacity: 4 qt. (add 1 qt. for new filter). Cooling system capacity: 15.5 qt. (with heater standard). Carburetor: Four-barrel. VIN code: T.

1973 CHRYSLER

Newport, Newport Custom base V-8
Overhead valve. Cast-iron block. Bore & stroke: 4.34 x 3.38 in. Displacement: 400 cid. Compression ratio: 8.20:1. Net brake hp: 185 at 3600 rpm. Taxable hp: 60.30. Net torque: 310 lbs.-ft. at 2400 rpm. Five main bearings. Hydraulic valve lifters. Crankcase capacity: 4 qt. (add 1 qt. for new filter). Cooling system capacity: 14.5 qt. (with heater standard). Carburetor: Carburetor: Two-barrel. VIN code: M.

New Yorker base V-8; Newport, Newport Custom optional V-8
Overhead valve. Cast-iron block. Bore & stroke: 4.32 x 3.75 in. Displacement: 440 cid. Compression ratio: 8.20:1. Net brake hp: 215 at 3600 rpm. Taxable hp: 59.70. Torque: 350 lbs.-ft. at 2400 rpm. Five main bearings. Hydraulic valve lifters. Crankcase capacity: 4 qt. (add 1 qt. for new filter). Cooling system capacity: 15.5 qt. (with heater standard). Carburetor: Four-barrel. VIN code: T.

Imperial base V-8
Overhead valve. Cast-iron block. Bore & stroke: 4.32 x 3.75 in. Displacement: 440 cid. Compression ratio: 8.20:1. Net brake hp: 215 at 3600 rpm. Taxable hp: 59.70. Torque: 350 lbs.-ft. at 2400 rpm. Five main bearings. Hydraulic valve lifters. Crankcase capacity: 4 qt. (add 1 qt. for new filter). Cooling system capacity: 15.5 qt. (with heater standard). Carburetor: Four-barrel. VIN code: T.

The 440-cid four-barrel V-8 was the engine used in the 1973 Chrysler Imperial. It had a standard electronic ignition, an 8.2:1 compression ratio, a single snorkel air cleaner, and a single exhaust system. By this time a flex-tube was attached between the left-hand valve cover and the air cleaner to vent gases back into the engine.

In 1973, the 440 V-8 was the largest engine offered by the Chrysler Corporation. The combination of its size, a four-barrel carburetor, and the electronic ignition system made it versatile and dependable. It had carburetor improvements and tri-metal connecting rod bearings. Four of the five main bearings on the 440's crankshaft were steel-back Babbitt, while the thrust bearing featured rugged aluminum-alloy-on-steel construction. The crankshaft was made of forged steel.

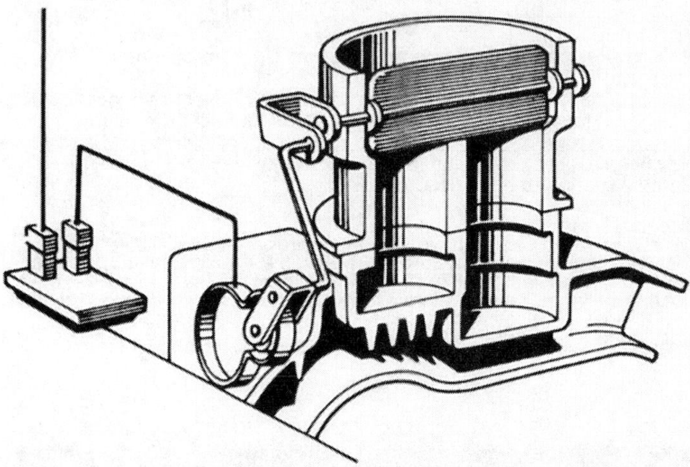

A new electric-assist choke was standard on all Chrysler V-8 engines for 1973. It incorporated an electric heating element that added heat to the choke's bimetallic spring when the heat around the engine rose above 63 degrees F. This stopped the choking action sooner, thereby enhancing emissions.

All 1973 versions of the Chrysler 440-4V V-8 had a carburetor that incorporated several improvements over previously used carburetors. These included better low-speed driveability and more precise control of fuel metering. In addition, cold-weather starting was assured because the "Thermo-Quad" design permitted the engine to operate well on lean fuel mixtures. This lowered hydrocarbon and carbon monoxide emissions. The Therm-Quad carburetor on the 440 4-barrel V-8 had a plastic main body that reduced heat transfer to the fuel.

1974 CHRYSLER

Newport, Newport Custom "no-cost" optional V-8
Overhead valve. Cast-iron block. Bore & stroke: 4.00 x 3.58 in. Displacement: 360 cid. Compression ratio: 8.40:1. Net brake hp: 180 at 4000 rpm. Taxable hp: 51.20. Net torque: 290 lbs.-ft. at 2400 rpm. Five main bearings. Hydraulic valve lifters. Crankcase capacity: 4 qt. (add 1 qt. for new filter). Cooling system capacity: 16 qt. (with heater). Carburetor: Carburetor: Two-barrel. VIN code: K.

Newport, Newport Custom base V-8
Overhead valve. Cast-iron block. Bore & stroke: 4.34 x 3.38 in. Displacement: 400 cid. Compression ratio: 8.20:1. Net brake hp: 185 at 4000 rpm. Taxable hp: 60.30. Net torque: 315 lbs.-ft. at 2400 rpm. Five main bearings. Hydraulic valve lifters. Crankcase capacity: 4 qt. (add 1 qt. for new filter). Cooling system capacity: 16.5 qt. (with heater). Carburetor: Carburetor: Two-barrel. VIN code: M.

Newport, Newport Custom optional V-8 (California)
Overhead valve. Cast-iron block. Bore & stroke: 4.34 x 3.38 in. Displacement: 400 cid. Compression ratio: 8.20:1. Net brake hp: 200 at 4400 rpm. Taxable hp: 60.30. Net torque: 310 lbs.-ft. at 2400 rpm. Five main bearings. Hydraulic valve lifters. Crankcase capacity: 4 qt. (add 1 qt. for new filter). Cooling system capacity: 16.5 qt. (with heater). Carburetor: Carburetor: Four-barrel. VIN code: N.

Newport, Newport Custom optional V-8 (outside California)
Overhead valve. Cast-iron block. Bore & stroke: 4.34 x 3.38 in. Displacement: 400 cid. Compression ratio: 8.20:1. Net brake hp: 205 at 4400 rpm. Taxable hp: 60.30. Net torque: 310 lbs.-ft. at 2400 rpm. Five main bearings. Hydraulic valve lifters. Crankcase capacity: 4 qt. (add 1 qt. for new filter). Cooling system capacity: 16.5 qt. (with heater). Carburetor: Carburetor: Four-barrel. VIN code: N.

New Yorker, Brougham, Town & Country base V-8; Newport, Custom base V-8 (California)
Overhead valve. Cast-iron block. Bore & stroke: 4.32 x 3.75 in. Displacement: 440 cid. Compression ratio: 8.20:1. Net brake hp: 220 at 4000 rpm. Taxable hp: 59.70. Torque: 345 lbs.-ft. at 3200 rpm. Five main bearings. Hydraulic valve lifters. Crankcase capacity: 4 qt. (add 1 qt. for new filter). Cooling system capacity: 16 qt. (with heater). Dual exhaust. Carburetor: Four-barrel. VIN code: T.

New Yorker, Brougham, T & C base V-8; Newport, Custom base V-8 (outside California)
Overhead valve. Cast-iron block. Bore & stroke: 4.32 x 3.75 in. Displacement: 440 cid. Compression ratio: 8.20:1. Net brake hp: 230 at 4000 rpm. Taxable hp: 59.70. Torque: 350 lbs.-ft. at 3200 rpm. Five main bearings. Hydraulic valve lifters. Crankcase capacity: 4 qt. (add 1 qt. for new filter). Cooling system capacity: 16 qt. (with heater). Dual exhaust. Carburetor: Four-barrel. VIN code: T.

Imperial LeBaron base V-8 (California)
Overhead valve. Cast-iron block. Bore & stroke: 4.32 x 3.75 in. Displacement: 440 cid. Compression ratio: 8.20:1. Net brake hp: 220 at 4000 rpm. Taxable hp: 59.70. Torque: 345 lbs.-ft. at 3200 rpm. Five main bearings. Hydraulic valve lifters. Crankcase capacity: 4 qt. (add 1 qt. for new filter). Cooling system capacity: 16 qt. (with heater). Dual exhaust. Carburetor: Four-barrel. VIN code: T.

Imperial LeBaron base V-8 (outside California)
Overhead valve. Cast-iron block. Bore & stroke: 4.32 x 3.75 in. Displacement: 440 cid. Compression ratio: 8.20:1. Net brake hp: 230 at 4000 rpm. Taxable hp: 59.70. Torque: 350 lbs.-ft. at 3200 rpm. Five main bearings. Hydraulic valve lifters. Crankcase capacity: 4 qt. (add 1 qt. for new filter). Cooling system capacity: 16 qt. (with heater). Dual exhaust. Carburetor: Four-barrel. VIN code: T.

As part of its "Extra Care in Engineering" promotion in 1974, Chrysler fitted all of its V-8s with an electronic voltage regulator, a coolant reserve system, an emissions control system, and an electronic ignition system. The latter provided up to 35 percent more starting voltage than conventional systems and reduced "tune-up" procedure to swapping spark plugs every 18,000 miles.

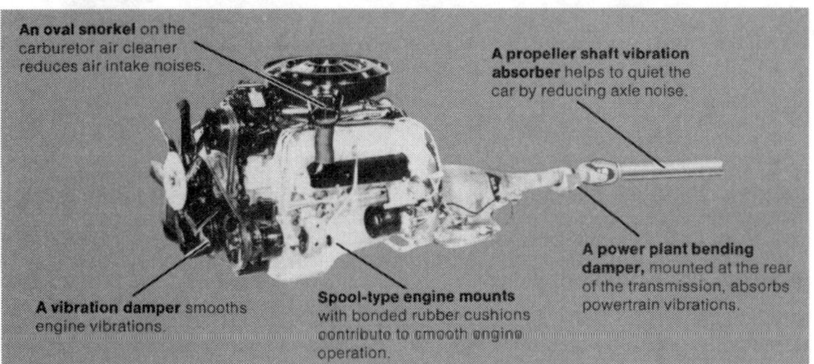

The 1974 Chrysler V-8 included an oval snorkel on the carburetor air cleaner to reduce air intake noises, a vibration damper to smooth engine vibrations, and spool-type engine mounts with rubber cushions to smooth engine operation.

1975 CHRYSLER

Cordoba base V-8 (California)
Overhead valve. Cast-iron block. Bore & stroke: 3.91 x 3.31 in. Displacement: 318 cid. Compression ratio: 8.50:1. Net brake hp: 135 at 3600 rpm. Taxable hp: 48.90. Torque: 245 lbs.-ft. at 1600 rpm. Five main bearings. Hydraulic valve lifters. Crankcase capacity: 4 qt. (add 1 qt. for new filter). Cooling system capacity: 16 qt. (with heater). Carburetor: Two-barrel. VIN code: G.

Cordoba base V-8 (outside California)
Overhead valve. Cast-iron block. Bore & stroke: 3.91 x 3.31 in. Displacement: 318 cid. Compression ratio: 8.50:1. Net brake hp: 150 at 4000 rpm. Taxable hp: 48.90. Torque: 255 lbs.-ft. at 1600 rpm. Five main bearings. Hydraulic valve lifters. Crankcase capacity: 4 qt. (add 1 qt. for new filter). Cooling system capacity: 16 qt. (with heater). Carburetor: Two-barrel. VIN code: G.

Newport, Newport Custom base V-8 (outside California)
Overhead valve. Cast-iron block. Bore & stroke: 4.34 x 3.38 in. Displacement: 400 cid. Compression ratio: 8.20:1. Net brake hp: 175 at 4000 rpm. Taxable hp: 60.30. Net torque: 300 lbs.-ft. at 2400 rpm. Five main bearings. Hydraulic valve lifters. Crankcase capacity: 4 qt. (add 1 qt. for new filter). Cooling system capacity: 16.5 qt. (with heater). Carburetor: Carburetor: Two-barrel. VIN code: M.

Cordoba, Newport, Newport Custom optional V-8 (outside California)
Overhead valve. Cast-iron block. Bore & stroke: 4.00 x 3.58 in. Displacement: 360 cid. Compression ratio: 8.40:1. Net brake hp: 180 at 4000 rpm. Taxable hp: 51.20. Net torque: 290 lbs.-ft. at 2400 rpm. Five main bearings. Hydraulic valve lifters. Crankcase capacity: 4 qt. (add 1 qt. for now filtor). Cooling oystom capacity: 16 qt. (with heater). Carburetor: Carburetor: Two-barrel. VIN code: K.

Newport, Newport Custom base V-8 (California)
Overhead valve. Cast-iron block. Bore & stroke: 4.34 x 3.38 in. Displacement: 400 cid. Compression ratio: 8.20:1. Net brake hp: 185 at 4000 rpm. Taxable hp: 60.30. Net torque: 285 lbs.-ft. at 3200 rpm. Five main bearings. Hydraulic valve lifters. Crankcase capacity: 4 qt. (add 1 qt. for new filter). Cooling system capacity: 16.5 qt. (with heater). Carburetor: Carburetor: Four-barrel. VIN code: P.

Cordoba, Newport, Newport Custom optional V-8 (California)
Overhead valve. Cast-iron block. Bore & stroke: 4.00 x 3.58 in. Displacement: 360 cid. Compression ratio: 8.40:1. Net brake hp: 190 at 4000 rpm. Taxable hp: 51.20. Net torque: 270 lbs.-ft. at 3200 rpm. Five main bearings. Hydraulic valve lifters. Crankcase capacity: 4 qt. (add 1 qt. for new filter). Cooling system capacity: 16 qt. (with heater). Carburetor: Carburetor: Four-barrel. VIN code: J.

New Yorker, Brougham, Town & Country base V-8
Overhead valve. Cast-iron block. Bore & stroke: 4.32 x 3.75 in. Displacement: 440 cid. Compression ratio: 8.20:1. Net brake hp: 215 at 4000 rpm. Taxable hp: 59.70. Torque: 330 lbs.-ft. at 3200 rpm. Five main bearings. Hydraulic valve lifters. Crankcase capacity: 4 qt. (add 1 qt. for new filter). Cooling system capacity: 16 qt. (with heater). Carburetor: Four-barrel. VIN code: T.

Imperial LeBaron base V-8 (California)
Overhead valve. Cast-iron block. Bore & stroke: 4.32 x 3.75 in.

In 1975, the 318-cid V-8 was an economy option for the Chrysler Cordoba. Notice the five-blade fan and the unique manifold casting. This engine was known for being reliable and rugged. It had hydraulic valve lifters that required no periodic adjustments.

The 1975 Chrysler 400-cid two-barrel V-8 was especially outstanding at low and middle speed ranges because of its displacement, large intake manifold, and large intake and exhaust valves, which contributed to easier breathing. It had five main bearings on its crankshaft and the distributor was located at the front of the engine.

The 360-cid V-8 was a larger displacement version 318 with a larger bore and longer stroke. In 1975, it was available in two- and four-barrel configurations. The 360 2-barrel was standard in the Cordoba and an economy option in Newport and Newport Custom models. The 360 4-barrel V-8 was available only with the E56 California Emission Control Package on the Cordoba, Newport, and Newport Custom.

Displacement: 440 cid. Compression ratio: 8.20:1. Net brake hp: 215 at 4000 rpm. Taxable hp: 59.70. Torque: 330 lbs.-ft. at 3200 rpm. Five main bearings. Hydraulic valve lifters. Crankcase capacity: 4 qt. (add 1 qt. for new filter). Cooling system capacity: 16 qt. (with heater). Carburetor: Four-barrel. VIN code: T.

The 1975 Chrysler 400-cid V-8 had a deep-skirted block and was an entirely different power plant than the 318- and 360-cid Chrysler V-8s. It was first available in two- and four-barrel versions. A high-performance version was released for sale late in 1975. It carried a four-barrel carburetor and a low-restriction dual exhaust system. It was an extra-cost option on the Cordoba only and was not available for sale in California.

The 1975 Chrysler 400-cid four-barrel V-8 came in two horsepower ratings. Listed as standard equipment for the Imperial, New Yorker Brougham, New Yorker, and Town & Country wagon—and optional for the Newport and Newport Custom models—was the 230-hp type. When equipped with the N95 emissions package, however, the engine was rated at 220 net horsepower.

1976 CHRYSLER

Cordoba optional V-8 (California)
Overhead valve. Cast-iron block. Bore & stroke: 3.91 x 3.31 in. Displacement: 318 cid. Compression ratio: 8.50:1. Net brake hp: 140 at 3600 rpm. Taxable hp: 48.90. Torque: 245 lbs.-ft. at 1600 rpm. Five main bearings. Hydraulic valve lifters. Crankcase capacity: 4 qt. (add 1 qt. for new filter). Cooling system capacity: 16 qt. (with heater). Carburetor: Carter BBD80695 or BBD80995 two-barrel. VIN code: G.

Cordoba optional V-8 (outside California)
Overhead valve. Cast-iron block. Bore & stroke: 3.91 x 3.31 in. Displacement: 318 cid. Compression ratio: 8.50:1. Net brake hp: 150 at 4000 rpm. Taxable hp: 48.90. Torque: 255 lbs.-ft. at 1600 rpm. Five main bearings. Hydraulic valve lifters. Crankcase capacity: 4 qt. (add 1 qt. for new filter). Cooling system capacity: 16 qt. (with heater). Carburetor: Carter BBD80695 or BBD80995 two-barrel. VIN code: G.

Cordoba, Newport optional V-8 (outside California)
Overhead valve. Cast-iron block. Bore & stroke: 4.00 x 3.58 in. Displacement: 360 cid. Compression ratio: 8.40:1. Net brake hp: 170 at 4000 rpm. Taxable hp: 51.20. Net torque: 280 lbs.-ft. at 2400 rpm. Five main bearings. Hydraulic valve lifters. Crankcase capacity: 4 qt. (add 1

"The Chrysler 318 V-8 engine has a long and enviable record for performance," pointed out the 1976 Chrysler Salesman's Guide. Quick and agile acceleration for safe passing on the highway was one result of the efficient design and engineering of this compact small-block V-8. For 1976, Chrysler eliminated the Orifice Spark Advance Control system. The EGR system was recalibrated, a new single-element catalytic converter was used, and the low-rpm spark advance mechanism was improved for better engine response.

On the 1976 Chrysler 360 two-barrel V-8, Chrysler shortened the Orifice Spark Advance Control system's time delay by 10 seconds. The temperature-sensing valve for the EGR system was relocated to the radiator header tank to provide more accurate control of the EGR system. Idle speed was reduced from 750 to 700 rpm to improve fuel economy. The four-barrel version of the same motor had smaller primary venturis, a reduced air pump speed, a recalibrated EGR system, and improvements to the OSAC device.

In 1976, the 440-cid four-barrel V-8 was again the largest engine available in Chrysler brand cars. This year it was rated for 205 net horsepower at 3600 rpm compared to 215 net horsepower at 4000 rpm the previous year. Tough aluminum-alloy-on-steel connecting rod and main bearings all contributed to the dependability and long life of the Chrysler 440-cid V-8.

The following changes were made to the Chrysler 400 two-barrel V-8 in 1976 to improve fuel economy and driveability: The vacuum spark advance time delay was shortened from 27 to 17 seconds in the OSAC system, the engine idle speed was reduced to 700 rpm, and the carburetor power mixture was set to operate at a higher road speed to improve low-speed and suburban fuel economy.

For 1976, the Chrysler 400-cid four-barrel V-8 was basically the same as the 400 two-barrel V-8, except for its larger carburetor. The high-performance version of this engine added a low-restriction dual exhaust system and a free-breathing four-barrel carburetor that combined gave a 55 hp boost in performance.

qt. for new filter). Cooling system capacity: 16 qt. (with heater). Carburetor: Holley R7364A two-barrel. VIN code: K.

Cordoba, Newport optional V-8 (California)
Overhead valve. Cast-iron block. Bore & stroke: 4.00 x 3.58 in. Displacement: 360 cid. Compression ratio: 8.40:1. Net brake hp: 175 at 4000 rpm. Taxable hp: 51.20. Net torque: 270 lbs.-ft. at 1600 rpm. Five main bearings. Hydraulic valve lifters. Crankcase capacity: 4 qt. (add 1 qt. for new filter). Cooling system capacity: 16 qt. (with heater). Carburetor: Carter TQ90555 two-barrel. VIN code: K.

Cordoba, Newport base V-8; New Yorker, Brougaham optional V-8 (outside California)
Overhead valve. Cast-iron block. Bore & stroke: 4.34 x 3.38 in. Displacement: 400 cid. Compression ratio: 8.20:1. Net brake hp: 175 at 4000 rpm. Taxable hp: 60.30. Net torque: 300 lbs.-ft. at 2400 rpm. Five main bearings. Hydraulic valve lifters. Crankcase capacity: 4 qt. (add 1 qt. for new filter). Cooling system capacity: 16.5 qt. (with heater). Carburetor: Holley R7366A two-barrel.

Cordoba, Newport base V-8; New Yorker, Brougaham optional V-8 (California)
Overhead valve. Cast-iron block. Bore & stroke: 4.34 x 3.38 in. Displacement: 400 cid. Compression ratio: 8.20:1. Net brake hp: 185 at 3600 rpm. Taxable hp: 60.30. Net torque: 285 lbs.-ft. at 3200 rpm. Five main bearings. Hydraulic valve lifters. Crankcase capacity: 4 qt. (add 1 qt. for new filter). Cooling system capacity: 16.5 qt. (with heater).

Carburetor: Holley R7366A two-barrel.

New Yorker, Brougham, Town & Country base V-8; Newport optional V-8 (California)
Overhead valve. Cast-iron block. Bore & stroke: 4.32 x 3.75 in. Displacement: 440 cid. Compression ratio: 8.20:1. Net brake hp: 200 at 3600 rpm. Taxable hp: 59.70. Torque: 310 lbs.-ft. at 2400 rpm. Five main bearings. Hydraulic valve lifters. Crankcase capacity: 4 qt. (add 1 qt. for new filter). Cooling system capacity: 16 qt. (with heater). Carburetor: Carter TQ9059-S four-barrel. VIN code: T.

New Yorker, Brougham, T & C base V-8; Newport optional V-8 (outside California)
Overhead valve. Cast-iron block. Bore & stroke: 4.32 x 3.75 in. Displacement: 440 cid. Compression ratio: 8.20:1. Net brake hp: 205 at 3600 rpm. Taxable hp: 59.70. Torque: 320 lbs.-ft. at 2000 rpm. Five main bearings. Hydraulic valve lifters. Crankcase capacity: 4 qt. (add 1 qt. for new filter). Cooling system capacity: 16 qt. (with heater). Carburetor: Carter TQ9058-S four-barrel. VIN code: T.

All models Lean-Burn V-8 (outside California)
Overhead valve. Cast-iron block. Bore & stroke: 4.34 x 3.38 in. Displacement: 400 cid. Compression ratio: 8.20:1. Net brake hp: 210 at 4000 rpm. Taxable hp: 60.30. Net torque: 305 lbs.-ft. at 3200 rpm. Five main bearings. Hydraulic valve lifters. Crankcase capacity: 4 qt. (add 1 qt. for new filter). Cooling system capacity: 16.5 qt. (with heater). Electronic Lean Burn System and Carter TQ9064S or TQ9097S four-barrel carburetor. VIN code: N.

Cordoba, Newport optional high-performance V-8 (outside California)
Overhead valve. Cast-iron block. Bore & stroke: 4.34 x 3.38 in. Displacement: 400 cid. Compression ratio: 8.20:1. Net brake hp: 240 at 4400 rpm. Taxable hp: 60.30. Net torque: 325 lbs.-ft. at 3200 rpm. Five main bearings. Hydraulic valve lifters. Crankcase capacity: 4 qt. (add 1 qt. for new filter). Cooling system capacity: 16.5 qt. (with heater). Dual exhaust. Carburetor: Carter TQ9054S four-barrel. VIN code: P.

1977 CHRYSLER

LeBaron base V-8; Cordoba optional V-8
Overhead valve. Cast-iron block and head. Bore & stroke: 3.91 x 3.31 in. Displacement: 318 cid (5.2 liters). Compression ratio: 8.5:1. Net brake hp: 145 at 4000 rpm. Taxable hp: 48.90. Net torque: 245 lbs.-ft. at 1600 rpm. Five main bearings. Hydraulic valve lifters. Carburetor: (Cordoba) Carter BBD81335 two-barrel; (LeBaron) Carter BBD81425 or R7990A two-barrel. VIN code: G.

Cordoba, Newport optional V-8
Overhead valve. Cast-iron block and head. Bore & stroke: 4.00 x 3.58 in. Displacement: 360 cid (5.9 liters). Compression ratio: 8.4:1. Net brake hp: 155 at 3600 rpm. Taxable hp: 51.20. Net torque: 275 lbs.-ft. at 2000 rpm. Five main bearings. Hydraulic valve lifters. Carburetor: Holley 2245 two-barrel. VIN code: K.

NOTE: A 360-cid 170-nhp four-barrel V-8 was available in California and high-altitude areas.

Cordoba, Newport base lean-burn V-8; New Yorker, Brougham, T&C optional lean-burn V-8
Overhead valve. Cast-iron block and head. Bore & stroke: 4.34 x 3.38 in. Displacement: 400 cid (6.6 liters). Compression ratio: 8.2:1. Net brake hp: 190 at 3600 rpm. Taxable hp: 60.30. Net torque: 305 lbs.-ft. at 3200 rpm. Five main bearings. Hydraulic valve lifters. Carburetor: Carter TQ91025 four-barrel. Electronic Lean Burn System. VIN code: N.

Town & Country, New Yorker Brougham base V-8; Newport optional V-8
Overhead valve. Cast-iron block and head. Bore & stroke: 4.32 x 3.75 in. Displacement: 440 cid (7.2 liters). Compression ratio: 8.2:1. Net brake hp: 195 at 3600 rpm. Taxable hp: 59.70. Net torque: 320 lbs.-ft. at 2000 rpm. Five main bearings. Hydraulic valve lifters. Carburetor: Carter TQ9078S four-barrel. Electronic Lean Burn System. VIN code: T.

1978 CHRYSLER

The 318-cid two-barrel V-8 was available on all 1978 Chrysler LeBarons and on Cordobas sold outside California. Electronic ignition circuitry was designed into the second-generation spark-control computer used with this engine's Lean-Burn system. It featured hydraulic valve lifters, induction-hardened valve seats, a cast ductile iron crankshaft, a single exhaust system with a catalytic converter, and an aluminized steel exhaust pipe and tailpipe.

The 360-cid two-barrel V-8 with the Lean-Burn system was available on all 1978 Chrysler LeBaron models, but was not available in California or in areas above 4,000 ft. altitude. Optional was a heavy-duty version with the code A36 Heavy-Duty package. It was also offered only in "federal" areas.

The 360-cid four-barrel V-8 came in two versions in 1978. The E56 version, for California, did not carry the Lean-Burn system and was specifically designed to meet California and high-altitude county emission-control regulations. It had an air pump and a catalytic converter. The E58 "heavy-duty" version of the same engine was offered only in California. It did not carry the Lean-Burn system either.

LeBaron, Cordoba optional V-8
Overhead valve. Cast-iron block and head. Bore & stroke: 3.91 x 3.31 in. Displacement: 318 cid (5.2 liters). Compression ratio: 8.5:1. Net brake hp: 140 at 4000 rpm. Taxable hp: 48.90. Net torque: 245 lbs.-ft. at 1600 rpm. Five main bearings. Hydraulic valve lifters. Carburetor: Carter BBD or Holley two-barrel. Electronic Lean Burn. VIN code: G.

The 400-cid four-barrel V-8 came in standard (E64) and heavy-duty (E68) versions in 1978. The heavy-duty version was offered for the Cordoba only and was not available in California or high-altitude counties. Features of both versions included a four-barrel carburetor, large intake and exhaust ports, and a large 2-1/4-in. diameter exhaust pipe.

Cordoba S base V-8; All other models optional V-8
Overhead valve. Cast-iron block and head. Bore & stroke: 4.00 x 3.58 in. Displacement: 360 cid (5.9 liters). Compression ratio: 8.4:1. Net brake hp: 155 at 3600 rpm. Taxable hp: 51.20. Net torque: 270 lbs.-ft. at 2400 rpm. Five main bearings. Hydraulic valve lifters. Carburetor: two-barrel Holley. Electronic Lean Burn. VIN code: K.
NOTE 1: A 360-cid 170-nhp four-barrel V-8 was available in California and high-altitude areas.
NOTE 2: A 360-cid 175-nhp "heavy-duty" four-barrel V-8 was also available.

Cordoba, Newport, New Yorker base V-8; Cordoba S optional V-8
Overhead valve. Cast-iron block and head. Bore & stroke: 4.34 x 3.38 in. Displacement: 400 cid (6.6 liters). Compression ratio: 8.2:1. Net brake hp: 190 at 3600 rpm. Taxable hp: 60.30. Net torque: 305 lbs.-ft. at 3200 rpm. Five main bearings. Hydraulic valve lifters. Carburetor: four-barrel Carter TQ. Electronic Lean Burn System. VIN code: N.

Newport, New Yorker optional V-8
Overhead valve. Cast-iron block and head. Bore & stroke: 4.32 x 3.75 in. Displacement: 440 cid (7.2 liters). Compression ratio: 8.2:1. Net brake hp: 195 at 3600 rpm. Taxable hp: 59.70. Net torque: 320 lbs.-ft. at 2000 rpm. Five main bearings. Hydraulic valve lifters. Carburetor: four-barrel Carter TQ. Electronic Lean Burn System. VIN code: T.

1979 CHRYSLER

The 318-cid two-barrel V-8 was standard in 1979 Cordobas and optional in LeBarons and Newports. It featured hydraulic valve lifters, induction-hardened valve seats, a cast ductile iron crankshaft, a single exhaust system with a catalytic converter, an aluminized steel exhaust pipe and tailpipe, and electronic ignition. Also available in California the same year was a 318-cid four-barrel V-8 with the ECS system.

The 360-cid four-barrel V-8 again came in two versions in 1979. The E56 version, for California, featured Electronic Spark Control or ESC. The E58 "heavy-duty" version of the same engine required the A36 Heavy-Duty package. On Cordobas, Newports, and New Yorkers, this engine was equipped with a complete dual exhaust system.

All models optional V-8
Overhead valve. Cast-iron block and head. Bore & stroke: 3.91 x 3.31 in. Displacement: 318 cid (5.2 liters). Compression ratio: 8.5:1. Net brake hp: 135 at 4000 rpm. Taxable hp: 48.90. Net torque: 250 lbs.-ft. at 1600 rpm. Five main bearings. Hydraulic valve lifters. Carburetor: Holley 2280 (R8448A) two-barrel. VIN code: G.

New Yorker base V-8; All other models optional V-8
Overhead valve. Cast-iron block and head. Bore & stroke: 4.00 x 3.58 in. Displacement: 360 cid (5.9 liters). Compression ratio: 8.4:1. Net brake hp: 150 at 3600 rpm. Taxable hp: 51.20. Net torque: 265 lbs.-ft. at 2400 rpm. Five main bearings. Hydraulic valve lifters. Carburetor: Holley 2245 two-barrel. VIN code: K.

All models optional high-output V-8
Overhead valve. Cast-iron block and head. Bore & stroke: 4.00 x 3.58 in. Displacement: 360 cid (5.9 liters). Compression ratio: 8.0:1. Net brake hp: 195 at 4000 rpm. Taxable hp: 51.20. Net torque: 280 lbs.-ft. at 2400 rpm. Five main bearings. Hydraulic valve lifters. Carburetor: Carter TQ two-barrel. VIN code: L.

1980 CHRYSLER

LeBaron, Corboba, Newport, New Yorker base V-8 (outside California)
Overhead valve. Cast-iron block and head. Bore & stroke: 3.91 x 3.31 in. Displacement: 318 cid (5.2 liters). Compression ratio: 8.5:1. Net brake hp: 120 at 3600 rpm. Taxable hp: 48.90. Net torque: 245 lbs.-ft. at 1600 rpm. Five main bearings. Hydraulic valve lifters. Carburetor: Carter BBD two-barrel. VIN code: G.
NOTE: A four-barrel 318 rated 155 horsepower was available in California.

Newport, New Yorker optional V-8
Overhead valve. Cast-iron block and head. Bore & stroke: 4.00 x 3.58 in. Displacement: 360 cid (5.9 liters). Compression ratio: 8.4:1. Net brake hp: 130 at 3200 rpm. Taxable hp: 51.20. Net torque: 255 lbs.-ft. at 2000 rpm. Five main bearings. Hydraulic valve lifters. Carburetor: Carter BBD two-barrel. VIN code: K.

LeBaron, Corboba, Newport, New Yorker base V-8 (California)
Overhead valve. Cast-iron block and head. Bore & stroke: 3.91 x 3.31 in. Displacement: 318 cid (5.2 liters). Compression ratio: 8.5:1. Net brake hp: 155. Taxable hp: 48.90. Five main bearings. Hydraulic valve lifters. Carburetor: Four-barrel. VIN code: G.

Cordoba "300" high-output V-8
Overhead valve. Cast-iron block and head. Bore & stroke: 4.00 x 3.58 in. Displacement: 360 cid (5.9 liters). Compression ratio: 8.0:1. Net brake hp: 185 at 4000 rpm. Taxable hp: 51.20. Net torque: 275 at 2000 rpm. Five main bearings. Hydraulic valve lifters. Carburetor: Carter TQ92445 four-barrel. VIN code: L.

1981 CHRYSLER

LeBaron, Corboba, Newport, New Yorker base V-8 (outside California)
Overhead valve. Cast-iron block and head. Bore & stroke: 3.91 x 3.31 in. Displacement: 318 cid (5.2-liters). Compression ratio: 8.5:1. Net brake hp: 130 at 4000 rpm. Taxable hp: 48.90. Net torque: 230 lbs.-ft. at 2000 rpm. Five main bearings. Hydraulic valve lifters. Carburetor: Carter BBD 8291S two-barrel. VIN code: K.

Newport, New Yorker optional V-8; LeBaron, Cordoba base V-8 (California only)
Overhead valve. Cast-iron block and head. Bore & stroke: 3.91 x 3.31 in. Displacement: 318 cid (5.2-liters). Compression ratio: 8.5:1. Net brake hp: 165 at 4000 rpm. Taxable hp: 48.90. Net torque: 240 lbs.-ft. at 2000 rpm. Five main bearings. Hydraulic valve lifters. Carburetor: Carter TQ9283S four-barrel. VIN code: M.

Imperial base V-8
Overhead valve. Cast-iron block and head. Bore & stroke: 3.91 x 3.31 in. Displacement: 318 cid (5.2-liters). Compression ratio: 8.5:1. Net brake hp: 140 at 4000 rpm. Taxable hp: 48.90. Net torque: 240 lbs.-ft. at 2000 rpm. Five main bearings. Hydraulic valve lifters. Induction: Electronic fuel injection. VIN code: J.

1982 CHRYSLER

Corboba, New Yorker base V-8 (Outside California)
Overhead valve. Cast-iron block and head. Bore & stroke: 3.91 x 3.31 in. Displacement: 318 cid (5.2-liters). Compression ratio: 8.5:1. Net brake hp: 130 at 4000 rpm. Taxable hp: 48.90. Net torque: 230 lbs.-ft. at 2000 rpm. Five main bearings. Hydraulic valve lifters. Carburetor: Carter BBD 8291S two-barrel. VIN code: K.

Corboba, New Yorker base V-8 (California)
Overhead valve. Cast-iron block and head. Bore & stroke: 3.91 x 3.31 in. Displacement: 318 cid (5.2-liters). Compression ratio: 8.5:1. Net brake hp: 165 at 4000 rpm. Taxable hp: 48.90. Net torque: 240 lbs.-ft. at 2000 rpm. Five main bearings. Hydraulic valve lifters. Carburetor: Carter TQ9283S four-barrel. VIN code: M.

Imperial base V-8
Overhead valve. Cast-iron block and head. Bore & stroke: 3.91 x 3.31 in. Displacement: 318 cid (5.2-liters). Compression ratio: 8.5:1. Net brake hp: 140 at 4000 rpm. Taxable hp: 48.90. Net torque: 245 lbs.-ft. at 2000 rpm. Five main bearings. Hydraulic valve lifters. Induction: Electronic fuel injection. VIN code: J.

1983 CHRYSLER

Corboba, Fifth Avenue base V-8
Overhead valve. Cast-iron block and head. Bore & stroke: 3.91 x 3.31 in. Displacement: 318 cid (5.2-liters). Compression ratio: 8.5:1. Net brake hp: 130 at 4000 rpm. Taxable hp: 48.90. Net torque: 230 lbs.-ft. at 2000 rpm. Five main bearings. Hydraulic valve lifters. Carburetor: Carter BBD two-barrel. VIN code: K.

Imperial base V-8
Overhead valve. Cast-iron block and head. Bore & stroke: 3.91 x 3.31 in. Displacement: 318 cid (5.2-liters). Compression ratio: 8.5:1. Net brake hp: 140 at 4000 rpm. Taxable hp: 48.90. Net torque: 245 lbs.-ft. at 2000 rpm. Five main bearings. Hydraulic valve lifters. Induction: Electronic fuel injection. VIN code: J.

1984 CHRYSLER

Fifth Avenue base V-8
Overhead valve. Cast-iron block and head. Bore & stroke: 3.91 x 3.31 in. Displacement: 318 cid (5.2-liters). Compression ratio: 8.7:1. Net brake hp: 130 at 4000 rpm. Taxable hp: 48.90. Net torque: 235 lbs.-ft. at 1600 rpm. Five main bearings. Hydraulic valve lifters. Carburetor: Carter BBD two-barrel. VIN code: P.

1985 CHRYSLER

Fifth Avenue base V-8
Overhead valve. Cast-iron block and head. Bore & stroke: 3.91 x 3.31 in. Displacement: 318 cid (5.2-liters). Compression ratio: 9.0:1. Net brake hp: 140 at 3600 rpm. Taxable hp: 48.90. Net torque: 265 lbs.-ft. at 1600 rpm. Five main bearings. Hydraulic valve lifters. Carburetor: Two-barrel. VIN code: P.

1986 CHRYSLER

Fifth Avenue base V-8
Overhead valve. Cast-iron block and head. Bore & stroke: 3.91 x 3.31 in. Displacement: 318 cid (5.2-liters). Compression ratio: 9.0:1. Net brake hp: 140 at 3600 rpm. Taxable hp: 48.90. Net torque: 265 lbs.-ft. at 1600 rpm. Five main bearings. Hydraulic valve lifters. Carburetor: Two-barrel. VIN code: P.

1987 CHRYSLER

Fifth Avenue base V-8
Overhead valve. Cast-iron block and head. Bore & stroke: 3.91 x 3.31 in. Displacement: 318 cid (5.2-liters). Compression ratio: 9.0:1. Net brake hp: 140 at 3600 rpm. Taxable hp: 48.90. Net torque: 265 lbs.-ft. at 1600 rpm. Five main bearings. Hydraulic valve lifters. Carburetor: Two-barrel. VIN code: P.

1988 CHRYSLER

Fifth Avenue base V-8
Overhead valve. Cast-iron block and head. Bore & stroke: 3.91 x 3.31 in. Displacement: 318 cid (5.2-liters). Compression ratio: 9.1:1. Net brake hp: 140 at 3600 rpm. Taxable hp: 48.90. Net torque: 265 lbs.-ft. at 2000 rpm. Five main bearings. Hydraulic valve lifters. Carburetor: Two-barrel. VIN code: P or 4.

1989 CHRYSLER

Fifth Avenue base V-8
Overhead valve. Cast-iron block and head. Bore & stroke: 3.91 x 3.31 in. Displacement: 318 cid (5.2-liters). Compression ratio: 9.1:1. Net brake hp: 140 at 3600 rpm. Taxable hp: 48.90. Net torque: 265 lbs.-ft. at 2000 rpm. Five main bearings. Hydraulic valve lifters. Carburetor: Two-barrel. VIN code: P.
NOTE: A 318-cid four-barrel V-8 Code S was offered in 1989, probably as a police package.

Selected Chrysler V-8 Tune-Up Specs 1951-1965

Year	Model	Spark Plugs Type	Spark Plugs Gap	Point Gap	Deg Cam Angle	Timing	Idle Speed	Compression Pressure	Battery (Delco)
1951	C52, C53, C54	AL-AR80	.035	.015-.018	NOTE 1	TC	425 rpm	150	Group 2
1952	C52-C55	AL-AR80	.035	.015-.018	NOTE 1	4B	475 rpm	150	Group 2
1953	Chrysler V-8	AL-AR80	.035	.015-.018	NOTE 1	4B	475 rpm	150	Group 2
	Imperial V-8	AL-AR80	.035	.015-.018	NOTE 1	4B	475 rpm	150	3SM/12H
1954	Chrysler V-8	AL-4GS150	.035	.015-.018	NOTE 1	4B	475 rpm	150	Group 2
	Imperial V-8	AL-4GS150	.035	.015-.018	NOTE 1	4B	475 rpm	150	3SM/12H
1955	Windsor V-8	AL-AR51	.035	.016-.018	NOTE 1	6B	475 rpm	155	Group 2
	Chrysler 300	AL-AGR41	.035	.016-.018	NOTE 1	10B	500 rpm	160	Group 2
	All other V-8	AL-AGR51	.035	.016-.018	NOTE 1	6B	475 rpm	160	Group 2
	Imperial V-8	AL-AGR51	.035	.016-.018	NOTE 1	6B	475 rpm	160	3SM/12H
1956	Windsor V-8	AL-AR52	.035	.015-.018	30.5	2B (*)	475 rpm	155	2SM/11HS
	Chrysler 300	AL-AGR41	.035	.015-.018	NOTE 1	8B	650 rpm	160	3SM/12H
	All other V-8	AL-AGR42	.035	.015-.018	NOTE 1	4B	475 rpm	160	3SM/12H

(*) Ignition timing for C71 Windsor with Power Pac V-8 is 4 degrees before top dead center (4B).

Year	Model	Spark Plugs Type	Spark Plugs Gap	Point Gap	Deg Cam Angle	Timing	Idle Speed	Compression Pressure	Battery (Delco)
1957	Windsor V-8	AL-AR42	.035	.015-.018	30.5	6B	475 rpm	155	3SM/12H
	Chrysler 300	AL-AGR41	.035	.015-.018	NOTE 1	6B	635 rpm	165	3SM/12H
	All other V-8	AL-AGR42	.035	.015-.018	NOTE 1	6B	475 rpm	165	3SM/12H

NOTE 1: These 1951-1957 Chrysler and Imperial V-8s have two sets of distributor points.
For 1951 the total cam angle is 34-36 degrees and for each point set the cam angle should be set at 27.5-30 degrees.
For 1952-55 the total cam angle is 32-36 degrees and for each point set the cam angle should be set at 26-28 degrees.
For 1956-57 the total cam angle is 36-39 degrees and for each point set the cam angle should be set at 29-32 degrees.

Year	Model	Spark Plugs Type	Spark Plugs Gap	Point Gap	Deg Cam Angle	Timing	Idle Speed	Compression Pressure	Battery (Delco)
1958	Windsor V-8	AL-AR42	.035	.015-.018	29	8B	500 rpm	NA	HO-11-60
	Saratoga V-8	AL-AR42	.035	.015-.018	29	6B	500 rpm	NA	HO-11-60
	New Yorker V-8	AL-AGR42	.035	.015-.018	34	6B	500 rpm	NA	HO-12-70
	"300-D" V-8	AL-AGR32	.035	.015-.018	34	6B	650 rpm	NA	HO-12-70
	Imperial V-8	AL-AGR42	.035	.015-.018	34	6B	500 rpm	NA	HO-12-70
1959	Windsor V-8	AL-AR42	.035	.015-.018	29	10B	500 rpm	150-180	12-H-70
	Saratoga V-8	AL-AR42	.035	.015-.018	29	10B	500 rpm	150-180	12-H-70
	New Yorker V-8	AL-A42	.035	.015-.018	34	10B	500 rpm	150-180	12-H-70
	"300-E" V-8	AL-A42	.035	.015-.018	34	10B	500 rpm	150-180	12-H-70
	Imperial V-	AL-A42	.035	.017	34	10B	500 rpm	150-180	12-H-70
1960	Windsor V-8	AL-AR42	.035	.014-.019	26-32	10B	500 rpm	150-180	12-H-70
	Saratoga V-8	AL-AR42	.035	.014-.019	26-32	10B	500 rpm	150-180	12-H-70
	New Yorker V-8	AL-A42	.035	.014-.019	26-32	10B	500 rpm	150-180	12-H-70
	"300-E" V-8	AL-A42	.035	.014-.019	26-32	5B	725 rpm	150-180	12-H-70
	Imperial V-8	AL-A42	.035	.014-.019	26-32	10B	500 rpm	150-180	12-H-70
1961	Newport V-8	AL-A42	.035	.015-.018	26-32	10B	500 rpm	135-165	11HS59B
	Windsor V-8	AL-A42	.035	.015-.018	26-32	10B	500 rpm	150-180	11HS59B
	New Yorker V-8	AL-A42	.035	.014-.019	27-32	10B	500 rpm	150-180	11H70B
	"300-F" V-8	AL-A32	.035	.015-.018	26-32	5B	500 rpm	150-180	12H-70
	Imperial V-8	AL-A42	.035	.015-.018	26-32	10B	500 rpm	150-180	12H-70

NOTE: Autolite A901 or Champion J9Y spark plugs used in Chrysler 300-G with optional Power-Pac V-8

Year	Model	Spark Plugs Type	Spark Plugs Gap	Point Gap	Deg Cam Angle	Timing	Idle Speed	Compression Pressure	Battery (Delco)
1962	Newport V-8	AL-A42	.035	.015-.018	26-32	10B	500 rpm	135-165	IIHS59B
	Sport 300 V-8	AL-A42	.035	.015-.018	26-32	10B	500 rpm	150-180	IIHS59B
	New Yorker V-8	AL-A42	.035	.014-.019	31-37	10B	500 rpm	150-180	11BHS59
	"300-H" V-8	AL-A32	.035	.015-.018	27-32	5B	500 rpm	150-180	12BHB70
	Imperial V-8	AL-A42	.035	.014-.019	27-32	10B	500 rpm	130-165	12H-70

1963	Newport V-8	CH-J12Y	.035	.017	28-33	10B	500 rpm	135-165	59 amp hr
	Sport 300 (383)	CH-J12Y	.035	.017	28-33	10B	500 rpm	150-180	59 amp hr
	Sport 300 (413)	CH-J12Y	.035	.017	NOTE 1	10B	500 rpm	150-180	70 amp hr
	New Yorker V-8	CH-J12Y	.035	.017	28-33	10B	500 rpm	150-180	70 amp hr
	"300-J" V-8	CH-J12Y	.035	.017	NOTE 1	12.5B	500 rpm	150-180	70 amp hr
	Imperial V-8	CH-J12Y	.035	.017	27-32	10B	500 rpm	130-165	70 amp hr

NOTE 1: The 1963 Chrysler Sport 300 "413" and the Chrysler 300-J had two sets of distributor points. The total dwell angle is 34-40 degrees and for each point set the dwell should be set at 26-28 degrees.

1964	361-cid V-8	CH-J12Y	.035	.017	30	10B	500 rpm	135-165	59 amp hr
	383-cid V-8	CH-J12Y	.035	.017	30	10B	500 rpm	150-180	59 amp h
	413-cid V-8	CH-J12Y	.035	.017	30	10B	500 rpm	150-180	70 amp hr
	413 Hi-Perf V-8	CH-XJ10Y	.035	.017	NOTE 1	10B	500 rpm	150-180	70 amp hr

NOTE 1: The 1964 Chrysler 300-K and other models with the 413-cid high-performance V-8 had two sets of distributor points. The total dwell angle is 34-40 degrees and for each point set the dwell should be set at 26-28 degrees.

1965	383-cid V-8	CH-J14Y	.035	.017	30	10B	500 rpm	150-180	59 amp hr
	383 Hi-Perf V-8	CH-J10Y	.035	.017	30	10B	500 rpm	150-180	70 amp hr 413
	cid V-8	CH-J12Y	.035	.017	30	10B	500 rpm	150-180	70 amp hr
	413 Hi-Perf V-8	CH-J10Y	.035	.017	NOTE 1	10B	500 rpm	150-180	70 amp hr

NOTE 1: The 1965 Chrysler 300-L and other models with the 413-cid high-performance V-8 had two sets of distributor points. The total dwell angle is 34-40 degrees and for each point set the dwell should be set at 26-28 degrees.

Chrysler and Imperial V-8 Firing Order

1951-1976
1-8-4-3-6-5-7-2

All 1951-1976 Chrysler and Imperial V-8s have the cylinders numbered from front to back. The cylinders on the left-hand bank are numbered 1-3-5-7 and the cylinders on the right-hand bank are numbered 2-4-6-8.

DESOTO
1952-1961

Arranged in order by horsepower ratings.

1952 DESOTO

The 1952 overhead valve DeSoto V-8 featured hemispherical combustion chambers and two rocker shafts per cylinder head. The cast-iron block had a bore and stroke of 3.625 x 3.344 inches. DeSoto advertised "more power from every drop of gas." The DeSoto V-8 was designed to opearate on regular gasoline.

DeSoto Firedome base V-8
Overhead valve V-8. Hemispherical combustion chambers. Two rocker shafts per cylinder head. Cast-iron block. Bore & stroke: 3.625 x 3.344 in. Displacement: 276.1 cid. Compression ratio: 7.1:1. Brake hp: 160 at 4000 rpm. Taxable hp: 42.1. Torque: 250 lbs.-ft. at 2000 rpm. Five main bearings. Hydraulic valve lifters. Crankcase capacity: 5 qt. (add 1 qt. if new filter installed). Cooling system capacity: 22 qt. (add 1 qt. for heater). Carburetor: Carter WCD-884-S two-barrel. Engine code: S17-1001 and up.

1953 DESOTO

DeSoto Firedome base V-8
Overhead valve V-8. Hemispherical combustion chambers. Two rocker shafts per cylinder head. Cast-iron block. Bore & stroke: 3.625 x 3.344 in. Displacement: 276.1 cid. Compression ratio: 7.1:1. Brake hp: 160 at 4000 rpm. Taxable hp: 42.1. Torque: 250 lbs.-ft. at 2000 rpm. Five main bearings. Hydraulic valve lifters. Crankcase capacity: 5 qt. (add 1 qt. if new filter installed). Cooling system capacity: 22 qt. (add 1 qt. for heater). Carburetor: Carter BBD-909-S two-barrel. Engine code: S16-1001 and up.

1954 DESOTO

Firedome base V-8
Overhead valve V-8. Hemispherical combustion chambers. Two rocker shafts per cylinder head. Cast-iron block. Bore & stroke: 3.625 x 3.344 in. Displacement: 276.1 cid. Compression ratio: 7.5:1. Brake hp: 170 at 4400 rpm. Taxable hp: 42.1. Torque: 255 lbs.-ft. at 2400 rpm. Five main bearings. Hydraulic valve lifters. Crankcase capacity: 5 qt. (add 1 qt. if new filter installed). Cooling system capacity: 22 qt. (add 1 qt. for heater). Carburetor: (early) Carter BB-2067-S two-barrel; (late) Carter BB 2120-S two-barrel. Engine code: S19-1001 and up.

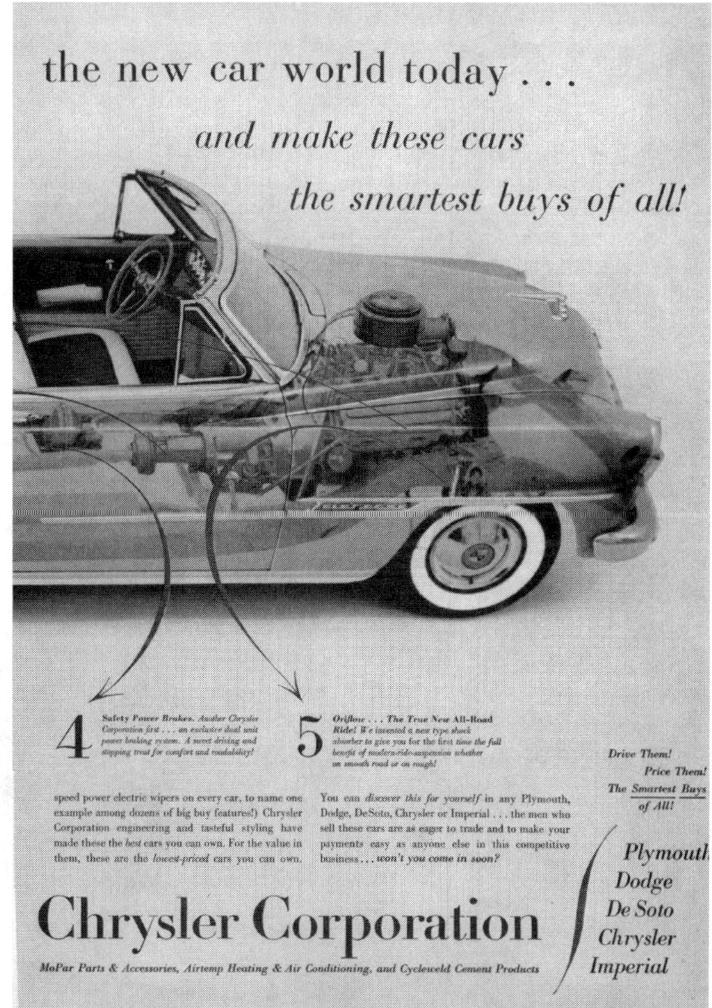

In 1954, the Firedome V-8 gained 10 hp from the same displacement block. The overhead hemi V-8 featured two rocker shafts per cylinder head. It had a 3.625 x 3.344 inch bore and stroke and displaced 276.1 cid. With a 7.5:1 compression ratio it cranked out 170 hp at 4400 rpm.

FireDome 8

It has the revolutionary new engine that's the talk of the engineering world...an engine with dome-shaped combustion chambers! Plus new, sensational Power Steering!

All over the country, De Soto-Plymouth Dealers are now displaying one of the most remarkable cars ever produced... the new De Soto FireDome Eight.

Its FireDome engine, with dome-shaped combustion chambers, develops more horsepower than conventional engines and can do it *on regular fuel.* It's America's most powerful engine design. It is the most important engine advance in more than a generation.

Delivering a mighty 160 horsepower, it will give you pick-up, acceleration and cruising performance far beyond anything you have ever known before.

In addition, this FireDome Eight brings you the magic of Power Steering. Now you can turn the wheel with one finger. Hydraulic power does the work for you. Road control is greater. Parking, even in tight spots, is child's play.

Want new super-safe Power Brakes? Want push-button Electric Window Lifts? Want Solex Heat-Resistant Glass? This new De Soto offers them all, plus the famous Tip-Toe Shift with Fluid Drive (*no-shift driving at its best!*), Oriflow Shock Absorbers, and many other major features for your comfort and safety.

This new De Soto has to be seen to be appreciated. And it must be driven to be believed. Plan to do both—soon!

DE SOTO DIVISION • CHRYSLER CORPORATION

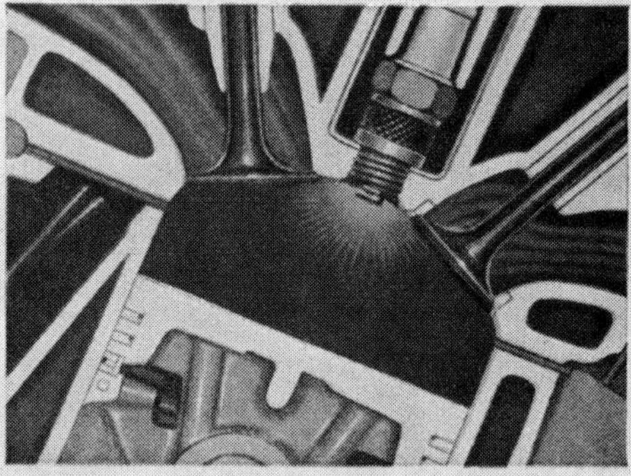

FIREDOME's spark plugs are centered in dome-shaped cylinder heads. More power from every drop of gas. Greater fuel economy. No need for premium fuel. Longer valve life. Smoother, quieter running.

POWER STEERING is the talk of the country. It makes steering as easy as dialing a telephone. You can turn the wheel with one finger... even when the car is at a standstill. Parking is easy... at last!

White sidewall tires, when available, are optional equipment.

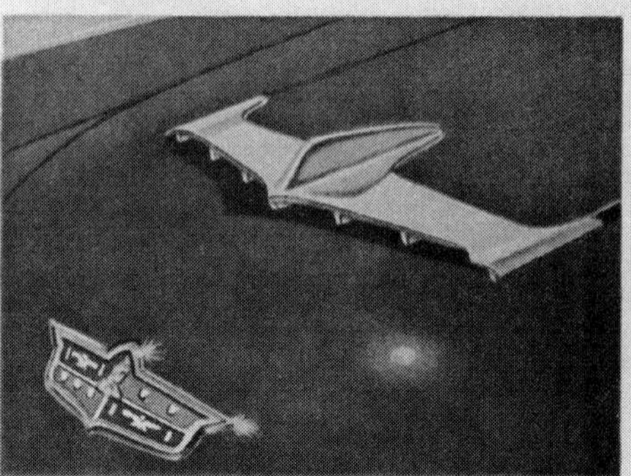

NEW AIR-VENT HOOD is typical of the way in which De Soto combines good engineering with smart design. It directs an added stream of cool air right to the carburetor... for greater engine power.

The spark plugs in the DeSoto "Firedome" V-8 were centered in dome-shaped cylinder heads. The hemi head design was said to squeeze more power from every drop of gas, give greater fuel economy, operate on regular gas, and create a smoother and quieter running engine. Valves lasted longer, too.

1955 DESOTO

DeSoto Firedome base V-8
Overhead valve V-8. Hemispherical combustion chambers. Two rocker shafts per cylinder head. Cast-iron block. Bore & stroke: 3.72 x 3.344 in. Displacement: 291.0 cid. Compression ratio: 7.5:1. Brake hp: 185 at 4400 rpm. Taxable hp: 44.3. Torque: 245 lbs.-ft. at 2800 rpm. Five main bearings. Hydraulic valve lifters. Crankcase capacity: 5 qt. (add 1 qt. if new filter installed). Cooling system capacity: 23 qt. (add 1 qt. for heater). Carburetor: Carter BBD-2117-S two-barrel. Engine code: S21-1001 and up.

DeSoto Fireflite base V-8
Overhead valve V-8. Hemispherical combustion chambers. Two rocker shafts per cylinder head. Cast-iron block. Bore & stroke: 3.72 x 3.344 in. Displacement: 291.0 cid. Compression ratio: 7.5:1. Brake hp: 200 at 4400 rpm. Taxable hp: 44.3. Torque: 274 lbs.-ft. at 2800 rpm. Five main bearings. Hydraulic valve lifters. Crankcase capacity: 5 qt. (add 1 qt. if new filter installed). Cooling system capacity: 23 qt. (add 1 qt. for heater). Carburetor: Carter WCFB-2210-S four-barrel. Engine code: S22-1001 and up.

1956 DESOTO

1DeSoto Firedome base V-8
Overhead valve V-8. Hemispherical combustion chambers. Two rocker shafts per cylinder head. Cast-iron block. Bore & stroke: 3.72 x 3.80 in. Displacement: 330 cid. Compression ratio: 8.5:1. Brake hp: 230 at 4400 rpm. Taxable hp: 44.3. Torque: 305 lbs.-ft. at 2800 rpm. Five main bearings. Hydraulic valve lifters. Crankcase capacity: 4 qt. (add 1 qt. if new filter installed). Cooling system capacity: 23 qt. (add 1 qt. for heater). Carburetor: Carter BBD-2308-S two-barrel. Engine code: S23-1001 and up.

DeSoto Fireflite base V-8
Overhead valve V-8. Hemispherical combustion chambers. Two rocker shafts per cylinder head. Cast-iron block. Bore & stroke: 3.72 x 3.80 in. Displacement: 330 cid. Compression ratio: 8.5:1. Brake hp: 255 at 4400 rpm. Taxable hp: 44.3. Torque: 350 lbs.-ft. at 3200 rpm. Five main bearings. Hydraulic valve lifters. Crankcase capacity: 4 qt. (add 1 qt. if new filter installed). Cooling system capacity: 23 qt. (add 1 qt. for heater). Carburetor: Carter WCFB-2311-S four-barrel. Engine code: S24-1001 and up.

DeSoto Adventurer base V-8
Overhead valve V-8. Hemispherical combustion chambers. Two rocker shafts per cylinder head. Cast-iron block. Bore & stroke: 3.78 x 3.80 in. Displacement: 341.1 cid. Compression ratio: 9.25:1. Brake hp: 320 at 5200 rpm. Taxable hp: 45.7. Torque: 356 lbs.-ft. at 4000 rpm. Five main bearings. Hydraulic valve lifters. Crankcase capacity: 4 qt. (add 1 qt. if new filter installed). Cooling system capacity: 23 qt. (add 1 qt. for heater). Carburetor: Carter WCFB-2311-S four-barrel. Engine code: S24A-1001 and up.

The DeSoto Adventurer's base V-8 was a 344.8-cid Hemi with two rocker shafts per cylinder head. With a 9.25:1 compression ratio it developed 345 hp at 5200 rpm. This was the first standard-equipment engine to produce one horsepower per cubic inch. (Jim DiGregorio photo)

DeSoto Firesweep base V-8
Overhead valve. Hemispherical combustion chambers. One rocker shaft per cylinder head. Cast-iron block. Bore & stroke: 3.69 x 3.80 in. Displacement: 325 cid. Compression ratio: 8.5:1. Brake hp: 245 at 4400 rpm. Taxable hp: 43.6. Torque: 320 lbs.-ft. at 2400 rpm. Five main bearings. Hydraulic valve lifters. Crankcase capacity: 5 qt. (add 1 qt. if new filter installed). Cooling system capacity: 20 qt. (add 1 qt. for heater). Carburetor: (manual transmission) Stromberg two-barrel Model WW-3-149; (automatic transmission) Stromberg two-barrel Model WW-3-150. Engine code: S27-1001 and up.

DeSoto Firesweep optional V-8
Overhead valve. Hemispherical combustion chambers. One rocker shaft per cylinder head. Cast-iron block. Bore & stroke: 3.69 x 3.80 in. Displacement: 325 cid. Compression ratio: 8.5:1. Brake hp: 260 at 4400 rpm. Taxable hp: 43.6. Torque: 335 lbs.-ft. at 2800 rpm. Five main bearings. Hydraulic valve lifters. Crankcase capacity: 5 qt. (add 1 qt. if new filter installed). Cooling system capacity: 20 qt. (add 1 qt. for heater). Carburetor: Four-barrel. Engine code: S27*-1001 and up.

1957 DESOTO

DeSoto Firedome base V-8
Overhead valve V-8. Hemispherical combustion chambers. Two rocker shafts per cylinder head. Cast-iron block. Bore & stroke: 3.78 x 3.80 in. Displacement: 341.1 cid. Compression ratio: 9.25:1. Brake hp: 270 at 4600 rpm. Taxable hp: 45.7. Torque: 350 lbs.-ft. at 2400 rpm. Five main bearings. Hydraulic valve lifters. Crankcase capacity: 4 qt. (add 1 qt. if new filter installed). Cooling system capacity: 23 qt. (add 1 qt. for heater). Carburetor: Carter BBD-2522-S two-barrel. Engine code: S25-1001 and up.

DeSoto Fireflite base V-8; optional Firedome V-8
Overhead valve V-8. Hemispherical combustion chambers. Two rocker shafts per cylinder head. Cast-iron block. Bore & stroke: 3.78 x 3.80 in. Displacement: 341.1 cid. Compression ratio: 9.25:1. Brake hp: 295 at 4600 rpm. Taxable hp: 45.7. Torque: 375 lbs.-ft. at 2800 rpm. Five main bearings. Hydraulic valve lifters. Crankcase capacity: 4 qt. (add 1 qt. if new filter installed). Cooling system capacity: 23 qt. (add 1 qt. for heater). Carburetor: Carter WCFB-2588-S four-barrel. Engine code: S26-1001 and up.

DeSoto Adventurer base V-8
Overhead valve V-8. Hemispherical combustion chambers. Two rocker shafts per cylinder head. Cast-iron block. Bore & stroke: 3.80 x 3.80 in. Displacement: 344.8 cid. Compression ratio: 9.25:1. Brake hp: 345 at 5200 rpm. Taxable hp: 46.7. Torque: 356 lbs.-ft. at 4000 rpm. Five main bearings. Hydraulic valve lifters. Crankcase capacity: 4 qt. (add 1 qt. if new filter installed). Cooling system capacity: 23 qt. (add 1 qt. for heater). Carburetor: Carter WCFB-2588-S four-barrel. Engine code: S26A-1001 and up.

1958 DESOTO

DeSoto Firesweep base V-8
Overhead valve. Cast-iron block. Bore & stroke: 4.06 x 3.39 in. Displacement: 350 cid. Compression ratio: 10.0:1. Brake hp: 280 at 4600 rpm. Taxable hp: 52.7. Torque: 380 lbs.-ft. at 2400 rpm. Five main bearings. Hydraulic valve lifters. Crankcase capacity: 4 qt. (add 1 qt. if new filter installed). Cooling system capacity: 16 qt. (add 1 qt. for heater). Carburetor: Carter BBD-2637-S two-barrel. Engine code: LS1-L. Engine number: L-350-1001 and up.

DeSoto Firedome base V-8;mFiresweep optional V-8
Overhead valve. Cast-iron block. Bore & stroke: 4.125 x 3.375 in. Displacement: 361 cid. Compression ratio: 10.0:1. Brake hp: 295 at 4600 rpm. Taxable hp: 54.30. Torque: 390 lbs.-ft. at 2400 rpm. Five main bearings. Hydraulic valve lifters. Crankcase capacity: 4 qt. (add 1 qt. if new filter installed). Cooling system capacity: 16 qt. (add 1 qt. for heater). Carburetor: Carter BBD-2637-S two-barrel. Engine code: (Firesweep) LS1-L; (Firedome) LS2-M. Engine number: (Firesweep) L-350-1001 and

DeSoto built only high-performance hemi-head V-8s from 1952 to 1957. The first of these came only in the top-rung Firedome models and was a 276-cid 160-hp version. This 1952 advertisement announced the new engine and labeled it, "America's most powerful engine design!" It developed more horsepower per cubic inch than any other 1952 American automobile engine.

For 1959, DeSoto ads told buyers that they could pick the engine they liked best. The engine lineup included mighty Turboflash V-8s, as well as the Adventurer V-8 with twin four-barrel carburetors. The Adventurer's Hemi V-8 was optional for all DeSotos.

up; (Firedome) L-360-1001 and up.

DeSoto Fireflite base V-8
Overhead valve. Cast-iron block. Bore & stroke: 4.125 x 3.375 in. Displacement: 361 cid. Compression ratio: 10.0:1. Brake hp: 305 at 4600 rpm. Taxable hp: 54.30. Torque: 400 lbs.-ft. at 2800 rpm. Five main bearings. Hydraulic valve lifters. Crankcase capacity: 4 qt. (add 1 qt. if new filter installed). Cooling system capacity: 16 qt. (add 1 qt. for heater). Carburetor: Carter AFB-2642-S four-barrel. Engine code: LS3-H. Engine number: L-360-1001 and up.

DeSoto Adventurer V-8
Overhead valve. Cast-iron block. Bore & stroke: 4.125 x 3.375 in. Displacement: 361 cid. Compression ratio: 10.25:1. Brake hp: 345 at 5000 rpm. Taxable hp: 54.30. Torque: 400 lbs.-ft. at 3600 rpm. Five main bearings. Hydraulic valve lifters. Crankcase capacity: 4 qt. (add 1 qt. if new filter installed). Cooling system capacity: 16 qt. (add 1 qt. for heater). Carburetor: Two Carter four-barrel (front) AFB-2652-S; (rear) AFB-2653-S. Engine code: LS1-L. Engine number: L-350-1001 and up. Engine code: LS3-S. Engine number: L-360-1001 and up.

DeSoto Adventurer [Engine code: LS1-L. Engine number: L-350-1001 and up.]optional EFI V-8
Overhead valve. Cast-iron block. Bore & stroke: 4.125 x 3.375 in. Displacement: 361 cid. Compression ratio: 10.0:1. Brake hp: 355 at 5000 rpm. Taxable hp: 54.30. Torque: unknown at unknown rpm. Five main bearings. Hydraulic valve lifters. Crankcase capacity: 4 qt. (add 1 qt. if new filter installed). Cooling system capacity: 16 qt. (add 1 qt. for heater). Fuel system: Bendix electronic fuel injection. Engine code: LS3-S. Engine number: L-360-1001 and up.

NOTE: Installed on a handful of cars before being recalled. Cars were later retro-fitted with dual four-barrel carburetors.

1959 DESOTO

DeSoto Firesweep base V-8
Overhead valve. Cast-iron block. Bore & stroke: 4.125 x 3.375 in. Displacement: 361 cid. Compression ratio: 10.00:1. Brake hp: 290 at 4600 rpm. Taxable hp: 54.30. Torque: 390 lbs.-ft. at 2400 rpm. Five main bearings. Hydraulic valve lifters. Crankcase capacity: 5 qt. (add 1 qt. if new filter installed). Cooling system capacity: 16 qt. (add 1 qt. for heater). Carburetor: Carter BBD-2870-S two-barrel. Engine code: MS-1. Engine number: ML-361-1001 and up.

DeSoto Firedome base V-8
Overhead valve. Cast-iron block. Bore & stroke: 4.25 x 3.375 in. Displacement: 383 cid. Compression ratio: 10.10:1. Brake hp: 305 at 4600 rpm. Taxable hp: 57.80. Torque: 410 lbs.-ft. at 2400 rpm. Five main bearings. Hydraulic valve lifters. Crankcase capacity: 5 qt. (add 1 qt. if new filter installed). Cooling system capacity: 15 qt. (add 1 qt. for heater). Carburetor: Carter BBD-2871-S two-barrel. Engine code: MS-2. Engine number: ML-383-1001 and up.

DeSoto Fireflite base V-8
Overhead valve. Cast-iron block. Bore & stroke: 4.25 x 3.375 in. Displacement: 383 cid. Compression ratio: 10.10:1. Brake hp: 325 at 4600 rpm. Taxable hp: 57.80. Torque: 425 lbs.-ft. at 2800 rpm. Five main bearings. Hydraulic valve lifters. Crankcase capacity: 5 qt. (add 1 qt. if new filter installed). Cooling system capacity: 15 qt. (add 1 qt. for heater). Carburetor: Carter AFB-2794-S four-barrel. Engine code: MS-3. Engine number: ML-383-1001 and up.

DeSoto Adventurer base V-8
Overhead valve. Cast-iron block. Bore & stroke: 4.25 x 3.375 in. Displacement: 383 cid. Compression ratio: 10.10:1. Brake hp: 350 at 5000 rpm. Taxable hp: 57.80. Torque: 425 lbs.-ft. at 3600 rpm. Five main bearings. Hydraulic valve lifters. Crankcase capacity: 5 qt. (add 1 qt. if new filter installed). Cooling system capacity: 15 qt. (add 1 qt. for heater). Carburetor: Two Carter four-barrel: (front) AFB-2790-S; (rear) AFB-2791-S. Engine code: MS-3. Engine number: ML-383-1001 and up.

1960 DESOTO

DeSoto Fireflite base V-8
Overhead valve. Cast-iron block. Bore & stroke: 4.125 x 3.375 in. Displacement: 361 cid. Compression ratio: 10.00:1. Brake hp: 295 at 4600 rpm. Taxable hp: 53.50. Torque: 390 lbs.-ft. at 2400 rpm. Five main bearings. Hydraulic valve lifters. Crankcase capacity: 5 qt. (add 1 qt. if new filter installed). Cooling system capacity: 16 qt. (add 1 qt. for heater). Carburetor: Carter BBD-2923-S two-barrel. Engine code: P-36.

DeSoto Adventurer base V-8
Overhead valve. Cast-iron block. Bore & stroke: 4.25 x 3.375 in. Displacement: 383 cid. Compression ratio: 10.10:1. Brake hp: 305 at 4600 rpm. Taxable hp: 57.80. Torque: 410 lbs.-ft. at 2400 rpm. Five main bearings. Hydraulic valve lifters. Crankcase capacity: 5 qt. (add 1 qt. if new filter installed). Cooling system capacity: 16 qt. (add 1 qt. for heater). Carburetor: Carter BBD-2923-S two-barrel. Engine code: P-38.

DeSoto Fireflite, Adventurer optional V-8
Overhead valve. Cast-iron block. Bore & stroke: 4.25 x 3.375 in. Displacement: 383 cid. Compression ratio: 10.00:1. Brake hp: 325 at 4600 rpm. Taxable hp: 57.80. Torque: 425 lbs.-ft. at 2800 rpm. Five main bearings. Hydraulic valve lifters. Crankcase capacity: 5 qt. (add 1 qt. if new filter installed). Cooling system capacity: 16 qt. (add 1 qt. for heater). Carburetor: Carter AFB-2903-S four-barrel. Engine code: P-38.

DeSoto Adventurer optional V-8
Overhead valve. Cast-iron block. Bore & stroke: 4.25 x 3.375 in. Displacement: 383 cid. Compression ratio: 10.00:1. Brake hp: 330 at 4800 rpm. Taxable hp: 57.80. Torque: 460 lbs.-ft. at 2800 rpm. Five main bearings. Hydraulic valve lifters. Crankcase capacity: 5 qt. (add 1 qt. if new filter installed). Cooling system capacity: 16 qt. (add 1 qt. for heater). Carburetor: Carter AFB-2903-S four-barrel. Engine code: P-38.

1961 DESOTO

DeSoto Adventurer base V-8
Overhead valve. Cast-iron block. Bore & stroke: 4.125 x 3.375 in. Displacement: 361 cid. Compression ratio: 10.00:1. Brake hp: 265 at 4400 rpm. Taxable hp: 54.40. Torque: 380 lbs.-ft. at 2400 rpm. Five main bearings. Hydraulic valve lifters. Crankcase capacity: 5 qt. (add 1 qt. if new filter installed). Cooling system capacity: 20 qt. (add 1 qt. for heater). Carburetor: Stromberg WWC-3-188 two-barrel. Engine code: R-36.

Selected DeSoto V-8 Tune-Up Specs 1952-1961

Year	Model	Spark Plugs Type	Gap	Point Gap	Deg Cam Angle	Timing	Idle Speed	Compression Pressure	Battery (Delco)
1952	Firedome V-8	AL-AR80	.035	.015-.018	NOTE 1	4B	475 rpm	150	Group 2

NOTE 1: This DeSoto V-8 has two sets of distributor points. The total cam angle is 32-36 degrees and for each point set the cam angle should be set at 26-28 degrees.

Year	Model	Spark Plugs Type	Gap	Point Gap	Deg Cam Angle	Timing	Idle Speed	Compression Pressure	Battery (Delco)
1953	Firedome V-8	AL-4S140	.035	.015-.018	NOTE 2	4B	475 rpm	150	Group 2

NOTE 2: This DeSoto V-8 has two sets of distributor points. The total cam angle is 32-36 degrees and for each point set the cam angle should be set at 26-28 degrees.

Year	Model	Spark Plugs Type	Gap	Point Gap	Deg Cam Angle	Timing	Idle Speed	Compression Pressure	Battery (Delco)
1954	Firedome V-8	AL-4S140	.035	.015-.018	NOTE 3	4B	475 rpm	150	Group 2

NOTE 3: This DeSoto V-8 has two sets of distributor points. The total cam angle is 32-36 degrees and for each point set the cam angle should be set at 26-28 degrees.

Year	Model	Spark Plugs Type	Gap	Point Gap	Deg Cam Angle	Timing	Idle Speed	Compression Pressure	Battery (Delco)
1955	Firedome V-8	AL-4S140	.035	.015-.018	NOTE 4	10B	475 rpm	150	Group 2
	Fireflite V-8	AL-4S165	.035	.015-.018	NOTE 4	4B	475 rpm	150	Group 2

NOTE 4: This DeSoto V-8 has two sets of distributor points. The total cam angle is 32-36 degrees and for each point set the cam angle should be set at 26-28 degrees.

Year	Model	Spark Plugs Type	Gap	Point Gap	Deg Cam Angle	Timing	Idle Speed	Compression Pressure	Battery (Delco)
1956	Firedome V-8	AL-AR51	.035	.015-.018	30	8B	475 rpm	155	2SM/11HS
	Fireflite V-8	AL-AR52	.035	.015-.018	30	4B	475 rpm	160	2SM/11HS
	Adventurer V-8	AL-4S52	.035	.015-.018	30	6B	600 rpm	165	2SM/11HS
1957	Firedome V-8	AL-AR42	.035	.015-.018	30	6B	475 rpm	160	2SM/11HS
	Fireflite V-8	AL-AR42	.035	.015-.018	30	6B	475 rpm	160	2SM/11HS
	Firesweep V-8	AL-AR42	.035	.015-.018	30	6B	475 rpm	140	2SM/11HS
1958	Firedome V-8	AL-AR42	.035	.015-.018	29	6B	500 rpm	160	HO-11-60
	Fireflite V-8	AL-AR42	.035	.015-.018	29	6B	500 rpm	160	HO-11-60
	Firesweep V-8	AL-AR42	.035	.015-.018	29	6B	500 rpm	140	HO-11-60
	Adventurer V-8	AL-AR42	.035	.015-.018	29	8B	500 rpm	160	HO-11-60
1959	Firedome V-8	AL-AR42	.035	.017	29	10B	500 rpm	160	12-H-70
	Fireflite V-8	AL-AR42	.035	.017	29	10B	500 rpm	160	12-H-70
	Firesweep V-8	AL-AR42	.035	.017	29	10B	500 rpm	140	12-H-70
	Adventurer V-8	AL-A42	.035	.017	29	10B	500 rpm	160	12-H-70
1960	Fireflite V-8	AL-AR42	.035	.014	29-32	10B	500 rpm	150-180	12-HS-60
	Adventurer V-8	AL-AR42	.035	.014	29-32	10B	500 rpm	150-180	12-HS-60
1961	Fireflite V-8	AL-A42	.035	.015-.018	29-34	10B	500 rpm	150-180	11-HS-59B
	Ram V-8	AL-A32	.035	.015-.018	29-34	7.5B	500 rpm	150-180	12-HS-60

DeSoto V-8 Firing Order

1952-1961
1-8-4-3-6-5-7-2

All DeSoto V-8s have the cylinders numbered from front to back. The cylinders on the left-hand bank are numbered 1-3-5-7 and the cylinders on the right-hand bank are numbered 2-4-6-8.

DODGE
1953-1989

Dodge built high-performance hemi-head V-8s 1953 to 1958, by which time up to 320 hp was available in the Super D-500 version. There had been an even more brutal 340-hp D-501 option in mid-1957, before automakers were forced to apply brakes to the horsepower race. Also interesting was the 330-hp V-8 with electronic fuel injection offered for a short time in early 1958. Cars ordered with this option were recalled and retro-fitted with "only" two four-barrel carbs.

A new, cheaper-to-build "wedge" V-8 arrived in 1959. It became well known in its 383-, 426- and 440-cid configurations. In the early '60s, it was available with wild-looking cross-ram intake manifolds in a 375-hp package. A few years later, the 426 Wedge made drag racing history in the hands of race teams like the Ramchargers and drivers like "Dandy" Dick Landy.

A 273-cid "small-block" V-8 was also developed in the early '60s for the compact Dodge Dart. This was quickly bored to 318 and 340 cid with the latter engine producing upwards of 290 hp when equipped with a "six pack" of three two-barrel carbs. Speed legend Andy Granatelli employed the small Mopar V-8 as the basis of his fuel-injected Indy Car engines in the late '60s.

Dodge re-introduced the Hemi V-8 in 1964, but this version was strictly for racing. It had a short tenure in NASCAR racing, but started tearing up America's drag strips and it remains a formidable force in quarter-mile competition today.

By the early '70s Dodge—like other automakers—was forced to join the trend towards smaller, less-powerful V-8s. Nevertheless, the big 440 was able to hang on until 1977, while a 400-cid V-8 clung on for a year past that. In 1982 the 360-cid V-8 disappeared and from then until 1989, the only V-8 was the 318. After that, there were none.

Arranged in order by horsepower ratings.

1953 DODGE

Dodge Coronet base "Red Ram" V-8
Overhead valve V-8. Hemispherical combustion chambers. Two rocker shafts per cylinder head. Cast-iron block. Bore & stroke: 3.44 x 3.25 in. Displacement: 241.3 cid. Compression ratio: 7.1:1. Brake hp: 140 at 4400 rpm. Taxable hp: 37.8. Torque: 220 lbs.-ft. at 2000 rpm. Five main bearings. Hydraulic valve lifters. Crankcase capacity: 5 qt. (add 1 qt. if new filter installed). Cooling system capacity: 19 qt. (add 1 qt. for heater). Carburetor: Stromberg Model WW-3-108 two-barrel. Engine code: D44 or D48.

1954 DODGE

Dodge base "Red Ram" V-8 (with manual transmission)
Overhead valve V-8. Hemispherical combustion chambers. Two rocker shafts per cylinder head. Cast-iron block. Bore & stroke: 3.44 x 3.25 in. Displacement: 241.3 cid. Compression ratio: 7.1:1. Brake hp: 140 at 4400 rpm. Taxable hp: 37.8. Torque: 220 lbs.-ft. at 2000 rpm. Five main bearings. Hydraulic valve lifters. Crankcase capacity: 5 qt. (add 1 qt. if new filter installed). Cooling system capacity: 19 qt. (add 1 qt. for heater). Carburetor: Stromberg Model WW-3-108 two-barrel. Engine code: (Meadowbrook) D50-1 or D50-1A.

Dodge optional "Red Ram" V-8 (with Gyro-Matic or Powerflite transmission)
Overhead valve V-8. Hemispherical combustion chambers. Two rocker shafts per cylinder head. Cast-iron block. Bore & stroke: 3.44 x 3.25 in.

The 140-hp Red Ram V-8 engine came out in the 1953 Dodge. Prior to that, only a flathead in-line six was available. The new V-8 was a 241-cid Hemi featuring the "triple power" advantages of a hemispherical combustion chamber, high-lift lateral valves, and a short-stroke design. It operated on regular gasoline.

Displacement: 241.3 cid. Compression ratio: 7.5:1. Brake hp: 150 at 4400 rpm. Taxable hp: 37.8. Torque: 222 lbs.-ft. at 2400 rpm. Five main bearings. Hydraulic valve lifters. Crankcase capacity: 5 qt. (add 1 qt. if new filter installed). Cooling system capacity: 19 qt. (add 1 qt. for heater). Carburetor: Stromberg Model WW-3-108 two-barrel. Engine code: (Coronet) D50-2; (Royal) D50-3 or D53-3; (Sierra) D53-2.

1955 DODGE

Dodge base "Red Ram" V-8
Overhead valve V-8. Hemispherical combustion chambers. One rocker shaft per cylinder head.Cast iron block. Bore & stroke: 3.63 x 3.26 in. Displacement: 270.2 cid. Compression ratio: 7.6:1. Brake hp: 175 at 4400 rpm. Taxable hp: 42.2. Torque: 246 lbs.-ft. at 2400 rpm. Five main bearings. Hydraulic valve lifters. Crankcase capacity: 5 qt. (add 1 qt. if new filter installed). Cooling system capacity: 19 qt. (add 1 qt. for heater). Carburetor: Stromberg Model WW-3-131 two-barrel. Engine code: (Coronet) D55-1; (Royal) D55-2.

Dodge optional "Super Red Ram" V-8
Overhead valve V-8. Hemispherical combustion chambers. Two rocker shafts per cylinder head. Cast-iron block. Bore & stroke: 3.63 x 3.26 in. Displacement: 270.2 cid. Compression ratio: 7.6:1. Brake hp: 183 at 4400 rpm. Taxable hp: 42.2. Torque: 245 lbs.-ft. at 2400 rpm. Five main bearings. Hydraulic valve lifters. Crankcase capacity: 5 qt. (add 1 qt. if new filter installed). Cooling system capacity: 19 qt. (add 1 qt. for heater). Carburetor: Stromberg Model WW-3-131 two-barrel. Engine code: (Custom Royal) D55-3; (Royal Lancer) D55-3.

Dodge optional "Super Red Ram" V-8 with Power-Pac
Overhead valve V-8. Hemispherical combustion chambers. Two rocker shafts per cylinder head. Cast-iron block. Bore & stroke: 3.63 x 3.26 in. Displacement: 270.2 cid. Compression ratio: 7.6:1. Brake hp: 193 at 4400 rpm. Taxable hp: 42.2. Torque: 245 lbs.-ft. at 2800 rpm. Five main bearings. Hydraulic valve lifters. Crankcase capacity: 5 qt. (add 1 qt. if new filter installed). Cooling system capacity: 19 qt. (add 1 qt. for heater). Dual exhaust. Carburetor: Carter four-barrel. Engine code: (Royal Lancer) D55-3.

1956 DODGE

Dodge base "Red Ram" V-8
Overhead valve V-8. Hemispherical combustion chambers. One rocker shaft per cylinder head. Cast-iron block. Bore & stroke: 3.63 x 3.26 in. Displacement: 270.2 cid. Compression ratio: 8.0:1. Brake hp: 189 at 4400 rpm. Taxable hp: 42.2. Torque: 266 lbs.-ft. at 2400 rpm. Five main bearings. Hydraulic valve lifters. Crankcase capacity: 5 qt. (add 1 qt. if new filter installed). Cooling system capacity: 19 qt. (add 1 qt. for heater). Carburetor: Stromberg two-barrel Model WW-3-135.Engine code: (Coronet) D63-1.

Dodge optional "Super Red Ram" V-8
Overhead valve V-8. Hemispherical combustion chambers. One rocker shaft per cylinder head. Cast-iron block. Bore & stroke: 3.63 x 3.80 in. Displacement: 315 cid. Compression ratio: 8.0:1. Brake hp: 218 at 4400 rpm. Taxable hp: 42.2. Torque: 309 lbs.-ft. at 2000 rpm. Five main bearings. Hydraulic valve lifters. Crankcase capacity: 5 qt. (add 1 qt. if new filter installed). Cooling system capacity: 19 qt. (add 1 qt. for heater). Carburetor: Stromberg two-barrel Model WW-3-148. Engine code: (Royal) D63-2; (Custom Royal) D63-3.

Dodge optional "Super Red Ram" V-8 with Power-Pac
Overhead valve V-8. Hemispherical combustion chambers. One rocker shaft per cylinder head. Cast-iron block. Bore & stroke: 3.63 x 3.80 in. Displacement: 315 cid. Compression ratio: 8.0:1. Brake hp: 230 at 4400 rpm. Taxable hp: 42.2. Torque: 316 lbs.-ft. at 2400 rpm. Five main bearings. Hydraulic valve lifters. Crankcase capacity: 5 qt. (add 1 qt. if new filter installed). Cooling system capacity: 19 qt. (add 1 qt. for heater). Dual exhaust. Carburetor: Carter four-barrel Model WCFB. Engine code: (Royal) D63-2; (Custom Royal) D63-2.

Dodge optional D-500 V-8 standard (midyear option)
Overhead valve. Hemispherical combustion chambers. Two rocker shafts per cylinder head. Cast-iron block. Bore & stroke: 3.63 x 3.80 in. Displacement: 315 cid. Compression ratio: 9.25:1. Brake hp: 260 at 4800 rpm. Taxable hp: 42.2. Torque: 330 lbs.-ft. at 3000 rpm. Five main bearings. Solid valve lifters. Crankcase capacity: 5 qt. (add 1 qt. if new filter installed). Cooling system capacity: 20 qt. (add 1 qt. for heater). Dual exhaust. Carburetor: Carter four-barrel Type WCFB. Engine code: (Dodge Coronet 500) D500.

Dodge optional D-500 V-8 dual four-barrel (midyear option)
Overhead valve. Hemispherical combustion chambers. Two rocker shafts per cylinder head. Cast-iron block. Bore & stroke: 3.63 x 3.80 in. Displacement: 315 cid. Compression ratio: 9.25:1. Brake hp: 295 at 4400 rpm. Taxable hp: unknown. Torque: unknown at unknown rpm. Five main bearings. Solid valve lifters. Crankcase capacity: 5 qt. (add 1 qt. if new filter installed). Cooling system capacity: 20 qt. (add 1 qt. for heater). Dual exhaust. Carburetor: Two Carter four-barrel Type WCFB. Engine code: (Dodge Coronet 500) D500.

1957 DODGE

The D-501 option available for the 1957 Dodge Coronet after mid model year included a 354-cid Hemi V-8 with twin Carter WCFB four-barrel carburetors. It had a 10.0:1 compression ratio and developed 340 hp at 5200 rpm. Note how each carburetor was fitted with its own air cleaner.

Dodge base "Red Ram" V-8
Overhead valve. Hemispherical combustion chambers. One rocker shaft per cylinder head. Cast-iron block. Bore & stroke: 3.69 x 3.80 in. Displacement: 325 cid. Compression ratio: 8.5:1. Brake hp: 245 at 4400 rpm. Taxable hp: 43.6. Torque: 320 lbs.-ft. at 2400 rpm. Five main bearings. Five main bearings. Hydraulic valve lifters. Crankcase capacity: 5 qt. (add 1 qt. if new filter installed). Cooling system capacity: 20 qt. (add 1 qt. for heater). Carburetor: Stromberg two-barrel Model WW-3-149. Engine code: (Coronet) D66-1; (Custom Coronet) D66-2; (Royal) D67-1; (Suburban/Sierra) D70; (Custom Sierra) D71.

Dodge optional "Super Red Ram" V-8
Overhead valve. Hemispherical combustion chambers. One rocker shaft per cylinder head. Cast-iron block. Bore & stroke: 3.69 x 3.80 in. Displacement: 325 cid. Compression ratio: 8.5:1. Brake hp: 260 at 4400 rpm. Taxable hp: 43.6. Torque: 335 lbs.-ft. at 2800 rpm. Five main bearings. Five main bearings. Hydraulic valve lifters. Crankcase capacity: 5 qt. (add 1 qt. if new filter installed). Cooling system capacity: 20 qt. (add 1 qt. for heater). Carburetor: Carter WCFB-2532S four-barrel. Engine code: (Coronet) D66-2; (Custom Coronet) D67-2; (Royal) D67-2; (Custom Royal) D67-2; (Suburban/Sierra) D67-2; (Custom Sierra) D67-2.

Dodge optional "D-500" V-8
Overhead valve. Hemispherical combustion chambers. Two rocker shafts per cylinder head. Cast-iron block. Bore & stroke: 3.69 x 3.80 in. Displacement: 325 cid. Compression ratio: 8.5:1. Brake hp: 285 at 4800 rpm. Taxable hp: 43.6. Torque: 345 lbs.-ft. at 2800 rpm. Five main bearings. Five main bearings. Hydraulic valve lifters. Crankcase capacity: 5 qt. (add 1 qt. if new filter installed). Cooling system capacity: 20 qt. (add 1 qt. for heater). Carburetor: Carter WCFB four-barrel. Engine code: (Coronet) D-500; (Custom Coronet) D-500; (Royal) D-500; (Custom Royal) D-500; (Suburban/Sierra) D-500; (Custom Sierra) D-500.

Dodge optional "Super D-500" V-8 with Power-Pac
Overhead valve. Hemispherical combustion chambers. Two rocker shafts per cylinder head. Cast-iron block. Bore & stroke: 3.69 x 3.80 in. Displacement: 325 cid. Compression ratio: 8.5:1. Brake hp: 310 at 4800 rpm. Taxable hp: 43.6. Torque: 350 lbs.-ft. at 3200 rpm. Five main bearings.Five main bearings. Hydraulic valve lifters. Crankcase capacity:

5 qt. (add 1 qt. if new filter installed). Cooling system capacity: 20 qt. (add 1 qt. for heater). Carburetor: Two Carter WCFB four-barrel. Engine code: (Coronet) D-500*; (Custom Coronet) D-500*; (Royal) D-500*; (Custom Royal) D-500*; (Suburban/Sierra) D-500*; (Custom Sierra) D-500*.

Dodge optional "D-501" V-8 (midyear option)
Overhead valve. Hemispherical combustion chambers. Two rocker shafts per cylinder head. Cast-iron block. Bore & stroke: 3.94 x 3.63 in. Displacement: 354 cid. Compression ratio: 10.0:1. Brake hp: 340 at 5200 rpm. Taxable hp: unknown. Torque: unknown at unknown rpm. Five main bearings. Five main bearings. Hydraulic valve lifters. Crankcase capacity: 5 qt. (add 1 qt. if new filter installed). Cooling system capacity: 20 qt. (add 1 qt. for heater). Carburetor: Two Carter WCFB four-barrel. Engine code: (Coronet) D-500-1; (Custom Coronet) D-500-1.

Mopar's 318-cid V-8 originally appeared as an option for the 1957 Plymouth Fury. Dodge got the 318 in 1960 and shown here is the 200-hp truck version from that model year. The Dodge passenger car version produced 230 hp at 4400 rpm. It had a two-barrel carburetor, while the trucks used a one-barrel carburetor.

1958 DODGE

Dodge Coronet "Red Ram" V-8
Overhead valve. Hemispherical combustion chambers. One rocker shaft per cylinder head. Cast-iron block. Bore & stroke: 3.69 x 3.80 in. Displacement: 325 cid. Compression ratio: 9.0:1. Brake hp: 252 at 4400 rpm. Taxable hp: 43.3. Torque: 345 lbs.-ft. at 2400 rpm. Five main bearings. Hydraulic valve lifters. Crankcase capacity: 5 qt. (add 1 qt. if new filter installed). Cooling system capacity: 20 qt. (add 1 qt. for heater). Carburetor: Stromberg two-barrel Model WW-3-163. Engine code: (Coronet) LD2-L1; (Custom Coronet) LD2-L2; (Royal) LD2-L.

Dodge Royal "Red Ram" V-8
Overhead valve. Hemispherical combustion chambers. One rocker shaft per cylinder head. Cast-iron block. Bore & stroke: 3.69 x 3.80 in. Displacement: 325 cid. Compression ratio: 9.0:1. Brake hp: 265 at 4400 rpm. Taxable hp: 43.3. Torque: 355 lbs.-ft. at 2800 rpm. Five main bearings. Hydraulic valve lifters. Crankcase capacity: 5 qt. (add 1 qt. if new filter installed). Cooling system capacity: 20 qt. (add 1 qt. for heater). Carburetor: Carter WCFB-2660-S four-barrel. Engine code: (Coronet) LD2-M; (Custom Coronet) LD2-M; (Royal) LD2-M.

Dodge Custom Royal "Ram Fire" V-8
Overhead valve. Cast-iron block. Bore & stroke: 4.06 x 3.39 in. Displacement: 350 cid. Compression ratio: 10.0:1. Brake hp: 295 at 4600 rpm. Taxable hp: 52.7. Torque: 385 lbs.-ft. at 2800 rpm. Five main bearings. Hydraulic valve lifters. Crankcase capacity: 4 qt. (add 1 qt. if new filter installed). Cooling system capacity: 16 qt. (add 1 qt. for heater). Carburetor: Carter AFB-2773-S four-barrel. Engine code: (Custom Royal) LD3-H; (Suburban/Sierra) LD3-L; (Custom Sierra) LD3-H.

Dodge optional "D-500" V-8
Overhead valve. Cast-iron block. Bore & stroke: 4.125 x 3.375 in. Displacement: 361 cid. Compression ratio: 10.0:1. Brake hp: 305 at 4600 rpm. Taxable hp: 54.30. Torque: 400 lbs.-ft. at 2800 rpm. Five main bearings. Hydraulic valve lifters. Crankcase capacity: 4 qt. (add 1 qt. if new filter installed). Cooling system capacity: 16 qt. (add 1 qt. for heater). Carburetor: Carter AFB-2870-S four-barrel. Engine code: (All) D-500.

Dodge optional "Super D-500" V-8
Overhead valve. Cast-iron block. Bore & stroke: 4.125 x 3.375 in. Displacement: 361 cid. Compression ratio: 10.0:1. Brake hp: 320 at 4600 rpm. Taxable hp: 54.30. Five main bearings. Hydraulic valve lifters. Crankcase capacity: 4 qt. (add 1 qt. if new filter installed). Cooling system capacity: 16 qt. (add 1 qt. for heater). Carburetor: Two Carter WCFB four-barrel. Engine code: (All) D-500.

Dodge optional EFI V-8
Overhead valve. Cast-iron block. Bore & stroke: 4.125 x 3.375 in. Displacement: 361 cid. Compression ratio: 10.0:1. Brake hp: 333 at 4800 rpm. Taxable hp: 54.30. Five main bearings. Hydraulic valve lifters. Crankcase capacity: 4 qt. (add 1 qt. if new filter installed). Cooling system capacity: 16 qt. (add 1 qt. for heater). Fuel system: Bendix electronic fuel injection. Engine code: (All) D-500.
NOTE: Installed on a handful of cars before being recalled. Cars were later retro-fitted with dual four-barrel carburetors.

1959 DODGE

Dodge Coronet base V-8
Overhead valve. Cast-iron block. Bore & stroke: 3.95 x 3.31 in. Displacement: 326 cid. Compression ratio: 9.20:1. Brake hp: 255 at 4400 rpm. Taxable hp: 49.90. Torque: 350 lbs.-ft. at 2400 rpm. Five main bearings. Hydraulic valve lifters. Crankcase capacity: 5 qt. (add 1 qt. if new filter installed). Cooling system capacity: 20 qt. (add 1 qt. for heater). Carburetor: Stromberg WW-3-164 two-barrel. Engine code: (Coronet) MD2-L.

Dodge Royal base "Ram Fire" V-8
Overhead valve. Cast-iron block. Bore & stroke: 4.125 x 3.375 in. Displacement: 361 cid. Compression ratio: 10.10:1. Brake hp: 295 at 4600 rpm. Taxable hp: 54.30. Torque: 390 lbs.-ft. at 2400 rpm. Five main bearings. Hydraulic valve lifters. Crankcase capacity: 5 qt. (add 1 qt. if new filter installed). Cooling system capacity: 16 qt. (add 1 qt. for heater). Carburetor: Carter BBD-2870-S two-barrel. Engine code: (Royal) MD3-M; (Station Wagon) MD3-L.

Dodge Custom Royal base "Ram Fire" V-8
Overhead valve. Cast-iron block. Bore & stroke: 4.125 x 3.375 in. Displacement: 361 cid. Compression ratio: 10.10:1. Brake hp: 305 at 4600 rpm. Taxable hp: 54.30. Torque: 400 lbs.-ft. at 2800 rpm. Five main bearings. Hydraulic valve lifters. Crankcase capacity: 5 qt. (add 1 qt. if new filter installed). Cooling system capacity: 16 qt. (add 1 qt. for heater). Carburetor: Carter AFB 2773-S four-barrel. Engine code: (Custom Royal) MD3-H; (Custom Station Wagon) MD3-H.

Dodge optional "D-500" V-8
Overhead valve. Cast-iron block. Bore & stroke: 4.25 x 3.375 in. Displacement: 383 cid. Compression ratio: 10.10:1. Brake hp: 320 at 4600 rpm. Taxable hp: 57.80. Torque: 420 lbs.-ft. at 2800 rpm. Five main bearings. Hydraulic valve lifters. Crankcase capacity: 5 qt. (add 1 qt. if new filter installed). Cooling system capacity: 16 qt. (add 1 qt. for heater). Carburetor: Carter four-barrel. Engine code: (All) D-500.

Dodge optional "Super D-500" V-8
Overhead valve. Cast-iron block. Bore & stroke: 4.25 x 3.375 in. Displacement: 383 cid. Compression ratio: 10.10:1. Brake hp: 345 at 5000 rpm. Taxable hp: 57.80. Torque: 420 lbs.-ft. at 3600 rpm. Five main bearings. Hydraulic valve lifters. Crankcase capacity: 5 qt. (add 1 qt. if new filter installed). Cooling system capacity: 16 qt. (add 1 qt. for heater). Carburetor: Two Carter four-barrel. Engine code: (All) D-500*.

1960 DODGE

Dodge Dart Seneca, Pioneer base V-8
Overhead valve. Cast-iron block. Bore & stroke: 3.91 x 3.31 in. Displacement: 318 cid. Compression ratio: 9.0:1. Brake hp: 230 at 4400 rpm. Taxable hp: 48.90. Torque: 340 lbs.-ft. at 2400. Five main bearings.

Hydraulic valve lifters. Crankcase capacity: 5 qt. (add 1 qt. if new filter installed). Cooling system capacity: 16 qt. (add 1 qt. for heater). Carburetor: Carter BBD-2921-S two-barrel. Engine code: P-318.

Dodge Dart Phoenix base V-8
Overhead valve. Cast-iron block. Bore & stroke: 3.91 x 3.31 in. Displacement: 318 cid. Compression ratio: 9.0:1. Brake hp: 255 at 4400 rpm. Taxable hp: 48.90. Torque: 345 lbs.-ft. at 2800. Five main bearings. Hydraulic valve lifters. Crankcase capacity: 5 qt. (add 1 qt. if new filter installed). Cooling system capacity: 16 qt. (add 1 qt. for heater). Carburetor: Carter AFB-2948-S four-barrel. Engine code: P-318.

Dodge Matador base V-8
Overhead valve. Cast-iron block. Bore & stroke: 4.125 x 3.375 in. Displacement: 361 cid. Compression ratio: 10.00:1. Brake hp: 295 at 4600 rpm. Taxable hp: 54.30. Torque: 390 lbs.-ft. at 2400 rpm. Five main bearings. Hydraulic valve lifters. Crankcase capacity: 5 qt. (add 1 qt. if new filter installed). Cooling system capacity: 16 qt. (add 1 qt. for heater). Carburetor: Stromberg WWC-3-188 two-barrel. Engine code: P-36.

Dodge Dart optional V-8
Overhead valve. Cast-iron block. Bore & stroke: 4.125 x 3.375 in. Displacement: 361 cid. Compression ratio: 10.00:1. Brake hp: 310 at 4800 rpm. Taxable hp: 54.30. Torque: 435 lbs.-ft. at 2800 rpm. Five main bearings. Hydraulic valve lifters. Crankcase capacity: 5 qt. (add 1 qt. if new filter installed). Cooling system capacity: 16 qt. (add 1 qt. for heater). Carburetor: Carter AFB-2903-S four-barrel. Engine code: P-36.

Dodge Polara base V-8
Overhead valve. Cast-iron block. Bore & stroke: 4.25 x 3.375 in. Displacement: 383 cid. Compression ratio: 10.00:1. Brake hp: 325 at 4600 rpm. Taxable hp: 57.80. Torque: 425 at 2800 rpm. Five main bearings. Hydraulic valve lifters. Crankcase capacity: 5 qt. (add 1 qt. if new filter installed). Cooling system capacity: 16 qt. (add 1 qt. for heater). Carburetor: Holley R-1971-A four-barrel. Engine code: P-38.

Chrysler Corporation engine engineers developed radical-looking ram manifolds for 1960 Dodge Dart models. These were available on the 381-cid V-8 that was part of the D-500 package. With two Carter four-barrel carburetors perched on top of criss-crossing 30-inch ram tubes, this engine had a really wild look. The fresh air inducted through the tubes at middle rpm ranges produced tremendous torque.

Dodge Dart, Dodge optional "D-500" V-8
Overhead valve. Cast-iron block. Bore & stroke: 4.25 x 3.375 in. Displacement: 383 cid. Compression ratio: 10.00:1. Brake hp: 330 at 4800 rpm. Taxable hp: 57.80. Torque: 460 lbs.-ft. at 2800 rpm. Five main bearings. Hydraulic valve lifters. Crankcase capacity: 5 qt. (add 1 qt. if new filter installed). Cooling system capacity: 16 qt. (add 1 qt. for heater). Carburetor: Carter AFB-2903-S four-barrel. Engine code: P-38.

1961 DODGE

Believe it or not, this Weiand tri-power setup was factory-approved as "special equipment" for the 1961 Dodge 383-cid V-8. The 383-cid was an option for Dodge Polara models that year. The factory decided to go the route of certifying aftermarket speed equipment to keep up with the Tri Power factory options from GM and Ford.

Dodge Dart Seneca, Pioneer base V-8
Overhead valve. Cast-iron block. Bore & stroke: 3.91 x 3.31 in. Displacement: 318 cid. Compression ratio: 9.0:1. Brake hp: 230 at 4400 rpm. Taxable hp: 48.90. Torque: 340 lbs.-ft. at 2400. Five main bearings. Hydraulic valve lifters. Crankcase capacity: 5 qt. (add 1 qt. if new filter installed). Cooling system capacity: 16 qt. (add 1 qt. for heater). Carburetor: Stromberg WW-1543 two-barrel. Engine code: R-318.

Dodge Polara base V-8
Overhead valve. Cast-iron block. Bore & stroke: 4.125 x 3.375 in. Displacement: 361 cid. Compression ratio: 10.00:1. Brake hp: 265 at 4400 rpm. Taxable hp: 54.40. Torque: 380 lbs.-ft. at 2400 rpm. Five main bearings. Hydraulic valve lifters. Crankcase capacity: 5 qt. (add 1 qt. if new filter installed). Cooling system capacity: 20 qt. (add 1 qt. for heater). Carburetor: Stromberg WWC-3-188 two-barrel. Engine code: R-36.

Dodge Dart Seneca, Pioneer, Phoenix optional V-8
Overhead valve. Cast-iron block. Bore & stroke: 4.125 x 3.375 in. Displacement: 361 cid. Compression ratio: 9.00:1. Brake hp: 305 at 4800 rpm. Taxable hp: 54.30. Torque: 395 lbs.-ft. at 3000 rpm. Five main bearings. Hydraulic valve lifters. Crankcase capacity: 5 qt. (add 1 qt. if new filter installed). Cooling system capacity: 16 qt. (add 1 qt. for heater). Carburetor: Carter AFB-3105-S four-barrel. Engine code: R-36.

Dodge Dart Seneca, Pioneer, Phoenix and Dodge Polara optional V-8
Overhead valve. Cast-iron block. Bore & stroke: 4.25 x 3.375 in. Displacement: 383 cid. Compression ratio: 10.00:1. Brake hp: 325 at 4600 rpm. Taxable hp: 57.80. Torque: 425 lbs.-ft. at 2800 rpm. Five main bearings. Hydraulic valve lifters. Crankcase capacity: 5 qt. (add 1 qt. if new filter installed). Cooling system capacity: (Polara) 20 qt.; (Other models) 16 qt. (add 1 qt. for heater). Carburetor: Carter AFB-2968-S four-barrel. Engine code: R-38.

Dodge Dart Seneca, Pioneer, Phoenix and Dodge Polara optional Cross Ram V-8
Overhead valve. Cast-iron block. Bore & stroke: 4.25 x 3.375 in. Displacement: 383 cid. Compression ratio: 10.00:1. Brake hp: 330 at 4800 rpm. Taxable hp: 57.80. Torque: 460 lbs.-ft. at 2800 rpm. Five main bearings. Hydraulic valve lifters. Crankcase capacity: 5 qt. (add 1 qt. if new filter installed). Cooling system capacity: (Polara) 20 qt.; (Other models) 16 qt. (add 1 qt. for heater). Carburetor: Two Carter AFB-2903-S four-barrel. Engine code: R-38.

Dodge Dart Seneca, Pioneer, Phoenix and Dodge Polara optional Cross Ram V-8
Overhead valve. Cast-iron block. Bore & stroke: 4.25 x 3.375 in. Displacement: 383 cid. Compression ratio: 10.00:1. Brake hp: 330 at 5200 rpm. Taxable hp: 57.80. Torque: 425 lbs.-ft. at 3600 rpm. Five main bearings. Hydraulic valve lifters. Crankcase capacity: 5 qt. (add 1 qt. if new filter installed). Cooling system capacity: (Polara) 20 qt.; (Other models) 16 qt. (add 1 qt. for heater). Carburetor: Two Carter AFB four-barrels: (front) 2970-S; (rear) 2791-S. Engine code: R-38.

Dodge Dart Seneca, Pioneer, Phoenix optional Cross Ram V-8 (D-500)
Overhead valve. Cast-iron block. Bore & stroke: 4.25 x 3.375 in. Displacement: 383 cid. Compression ratio: 10.00:1. Brake hp: 340 at 5000 rpm. Taxable hp: 57.80. Torque: 440 lbs.-ft. at 2800 rpm. Five main bearings. Hydraulic valve lifters. Crankcase capacity: 5 qt. (add 1 qt. if new filter installed). Cooling system capacity: 16 qt. (add 1 qt. for heater). Carburetor: Two Carter AFB 3084-S four-barrel. Engine code: R-38.

Dodge Dart Seneca, Pioneer, Phoenix optional "D-500" V-8
Overhead valve. Cast-iron block. Bore & stroke: 4.19 x 3.75 in. Displacement: 413 cid. Compression ratio: 10.00:1. Brake hp: 350 at 4600 rpm. Torque: 470 lbs.-ft. at 2800 rpm. Five main bearings. Hydraulic valve lifters. Crankcase capacity: 5 qt. (add 1 qt. if new filter installed). Cooling system capacity: 16 qt. (add 1 qt. for heater). Carburetor: Carter AFB 3108-S four-barrel. Engine code: R-41.

Dodge Dart Seneca, Pioneer, Phoenix optional "Super D-500" Cross-Ram V-8
Overhead valve. Cast-iron block. Bore & stroke: 4.19 x 3.75 in. Displacement: 413 cid. Compression ratio: 10.00:1. Brake hp: 375 at 5000 rpm. Torque: 465 lbs.-ft. at 2800 rpm. Five main bearings. Solid valve lifters. Crankcase capacity: 5 qt. (add 1 qt. if new filter installed). Cooling system capacity: 16 qt. (add 1 qt. for heater). Carburetor: Two Carter AFB 3084-S four-barrel. (Option added at midyear; includes "long" ram-tuned induction system.) Engine code: R-41.

Dodge Dart Seneca, Pioneer, Phoenix optional "Super Ram-Tuned D-500" Cross-Ram V-8
Overhead valve. Cast-iron block. Bore & stroke: 4.19 x 3.75 in. Displacement: 413 cid. Compression ratio: 10.00:1. Brake hp: 375 at 5200 rpm. Torque: 450 lbs.-ft. at 2800 rpm. Five main bearings. Solid valve lifters. Crankcase capacity: 5 qt. (add 1 qt. if new filter installed). Cooling system capacity: 16 qt. (add 1 qt. for heater). Carburetor: Two Carter AFB four-barrels: (front) 2970-S; (rear) 2791-S. (Option added at midyear; includes "short" ram-tuned induction system.) Engine code: R-41.

1962 DODGE

Dodge Dart base V-8
Overhead valve. Cast-iron block. Bore & stroke: 3.91 x 3.31 in. Displacement: 318 cid. Compression ratio: 9.0:1. Brake hp: 230 at 4400 rpm. Taxable hp: 48.90. Torque: 340 lbs.-ft. at 2400. Five main bearings. Hydraulic valve lifters. Crankcase capacity: 4 qt. (add 1 qt. if new filter installed). Cooling system capacity: 20 qt. (add 1 qt. for heater). Carburetor: Ball & Ball BBD-3240-S or Bendix WW-3-198 two-barrel. Engine code: S-318.

Dodge Dart optional V-8
Overhead valve. Cast-iron block. Bore & stroke: 3.91 x 3.31 in. Displacement: 318 cid. Compression ratio: 9.0:1. Brake hp: 260 at 4400 rpm. Taxable hp: 48.90. Torque: 345 lbs.-ft. at 2800. Five main bearings. Hydraulic valve lifters. Crankcase capacity: 4 qt. (add 1 qt. if new filter installed). Cooling system capacity: 20 qt. (add 1 qt. for heater). Carburetor: Carter AFB 3247-S four-barrel. Engine code: S-318.

Dodge Polara base V-8
Overhead valve. Cast-iron block. Bore & stroke: 4.125 x 3.375 in. Displacement: 361 cid. Compression ratio: 9.00:1. Brake hp: 305 at 4800 rpm. Taxable hp: 54.30. Torque: 395 lbs.-ft. at 3000 rpm. Five main bearings. Hydraulic valve lifters. Crankcase capacity: 4 qt. (add 1 qt. if new filter installed). Cooling system capacity: 16 qt. (add 1 qt. for heater). Carburetor: Carter AFB-3252-S four-barrel. Engine code: S-36.

Dodge Dart, Dodge Polara optional Cross Ram V-8
Overhead valve. Cast-iron block. Bore & stroke: 4.125 x 3.375 in. Displacement: 361 cid. Compression ratio: 9.00:1. Brake hp: 310 at 5200 rpm. Taxable hp: 54.30. Torque: 390 lbs.-ft. at 3400 rpm. Five main bearings. Hydraulic valve lifters. Crankcase capacity: 4 qt. (add 1 qt. if new filter installed). Cooling system capacity: 16 qt. (add 1 qt. for heater). Carburetor: Two Carter AFB four-barrels: (front) 2970-S; (rear) 2971-S. Engine code: S-36.

Dodge Dart, Dodge Polara optional V-8
Overhead valve. Cast-iron block. Bore & stroke: 4.25 x 3.375 in. Displacement: 383 cid. Compression ratio: 10.00:1. Brake hp: 330 at 4600 rpm. Taxable hp: 57.80. Torque: 452 lbs.-ft. at 2800 rpm. Five main bearings. Hydraulic valve lifters. Crankcase capacity: 4 qt. (add 1 qt. if new filter installed). Cooling system capacity: 16 qt. (add 1 qt. for heater). Carburetor: Carter AFB-3438-S four-barrel. Engine code: S-38.

Dodge Dart, Dodge Polara optional Cross Ram V-8
Overhead valve. Cast-iron block. Bore & stroke: 4.25 x 3.375 in. Displacement: 383 cid. installed Compression ratio: 10.00:1. Brake hp: 335 at 5200 rpm. Taxable hp: 57.80. Torque: 420 lbs.-ft. at 3600 rpm. Five main bearings. Hydraulic valve lifters. Crankcase capacity: 4 qt. (add 1 qt. if new filter). Cooling system capacity: 16 qt. (add 1 qt. for heater). Carburetor: Two Carter AFB four-barrels: (front) 2970-S; (rear) 2971-S. Engine code: S-38.

Dodge Dart, Dodge Polara optional V-8
Overhead valve. Cast-iron block. Bore & stroke: 4.19 x 3.75 in. Displacement: 413 cid. Compression ratio: 11.00:1. Brake hp: 365 at 4600 rpm. Taxable hp: 55.90. Torque: 460 lbs.-ft. at 2800 rpm. Five main bearings. Hydraulic valve lifters. Crankcase capacity: 4 qt. (add 1 qt. if new filter installed). Cooling system capacity: 16 qt. (add 1 qt. for heater). Carburetor: Carter AFB 3251-S four-barrel. Engine code: S-41.

Dodge Dart, Dodge Polara optional Cross Ram V-8
Overhead valve. Cast-iron block. Bore & stroke: 4.19 x 3.75 in. Displacement: 413 cid. Compression ratio: 11.00:1. Brake hp: 385 at 5200 rpm. Taxable hp: 55.90. Torque: 455 lbs.-ft. at 3600 rpm. Five main bearings. Hydraulic valve lifters. Crankcase capacity: 4 qt. (add 1 qt. if new filter installed). Cooling system capacity: 16 qt. (add 1 qt. for heater). Carburetor: Two Carter AFB four-barrels: (front) 2970-S; (rear) 2971-S. Engine code: S-41.

Dodge Dart, Dodge Polara Ram Charger "Max Wedge" V-8
Overhead valve. Cast-iron block. Bore & stroke: 4.19 x 3.75 in. Displacement: 413 cid. Compression ratio: 11.00:1. Brake hp: 410 at 5200 rpm. Taxable hp: 55.90. Torque: 460 lbs.-ft. at 4400 rpm. Five main bearings. Solid valve lifters. Crankcase capacity: 4 qt. (add 1 qt. if new filter installed). Cooling system capacity: 16 qt. (add 1 qt. for heater). Carburetor: Two Carter AFB 2903-S four-barrels. Engine code: S-41.

Dodge Dart, Dodge Polara Ram Charger "Max Wedge" V-8
Overhead valve. Cast-iron block. Bore & stroke: 4.19 x 3.75 in. Displacement: 413 cid. Compression ratio: 13.50:1. Brake hp: 420 at 5400 rpm. Five main bearings. Solid valve lifters. Crankcase capacity: 4 qt. (add 1 qt. if new filter installed). Cooling system capacity: 16 qt. (add 1 qt. for heater). Carburetor: Two Carter AFB 3084-S four-barrels. Engine code: S-41. (Midyear option.)

1963 DODGE

Dodge 330, 440, Polara base V-8
Overhead valve. Cast-iron block. Bore & stroke: 3.91 x 3.31 in. Displacement: 318 cid. Compression ratio: 9.0:1. Brake hp: 230 at 4400 rpm. Taxable hp: 48.90. Torque: 340 lbs.-ft. at 2400. Five main bearings. Hydraulic valve lifters. Crankcase capacity: (880 models) 5 qt.; (Other models) 4 qt. (add 1 qt. for new filter). Cooling system capacity: (330, 440, Polara) 20 qt.; (other models) 16 qt. (add 1 qt. for heater). Carburetor: Carter two-barrel BBD-3472-S or Stromberg WW-3-222A two-barrel. Engine code: T-318.

Dodge 880, Dodge Custom 880 base V-8
Overhead valve. Cast-iron block. Bore & stroke: 4.13 x 3.38 in. Displacement: 361 cid. Compression ratio: 9.0:1. Brake hp: 265 at 4400 rpm. Taxable hp: 54.30. Torque: 380 lbs.-ft. at 2400. Five main bearings. Hydraulic valve lifters. Crankcase capacity: (880 models) 5 qt.; (Other models) 4 qt. (add 1 qt. for new filter). Cooling system capacity: (330, 440, Polara) 20 qt.; (other models) 16 qt. (add 1 qt. for heater). Carburetor: Carter BBD-34763 two-barrel. Engine code: T-36.

Dodge Polara 500 base V-8; Dodge 330, 440, 880, Custom 880, Polara optional V-8
Overhead valve. Cast-iron block. Bore & stroke: 4.25 x 3.38 in. Displacement: 383 cid. Compression ratio: 10.00:1. Brake hp: 305 at 4600 rpm. Taxable hp: 57.80. Torque: 410 lbs.-ft. at 2400 rpm. Five main bearings. Hydraulic valve lifters. Crankcase capacity: (880 models) 5 qt.;

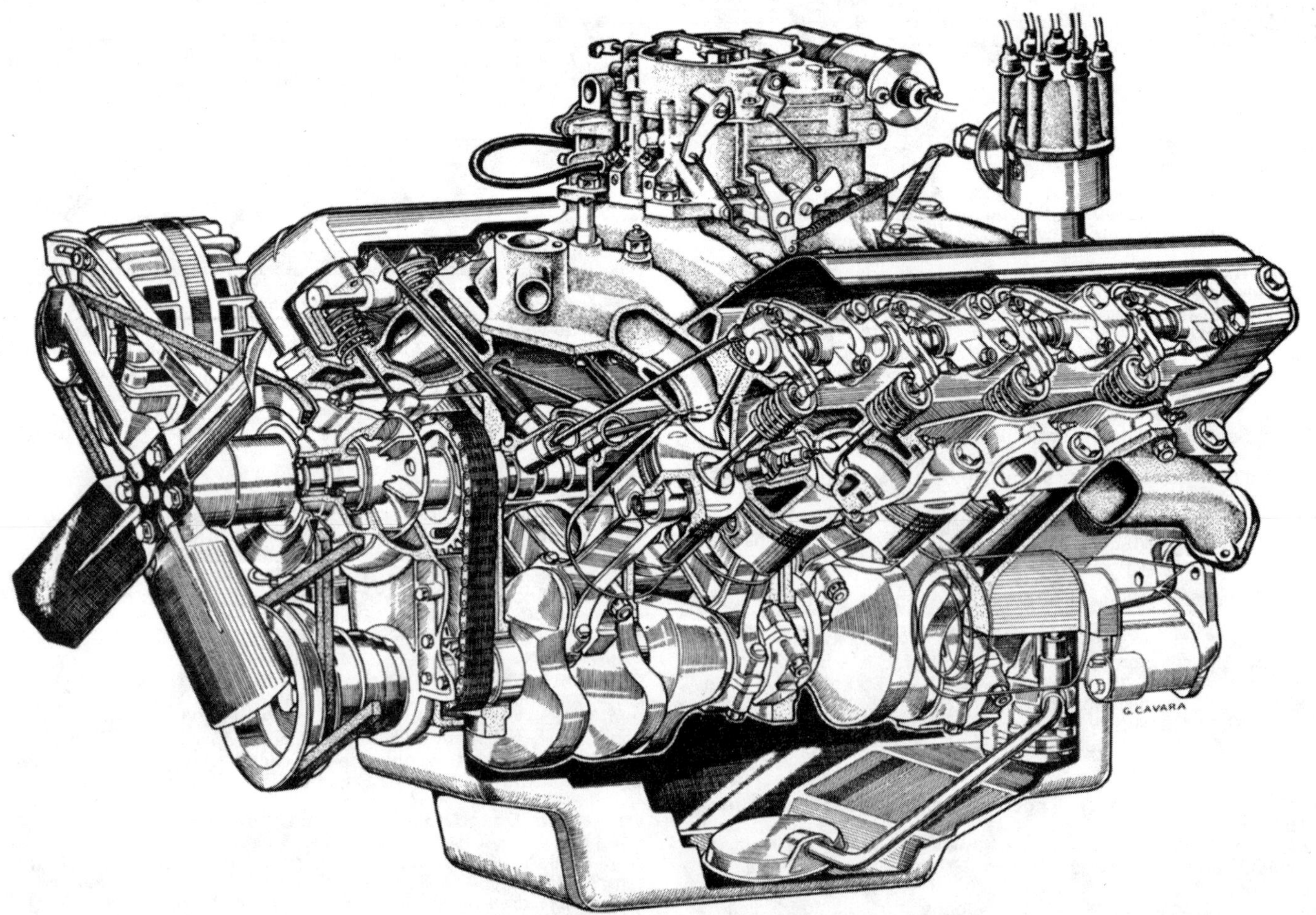

Here's a factory cutaway drawing of the 1963 Dodge 318-cid 230-hp V-8 with a two-barrel carburetor. In 1953-1954, a Hemi was the only Dodge V-8. The 270-cid 175-hp "polyspheric" V-8 of 1955 was the granddaddy of the 318, which became the standard V-8 in lower-priced Chrysler products by the early 1960s.

(Other models) 4 qt. (add 1 qt. for new filter). Cooling system capacity: (330, 440, Polara) 20 qt.; (other models) 16 qt. (add 1 qt. for heater). Carburetor: Carter BBD-3475-S two-barrel. Engine code: T-38.

Dodge 330, 440, 880, Polara, Polara 500 optional V-8
Overhead valve. Cast-iron block. Bore & stroke: 4.25 x 3.38 in. Displacement: 383 cid. Compression ratio: 10.0:1. Brake hp: 330 at 4600 rpm. Taxable hp: 57.80. Torque: 425 lbs.-ft. at 2800 rpm. Five main bearings. Hydraulic valve lifters. Crankcase capacity: (880 models) 5 qt.; (Other models) 4 qt. (add 1 qt. for new filter). Cooling system capacity: (330, 440, Polara) 20 qt.; (other models) 16 qt. (add 1 qt. for heater). Carburetor: Carter AFB -3437-S four-barrel. Engine code: T-38.

Dodge optional V-8
Overhead valve. Cast-iron block. Bore & stroke: 4.19 x 3.75 in. Displacement: 413 cid. Compression ratio: 10.10:1. Brake hp: 360 at 4800 rpm. Five main bearings. Hydraulic valve lifters. Crankcase capacity: (880 models) 5 qt.; (Other models) 4 qt. (add 1 qt. for new filter). Cooling system capacity: (330, 440, Polara) 20 qt.; (other models) 16 qt. (add 1 qt. for heater). Carburetor: Carter AFB four-barrel. Engine code: 5-41

Dodge optional "Max Wedge Stage II" Ramcharger V-8
Overhead valve. Cast-iron block. Bore & stroke: 4.25 x 3.75 in. Displacement: 426 cid. Compression ratio: 11.0:1. Brake hp: 415 at 5800 rpm. Solid valve lifters. Crankcase capacity: (880 models) 5 qt.; (Other models) 5 qt. (add 1 qt. for new filter). Cooling system capacity: (330, 440, Polara) 20 qt.; (other models) 16 qt. (add 1 qt. for heater). Carburetor: Two Carter four-barrel. Engine code: T-42.

Dodge optional "Max Wedge Stage II" Ramcharger V-8
Overhead valve. Cast-iron block. Bore & stroke: 4.25 x 3.75 in. Displacement: 426 cid. Compression ratio: 13.5:1. Brake hp: 425 at 5600 rpm. Five main bearings. Solid valve lifters. Crankcase capacity: (880 models) 5 qt.; (Other models) 5 qt. (add 1 qt. for new filter). Cooling system capacity: (330, 440, Polara) 20 qt.; (other models) 16 qt. (add 1 qt. for heater). Carburetor: Two Carter four-barrel. Engine code: T-42.
NOTE: Dodge changed the spelling of "Ram-Charger" to "Ramcharger" for identification of Chrysler's new 426-cid Max Wedge Stage II engine. This was an all-out racing mill. The Plymouth version had a black seven-blade fan, while Dodge's had a chrome fan in sales literature.

1964 DODGE

The awesome 426-cid "Hemi" that arrived in 1964 Dodges (and Plymouths) was available in two for-racing-only versions. Both had dual Carter AFB four-barrel carburetors. The first had an 11.0:1 compression ratio and developed 415 hp. The second had a 12.5:1 compression ratio and cranked up 425 hp. A third Hemi, with a single four-barrel carburetor and 400 hp was used only for NASCAR racing, until the Hemi was banned from stock-car competition. A handful were sold for street use on an "as is" basis, meaning that Chrysler's normal warranty did not apply. The company warned drag racers not to run their Hemi at full bore for more than 15 seconds.

Dodge Dart optional V-8
Overhead valve. Cast-iron block. Bore & stroke: 3.63 x 3.31 in. Displacement: 273 cid. Compression ratio: 8.8:1. Brake hp: 180 at 4200 rpm. Taxable hp: 42.20. Torque: 260 lbs.-ft. at 1600 rpm. Five main bearings. Hydraulic valve lifters. Crankcase capacity: 4 qt. (add 1 qt. for new filter). Cooling system capacity: 16 qt. (add 1 qt. for heater). Carburetor: Carter BBD-3843S two-barrel. Engine code: V-273.

Dodge 330, 440, Polara base V-8
Overhead valve. Cast-iron block. Bore & stroke: 3.91 x 3.31 in. Displacement: 318 cid. Compression ratio: 9.0:1. Brake hp: 230 at 4400 rpm. Taxable hp: 48.90. Torque: 340 lbs.-ft. at 2400 rpm. Five main bearings. Hydraulic valve lifters. Crankcase capacity: 4 qt. (add 1 qt. for new filter). Cooling system capacity: 20 qt. (add 1 qt. for heater). Carburetor: Carter BBD-3682-S two-barrel. Engine code: V-318.

Dodge 880, Custom 880 base V-8
Overhead valve. Cast-iron block. Bore & stroke: 4.13 x 3.38 in. Displacement: 361 cid. Compression ratio: 9.0:1. Brake hp: 265 at 4400 rpm. Taxable hp: 54.30. Torque: 380 lbs.-ft. at 2400 rpm. Five main bearings. Hydraulic valve lifters. Crankcase capacity: 5 qt. (add 1 qt. for new filter). Cooling system capacity: 16 qt. (add 1 qt. for heater). Carburetor: Stromberg WWC-3-244 two-barrel. Engine code: V-36.

Dodge Polara 500 base V-8; Dodge 330, 440, 880, Custom 880 optional V-8
Overhead valve. Cast-iron block. Bore & stroke: 4.25 x 3.38 in. Displacement: 383 cid. Compression ratio: 10.0:1. Brake hp: 305 at 4600 rpm. Taxable hp: 57.80. Torque: 410 lbs.-ft. at 2400 rpm. Five main bearings. Hydraulic valve lifters. Crankcase capacity: 4 qt. (add 1 qt. for new filter). Cooling system capacity: 16 qt. (add 1 qt. for heater). Carburetor: Carter BBD-3476-S two-barrel. Engine code: V-38.

Dodge 330, 440, Polara optional V-8
Overhead valve. Cast-iron block. Displacement: 383 cid. Bore & stroke: 4.25 x 3.38 in. Compression ratio: 10.0:1. Brake hp: 330 at 4600 rpm. Taxable hp: 57.80. Torque: 425 lbs.-ft. at 2800 rpm. Five main bearings. Hydraulic valve lifters. Crankcase capacity: 4 qt. (add 1 qt. for new filter). Cooling system capacity: 16 qt. (add 1 qt. for heater). Carburetor: Carter BBD-3684-S four-barrel. Engine code: V-38.

Dodge 330, 440, Polara optional Ramcharger (Wedge) V-8
Overhead valve. Cast-iron block. Displacement: 426 cid. Bore & stroke: 4.25 x 3.75 in. Compression ratio: 10.30:1. Brake hp: 365 at 4800 rpm. Taxable hp: 57.80. Torque: 475 lbs.-ft. at 3200 rpm. Five main bearings. Hydraulic valve lifters. Crankcase capacity: 5 qt. (add 1 qt. for new filter). Cooling system capacity: 16 qt. (add 1 qt. for heater). Carburetor: Carter AFB 3859-S four-barrel. Engine code: V-42.

Dodge 330, 440, Polara optional Ramcharger "Eight Barrel" (Wedge) V-8
Overhead valve. Cast-iron block. Displacement: 426 cid. Bore & stroke: 4.25 x 3.75 in. Compression ratio: 11.00:1. Brake hp: 415 at 5600 rpm. Taxable hp: 57.80. Torque: 470 lbs.-ft. at 4400 rpm. Five main bearings. Solid valve lifters. Crankcase capacity: 5 qt. (add 1 qt. for new filter). Cooling system capacity: 16 qt. (add 1 qt. for heater). Carburetor: Two Carter AFB four-barrel. Engine code: 426VMP.

Dodge 330, 440, Polara optional Ramcharger "Eight Barrel" (Wedge) V-8
Overhead valve. Cast-iron block. Displacement: 426 cid. Bore & stroke: 4.25 x 3.75 in. Compression ratio: 13.50:1. Brake hp: 425 at 5600 rpm. Taxable hp: 57.80. Torque: 480 lbs.-ft. at 4400 rpm. Five main bearings. Solid valve lifters. Crankcase capacity: 5 qt. (add 1 qt. for new filter). Cooling system capacity: 16 qt. (add 1 qt. for heater). Carburetor: Two Carter AFB 3084-S four-barrel. Engine code: 426VMPHC.

Dodge 330, 440, Polara optional "NASCAR" Hemi V-8
Overhead valve with hemispherical segment combustion chambers. Cast-iron block. Bore & stroke: 4.25 x 3.75 in. Displacement: 426 cid. Compression ratio: 12.5:1. Brake hp: 400 at 5600 rpm. Taxable hp: unknown. Torque: unknown at unknown rpm. Five main bearings. Solid valve lifters. Crankcase capacity: 5 qt. (add 1 qt. for new filter). Cooling system capacity: 16 qt. (add 1 qt. for heater). Carburetor: Holley four-barrel.

Dodge 330, 440, Polara optional "Race Hemi" V-8
Overhead valve with hemispherical segment combustion chambers. Cast-iron block. Displacement: 426 cid. Bore & stroke: 4.25 x 3.75 in. Compression ratio: 11.0:1. Brake hp: 415 at 5600 rpm. Taxable hp: 57.80. Torque: 470 lbs.-ft. at 4400 rpm. Five main bearings. Solid valve lifters. Crankcase capacity: 5 qt. (add 1 qt. for new filter). Cooling system capacity: 16 qt. (add 1 qt. for heater). Carburetor: Two Carter AFB 3859-S four-barrel.

Dodge 330, 440, Polara optional "Race Hemi" V-8
Overhead valve with hemispherical segment combustion chambers. Cast-iron block. Displacement: 426 cid. Bore & stroke: 4.25 x 3.75 in. Compression ratio: 12.5:1. Brake hp: 425 at 5600 rpm. Taxable hp:

57.80. Torque: 480 lbs.-ft. at 4600 rpm. Five main bearings. Solid valve lifters. Crankcase capacity: 5 qt. (add 1 qt. for new filter). Cooling system capacity: 16 qt. (add 1 qt. for heater). Carburetor: Two Carter AFB 3084-S four-barrel.

1965 DODGE

Every 1965 Dodge, Chrysler, or Plymouth V-8, including engine no. A2731219, was backed by a 5-year 50,000-mile warranty.

Dodge Dart 270, GT base V-8; Dodge Coronet, Coronet 440, Coronet 500 base V-8
Overhead valve. Cast-iron block. Bore & stroke: 3.63 x 3.31 in. Displacement: 273 cid. Compression ratio: 8.80:1. Brake hp: 180 at 4200 rpm. Taxable hp: 42.20. Torque: 260 lbs.-ft. at 1600 rpm. Five main bearings. Hydraulic valve lifters. Crankcase capacity: 4 qt. (add 1 qt. for new filter). Cooling system capacity: 17 qt. (add 1 qt. for heater). Carburetor: Carter BBD-3943-S two-barrel. Engine code: A-273.

Dodge Coronet, Coronet 440, Coronet 500 optional V-8
Overhead valve. Cast-iron block. Bore & stroke: 3.91 x 3.31 in. Displacement: 318 cid. Compression ratio: 9.0:1. Brake hp: 230 at 4400 rpm. Taxable hp: 48.90. Torque: 340 lbs.-ft. at 2400 rpm. Five main bearings. Hydraulic valve lifters. Crankcase capacity: 4 qt. (add 1 qt. for new filter). Cooling system capacity: 20 qt. (add 1 qt. for heater). Carburetor: Carter BBD-3947-S two-barrel. Engine code: A-318.

Dodge Coronet, Coronet 440, Coronet 500 optional V-8
Overhead valve. Cast-iron block. Bore & stroke: 4.13 x 3.38 in. Displacement: 361 cid. Compression ratio: 9.0:1. Brake hp: 265 at 4400 rpm. Taxable hp: 54.30. Torque: 380 lbs.-ft. at 2400 rpm. Five main bearings. Hydraulic valve lifters. Crankcase capacity: 4 qt. (add 1 qt. for new filter). Cooling system capacity: 16 qt. (add 1 qt. for heater). Carburetor: Carter BBD-3849S two-barrel. Engine code: A-361.

Dodge Polara, Custom 880 base V-8
Overhead valve. Cast-iron block. Displacement: 383 cid. Bore & stroke: 4.25 x 3.38 in. Compression ratio: 9.20:1. Brake hp: 270 at 4400 rpm. Taxable hp: 57.80. Torque: 390 lbs.-ft. at 2800 rpm. Five main bearings. Hydraulic valve lifters. Crankcase capacity: 4 qt. (add 1 qt. for new filter). Cooling system capacity: 16 qt. (add 1 qt. for heater). Carburetor: Carter BBD-4125-S two-barrel. Engine code: A-383.

Dodge Monaco base V-8; Polara, Custom 880 optional V-8
Overhead valve. Cast-iron block. Displacement: 383 cid. Bore & stroke: 4.25 x 3.38 in. Compression ratio: 10.00:1. Brake hp: 315 at 4400 rpm. Taxable hp: 57.80. Torque: 420 lbs.-ft. at 2800 rpm. Five main bearings. Hydraulic valve lifters. Crankcase capacity: 4 qt. (add 1 qt. for new filter). Cooling system capacity: 16 qt. (add 1 qt. for heater). Carburetor: Carter AFB-3855-S four-barrel. Engine code: A-383.

Dodge Coronet, Coronet 440, Coronet 500 optional V-8
Overhead valve. Cast-iron block. Displacement: 383 cid. Bore & stroke: 4.25 x 3.38 in. Compression ratio: 10.0:1. Brake hp: 330 at 4600 rpm. Taxable hp: 57.80. Torque: 425 lbs.-ft. at 2800 rpm. Five main bearings. Hydraulic valve lifters. Crankcase capacity: 4 qt. (add 1 qt. for new filter). Cooling system capacity: 16 qt. (add 1 qt. for heater). Carburetor: Carter BBD-3855-S four-barrel. Engine code: A-383.

Dodge Polara, Custom 880, Monaco optional V-8
Overhead valve. Cast-iron block. Bore & stroke: 4.19 x 3.75 in. Displacement: 413 cid. Compression ratio: 10.10:1. Brake hp: 340 at 4600 rpm. Taxable hp: 57.80. Torque: 470 lbs.-ft. at 2800 rpm. Five main bearings. Hydraulic valve lifters. Crankcase capacity: (880 models) 5 qt.; (Other models) 4 qt. (add 1 qt. for new filter). Cooling system capacity: (330, 440, Polara) 20 qt.; (other models) 16 qt. (add 1 qt. for heater). Carburetor: Carter 3858-S AFB four-barrel. Engine code: A-413.

Dodge Polara, Custom 880, Monaco optional V-8
Overhead valve. Cast-iron block. Bore & stroke: 4.19 x 3.75 in. Displacement: 413 cid. Compression ratio: 10.30:1. Brake hp: 360 at 4800 rpm. Taxable hp: 57.80. Torque: 470 lbs.-ft. at 3200 rpm. Five main bearings. Hydraulic valve lifters. Crankcase capacity: (880 models) 5 qt.; (Other models) 4 qt. (add 1 qt. for new filter). Cooling system capacity: (330, 440, Polara) 20 qt.; (other models) 16 qt. (add 1 qt. for heater). Carburetor: Carter 3858-S AFB four-barrel.

Dodge Polara, Custom 880, Monaco V-8; Coronet 440, Coronet 500 optional "Wedge" V-8
Overhead valve. Cast-iron block. Displacement: 426 cid. Bore & stroke: 4.25 x 3.75 in. Compression ratio: 10.10:1. Brake hp: 365 at 4800 rpm. Taxable hp: 57.80. Torque: 470 lbs.-ft. at 3200 rpm. Five main bearings. Hydraulic valve lifters. Crankcase capacity: 4 qt. (add 1 qt. for new filter). Cooling system capacity: 16 qt. (add 1 qt. for heater). Carburetor: Carter AFB 3959-S four-barrel. Engine code: A-426.

Dodge 330, 440, Polara optional Ramcharger "Eight Barrel" (Wedge) V-8
Overhead valve. Cast-iron block. Displacement: 426 cid. Bore & stroke: 4.25 x 3.75 in. Compression ratio: 13.50:1. Brake hp: 425 at 5600 rpm. Taxable hp: 57.80. Torque: 480 lbs.-ft. at 4400 rpm. Five main bearings. Solid valve lifters. Crankcase capacity: 5 qt. (add 1 qt. for new filter). Cooling system capacity: 16 qt. (add 1 qt. for heater). Carburetor: Two Carter AFB 3084-S four-barrel. Engine code: AH-426

Dodge optional "Race Hemi" V-8
Overhead valve. Cast-iron block. Displacement: 426 cid. Bore & stroke: 4.25 x 3.75 in. Compression ratio: 11.00:1. Brake hp: 415 at 5600 rpm. Taxable hp: 57.80. Torque: 470 lbs.-ft. at 4400 rpm. Five main bearings. Solid valve lifters. Crankcase capacity: 4 qt. (add 1 qt. for new filter). Cooling system capacity: 16 qt. (add 1 qt. for heater). Carburetor: Two Carter AFB 3859-S four-barrel. Engine code: AH-426.

Dodge optional "Race Hemi" V-8
Overhead valve with hemispherical segment combustion chambers. Cast-iron block. Displacement: 426 cid. Bore & stroke: 4.25 x 3.75 in. Compression ratio: 12.0:1. Brake hp: 425 at 5600 rpm. Taxable hp:

The 426-cid "Hemi" was back for 1965, when Dodge promoted it as the "Hemi Charger Acceleration Engine." This engine option had a price tag in the $1,800 range. Specifications were similar to 1964. The 426-powered Dodge Hemi Chargers continued to dominate drag strip action across the United States.

57.80. Torque: 480 lbs.-ft. at 4600 rpm. Five main bearings. Solid valve lifters. Crankcase capacity: 4 qt. (add 1 qt. for new filter). Cooling system capacity: 16 qt. (add 1 qt. for heater). Carburetor: Two Carter AFB 3084-S four-barrel. Engine code: AH-426HC.

1966 DODGE

Dodge Dart 270, GT base V-8; Dodge Coronet, Deluxe, Coronet 440, Coronet 500 base V-8
Overhead valve. Cast-iron block. Bore & stroke: 3.63 x 3.31 in. Displacement: 273 cid. Compression ratio: 8.80:1. Brake hp: 180 at 4200 rpm. Taxable hp: 42.20. Torque: 260 lbs.-ft. at 1600 rpm. Five main bearings. Hydraulic valve lifters. Crankcase capacity: 4 qt. (add 1 qt. for new filter). Cooling system capacity: 18 qt. (With heater standard). Carburetor: Carter BBD-4113-S two-barrel. Engine code: B-273. VIN code: D.

Dodge Dart 270, GT optional V-8
Overhead valve. Cast-iron block. Bore & stroke: 3.63 x 3.31 in. Displacement: 273 cid. Compression ratio: 10.50:1. Brake hp: 235 at 5200 rpm. Taxable hp: 42.20. Torque: 280 lbs.-ft. at 4000 rpm. Five main bearings. Hydraulic valve lifters. Crankcase capacity: 4 qt. (add 1 qt. for new filter). Cooling system capacity: 18 qt. (With heater standard). Carburetor: Carter AFB-4119-S four-barrel. Engine code: B-273. VIN code: D.

Dodge Coronet, Deluxe, Coronet 440, Coronet 500, Charger, Polara base V-8
Overhead valve. Cast-iron block. Bore & stroke: 3.91 x 3.31 in. Displacement: 318 cid. Compression ratio: 9.0:1. Brake hp: 230 at 4400 rpm. Taxable hp: 48.90. Torque: 340 lbs.-ft. at 2400 rpm. Five main bearings. Hydraulic valve lifters. Crankcase capacity: 4 qt. (add 1 qt. for new filter). Cooling system capacity: 21 qt. (With heater standard). Carburetor: Stromberg WW3-258 two-barrel. Engine code: B-318. VIN code: E.

Dodge Coronet, Deluxe, Coronet 440, Coronet 500, Charger optional V-8
Overhead valve. Cast-iron block. Bore & stroke: 4.13 x 3.38 in. Displacement: 361 cid. Compression ratio: 9.0:1. Brake hp: 265 at 4400 rpm. Taxable hp: 54.30. Torque: 380 lbs.-ft. at 2400 rpm. Five main bearings. Hydraulic valve lifters. Crankcase capacity: 4 qt. (add 1 qt. for new filter). Cooling system capacity: 17 qt. (With heater standard). Carburetor: Carter BBD-3849-S two-barrel. Engine code: B-361. VIN code: F.

Dodge Polara, Monaco base V-8; Monaco 500 optional V-8
Overhead valve. Cast-iron block. Displacement: 383 cid. Bore & stroke: 4.25 x 3.38 in. Compression ratio: 9.20:1. Brake hp: 270 at 4400 rpm. Taxable hp: 57.80. Torque: 390 lbs.-ft. at 2800 rpm. Five main bearings. Hydraulic valve lifters. Crankcase capacity: 4 qt. (add 1 qt. for new filter). Cooling system capacity: 17 qt. (With heater standard). Carburetor: Carter BBD-4125-S two-barrel. Engine code: B-383. VIN code: G.

Shown here with the air cleaner and valve covers removed is a prototype Street Hemi V-8 in a 1966 Dodge convertible that turned some 500 miles at Riverside Raceway during a magazine preview. The original photo caption stated, "The 426-cu.-in. engine will enter production next month and the units should be on the street before Christmas." (Dodge News Photo)

Dodge Coronet, Deluxe, Coronet 440, Coronet 500, Charger, Monaco 500 base V-8
Dodge Monaco, Polara optional V-8
Overhead valve. Cast-iron block. Displacement: 383 cid. Bore & stroke: 4.25 x 3.38 in. Compression ratio: 10.0:1. Brake hp: 325 at 4800 rpm. Taxable hp: 57.80. Torque: 425 lbs.-ft. at 2800 rpm. Five main bearings. Hydraulic valve lifters. Crankcase capacity: 4 qt. (add 1 qt. for new filter). Cooling system capacity: 17 qt. (With heater standard). Carburetor: Carter BBD-3855-S four-barrel. Engine code: B-383. VIN code: G.

Dodge Polara, Monaco, Monaco 500 optional "440 TNT" (Wedge) V-8
Overhead valve. Cast-iron block. Bore & stroke: 4.32 x 3.75 in. Displacement: 440 cid. Compression ratio: 10.10:1. Brake hp: 350 at 4400 rpm. Taxable hp: 59.70. Torque: 480 lbs.-ft. at 2800 rpm. Five main bearings. Hydraulic valve lifters. Crankcase capacity: 4 qt. (add 1 qt. for new filter). Cooling system capacity: 17 qt. (With heater standard). Carburetor: Carter AFB 3959-S four-barrel. Engine code: B-440. VIN code: J.

Dodge optional "Street Hemi" Eight-Barrel V-8
Overhead valve with hemispherical segment combustion chambers. Cast-iron block. Displacement: 426 cid. Bore & stroke: 4.25 x 3.75 in. Compression ratio: 10.25:1. Brake hp: 425 at 5000 rpm. Taxable hp: 57.80. Torque: 490 lbs.-ft. at 4000 rpm. Five main bearings. Hydraulic valve lifters. Crankcase capacity: 4 qt. (add 1 qt. for new filter). Cooling system capacity: 17 qt. (With heater). Carburetor: Two Carter AFB 3084-S four-barrel. Engine code: B-426. VIN code: H.

Dodge optional "Race Hemi" Eight-Barrel V-8
Overhead valve with hemispherical segment combustion chambers. Cast-iron block. Displacement: 426 cid. Bore & stroke: 4.25 x 3.75 in. Compression ratio: 10.25:1. Compression ratio: 12.0:1. Brake hp: 425 at 5600 rpm. Taxable hp: 57.80. Torque: 480 lbs.-ft. at 4600 rpm. Five main bearings. Solid valve lifters. Crankcase capacity: 6 qt. (add 1 qt. for new filter). Cooling system capacity: 17 qt. (With heater). Carburetor: Two Carter AFB 3084-S four-barrel. Engine code: B-426. VIN code: K.

A new "Street Hemi" engine arrived for 1966 Dodges. Cars with this option wore badges like the one shown here. The Street Hemi was essentially the same as the Race Hemi, except that it had a 10.25:1 compression ratio and hydraulic valve lifters.

1967 DODGE

Dodge Dart 270, GT base V-8; Dodge Coronet, Deluxe, Coronet 440, Coronet 500 base V-8
Overhead valve. Cast-iron block. Bore & stroke: 3.63 x 3.31 in. Displacement: 273 cid. Compression ratio: 8.80:1. Brake hp: 180 at 4200 rpm. Taxable hp: 42.20. Torque: 260 lbs.-ft. at 1600 rpm. Five main bearings. Hydraulic valve lifters. Crankcase capacity: 4 qt. (add 1 qt. for new filter). Cooling system capacity: 19 qt. (add 1 qt. for heater). Carburetor: Carter BBD-4113-S two-barrel. Engine code: C-273. VIN code: D.

Dodge Dart 270, GT optional high-performance V-8
Overhead valve. Cast-iron block. Bore & stroke: 3.63 x 3.31 in. Displacement: 273 cid. Compression ratio: 10.50:1. Brake hp: 235 at

This factory cutaway illustration shows the guts of the 1967 Hemi V-8. By this time, only one version of the monster "mill" was available. It had a 10.25:1 compression ratio and twin Carter AFBs. The official horsepower rating was again 425 hp. In Hemis, the spark plugs were slightly tilted, instead of in the same line as the cylinder bore. Dual valve springs were required.

This 1967 Dodge Charger that appeared in *Car Exchange* magazine was fitted with the big-block 383-cid V-8. The 383 came in two-barrel (270-hp) and four-barrel (325-hp) versions that season. The later 383 with a 4.25 x 3.38-inch bore and stroke was not the same as the 383-cid V-8 used in 1959-1960 Chryslers, which had 4.03 x 3.75-inch bore and stroke dimensions.

5200 rpm. Taxable hp: 42.20. Torque: 280 lbs.-ft. at 4000 rpm. Five main bearings. Hydraulic valve lifters. Crankcase capacity: 4 qt. (add 1 qt. for new filter). Cooling system capacity: 19 qt. (add 1 qt. for heater). Carburetor: Carter AFB-3855-S four-barrel. Engine code: C-273. VIN code: E.

Charger, Polara base V-8; Dodge Coronet, Deluxe, Coronet 440, Coronet 500 optional V-8
Overhead valve. Cast-iron block. Bore & stroke: 3.91 x 3.31 in. Displacement: 318 cid. Compression ratio: 9.0:1. Brake hp: 230 at 4400 rpm. Taxable hp: 48.90. Torque: 340 lbs.-ft. at 2400 rpm. Five main bearings. Hydraulic valve lifters. Crankcase capacity: 4 qt. (add 1 qt. for new filter). Cooling system capacity: 18 qt. (add 1 qt. for heater). Carburetor: Stromberg WW3-258 two-barrel. Engine code: C-318. VIN code: F.

Dodge Polara, Polara 500, Monaco base V-8
Dodge Coronet, Deluxe, Coronet 440, Coronet 500, Charger, Monaco 500 optional V-8
Overhead valve. Cast-iron block. Displacement: 383 cid. Bore & stroke: 4.25 x 3.38 in. Compression ratio: 9.20:1. Brake hp: 270 at 4400 rpm. Taxable hp: 57.80. Torque: 390 lbs.-ft. at 2800 rpm. Five main bearings. Hydraulic valve lifters. Crankcase capacity: 4 qt. (add 1 qt. for new filter). Cooling system capacity: 17 qt. (add 1 qt. for heater). Carburetor: Carter BBD-4125-S two-barrel. Engine code: C-383. VIN code: G.

Dodge Monaco 500 base V-8
Dodge Coronet, Deluxe, 440, 500, Charger, Polara 500, Monaco optional V-8
Overhead valve. Cast-iron block. Displacement: 383 cid. Bore & stroke: 4.25 x 3.38 in. Compression ratio: 10.0:1. Brake hp: 325 at 4800 rpm. Taxable hp: 57.80. Torque: 425 lbs.-ft. at 2800 rpm. Five main bearings. Hydraulic valve lifters. Crankcase capacity: 4 qt. (add 1 qt. for new filter). Cooling system capacity: 17 qt. (add 1 qt. for heater). Carburetor: Carter BBD-3855-S four-barrel. Engine code: C-383. VIN code: H.

Dodge Polara, Monaco, Monaco 500 optional 440 "Wedge" V-8
Overhead valve. Cast-iron block. Bore & stroke: 4.32 x 3.75 in. Displacement: 440 cid. Compression ratio: 10.10:1. Brake hp: 350 at

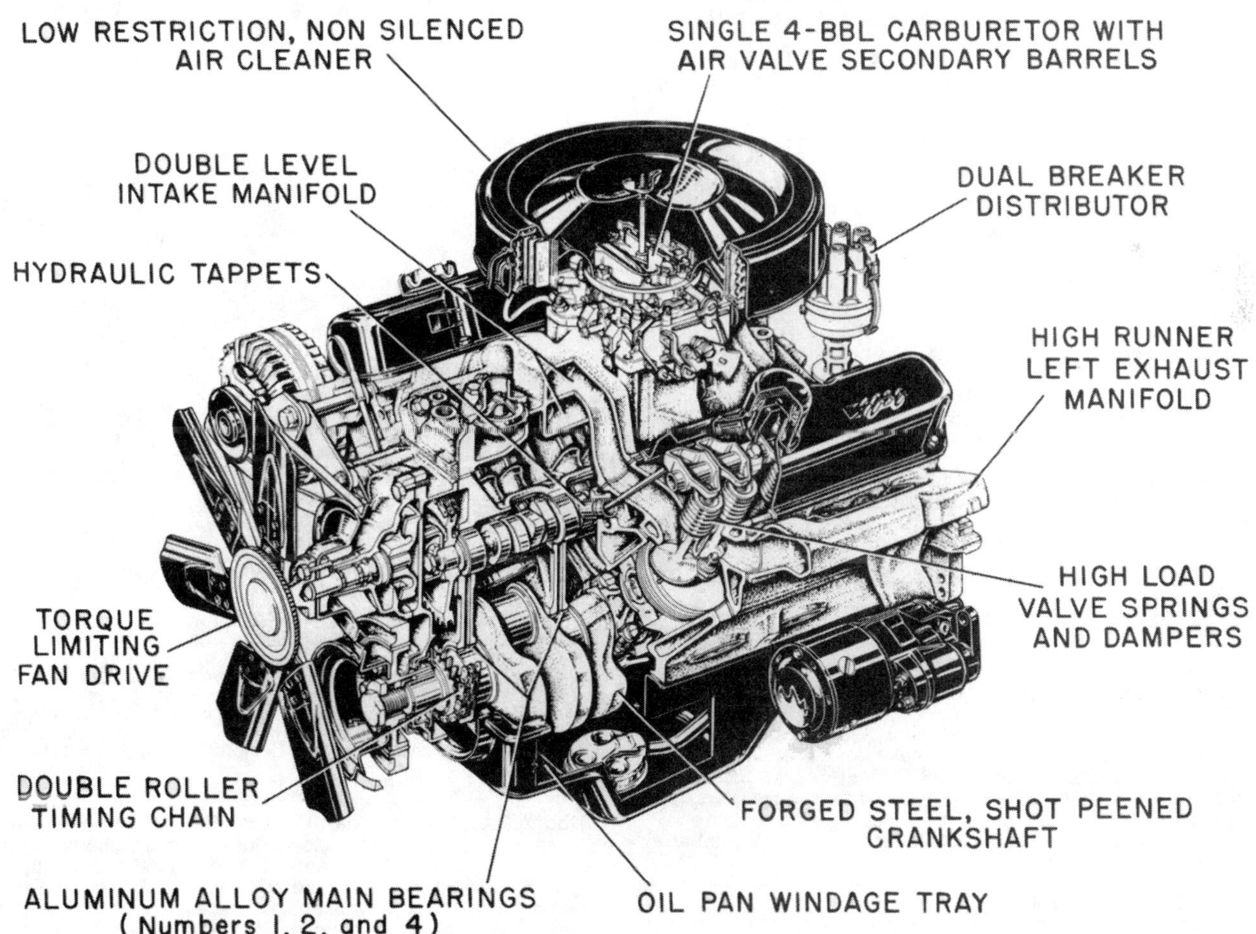

This factory illustration clearly labeled all of the important features of the 340-cid Mopar V-8. This engine first appeared as base V-8 for the 1968 Dodge Dart GTS and 1968 Plymouth Barracuda Formula 'S.' Dodge offered it through 1973, Plymouth through 1974.

4400 rpm. Taxable hp: 59.70. Torque: 480 lbs.-ft. at 2800 rpm. Five main bearings. Hydraulic valve lifters. Crankcase capacity: 4 qt. (add 1 qt. for new filter). Cooling system capacity: 18 qt. (add 1 qt. for heater). Carburetor: Carter AFB 3959-S four-barrel. Engine code: C-440. VIN code: K.

Dodge Coronet R/T base V-8; Polara, Polara 500, Monaco, Monaco 500 optional 440 V-8
Overhead valve. Cast-iron block. Bore & stroke: 4.32 x 3.75 in. Displacement: 440 cid. Compression ratio: 10.10:1. Brake hp: 375 at 4600 rpm. Taxable hp: 59.70. Torque: 480 lbs.-ft. at 3200 rpm. Five main bearings. Hydraulic valve lifters. Crankcase capacity: 4 qt. (add 1 qt. for new filter). Cooling system capacity: 18 qt. (add 1 qt. for heater). Carburetor: Carter AFB 3959-S four-barrel. Engine code: C-440. VIN code: L.

Dodge Coronet, Deluxe, 440, 500, Charger, R/T optional "Street Hemi" Eight-Barrel V-8
Overhead valve with hemispherical segment combustion chambers. Cast-iron block. Displacement: 426 cid. Bore & stroke: 4.25 x 3.75 in. Compression ratio: 10.25:1. Brake hp: 425 at 5000 rpm. Taxable hp: 57.80. Torque: 490 lbs.-ft. at 4000 rpm. Five main bearings. Hydraulic valve lifters. Crankcase capacity: 6 qt. (add 1 qt. for new filter). Cooling system capacity: 17 qt. (add 1 qt. for heater). Carburetor: Two Carter AFB 3084-S four-barrel. Engine code: C-426. VIN code: J.

1968 DODGE

Dodge Dart 270, GT base V-8; Dodge Coronet, Deluxe, Coronet 440 base V-8
Overhead valve. Cast-iron block. Bore & stroke: 3.63 x 3.31 in. Displacement: 273 cid. Compression ratio: 9.00:1. Brake hp: 190 at 4400 rpm. Taxable hp: 42.20. Torque: 260 lbs.-ft. at 2000 rpm. Five main bearings. Hydraulic valve lifters. Crankcase capacity: 4 qt. (add 1 qt. for new filter). Cooling system capacity: 19 qt. (with heater). Carburetor: Carter BBD-4116-S two-barrel. Engine code: PT-273. VIN code: D.

Dodge Coronet 500, Charger, Polara, Polara 500 base V-8
Dodge Dart 270, GT, Coronet Deluxe, Coronet 440 optional high-performance V-8
Overhead valve. Cast-iron block. Bore & stroke: 3.91 x 3.31 in. Displacement: 318 cid. Compression ratio: 9.0:1. Brake hp: 230 at 4400 rpm. Taxable hp: 48.90. Torque: 340 lbs.-ft. at 2400 rpm. Five main bearings. Hydraulic valve lifters. Crankcase capacity: 4 qt. (add 1 qt. for new filter). Cooling system capacity: 18 qt. (with heater). Carburetor: Carter BBD-4420-S two-barrel. Engine code: PT-318. VIN code: F.

Dodge Polara, Polara 500, Monaco base V-8
Dodge Coronet, Deluxe, Coronet 440, Coronet 500, Charger, Monaco 500 optional V-8
Overhead valve. Cast-iron block. Displacement: 383 cid. Bore & stroke: 4.25 x 3.38 in. Compression ratio: 9.20:1. Brake hp: 270 at 4400 rpm. Taxable hp: 57.80. Torque: 390 lbs.-ft. at 2800 rpm. Five main bearings. Hydraulic valve lifters. Crankcase capacity: 4 qt. (add 1 qt. for new filter). Cooling system capacity: 17 qt. (with heater). Carburetor: Carter BBD-4422-S two-barrel. Engine code: PT-383. VIN code: G.

Dodge Dart GTS base V-8
Overhead valve. Cast-iron block. Bore & stroke: 4.04 x 3.31 in. Displacement: 340 cid. Compression ratio: 10.5:1. Brake hp: 275 at 5000 rpm. Taxable hp: 52.20. Torque: 340 lbs.-ft. at 3200 rpm. Five main bearings. Hydraulic valve lifters. Crankcase capacity: 4 qt. (add 1 qt. for new filter). Cooling system capacity: 17 qt. (with heater). Carburetor: Carter Thermo-Quad AVS-4424-S four-barrel. Engine code: PT-340. VIN code: H.

Dodge Monaco, Monaco 500 base V-8
Dodge Coronet Deluxe, Coronet 440, Coronet 500, Charger, Polara, Polara 500 optional V-8
Overhead valve. Cast-iron block. Displacement: 383 cid. Bore & stroke: 4.25 x 3.38 in. Compression ratio: 9.20:1. Brake hp: 290 at 4400 rpm. Taxable hp: 57.80. Torque: 390 lbs.-ft. at 2800 rpm (except Polara 500 is 380 ft.-lbs. at 2400 rpm). Five main bearings. Hydraulic valve lifters. Crankcase capacity: 4 qt. (add 1 qt. for new filter). Cooling system capacity: 17 qt. (with heater). Carburetor: Carter BBD-4422-S two-barrel. Engine code: PT-383. VIN code: G.

Dodge Dart GTS optional V-8
Overhead valve. Cast-iron block. Displacement: 383 cid. Bore & stroke: 4.25 x 3.38 in. Compression ratio: 10.00:1. Brake hp: 300 at 4400 rpm. Taxable hp: 57.80. Torque: 400 lbs.-ft. at 2400 rpm. Five main bearings. Hydraulic valve lifters. Crankcase capacity: 4 qt. (add 1 qt. for new filter). Cooling system capacity: 17 qt. (with heater). Carburetor: Carter BBD-4422-S two-barrel. Engine code: PT-383. VIN code: H.

Dodge Monaco 500 base V-8
Dodge Polara and Monaco station wagons optional V-8
Overhead valve. Cast-iron block. Displacement: 383 cid. Bore & stroke: 4.25 x 3.38 in. Compression ratio: 10.10:1. Brake hp: 330 at 5000 rpm. Taxable hp: 57.80. Torque: 425 lbs.-ft. at 3200 rpm. Five main bearings. Hydraulic valve lifters. Crankcase capacity: 4 qt. (add 1 qt. for new filter). Cooling system capacity: 17 qt. (with heater). Carburetor: Carter four-barrel. Engine code: PT-383. VIN code: H.

Dodge Polara, Polara 500, Monaco, Monaco 500 optional 440 "Wedge" V-8
Overhead valve. Cast-iron block. Bore & stroke: 4.32 x 3.75 in. Displacement: 440 cid. Compression ratio: 10.10:1. Brake hp: 350 at 4400 rpm. Taxable hp: 59.70. Torque: 480 lbs.-ft. at 2800 rpm. Five main bearings. Hydraulic valve lifters. Crankcase capacity: 4 qt. (add 1 qt. for new filter). Cooling system capacity: 18 qt. (with heater). Carter AVS-4428-S four-barrel. Engine code: PT-440. VIN code: K.

Dodge Coronet R/T, Charger R/T base V-8
Overhead valve. Cast-iron block. Bore & stroke: 4.32 x 3.75 in. Displacement: 440 cid. Compression ratio: 10.10:1. Brake hp: 375 at 4600 rpm. Taxable hp: 59.70. Torque: 480 lbs.-ft. at 3200 rpm. Five main bearings. Hydraulic valve lifters. Crankcase capacity: 4 qt. (add 1 qt. for new filter). Cooling system capacity: 17 qt. (with heater). Carburetor: Carter AVS-4428-S four-barrel. Engine code: PT-440. VIN code: L.

Dodge Coronet, Deluxe, 440, 500, Charger, R/T optional "Street Hemi" Eight-Barrel V-8
Overhead valve with hemispherical segment combustion chambers. Cast-iron block. Displacement: 426 cid. Bore & stroke: 4.25 x 3.75 in. Compression ratio: 10.25:1. Brake hp: 425 at 5000 rpm. Taxable hp: 57.80. Torque: 490 lbs.-ft. at 4000 rpm. Five main bearings. Hydraulic valve lifters. Crankcase capacity: 6 qt. (add 1 qt. for new filter). Cooling system capacity: 18 qt. (with heater). Carburetor: Two Carter AFB 3084-S four-barrel. Engine code: PT-426. VIN code: J.

1969 DODGE

A favored setup of drag racers in 1969 was the Dodge Magnum 440 V-8 with four-speed manual transmission attachment. This big-block V-8 was in the same engine family as the 383 and came in three horsepower ranges: 250, 375, and 390. The first two had four-barrel carburetors and the last had three two-barrel carburetors, which Mopar called a "Six Pack." The "440 TNT" wedge-head engine was first seen in a Dodge in 1966 and last seen in a Dodge in 1977. The engine displacement had no conection to the 440 model designation used earlier by Dodge Division.

Dodge Dart, Swinger, dart Custom, Dart GT optional V-8
Overhead valve. Cast-iron block. Bore & stroke: 3.63 x 3.31 in. Displacement: 273 cid. Compression ratio: 9.00:1. Brake hp: 190 at 4400 rpm. Taxable hp: 42.20. Torque: 260 lbs.-ft. at 2000 rpm. Five main bearings. Hydraulic valve lifters. Crankcase capacity: 4 qt. (add 1 qt. for new filter). Cooling system capacity: 19 qt. (with heater). Carburetor: Carter BBD-4605-S two-barrel. Engine code: PT-273. VIN code: D.

Dodge Coronet 500, Charger, Polara, Polara 500 base V-8
Overhead valve. Cast-iron block. Bore & stroke: 3.91 x 3.31 in.

Displacement: 318 cid. Compression ratio: 9.0:1. Brake hp: 230 at 4400 rpm. Taxable hp: 48.90. Torque: 340 lbs.-ft. at 2400 rpm. Five main bearings. Hydraulic valve lifters. Crankcase capacity: 4 qt. (add 1 qt. for new filter). Cooling system capacity: 18 qt. (with heater). Carburetor: Carter BBD-4607-S two-barrel. Engine code: PT-318. VIN code: F.

Dodge Dart GTS base V-8
Overhead valve. Cast-iron block. Bore & stroke: 4.04 x 3.31 in. Displacement: 340 cid. Compression ratio: 10.5:1. Brake hp: 275 at 5000 rpm. Taxable hp: 52.20. Torque: 340 lbs.-ft. at 3200 rpm. Five main bearings. Hydraulic valve lifters. Crankcase capacity: 4 qt. (add 1 qt. for new filter). Cooling system capacity: 17 qt. (with heater). Carburetor: Carter Thermo-Quad AVS-4611-S four-barrel. Engine code: PT-340. VIN code: P.

Dodge Monaco base V-8
Overhead valve. Cast-iron block. Displacement: 383 cid. Bore & stroke: 4.25 x 3.38 in. Compression ratio: 9.20:1. Brake hp: 290 at 4400 rpm. Taxable hp: 57.80. Torque: 390 lbs.-ft. at 2800 rpm (except Polara 500 is 380 ft.-lbs. at 2400 rpm). Five main bearings. Hydraulic valve lifters. Crankcase capacity: 4 qt. (add 1 qt. for new filter). Cooling system capacity: 17 qt. (with heater). Carburetor: Carter BBD-4613-S two-barrel. Engine code: PT-383. VIN code: G.

Dodge Coronet 500, Charger, Polara, Polara 500, Monaco optional V-8
Overhead valve. Cast-iron block. Displacement: 383 cid. Bore & stroke: 4.25 x 3.38 in. Compression ratio: 10.00:1. Brake hp: 330 at 5200 rpm. Taxable hp: 57.80. Torque: 425 lbs.-ft. at 3200 rpm. Five main bearings. Hydraulic valve lifters. Crankcase capacity: 4 qt. (add 1 qt. for new filter). Cooling system capacity: 17 qt. (with heater). Carburetor: Carter four-barrel. Engine code: PT-383. VIN code: H.

Dodge Super Bee base "Magnum" V-8
Overhead valve. Cast-iron block. Displacement: 383 cid. Bore & stroke: 4.25 x 3.38 in. Compression ratio: 10.00:1. Brake hp: 335 at 5000 rpm. Taxable hp: 57.80. Torque: 425 lbs.-ft. at 3400 rpm. Five main bearings. Hydraulic valve lifters. Crankcase capacity: 4 qt. (add 1 qt. for new filter). Cooling system capacity: 17 qt. (with heater). Carburetor: Carter Thermo-Quad AVS-4615-S four-barrel. Engine code: PT-383. VIN code: H.

Dodge optional 440 "Wedge" V-8
Overhead valve. Cast-iron block. Bore & stroke: 4.32 x 3.75 in. Displacement: 440 cid. Compression ratio: 10.10:1. Brake hp: 350 at 4400 rpm. Taxable hp: 59.70. Torque: 480 lbs.-ft. at 2800 rpm. Five main bearings. Hydraulic valve lifters. Crankcase capacity: 4 qt. (add 1 qt. for new filter). Cooling system capacity: 18 qt. (with heater). Carter AVS-4428-S four-barrel. Engine code: PT-440. VIN code: K.

Dodge Coronet R/T, Charger R/T base V-8
Overhead valve. Cast-iron block. Bore & stroke: 4.32 x 3.75 in. Displacement: 440 cid. Compression ratio: 10.10:1. Brake hp: 375 at 4600 rpm. Taxable hp: 59.70. Torque: 480 lbs.-ft. at 3200 rpm. Five main bearings. Hydraulic valve lifters. Crankcase capacity: 4 qt. (add 1 qt. for new filter). Cooling system capacity: 17 qt. (with heater). Carburetor: Carter AVS-4617-S four-barrel. Engine code: PT-440. VIN code: L.

Starting with a 273-cid displacement, the small-block Chrysler V-8 was developed first for the Plymouth Valiant and Dodge Dart late in 1964, to help them compete with the Mustang for sales. Four years later, it became a 340-cid V-8, as seen here in its 1969 configuration. The 340 filled a gap in Chrysler's engine lineup between the 318 and the 383.

Dodge Super Bee optional "6-Pak" V-8
Overhead valve. Cast-iron block. Bore & stroke: 4.32 x 3.75 in. Displacement: 440 cid. Compression ratio: 10.50:1. Brake hp: 390 at 4700 rpm. Taxable hp: 59.70. Torque: 490 lbs.-ft. at 3200 rpm. Five main bearings. Hydraulic valve lifters. Crankcase capacity: 4 qt. (add 1 qt. for new filter). Cooling system capacity: 15.5 qt. (with heater). Carburetor: three Holley two-barrel. Engine code: PT-440. VIN code: M.

Dodge Coronet 500, Coronet R/T, Charger, Charger R/T optional "Street Hemi" V-8
Overhead valve with hemispherical segment combustion chambers. Cast-iron block. Displacement: 426 cid. Bore & stroke: 4.25 x 3.75 in. Compression ratio: 10.25:1. Brake hp: 425 at 5000 rpm. Taxable hp: 57.80. Torque: 490 lbs.-ft. at 4000 rpm. Five main bearings. Hydraulic valve lifters. Crankcase capacity: 6 qt. (add 1 qt. for new filter). Cooling system capacity: 18 qt. (with heater). Carburetor: Two Carter AFB 3084-S four-barrel. Engine code: PT-426. VIN code: J.

1970 DODGE

Dodge Coronet 500, Polara, Polara Custom base V-8
Overhead valve. Cast-iron block. Bore & stroke: 3.91 x 3.31 in. Displacement: 318 cid. Compression ratio: 8.80:1. Brake hp: 230 at 4400 rpm. Taxable hp: 48.90. Torque: 320 lbs.-ft. at 2000 rpm. Five main bearings. Hydraulic valve lifters. Crankcase capacity: 4 qt. (add 1 qt. for new filter). Cooling system capacity: 16 qt. (with heater). Carburetor: Carter BBD-4721-S two-barrel. VIN code: G.

In 1970, Dodge also offered Six Pack carburetion for the 340-cid small-block V-8, as seen here under the hood of a Dodge Dart. This engine had a 10.0:1 compression ratio. It produced 290 hp at 4400 rpm and 340 foot-pounds of torque at 3200 rpm.

Dodge Dart Swinger base V-8
Overhead valve. Cast-iron block. Bore & stroke: 4.04 x 3.31 in. Displacement: 340 cid. Compression ratio: 8.80:1. Brake hp: 275 at 5000 rpm. Taxable hp: 52.20. Torque: 340 lbs.-ft. at 3200 rpm. Five main bearings. Hydraulic valve lifters. Crankcase capacity: 4 qt. (add 1 qt. for new filter). Cooling system capacity: 16 qt. (with heater). Carburetor: Carter Thermo-Quad AVS-4933-S four-barrel. VIN code: P.

Dodge Challenger, Monaco base V-8
Overhead valve. Cast-iron block. Displacement: 383 cid. Bore & stroke: 4.25 x 3.38 in. Compression ratio: 8.70:1. Brake hp: 290 at 4400 rpm. Taxable hp: 57.80. Torque: 390 lbs.-ft. at 2800 rpm (except Polara 500 is 380 lbs.-ft. at 2400 rpm). Five main bearings. Hydraulic valve lifters. Crankcase capacity: 4 qt. (add 1 qt. for new filter). Cooling system capacity: 16 qt. (with heater). Carburetor: Carter BBD-4613-S two-barrel. VIN code: G.

Dodge Challenger R/T, Coronet Super Bee base V-8
Dodge Challenger, Coronet Deluxe, Coronet 440, Coronet 500 optional V-8
Overhead valve. Cast-iron block. Displacement: 383 cid. Bore & stroke: 4.25 x 3.38 in. Compression ratio: 9.50:1. Brake hp: 330 at 5000 rpm. Taxable hp: 57.80. Torque: 425 lbs.-ft. at 3200 rpm. Five main bearings. Hydraulic valve lifters. Crankcase capacity: 4 qt. (add 1 qt. for new filter). Cooling system capacity: 16 qt. (with heater). Carburetor: Carter AVS-4376-S four-barrel. VIN code: G.

Dodge Challenger optional high-performance V-8
Overhead valve. Cast-iron block. Displacement: 383 cid. Bore & stroke: 4.25 x 3.38 in. Compression ratio: 9.50:1. Brake hp: 335 at 5200 rpm. Taxable hp: 57.80. Torque: 425 lbs.-ft. at 3400 rpm. Five main bearings. Hydraulic valve lifters. Crankcase capacity: 4 qt. (add 1 qt. for new filter). Cooling system capacity: 16 qt. (with heater). Carburetor: Carter AVS-4376-S four-barrel. VIN code: H.

Dodge Polara, Monaco optional 440 "Wedge" V-8
Overhead valve. Cast-iron block. Bore & stroke: 4.32 x 3.75 in. Displacement: 440 cid. Compression ratio: 9.70:1. Brake hp: 350 at 4400 rpm. Taxable hp: 59.70. Torque: 480 lbs.-ft. at 2800 rpm. Five main bearings. Hydraulic valve lifters. Crankcase capacity: 4 qt. (add 1 qt. for new filter). Cooling system capacity: 15.5 qt. (with heater). Carburetor: Carter AVS-4737-S four-barrel. VIN code: K.

Dodge Coronet R/T, Charger R/T, Charger Daytona base V-8; Challenger R/T optional V-8
Overhead valve. Cast-iron block. Bore & stroke: 4.32 x 3.75 in.

The 318-cid Dodge V-8 was part of Chrysler Corporation's "small-block" engine family which was comprised of the earlier 273, the 318, and the 340. Notice the Dodge name on the manifold heat shield on this 1970 version of the 318. In 1970, the small-block V-8s gained great importance as the basis of Dan Gurney's Trans-Am racing team. The Trans-Am cars used a bored version of the 273 block to achieve a 305-cid V-8 that was legal under Trans-Am racing formulas.

Displacement: 440 cid. Compression ratio: 9.70:1. Brake hp: 375 at 4600 rpm. Taxable hp: 59.70. Torque: 480 lbs.-ft. at 2800 rpm. Five main bearings. Hydraulic valve lifters. Crankcase capacity: 4 qt. (add 1 qt. for new filter). Cooling system capacity: 15.5 qt. (with heater). Carburetor: Carter AVS-4737-S four-barrel. VIN code: L.

Dodge Challenger R/T, Coronet Super Bee, Coronet R/T, Charger R/T optional "6-Pak" V-8
Overhead valve. Cast-iron block. Bore & stroke: 4.32 x 3.75 in. Displacement: 440 cid. Compression ratio: 10.50:1. Brake hp: 390 at 4700 rpm. Taxable hp: 59.70. Torque: 490 lbs.-ft. at 3200 rpm. Five main bearings. Hydraulic valve lifters. Crankcase capacity: 4 qt. (add 1 qt. for new filter). Cooling system capacity: 15.5 qt. (with heater). Carburetor: three Holley two-barrel. VIN code: M.

Dodge Challenger R/T, Coronet Super Bee, Coronet R/T, Charger R/T, Charger Daytona optional "Street Hemi" V-8
Overhead valve with hemispherical segment combustion chambers. Cast-iron block. Displacement: 426 cid. Bore & stroke: 4.25 x 3.75 in. Compression ratio: 10.20:1. Brake hp: 425 at 5000 rpm. Taxable hp: 57.80. Torque: 490 lbs.-ft. at 4000 rpm. Five main bearings. Hydraulic valve lifters. Crankcase capacity: 6 qt. (add 1 qt. for new filter). Cooling system capacity: 15.5 qt. (with heater). Carburetor: Two Carter AFB 3084-S four-barrel. VIN code: J.

The 1970 Dodge 440 Six-Pack V-8 featured a dual-breaker point ignition system, heavy-duty valve springs, and especially selected, magnafluxed connecting rods. Three Holley two-barrel carburetors were fitted to its special intake manifold. A special low-restriction air cleaner covered all three carburetors and carried the Dodge name at its front end.

The Dodge name was carried on the original oil filter attached to this 1970 Dodge 383-cid V-8. This version of the Chrysler "B-block" wore a slimline dual-snorkel air cleaner clearly marked to indicate which end faced the front.

Shown without its air cleaner is the 1970 Dodge 340-cid 275-hp V-8. When fitted in a Dodge, this V-8 carried the Dodge name on the left-hand valve cover. Other than such minor differences, the 318- and the 340-cid Mopar V-8s looked very much alike.

1971 DODGE

Dodge Polara optional V-8
Overhead valve. Cast-iron block. Bore & stroke: 4.00 x 3.58 in. Displacement: 360 cid. Compression ratio: 8.70:1. Brake hp: 225 at 4400 rpm. Taxable hp: 51.0. Torque: 360 lbs.-ft. at 2400 rpm. Five main bearings. Hydraulic valve lifters. Crankcase capacity: 6 qt. (add 1 qt. for new filter). Cooling system capacity: 16 qt. (with heater). Carburetor: two-barrel. VIN code: K.

Dodge Charger base V-8
Overhead valve. Cast-iron block. Bore & stroke: 3.91 x 3.31 in. Displacement: 318 cid. Compression ratio: 8.60:1. Brake hp: 230 at 4400 rpm. Taxable hp: 48.90. Torque: 320 lbs.-ft. at 2000 rpm. Five main bearings. Hydraulic valve lifters. Crankcase capacity: 4 qt. (add 1 qt. for new filter). Cooling system capacity: 16 qt. (with heater). Carburetor: Carter BBD-4957-S two-barrel. VIN code: G.

Dodge Dart base V-8
Overhead valve. Cast-iron block. Bore & stroke: 4.04 x 3.31 in. Displacement: 340 cid. Compression ratio: 10.20:1. Brake hp: 275 at 5000 rpm. Taxable hp: 52.20. Torque: 340 lbs.-ft. at 3200 rpm. Five main bearings. Hydraulic valve lifters. Crankcase capacity: 4 qt. (add 1 qt. for new filter). Cooling system capacity: 16 qt. (with heater). Carburetor: Carter Thermo-Quad TQ-4972-S four-barrel. VIN code: H.

Dodge Polara, Monaco base V-8
Overhead valve. Cast-iron block. Displacement: 383 cid. Bore & stroke: 4.25 x 3.38 in. Compression ratio: 8.50:1. Brake hp: 275 at 4400 rpm. Taxable hp: 57.80. Torque: 375 lbs.-ft. at 2800 rpm (except Polara 500 is 380 lbs.-ft. at 2400 rpm). Five main bearings. Hydraulic valve lifters. Crankcase capacity: 4 qt. (add 1 qt. for new filter). Cooling system capacity: 14.5 qt. (with heater). Carburetor: Carter BBD-4961-S two-barrel. VIN code: L.

Dodge Challenger optional "340 6-Pak" V-8
Overhead valve. Cast-iron block. Bore & stroke: 4.04 x 3.31 in. Displacement: 340 cid. Compression ratio: 10.20:1. Brake hp: 290 at

5000 rpm. Taxable hp: 52.20. Torque: 340 lbs.-ft. at 3200 rpm. Five main bearings. Hydraulic valve lifters. Crankcase capacity: 4 qt. (add 1 qt. for new filter). Cooling system capacity: 16 qt. (with heater). Carburetor: three Holley two-barrel. VIN code: J.

Dodge Challenger, Monaco base V-8
Overhead valve. Cast-iron block. Displacement: 383 cid. Bore & stroke: 4.25 x 3.38 in. Compression ratio: 8.50:1. Brake hp: 300 at 4800 rpm. Taxable hp: 57.80. Torque: 410 at 3400 rpm. Five main bearings. Hydraulic valve lifters. Crankcase capacity: 4 qt. (add 1 qt. for new filter). Cooling system capacity: (Challenger) 16 qt.; (Monaco) 14.5 qt. (with heater). Carburetor: Holley R-4668A four-barrel. VIN code: L.

Dodge Polara, Monaco optional V-8
Overhead valve. Cast-iron block. Bore & stroke: 4.32 x 3.75 in. Displacement: 440 cid. Compression ratio: 8.80:1. Brake hp: 335 at 4400 rpm. Taxable hp: 59.70. Torque: 460 at 3200 rpm. Five main bearings. Hydraulic valve lifters. Crankcase capacity: 4 qt. (add 1 qt. for new filter). Cooling system capacity: 17 qt. (with heater). Carburetor: Carter AVS-4967-S four-barrel. VIN code: T.

Dodge Charger R/T base "Magnum" high-performance V-8; optional Challenger V-8
Overhead valve. Cast-iron block. Bore & stroke: 4.32 x 3.75 in. Displacement: 440 cid. Compression ratio: 9.70:1. Brake hp: 370 at 4600 rpm. Taxable hp: 59.70. Torque: 480 lbs.-ft. at 3200 rpm. Five main bearings. Hydraulic valve lifters. Crankcase capacity: 4 qt. (add 1 qt. for new filter). Cooling system capacity: (Challenger) 15.5 qt.; (Charger) 17 qt.

Charger optional "Street Hemi" V-8
Overhead valve with hemispherical segment combustion chambers. Cast-iron block. Displacement: 426 cid. Bore & stroke: 4.25 x 3.75 in. Compression ratio: 10.20:1. Brake hp: 425 at 5000 rpm. Taxable hp: 57.80. Torque: 490 lbs.-ft. at 4000 rpm. Five main bearings. Hydraulic valve lifters. Crankcase capacity: 6 qt. (add 1 qt. for new filter). Cooling system capacity: (Challenger) 15.5 qt.; (Charger) 17 qt. (with heater). Carburetor: Two Carter AFB 3084-S four-barrel. VIN code: R.

1972 DODGE

Dodge Dart, Challenger, Charger, Coronet, Polara base V-8
Overhead valve. Cast-iron block. Bore & stroke: 3.91 x 3.31 in. Displacement: 318 cid. Compression ratio: 8.60:1. Net brake hp: 150 at 4000 rpm. Net torque: 260 lbs.-ft. at 1600 rpm. Four main bearings. Hydraulic valve lifters. Crankcase capacity: 4 qt. (add 1 qt. for new filter). Cooling system capacity: (Dart) 16 qt.; (Other models) 16.5 qt. (with air conditioning). Carburetor: Carter BBD-6150-S two-barrel. VIN code: G.

This 318-cid Dodge engine was base V-8 in the 1973 Dart, Challenger, Coronet, Charger, and Polara. Shown with the air cleaner removed, it was now equipped with Electric Choke Assist to help reduce hydrocarbon and carbon monoxide emissions during starting and warm up. An exhaust gas recirculation (EGR) system was used to introduce controlled amounts of exhaust gases into the fuel-air mixture to help reduce Nox emissions.

Dodge Monaco base V-8
Overhead valve. Cast-iron block. Bore & stroke: 4.00 x 3.58 in. Displacement: 360 cid. Compression ratio: 8.80:1. Net brake hp: 175 at 4000 rpm. Net torque: 285 at 2400 rpm. Five main bearings. Hydraulic valve lifters. Crankcase capacity: 4 qt. (add 1 qt. for new filter). Cooling system capacity: 15 qt. (with air conditioning). Carburetor: Holley R-6161A two-barrel. VIN code: K.

Dodge Charger, Coronet, Polara, Monaco optional V-8
Overhead valve. Cast-iron block. Bore & stroke: 4.34 x 3.38 in. Displacement: 400 cid. Compression ratio: 8.20:1. Net brake hp: 190 at 4400 rpm. Net torque: 310 lbs.-ft. at 2400 rpm. Five main bearings. Hydraulic valve lifters. Crankcase capacity: 4 qt. (add 1 qt. for new filter). Cooling system capacity: (Coronet, Charger) 15 qt.; (Polara, Monaco) 14 qt. (with air conditioning). Carburetor: Holley ZR-6368A two-barrel. VIN code: M.

Dodge Polara, Monaco optional V-8
Overhead valve. Cast-iron block. Bore & stroke: 4.32 x 3.75 in. Displacement: 440 cid. Compression ratio: 8.20:1. Net brake hp: 225 at 4400 rpm. Net torque: 345 lbs.-ft. at 3200 rpm. Five main bearings. Hydraulic valve lifters. Crankcase capacity: 4 qt. (add 1 qt. for new filter). Cooling system capacity: 16 qt. (with air conditioning). Carburetor: Holley R-6160A four-barrel. VIN code: T.

Dodge Dart base V-8
Overhead valve. Cast-iron block. Bore & stroke: 4.04 x 3.31 in. Displacement: 340 cid. Compression ratio: 8.50:1. Net brake hp: 240 at 4800 rpm. Net torque: 290 lbs.-ft. at 3600 rpm. Five main bearings. Hydraulic valve lifters. Crankcase capacity: 4 qt. (add 1 qt. for new filter). Cooling system capacity: (Dart) 15 qt.; (Other models) 14.5 qt. (with air conditioning). Carburetor: Carter Thermo-Quad TQ-6139-S four-barrel. VIN code: H.

Charger, Coronet optional V-8
Overhead valve. Cast-iron block. Bore & stroke: 4.32 x 3.75 in. Displacement: 440 cid. Compression ratio: 8.20:1. Net brake hp: 245 at 4400 rpm. Net torque: 360 lbs.-ft. at 3200 rpm. Five main bearings. Hydraulic valve lifters. Crankcase capacity: 4 qt. (add 1 qt. for new filter). Cooling system capacity: 15 qt. (with air conditioning). Carburetor: Holley R-6160A four-barrel. VIN code: U.

Dodge Dodge Charger, Coronet optional V-8
Overhead valve. Cast-iron block. Bore & stroke: 4.34 x 3.38 in. Displacement: 400 cid. Compression ratio: 8.2:1. Net brake hp: 255 at 4400 rpm. Net torque: 340 lbs.-ft. at 3200 rpm. Five main bearings. Hydraulic valve lifters. Crankcase capacity: 4 qt. (add 1 qt. for new filter). Cooling system capacity: 14.5 qt. (with air conditioning). Carburetor: Carter Thermo-Quad TQ-6090-S four-barrel. VIN code: P.

Charger, Coronet optional "440 6-Pak" V-8
Overhead valve. Cast-iron block. Displacement: 440 cid. Bore & stroke: 4.32 x 3.75 in. Compression ratio: 10.30:1. Net brake hp: 330 at 4800 rpm. Net torque: 410 lbs.-ft. at 3600 rpm. Five main bearings. Hydraulic valve lifters. Crankcase capacity: 4 qt. (add 1 qt. for new filter). Cooling system capacity: 17 qt. (with air conditioning). Carburetor: Three Holley two-barrel. VIN code: V.

1973 DODGE

Dodge Dart, Challenger, Coronet, Charger, Polara base V-8
Overhead valve. Cast-iron block. Bore & stroke: 3.91 x 3.31 in. Displacement: 318 cid. Compression ratio: 8.60:1. Net brake hp: 150 at 4000 rpm. Net torque: 265 lbs.-ft. at 2000 rpm. Four main bearings. Hydraulic valve lifters. Crankcase capacity: 4 qt. (add 1 qt. for new filter). Cooling system capacity: (Polara, Monaco) 19 qt.; (Other models) 18 qt. Carburetor: Carter Two-barrel. VIN code: G.

Dodge Polara station wagon base, Monaco base V-8; Polara (car) optional V-8
Overhead valve. Cast-iron block. Bore & stroke: 4.00 x 3.58 in. Displacement: 360 cid. Compression ratio: 8.40:1. Net brake hp: 170 at 4000 rpm. Net torque: 285 lbs.-ft. at 2400 rpm. Five main bearings. Hydraulic valve lifters. Crankcase capacity: 4 qt. (add 1 qt. for new filter). Cooling system capacity: 16 qt. (with air conditioning). Carburetor: Two-barrel. VIN code: K.

Dodge Coronet, Charger optional V-8
Overhead valve. Cast-iron block. Bore & stroke: 4.34 x 3.38 in. Displacement: 400 cid. Compression ratio: 8.20:1. Net brake hp: 175 at 3600 rpm. Net torque: 305 lbs.-ft. at 2400 rpm. Five main bearings. Hydraulic valve lifters. Crankcase capacity: 4 qt. (add 1 qt. for new filter). Cooling system capacity: 17 qt. (with air conditioning). Carburetor: Two-barrel. VIN code: M.

Dodge Polara, Monaco optional V-8
Overhead valve. Cast-iron block. Bore & stroke: 4.34 x 3.38 in. Displacement: 400 cid. Compression ratio: 8.20:1. Net brake hp: 185 at 3600 rpm. Net torque: 310 lbs.-ft. at 2400 rpm. Five main bearings. Hydraulic valve lifters. Crankcase capacity: 4 qt. (add 1 qt. for new filter). Cooling system capacity: 17 qt. (with air conditioning). Carburetor: Two-barrel. VIN code: M.

Dodge Polara, Monaco optional "Magnum" V-8
Overhead valve. Cast-iron block. Bore & stroke: 4.32 x 3.75 in. Displacement: 440 cid. Compression ratio: 8.20:1. Net brake hp: 220 at 3600 rpm. Net torque: 350 lbs.-ft. at 2400 rpm. Five main bearings. Hydraulic valve lifters. Crankcase capacity: 4 qt. (add 1 qt. for new filter). Cooling system capacity: 17 qt. (with air conditioning). Carburetor: Four-barrel. VIN code: T.

Dodge Dart "340" Sport base V-8; Challenger, Coronet, Charger optional V-8
Overhead valve. Cast-iron block. Bore & stroke: 4.04 x 3.31 in. Displacement: 340 cid. Compression ratio: 8.50:1. Net brake hp: 240 at 4800 rpm. Net torque: 290 lbs.-ft. at 3600 rpm. Five main bearings. Hydraulic valve lifters. Crankcase capacity: 4 qt. (add 1 qt. for new filter). Cooling system capacity: 16 qt. (with air conditioning). Carburetor: Four-barrel. VIN code: H.

Dodge Coronet, Charger optional V-8
Overhead valve. Cast-iron block. Bore & stroke: 4.34 x 3.38 in. Displacement: 400 cid. Compression ratio: 8.20:1. Net brake hp: 260 at 4800 rpm. Net torque: 335 lbs.-ft. at 3600 rpm. Five main bearings. Hydraulic valve lifters. Crankcase capacity: 4 qt. (add 1 qt. for new filter). Cooling system capacity: 17 qt. (with air conditioning). Carburetor: Four-barrel. VIN code: P.

Dodge Coronet, Charger optional "Magnum" V-8
Overhead valve. Cast-iron block. Bore & stroke: 4.32 x 3.75 in. Displacement: 440 cid. Compression ratio: 8.20:1. Net brake hp: 280 at 4800 rpm. Net torque: 380 lbs.-ft. at 3200 rpm. Five main bearings. Hydraulic valve lifters. Crankcase capacity: 4 qt. (add 1 qt. for new filter). Cooling system capacity: 17 qt. (with air conditioning). Carburetor: Four-barrel. VIN code: U.

1974 DODGE

Dodge Dart, Challenger, Coronet, Charger base V-8
Overhead valve. Cast-iron block. Bore & stroke: 3.91 x 3.31 in. Displacement: 318 cid. Compression ratio: 8.60:1. Net brake hp: 150 at 4000 rpm. Net torque: 265 lbs.-ft. at 2000 rpm. Four main bearings. Hydraulic valve lifters. Crankcase capacity: 4 qt. (add 1 qt. for new filter). Cooling system capacity: 16 qt. Carburetor: Carter Two-barrel. VIN code: G.

Dodge Monaco base V-8; Charger optional V-8 (Federal)
Overhead valve. Cast-iron block. Bore & stroke: 4.00 x 3.58 in. Displacement: 360 cid. Compression ratio: 8.40:1. Net brake hp: 180 at 4000 rpm. Net torque: 290 lbs.-ft. at 2400 rpm. Five main bearings. Hydraulic valve lifters. Crankcase capacity: 4 qt. (add 1 qt. for new filter). Cooling system capacity: 16 qt. (with air conditioning). Carburetor: Two-barrel. VIN code: K.

Dodge Monaco station wagon, Brougham base V-8 (Federal)
Overhead valve. Cast-iron block. Bore & stroke: 4.34 x 3.38 in. Displacement: 400 cid. Compression ratio: 8.20:1. Net brake hp: 185 at 4000 rpm. Net torque: 315 lbs.-ft. at 2400 rpm. Five main bearings. Hydraulic valve lifters. Crankcase capacity: 4 qt. (add 1 qt. for new filter). Cooling system capacity: 16 qt. (with air conditioning). Carburetor: Two-barrel. VIN code: M.

Dodge Dart "360" Sport base V-8; Challenger, Charger, Monaco optional V-8 (California)
Overhead valve. Cast-iron block. Bore & stroke: 4.00 x 3.58 in. Displacement: 360 cid. Compression ratio: 8.40:1. Net brake hp: 200 at 4000 rpm. Net torque: 290 lbs.-ft. at 3200 rpm. Five main bearings. Hydraulic valve lifters. Crankcase capacity: 4 qt. (add 1 qt. for new filter). Cooling system capacity: (Coronet, Charger) 16.5 qt.; (Other models) 16 qt. (with air conditioning). Carburetor: Four-barrel. VIN code: J.

Dodge Monaco optional V-8 (California)
Overhead valve. Cast-iron block. Bore & stroke: 4.34 x 3.38 in. Displacement: 400 cid. Compression ratio: 8.20:1. Net brake hp: 200 at 4400 rpm. Net torque: 310 lbs.-ft. at 2400 rpm. Five main bearings. Hydraulic valve lifters. Crankcase capacity: 4 qt. (add 1 qt. for new filter).

Cooling system capacity: 16.5 qt. (with air conditioning). Carburetor: Four-barrel. VIN code: N.

Dodge Coronet, Charger optional V-8 (Federal)
Overhead valve. Cast-iron block. Bore & stroke: 4.34 x 3.38 in. Displacement: 400 cid. Compression ratio: 8.20:1. Net brake hp: 205 at 4400 rpm. Net torque: 310 lbs.-ft. at 2400 rpm. Five main bearings. Hydraulic valve lifters. Crankcase capacity: 4 qt. (add 1 qt. for new filter). Cooling system capacity: 16.5 qt. (with air conditioning). Carburetor: Four-barrel. VIN code: N.

Monaco station wagon base V-8 (California)
Overhead valve. Cast-iron block. Bore & stroke: 4.32 x 3.75 in. Displacement: 440 cid. Compression ratio: 8.20:1. Net brake hp: 220 at 4000 rpm. Net torque: 345 lbs.-ft. at 3200 rpm. Five main bearings. Hydraulic valve lifters. Crankcase capacity: 4 qt. (add 1 qt. for new filter). Cooling system capacity: 16 qt. (with air conditioning). Carburetor: Four-barrel. VIN code: T.

Dodge Monaco optional V-8 (Federal)
Overhead valve. Cast-iron block. Bore & stroke: 4.32 x 3.75 in. Displacement: 440 cid. Compression ratio: 8.20:1. Net brake hp: 230 at 4000 rpm. Net torque: 350 lbs.-ft. at 3200 rpm. Five main bearings. Hydraulic valve lifters. Crankcase capacity: 4 qt. (add 1 qt. for new filter). Cooling system capacity: 16 qt. (with air conditioning). Carburetor: Four-barrel. VIN code: T.

Dodge Dart "360" Sport base V-8; Challenger, Charger optional V-8 (Federal)
Overhead valve. Cast-iron block. Bore & stroke: 4.00 x 3.58 in. Displacement: 360 cid. Compression ratio: 8.40:1. Net brake hp: 245 at 4800 rpm. Net torque: 320 lbs.-ft. at 3600 rpm. Five main bearings. Hydraulic valve lifters. Crankcase capacity: 4 qt. (add 1 qt. for new filter). Cooling system capacity: (Coronet, Charger) 16.5 qt.; (Other models) 16 qt. Dual exhaust. Carburetor: Four-barrel. VIN code: L.

Dodge Coronet optional V-8 (California)
Overhead valve. Cast-iron block. Bore & stroke: 4.34 x 3.38 in. Displacement: 400 cid. Compression ratio: 8.20:1. Net brake hp: 250 at 4800 rpm. Net torque: 330 lbs.-ft. at 3400 rpm. Five main bearings. Hydraulic valve lifters. Crankcase capacity: 4 qt. (add 1 qt. for new filter). Cooling system capacity: 16.5 qt. (with air conditioning). Dual exhaust. Carburetor: Four-barrel. VIN code: P.

Dodge Coronet, Charger optional V-8
Overhead valve. Cast-iron block. Bore & stroke: 4.32 x 3.75 in. Displacement: 440 cid. Compression ratio: 8.20:1. Net brake hp: 275 at 4400 rpm. Net torque: 375 lbs.-ft. at 3200 rpm. Five main bearings. Hydraulic valve lifters. Crankcase capacity: 4 qt. (add 1 qt. for new filter). Cooling system capacity: 16 qt. (with air conditioning). Dual exhaust. Carburetor: Four-barrel. VIN code: U.

1975 DODGE

Dodge Charger base V-8 (California)
Overhead valve. Cast-iron block. Bore & stroke: 3.91 x 3.31 in. Displacement: 318 cid. Compression ratio: 8.5:1. Net brake hp: 135 at 3600 rpm. Net torque: 245 lbs.-ft. at 1600 rpm. Five main bearings. Hydraulic valve lifters. Crankcase capacity: 4 qt. (add 1 qt. for new filter). Cooling system capacity: 16 qt. (with air conditioning). Carburetor: Two-barrel. VIN code: G.

Dodge base V-8 (California)
Overhead valve. Cast-iron block. Bore & stroke: 3.91 x 3.31 in. Displacement: 318 cid. Compression ratio: 8.5:1. Net brake hp: 140 at 3600 rpm. Net torque: 255 lbs.-ft. at 1600 rpm. Five main bearings. Hydraulic valve lifters. Crankcase capacity: 4 qt. (add 1 qt. for new filter). Cooling system capacity: 16 qt. (with air conditioning). Carburetor: Two-barrel. VIN code: G.

Dodge Charger base V-8 (Federal)
Overhead valve. Cast-iron block. Bore & stroke: 3.91 x 3.31 in. Displacement: 318 cid. Compression ratio: 8.5:1. Net brake hp: 145 at 4000 rpm. Net torque: 255 lbs.-ft. at 1600 rpm. Five main bearings. Hydraulic valve lifters. Crankcase capacity: 4 qt. (add 1 qt. for new filter). Cooling system capacity: 16 qt. (with air conditioning). Carburetor: Two-barrel. VIN code: G.

Dodge Coronet base V-8 Dodge Charger, Monaco, Monaco Royal Brougham optional V-8 (Federal)
Overhead valve. Cast-iron block. Bore & stroke: 3.91 x 3.31 in. Displacement: 318 cid. Compression ratio: 8.5:1. Net brake hp: 150 at 4000 rpm. Net torque: 255 lbs.-ft. at 1600 rpm. Five main bearings. Hydraulic valve lifters. Crankcase capacity: 4 qt. (add 1 qt. for new filter). Cooling system capacity: 16 qt. (with air conditioning). Carburetor: Two-barrel. VIN code: G.

Dodge Coronet, Charger optional V-8 (Federal)
Overhead valve. Cast-iron block. Bore & stroke: 4.34 x 3.38 in. Displacement: 400 cid. Compression ratio: 8.2:1. Net brake hp: 165 at 4000 rpm. Net torque: 295 lbs.-ft. at 3200 rpm. Five main bearings. Hydraulic valve lifters. Crankcase capacity: 4 qt. (add 1 qt. for new filter). Cooling system capacity: 16 qt. (with air conditioning). Carburetor: Two-barrel. VIN code: M.

Dodge Monaco wagon/Brougham base V-8; Monaco (car) optional V-8 (Federal)
Overhead valve. Cast-iron block. Bore & stroke: 4.34 x 3.38 in. Displacement: 400 cid. Compression ratio: 8.2:1. Net brake hp: 175 at 4000 rpm. Net torque: 300 lbs.-ft. at 2400 rpm. Five main bearings. Hydraulic valve lifters. Crankcase capacity: 4 qt. (add 1 qt. for new filter). Cooling system capacity: 16 qt. (with air conditioning). Carburetor: Two-barrel. VIN code: M.

Dodge Charger, Monaco base V-8 Dodge Coronet, Monaco Royal Brougham optional V-8 (Federal)
Overhead valve. Cast-iron block. Bore & stroke: 4.00 x 3.58 in. Displacement: 360 cid. Compression ratio: 8.4:1. Net brake hp: 180 at 4000 rpm. Net torque: 290 lbs.-ft. at 2400 rpm. Five main bearings. Hydraulic valve lifters. Crankcase capacity: 4 qt. (add 1 qt. for new filter). Cooling system capacity: 16 qt. (with air conditioning). Carburetor: Two-barrel. VIN code: K.

Dodge Charger optional V-8 (California)
Overhead valve. Cast-iron block. Bore & stroke: 4.34 x 3.38 in. Displacement: 400 cid. Compression ratio: 8.2:1. Net brake hp: 185 at 4000 rpm. Net torque: 285 lbs.-ft. at 3200 rpm. Five main bearings. Hydraulic valve lifters. Crankcase capacity: 4 qt. (add 1 qt. for new filter). Cooling system capacity: 16 qt. (with air conditioning). Carburetor: Four-barrel. VIN code: N.

Dodge Coronet, Charger optional V-8 (Federal)
Overhead valve. Cast-iron block. Bore & stroke: 4.00 x 3.58 in. Displacement: 360 cid. Compression ratio: 8.4:1. Net brake hp: 190 at 4000 rpm. Net torque: 270 lbs.-ft. at 3200 rpm. Five main bearings. Hydraulic valve lifters. Crankcase capacity: 4 qt. (add 1 qt. for new filter). Cooling system capacity: 16 qt. (with air conditioning). Carburetor: Four-barrel. VIN code: J.

Dodge Coronet, Charger optional V-8 (Federal)
Overhead valve. Cast-iron block. Bore & stroke: 4.34 x 3.38 in. Displacement: 400 cid. Compression ratio: 8.2:1. Net brake hp: 190 at 4000 rpm. Net torque: 290 lbs.-ft. at 3200 rpm. Five main bearings. Hydraulic valve lifters. Crankcase capacity: 4 qt. (add 1 qt. for new filter). Cooling system capacity: 16 qt. (with air conditioning). Carburetor: Four-barrel. VIN code: N.

Dodge Monaco, Monaco Royal Brougham base V-8 (California)
Overhead valve. Cast-iron block. Bore & stroke: 4.34 x 3.38 in. Displacement: 400 cid. Compression ratio: 8.2:1. Net brake hp: 195 at 4000 rpm. Net torque: 285 lbs.-ft. at 3200 rpm. Five main bearings. Hydraulic valve lifters. Crankcase capacity: 4 qt. (add 1 qt. for new filter). Cooling system capacity: 16 qt. (with air conditioning). Carburetor: Four-barrel. VIN code: N.

Dodge Coronet, Monaco, Royal Brougham optional "Magnum" V-8 (Federal)
Overhead valve. Cast-iron block. Bore & stroke: 4.32 x 3.75 in. Displacement: 440 cid. Compression ratio: 8.20:1. Net brake hp: 215 at 4000 rpm. Net torque: 330 lbs.-ft. at 3200 rpm. Five main bearings. Hydraulic valve lifters. Crankcase capacity: 4 qt. (add 1 qt. for new filter). Cooling system capacity: 16 qt. (with air conditioning). Carburetor: Four-barrel. VIN code: T.

Dodge Dart "360" Sport base V-8(Federal)
Overhead valve. Cast-iron block. Bore & stroke: 4.00 x 3.58 in. Displacement: 360 cid. Compression ratio: 8.4:1. Net brake hp: 230 at 4200 rpm. Net torque: 320 lbs.-ft. at 3200 rpm. Five main bearings. Hydraulic valve lifters. Crankcase capacity: 4 qt. (add 1 qt. for new filter). Cooling system capacity: 16 qt. (with air conditioning). Dual exhaust. Carburetor: Four-barrel. VIN code: N.

Dodge Coronet, Charger optional V-8 (Federal)
Displacement: 400 cid. Compression ratio: 8.2:1. Net brake hp: 240 at 4400 rpm. Net torque: 325 lbs.-ft. at 3200 rpm. Five main bearings. Hydraulic valve lifters. Crankcase capacity: 4 qt. (add 1 qt. for new filter). Cooling system capacity: 16 qt. (with air conditioning). Dual exhaust. Carburetor: Four-barrel. VIN code: P.

Dodge Coronet, Monaco optional "Magnum" V-8 (California)
Overhead valve. Cast-iron block. Bore & stroke: 4.32 x 3.75 in.

Displacement: 440 cid. Compression ratio: 8.20:1. Net brake hp: 250 at 4000 rpm. Net torque: 350 lbs.-ft. at 3200 rpm. Five main bearings. Hydraulic valve lifters. Crankcase capacity: 5 qt. (add 1 qt. for new filter). Cooling system capacity: 16 qt. (with air conditioning). Dual exhaust. Carburetor: Four-barrel. VIN code: U.

Dodge Coronet, Monaco optional "Magnum" V-8 (Federal)
Overhead valve. Cast-iron block. Bore & stroke: 4.32 x 3.75 in. Displacement: 440 cid. Compression ratio: 8.20:1. Net brake hp: 260 at 4000 rpm. Net torque: 355 lbs.-ft. at 3200 rpm. Five main bearings. Hydraulic valve lifters. Crankcase capacity: 5 qt. (add 1 qt. for new filter). Cooling system capacity: 16 qt. (with air conditioning). Dual exhaust. Carburetor: Four-barrel. VIN code: U.

1976 DODGE

Dodge Dart, Aspen, Coronet, Charger, Monaco base V-8 (Federal)
Overhead valve. Cast-iron block and head. Bore & stroke: 3.91 x 3.31 in. Displacement: 318 cid. (5.2 liters). Compression ratio: 8.5:1. Net brake hp: 140 at 3600 rpm. Net torque: 250 lbs.-ft. at 2000 rpm. Five main bearings. Hydraulic valve lifters. Crankcase capacity: 4 qt. (add 1 qt. for new filter). Cooling system capacity: 16 qt. (with air conditioning). Carburetor: Carter BBD 8068S or 8069S two-barrel. VIN Code: G.

Dodge Dart, Aspen, Coronet, Charger, Monaco base V-8 (California)
Overhead valve. Cast-iron block and head. Bore & stroke: 3.91 x 3.31 in. Displacement: 318 cid. (5.2 liters). Compression ratio: 8.5:1. Net brake hp: 150 at 4000 rpm. Net torque: 255 lbs.-ft. at 1600 rpm. Five main bearings. Hydraulic valve lifters. Crankcase capacity: 4 qt. (add 1 qt. for new filter). Cooling system capacity: 16 qt. (with air conditioning). Carburetor: Carter BBD 8068S or 8069S two-barrel. VIN Code: G.

Dodge Coronet wagon, Royal Monaco base V-8 (Federal) Dodge Aspen, Coronet (car), Charger optional V-8 (Federal)
Overhead valve. Cast-iron block and head. Bore & stroke: 4.00 x 3.58 in. Displacement: 360 cid. (5.9 liters). Compression ratio: 8.4:1. Net brake hp: 170 at 4000 rpm. Net torque: 280 lbs.-ft. at 2400 rpm. Five main bearings. Hydraulic valve lifters. Crankcase capacity: 4 qt. (add 1 qt. for new filter). Cooling system capacity: 16 qt. (with air conditioning). Carburetor: Holley 2245 (R7364A) two-barrel. VIN Code: K.

Dodge Coronet wagon base V-8 (California) Dodge Aspen, Coronet (car), Charger optional V-8 (California)
Overhead valve. Cast-iron block and head. Bore & stroke: 4.00 x 3.58 in. Displacement: 360 cid. (5.9 liters). Compression ratio: 8.4:1. Net brake hp: 175 at 4000 rpm. Net torque: 270 lbs.-ft. at 1600 rpm. Five main bearings. Hydraulic valve lifters. Crankcase capacity: 4 qt. (add 1 qt. for new filter). Cooling system capacity: 16 qt. (with air conditioning). Carburetor: Carter TQ four-barrel carburetor. VIN Code: J.

Dodge Royal Monaco Brougham base V-8 Coronet, Charger, Monaco optional V-8 (Federal)
Overhead valve. Cast-iron block and head. Bore & stroke: 4.34 x 3.38 in. Displacement: 400 cid. (6.6 liters). Compression ratio: 8.2:1. Net brake hp: 175 at 4000 rpm. Net torque: 300 lbs.-ft. at 2400 rpm. Five main bearings. Hydraulic valve lifters. Crankcase capacity: 4 qt. (add 1 qt. for new filter). Cooling system capacity: 16 qt. (with air conditioning). Carburetor: Holley 2245 (R7366A) two-barrel. VIN Code: M.

Dodge Coronet, Charger optional V-8 (California)
Overhead valve. Cast-iron block and head. Bore & stroke: 4.34 x 3.38 in. Displacement: 400 cid. (6.6 liters). Compression ratio: 8.2:1. Net brake hp: 185 at 3600 rpm. Net torque: 285 lbs.-ft. at 3200 rpm. Five main bearings. Hydraulic valve lifters. Crankcase capacity: 4 qt. (add 1 qt. for new filter). Cooling system capacity: 16 qt. (with air conditioning). Carburetor: Carter TQ-9056-S four-barrel. VIN Code: N.

Dodge Monaco, Royal Monaco Brougham optional V-8 (California)
Overhead valve. Cast-iron block and heads. Bore & stroke: 4.32 x 3.75 in. Displacement: 440 cid. (7.2 liters). Compression ratio: 8.2:1. Net brake hp: 200 at 3600 rpm. Net torque: 310 lbs.-ft. at 2400 rpm. Five main bearings. Hydraulic valve lifters. Crankcase capacity: 5 qt. (add 1 qt. for new filter). Cooling system capacity: 16 qt. (with air conditioning). Carburetor: Carter TQ-9058-S four-barrel. VIN Code: T.

Dodge Monaco, Royal Monaco Brougham optional V-8 (Federal)
Overhead valve. Cast-iron block and heads. Bore & stroke: 4.32 x 3.75 in. Displacement: 440 cid. (7.2 liters). Compression ratio: 8.2:1. Net brake hp: 205 at 3600 rpm. Net torque: 320 lbs.-ft. at 2000 rpm. Five main bearings. Hydraulic valve lifters. Crankcase capacity: 5 qt. (add 1 qt. for new filter). Cooling system capacity: 16 qt. (with air conditioning). Carburetor: Carter TQ-9058-S four-barrel. VIN Code: T.

Dodge Monaco, Royal Monaco Brougham optional V-8 (Federal)
Overhead valve. Cast-iron block and head. Bore & stroke: 4.34 x 3.38 in. Displacement: 400 cid. (6.6 liters). Compression ratio: 8.2:1. Net brake hp: 210 at 4000 rpm. Net torque: 305 lbs.-ft. at 3200 rpm. Five main bearings. Hydraulic valve lifters. Crankcase capacity: 4 qt. (add 1 qt. for new filter). Cooling system capacity: 16 qt. (with air conditioning). Carburetor: Carter TQ-9064-S four-barrel. VIN Code: N.

Dodge Dart "360" sport base V-8; Coronet optional V-8 (Federal)
Overhead valve. Cast-iron block and head. Bore & stroke: 4.00 x 3.58 in. Displacement: 360 cid. (5.9 liters). Compression ratio: 8.4:1. Net brake hp: 220 at 4400 rpm. Net torque: 280 lbs.-ft. at 3200 rpm. Five main bearings. Hydraulic valve lifters. Crankcase capacity: 4 qt. (add 1 qt. for new filter). Cooling system capacity: 16 qt. (with air conditioning). Carburetor: Carter TQ-9002-S four-barrel carburetor. VIN Code: L.

Dodge Coronet, Charger S/E, Royal Monaco Brougham optional V-8 (Federal)
Overhead valve. Cast-iron block and head. Bore & stroke: 4.34 x 3.38 in. Displacement: 400 cid. (6.6 liters). Compression ratio: 8.2:1. Net brake hp: 240 at 4400 rpm. Net torque: 325 lbs.-ft. at 3200 rpm. Five main bearings. Hydraulic valve lifters. Crankcase capacity: 5 qt. (add 1 qt. for new filter). Cooling system capacity: 16 qt. (with air conditioning). Dual exhaust. Carburetor: Carter TQ-9054-S four-barrel. VIN Code: P.

Dodge Monaco, Royal Monaco Brougham optional V-8 (California)
Overhead valve. Cast-iron block and heads. Bore & stroke: 4.32 x 3.75 in. Displacement: 440 cid. (7.2 liters). Compression ratio: 8.2:1. Net brake hp: 250 at 4000 rpm. Net torque: 350 lbs.-ft. at 3200 rpm. Five main bearings. Hydraulic valve lifters. Crankcase capacity: 5 qt. (add 1 qt. for new filter). Cooling system capacity: 16 qt. (with air conditioning). Dual exhaust. Carburetor: Carter TQ-9058-S four-barrel. VIN Code: T.

Dodge Monaco, Royal Monaco Brougham optional V-8 (Federal)
Overhead valve. Cast-iron block and heads. Bore & stroke: 4.32 x 3.75 in. Displacement: 440 cid. (7.2 liters). Compression ratio: 8.2:1. Net brake hp: 255 at 4400 rpm. Net torque: 355 lbs.-ft. at 3200 rpm. Five main bearings. Hydraulic valve lifters. Crankcase capacity: 5 qt. (add 1 qt. for new filter). Cooling system capacity: 16 qt. (with air conditioning). Dual exhaust. Carburetor: Carter TQ-9058-S four-barrel. VIN Code: T.

1977 DODGE

Aspen, Monaco, Diplomat, Charger SE, Royal Monaco base V-8
Overhead valve. Cast-iron block and head. Displacement: 318 cid (5.2 liters). Bore & stroke: 3.91 x 3.31 in. Compression ratio: 8.5:1. Net brake hp: 145 at 4000 rpm. Net torque: 245 lbs.-ft. at 1600 rpm. Five main bearings. Hydraulic valve lifters. Carburetor: Carter BBD 8093S or 8094S two-barrel. VIN Code: G.

Monaco wagon, Royal Monaco Brougham base V-8; Aspen, Monaco, Charger SE, Royal Monaco optional V-8
Overhead valve. Cast-iron block and head. Bore & stroke: 4.00 x 3.58 in. Displacement: 360 cid (5.9 liters). Compression ratio: 8.4:1. Net brake hp: 155 at 3600 rpm. Net torque: 275 lbs.-ft. at 2000 rpm. Five main bearings. Hydraulic valve lifters. Carburetor: Holley 2245 (R7671A) two-barrel. VIN Code: K.

Aspen optional V-8
Overhead valve. Cast-iron block and head. Bore & stroke: 4.00 x 3.58 in. Displacement: 360 cid (5.9 liters). Compression ratio: 8.4:1. Net brake hp: Horsepower: 175 at 4000 rpm. Net torque: 275 lbs.-ft. at 2000 rpm. Five main bearings. Hydraulic valve lifters. Carburetor: Carter TQ four-barrel carburetor and Lean Burn system. VIN Code: L.

Royal Monaco station wagon base V-8; Charger SE, Monaco, Royal Monaco optional V-8
Overhead valve. Cast-iron block and head. Bore & stroke: 4.34 x 3.38 in. Displacement: 400 cid (6.6 liters). Compression ratio: 8.2:1. Net brake hp: 190 at 3600 rpm. Net torque: 305 lbs.-ft. at 3200 rpm. Five main bearings. Hydraulic valve lifters. Carburetor: Carter TQ9102S four-barrel. VIN Code: N.

Royal Monaco optional V-8
Overhead valve. Cast-iron block and head. Bore & stroke: 4.32 x 3.75 in. Displacement: 440 cid (7.2 liters). Compression ratio: 8.2:1. Net brake hp: 195 at 3600 rpm. Net torque: 320 lbs.-ft. at 2000 rpm. Five main bearings. Hydraulic valve lifters. Carburetor: Carter TQ9078S four-barrel. VIN Code: T.

1978 DODGE

Charger SE, Magnum XE, Aspen, Diplomat, Monaco base V-8
Overhead valve. Cast-iron block and head. Bore & stroke: 3.91 x 3.31 in. Displacement: 318 cid (5.2 liters). Compression ratio: 8.5:1. Net brake hp: 140 at 4000 rpm. Net torque: 245 lbs.-ft. at 1600 rpm. Five main bearings. Hydraulic valve lifters. Carburetor: Carter BBD two-barrel. VIN Code: G.

Monaco wagon base V-8; Aspen, Diplomat, Charger SE, Magnum XE, Monaco optional V-8
Overhead valve. Cast-iron block and head. Bore & stroke: 4.00 x 3.58 in. Displacement: 360 cid (5.9 liters). Compression ratio: 8.4:1. Net brake hp: 155 at 3600 rpm. Net torque: 270 lbs.-ft. at 2400 rpm. Five main bearings. Hydraulic valve lifters. Carburetor: Holley 2245 two-barrel. VIN Code: K.

Aspen optional V-8
Overhead valve. Cast-iron block and head. Bore & stroke: 4.00 x 3.58 in. Displacement: 360 cid (5.9 liters). Compression ratio: 8.4:1. Net brake hp: 175 at 4000 rpm. Net torque: 260 lbs.-ft. at 2400 rpm. Five main bearings. Hydraulic valve lifters. Carburetor: Carter TQ four-barrel. VIN Code: J.

Charger SE, Magnum XE, Monaco optional V-8
Overhead valve. Cast-iron block and head. Bore & stroke: 4.34 x 3.38 in. Displacement: 400 cid (6.6 liters). Compression ratio: 8.2:1. Net brake hp: 190 at 3600 rpm. Net torque: 305 lbs.-ft. at 3200 rpm. Five main bearings. Hydraulic valve lifters. Carburetor: Carter TQ four-barrel. VIN Code: N.

Charger SE, Magnum XE, Monaco optional "heavy-duty" V-8 (police package)
Overhead valve. Cast-iron block and head. Bore & stroke: 4.34 x 3.38 in. Displacement: 400 cid (6.6 liters). Compression ratio: unknown. Net brake hp: unknown at unknown rpm. Net torque: unknown at unknown rpm. Five main bearings. Hydraulic valve lifters. Carburetor: unknown. VIN Code: P.

1979 DODGE

Magnum XE base V-8; Aspen, Diplomat, St. Regis optional V-8 (Federal)
Overhead valve. Cast-iron block and head. Bore & stroke: 3.91 x 3.31 in. Displacement: 318 cid. (5.2 liters). Compression ratio: 8.5:1. Net brake hp: 135 at 4000 rpm. Net torque: 250 lbs.-ft. at 1600 rpm. Five main bearings. Hydraulic valve lifters. Carburetor: Holley 2280 (R8448A) two-barrel. VIN Code: G. Sales code: E44

A factory illustration shows the 1967 Chrysler Corporation 318-cid two-barrel V-8, which was rated at 230 hp. This was the base V-8 in the Dodge Charger and Polara. It was optional equipment in the Dodge Coronet, Coronet Deluxe, Coronet 440, and Coronet 500 models. The 318, first used by Dodge Division in 1960 Dart models, was used through the 1970's, and was the last V-8 available in a Dodge in 1989.

Diplomat, Magnum XE, St. Regis optional V-8 (Federal)
Overhead valve. Cast-iron block and head. Bore & stroke: 4.00 x 3.58 in. Displacement: 360 cid. (5.9 liters). Compression ratio: 8.4:1. Net brake hp: 150 at 3600 rpm. Net torque: 265 lbs.-ft. at 2400 rpm. Five main bearings. Hydraulic valve lifters. Carburetor: Holley 2245 two-barrel. VIN Code K. Sales code: E57

Magnum XE base V-8; Aspen, Diplomat, St. Regis optional V-8 (California)
Overhead valve. Cast-iron block and head. Bore & stroke: 3.91 x 3.31 in. Displacement: 318 cid. (5.2 liters). Compression ratio: 8.5:1. Net brake hp: 155 at 4000 rpm. Net torque: 245 lbs.-ft. at 1600 rpm. Five main bearings. Hydraulic valve lifters. Carburetor: Four-barrel. Electronic Spark Control. VIN Code: G. Sales code: E46.

Diplomat, Magnum XE, St. Regis optional V-8 (California)
Overhead valve. Cast-iron block and head. Bore & stroke: 4.00 x 3.58 in. Displacement: 360 cid. (5.9 liters). Compression ratio: 8.4:1. Net brake hp: 170 at 4000 rpm. Net torque: 270 lbs.-ft. at 1600 rpm. Five main bearings. Hydraulic valve lifters. Carburetor: Four-barrel. Electronic Spark Control. VIN Code K. Sales code: E56

The 360-cid four-barrel V-8 with Electronic Spark Control was available in Dodge Magnum XE, St. Regis, and Diplomat models sold in California in 1979. This engine came equipped with an air pump, a catalytic converter, and a single exhaust system. It required the N95 California Emissions Control Package at extra cost.

The 360-cid two-barrel V-8 with Electronic Spark Control was available in Dodge Magnum XE, St. Regis, and Diplomat models sold outside of California in 1979. The 360 was a larger-displacement version of the compact 318-cid V-8. Its valve timing, valve lift, and duration were carefully engineered for low emissions. The camshaft was designed to seat the valves smoothly to decrease the possibility of valve bounce and the stress it causes in valve stems.

Aspen coupe, Diplomat, Magnum XE, St. Regis optional "heavy-duty" V-8 (police package)
Overhead valve. Cast-iron block and head. Bore & stroke: 4.00 x 3.58 in. Displacement: 360 cid. (5.9 liters). Compression ratio: 8.0:1. Net brake hp: 190 at 4000 rpm. Net torque: 275 lbs.-ft. at 2000 rpm. Five main bearings. Hydraulic valve lifters. Carburetor: Carter TQ four-barrel. Dual exhaust on Magnum XE and St. Regis models; other models utilize dual mufflers and catalysts converging into a single tailpipe. VIN Code L. Sales code E58.

1980 DODGE

Dodge Aspen, Diplomat, Mirada, St. Regis base V-8
Overhead valve. Cast-iron block and heads. Bore & stroke: 3.91 x 3.31 in. Displacement: 318 cid. (5.2 liters). Compression ratio: 8.5:1. Net brake hp: 120 at 3600 rpm. Net torque: 245 lbs.-ft. at 1600 rpm. Five main bearings. Hydraulic valve lifters. Carburetor: Carter BBD two-barrel. VIN Code: G or H.

Dodge St. Regis optional V-8
Overhead valve. Cast-iron block and head. Bore & stroke: 4.00 x 3.58 in. Displacement: 360 cid. (5.9 liters). Compression ratio: 8.4:1. Net brake hp: 130 at unknown rpm. Net torque: unknown at unknown rpm. Five main bearings. Hydraulic valve lifters. Carburetor: Carter BBD two-barrel. VIN Code: K or J.

Dodge Mirada optional V-8
Overhead valve. Cast-iron block and head. Bore & stroke: 4.00 x 3.58 in. Displacement: 360 cid. (5.9 liters). Compression ratio: 8.4:1. Net brake hp: 185 at 4000 rpm. Net torque: 275 lbs.-ft. at 2000 rpm. Five main bearings. Hydraulic valve lifters. Carburetor: Carter TQ four-barrel. VIN Code: L.

1981 DODGE

Dodge Diplomat, Mirada, St. Regis base V-8
Overhead valve. Cast-iron block and heads. Bore & stroke: 3.91 x 3.31 in. Displacement: 318 cid. (5.2 liters). Compression ratio: 8.5:1. Net brake hp: 130 at 4000 rpm. Net torque: 230 lbs.-ft. at 2000 rpm. Five main bearings. Hydraulic valve lifters. Carburetor: Carter BBD 8291S two-barrel. VIN Code: K.

Dodge Diplomat, Mirada, St. Regis optional V-8
Overhead valve. Cast-iron block and heads. Bore & stroke: 3.91 x 3.31 in. Displacement: 318 cid. (5.2 liters). Compression ratio: 8.5:1. Net brake hp: 165 at 4000 rpm. Net torque: 240 lbs.-ft. at 2000 rpm. Five main bearings. Hydraulic valve lifters. Carburetor: Carter BBD 8291S two-barrel. VIN Code: L.

1982 DODGE

Dodge Diplomat, Mirada base V-8
Overhead valve. Cast-iron block and heads. Bore & stroke: 3.91 x 3.31 in. Displacement: 318 cid. (5.2 liters). Compression ratio: 8.5:1. Net brake hp: 130 at 4000 rpm. Net torque: 230 lbs.-ft. at 2000 rpm. Five main bearings. Hydraulic valve lifters. Carburetor: Carter BBD two-barrel. VIN Code: K.

Dodge Diplomat, Mirada optional V-8
Overhead valve. Cast-iron block and heads. Bore & stroke: 3.91 x 3.31 in. Displacement: 318 cid. (5.2 liters). Compression ratio: 8.5:1. Net brake hp: 165 at 4000 rpm. Net torque: 240 lbs.-ft. at 2000 rpm. Five main bearings. Hydraulic valve lifters. Carburetor: Carter BBD 8291S two-barrel. VIN Code: L.

1983 DODGE

Dodge Diplomat, Mirada base V-8
Overhead valve. Cast-iron block and heads. Bore & stroke: 3.91 x 3.31 in. Displacement: 318 cid. (5.2 liters). Compression ratio: 8.5:1. Net brake hp: 130 at 4000 rpm. Net torque: 230 lbs.-ft. at 1600 rpm. Five main bearings. Hydraulic valve lifters. Carburetor: Carter BBD two-barrel. VIN Code: P.

1984 DODGE

Dodge Diplomat base V-8
Overhead valve. Cast-iron block and heads. Bore & stroke: 3.91 x 3.31 in. Displacement: 318 cid. (5.2 liters). Compression ratio: 8.5:1. Net brake hp: 130 at 4000 rpm. Net torque: 235 lbs.-ft. at 1600 rpm. Five main bearings. Hydraulic valve lifters. Carburetor: Carter BBD two-barrel. VIN Code: P.

Dodge Diplomat police V-8
Overhead valve. Cast-iron block and heads. Bore & stroke: 3.91 x 3.31 in. Displacement: 318 cid. (5.2 liters). Compression ratio: 8.5:1. Net brake hp: 165 at 4000 rpm. Net torque: 240 lbs.-ft. at 2000 rpm. Five main bearings. Hydraulic valve lifters. Carburetor: Carter TQ 9295 four-barrel. VIN Code: P.

1985 DODGE

Dodge Diplomat base V-8
Overhead valve. Cast-iron block and heads. Bore & stroke: 3.91 x 3.31 in. Displacement: 318 cid. (5.2 liters). Compression ratio: 8.7:1. Net brake hp: 140 at 3600 rpm. Net torque: 265 lbs.-ft. at 1600 rpm. Five main bearings. Hydraulic valve lifters. Carburetor: Holley 6280 two-barrel. VIN Code: P.

Dodge Diplomat police V-8
Overhead valve. Cast-iron block and heads. Bore & stroke: 3.91 x 3.31 in. Displacement: 318 cid. (5.2 liters). Compression ratio: 8.5:1. Net brake hp: 175 at unknown rpm. Net torque: unknown at unknown rpm. Five main bearings. Hydraulic valve lifters. Carburetor: Four-barrel. VIN Code: P.

1986 DODGE

Dodge Diplomat base V-8
Overhead valve. Cast-iron block and heads. Bore & stroke: 3.91 x 3.31 in. Displacement: 318 cid. (5.2 liters). Compression ratio: 9.0:1. Net brake hp: 140 at 3600 rpm. Net torque: 265 lbs.-ft. at 1600 rpm. Five main bearings. Hydraulic valve lifters. Carburetor: Holley 6280 two-barrel. VIN Code: P.

Dodge Diplomat police V-8
Overhead valve. Cast-iron block and heads. Bore & stroke: 3.91 x 3.31 in. Displacement: 318 cid. (5.2 liters). Compression ratio: 8.5:1. Net brake hp: 175 at unknown rpm. Net torque: unknown at unknown rpm. Five main bearings. Hydraulic valve lifters. Carburetor: Four-barrel. VIN Code: P.

1987 DODGE

Dodge Diplomat base V-8
Overhead valve. Cast-iron block and heads. Bore & stroke: 3.91 x 3.31 in. Displacement: 318 cid (5.2 liters). Compression ratio: 9.0:1. Net brake hp: 140 at 3600 rpm. Net torque: 265 lbs.-ft. at 1600 rpm. Five main bearings. Hydraulic valve lifters. Carburetor: Four-barrel. VIN Code: P.

1988 DODGE

Dodge Diplomat base V-8
Overhead valve. Cast-iron block and heads. Bore & stroke: 3.91 x 3.31 in. Displacement: 318 cid. (5.2 liters). Compression ratio: 9.1:1. Net brake hp: 140 at 3600 rpm. Net torque: 265 lbs.-ft. at 1600 rpm. Five main bearings. Hydraulic valve lifters. Carburetor: Two-barrel. VIN Code: P.

Dodge Diplomat police V-8
Overhead valve. Cast-iron block and heads. Bore & stroke: 3.91 x 3.31 in. Displacement: 318 cid. (5.2 liters). Compression ratio: 8.5:1. Net brake hp: 175 at unknown rpm. Net torque: unknown at unknown rpm. Five main bearings. Hydraulic valve lifters. Carburetor: Four-barrel. VIN Code: P.

1989 DODGE

Dodge Diplomat base V-8
Overhead valve. Cast-iron block and heads. Bore & stroke: 3.91 x 3.31 in. Displacement: 318 cid. (5.2 liters). Compression ratio: 9.1:1. Net brake hp: 140 at 3600 rpm. Net torque: 265 lbs.-ft. at 1600 rpm. Five main bearings. Hydraulic valve lifters. Carburetor: Two-barrel. VIN Code: P.

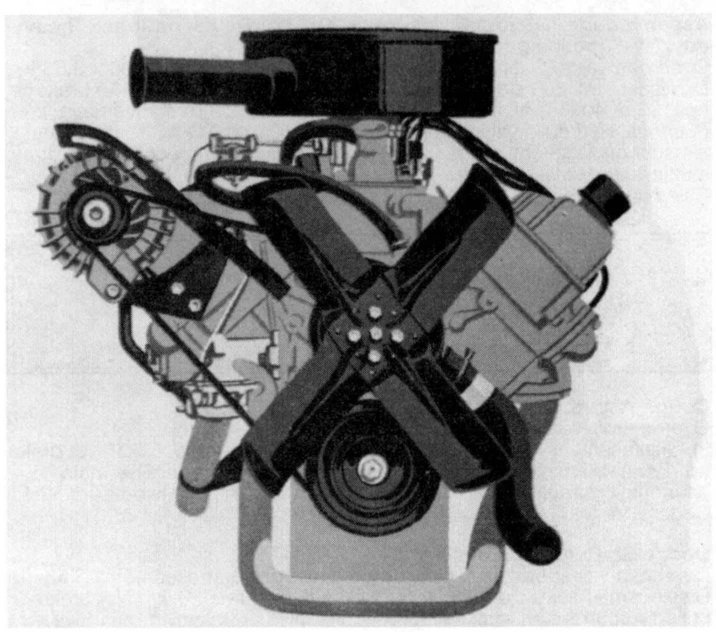

The 318-cid V-8 was available with 200 hp in Dodge trucks and with a choice of 230 or 260 hp in cars. The more powerful car engine carried a four-barrel carburetor instead of the base two-barrel carburetor. Features of the 318 included high-compression combustion chambers, a rotary-type oil pump, silent-chain camshaft drive, free-turning exhaust valves, and a drop-forged crankshaft.

Selected Dodge V-8 Tune-Up Specs 1953-1965

Year	Model	Spark Plugs Type	Gap	Point Gap	Deg Cam Angle	Timing	Idle Speed	Compression Pressure	Battery (Delco)
1953	Red Ram V-8	AL-4S140	.035	.015-.018	NOTE 1	4B	475 rpm	150	Group 1
1954	Red Ram V-8	AL-4S140	.035	.015-.018	NOTE 1	4B	475 rpm	155	Group 1
1955	D-55-1, D-55-2	AL-4S165	.035	.015-.018	NOTE 1	4B	475 rpm	155	Group 1
	D-55-3	AL-4S165	.035	.015-.018	NOTE 1	4B	475 rpm	155	Group 1

NOTE 1: These 1953-1955 Dodge "Red Ram" V-8s have two sets of distributor points. The total cam angle is 32-36 degrees and for each point set the cam angle should be set at 26-28 degrees.
NOTE 2: D-55-1 and D-55-2 have single rocker shaft; the D-55-3 has double rocker shafts.

Year	Model	Spark Plugs Type	Gap	Point Gap	Deg Cam Angle	Timing	Idle Speed	Compression Pressure	Battery (Delco)
1956	D-63-1	AL-AR52	.035	.015-.018	31	4B	475 rpm	155	2SM/11HS
	D-63-2, D-63-3	AL-AR52	.035	.015-.018	31	6B	475 rpm	155	2SM/11HS
	D-500	AL-4S250	.035	.015-.018	38	2B	475 rpm	165	2SM/11HS

NOTE 3: D-63-1, D-63-2 and D-63-3 have single rocker shaft; D-500 has double rocker shafts.

Year	Model	Spark Plugs Type	Gap	Point Gap	Deg Cam Angle	Timing	Idle Speed	Compression Pressure	Battery (Delco)
1957	D66 to D71	AL-AR42	.035	.015-.018	31	6B	475 rpm	155	2SM/11HS
	D-500	AL-AR42	.035	.015-.018	NOTE 4	6B	475 rpm	165	2SM/11HS

NOTE 4: The 1957 Dodge D-500 with a single four-barrel carburetor had a single set of points and cam angle should be set to 29-32 degrees. The 1957 Dodge D-500 with dual four-barrel carburetors had a dual points distributor and the cam angle for each set of points should be set to 29-32 degrees to give an overall cam angle setting of 36-39 degrees. The D-500 also had double rocker shafts, while the D-66-1, D-66-2, D-67-1, D-70 and D-71 V-8s had a single rocker shaft setup.

Year	Model	Spark Plugs Type	Gap	Point Gap	Deg Cam Angle	Timing	Idle Speed	Compression Pressure	Battery (Delco)
1958	LD2	AL-AGR42	.035	.015-.018	29	6B	500 rpm	155	11-HS-60
	LD3	AL-AR42	.035	.015-.018	29	6B	500 rpm	165	11-HS-60
1959	Royal (MD3-H)	AL-AR42	.035	.015-.018	29	10B	500 rpm	155	HO-11-50
	Royal (MD3-M)	AL-A42	.035	.015	29	10B	500 rpm	155	HO-11-60
	Custom Royal	AL-A42	.035	.015	29	10B	500 rpm	155	HO-11-60
	Sta. Wag., D-500	AL-A42	.035	.015	29	10B	500 rpm	165	HO-11-60
	Coronet, D-500	AL-A32	.035	.015-.018	29-34	10B	500 rpm	165	HO-11-50
1960	Dodge Dart (man.)	AL-AR42	.035	.015-.018	29-34	5B	500 rpm	125-165	HO-11-50
	Dodge Dart (auto.)	AL-AR42	.035	.015-.018	29-34	10B	500 rpm	125-165	HO-11-50
	Matador	AL-A42	.035	.014-.019	26-32	10B	500 rpm	125-165	HO-11-60
	Polara	AL-A42	.035	.014-.019	26-32	10B	500 rpm	125-165	HO-11-60
	D-500	AL-A32	.035	.014-.019	26-32	7.5B	725 rpm	125-165	HO-11-60
	Dart (Hi-Perf.)	AL-A32	.035	.014-.019	26-32	10B	500 rpm	150-180	HO-11-50
	Dodge (Hi-Perf.)	AL-A32	.035	.014-.019	26-32	10B	500 rpm	150-180	HO-11-50
1961	Dart (man.)	AL-AR42	.035	.014-.019	27-32	5B	500 rpm	135-165	MB-24-50
	Dart (auto.)	AL-AR42	.035	.014-.019	27-32	10B	500 rpm	135-165	MB-24-50
	Polara	AL-A42	.035	.015-.018	29-34	10B	500 rpm	150-180	11-HS-59B
	D-500	AL-A32	.035	.015-.018	29-34	7.5B	725 rpm	150-180	11-HS-59B
	Dart (Hi-Perf.)	AL-A32	.035	.015-.018	31-37	7.5B/10B	500 rpm	150-180	HO-11-60
	Dodge (Hi-Perf.)	AL-A32	.035	.015-.018	31-37	7.5B/10B	500 rpm	150-180	HO-11-60

Selected Dodge V-8 Tune-Up Specs 1953-1965

Year	Model	Spark Plugs Type	Gap	Point Gap	Deg Cam Angle	Timing	Idle Speed	Compression Pressure	Battery (Delco)
1962	Dart (man.)	AL-AR42	.035	.014-.019	27-32	5B	500 rpm	135-165	11-HS-48B
	Dart (auto.)	AL-AR42	.035	.014-.019	27-32	10B	500 rpm	135-165	11-HS-48B
	Polara	AL-A42	.035	.014-.019	29	10B	500 rpm	125-155	11-HS-59B
	D-500	CH-J12Y	.035	.014-.019	29	10B	500 rpm	125-155	11-HS-59B
1963	318 (man.)	CH-J12Y	.035	.017	28-33	5B	500 rpm	135	48 amp hr
	318 (auto.)	CH-J12Y	.035	.017	28-33	10B	500 rpm	135	48 amp hr
	361, 383 2-barrel	AL-A42	.035	.017	28-33	10B	500 rpm	140	59 amp hr
	383 4-barrel	CH-J12Y	.035	.017	NOTE 5	10B	500 rpm	145	59 amp hr
	426	CH-J12Y	.035	.017	NOTE 5	10B	500 rpm	145	70 amp hr

NOTE 5: For 1961 to 1963 dual point distributor set at 34-40 degrees overall, 27-32 degrees for each set.

Year	Model	Spark Plugs Type	Gap	Point Gap	Deg Cam Angle	Timing	Idle Speed	Compression Pressure	Battery (Delco)
1964	318 (man.)	CH-J12Y	.035	.017	30	5B	550 rpm	135	z8 amp hr
	318 (auto.)	CH-J12Y	.035	.017	30	10B	475 rpm	135	48 amp hr
	361	CH-J12Y	.035	.017	30	10B	500 rpm	140	59 amp hr
	383 2-barrel	CH-J12Y	.035	.017	30	10B	500 rpm	145	59 amp hr
	383 4-barrel	CH-J10Y	.035	.017	NOTE 6	10B	500 rpm	145	59 amp hr
	413	CH-J10Y	.035	.017	30	10B	500 rpm	145	59 amp hr
	426	CH-N14Y	.035	.017	NOTE 6	10B	500 rpm	145	70 amp hr
	426 Hemi	CH-N61Y	.020	.012	NOTE 6	31B	1400 rpm	145	70 amp hr
1965	318 (man.)	CH-J14Y	.035	.017	30	5B	550 rpm	140	48 amp hr
	318 (auto.)	CH-J14Y	.035	.017	30	105B	475 rpm	140	48 amp hr
	361	CH-J14Y	.035	.017	30	10B	500 rpm	140	59 amp hr
	383 2-barrel	CH-J14Y	.035	.017	NOTE 6	10B	550 rpm	145	59 amp hr
	383 4-barrel	CH-J10Y	.035	.017	NOTE 6	10B	500 rpm	145	59 amp hr
	413 Standard	CH-J14Y	.035	.017	30	12.5B	500 rpm	145	59 amp hr
	413 Spec. Cam.	CH-J10Y	.035	.017	NOTE 6	10B	500 rpm	145	59 amp hr
	426 Hemi	CH-J10Y	.035	.017	NOTE 6	10B	500 rpm	145	70 amp hr

NOTE 6: For 1964 to 1965 dual point distributor set at 37-42 degrees overall, 27-32 degrees for each set.

Dodge and Dodge Dart V-8 Firing Order

1953-1976
1-8-4-3-6-5-7-2

All Dodge and Dodge Dart V-8s have the cylinders numbered from front to back. The cylinders on the left-hand bank are numbered 1-3-5-7 and the cylinders on the right-hand bank are numbered 2-4-6-8.

EDSEL
1958-1960

Henry Ford II knew that Ford Motor Co. was lacking the same type of "model ladder" that GM benefited from. This was the seed for the Edsel. The Korean War interrupted efforts to develop this market-niche marque, but in 1955, a Special Products Div. was formed to develop it.

The Special Products Div. was not an arm of Ford or Mercury, but shared basic components with both. Under the leadership of R.E. Krafve, it developed the E-car ("E" meant experimental). The Foote, Cone & Belding ad agency created a list of 18,000 possible names, but on Nov. 19, 1956, after much negotiating with the Ford family, Edsel was picked.

There were four Edsel series. The Ranger was the price leader, followed by the Pacer. Both were Ford Fairlane based. The Corsair and Citation models were based on the 1957-1958 Mercury with a special 124-in. wheelbase. Production began July 15, 1957. Ford-based models were made in Mahwah, N.J., Louisville, Ky. and San Jose, Calif. Mercury-based cars were built at Somerville, Mass. By Aug. 19, 1957 the Wayne, Mich. and Los Angeles, Calif. plants were added.

The Edsel was billed as "The Newest Thing on Wheels." The junior Edsels were powered by the E-400 V-8, with 361 cid and a 303-hp rating. It was part of the FE series of Ford engines introduced in 1958. For the senior Corsair and Citation models, the E-475 with 410 cid and 345 hp kept them moving. There were no optional V-8s for 1958. By October 1957, the sales picture was bleak. By January 1958, the Edsel Div. was reorganized and placed in the new Mercury-Edsel-Lincoln Div. of Ford Motor Co.

The Edsel lineup was cut back to 10 models for 1959. Production was limited to the Louisville plant. Standard power for the Ranger series was now the 292-cid Y-block V-8 available in Fords since 1955. Standard in Villager station wagons and the Corsair series was the 332-cid V-8, also a Ford engine. An extra-cost option was the 361-cid V-8 carried over from 1958.

On Oct. 15, 1959, the 1960 Edsels were put on sale. They were little more than face-lifted Fords. The standard engine for all models was the 292-cid V-8. A 300-hp version of the Ford 352-cid V-8 was available as an extra-cost option in. Sales remained dismal and on Nov. 19, 1959, Ford announced the end of the Edsel.

1958 EDSEL

Ranger, Pacer base V-8
Overhead valve. Cast-iron block. Bore & stroke: 4.05 x 3.50 in. Displacement: 361 cid. Compression ratio: 10.50:1. Brake hp: 303 at 4600 rpm. Taxable hp: 52.49. Torque: 400 lbs.-ft. at 2800 rpm. Five main bearings. Solid valve lifters. Crankcase capacity: 5 qt. (add 1 qt. with new oil filter). Cooling system capacity (add 1 qt. for heater): 18.5 qt. Carburetor: (Manual transmission) Holley R-1482-A four-barrel; (Automatic transmission) Holley R-1483-A four-barrel. Engine code: W.

Corsair, Citation base V-8
Overhead valve. Cast-iron block. Bore & stroke: 4.20 x 3.70 in. Displacement: 410 cid. Compression ratio: 10.50:1. Brake hp: 345 at 4600 rpm. Taxable hp: 56.45. Torque: 475 lbs.-ft. at 2900 rpm. Five main bearings. Solid valve lifters. Crankcase capacity: 5 qt. (add 1 qt. with new oil filter). Cooling system capacity (add 1 qt. for heater): 22 qt. Carburetor: (Automatic transmission only) Holley R-1619-A four-barrel. Engine code: X.

1959 EDSEL

Ranger, Villager base V-8
Overhead valve. Cast-iron block. Bore & stroke: 3.75 x 3.30 in. Displacement: 292 cid. Compression ratio: 8.80:1. Brake hp: 200 at 4400 rpm. Taxable hp: 45.0. Torque: 285 lbs.-ft. at 2200 rpm. Five main bearings. Solid valve lifters. Crankcase capacity: 5 qt. (add 1 qt. with new oil filter). Cooling system capacity (add 1 qt. for heater): 19 qt. Carburetor: Ford B9A-9510-A two-barrel Model 5752306 or Model 5752307. Engine code: C.

Corsair base V-8; Ranger optional V-8
Overhead valve. Cast-iron block. Bore & stroke: 4.00 x 3.30 in. Displacement: 332 cid. Compression ratio: 8.90:1. Brake hp: 225 at 4400 rpm. Taxable hp: 51.2. Torque: 325 lbs.-ft. at 2200 rpm. Five main bearings. Hydraulic valve lifters. Crankcase capacity: 5 qt. (add 1 qt. with new oil filter). Cooling system capacity (add 1 qt. for heater): 19 qt. Carburetor: Holley two-barrel Model R-1843A with standard transmission, R-1844A with automatic transmission or R-1929-AAS as a field service unit. Engine code: B.

Ranger, Corsair, Villager optional V-8
Overhead valve. Cast-iron block. Bore & stroke: 4.05 x 3.50 in. Displacement: 361 cid. Compression ratio: 9.60:1. Brake hp: 303 at 4600 rpm. Taxable hp: 42. Torque: 390 lbs.-ft. at 2900 rpm. Five main bearings. Solid valve lifters. Crankcase capacity: 5 qt. (add 1 qt. with new oil filter). Cooling system capacity (add 1 qt. for heater): 19 qt. Carburetor: (Manual transmission and Canada) Holley R-1847A four-barrel; (Automatic transmission and filed service unit) Holley 1848A four-barrel. Engine code: W.

1960 EDSEL

Ranger, Villager base V-8
Overhead valve. Cast-iron block. Bore & stroke: 3.75 x 3.30 in. Displacement: 292 cid. Compression ratio: 8.80:1. Brake hp: 185 at 4200 rpm. Taxable hp: 45.0. Torque: 292 lbs.-ft. at 2200 rpm. Five main bearings. Solid valve lifters. Crankcase capacity: 5 qt. (add 1 qt. with new oil filter). Cooling system capacity (add 1 qt. for heater): 19 qt. Carburetor: (Manual transmission) Ford COAE-9510-AF two-barrel; (Automatic transmission) COAE-9510-AG. Engine code: W.

Ranger, Villager optional V-8
Overhead valve. Cast-iron block. Bore & stroke: 4.00 x 3.50 in. Displacement: 352 cid. Compression ratio: 9.60:1. Brake hp: 300 at 4600 rpm. Torque: 381 lbs.-ft. at 2800 rpm. Five main bearings. Hydraulic valve lifters. Carburetor: Crankcase capacity: 5 qt. (add 1 qt. with new oil filter). Cooling system capacity (add 1 qt. for heater): 19.5 qt. Dual exhaust. Carburetor: Ford-Holley four-barrel Model COAE-9510-K. Engine code: Y.

Selected Edsel V-8 Tune-Up Specs 1958-1960

	Spark Plugs	Point	Cam		Idle	Compression		Battery	
Year	Model	Type	Gap	Gap	Angle	Timing	Speed	Pressure	(Delco)
1958	V-8 (361)	F-11-Y	.032-.036	.014-.016	26-28.5	3/6B	Note 3	160 lb.	55/65 amp-hr
	V-8 (410)	F-11-Y	.032-.036	.014-.016	26-28.5	3/6B	Note 3	160 lb.	55/65 amp-hr

Note 1: Timing is 3B for manual transmission; 6B for automatic transmission.
Note 2: 55 amp-hr battery for manual transmission; 65 amp-hr battery for automatic transmission.
Note 3: Idle speed 475-500 rpm with manual transmission; 450-475 rpm with automatic transmission (in drive).

1959	V-8 (292)	FEH-12405A	.035	.014-.016	26-28.5	3/6B	Note 5	160 lb.	21-66-55AH
	V-8 (332)	FEV-12405A	.035	.014-.016	26-28.5	3/6B	Note 5	160 lb.	21-66-55AH
	V-8 (361)	FEV-12405A	.035	.014-.016	26-28.5	3/6B	Note 5	160 lb.	21-66-55AH

Note 4: Timing is 3B for manual transmission; 6B for automatic transmission.
Note 5: Idle speed 475-500 rpm with manual transmission; 450-475 rpm with automatic transmission (in drive).

1960	V-8 (292)	F-14-Y	.032	.014-.016	26-28.5	3/6B	Note 7	160 lb.	B9A-10654D-E-F
	V-8 (352)	F-11-Y	.032	.014-.016	26-28.5	3/6B	Note 7	180 lb.	B9A-10654D-E-F

Note 6: Timing is 3B for manual transmission; 6B for automatic transmission; 2-degree minimum advance, 10-degree maximum advance both engines.
Note 7: Idle speed 500-525 rpm with manual transmission in neutral; 450 rpm with automatic transmission in drive.

Edsel V-8 Firing Order

1958-1960 (Overhead valve V-8s)

292-cid: 1-5-4-8-6-3-7-2 with RH bank numbered 1-2-3-4 and LH bank numbered 5-6-7-8 front to rear.

Others: 1-5-4-2-6-3-7-8 with RH bank numbered 1-2-3-4 and LH bank numbered 5-6-7-8 front to rear.

FORD
1932-2002

1932 FORD

Model V-8-106/Model 18
L-head. Cast-iron block. Bore & stroke: 3-1/16 x 3-3/4 in. Displacement: 221 cid. Compression ratio: 5.50:1. Brake hp: 65 at 3400 rpm. Taxable hp: 30. Torque: 114 lbs.-ft. at 1800 rpm. Three main bearings. Solid valve lifters. Crankcase capacity: 5 qt. Cooling system capacity: 22 qt. Carburetor: Special Ford Detroit Lubricator Model 18-9510 single-barrel downdraft (1.50-in. throat). Engine number same as serial number. Serial number range approximately 18-1 to 203127.

1933 FORD

Model V-8-112/Model 40
L-head. Cast-iron block. Bore & stroke: 3-1/16 x 3-3/4 in. Displacement: 221 cid. Compression ratio: 6.33:1. Brake hp: 75 at 3800 rpm. Taxable hp: 30. Torque: 138 lbs.-ft. at 1800 rpm. Three main bearings. Solid valve lifters. Crankcase capacity: 5 qt. Cooling system capacity: 22 qt. Carburetor: Special Ford Detroit Lubricator Model 40-9510 single-barrel downdraft (1.25-in. throat). Engine number same as serial number. Serial number range 18-203127 to 457447.

1934 FORD

Model V-8-112/Model 40 (V-8 85)
L-head. Cast-iron block. Bore & stroke: 3-1/16 x 3-3/4 in. Displacement: 221 cid. Compression ratio: 6.33:1. Brake hp: 85 at 3800 rpm. Taxable hp: 30. Torque: 150 lbs.-ft. at 2200 rpm. Three main bearings. Solid valve lifters. Crankcase capacity: 5 qt. Cooling system capacity: 22 qt. Carburetor: Stromberg EE-1 two-barrel downdraft. Engine number same as serial number. Serial number range 18-457478 to 18-1234356.

1935 FORD

Model 48 V-8 (V-8 85)
L-head. Cast-iron block. Bore & stroke: 3-1/16 x 3-3/4 in. Displacement: 221 cid. Compression ratio: 6.33:1. Brake hp: 85 at 3800 rpm. Taxable hp: 30. Torque: 148 lbs.-ft. at 2200 rpm. Three main bearings. Solid valve lifters. Crankcase capacity: 5 qt. Cooling system capacity: 20 qt. Carburetor: Stromberg EE-1 two-barrel downdraft. Engine number same as serial number. Serial number range 18-1234357 to approximately 18-2207110.

1936 FORD

Model 68 V-8 (V-8 85)
L-head. Cast-iron block. Bore & stroke: 3-1/16 x 3-3/4 in. Displacement: 221 cid. Compression ratio: 6.3:1. Brake hp: 85 at 3800 rpm. Taxable hp: 30. Torque: 148 lbs.-ft. at 2200 rpm. Three main bearings. Solid valve lifters. Crankcase capacity: 5 qt. Cooling system capacity: 22 qt. Carburetor: Stromberg EE-1 two-barrel downdraft. Engine number same as serial number. Serial number range 18-2207111 to approximately 18-3331856.

1937 FORD

Model 78 (V-8 85)
L-head. Cast-iron block. Bore & stroke: 3-1/16 x 3-3/4 in. Displacement: 221 cid. Compression ratio: 6.3:1. Brake hp: 85 at 3800 rpm. Taxable hp: 30. Torque: 154 lbs.-ft. at 2200 rpm. Three main bearings. Solid valve lifters. Crankcase capacity: 5 qt. Cooling system capacity: 22 qt. Carburetor: Stromberg 67-9510A two-barrel downdraft. Engine number same as serial number. Serial number range 18-4186447 to approximately 18-4661000.

Model 74 (V-8 60)
L-head. Cast-iron block. Bore & stroke: 2.6 x 3.2 in. Displacement: 136 cid. Compression ratio: 6.6:1. Brake hp: 60 at 3500 rpm. Taxable hp: 21.6. Torque: 94 lbs.-ft. at 2500 rpm. Three main bearings. Solid valve lifters. Crankcase capacity: 4 qt. Cooling system capacity: 15 qt. Carburetor: Stromberg 922A-9510A two-barrel downdraft. Engine number same as serial number. Serial number range 54-358335 and up. **NOTE:** Sheet metal water jacket; 1937 only.

1938 FORD

Model 81A (V-8 85)
L-head. Cast-iron block. Bore & stroke: 3-1/16 x 3-3/4 in. Displacement: 221 cid. Compression ratio: 6.12:1. Brake hp: 85 at 3800 rpm. Taxable hp: 30. Torque: 155 lbs.-ft. at 2200 rpm. Three main bearings. Solid valve lifters. Crankcase capacity: 5 qt. Cooling system capacity: 22 qt. Carburetor: Stromberg 21A-9510A two-barrel downdraft. Engine number same as serial number. Serial number range 18-4186447 and up.

Model 82A (V-8 60)
L-head. Cast-iron block. Bore & stroke: 2.6 x 3.2 in. Displacement: 136 cid. Compression ratio: 6.6:1. Brake hp: 60 at 3500 rpm. Taxable hp: 21.6. Torque: 94 lbs.-ft. at 2500 rpm. Three main bearings. Solid valve lifters. Crankcase capacity: 4 qt. Cooling system capacity: 15.25 qt. Carburetor: Stromberg 922A-9510A two-barrel downdraft. Engine number same as serial number. Serial number range 54-358335 and up.

1939 FORD

Model 91A
L-head. Cast-iron block. Bore & stroke: 3-1/16 x 3-3/4 in. Displacement: 221 cid. Compression ratio: 6.20:1. Brake hp: 90 at 3800 rpm. Taxable hp: 30. Torque: 154 lbs.-ft. at 2200 rpm. Three main bearings. Solid valve lifters. Crankcase capacity: 5 qt. Cooling system capacity: 22 qt. Carburetor: Holley (Chandler-Groves) Model AA-1 or Ford Model 91A-9510-A two-barrel downdraft. Engine number same as serial number. Serial number range: 18-4661001 to approximately 18-5210700.

Model 922A (V-8 60)
L-head. Cast-iron block. Bore & stroke: 2.6 x 3.2 in. Displacement: 136 cid. Compression ratio: 6.6:1. Brake hp: 60 at 3500 rpm. Taxable hp: 21.6. Torque: 94 lbs.-ft. at 2500 rpm. Three main bearings. Solid valve lifters. Crankcase capacity: 4 qt. Cooling system capacity: 15 qt. Carburetor: Stromberg EE-7/8 or Holley (Chandler-Groves) Model AA-7/8 two-barrel downdraft. Engine number same as serial number. Serial number range: continued from 1938 to approximately 54-506500.

1940 FORD

Model 01A
L-head. Cast-iron block. Bore & stroke: 3-1/16 x 3-3/4 in. Displacement: 221 cid. Compression ratio: 6.20:1. Brake hp: 90 at 3800 rpm. Torque: 154 lbs.-ft. at 2200 rpm. Taxable hp: 30. Three main bearings. Solid valve lifters. Cooling system capacity: 22 qt. Crankcase capacity: 5 qt. Carburetor: Holley (Chandler-Groves) Model AA-1 or Ford Model 21A-9510-A two-barrel downdraft. Engine number same as serial number. Serial number range: 18-5210701 to approximately 18-5896294.

Model 02A (V-8 60)
L-head. Cast-iron block. Bore & stroke: 2.6 x 3.2 in. Displacement: 136 cid. Compression ratio: 6.6:1. Brake hp: 60 at 3500 rpm. Taxable hp: 21.6. Torque: 94 lbs.-ft. at 2500 rpm. Three main bearings. Solid valve lifters. Crankcase capacity: 4 qt. Cooling system capacity: 15 qt. Carburetor: Stromberg EE-7/8 or Holley (Chandler-Groves) Model AA-7/8 or Ford 922A-9510-A two-barrel downdraft. Engine number same as serial number. Serial number range approximately 54-506501 and up.

1941 FORD

Model 11A
L-head. Cast-iron block. Bore & stroke: 3-1/16 x 3-3/4 in. Displacement: 221 cid. Compression ratio: 6.20:1. Brake hp: 90 at 3800 rpm. Torque: 154 at 2200 rpm. Taxable hp: 30. Three main bearings. Solid valve lifters. Crankcase capacity: 5 qt. Cooling system capacity: 22 qt. Carburetor: Holley (Chandler-Groves) Ford Model 91A-9510-A two-barrel downdraft. Engine number same as serial number. Serial number range approximately 18-5896295 and up.

1942 FORD

Model 21A
L-head. Cast-iron block. Bore & stroke: 3-1/16 x 3-3/4 in. Displacement: 221 cid. Compression ratio: 6.20:1. Brake hp: 90 at 3800 rpm. Torque: 156 at 2200 rpm. Taxable hp: 30. Three main bearings. Solid valve lifters. Crankcase capacity: 5 qt. Cooling system capacity: 22 qt. Carburetor:

"More power to you" heralded a 1946 *Life* magazine advertisement promoting Ford's new 100-hp V-8. The 1946 engine featured new four-ring aluminum pistons that made it thrifty on gas and oil. Another new feature of this engine was its "balanced carburetion." It used a Holley model 94 two-barrel carburetor.

"Ford's out front with a famous V-8. And a brilliant, new 6 Folks say is great!" was the theme of this 1946 *Life* magazine advertisement showing both engines available in Fords of that vintage. The V-8 had a 3.19 x 3.57-inch bore and stroke. It had 130 more cubic inches and 10 more horsepower than the six. Other specifications included a 6.6:1 compression ratio and three main bearings.

Holley (Chandler-Groves) Ford Model 21A-9510-A two-barrel downdraft. Engine number same as serial number. Serial number range approximately 18-6769036 and up.

1946 FORD

Model 69A
L-head. Cast-iron block. Bore & stroke: 3.19 x 3.75 in. Displacement: 239 cid. Compression ratio: 6.75:1. Brake hp: 100 at 3800 rpm. Torque: 180 at 2000 rpm. Taxable hp: 32.5. Three main bearings. Solid valve lifters. Crankcase capacity: 5 qt. (add 1 qt. with new oil filter). Cooling system capacity: 22 qt. Carburetor: Holley (Chandler-Groves) Ford No. 59A-9510-A two-barrel downdraft. Engine number same as serial number. Serial number range approximately 99A-650280 and up.

1947 FORD

Model 69A
L-head. Cast-iron block. Bore & stroke: 3.19 x 3.75 in. Displacement: 239 cid. Compression ratio: 6.75:1. Brake hp: 100 at 3800 rpm. Torque: 180 lbs.-ft. at 2000 rpm. Taxable hp: 32.5. Three main bearings. Solid valve lifters. Crankcase capacity: 5 qt. (add 1 qt. with new oil filter). Cooling system capacity: 22 qt. Carburetor: Holley (Chandler-Groves) Ford No. 59A-9510-A two-barrel downdraft. Engine number same as serial number. Serial number range approximately 799A-1412708 and up.

1948 FORD

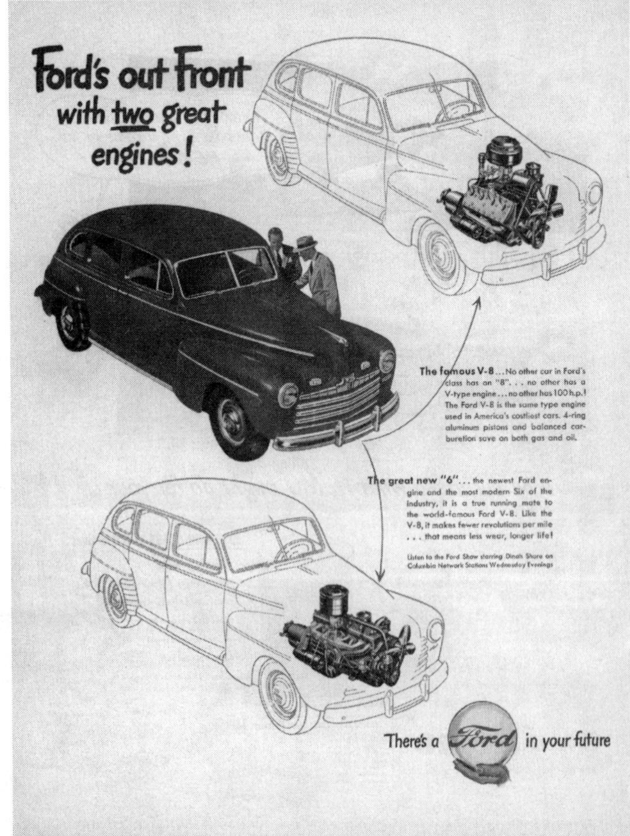

There were no important changes to the flathead Ford V-8 in 1947 or 1948. This advertising art comes from 1948, when the Dearborn automaker was sponsoring the Ford show, starring Dinah Shore, on Columbia Network radio stations on Wednesday evenings.

Model 89A
L-head. Cast-iron block. Bore & stroke: 3.19 x 3.75 in. Displacement: 239 cid. Compression ratio: 6.75:1. Brake hp: 100 at 3800 rpm. Taxable hp: 32.5. Torque: 180 lbs.-ft. at 2000 rpm. Three main bearings. Solid valve lifters. Crankcase capacity: 5 qt. (add 1 qt. with new oil filter). Cooling system capacity: 22 qt. Carburetor: Holley (Chandler-Groves) Ford No. 59A-9510-A two-barrel downdraft. Engine number same as serial number. Serial number range approximately 899A-198459 and up.

1949 FORD

Model 98BA
L-head. Cast-iron block. Bore & stroke: 3.19 x 3.75 in. Displacement: 239 cid. Compression ratio: 6.8:1. Brake hp: 100 at 3600 rpm. Taxable hp: 32.5. Torque: 181 lbs.-ft. at 2000 rpm. Three main bearings. Solid valve lifters. Crankcase capacity: 5 qt. (add 1 qt. with new oil filter). Cooling system capacity: 22 qt. Code: BA. Carburetor: Holley-Ford No. 8BA-9510-A two-barrel downdraft.

1950 FORD

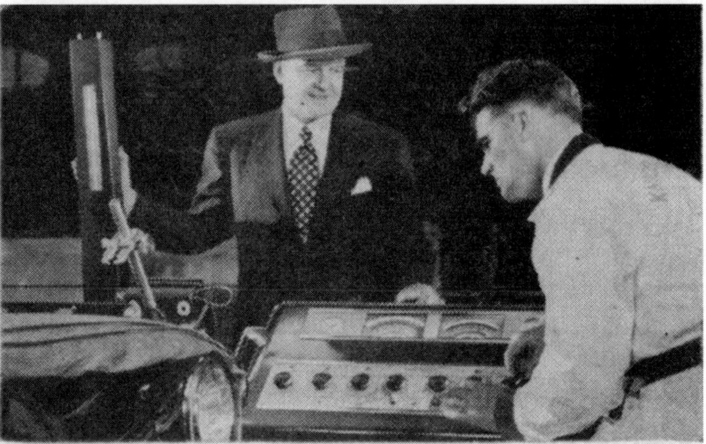

This image from a 1950 advertisement shows the type of tune-up equipment that Ford factory service technicians used to tune-up the flathead V-8 engines of that era. The 1949-1950 V-8s got a new Holley AA-1 two-barrel carburetor, but little else changed.

Model 0BA
L-head. Cast-iron block. Bore & stroke: 3.19 x 3.75 in. Displacement: 239 cid. Compression ratio: 6.8:1. Brake hp: 100 at 3600 rpm. Taxable hp: 32.5. Torque: 181 lbs.-ft. at 2000 rpm. Three main bearings. Solid valve lifters. Crankcase capacity: 5 qt. (add 1 qt. with new oil filter). Cooling system capacity: 22 qt. Carburetor: Ford AA-1 or Stromberg EE-1 two-barrel downdraft. Code: BA.

1951 FORD

Model 1BA
L-head. Cast-iron block. Bore & stroke: 3.19 x 3.75 in. Displacement: 239 cid. Compression ratio: 6.8:1. Brake hp: 100 at 3600 rpm. Taxable hp: 32.5. Torque: 181 lbs.-ft. at 2000 rpm. Three main bearings. Solid valve lifters. Crankcase capacity: 5 qt. (add 1 qt. with new oil filter). Cooling system capacity: 22 qt. Carburetor: Ford 8BA or Stromberg EE-1 two-barrel downdraft. Code: BA.

Six features of the 1951 Ford flathead V-8 were pointed out in this advertisement: 1) Silent-spin fan; 2) New waterproof ignition system; 3) Automatic Mileage Maker for gas savings that placed first in class in the Mobilgas Economy Run; 4) The only V-8 in the low-priced field (Plymouth and Chevy didn't offer V-8s yet); 5) Rota-quiet Valves; and 6) Super-fitted pistons. The Holley carburetor used in the 1949-1950 models was replaced with a new Ford 8BA two-barrel.

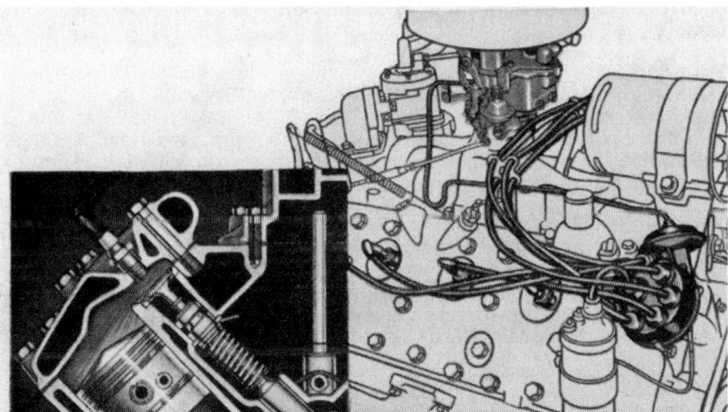

In 1951, Ford promoted its flathead V-8 with "Automatic Mileage Maker." This early engine management system was said to match ignition timing to fuel charges so that "every drop of gas is used – none wasted!" It promised "high-compression performance on regular gas." Also new was a waterproof ignition system that prevented moisture from shorting out the spark.

1952 FORD

Model B2
L-head. Cast-iron block. Bore & stroke: 3.19 x 3.75 in. Displacement: 239 cid. Compression ratio: 7.20:1. Brake hp: 110 at 3800 rpm. Taxable hp: 32.5. Torque: 196 lbs.-ft. at 2000 rpm. Three main bearings. Solid valve lifters. Crankcase capacity: 5 qt. (add 1 qt. with new oil filter). Cooling system capacity: 22 qt. Carburetor: Holley 2100 two-barrel downdraft. Code: B.

Ford called its 1952 engines the "Mileage Maker Six" and the "Strato-Star" V-8. The latter power plant was still of flathead design, although the industry was trending towards overhead-valve V-8s. To make the V-8 Ford more competitive in the sales race, the flathead was treated to a higher 7.2:1 compression ratio, which helped boost its output to 110 hp at 3800 rpm.

1953 FORD

Model B3
L-head. Cast-iron block. Bore & stroke: 3.19 x 3.75 in. Displacement: 239 cid. Compression ratio: 7.20:1. Brake hp: 110 at 3800 rpm. Taxable hp: 32.5. Torque: 196 lbs.-ft. at 2000 rpm. Three main bearings. Solid valve lifters. Crankcase capacity: 5 qt. (add 1 qt. with new oil filter). Cooling system capacity: 22 qt. Carburetor: Holley 2100 two-barrel downdraft. Code: B.

1954 FORD

The 1954 Ford Y-block V-8 was a rugged engine. It had new 3.50 x 3.10-inch bore and stroke dimensions, but displacement came out to be the same 239 cid as the flathead V-8. It had a 7.2:1 compression ratio and developed 130 hp at 4200 rpm. Of 1,165,942 Fords built in 1954, 863,096 had the V-8 engine installed.

Model U4
Overhead valve. Cast-iron block. Bore & stroke: 3.50 x 3.10 in. Displacement: 239 cid. Compression ratio: 7.20:1. Brake hp: 130 at 4200 rpm. Taxable hp: 39.2. Torque: 214 lbs.-ft. at 2000 rpm. Five main bearings. Solid valve lifters. Crankcase capacity: 5 qt. (add 1 qt. with new oil filter). Cooling system capacity: 20 qt. Carburetor: Holley two-barrel Model AA-1. Code: U.

The low-mounted generator introduced on the 1954 Ford Y-block V-8 remained a characteristic of Ford V-8s for many years. In Y-block engines the porting line is below the centerline of the crankshaft, which allows the use of main bearing webs with a much deeper cross section. Deep cast iron "skirts" surrounded the lower portion of the crankcase, making for a stronger, long-lasting engine.

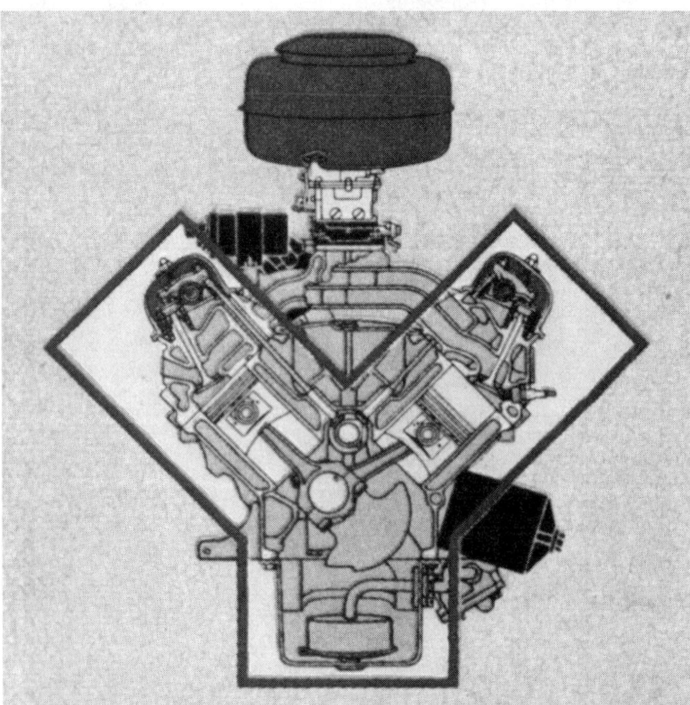

In 1954, Ford introduced a new overhead-valve V-8. It had the same displacement and compression ratio as the flathead, but the OHV design and a new Holley model AA-1 carburetor combined to raise horsepower to 130. In an attempt to differentiate its V-8 from the pack, Ford characterized its deep-block construction as the letter "Y" and referred to the engine as a Y-block.

Ford's bright-metal V-8 emblem was even placed against a red "Y" to emphasize that it was different than other V-8s. The 239-cid Y-block V-8 introduced in 1954 was a modern high-compression, low-friction design with deep-block construction that made it quieter and smoother running.

1955 FORD

Model U
Overhead valve. Cast-iron block. Bore & stroke: 3.62 x 3.30 in. Displacement: 272 cid. Compression ratio: 7.60:1. Brake hp: 162 at 4400 rpm. Taxable hp: 42.0. Torque: 258 lbs.-ft. at 2200 rpm. Five main bearings. Solid valve lifters. Crankcase capacity: 5 qt. (add 1 qt. with new oil filter). Cooling system capacity: 19 qt. Carburetor: Holley two-barrel. Code: U.

Model M (Special with Power Pack)
Overhead valve. Cast-iron block. Bore & stroke: 3.62 x 3.30 in. Displacement: 272 cid. Compression ratio: 7.60:1. Brake hp: 182 at 4400 rpm. Taxable hp: 42.0. Torque: 268 lbs.-ft. at 2600 rpm. Five main bearings. Solid valve lifters. Crankcase capacity: 5 qt. (add 1 qt. with new oil filter). Cooling system capacity: 19 qt. Carburetor: Holley four-barrel. Code: M.

Thunderbird
Overhead valve. Cast-iron block. Bore & stroke: 3.75 x 3.30 in. Displacement: 292 cid. Compression ratio: 8.1:1. Brake hp: 193 at 4400 rpm. Taxable hp: 45. Torque: 280 lbs.-ft. at 2600 rpm. Five main bearings. Solid valve lifters. Crankcase capacity: 5 qt. (add 1 qt. with new oil filter). Cooling system capacity: 19 qt. Carburetor: Holley four-barrel. Code: P.

Thunderbird (Ford-O-Matic)
Overhead valve. Cast-iron block. Bore & stroke: 3.75 x 3.30 in. Displacement: 292 cid. Compression ratio: 8.50:1. Brake hp: 198 at 4400 rpm. Taxable hp: 45. Torque: 286 lbs.-ft. at 2500 rpm. Five main bearings. Solid valve lifters. Crankcase capacity: 5 qt. (add 1 qt. with new oil filter). Cooling system capacity: 19 qt. Carburetor: Holley four-barrel. Code: P.

1956 FORD

Model U V-8 ((Manual transmission)
Overhead valve. Cast-iron block. Bore & stroke: 3.62 x 3.30 in. Displacement: 272 cid. Compression ratio: 8.40:1. Brake hp: 173 at 4400 rpm. Taxable hp: 42.0. Torque: 260 lbs.-ft. at 2400 rpm. Taxable hp: 42. Five main bearings. Solid valve lifters. Crankcase capacity: 5 qt. (add 1 qt. with new oil filter). Cooling system capacity: 19 qt. Carburetor: Two-barrel. Code: U.

Model U V-8 (Ford-O-Matic)
Overhead valve. Cast-iron block. Bore & stroke: 3.62 x 3.30 in. Displacement: 272 cid. Compression ratio: 8.40:1. Brake hp: 176 at 4400 rpm. Taxable hp: 42.0. Five main bearings. Solid valve lifters. Crankcase

Here's a good look at the stock, as-installed 1956 Thunderbird V-8. As you can see, the spark plugs are a little hard to get to and difficult to change. The base Thunderbird V-8 was a 292-cid 202-hp job. A 312-cid 225-hp engine was optional. Late in the year, a number of performance options were added, including dual four-barrel carburetors for the 312.

capacity: 5 qt. (add 1 qt. with new oil filter). Cooling system capacity: 19 qt. Carburetor: Two-barrel. Code: U.

Model M Fairlane V-8
Overhead valve. Cast-iron block. Bore & stroke: 3.75 x 3.30 in. Displacement: 292 cid. Compression ratio: 8.40:1. Brake hp: 200 at 4600 rpm. Taxable hp: 45. Torque: 285 lbs.-ft. at 2600 rpm. Taxable hp: 45.0. Five main bearings. Solid valve lifters. Crankcase capacity: 5 qt. (add 1 qt. with new oil filter). Cooling system capacity: 19 qt. Carburetor: Holley 4000 four-barrel. Code: M.

Thunderbird V-8
Overhead valve. Cast-iron block. Bore & stroke: 3.75 x 3.30 in. Displacement: 292 cid. Compression ratio: 8.40:1. Brake hp: 202 at 4600 rpm. Taxable hp: 45. Torque: 289 lbs.-ft. at 2600 rpm. Taxable hp: 45.0. Five main bearings. Solid valve lifters. Crankcase capacity: 5 qt. (add 1 qt. with new oil filter). Cooling system capacity: 19 qt. Carburetor: Holley

"The Thunderbird Special 225-hp V-8 engine can now be yours in most Fords," said this 1956 Ford advertisement. "And the 202-hp Thunderbird V-8 is available in all Fords." As you can see, the Fords did not get the chrome dress up items, but the "Thunderbird" name was easy to spot on the valve covers. The air cleaner is removed here.

Advertising artists depicted a well-dressed Ford salesman showing off the 292-cid 202-hp Thunderbird V-8, which was also available in Ford Fairlanes with Fordomatic transmission and in Ford station wagons. This engine carried a large air cleaner.

4000 four-barrel. Code: M.

Thunderbird Special V-8 (with manual transmission or overdrive)
Overhead valve. Cast-iron block. Bore & stroke: 3.60 x 3.44 in. Displacement: 312 cid. Compression ratio: 8.4:1. Brake hp: 215 at 4600 rpm. Taxable hp: 45. Torque: 317 lbs.-ft. at 2600 rpm. Five main bearings. Crankcase capacity: 5 qt. (add 1 qt. with new oil filter). Cooling system capacity: 19 qt. Carburetor: Holley 4000 four-barrel. Code: P.

Thunderbird Special V-8 (with Ford-O-Matic)
Overhead valve. Cast-iron block. Bore & stroke: 3.60 x 3.44 in. Displacement: 312 cid. Compression ratio: 9.0:1. Brake hp: 225 at 4600 rpm. Taxable hp: 45. Torque: 324 lbs.-ft. at 2600 rpm. Five main bearings. Crankcase capacity: 5 qt. (add 1 qt. with new oil filter). Cooling system capacity: 19 qt. Carburetor: Holley 4000 four-barrel. Code: L.

Thunderbird Special V-8 (with dual four-barrel carburetion)
Overhead valve. Cast-iron block. Bore & stroke: 3.60 x 3.44 in. Displacement: 312 cid. Compression ratio: 9.5:1. Brake hp: 260 at unknown rpm. Taxable hp: 45. Torque: unknown. Five main bearings. Crankcase capacity: 5 qt. (add 1 qt. with new oil filter). Cooling system capacity: 19 qt. Carburetor: Two Holley four-barrel. Code: P.

1957 FORD

Base V-8
Overhead valve. Cast-iron block. Bore & stroke: 3.62 x 3.30 in. Displacement: 272 cid. Compression ratio: 8.60:1. Brake hp: 190 at 4600 rpm. Torque: 270 lbs.-ft. at 2400 rpm. Taxable hp: 42. Five main bearings. Solid valve lifters. Crankcase capacity: 5 qt. (add 1 qt. with new oil filter). Cooling system capacity: 19 qt. Carburetor: Holley two-barrel. Code: B.

Base Thunderbird V-8; Fairlane V-8 (Manual)
Overhead valve. Cast-iron block. Bore & stroke: 3.75 x 3.30 in. Displacement: 292 cid. Compression ratio: 9.10:1. Brake hp: 206 at 4500 rpm. Torque: 297 lbs.-ft. at 2700 rpm. Taxable hp: 45.0. Five main bearings. Solid valve lifters. Crankcase capacity: 5 qt. (add 1 qt. with new oil filter). Cooling system capacity: 19 qt. Carburetor: Holley 4000 four-barrel. Code: C.

Base Thunderbird V-8; Fairlane V-8 (Ford-O-Matic)
Overhead valve. Cast-iron block. Bore & stroke: 3.75 x 3.30 in. Displacement: 292 cid. Compression ratio: 9.10:1. Brake hp: 212 at 4500 rpm. Torque: 297 lbs.-ft. at 2700 rpm. Taxable hp: 45.0. Five main bearings. Solid valve lifters. Crankcase capacity: 5 qt. (add 1 qt. with new

This fullhouse flathead Ford V-8 sports a dual-coil ignition system and "three deuces" (triple two-barrel carburetors). Note the front mounting of the dual-point distributor and the finned aluminum cylinder heads. Today, early speed equipment for flathead Ford V-8s is a hot-selling and pricey item at old-car swap meets.

oil filter). Cooling system capacity: 19 qt. Carburetor: Holley two-barrel. Code: C.

Thunderbird Special V-8
Overhead valve. Cast-iron block. Bore & stroke: 3.60 x 3.44 in. Displacement: 312 cid. Compression ratio: 9.70:1. Brake hp: 245 at 4500 rpm. Taxable hp: 45.0. Torque: 332 lbs.-ft. at 3200 rpm. Five main bearings. Crankcase capacity: 5 qt. (add 1 qt. with new oil filter). Cooling system capacity: 19 qt. Carburetor: Single Holley four-barrel. Code: D.

Thunderbird Super V-8 (with dual four-barrel carburetion)
Overhead valve. Cast-iron block. Bore & stroke: 3.60 x 3.44 in. Displacement: 312 cid. Compression ratio: 9.70:1 (10.0:1 with racing kit installed). Brake hp: 270 at 4500 rpm. (285 hp with racing kit installed). Taxable hp: 45.0. Torque: 332 lbs.-ft. at 2300 rpm. Five main bearings. Solid valve lifters. Crankcase capacity: 5 qt. (add 1 qt. with new oil filter). Cooling system capacity: 19 qt. Carburetor: Two Holley four-barrel. Code: E.

Thunderbird Special Supercharged V-8
Overhead valve. Cast-iron block. Bore & stroke: 3.60 x 3.44 in. Displacement: 312 cid. Compression ratio: 8.50:1. Brake hp: 300 at 4800 rpm. Taxable hp: 45.0. Torque: 340 lbs.-ft. at 5300 rpm. Five main bearings. Solid valve lifters. Crankcase capacity: 5 qt. (add 1 qt. with new oil filter). Cooling system capacity: 19 qt. Carburetor: Holley four-barrel with McCulloch/Paxton centrifugal supercharger. Code F.
NOTE: A 340-hp competition version of this engine was available for NASCAR racing.

1958 FORD

Custom 300, Fairlane Base V-8
Overhead valve. Cast-iron block. Bore & stroke: 3.75 x 3.30 in. Displacement: 292 cid. Compression ratio: 9.10:1. Brake hp: 205 at 4500 rpm. Taxable hp: 45.0. Torque: 295 lbs.-ft. at 2400 rpm. Five main bearings. Solid valve lifters. Crankcase capacity: 5 qt. (add 1 qt. with new oil filter). Cooling system capacity: 19 qt. Carburetor: Holley four-barrel Model R-1281-3A with standard transmission or R-1282-3A with automatic transmission. Code: C.

Fairlane 500 Interceptor V-8
Overhead valve. Cast-iron block. Bore & stroke: 4.00 x 3.30 in. Displacement: 332 cid. Compression ratio: 9.50:1. Brake hp: 240 at 4600 rpm. Taxable hp: 51.2. Torque: 360 lbs.-ft. at 2800 rpm. Five main bearings. Valve lifters: (early) solid; (late) hydraulic. Crankcase capacity: 5 qt. (add 1 qt. with new oil filter). Cooling system capacity: 19 qt. Dual exhaust. Carburetor: Holley four-barrel Model R-1406-A with standard transmission or R-1552-A with automatic transmission. Code: G.

Fairlane 500 Interceptor V-8 (solid lifters)
Overhead valve. Cast-iron block. Bore & stroke: 4.00 x 3.30 in. Displacement: 332 cid. Compression ratio: 9.60:1. Brake hp: 265 at 4600 rpm. Taxable hp: 51.2. Torque: 360 lbs.-ft. at 2800 rpm. Five main bearings. Solid valve lifters. Crankcase capacity: 5 qt. (add 1 qt. with new oil filter). Cooling system capacity: 19 qt. Dual exhaust. Carburetor: Holley two-barrel Model R-1406-A with standard transmission or R-1552-A with automatic transmission. Code: G.

Thunderbird V-8
Overhead valve. Cast-iron block. Bore & stroke: 4.00 x 3.50 in. Displacement: 352 cid. Compression ratio: 10.2:1. Brake hp: 300 at 4600 rpm. Taxable hp: 51.2. Torque: 395 lbs.-ft. at 2800 rpm. Five main bearings. Valve lifters: (early) solid; (late) hydraulic. Crankcase capacity: 5 qt. (add 1 qt. with new oil filter). Cooling system capacity: 19.5 qt. Dual exhaust. Carburetor: Ford-Holley Model B8A-9510-E four-barrel or Carter nos. 2640S-SA-SC four-barrel. Code: H.

Police Special V-8
Overhead valve. Cast-iron block. Bore & stroke: 4.05 x 3.50 in. Displacement: 361 cid. Compression ratio: 10.50:1. Brake hp: 303 at 4600 rpm. Taxable hp: 52.49. Torque: 400 lbs.-ft. at 2800 rpm. Five main bearings. Solid valve lifters. Crankcase capacity: 5 qt. (add 1 qt. with new oil filter). Cooling system capacity (Add 1 qt. for heater): 18.5 qt. Carburetor: (Manual transmission) Holley R-1482-A four-barrel; (Automatic transmission) Holley R-1483-A four-barrel. Engine code: W.

1959 FORD

Base V-8
Overhead valve. Cast-iron block. Bore & stroke: 3.75 x 3.30 in. Displacement: 292 cid. Compression ratio: 8.80:1. Brake hp: 200 at 4400 rpm. Taxable hp: 45.0. Torque: 285 lbs.-ft. at 2200 rpm. Five main bearings. Solid valve lifters. Crankcase capacity: 5 qt. (add 1 qt. with new oil filter). Cooling system capacity: 19 qt. Carburetor: Ford two-barrel Model 5752306 and 5752307. Code: C.

332 Special V-8
Overhead valve. Cast-iron block. Bore & stroke: 4.00 x 3.30 in. Displacement: 332 cid. Compression ratio: 8.90:1. Brake hp: 225 at 4400 rpm. Taxable hp: 51.2. Torque: 325 lbs.-ft. at 2200 rpm. Five main bearings. Hydraulic valve lifters. Crankcase capacity: 5 qt. (add 1 qt. with new oil filter). Cooling system capacity: 19 qt. Carburetor: Holley two-barrel Model R-1843A with standard transmission, R-1844A with automatic transmission or R-1929-AAS as a field service unit. Code: B.

Thunderbird 352 Special V-8
Overhead valve. Cast-iron block. Bore & stroke: 4.00 x 3.50 in. Displacement: 352 cid. Compression ratio: 10.2:1. Brake hp: 300 at 4600 rpm. Taxable hp: 51.2. Torque: 395 lbs.-ft. at 2800 rpm. Five main bearings. Hydraulic valve lifters. Crankcase capacity: 5 qt. (add 1 qt. with new oil filter). Cooling system capacity: 19.5 qt. Dual exhaust. Carburetor: Ford four-barrel Model 5752304 and 5752305 or Holley four-barrel. Code: H.

Thunderbird 430 Special V-8 (Thunderbird with Cruise-O-Matic only)
Overhead valve. Cast-iron block. Bore & stroke: 4.30 x 3.70 in. Displacement: 430 cid. Compression ratio: 10.0:1. Brake hp: 350 at 4600 rpm. Taxable hp: 51.2. Torque: 490 lbs.-ft. at 2800 rpm. Five main bearings. Hydraulic valve lifters. Crankcase capacity: 5 qt. (add 1 qt. with new oil filter). Cooling system capacity: 19.5 qt. Dual exhaust. Carburetor: Holley four-barrel Model 4160-C. Code: J.

1960 FORD

Base V-8
Overhead valve. Cast-iron block. Bore & stroke: 3.75 x 3.30 in. Displacement: 292 cid. Compression ratio: 8.80:1. Brake hp: 185 at 4200

rpm. Taxable hp: 45.0. Torque: 292 lbs.-ft. at 2200 rpm. Five main bearings. Solid valve lifters. Crankcase capacity: 5 qt. (add 1 qt. with new oil filter). Cooling system capacity: 19 qt. Carburetor: Ford-Holley two-barrel Model COAE-9510-M with standard transmission or COAE-9510-H with automatic transmission. Code: W.

Interceptor V-8
Overhead valve. Cast-iron block. Bore & stroke: 4.00 x 3.50 in. Displacement: 352 cid. Compression ratio: 8.90:1. Brake hp: 235 at 4400 rpm. Taxable hp: 51.2. Torque: 350 lbs.-ft. at 2400 rpm. Five main bearings. Hydraulic valve lifters. Crankcase capacity: 5 qt. (add 1 qt. with new oil filter). Cooling system capacity: 19.5 qt. Dual exhaust. Carburetor: Ford-Holley two-barrel Model COAE-9510-R with standard transmission or COAE-9510-S with automatic transmission. Code: X (Export code G).

Interceptor Special V-8; Base Thunderbird V-8
Overhead valve. Cast-iron block. Bore & stroke: 4.00 x 3.50 in. Displacement: 352 cid. Compression ratio: 9.60:1. Brake hp: 300 at 4600 rpm. Taxable hp: 51.2. Torque: 381 lbs.-ft. at 2800 rpm. Five main bearings. Hydraulic valve lifters. Crankcase capacity: 5 qt. (add 1 qt. with new oil filter). Cooling system capacity: 19.5 qt. Dual exhaust. Carburetor: Ford-Holley four-barrel Model COAE-9510-J with standard transmission or COAE-9510-K with automatic transmission. Code: Y.
NOTE: Not mentioned in early 1960 Ford literature. Aluminum intake. Holley carburetor. Not available in cars with power steering or power brakes.

Thunderbird Special V-8 (Thunderbird only)
Overhead valve. Cast-iron block. Bore & stroke: 4.30 x 3.70 in. Displacement: 430 cid. Compression ratio: 10.0:1. Brake hp: 350 at 4600 rpm. Taxable hp: 59.15. Torque: 490 lbs.-ft. at 2800 rpm. Taxable hp: 51.2. Five main bearings. Hydraulic valve lifters. Crankcase capacity: 5 qt. (add 1 qt. with new oil filter). Cooling system capacity: 20 qt. with heater (16.5 qt. without heater). Dual exhaust. Carburetor: Carter AFB No. 2992S. Code: J.

Interceptor Special V-8; Optional V-8
Overhead valve. Cast-iron block. Bore & stroke: 4.00 x 3.50 in. Displacement: 352 cid. Compression ratio: 10.60:1. Brake hp: 360. Taxable hp: 51.2. Five main bearings. Hydraulic valve lifters. Crankcase capacity: 5 qt. (add 1 qt. with new oil filter). Cooling system capacity: 19.5 qt. Dual exhaust with special header-type exhaust manifolds. Aluminum intake manifold. Dual point distributor. Carburetor: Holley four-barrel. Code: Y.

1961 FORD

Base V-8
Overhead valve. Cast-iron block. Bore & stroke: 3.75 x 3.30 in. Displacement: 292 cid. Compression ratio: 8.80:1. Brake hp: 175 at 4200 rpm. Taxable hp: 45.0. Torque: 279 lbs.-ft. at 2200 rpm. Five main bearings. Solid valve lifters. Crankcase capacity: 5 qt. (add 1 qt. with new oil filter). Cooling system capacity: 19 qt. Carburetor: (with PCV) Ford two-barrel Model CIAE-9510-AA with standard transmission or CIAE-9510-AB with automatic transmission; (without PCV) Ford two-barrel Model CIAE-9510-Y with standard transmission or CIAE-9510-Z with automatic transmission. Code: W (Export code: T).

Interceptor V-8
Overhead valve. Cast-iron block. Bore & stroke: 4.00 x 3.50 in. Displacement: 352 cid. Compression ratio: 8.90:1. Brake hp: 220 at 4400 rpm. Taxable hp: 51.2. Torque: 336 lbs.-ft. at 2400 rpm. Five main bearings. Hydraulic valve lifters. Crankcase capacity: 5 qt. (add 1 qt. with new oil filter). Cooling system capacity: 19.5 qt. Carburetor: (with PCV) Ford two-barrel Model CIAE-9510-AE with standard transmission or CIAE-9510-AF with automatic transmission; (without PCV) Ford two-barrel Model CIAE-9510-AC with standard transmission or CIAE-9510-AD with automatic transmission. Code: X (Export code: G).

Thunderbird V-8
Overhead valve. Cast-iron block. Bore & stroke: 4.05 x 3.78 in. Displacement: 390 cid. Compression ratio: 9.60:1. Brake hp: 300 at 4600 rpm. Taxable hp: 52.49. Torque: 427 lbs.-ft. at 2800 rpm. Five main bearings. Hydraulic valve lifters. Crankcase capacity: 5 qt. (add 1 qt. with new oil filter). Cooling system capacity: 19 qt. Carburetor: Ford four-barrel Model CIAE-9510-AG with standard transmission or CIAE-9510-AH with automatic transmission. Code: Z.

Interceptor 390 V-8
Overhead valve. Cast-iron block. Displacement: 390 cid. Bore & stroke: 4.05 x 3.78 in. Compression ratio: 9.60:1. Brake hp: 330 at 5000 rpm. Torque: 427 lbs.-ft. at 3200 rpm. Five main bearings. Hydraulic valve lifters. Carburetor: (with PCV) Ford four-barrel Model C2AE-9510-AJ with standard transmission or C2AE-9510-AK with automatic transmission; (without PCV) Ford two-barrel Model C2AE-9510-AG with standard transmission or C2AE-9510-AR with automatic transmission. Cooling system capacity: 20.5 qt. with heater. Crankcase capacity: 5 qt. (add 1 qt. with new oil filter). Dual exhaust. Code: Z or Code: P (Police).

Thunderbird Special V-8
Overhead valve. Cast-iron block. Bore & stroke: 4.05 x 3.78 in. Displacement: 390 cid. Compression ratio: 10.60:1. Brake hp: 375 at 6000 rpm. Taxable hp: 52.49. Torque: 427 lbs.-ft. at 3400 rpm. Five main bearings. Solid valve lifters. Crankcase capacity: 5 qt. (add 1 qt. with new oil filter). Cooling system capacity: 19 qt. Carburetor: Ford four-barrel Model CIAE-9510-AG with standard transmission or CIAE-9510-AH with automatic transmission. Code: Q (Early versions may use Code: Z and export models used Code: R).
NOTE: Some cars were delivered with the intake manifold and carburetors in the trunk.

Thunderbird Special 6V V-8
Overhead valve. Cast-iron block. Bore & stroke: 4.05 x 3.78 in. Displacement: 390 cid. Compression ratio: 10.60:1. Brake hp: 401 at 6000 rpm. Taxable hp: 52.49. Torque: 430 lbs.-ft. at 3500 rpm. Five main bearings. Solid valve lifters. Crankcase capacity: 5 qt. (add 1 qt. with new oil filter). Cooling system capacity: 19 qt. Carburetor: Three two-barrel carburetors. Code: Q (Early versions may use Code: Z).

1962 FORD

A factory cutaway illustration reveals the inner workings of the 1962 Ford 406-cid Super Stock engine with three two-barrel carburetors. This 405-hp V-8 featured large-port cylinder heads, smooth-flowing exhaust headers, and a beefed-up bottom end. Cars using this engine dominated drag racing in the early '60s.

Base Fairlane V-8
Overhead valve. Cast-iron block. Displacement: 221 cid. Bore & stroke: 3.50 x 2.87 in. Compression ratio: 8.70:1. Brake hp: 145 at 4400 rpm. Torque: 216 lbs.-ft. at 2200 rpm. Five main bearings. Hydraulic valve lifters. Carburetor: Ford two-barrel Model C20E-9510-N with standard transmission or C20E-9510-T with automatic transmission. Cooling system capacity: 14.5 qt. with heater. Crankcase capacity: 4 qt. (add 1 qt. with new oil filter). Code: L.

Callenger 260 V-8 (midyear release)
Overhead valve. Cast-iron block. Displacement: 260 cid. Bore & stroke: 3.80 x 2.87 in. Compression ratio: 8.70:1. Brake hp: 164 at 4400 rpm. Torque: 258 lbs.-ft. at 2200 rpm. Five main bearings. Hydraulic valve lifters. Carburetor: Ford two-barrel Model C20F-9510-A. Cooling system capacity: 14.5 qt. with heater. Crankcase capacity: 4 qt. (add 1 qt. with new oil filter). Code: F.

Base Ford V-8
Overhead valve. Cast-iron block. Displacement: 292 cid. Bore & stroke: 3.75 x 3.30 in. Compression ratio: 8.80:1. Brake hp: 170 at 4200 rpm.

Torque: 279 lbs.-ft. at 2200 rpm. Five main bearings. Solid valve lifters. Carburetor: (with PCV) Ford two-barrel Model C2AE-9510-AA with standard transmission or C2AE-9510-AB with automatic transmission; (without PCV) Ford two-barrel Model C2AE-9510-Y with standard transmission or C2AE-9510-Z with automatic transmission. Cooling system capacity: 16 qt. with heater. Crankcase capacity: 5 qt. (add 1 qt. with new oil filter). Code: W.

Interceptor 352 V-8
Overhead valve. Cast-iron block. Displacement: 352 cid. Bore & stroke: 4.00 x 3.50 in. Compression ratio: 8.90:1. Brake hp: 220 at 4300 rpm. Torque: 336 lbs.-ft. at 2600 rpm. Five main bearings. Hydraulic valve lifters. Carburetor: (with PCV) Ford two-barrel Model C2AE-9510-AE with standard transmission or C2AE-9510-AF with automatic transmission; (without PCV) Ford two-barrel Model C2AE-9510-AC with standard transmission or C2AE-9510-AD with automatic transmission. Cooling system capacity: 20.5 qt. with heater. Crankcase capacity: 5 qt. (add 1 qt. with new oil filter). Code: X.

Interceptor 390 V-8
Overhead valve. Cast-iron block. Displacement: 390 cid. Bore & stroke: 4.05 x 3.78 in. Compression ratio: 9.60:1. Brake hp: 300 at 4600 rpm. Torque: 427 lbs.-ft. at 2800 rpm. Five main bearings. Hydraulic valve lifters. Carburetor: (with PCV) Ford four-barrel Model C2AE-9510-AJ with standard transmission or C2AE-9510-AK with automatic transmission. Cooling system capacity: 20.5 qt. with heater. Crankcase capacity: 5 qt. (add 1 qt. with new oil filter). Dual exhaust. Code: Z.

Interceptor 390 V-8
Overhead valve. Cast-iron block. Displacement: 390 cid. Bore & stroke: 4.05 x 3.78 in. Compression ratio: 9.60:1. Brake hp: 330 at 5000 rpm. Torque: 427 lbs.-ft. at 3200 rpm. Five main bearings. Solid valve lifters. Carburetor: (with PCV) Ford four-barrel Model C2AE-9510-AJ with standard transmission or C2AE-9510-AK with automatic transmission; (without PCV) Ford two-barrel Model C2AE-9510-AG with standard transmission or C2AE-9510-AR with automatic transmission. Cooling system capacity: 20.5 qt. with heater. Crankcase capacity: 5 qt. (add 1 qt. with new oil filter). Dual exhaust. Code: P.

Thunderbird 390 V-8 (Available only in Thunderbird)
Overhead valve. Cast-iron block. Displacement: 390 cid. Bore & stroke: 4.05 x 3.78 in. Compression ratio: 10.50:1. Brake hp: 340 at 5000 rpm. Torque: 430 lbs.-ft. at 3200 rpm. Five main bearings. Hydraulic valve lifters. Carburetor: Holley three two-barrels. Cooling system capacity: 20.5 qt. with heater. Crankcase capacity: 5 qt. (add 1 qt. with new oil filter). Dual exhaust. Code: M.

390-375 optional Ford V-8 (Not available in Thunderbird)
Overhead valve. Cast-iron block. Displacement: 390 cid. Bore & stroke: 4.05 x 3.78 in. Compression ratio: 10.60:1. Brake hp: 375 at 6000 rpm. Torque: 427 lbs.-ft. at 3400 rpm. Five main bearings. Solid valve lifters. Carburetor: Holley Model R-3228A four-barrel. Cooling system capacity: 20.5 qt. with heater. Crankcase capacity: 5 qt. (add 1 qt. with new oil filter). Dual exhaust. Code: Q.

Thunderbird 406 V-8
Overhead valve. Cast-iron block. Displacement: 406 cid. Bore & stroke: 4.13 x 3.78 in. Compression ratio: 11.40:1. Brake hp: 385 at 5800 rpm. Torque: 444 lbs.-ft. at 3400 rpm. Five main bearings. Solid valve lifters. Carburetor: Ford four-barrel Model CIAE-9510-AG with standard transmission or CIAE-9510-AH with automatic transmission. Cooling system capacity: 20.5 qt. with heater. Crankcase capacity: 5 qt. (add 1 qt. with new oil filter). Dual exhaust. Code: B.

390-401 optional Ford V-8 (Not available in Thunderbird)
Overhead valve. Cast-iron block. Displacement: 390 cid. Bore & stroke: 4.05 x 3.78 in. Compression ratio: 10.60:1. Brake hp: 401 at 6000 rpm. Torque: 430 lbs.-ft. at 3500 rpm. Five main bearings. Solid valve lifters. Carburetor: Holley three two barrels. Cooling system capacity: 20.5 qt. with heater. Crankcase capacity: 5 qt. (add 1 qt. with new oil filter). Dual exhaust. Code: M.

Thunderbird Special 406 V-8
Overhead valve. Cast-iron block. Displacement: 406 cid. Bore & stroke: 4.13 x 3.78 in. Compression ratio: 11.40:1. Brake hp: 405 at 5800 rpm. Torque: 448 lbs.-ft. at 3500 rpm. Five main bearings. Solid valve lifters. Carburetor: Holley three two-barrel. Cooling system capacity: 20.5 qt. with heater. Crankcase capacity: 5 qt. (add 1 qt. with new oil filter). Dual exhaust. Code: G.

1963 FORD

Base Fairlane V-8
Overhead valve. Cast-iron block. Displacement: 221 cid. Bore & stroke: 3.50 x 2.87 in. Compression ratio: 8.70:1. Brake hp: 145 at 4400 rpm. Torque: 216 lbs.-ft. at 2200 rpm. Five main bearings. Hydraulic valve lifters. Carburetor: Ford two-barrel C30F-9510-C. Cooling system capacity: 14.5 qt. with heater. Crankcase capacity: 4 qt. (add 1 qt. with new oil filter). Code: L. (Export code 3).

Challenger 260 V-8; Optional in Fairlane and full-size Fords
Overhead valve. Cast-iron block. Displacement: 260 cid. Bore & stroke: 3.80 x 2.87 in. Compression ratio: 8.70:1. Brake hp: 164 at 4400 rpm. Torque: 258 lbs.-ft. at 2200 rpm. Five main bearings. Hydraulic valve lifters. Carburetor: Ford two-barrel C30F-9510-E. Cooling system capacity: 14.5 qt. with heater. Crankcase capacity: 4 qt. (add 1 qt. with new oil filter). Code: F. (Export code 6).

Challenger 289 V-8 (1963-1/2 release for Fairlanes except station wagons)
Overhead valve. Cast-iron block. Displacement: 289 cid. Bore & stroke: 4.00 x 2.87 in. Compression ratio: 9.00:1. Brake hp: 195 at 4400 rpm. Torque: 282 lbs.-ft. at 2200 rpm. Five main bearings. Hydraulic valve lifters. Carburetor: Ford two-barrel C30F-9510. Cooling system capacity: 14.5 qt. with heater. Crankcase capacity: 4 qt. (add 1 qt. with new oil

The Challenger 289 V-8, introduced in 1963, was a completely different animal than the Y-block. With this new small-block, the porting line was on the centerline of the crankshaft and the main bearing webs were not as deep, but were made thicker to give them the necessary strength. This is the high-performance 289 from a 1963 Fairlane. The 289 was produced through 1968, when it was replaced by a 302-cid V-8 of similar overall design.

This Ford 427 "Wedge" V-8 is a medium-riser version with dual four-barrel Holley carburetors perched on a special dual-plane aluminum intake manifold. This produced good low-end torque, as opposed to the single-plane manifold, which was better for high-rpm drag strip use. A tunnel-port version was also available. To spot a 427, look for the heads of three tie bolts on each side of the block, just above the oil pan rail. This engine was offered from 1963 through 1968.

filter). Code: C.

Interceptor V-8; Base Galaxie 500XL V-8; optional in other full-size Fords
Overhead valve. Cast-iron block. Displacement: 352 cid. Bore & stroke: 4.00 x 3.50 in. Compression ratio: 8.90:1. Brake hp: 220 at 4300 rpm. Torque: 336 lbs.-ft. at 2600 rpm. Five main bearings. Hydraulic valve lifters. Carburetor: Two-barrel. Cooling system capacity: 20.5 qt. with heater. Crankcase capacity: 5 qt. (add 1 qt. with new oil filter). Code: X.

High-Perf. Challenger 289 V-8 (1963-1/2 release for Fairlanes except station wagons)
Overhead valve. Cast-iron block. Displacement: 289 cid. Bore & stroke: 4.00 x 2.87 in. Compression ratio: 10.50:1. Brake hp: 271 at 6000 rpm. Torque: 312 lbs.-ft. at 3400 rpm. Five main bearings. Solid valve lifters. Carburetor: Four-barrel. Cooling system capacity: 14.5 qt. with heater. Crankcase capacity: 4 qt. (add 1 qt. with new oil filter). Code: K.

Interceptor 390 V-8; optional in full-size Fords; standard in Thunderbird
Overhead valve. Cast-iron block. Displacement: 390 cid. Bore & stroke: 4.05 x 3.78 in. Compression ratio: 9.60:1. Brake hp: 300 at 4600 rpm. Torque: 427 lbs.-ft. at 2800 rpm. Five main bearings. Hydraulic valve lifters. Carburetor: Ford four-barrel C2AF-9510. Cooling system capacity: 20.5 qt. with heater. Crankcase capacity: 5 qt. (add 1 qt. with new oil filter). Dual exhaust. Code: Z.

Interceptor 390 V-8
Overhead valve. Cast-iron block. Displacement: 390 cid. Bore & stroke: 4.05 x 3.78 in. Compression ratio: 9.60:1. Brake hp: 330 at 5000 rpm. Torque: 427 lbs.-ft. at 3200 rpm. Five main bearings. Solid valve lifters. Carburetor: Ford four-barrel. Cooling system capacity: 20.5 qt. with heater. Crankcase capacity: 5 qt. (add 1 qt. with new oil filter). Dual exhaust. Code: P. (Export code 9).

Thunderbird Special 390 V-8 (Available only in Thunderbird)
Overhead valve. Cast-iron block. Displacement: 390 cid. Bore & stroke: 4.05 x 3.78 in. Compression ratio: 10.50:1. Brake hp: 340 at 5000 rpm. Torque: 430 lbs.-ft. at 3200 rpm. Five main bearings. Hydraulic valve lifters. Carburetor: Holley three Model 2300 two-barrel. Cooling system capacity: 20.5 qt. with heater. Crankcase capacity: 5 qt. (add 1 qt. with new oil filter). Dual exhaust. Code: M.

Thunderbird 406 V-8
Overhead valve. Cast-iron block. Displacement: 406 cid. Bore & stroke: 4.13 x 3.78 in. Compression ratio: 11.40:1. Brake hp: 385 at 5800 rpm. Torque: 444 lbs.-ft. at 3400 rpm. Five main bearings. Solid valve lifters. Carburetor: Holley four-barrel. Cooling system capacity: 20.5 qt. with heater. Crankcase capacity: 5 qt. (add 1 qt. with new oil filter). Dual exhaust. Code: B.

Thunderbird Special 406 V-8
Overhead valve. Cast-iron block. Displacement: 406 cid. Bore & stroke: 4.13 x 3.78 in. Compression ratio: 11.40:1. Brake hp: 405 at 5800 rpm. Torque: 448 lbs.-ft. at 3500 rpm. Five main bearings. Solid valve lifters. Carburetor: Holley three two-barrel. Cooling system capacity: 20.5 qt. with heater. Crankcase capacity: 5 qt. (add 1 qt. with new oil filter). Dual exhaust. Code: G.
NOTE: Export engines have lower compression and less horsepower.

1964 FORD

Challenger 260 V-8; Optional in Fairlane and full-size Fords
Overhead valve. Cast-iron block. Displacement: 260 cid. Bore & stroke: 3.80 x 2.87 in. Compression ratio: 8.8:1. Brake hp: 164 at 4400 rpm. Torque: 258 lbs.-ft. at 2200 rpm. Five main bearings. Hydraulic valve lifters. Carburetor: Two-barrel. Cooling system capacity: 15.5 qt. with heater. Crankcase capacity: 4 qt. (add 1 qt. with new oil filter). Code: F.

Challenger 289 V-8
Overhead valve. Cast-iron block. Displacement: 289 cid. Bore & stroke: 4.00 x 2.87 in. Compression ratio: 9.00:1. Brake hp: 195 at 4400 rpm. Torque: 282 lbs.-ft. at 2200 rpm. Five main bearings. Hydraulic valve lifters. Carburetor: Ford two-barrel C30F-9510. Cooling system capacity: 16 qt. with heater. Crankcase capacity: 4 qt. (add 1 qt. with new oil filter). Code C.

Challenger 289 V-8 Four-Barrel
Overhead valve. Cast-iron block. Displacement: 289 cid. Bore & stroke: 4.00 x 2.87 in. Compression ratio: 9.0:1. Brake hp: 210 at 4400 rpm. Torque: 300 lbs.-ft. at 2800 rpm. Five main bearings. Hydraulic valve lifters. Carburetor: Four-barrel. Cooling system capacity: 16 qt. with heater. Crankcase capacity: 4 qt. (add 1 qt. with new oil filter). Code: C.

Interceptor V-8; Base Galaxie 500XL V-8; optional in other full-size Fords
Overhead valve. Cast-iron block. Displacement: 352 cid. Bore & stroke: 4.00 x 3.50 in. Compression ratio: 9.30:1. Brake hp: 250 at 4400 rpm. Torque: 352 lbs.-ft. at 2800 rpm. Five main bearings. Hydraulic valve lifters. Carburetor: Four-barrel. Cooling system capacity: 20 qt. with heater. Crankcase capacity: 5 qt. (add 1 qt. with new oil filter). Code: X.

High-Perf. Challenger 289 V-8
Overhead valve. Cast-iron block. Displacement: 289 cid. Bore & stroke: 4.00 x 2.87 in. Compression ratio: 10.50:1. Brake hp: 271 at 6000 rpm. Torque: 312 lbs.-ft. at 3400 rpm. Five main bearings. Solid valve lifters. Carburetor: Four-barrel. Cooling system capacity: 16 qt. with heater. Crankcase capacity: 4 qt. (add 1 qt. with new oil filter). Code: K.

Thunderbird 390 V-8; optional in full-size Fords; standard in Thunderbird
Overhead valve. Cast-iron block. Displacement: 390 cid. Bore & stroke: 4.05 x 3.78 in. Compression ratio: 10.0:1. Brake hp: 300 at 4600 rpm. Torque: 427 lbs.-ft. at 2800 rpm. Five main bearings. Hydraulic valve lifters. Carburetor: Four-barrel. Cooling system capacity: 20 qt. with heater. Crankcase capacity: 5 qt. (add 1 qt. with new oil filter). Dual exhaust. Code: Z.

Thunderbird Police Special 390 V-8
Overhead valve. Cast-iron block. Displacement: 390 cid. Bore & stroke: 4.05 x 3.78 in. Compression ratio: 10.0:1. Brake hp: 330 at 5000 rpm. Torque: 427 lbs.-ft. at 3200 rpm. Five main bearings. Solid valve lifters. Carburetor: Ford four-barrel. Cooling system capacity: 20 qt. with heater. Crankcase capacity: 5 qt. (add 1 qt. with new oil filter). Dual exhaust. Code: P.

Thunderbird High-Performance V-8
Overhead valve. Cast-iron block. Displacement: 427 cid. Bore & stroke: 4.23 x 3.78 in. Compression ratio: 11.50:1. Brake hp: 410 at 5600 rpm. Torque: 476 lbs.-ft. at 3400 rpm. Five main bearings. Solid valve lifters. Carburetor: Holley four-barrel. Cooling system capacity: 20 qt. with heater. Crankcase capacity: 5 qt. (add 1 qt. with new oil filter). Dual exhaust. Code: Q.

Thunderbird Super High-Performance V-8
Overhead valve. Cast-iron block. Displacement: 427 cid. Bore & stroke: 4.23 x 3.78 in. Compression ratio: 11.50:1. Brake hp: 425 at 6000 rpm. Torque: 480 lbs.-ft. at 3700 rpm. Five main bearings. Solid valve lifters. Carburetor: Two Holley four-barrel. Cooling system capacity: 20 qt. with heater. Crankcase capacity: 5 qt. (add 1 qt. with new oil filter). Dual exhaust. Code: R.

1965 FORD

A prototype 1965 Ford 427-cid Single Overhead Cam "Hemi" V-8. This engine evolved from the 427 "hi-riser" push rod engine and was used only in race cars. It was never installed in a production vehicle. The more powerful version shown here was good for 657 hp at 7500 rpm. These engines cost close to $3,000.

Challenger 289 V-8
Overhead valve. Cast-iron block. Displacement: 289 cid. Bore & stroke: 4.00 x 2.87 in. Compression ratio: 9.30:1. Brake hp: 200 at 4400 rpm. Torque: 282 lbs.-ft. at 2400 rpm. Five main bearings. Hydraulic valve lifters. Carburetor: Two-barrel. Cooling system capacity: 16 qt. with heater. Crankcase capacity: 4 qt. (add 1 qt. with new oil filter). Code: C.

Challenger 289 Four-Barrel V-8
Overhead valve. Cast-iron block. Displacement: 289 cid. Bore & stroke: 4.00 x 2.87 in. Compression ratio: 10.0:1. Brake hp: 225 at 4800 rpm. Torque: 305 lbs.-ft. at 3200 rpm. Five main bearings. Hydraulic valve lifters. Carburetor: Four-barrel. Cooling system capacity: 16 qt. with heater. Crankcase capacity: 4 qt. (add 1 qt. with new oil filter). Code: A.

Interceptor V-8; Base Galaxie 500XL V-8; optional in other full-size Fords
Overhead valve. Cast-iron block. Displacement: 352 cid. Bore & stroke: 4.00 x 3.50 in. Compression ratio: 9.30:1. Brake hp: 250 at 4400 rpm. Torque: 352 lbs.-ft. at 2800 rpm. Five main bearings. Hydraulic valve lifters. Carburetor: Four-barrel. Cooling system capacity: 20.5 qt. with heater. Crankcase capacity: 5 qt. (add 1 qt. with new oil filter). Code: X.

Challenger 289 High-Performance Four-Barrel V-8
Overhead valve. Cast-iron block. Displacement: 289 cid. Bore & stroke: 4.00 x 2.87 in. Compression ratio: 10.5:1. Brake hp: 271 at 6000 rpm. Torque: 312 lbs.-ft. at 3400 rpm. Five main bearings. Solid valve lifters. Carburetor: Four-barrel. Cooling system capacity: 16 qt. with heater. Crankcase capacity: 4 qt. (add 1 qt. with new oil filter). Code: K.

Thunderbird 390 V-8; optional in full-size Fords; standard in Thunderbird
Overhead valve. Cast-iron block. Displacement: 390 cid. Bore & stroke: 4.05 x 3.78 in. Compression ratio: 10.0:1. Brake hp: 300 at 4600 rpm. Torque: 427 lbs.-ft. at 2800 rpm. Five main bearings. Hydraulic valve lifters. Carburetor: Four-barrel. Cooling system capacity: 20.5 qt. with heater. Crankcase capacity: 5 qt. (add 1 qt. with new oil filter). Dual exhaust. Code: Z.

Thunderbird Interceptor Special 390 V-8
Overhead valve. Cast-iron block. Displacement: 390 cid. Bore & stroke: 4.05 x 3.78 in. Compression ratio: 10.0:1. Brake hp: 330 at 5000 rpm. Torque: 427 lbs.-ft. at 3200 rpm. Five main bearings. Solid valve lifters. Carburetor: Ford four-barrel. Cooling system capacity: 20.5 qt. with heater. Crankcase capacity: 5 qt. (add 1 qt. with new oil filter). Dual exhaust. Code: P.

Thunderbird Super High-Performance V-8
Overhead valve. Cast-iron block. Displacement: 427 cid. Bore & stroke: 4.23 x 3.78 in. Compression ratio: 11.50:1. Brake hp: 425 at 6000 rpm. Torque: 480 lbs.-ft. at 3700 rpm. Five main bearings. Solid valve lifters. Carburetor: Two Holley four-barrel. Cooling system capacity: 20.5 qt. with heater. Crankcase capacity: 5 qt. (add 1 qt. with new oil filter). Dual exhaust. Code: R.

"SOHC 427" 4V V-8
Overhead valve. Cast-iron block. Hemispherical combustion chambers with overhead valves and overhead camshafts for each engine bank. Displacement: 427 cid. Bore & stroke: 4.23 x 3.78 in. Compression ratio: 12.10:1. Brake hp: 616 at 7000 rpm. Five main bearings. Carburetor: Holley four-barrel. Cooling system capacity: 20.5 qt. with heater. Crankcase capacity: 5 qt. (add 1 qt. with new oil filter). Dual exhaust. Code: L ($2,500).
NOTE: Not installed in any production vehicles.

"SOHC 427' 4V V-8
Overhead valve. Cast-iron block. Hemispherical combustion chambers with overhead valves and overhead camshafts for each engine bank. Displacement: 427 cid. Bore & stroke: 4.23 x 3.78 in. Compression ratio: 12.10:1. Brake hp: 657 at 7500 rpm. Five main bearings. Hydraulic valve lifters. Carburetor: Two Holley four-barrel. Cooling system capacity: 20.5 qt. with heater. Crankcase capacity: 5 qt. (add 1 qt. with new oil filter). Dual exhaust. Code: M.
NOTE: Not installed in any production vehicles.

1966 FORD

Challenger 289 V-8
Overhead valve. Cast-iron block. Displacement: 289 cid. Bore & stroke: 4.00 x 2.87 in. Compression ratio: 9.30:1. Brake hp: 200 at 4400 rpm. Torque: 282 lbs.-ft. at 2400 rpm. Five main bearings. Hydraulic valve lifters. Carburetor: Two-barrel. Cooling system capacity: 15 qt. with heater. Crankcase capacity: 4 qt. (add 1 qt. with new oil filter). Code: C.

Ford's 289-cid small-block V-8 was unchanged in 1966. Seen here is the basic two-barrel 200-hp version that was an option in the Falcon, Fairlane, Galaxie, Mustang, and Mercury Comet. This was FoMoCo's bread-and-butter regular-fuel V-8.

Challenger 289 Four-Barrel V-8
Overhead valve. Cast-iron block. Displacement: 289 cid. Bore & stroke: 4.00 x 2.87 in. Compression ratio: 10.0:1. Brake hp: 225 at 4800 rpm. Torque: 305 lbs.-ft. at 3200 rpm. Five main bearings. Hydraulic valve lifters. Carburetor: Four-barrel. Cooling system capacity: 15 qt. with heater. Crankcase capacity: 4 qt. (add 1 qt. with new oil filter). Code: A.

Interceptor V-8; Base Galaxie 500XL V-8; optional in other full-size Fords
Overhead valve. Cast-iron block. Displacement: 352 cid. Bore & stroke: 4.00 x 3.50 in. Compression ratio: 9.30:1. Brake hp: 250 at 4400 rpm. Torque: 352 lbs.-ft. at 2800 rpm. Five main bearings. Hydraulic valve lifters. Carburetor: Four-barrel. Cooling system capacity: 20.5 qt. with heater. Crankcase capacity: 5 qt. (add 1 qt. with new oil filter). Code: X.

Challenger 289 High-Performance Four-Barrel V-8
Overhead valve. Cast-iron block. Displacement: 289 cid. Bore & stroke: 4.00 x 2.87 in. Compression ratio: 10.5:1. Brake hp: 271 at 6000 rpm. Torque: 312 lbs.-ft. at 3400 rpm. Five main bearings. Hydraulic valve lifters. Carburetor: Four-barrel. Cooling system capacity: 15 qt. with heater. Crankcase capacity: 4 qt. (add 1 qt. with new oil filter). Code: K.

Thunderbird 390 V-8; optional in full-size Fords; standard in Thunderbird
Overhead valve. Cast-iron block. Displacement: 390 cid. Bore & stroke: 4.05 x 3.78 in. Compression ratio: 9.5:1. Brake hp: 275 at 4400 rpm. Torque: 405 lbs.-ft. at 2600 rpm. Five main bearings. Solid valve lifters. Carburetor: Two-barrel. Cooling system capacity: 20.5 qt. with heater. Crankcase capacity: 4 qt. (add 1 qt. with new oil filter). Code: Y.

Thunderbird 390 Four-barrel V-8; optional in full-size Fords and Thunderbirds
Overhead valve. Cast-iron block. Displacement: 390 cid. Bore & stroke: 4.05 x 3.78 in. Compression ratio: 10.5:1. Brake hp: 315 at 4600 rpm. Torque: 427 lbs.-ft. at 2800 rpm. Five main bearings. Hydraulic valve lifters. Carburetor: Four-barrel. Cooling system capacity: 20.5 qt. with heater. Crankcase capacity: 4 qt. (add 1 qt. with new oil filter). Dual exhaust. Code: Z.

Optional Fairlane GT 390 V-8
Overhead valve. Cast-iron block. Displacement: 390 cid. Bore & stroke: 4.05 x 3.78 in. Compression ratio: 11.0:1. Brake hp: 335 at 4800 rpm. Torque: 427 lbs.-ft. at 3200 rpm. Five main bearings. Hydraulic valve lifters. Carburetor: Holley four-barrel. Cooling system capacity: 20.5 qt. with heater. Crankcase capacity: 4 qt. (add 1 qt. with new oil filter). Dual exhaust. Code: S.

Thunderbird Special V-8
Overhead valve. Cast-iron block. Displacement: 428 cid. Bore & stroke: 4.13 x 3.98 in. Compression ratio: 10.50:1. Brake hp: 345 at 4600 rpm. Torque: 462 lbs.-ft. at 2800 rpm. Five main bearings. Hydraulic valve lifters. Carburetor: Four-barrel. Cooling system capacity: 20.5 qt. with heater. Crankcase capacity: 4 qt. (add 1 qt. with new oil filter). Dual exhaust. Code: Q.

Police Interceptor V-8
Overhead valve. Cast-iron block. Displacement: 428 cid. Bore & stroke: 4.13 x 3.98 in. Compression ratio: 10.50:1. Brake hp: 360 at 5400 rpm. Torque: 459 lbs.-ft. at 3200 rpm. Five main bearings. Solid valve lifters. Carburetor: Four-barrel. Cooling system capacity: 20.5 qt. with heater. Crankcase capacity: 4 qt. (add 1 qt. with new oil filter). Dual exhaust. Code: P.

Thunderbird High-Performance V-8
Overhead valve. Cast-iron block. Displacement: 427 cid. Bore & stroke: 4.23 x 3.78 in. Compression ratio: 11.00:1. Brake hp: 410 at 5600 rpm. Torque: 476 lbs.-ft. at 3400 rpm. Five main bearings. Solid valve lifters. Carburetor: Holley four-barrel. Cooling system capacity: 20.5 qt. with heater. Crankcase capacity: 5 qt. (add 1 qt. with new oil filter). Dual exhaust. Code: W.

Thunderbird Super High-Performance V-8
Overhead valve. Cast-iron block. Displacement: 427 cid. Bore & stroke: 4.23 x 3.78 in. Compression ratio: 11.50:1. Brake hp: 425 at 6000 rpm. Torque: 480 lbs.-ft. at 3700 rpm. Five main bearings. Hydraulic valve lifters. Carburetor: Two Holley four-barrel. Cooling system capacity: 20.5 qt. with heater. Crankcase capacity: 5 qt. (add 1 qt. with new oil filter). Dual exhaust. Code: R.

"SOHC 427" 4V V-8
Overhead valve. Cast-iron block. Hemispherical combustion chambers with overhead valves and overhead camshafts for each engine bank. Displacement: 427 cid. Bore & stroke: 4.23 x 3.78 in. Compression ratio: 12.10:1. Brake hp: 616 at 7000 rpm. Five main bearings. Carburetor: Holley four-barrel. Cooling system capacity: 20.5 qt. with heater. Crankcase capacity: 5 qt. (add 1 qt. with new oil filter). Dual exhaust. Code: L. Engine available only "over the counter" for $2,500.

"SOHC 427" 4V V-8
Overhead valve. Cast-iron block. Hemispherical combustion chambers with overhead valves and overhead camshafts for each engine bank. Displacement: 427 cid. Bore & stroke: 4.23 x 3.78 in. Compression ratio: 12.10:1. Brake hp: 657 at 7500 rpm. Five main bearings. Hydraulic valve lifters. Carburetor: Two Holley four-barrel. Cooling system capacity: 20.5 qt. with heater. Crankcase capacity: 5 qt. (add 1 qt. with new oil filter). Dual exhaust. Code: M.

1967 FORD

Challenger 289 V-8
Overhead valve. Cast-iron block. Displacement: 289 cid. Bore & stroke: 4.00 x 2.87 in. Compression ratio: 9.30:1. Brake hp: 200 at 4400 rpm. Torque: 282 lbs.-ft. at 2400 rpm. Five main bearings. Hydraulic valve lifters. Carburetor: Two-barrel. Cooling system capacity: 15 qt. with heater. Crankcase capacity: 4 qt. (add 1 qt. with new oil filter). Code: C.

Challenger 289 Four-Barrel V-8
Overhead valve. Cast-iron block. Displacement: 289 cid. Bore & stroke: 4.00 x 2.87 in. Compression ratio: 10.0:1. Brake hp: 225 at 4800 rpm. Torque: 305 lbs.-ft. at 3200 rpm. Five main bearings. Hydraulic valve lifters. Carburetor: Four-barrel. Cooling system capacity: 15 qt. with heater. Crankcase capacity: 4 qt. (add 1 qt. with new oil filter). Code: A.

In addition to the base two-barrel version and the high-performance version, the 1967 Ford 289-cid V-8 was available in regular four-barrel version. This engine had a 10.0:1 compression ratio, and a Holley four-barrel, and was called the C code engine. It developed 225 hp at 4800 rpm.

From the same family of engines as the 289-cid Ford small-block, the new 302 was introduced in 1968. It came in the normal two-barrel (210-hp) and four-barrel (230-hp) variations. Interestingly, the hydraulic valve lifters used in each of these engines did not look identical, although they were interchangeable. The 302 had a new emissions control setup with hoses and a canister that took air from the air cleaner and sent it to the exhaust manifold.

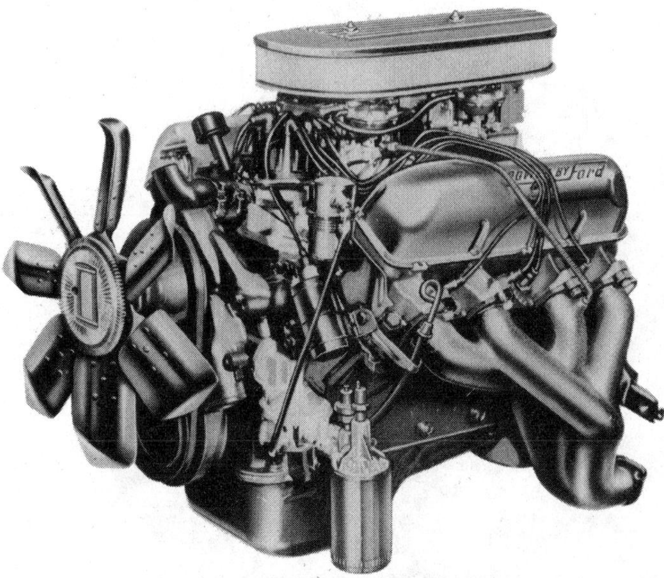

Ford offered two versions of the ultra-high-performance 427-cid big-block in 1967. Shown here is the most powerful one, which had two four-barrel carburetors under a neat-looking chrome finned air cleaner. It produced 425 hp. This engine evolved from the 352-cid Ford V-8 that was first introduced in 1958.

Challenger 289 High-Performance Four-Barrel V-8
Overhead valve. Cast-iron block. Displacement: 289 cid. Bore & stroke: 4.00 x 2.87 in. Compression ratio: 10.5:1. Brake hp: 271 at 6000 rpm. Torque: 312 lbs.-ft. at 3400 rpm. Five main bearings. Solid valve lifters. Carburetor: Four-barrel. Cooling system capacity: 15 qt. with heater. Crankcase capacity: 4 qt. (add 1 qt. with new oil filter). Code: K.

Thunderbird 390 V-8; optional in full-size Fords; standard in Thunderbird
Overhead valve. Cast-iron block. Displacement: 390 cid. Bore & stroke: 4.05 x 3.78 in. Compression ratio: 9.5:1. Brake hp: 275 at 4400 rpm. Torque: 405 lbs.-ft. at 2600 rpm. Five main bearings. Hydraulic valve lifters. Carburetor: Two-barrel. Cooling system capacity: 20.5 qt. with heater. Crankcase capacity: 4 qt. (add 1 qt. with new oil filter). Code: Y.

Thunderbird 390 Four-barrel V-8; optional in full-size Fords and Thunderbirds
Overhead valve. Cast-iron block. Displacement: 390 cid. Bore & stroke: 4.05 x 3.78 in. Compression ratio: 10.5:1. Brake hp: 315 at 4600 rpm. Torque: 427 lbs.-ft. at 2800 rpm. Five main bearings. Hydraulic valve lifters. Carburetor: Four-barrel. Cooling system capacity: 20.5 qt. with heater. Crankcase capacity: 4 qt. (add 1 qt. with new oil filter). Dual exhaust. Code: Z.

Optional Fairlane GT 390 V-8; Optional Mustang GT/GTS 320-hp V-8
Overhead valve. Cast-iron block. Displacement: 390 cid. Bore & stroke: 4.05 x 3.78 in. Compression ratio: 11.0:1. Brake hp: 335 at 4800 rpm. Torque: 427 lbs.-ft. at 3200 rpm. Five main bearings. Hydraulic valve lifters. Carburetor: Holley four-barrel. Cooling system capacity: 20.5 qt. with heater. Crankcase capacity: 4 qt. (add 1 qt. with new oil filter). Dual exhaust. Code: S.

Optional Mustang GT/GTS V-8
Overhead valve. Cast-iron block. Displacement: 390 cid. Bore & stroke: 4.05 x 3.78 in. Compression ratio: 10.5:1. Brake hp: 320 at 4600 rpm. Torque: 427 lbs.-ft. at 3200 rpm. Five main bearings. Hydraulic valve lifters. Carburetor: Holley four-barrel. Cooling system capacity: 20.5 qt. with heater. Crankcase capacity: 4 qt. (add 1 qt. with new oil filter). Dual exhaust. Code: S.

Thunderbird Special V-8
Overhead valve. Cast-iron block. Displacement: 428 cid. Bore & stroke: 4.13 x 3.98 in. Compression ratio: 10.50:1. Brake hp: 345 at 4600 rpm. Torque: 462 lbs.-ft. at 2800 rpm. Five main bearings. Hydraulic valve lifters. Carburetor: Four-barrel. Cooling system capacity: 20.5 qt. with heater. Crankcase capacity: 4 qt. (add 1 qt. with new oil filter). Dual exhaust. Code: Q.

Police Interceptor V-8
Overhead valve. Cast-iron block. Displacement: 428 cid. Bore & stroke: 4.13 x 3.98 in. Compression ratio: 10.50:1. Brake hp: 360 at 5400 rpm. Torque: 459 lbs.-ft. at 3200 rpm. Five main bearings. Solid valve lifters. Carburetor: Holley four-barrel. Cooling system capacity: 20.5 qt. with heater. Crankcase capacity: 4 qt. (add 1 qt. with new oil filter). Dual exhaust. Code: P.

Thunderbird High-Performance V-8
Overhead valve. Cast-iron block. Displacement: 427 cid. Bore & stroke: 4.23 x 3.78 in. Compression ratio: 11.00:1. Brake hp: 410 at 5600 rpm. Torque: 476 lbs.-ft. at 3400 rpm. Five main bearings. Solid valve lifters. Carburetor: Holley four-barrel. Cooling system capacity: 19.5 qt. with heater. Crankcase capacity: 5 qt. (add 1 qt. with new oil filter). Dual exhaust. Code: W.

Thunderbird Super High-Performance V-8
Overhead valve. Cast-iron block. Displacement: 427 cid. Bore & stroke: 4.23 x 3.78 in. Compression ratio: 11.50:1. Brake hp: 425 at 6000 rpm. Torque: 480 lbs.-ft. at 3700 rpm. Five main bearings. Solid valve lifters. Carburetor: Two Holley four-barrel. Cooling system capacity: 19.5 qt. with heater. Crankcase capacity: 5 qt. (add 1 qt. with new oil filter). Dual exhaust. Code: R.

"SOHC 427" 4V V-8
Overhead valve. Cast-iron block. Hemispherical combustion chambers with overhead valves and overhead camshafts for each engine bank. Displacement: 427 cid. Bore & stroke: 4.23 x 3.78 in. Compression ratio: 12.10:1. Brake hp: 616 at 7000 rpm. Five main bearings. Carburetor: Holley four-barrel. Cooling system capacity: 20.5 qt. with heater. Crankcase capacity: 5 qt. (add 1 qt. with new oil filter). Dual exhaust. Code: L. Engine available only "over the counter" for $2,500.

"SOHC 427" 4V V-8
Overhead valve. Cast-iron block. Hemispherical combustion chambers with overhead valves and overhead camshafts for each engine bank. Displacement: 427 cid. Bore & stroke: 4.23 x 3.78 in. Compression ratio: 12.10:1. Brake hp: 657 at 7500 rpm. Five main bearings. Hydraulic valve lifters. Carburetor: Two Holley four-barrel. Cooling system capacity: 20.5 qt. with heater. Crankcase capacity: 5 qt. (add 1 qt. with new oil filter). Dual exhaust. Code: M.
NOTE: A tunnel-port 427 was available as an over-the-counter kit. It included a special tunnel-port intake manifold and special cylinder heads.

1968 FORD

Challenger 289 V-8
Overhead valve. Cast-iron block. Displacement: 289 cid. Bore & stroke: 4.00 x 2.87 in. Compression ratio: 8.70:1. Brake hp: 195 at 4600 rpm. Torque: 288 lbs.-ft. at 2600 rpm. Five main bearings. Hydraulic valve lifters. Carburetor: Two-barrel. Cooling system capacity: 15 qt. with heater. Crankcase capacity: 4 qt. (add 1 qt. with new oil filter). Code: C.

302 Two-Barrel V-8
Overhead valve. Cast-iron block. Displacement: 302 cid. Bore & stroke: 4.00 x 3.00 in. Compression ratio: 9.00:1. Brake hp: 210 at 4400 rpm.

A new cast-iron-headers exhaust system was part of the Cobra Jet package seen on this 1968 CJ-428 V-8. It also featured a chrome dress-up kit and a de-clutching fan. The deep-skirted block made the engine very rugged and the spark plugs were arranged in a specific manner that prevented crossfire from adjacent wires. A Holley 4150C four-barrel carburetor was employed.

Looking much like the 390 GT V-8 on the outside, the 1968 Ford 428-cid high-performance engine was fitted with a chrome accessory kit. The 428-cid big-block V-8 first bowed in the 1966 Thunderbird. It was a bored and stroked 390 and was offered in 340- and 360-hp editions. There was also a Cobra Jet version.

Torque: 300 lbs.-ft. at 2600 rpm. Five main bearings. Hydraulic valve lifters. Carburetor: Two-barrel. Cooling system capacity: 15 qt. with heater. Crankcase capacity: (Mid-size/Mustang) 4 qt.; (Full-Size/T-Bird) 5 qt. (add 1 qt. with new oil filter). Code: F (Export code 6).

302 Four-Barrel V-8
Overhead valve. Cast-iron block. Displacement: 302 cid. Bore & stroke: 4.00 x 3.00 in. Compression ratio: 10.00:1. Brake hp: 230 at 4800 rpm. Torque: 310 lbs.-ft. at 2800 rpm. Five main bearings. Hydraulic valve lifters. Carburetor: Two-barrel. Cooling system capacity: 15 qt. with heater. Crankcase capacity: (Mid-size/Mustang) 4 qt.; (Full-Size/T-Bird) 5 qt. (add 1 qt. with new oil filter). Code: J.

Thunderbird 390 V-8; optional in full-size Fords; standard in Thunderbird
Overhead valve. Cast-iron block. Displacement: 390 cid. Bore & stroke: 4.05 x 3.78 in. Compression ratio: 9.5:1. Brake hp: 265 at 4400 rpm. Torque: 390 lbs.-ft. at 2600 rpm. Five main bearings. Hydraulic valve lifters. Carburetor: Two-barrel. Cooling system capacity: 20.5 qt. with heater. Crankcase capacity: 5 qt. (add 1 qt. with new oil filter). Code: Y.

Thunderbird 390 V-8; optional in full-size Fords and Thunderbirds
Overhead valve. Cast-iron block. Displacement: 390 cid. Bore & stroke: 4.05 x 3.78 in. Compression ratio: 10.5:1. Brake hp: 280 at 4400 rpm. Torque: 403 lbs.-ft. at 2600 rpm. Five main bearings. Hydraulic valve

Chromed valve covers and a chromed slim-line air cleaner dressed up the high-performance version of the Ford 427-cid V-8 in 1968. This 90-degree V-8 had a cast-iron block, cast-iron cylinder heads. The valves, springs, and rockers were in the heads, which had machined combustion chambers.

Although it was part of the 427 big-block V-8 engine family, the 1968 Ford 390 GT engine had a couple of obvious distinctions including a de-clutching type radiator fan and header-type exhaust manifolds. The 390 first appeared in the 1961 Thunderbird. In 1968, the GT version was exclusive to Mustangs and Fairlanes.

lifters. Carburetor: Two-barrel. Cooling system capacity: 20.5 qt. with heater. Crankcase capacity: (Mid-size/Mustang) 4 qt.; (Full-Size/T-Bird) 5 qt. (add 1 qt. with new oil filter). Dual exhaust. Code: X.

Thunderbird 390 Four-barrel V-8; optional in full-size Fords and Thunderbirds
Overhead valve. Cast-iron block. Displacement: 390 cid. Bore & stroke: 4.05 x 3.78 in. Compression ratio: 10.5:1. Brake hp: 315 at 4600 rpm. Torque: 427 lbs.-ft. at 2800 rpm. Five main bearings. Hydraulic valve lifters. Carburetor: Four-barrel. Cooling system capacity: 20.5 qt. with heater. Crankcase capacity: (Mid-size/Mustang) 4 qt.; (Full-Size/T-Bird) 5 qt. (add 1 qt. with new oil filter). Dual exhaust. Code: Z.

Optional Mustang, Fairlane GT 390 V-8
Overhead valve. Cast-iron block. Displacement: 390 cid. Bore & stroke: 4.05 x 3.78 in. Compression ratio: 10.5:1. Brake hp: 325 at 4800 rpm. Torque: 427 lbs.-ft. at 3200 rpm. Five main bearings. Hydraulic valve lifters. Carburetor: Holley four-barrel. Cooling system capacity: 20.5 qt.

The 1968 Ford Cobra Jet version of the 428 (a.k.a. 428-CJ) was optional equipment in the Ford Mustang, Ford Fairlane, Mercury Cougar, and new Mercury Montego. The special through-the-hood air cleaner doubled as a functional ram-induction hood scoop. The 428-CJ did not share the 427's cross-bolted main bearing caps and could not be revved as high. It topped out at 5000-6000 rpm.

An all-new lightweight 429-cid V-8 was used in the 1968 Ford Thunderbird. It featured a thin wall cast-iron block and wider bore spacing than other big-block Ford V-8s. An Autolite four-barrel carburetor was used. This motor was derived from the 460-cid Lincoln V-8.

with heater. Crankcase capacity: (Mid-size/Mustang) 4 qt.; (Full-Size/T-Bird) 5 qt. (add 1 qt. with new oil filter). Dual exhaust. Code: S.
NOTE: Some literature listed the above engine with a 335-hp rating.

Mustang, Cobra Jet 428 V-8
Overhead valve. Cast-iron block. Displacement: 428 cid. Bore & stroke: 4.13 x 3.98 in. Compression ratio: 10.70:1. Brake hp: 335 at 5600 rpm. Torque: 445 lbs.-ft. at 3400 rpm. Five main bearings. Hydraulic valve lifters. Carburetor: Holley four-barrel. Cooling system capacity: 20 qt. with heater. Crankcase capacity: 5 qt. (add 1 qt. with new oil filter). Dual exhaust. Code: Q or R.

Thunderbird 428 V-8
Overhead valve. Cast-iron block. Displacement: 428 cid. Bore & stroke: 4.13 x 3.98 in. Compression ratio: 10.50:1. Brake hp: 340 at 5400 rpm. Torque: 462 lbs.-ft. at 2800 rpm. Five main bearings. Hydraulic valve lifters. Carburetor: Holley four-barrel. Cooling system capacity: 20 qt. with heater. Crankcase capacity: 5 qt. (add 1 qt. with new oil filter). Dual exhaust. Code: Q.

Cobra Jet 428 V-8
Overhead valve. Cast-iron block. Displacement: 428 cid. Bore & stroke: 4.13 x 3.98 in. Compression ratio: 10.50:1. Brake hp: 360 at 5400 rpm. Torque: 460 lbs.-ft. at 3200 rpm. Five main bearings. Solid valve lifters. Carburetor: Holley four-barrel. Cooling system capacity: 20 qt. with heater. Crankcase capacity: 5 qt. (add 1 qt. with new oil filter). Dual exhaust. Code: P.

Thunder Jet 429 V-8
Overhead valve. Cast-iron block. Displacement: 429 cid. Bore & stroke: 4.36 x 3.59 in. Compression ratio: 10.50:1. Brake hp: 360 at 4600 rpm. Torque: 480 lbs.-ft. at 2800 rpm. Five main bearings. Hydraulic valve lifters. Carburetor: Motorcraft four-barrel. Cooling system capacity: 18.6 qt. with heater (T-bird 19.4 qt.). Crankcase capacity: 5 qt. (add 1 qt. with new oil filter). Dual exhaust. Code: N.

Mustang, Thunderbird High-Performance V-8
Overhead valve. Cast-iron block. Displacement: 427 cid. Bore & stroke: 4.23 x 3.78 in. Compression ratio: 10.90:1. Brake hp: 390 at 4600 rpm. Torque: 460 lbs.-ft. at 3200 rpm. Five main bearings. Hydraulic valve lifters. Carburetor: Holley four-barrel. Cooling system capacity: 20 qt. with heater. Crankcase capacity: 4 qt. (add 1 qt. with new oil filter). Dual exhaust. Code: W.

1969 FORD

302 Two-Barrel V-8
Overhead valve. Cast-iron block. Displacement: 302 cid. Bore & stroke: 4.00 x 3.00 in. Compression ratio: 9.50:1. Brake hp: 220 at 4600 rpm. Torque: 300 lbs.-ft. at 2600 rpm. Five main bearings. Hydraulic valve lifters. Carburetor: Two-barrel. Cooling system capacity: (Mid-size/Mustang)15 qt. with heater; (Full-size) 15.4 qt. with heater. Crankcase capacity: (Mid-size/Mustang) 4 qt.; (Full-Size/T-Bird) 5 qt. (add 1 qt. with new oil filter). Serial number code: F (police/taxi code D).

351 Two-barrel V-8
Overhead valve. Cast-iron block. Displacement: 351 cid. Bore & stroke: 4.00 x 3.50 in. Compression ratio: 9.50:1. Brake hp: 250 at 4600 rpm. Torque: 355 lbs.-ft. at 2600 rpm. Five main bearings. Hydraulic valve lifters. Carburetor: Motorcraft two-barrel. Cooling system capacity: (Mid-size/Mustang) 15 qt.; (Full-size) 16.5 qt. with heater. Crankcase capacity: (Mid-size/Mustang) 4 qt.; (Full-Size/T-Bird) 5 qt. (add 1 qt. with new oil filter). Serial number code: H.

Thunderbird 390 V-8; optional in full-size Fords; standard in Thunderbird
Overhead valve. Cast-iron block. Displacement: 390 cid. Bore & stroke: 4.05 x 3.78 in. Compression ratio: 9.5:1. Brake hp: 265 at 4400 rpm. Torque: 390 lbs.-ft. at 2600 rpm. Five main bearings. Hydraulic valve lifters. Carburetor: Two-barrel. Cooling system capacity: (Mid-size/Mustang) 20.5 qt.; (Full-size) 20.1 qt. with heater. Crankcase capacity: (Mid-size/Mustang) 4 qt.; (Full-Size/T-Bird) 5 qt. (add 1 qt. with new oil filter). Code: Y.

Thunderbird 390 High-Compression V-8; optional in full-size Fords and Thunderbirds
Overhead valve. Cast-iron block. Displacement: 390 cid. Bore & stroke: 4.05 x 3.78 in. Compression ratio: 10.5:1. Brake hp: 280 at 4400 rpm. Torque: 403 lbs.-ft. at 2600 rpm. Five main bearings. Hydraulic valve lifters. Carburetor: Holley two-barrel. Cooling system capacity: (Mid-size/Mustang) 20.5 qt.; (Full-size) 20.1 qt. with heater. Crankcase capacity: (Mid-size/Mustang) 4 qt.; (Full-Size/T-Bird) 5 qt. (add 1 qt. with new oil filter). Dual exhaust. Code: X.

Boss 302 V-8
Overhead valve. Cast-iron block. Displacement: 302 cid. Bore & stroke: 4.00 x 3.00 in. Compression ratio: 10.50:1. Brake hp: 290 at 5800 rpm. Torque: 290 lbs.-ft. at 4300 rpm. Five main bearings. Solid valve lifters. Carburetor: Holley four-barrel. Cooling system capacity: (Mid-size/Mustang) 15 qt. with heater; (Full-size) 15.4 qt. with heater. Crankcase capacity: (Mid-size/Mustang) 4 qt.; (Full-Size/T-Bird) 5 qt. (add 1 qt. with new oil filter). Serial number code: G.

351 Four-barrel V-8
Overhead valve. Cast-iron block. Displacement: 351 cid. Bore & stroke: 4.00 x 3.50 in. Compression ratio: 10.70:1. Brake hp: 290 at 4800 rpm. Torque: 385 lbs.-ft. at 3200 rpm. Five main bearings. Hydraulic valve lifters. Carburetor: Motorcraft four-barrel. Cooling system capacity: (Mid-size/Mustang) 20.5 qt.; (Full-size) 20.1 qt. with heater. Crankcase capacity: (Mid-size/Mustang) 4 qt.; (Full-Size/T-Bird) 5 qt. (add 1 qt. with new oil filter). Serial number code: M.

The 1969 Ford "Cyclone" 429 V-8 was a de-stroked version of the Lincoln 460 with identical bore measurements. This premium fuel engine had wedge-shaped combustion chambers and "porcupine" valves that were positioned at an angle so that the valve seats and faces conformed to the contours of the combustion chambers. This engine was originally intended for the Mercury Cyclone, but later was released for other Fords and Mercurys. It is not the same as the "Boss 429" made famous in the Mustang.

The 428 Cobra Jet "Ram Air" engine was a highly-competitive 1969 Mustang option and came as part of a package including the ram-air induction system, a competition suspension, a heavy-duty 80-amp battery and a 3.50:1 non-locking rear axle. The Ram-Air equipment included a through-the-hood "shaker" type air scoop and a special air cleaner assembly.

Thunderbird 390 Four-barrel V-8; optional in, Mustangs, full-size Fords and Thunderbirds
Overhead valve. Cast-iron block. Displacement: 390 cid. Bore & stroke: 4.05 x 3.78 in. Compression ratio: 10.5:1. Brake hp: 320 at 4600 rpm. Torque: 427 lbs.-ft. at 3200 rpm. Five main bearings. Hydraulic valve lifters. Carburetor: Four-barrel. Cooling system capacity: (Mid-size/Mustang) 20.5 qt.; (Full-size) 20.1 qt. with heater. Crankcase capacity: (Mid-size/Mustang) 4 qt.; (Full-Size/T-Bird) 5 qt. (add 1 qt. with new oil filter). Dual exhaust. Code: S.

Thunder Jet 429 Two-Barrel V-8
Overhead valve. Cast-iron block. Displacement: 429 cid. Bore & stroke: 4.36 x 3.59 in. Compression ratio: 10.50:1. Brake hp: 320 at 4400 rpm. Torque: 460 lbs.-ft. at 2200 rpm. Five main bearings. Hydraulic valve lifters. Carburetor: Motorcraft two-barrel. Cooling system capacity: 18.6 qt. with heater (T-bird 19.4 qt.). Crankcase capacity: 5 qt. (add 1 qt. with new oil filter). Dual exhaust. Code: K.

Cobra Jet 428 V-8
Overhead valve. Cast-iron block. Displacement: 428 cid. Bore & stroke: 4.13 x 3.98 in. Compression ratio: 10.60:1. Brake hp: 335 at 5200 rpm. Torque: 440 lbs.-ft. at 3400 rpm. Five main bearings. Hydraulic valve lifters. Carburetor: Holley four-barrel. Cooling system capacity: (Mid-size/Mustang) 20.5 qt.; (Full-size) 20.1 qt. with heater. Crankcase capacity: (Mid-size/Mustang) 4 qt.; (Full-Size/T-Bird) 5 qt. (add 1 qt. with new oil filter). Dual exhaust. Code: Q.

Cobra Jet 428 V-8
Overhead valve. Cast-iron block. Displacement: 428 cid. Bore & stroke: 4.13 x 3.98 in. Compression ratio: 10.50:1. Brake hp: 360 at 5400 rpm. Torque: 460 lbs.-ft. at 3200 rpm. Five main bearings. Solid valve lifters. Carburetor: Holley four-barrel. Cooling system capacity: (Mid-size/Mustang) 20.5 qt.; (Full-size) 20.1 qt. with heater. Crankcase capacity: (Mid-size/Mustang) 4 qt.; (Full-Size/T-Bird) 5 qt. (add 1 qt. with new oil filter). Dual exhaust. Code: P.

Thunder Jet 429 Four-Barrel V-8
Overhead valve. Cast-iron block. Displacement: 429 cid. Bore & stroke: 4.36 x 3.59 in. Compression ratio: 10.50:1. Brake hp: 360 at 4600 rpm. Torque: 480 lbs.-ft. at 2800 rpm. Five main bearings. Hydraulic valve lifters. Carburetor: Motorcraft four-barrel. Cooling system capacity: 18.6 qt. with heater (T-bird 19.4 qt.). Crankcase capacity: 5 qt. (add 1 qt. with new oil filter). Dual exhaust. Code: N.

Boss 429 V-8
Overhead valve. Cast-iron block. Displacement: 429 cid. Bore & stroke: 4.36 x 3.59 in. Compression ratio: 11.30:1. Brake hp: 375 at 5600 rpm. Torque: 450 lbs.-ft. at 3400 rpm. Five main bearings. Solid or hydraulic valve lifters. Carburetor: Holley four-barrel. Cooling system capacity: 18.6 qt. with heater (T-bird 19.4 qt.). Crankcase capacity: 5 qt. (add 1 qt. with new oil filter). Dual exhaust. Code: Z.

1970 FORD

302 Two-Barrel V-8
Overhead valve. Cast-iron block. Displacement: 302 cid. Bore & stroke: 4.00 x 3.00 in. Compression ratio: 9.50:1. Brake hp: 220 at 4600 rpm. Torque: 300 lbs.-ft. at 2600 rpm. Five main bearings. Hydraulic valve lifters. Carburetor: Motorcraft two-barrel. Cooling system capacity: (Mid-size/Mustang) 13.5 qt. with heater; (Full-size) 15.4 qt. with heater. Crankcase capacity: (Mid-size/Mustang) 4 qt.; (Full-Size) 5 qt. (add 1 qt. with new oil filter). Serial number code: F.

351 Two-Barrel "Windsor" V-8
Overhead valve. Cast-iron block. Displacement: 351 cid. Bore & stroke: 4.00 x 3.50 in. Compression ratio: 9.50:1. Brake hp: 250 at 4600 rpm. Torque: 355 lbs.-ft. at 2600 rpm. Five main bearings. Hydraulic valve lifters. Carburetor: Motorcraft two-barrel. Cooling system capacity: (Mid-size/Mustang) 14.6 qt. with heater; (Full-size) 16.5 qt. with heater.

Ford offered two Thunderjet versions, one Cobra-Jet version and one Super-Cobra-Jet version of the 429-cid V-8 in 1970. This Cobra-Jet edition fills the engine bay of a rare 1970-1/2 Ford Falcon. This Cobra-Jet version produced 370 hp at 5400 rpm, while the Super-Cobra-Jet version pumped up 375 hp at 5600 rpm.

The last year that the 428-cid V-8 was offered in a factory-built Ford product was 1970. As you can see, it was used in Mercurys and Fords in its final season. This motor was in the 352-390-428 family and was not an all-out racing engine like the 406 or the 427. However, it could easily be built up for racing and some of the heavy-duty 427 parts were useable.

The 351 Cleveland (351C) V-8 used in Ford and Mercury products got its name from the fact that it was produced at Ford engine plant No. 2 in Cleveland, Ohio. The 351C and 351W had the same bore and center-to-center spacing, but the Cleveland-built engine had a lower deck height and was the hotter of the two. The C block was also longer with the front portion stretched to cover the timing chain and gears. The 1970 351C shown here came in two-barrel (250-hp) and four-barrel (300-hp) options.

Crankcase capacity: (Mid-size/Mustang) 4 qt.; (Full-Size) 5 qt. (add 1 qt. with new oil filter). Serial number code: H.

Thunderbird 390 V-8; optional in full-size Fords; standard in Thunderbird
Overhead valve. Cast-iron block. Displacement: 390 cid. Bore & stroke: 4.05 x 3.78 in. Compression ratio: 9.5:1. Brake hp: 270 at 4400 rpm. Torque: 390 lbs.-ft. at 2600 rpm. Five main bearings. Hydraulic valve lifters. Carburetor: Motorcraft two-barrel. Cooling system capacity: (Mid-size/Mustang) 14.6 qt. with heater; (Full-size) 20.1 qt. with heater. Crankcase capacity: (Mid-size/Mustang) 4 qt.; (Full-Size) 5 qt. (add 1 qt. with new oil filter). Code: X.

Boss 302 V-8
Overhead valve. Cast-iron block. Displacement: 302 cid. Bore & stroke: 4.00 x 3.00 in. Compression ratio: 10.60:1. Brake hp: 290 at 5800 rpm. Torque: 290 lbs.-ft. at 4300 rpm. Five main bearings. Solid valve lifters. Carburetor: Holley four-barrel. Cooling system capacity: (Mid-size/Mustang) 13.5 qt. with heater; (Full-size) 15.4 qt. with heater. Crankcase capacity: (Mid-size/Mustang) 4 qt.; (Full-Size) 5 qt. (add 1 qt. with new oil filter). Serial number code: G.

351 "Cleveland" Four-Barrel V-8
Overhead valve. Cast-iron block. Displacement: 351 cid. Bore & stroke: 4.00 x 3.50 in. Compression ratio: 11.0:1. Brake hp: 300 at 5400 rpm. Torque: 380 lbs.-ft. at 3400 rpm. Five main bearings. Hydraulic valve lifters. Carburetor: Motorcraft four-barrel. Cooling system capacity: (Mid-size/Mustang) 14.6 qt. with heater; (Full-size) 16.5 qt. with heater. Crankcase capacity: (Mid-size/Mustang) 4 qt.; (Full-Size) 5 qt. (add 1 qt. with new oil filter). Serial number code: M.

Thunder Jet 429 Two-Barrel V-8
Overhead valve. Cast-iron block. Displacement: 429 cid. Bore & stroke: 4.36 x 3.59 in. Compression ratio: 10.50:1. Brake hp: 320 at 4400 rpm. Torque: 460 lbs.-ft. at 2200 rpm. Five main bearings. Hydraulic valve lifters. Carburetor: Motorcraft two-barrel. Cooling system capacity: (Mid-size/Mustang) 19 qt. with heater; (Full-size) 18.6 qt. with heater. Crankcase capacity: (Mid-size/Mustang) 4 qt.; (Full-Size) 5 qt. (add 1 qt. with new oil filter). Dual exhaust. Code: K.

Cobra Jet 428 V-8
Overhead valve. Cast-iron block. Displacement: 428 cid. Bore & stroke: 4.13 x 3.98 in. Compression ratio: 10.60:1. Brake hp: 335 at 5200 rpm. Torque: 440 lbs.-ft. at 3400 rpm. Five main bearings. Hydraulic valve lifters. Carburetor: Holley four-barrel. Cooling system capacity: 19.3 qt. with heater. Crankcase capacity: 4 qt. (add 1 qt. with new oil filter). Dual exhaust. Code: Q or R.

Super Cobra Jet 428 V-8
Overhead valve. Cast-iron block. Displacement: 428 cid. Bore & stroke: 4.13 x 3.98 in. Compression ratio: 10.50:1. Brake hp: 360 at 5400 rpm. Torque: 459 lbs.-ft. at 3200 rpm. Five main bearings. Hydraulic valve lifters. Carburetor: Holley four-barrel. Cooling system capacity: 19.3 qt. with heater. Crankcase capacity: 6 qt. (add 1 qt. with new oil filter). Dual exhaust. Code: R.
NOTE: Some Ford literature listed higher horsepower ratings for Super CJ engines, but most did not. Researcher Bruce K. Dodds believes the 360-hp version was made for police use.

Thunder Jet 429 Four-Barrel V-8
Overhead valve. Cast-iron block. Displacement: 429 cid. Bore & stroke: 4.36 x 3.59 in. Compression ratio: 10.50:1. Brake hp: 360 at 4600 rpm. Torque: 480 lbs.-ft. at 3400 rpm. Five main bearings. Hydraulic valve lifters. Carburetor: Motorcraft four-barrel. Cooling system capacity: (Mid-size/Mustang) 19 qt. with heater; (Full-size) 18.6 with heater. Crankcase capacity: (Mid-size/Mustang) 4 qt.; (Full-Size) 5 qt. (add 1 qt. with new oil filter). Dual exhaust. Code: N.

1971 FORD

The new-for-1971 400-cid V-8 was big in cubes, if not in high performance. It had a 4.00 x 4.00 bore and stroke and a 9.5:1 compression ratio. The cast-iron V-8 produced 260 hp at 4400 rpm and 400 foot-pounds of torque at 2200 rpm. It also featured hydraulic valve lifters and a Motorcraft two-barrel carburetor.

Police Interceptor 429 V-8
Overhead valve. Cast-iron block. Displacement: 429 cid. Bore & stroke: 4.36 x 3.59 in. Compression ratio: 11.30:1. Brake hp: 370 at 5400 rpm. Torque: 450 lbs.-ft. at 3400 rpm. Five main bearings. Solid valve lifters. Carburetor: Four-barrel. Cooling system capacity: (Mid-size/Mustang) 19 qt. with heater; (Full-size) 18.6 with heater. Crankcase capacity: (Mid-size/Mustang) 4 qt.; (Full-Size) 5 qt. (add 1 qt. with new oil filter). Dual exhaust. Code: P.

Boss 429 Ram Air V-8 (Aluminum cylinder head; crescent-shaped combustion chambers)
Overhead valve. Cast-iron block. Displacement: 429 cid. Bore & stroke: 4.36 x 3.59 in. Compression ratio: 10.50:1. Brake hp: 375 at 5200 rpm. Torque: 450 lbs.-ft. at 3400 rpm. Five main bearings. Hydraulic valve lifters. Carburetor: Holley four-barrel. Cooling system capacity: (Mid-size/Mustang) 19 qt. with heater; (Full-size) 18.6 qt. with heater. Crankcase capacity: (Mid-size/Mustang) 4 qt.; (Full-Size) 5 qt. (add 1 qt. with new oil filter). Dual exhaust. Code: Z.

Boss 429 Ram Air V-8 (Aluminum cylinder head; crescent-shaped combustion chambers)
Overhead valve. Cast-iron block. Displacement: 429 cid. Bore & stroke: 4.36 x 3.59 in. Compression ratio: 11.30:1. Brake hp: 375 at 5600 rpm. Torque: 450 lbs.-ft. at 3400 rpm. Five main bearings. Hydraulic valve lifters. Carburetor: Holley four-barrel. Cooling system capacity: (Mid-size/Mustang) 19 qt. with heater; (Full-size) 18.6 qt. with heater. Crankcase capacity: (Mid-size/Mustang) 4 qt.; (Full-Size) 5 qt. (add 1 qt. with new oil filter). Dual exhaust. Code: Z.

1971 302 Two-Barrel V-8; Base Mustang V-8
Overhead valve. Cast-iron block. Displacement: 302 cid. Bore & stroke: 4.00 x 3.00 in. Compression ratio: 9.00:1. Brake hp: 210 at 4600 rpm. Torque: 296 lbs.-ft. at 2600 rpm. Five main bearings. Hydraulic valve lifters. Carburetor: Motorcraft two-barrel. Cooling system capacity: (Mid-size/Mustang) 13.5 qt. with heater; (Full-size) 15.4 qt. with heater. Crankcase capacity: (Mid-size/Mustang) 4 qt.; (Full-Size) 5 qt. (add 1 qt. with new oil filter). Serial number code: F.

351 "Windsor" V-8; Base Torino/Ford V-8
Overhead valve. Cast-iron block. Displacement: 351 cid. Bore & stroke: 4.00 x 3.50 in. Compression ratio: 9.00:1. Brake hp: 240 at 4600 rpm. Torque: 350 lbs.-ft. at 2600 rpm. Hydraulic valve lifters. Carburetor: Motorcraft two-barrel. Cooling system capacity: (Mid-size/Mustang) 14.6 qt. with heater; (Full-size) 16.5 qt. with heater. Crankcase capacity: (Mid-size/Mustang) 4 qt.; (Full-Size) 5 qt. (add 1 qt. with new oil filter). Serial number code: H.

351 "Cleveland" Two-Barrel V-8
Overhead valve. Cast-iron block. Displacement: 351 cid. Bore & stroke: 4.00 x 3.50 in. Compression ratio: 9.50:1. Brake hp: 250 at 4600 rpm. Torque: 350 lbs.-ft. at 2600 rpm. Five main bearings. Hydraulic valve lifters. Carburetor: Motorcraft two-barrel. Cooling system capacity: (Mid-size/Mustang) 14.6 qt. with heater; (Full-size) 16.5 qt. with heater. Crankcase capacity: (Mid-size/Mustang) 4 qt.; (Full-Size) 5 qt. (add 1 qt. with new oil filter). Serial number code: H.

390 V-8; optional in full-size Fords
Overhead valve. Cast-iron block. Displacement: 390 cid. Bore & stroke: 4.05 x 3.78 in. Compression ratio: 8.6:1. Brake hp: 255 at 4400 rpm. Torque: 376 lbs.-ft. at 2600 rpm. Five main bearings. Hydraulic valve lifters. Carburetor: Two-barrel. Crankcase capacity: (Mid-size/Mustang) 4 qt.; (Full-Size) 5 qt. (add 1 qt. with new oil filter). Code: X.

400 V-8
Overhead valve. Cast-iron block. Displacement: 400 cid. Bore & stroke: 4.00 x 4.00 in. Compression ratio: 9.5:1. Brake hp: 260 at 4400 rpm. Torque: 400 lbs.-ft. at 2200 rpm. Hydraulic valve lifters. Carburetor: Motorcraft two-barrel. Cooling system capacity: (Mid-size/Mustang) 14.6 qt. with heater; (Full-size) 20.1 qt. with heater. Crankcase capacity: (Mid-size/Mustang) 4 qt.; (Full-Size) 5 qt. (add 1 qt. with new oil filter). Code: S.

351 "Cleveland" Four-Barrel V-8 (Cobra Jet after May 1971)
Overhead valve. Cast-iron block. Displacement: 351 cid. Bore & stroke: 4.00 x 3.50 in. Compression ratio: 10.7:1. Brake hp: 285 at 5400 rpm. (280 hp after May 1971) Torque: 370 lbs.-ft. at 3400 rpm. Five main bearings. Hydraulic valve lifters. Carburetor: Four-barrel. Cooling system capacity: (Mid-size/Mustang) 14.6 qt. with heater; (Full-size) 16.5 qt. with heater. Crankcase capacity: (Mid-size/Mustang) 4 qt.; (Full-Size) 5 qt. (add 1 qt. with new oil filter). Serial number code: M.

429 Two-barrel V-8; optional in Fords
Overhead valve. Cast-iron block. Displacement: 429 cid. Bore & stroke: 4.36 x 3.59 in. Compression ratio: 10.50:1. Brake hp: 320 at 4400 rpm. Torque: 460 lbs.-ft. at 2200 rpm. Five main bearings. Hydraulic valve lifters. Carburetor: Motorcraft two-barrel. Cooling system capacity: (Mid-size/Mustang) 19 qt. with heater; (Full-size) 18.6 qt. with heater. Crankcase capacity: (Mid-size/Mustang) 4 qt.; (Full-Size) 5 qt. (add 1 qt. with new oil filter). Dual exhaust. Code: K.

Boss 351 H.O. Four-barrel (Mustang) V-8
Overhead valve. Cast-iron block. Displacement: 351 cid. Bore & stroke: 4.00 x 3.50 in. Compression ratio: 11.7:1. Brake hp: 330 at 5400 rpm. Torque: 370 lbs.-ft. at 4000 rpm. Five main bearings. Hydraulic valve lifters. Carburetor: Holley four-barrel. Cooling system capacity: (Mid-size/Mustang) 14.6 qt. with heater; (Full-size) 16.5 qt. with heater. Crankcase capacity: (Mid-size/Mustang) 4 qt.; (Full-Size) 5 qt. (add 1 qt. with new oil filter). Serial number code: R.

429 Four-barrel V-8; Base Thunderbird V-8
Overhead valve. Cast-iron block. Displacement: 429 cid. Bore & stroke: 4.36 x 3.59 in. Compression ratio: 10.50:1. Brake hp: 360 at 4600 rpm. Torque: 480 lbs.-ft. at 2800 rpm. Five main bearings. Hydraulic valve lifters. Carburetor: Four-barrel. Cooling system capacity: (Mid-size/Mustang) 19 qt. with heater; (Full-size) 18.6 qt. with heater. Crankcase capacity: (Mid-size/Mustang) 4 qt.; (Full-Size) 5 qt. (add 1 qt. with new oil filter). Dual exhaust. Code: N.

429 Cobra Jet V-8; Optional Mustang V-8
Overhead valve. Cast-iron block. Displacement: 429 cid. Bore & stroke: 4.36 x 3.59 in. Compression ratio: 11.30:1. Brake hp: 370 at 5400 rpm. Torque: 450 lbs.-ft. at 3400 rpm. Five main bearings. Hydraulic valve lifters. Carburetor: Four-barrel. Cooling system capacity: (Mid-size/Mustang) 19 qt. with heater; (Full-size) 18.6 qt. with heater. Crankcase capacity: (Mid-size/Mustang) 4 qt.; (Full-Size) 5 qt. (add 1 qt. with new oil filter). Dual exhaust. Code: C.

429 Super Cobra Jet V-8; Optional Mustang V-8
Overhead valve. Cast-iron block. Displacement: 429 cid. Bore & stroke: 4.36 x 3.59 in. Compression ratio: 11.50:1. Brake hp: 375 at 5600 rpm. Torque: 450 lbs.-ft. at 3400 rpm. Five main bearings. Solid valve lifters. Carburetor: Holley four-barrel. Cooling system capacity: (Mid-size/Mustang) 19 qt. with heater; (Full-size) 18.6 qt. with heater. Crankcase capacity: (Mid-size/Mustang) 4 qt.; (Full-Size) 5 qt. (add 1 qt. with new oil filter). Dual exhaust. Code: J.

1972 FORD

Starting in 1972 horsepower and torque ratings are net, as installed in the vehicle.

302 Two-Barrel V-8; Compact, Intermediate, Full-Size Ford and Mustang
Overhead valve. Cast-iron block. Displacement: 302 cid. Bore & stroke: 4.00 x 3.00 in. Compression ratio: 8.50:1. Brake hp: 140 at 4000 rpm. Torque: 239 lbs.-ft. at 2000 rpm. Five main bearings. Hydraulic valve lifters. Carburetor: Motorcraft two-barrel. Cooling system capacity with air conditioning: (Maverick) 14.25 qt.; (Mustang) 15.5 qt.; (Torino) 16.25 qt. Crankcase capacity: (All) 4 qt. (add 1 qt. with new oil filter). Serial number code: F.

302 Two-Barrel V-8; Compact and Intermediate Ford
Overhead valve. Cast-iron block. Displacement: 302 cid. Bore & stroke: 4.00 x 3.00 in. Compression ratio: 8.50:1. Brake hp: 141 at 4000 rpm. Torque: 242 lbs.-ft. at 2000 rpm. Five main bearings. Hydraulic valve lifters. Carburetor: Motorcraft two-barrel. Cooling system capacity with air conditioning: (Maverick) 14.25 qt.; (Mustang) 15.5 qt.; (Torino) 16.25 qt. Crankcase capacity: (All) 4 qt. (add 1 qt. with new oil filter). Serial number code: F.

302 Two-Barrel V-8; Compact and Intermediate Ford
Overhead valve. Cast-iron block. Displacement: 302 cid. Bore & stroke: 4.00 x 3.00 in. Compression ratio: 8.50:1. Brake hp: 143 at 4200 rpm. Torque: 242 lbs.-ft. at 2000 rpm. Five main bearings. Hydraulic valve lifters. Carburetor: Motorcraft two-barrel. Cooling system capacity with air conditioning: (Maverick) 14.25 qt.; (Mustang) 15.5 qt.; (Torino) 16.25 qt. Crankcase capacity: (All) 4 qt. (add 1 qt. with new oil filter). Serial number code: F.

351 "Windsor" Two-Barrel V-8; Full-Size Ford
Overhead valve. Cast-iron block. Displacement: 351 cid. Bore & stroke: 4.00 x 3.50 in. Compression ratio: 8.30:1. Brake hp: 153 at 3800 rpm. Torque: 266 lbs.-ft. at 2000 rpm. Five main bearings. Hydraulic valve lifters. Carburetor: Motorcraft two-barrel. Cooling system capacity with air conditioning: 16.75 qt. Crankcase capacity: 4 qt. (add 1 qt. with new oil filter). Serial number code: H.

351 "Cleveland" Two-Barrel V-8; Compact and Intermediate Ford
Overhead valve. Cast-iron block. Displacement: 351 cid. Bore & stroke: 4.00 x 3.50 in. Compression ratio: 8.60:1. Brake hp: 161 at 4000 rpm. Torque: 276 lbs.-ft. at 2000 rpm. Five main bearings. Hydraulic valve

lifters. Carburetor: Motorcraft two-barrel. Cooling system capacity with air conditioning: 15.75 qt. Crankcase capacity: 4 qt. (add 1 qt. with new oil filter). Serial number code: H.

351 "Cleveland" Two-Barrel V-8; Full-Size Ford
Overhead valve. Cast-iron block. Displacement: 351 cid. Bore & stroke: 4.00 x 3.50 in. Compression ratio: 8.60:1. Brake hp: 163 at 3800 rpm. Torque: 277 lbs.-ft. at 2000 rpm. Five main bearings. Hydraulic valve lifters. Carburetor: Motorcraft two-barrel. Cooling system capacity with air conditioning: 16.75 qt. Crankcase capacity: 4 qt. (add 1 qt. with new oil filter). Serial number code H. Serial number code: H.

351 "Cleveland" Two-Barrel V-8; Compact and Intermediate Ford
Overhead valve. Cast-iron block. Displacement: 351 cid. Bore & stroke: 4.00 x 3.50 in. Compression ratio: 8.60:1. Brake hp: 164 at 4000 rpm. Torque: 276 lbs.-ft. at 2000 rpm. Five main bearings. Hydraulic valve lifters. Carburetor: Motorcraft two-barrel. Cooling system capacity with air conditioning: 15.75 qt. Crankcase capacity: 4 qt. (add 1 qt. with new oil filter). Serial number code: H.

351 Two-Barrel V-8; Mustang
Overhead valve. Cast-iron block. Displacement: 351 cid. Bore & stroke: 4.00 x 3.50 in. Compression ratio: 8.60:1. Brake hp: 168. Five main bearings. Hydraulic valve lifters. Carburetor: Motorcraft two-barrel. Cooling system capacity with air conditioning: 15.75 qt. Crankcase capacity: 4 qt. (add 1 qt. with new oil filter). Serial number code: H.

400 Two-Barrel V-8; Compact and Intermediate Ford
Overhead valve. Cast-iron block. Displacement: 400 cid. Bore & stroke: 4.00 x 4.00 in. Compression ratio: 8.40:1. Brake hp: 168 at 4200 rpm. Torque: 297 lbs.-ft. at 2200 rpm. Five main bearings. Hydraulic valve lifters. Carburetor: Motorcraft two-barrel. Cooling system capacity with air conditioning: 18.3 qt. Crankcase capacity: 4 qt. (add 1 qt. with new oil filter). Serial number code: S.

400 Two-Barrel V-8; Full-Size Ford; Delete option Thunderbird
Overhead valve. Cast-iron block. Displacement: 400 cid. Bore & stroke: 4.00 x 4.00 in. Compression ratio: 8.40:1. Brake hp: 172 at 4000 rpm. Torque: 298 lbs.-ft. at 2200 rpm. Five main bearings. Hydraulic valve lifters. Carburetor: Motorcraft two-barrel. Cooling system capacity with air conditioning: 17.75 qt. Crankcase capacity: 4 qt. (add 1 qt. with new oil filter). Serial number code: S.

351 "Cleveland" Two-Barrel V-8; Optional Compact and Intermediate Ford and Mustang
Overhead valve. Cast-iron block. Displacement: 351 cid. Bore & stroke: 4.00 x 3.50 in. Compression ratio: 8.60:1. Brake hp: 177 at 4000 rpm. Torque: 284 lbs.-ft. at 2000 rpm. Five main bearings. Hydraulic valve lifters. Carburetor: Motorcraft two-barrel. Cooling system capacity with air conditioning: 15.75 qt. Crankcase capacity: 4 qt. (add 1 qt. with new oil filter). Serial number code: H.

429 V-8; Compact and Intermediate Ford
Overhead valve. Cast-iron block. Displacement: 429 cid. Bore & stroke: 4.36 x 3.59 in. Compression ratio: 8.50:1. Brake hp: 205 at 4400 rpm. Torque: 322 lbs.-ft. at 2600 rpm. Five main bearings. Hydraulic valve lifters. Carburetor: Motorcraft two-barrel. Cooling system capacity with air conditioning: 19.5 qt. Crankcase capacity: 4 qt. (add 1 qt. with new oil filter). Serial number code: N.

429 V-8; Full-size Ford
Overhead valve. Cast-iron block. Displacement: 429 cid. Bore & stroke: 4.36 x 3.59 in. Compression ratio: 8.50:1. Brake hp: 208 at 4400 rpm. Torque: 322 lbs.-ft. at 2800 rpm. Five main bearings. Hydraulic valve lifters. Carburetor: Motorcraft two-barrel. Cooling system capacity with air conditioning: 20 qt. Crankcase capacity: 4 qt. (add 1 qt. with new oil filter). Serial number code: N.

429 V-8; Base Thunderbird
Overhead valve. Cast-iron block. Displacement: 429 cid. Bore & stroke: 4.36 x 3.59 in. Compression ratio: 8.50:1. Brake hp: 212 at 4400 rpm. Torque: 327 lbs.-ft. at 2600 rpm. Five main bearings. Hydraulic valve lifters. Carburetor: Motorcraft four-barrel. Cooling system capacity with air conditioning: 20 qt. Crankcase capacity: 4 qt. (add 1 qt. with new oil filter). Serial number code: N.

460 V-8; Optional Thunderbird
Overhead valve. Cast-iron block. Displacement: 429 cid. Bore & stroke: 4.36 x 3.85 in. Compression ratio: 8.00:1. Brake hp: 224 at 4400 rpm. Torque: 342 lbs.-ft. at 2600 rpm. Five main bearings. Hydraulic valve lifters. Carburetor: Motorcraft four-barrel. Cooling system capacity with air conditioning: 20 qt. Crankcase capacity: 4 qt. (add 1 qt. with new oil filter). Serial number code: N.

351 "Cleveland" CJ Four-Barrel V-8; Compact and Intermediate Ford
Overhead valve. Cast-iron block. Displacement: 351 cid. Bore & stroke: 4.00 x 3.50 in. Compression ratio: 8.60:1. Brake hp: 248 at 5400 rpm. Torque: 299 lbs.-ft. at 3600 rpm. Five main bearings. Hydraulic valve lifters. Carburetor: Motorcraft four-barrel. Cooling system capacity with air conditioning: 15.75 qt. Crankcase capacity: (All) 4 qt. (add 1 qt. with new oil filter). Serial number code: Q.

351 "Cleveland" Four-Barrel V-8; Compact and Intermediate Ford
Overhead valve. Cast-iron block. Displacement: 351 cid. Bore & stroke: 4.00 x 3.50 in. Compression ratio: 8.60:1. Brake hp: 262 at 5400 rpm. Torque: 299 lbs.-ft. at 3600 rpm. Five main bearings. Hydraulic valve lifters. Carburetor: Motorcraft four-barrel. Carburetor: Motorcraft four-barrel. Cooling system capacity with air conditioning: 15.75 qt. Crankcase capacity: (All) 4 qt. (add 1 qt. with new oil filter). Serial number code: Q.

351 "Cleveland" Four-Barrel HO V-8; Optional Compact, Intermediate Ford and Mustang
Overhead valve. Cast-iron block. Displacement: 351 cid. Bore & stroke: 4.00 x 3.50 in. Compression ratio: 8.60:1. Brake hp: 266 at 5400 rpm. Torque: 301 lbs.-ft. at 3600 rpm. Five main bearings. Solid valve lifters. Carburetor: Motorcraft four-barrel. Cooling system capacity with air conditioning: 15.75 qt. Crankcase capacity: (Mustang H.O.) 5 qt.; (Others) 4 qt. (add 1 qt. with new oil filter). Serial number code: R.

351 Four-Barrel H.O. V-8; Optional Mustang
Overhead valve. Cast-iron block. Displacement: 351 cid. Bore & stroke: 4.00 x 3.50 in. Compression ratio: 8.60:1. Brake hp: 275 at 6000 rpm. Torque: 286 lbs.-ft. at 3800 rpm. Five main bearings. Hydraulic valve lifters. Carburetor: Motorcraft four-barrel. Cooling system capacity with air conditioning: 15.75 qt. Crankcase capacity: 5 qt. (add 1 qt. with new oil filter). Serial number code: Q.

1973 FORD

302 Two-Barrel V-8; Compact and Intermediate Ford
Overhead valve. Cast-iron block. Displacement: 302 cid. Bore & stroke: 4.00 x 3.00 in. Compression ratio: 8.0:1. Net hp: 135 at 4200 rpm. Net torque: 228 lbs.-ft. at 2200 rpm. Carburetor: Motorcraft two-barrel. Five main bearings. Cooling system capacity with air conditioning: (Maverick) 14.25 qt.; (Mustang) 15.25 qt.; (Torino) 15.75 qt. Crankcase capacity: (All) 4 qt. (add 1 qt. with new oil filter). Serial number code: F.

302 Two-Barrel V-8; Compact and Intermediate Ford
Overhead valve. Cast-iron block. Displacement: 302 cid. Bore & stroke: 4.00 x 3.00 in. Compression ratio: 8.0:1. Net hp: 137 at 4200 rpm. Net torque: 230 lbs.-ft. at 2200 rpm. Carburetor: Motorcraft two-barrel. Five main bearings. Cooling system capacity with air conditioning: (Maverick) 14.25 qt.; (Mustang) 15.25 qt.; (Torino) 15.75 qt. Crankcase capacity: (All) 4 qt. (add 1 qt. with new oil filter). Serial number code: F.

302 Two-Barrel V-8; Mustang
Overhead valve. Cast-iron block. Displacement: 302 cid. Bore & stroke: 4.00 x 3.00 in. Compression ratio: 8.0:1. Net hp: 140 at 4000 rpm. Net torque: 232 lbs.-ft. at 2200 rpm. Carburetor: Motorcraft two-barrel. Five main bearings. Cooling system capacity with air conditioning: 15.25 qt. Crankcase capacity: 4 qt. (add 1 qt. with new oil filter). Serial number code: F.

351 "Windsor" V-8
Overhead valve. Cast-iron block. Displacement: 351 cid. Bore & stroke: 4.00 x 3.50 in. Compression ratio: 8.0:1. Net hp: 156 at 3800 rpm. Net torque: 234 lbs.-ft. at 2200 rpm. Carburetor: Motorcraft two-barrel. Five main bearings. Cooling system capacity with air conditioning: 17.0 qt. Crankcase capacity: 4 qt. (add 1 qt. with new oil filter). Serial number code: H.

351 "Cleveland" V-8
Overhead valve. Cast-iron block. Displacement: 351 cid. Bore & stroke: 4.00 x 3.50 in. Compression ratio: 8.0:1. Net hp: 154 at 4000 rpm. Net torque: 246 lbs.-ft. at 2400 rpm. Carburetor: Motorcraft two-barrel. Five main bearings. Cooling system capacity with air conditioning: (Torino) 16.75 qt.; (Full-size) 16.25 qt. Crankcase capacity: 4 qt. (add 1 qt. with new oil filter). Serial number code: H.

351 "Cleveland" V-8
Overhead valve. Cast-iron block. Displacement: 351 cid. Bore & stroke: 4.00 x 3.50 in. Compression ratio: 8.0:1. Net hp: 159 at 4000 rpm. Net torque: 250 lbs.-ft. at 2400 rpm. Carburetor: Motorcraft two-barrel. Five main bearings. Cooling system capacity with air conditioning: (Torino) 16.75 qt.; (Full-size) 16.25 qt. Crankcase capacity: 4 qt. (add 1 qt. with new oil filter). Serial number code: H.

351 "Cleveland" V-8
Overhead valve. Cast-iron block. Displacement: 351 cid. Bore & stroke: 4.00 x 3.50 in. Compression ratio: 8.6:1. Net hp: 161 at 4000 rpm. Net torque: 254 lbs.-ft. at 2400 rpm. Carburetor: Motorcraft two-barrel. Five

main bearings. Cooling system capacity with air conditioning: (Torino) 16.75 qt.; (Full-size) 16.25 qt. Crankcase capacity: 4 qt. (add 1 qt. with new oil filter). Serial number code: H.

400 V-8
Overhead valve. Cast-iron block. Displacement: 400 cid. Bore & stroke: 4.00 x 4.00 in. Compression ratio: 8.0:1. Net hp: 163 at 3800 rpm. Net torque: 300 lbs.-ft. at 2000 rpm. Carburetor: Motorcraft two-barrel. Five main bearings. Cooling system capacity with air conditioning: (Torino) 17.75 qt.; (Full-size) 18.00 qt. Crankcase capacity: 4 qt. (add 1 qt. with new oil filter). Serial number code: S.

400 V-8
Overhead valve. Cast-iron block. Displacement: 400 cid. Bore & stroke: 4.00 x 4.00 in. Compression ratio: 8.0:1. Net hp: 168 at 3800 rpm. Net torque: 310 lbs.-ft. at 2000 rpm. Carburetor: Motorcraft two-barrel. Five main bearings. Cooling system capacity with air conditioning: (Torino) 17.75 qt.; (Full-size) 18.00 qt. Crankcase capacity: 4 qt. (add 1 qt. with new oil filter). Serial number code: S.

400 V-8
Overhead valve. Cast-iron block. Displacement: 400 cid. Bore & stroke: 4.00 x 4.00 in. Compression ratio: 8.4:1. Net hp: 171 at 3600 rpm. Net torque: 314 lbs.-ft. at 2000 rpm. Carburetor: Motorcraft two-barrel. Five main bearings. Cooling system capacity with air conditioning: (Torino) 17.75 qt.; (Full-size) 18.00 qt. Crankcase capacity: 4 qt. (add 1 qt. with new oil filter). Serial number code: S.

351 "Cleveland" Mustang V-8
Overhead valve. Cast-iron block. Displacement: 351 cid. Bore & stroke: 4.00 x 3.50 in. Compression ratio: 8.6:1. Net hp: 177 at 3800 rpm. Net torque: 256 lbs.-ft. at 2400 rpm. Carburetor: Motorcraft two-barrel. Five main bearings. Cooling system capacity with air conditioning: 15.75 qt. Crankcase capacity: 4 qt. (add 1 qt. with new oil filter). Serial number code: H.

Thunderbird 429 V-8
Overhead valve. Cast-iron block. Displacement: 429 cid. Bore & stroke: 4.36 x 3.59 in. Compression ratio: 8.0:1. Net hp: 197 at 4400 rpm. Net torque: 320 lbs.-ft. at 2600 rpm. Carburetor: Motorcraft four-barrel. Five main bearings. Cooling system capacity with air conditioning: (Torino) 18.75 qt.; (Full-size) 19.50 qt. Crankcase capacity: (Torino) 6 qt.; (Others) 4 qt. (add 1 qt. with new oil filter). Serial number code: N.

Thunderbird 429 V-8
Overhead valve. Cast-iron block. Displacement: 429 cid. Bore & stroke: 4.36 x 3.59 in. Compression ratio: 8.0:1. Net hp: 198 at 4400 rpm. Net torque: 320 lbs.-ft. at 2800 rpm. Carburetor: Motorcraft four-barrel. Five main bearings. Cooling system capacity with air conditioning: (Torino) 18.75 qt.; (Full-size) 19.50 qt. Crankcase capacity: (Torino) 6 qt.; (Others) 4 qt. (add 1 qt. with new oil filter). Serial number code: N.

Thunderbird 429 V-8
Overhead valve. Cast-iron block. Displacement: 429 cid. Bore & stroke: 4.36 x 3.59 in. Compression ratio: 8.0:1. Net hp: 201 at 4400 rpm. Net torque: 322 lbs.-ft. at 2600 rpm. Carburetor: Motorcraft four-barrel. Five main bearings. Cooling system capacity with air conditioning: (Torino) 18.75 qt.; (Full-size) 19.50 qt. Crankcase capacity: (Torino) 6 qt.; (Others) 4 qt. (add 1 qt. with new oil filter). Serial number code: N.

Thunderbird 460 V-8
Overhead valve. Cast-iron block. Displacement: 460 cid. Bore & stroke: 4.36 x 3.85 in. Compression ratio: 8.0:1. Net hp: 202 at 4400 rpm. Net torque: 330 lbs.-ft. at 2800 rpm. Carburetor: Motorcraft four-barrel. Five main bearings. Cooling system capacity with air conditioning: (Torino) 19.50 qt.; (Full-size) 19.50 qt. Crankcase capacity: 4 qt. (add 1 qt. with new oil filter). Serial number code: A.

Thunderbird 460 V-8
Overhead valve. Cast-iron block. Displacement: 460 cid. Bore & stroke: 4.36 x 3.85 in. Compression ratio: 8.0:1. Net hp: 208 at 4400 rpm. Net torque: 338 lbs.-ft. at 2800 rpm. Carburetor: Motorcraft four-barrel. Five main bearings. Cooling system capacity with air conditioning: (Torino) 19.50 qt.; (Full-size) 19.50 qt. Crankcase capacity: 4 qt. (add 1 qt. with new oil filter). Serial number code: A.

Thunderbird 460 V-8
Overhead valve. Cast-iron block. Displacement: 460 cid. Bore & stroke: 4.36 x 3.85 in. Compression ratio: 8.0:1. Net hp: 219 at 4400 rpm. Net torque: 360 lbs.-ft. at 2800 rpm. Carburetor: Motorcraft four-barrel. Five main bearings. Cooling system capacity with air conditioning: (Torino) 19.50 qt.; (Full-size) 19.50 qt. Crankcase capacity: 4 qt. (add 1 qt. with new oil filter). Serial number code: A.

351 "Cobra Jet Cleveland" V-8
Overhead valve. Cast-iron block. Displacement: 351 cid. Bore & stroke: 4.00 x 3.50 in. Compression ratio: 8.0:1. Net hp: 246 at 5400 rpm (248 at 5400 rpm in Mustang). Net torque: 312 lbs.-ft. at 3600 rpm. Carburetor: Holley four-barrel. Five main bearings. Cooling system capacity with air conditioning: (Mustang) 15.75 qt.; (Torino) 16.75 qt.; (Full-size) 16.25 qt. Crankcase capacity: 4 qt. (add 1 qt. with new oil filter). Serial number code: Q.

351 "Cobra Jet Cleveland" V-8
Overhead valve. Cast-iron block. Displacement: 351 cid. Bore & stroke: 4.00 x 3.50 in. Compression ratio: 7.9:1. Net hp: 266 at 5400 rpm. Net torque: 301 lbs.-ft. at 3600 rpm. Carburetor: Holley four-barrel. Five main bearings. Cooling system capacity with air conditioning: (Mustang) 15.75 qt.; (Torino) 16.75 qt.; (Full-size) 16.25 qt. Crankcase capacity: 4 qt. (add 1 qt. with new oil filter). Serial number code: Q.
Note: Ford rated each 1973 engine with two or three different horsepower ratings, depending on the model each engine was installed in. As body size and weight increased between models, horsepower ratings increased correspondingly (i.e.: 302-cid V-8 rated at 135 hp in Maverick and 138 hp in Gran Torino). Most engine ratings varied between one and five horsepower, with the 460-cid V-8 varying 17 hp.

1974 FORD

302 V-8
Overhead valve. Cast-iron block. Displacement: 302 cid. Bore & stroke: 4.00 x 3.00 in. Compression ratio: 8.0:1. Net hp: 140 at 3800 rpm. Net torque: 230 lbs.-ft. at 2600 rpm. Carburetor: Motorcraft two-barrel. Five main bearings. Cooling system capacity with air conditioning: (Maverick) 14.25 qt.; (Torino) 15.5 qt. Crankcase capacity: (All) 4 qt. (add 1 qt. with new oil filter). Serial number code: F.

351 "Windsor" V-8
Overhead valve. Cast-iron block. Displacement: 351 cid. Bore & stroke: 4.00 x 3.50 in. Compression ratio: 8.2:1. Net hp: 163 at 4200 rpm. Net torque: 278 lbs.-ft. at 2000 rpm. Carburetor: Motorcraft two-barrel. Five main bearings. Cooling system capacity with air conditioning: 16.4 qt. Crankcase capacity: 4 qt. (add 1 qt. with new oil filter). Serial number code: H.

351 "Cleveland" V-8
Overhead valve. Cast-iron block. Displacement: 351 cid. Bore & stroke: 4.00 x 3.50 in. Compression ratio: 8.0:1. Net hp: 162 at 4000 rpm. Net torque: 278 lbs.-ft. at 2000 rpm. Carburetor: Motorcraft two-barrel. Five main bearings. Cooling system capacity with air conditioning: (Torino) 16.5 qt.; (full-size Ford) 17 qt. Crankcase capacity: (All) 4 qt. (add 1 qt. with new oil filter). Serial number code: Q.

351 "Cleveland" Four-Barrel V-8
Overhead valve. Cast-iron block. Displacement: 351 cid. Bore & stroke: 4.00 x 3.50 in. Compression ratio: 7.9:1. Net hp: 255 at 5600 rpm. Net torque: 290 lbs.-ft. at 3400 rpm. Carburetor: Motorcraft four-barrel. Five main bearings. Cooling system capacity with air conditioning: (Torino) 16.5 qt.; (full-size Ford) 17 qt. Crankcase capacity: (All) 4 qt. (add 1 qt. with new oil filter). Serial number code: Q.

400 V-8
Overhead valve. Cast-iron block. Displacement: 400 cid. Bore & stroke: 4.00 x 4.00 in. Compression ratio: 8.0:1. Net hp: 170 at 3400 rpm. Net torque: 330 lbs.-ft. at 2000 rpm. Carburetor: Motorcraft two-barrel. Five main bearings. Cooling system capacity with air conditioning: (Torino) 18.3 qt.; (full-size Ford) 18.5 qt. Crankcase capacity: (All) 4 qt. (add 1 qt. with new oil filter). Serial number code: S.

460 V-8; Optional in Fords
Overhead valve. Cast-iron block. Displacement: 460 cid. Bore & stroke: 4.36 x 3.85 in. Compression ratio: 8.0:1. Net hp: 215 at 4000 rpm. Net torque: 350 lbs.-ft. at 2600 rpm. Carburetor: Carter four-barrel. Five main bearings. Cooling system capacity with air conditioning: (Torino) 19.5 qt.; (full-size Ford) 19.5 qt. Crankcase capacity: (All) 4 qt. (add 1 qt. with new oil filter). Serial number code: A.

Thunderbird Base 460 V-8
Overhead valve. Cast-iron block. Displacement: 460 cid. Bore & stroke: 4.36 x 3.85 in. Compression ratio: 8.0:1. Net hp: 220 at 4000 rpm. Net torque: 355 lbs.-ft. at 2600 rpm. Carburetor: Carter four-barrel. Five main bearings. Cooling system capacity with air conditioning: (Torino) 19.5 qt.; (full-size Ford) 19.5 qt. Crankcase capacity: (All) 4 qt. (add 1 qt. with new oil filter). Serial number code: A.
Note: This was the first year since it was introduced that the Mustang did not offer a V-8.

1975 FORD

Maverick optional 302 V-8
Overhead valve. Cast-iron block. Displacement: 302 cid. Bore & stroke: 4.00 x 3.00 in. Compression ratio: 8.0:1. Net hp: 129 at 3800 rpm (122 hp in Mustang). Net torque: 220 lbs.-ft. at 1800 rpm. Carburetor: Motorcraft two-barrel. Five main bearings. Cooling system capacity with air conditioning: 14.1 qt. Crankcase capacity: 4 qt. (add 1 qt. with new oil filter). Serial number code: F.

351 "Windsor" V-8
Overhead valve. Cast-iron block. Displacement: 351 cid. Bore & stroke: 4.00 x 3.50 in. Compression ratio: 8.2:1. Net hp: 143 at 3600 rpm. Net torque: 255 lbs.-ft. at 2200 rpm. Carburetor: Motorcraft two-barrel. Five main bearings. Cooling system capacity with air conditioning: 16.2 qt. Crankcase capacity: 4 qt. (add 1 qt. with new oil filter). Serial number code: H.

351 "Modified" V-8
Overhead valve. Cast-iron block. Displacement: 351 cid. Bore & stroke: 4.00 x 3.50 in. Compression ratio: 8.0:1. Net hp: 148 at 3800 rpm. Net torque: 243 lbs.-ft. at 2400 rpm. Carburetor: Motorcraft two-barrel. Five main bearings. Cooling system capacity with air conditioning: 18.0 qt. Crankcase capacity: 4 qt. (add 1 qt. with new oil filter). Serial number code: H.

400 V-8
Overhead valve. Cast-iron block. Displacement: 400 cid. Bore & stroke: 4.00 x 4.00 in. Compression ratio: 8.0:1. Net hp: 158 at 3800 rpm. Net torque: 276 lbs.-ft. at 2000 rpm. Carburetor: Motorcraft two-barrel. Five main bearings. Cooling system capacity with air conditioning: (Torino) 17.5 qt.; (Full-size) 17.6 qt. Crankcase capacity: 4 qt. (add 1 qt. with new oil filter). Serial number code: S.

460 Police Interceptor V-8
Overhead valve. Cast-iron block. Displacement: 460 cid. Bore & stroke: 4.36 x 3.85 in. Compression ratio: 8.0:1. Net hp: 217 at 4000 rpm. Net torque: 365 lbs.-ft. at 2600 rpm. Carburetor: Carter four-barrel. Five main bearings. Cooling system capacity with air conditioning: (Torino) 19.7 qt.; (Full-size Ford standard) 18.5 qt.; (Full-size Ford medium-duty) 19 qt.; (Full-size Ford heavy-duty and police) 20 qt. Crankcase capacity: (Torino) 6 qt.; (Police) 6.5 qt.; (Others) 4 qt. (add 1 qt. with new oil filter). Serial number code: A.

460 V-8
Overhead valve. Cast-iron block. Displacement: 460 cid. Bore & stroke: 4.36 x 3.85 in. Compression ratio: 8.0:1. Net hp: 218 at 4000 rpm. Net torque: 369 lbs.-ft. at 2600 rpm. Carburetor: Carter four-barrel. Five main bearings. Cooling system capacity with air conditioning: (Torino) 19.7 qt.; (Full-size Ford standard) 18.5 qt.; (Full-size Ford medium-duty) 19 qt.; (Full-size Ford heavy-duty) 20 qt. Crankcase capacity: (Torino) 6 qt.; (Police) 6.5 qt.; (Others) 4 qt. (add 1 qt. with new oil filter). Serial number code: A.

1976 FORD

302 V-8
Overhead valve. Cast-iron block. Displacement: 302 cid (5.0 liters). Bore & stroke: 4.00 x 3.00 in. Compression ratio: 8.0:1. Net hp: 133 at 3600 rpm. Net torque: 243 lbs.-ft. at 1800 rpm. Carburetor: Motorcraft two-barrel. Five main bearings. Cooling system capacity with air conditioning: 14.1 qt. Crankcase capacity: 4 qt. (add 1 qt. with new oil filter). Serial number code: F.

302 V-8
Overhead valve. Cast-iron block. Displacement: 302 cid (5.0 liters). Bore & stroke: 4.00 x 3.00 in. Compression ratio: 8.0:1. Net hp: 134 at 3600 rpm. Net torque: 242 lbs.-ft. at 2000 rpm. Carburetor: Motorcraft two-barrel. Five main bearings. Cooling system capacity with air conditioning: 14.1 qt. Crankcase capacity: 4 qt. (add 1 qt. with new oil filter). Serial number code: F.

302 V-8; Mustang
Overhead valve. Cast-iron block. Displacement: 302 cid (5.0 liters). Bore & stroke: 4.00 x 3.00 in. Compression ratio: 8.0:1. Net hp: 139 at 3600 rpm. Net torque: 250 lbs.-ft. at 1600 rpm. Carburetor: Motorcraft two-barrel. Five main bearings. Cooling system capacity with air conditioning: 14.1 qt. Crankcase capacity: 4 qt. (add 1 qt. with new oil filter). Serial number code: F.

Base Elite, LTD V-8; Optional Granada V-8
90-degree, overhead valve V-8. Cast-iron block and head. Displacement: 351 cid (5.8 liters). Bore & stroke: 4.00 x 3.50 in. Compression ratio: 8.0:1. Brake horsepower: 143 at 3200 rpm. Torque: 285 lbs.-ft. at 1600 rpm. Five main bearings. Hydraulic valve lifters. Carburetor: Ford 2150A two-barrel. Cooling system capacity with air conditioning: 17.5 qt. Crankcase capacity: 4 qt. (add 1 qt. with new oil filter). Serial number code: H.

Modified 351 V-8
90-degree, overhead valve V-8. Cast-iron block and head. Displacement: 351 cid (5.8 liters). Bore & stroke: 4.00 x 3.50 in. Compression ratio: 8.0:1. Brake horsepower: 152 at 3800 rpm. Torque: 274 lbs.-ft. at 1600 rpm. Five main bearings. Hydraulic valve lifters. Carburetor: Ford 2150A two-barrel. Cooling system capacity with air conditioning: 17.5 qt. Crankcase capacity: 4 qt. (add 1 qt. with new oil filter). Serial number code: H.

Base Torino 351 Windsor V-8;
90-degree, overhead valve V-8. Cast-iron block and head. Displacement: 351 cid (5.8 liters). Bore & stroke: 4.00 x 3.50 in. Compression ratio: 8.0:1. Brake horsepower: 154 at 3400 rpm. Torque: 286 lbs.-ft. at 1800 rpm. Five main bearings. Hydraulic valve lifters. Carburetor: Ford 2150A two-barrel. Serial number code: H.

Base 500/LTD Wagon V-8; Optional Torino, Elite, LTD V-8
90-degree, overhead valve V-8. Cast-iron block and head. Displacement: 400 cid (6.6 liters). Bore & stroke: 4.00 x 4.00 in. Compression ratio: 8.0:1. Brake horsepower: 180 at 3800 rpm. Torque: 336 lbs.-ft. at 1800 rpm. Five main bearings. Hydraulic valve lifters. Carburetor: Ford 2150A two-barrel. Cooling system capacity with air conditioning: (Torino) 17.6 qt.; (Full-size Ford) 17.6 qt. Crankcase capacity: 4 qt. (add 1 qt. with new oil filter). Serial number code: S.

Base Thunderbird V-8; Optional Torino, Elite, LTD V-8
90-degree, overhead valve V-8. Cast-iron block and head. Displacement: 460 cid (7.5 liters). Bore & stroke: 4.36 x 3.85 in. Compression ratio: 8.0:1. Brake horsepower: 202 at 3800 rpm. Torque: 352 lbs.-ft. at 1600 rpm. Five main bearings. Hydraulic valve lifters. Carburetor: Motorcraft 9510 or Ford 4350A9510 four-barrel. Cooling system capacity with air conditioning: (Torino) 19.2 qt.; (Full-size Ford standard) 18.5 qt.; (Full-size Ford medium-duty) 19 qt.; (Full-size Ford heavy-duty) 20 qt. Crankcase capacity: (Torino) 6 qt.; (Police) 6.5 qt.; (Others) 4 qt. (add 1 qt. with new oil filter). Serial number code: A.

1977 FORD

Base Granada V-8
90-degree, overhead valve V-8. Cast-iron block and head. Displacement: 302 cid (5.0 liters). Bore & stroke: 4.00 x 3.00 in. Compression ratio: 8.4:1. Brake horsepower: 122 at 3200 rpm. Torque: 237 lbs.-ft. at 1600 rpm. Five main bearings. Hydraulic valve lifters. Carburetor: Motorcraft 2150 two-barrel. Serial number code: F.

Base LTD II, Thunderbird V-8; Optional Maverick, Granada V-8
90-degree, overhead valve V-8. Cast-iron block and head. Displacement: 302 cid (5.0 liters). Bore & stroke: 4.00 x 3.00 in. Compression ratio: 8.4:1. Brake horsepower: 130-137 at 3400-3600 rpm. Torque: 243-245 lbs.-ft. at 1600-1800 rpm. Five main bearings. Hydraulic valve lifters. Carburetor: Motorcraft 2150 two-barrel. Serial number code: F.

302 V-8; Mustang
Overhead valve. Cast-iron block. Displacement: 302 cid (5.0 liters). Bore & stroke: 4.00 x 3.00 in. Compression ratio: 8.0:1. Net hp: 139 at 3600 rpm. Net torque: 250 lbs.-ft. at 1600 rpm. Carburetor: Motorcraft two-barrel. Five main bearings. Cooling system capacity with air conditioning: 14.1 qt. Crankcase capacity: 4 qt. (add 1 qt. with new oil filter). Serial number code: F.

Optional LTD II, Thunderbird V-8
90-degree, overhead valve V-8. Cast-iron block and head. Displacement: 351 cid (5.8 liters). Bore & stroke: 4.00 x 3.50 in. Compression ratio: 8.3:1. Brake horsepower: 149 at 3200 rpm. Torque: 291 lbs.-ft. at 1600 rpm. Five main bearings. Hydraulic valve lifters. Carburetor: Motorcraft 2150 two-barrel. Windsor engine. Serial number code: H.

LTD II Wagon V-8; Optional Granada V-8
90-degree, overhead valve V-8, Cast-iron block and head. Displacement: 351 cid (5.8 liters). Bore & stroke: 4.00 x 3.50 in. Compression ratio: 8.3:1. Brake horsepower: 135 at 3200 rpm. Torque: 275 lbs.-ft. at 1600

rpm. Five main bearings. Hydraulic valve lifters. Carburetor: Motorcraft 2150 two-barrel. Windsor engine. Serial number code: H.

Optional Granada Ghia, LTD II, Thunderbird V-8
90-degree, overhead valve V-8, Cast-iron block and head. Displacement: 351 cid (5.8 liters). Bore & stroke: 4.00 x 3.50 in. Compression ratio: 8.0:1. Brake horsepower: 161 at 3600 rpm. Torque: 285 lbs.-ft. at 1800 rpm. Five main bearings. Hydraulic valve lifters. Carburetor: Motorcraft 2150 two-barrel. Windsor engine. Serial number code: H.

Base LTD wagon V-8: Optional LTD II, Thunderbird, LTD V-8
90-dgree, overhead valve V-8. Cast-iron block and head. Displacement: 400 cid (6.6 liters). Bore & stroke: 4.00 x 4.00 in. Compression ratio: 8:0.1. Brake horsepower: 173 at 3800 rpm. Torque: 326 lbs.-ft. at 1600 rpm. Five main bearings. Hydraulic valve lifters. Carburetor: Motorcraft 2150 two-barrel. Serial number code: S.

Optional LTD V-8
90-degree, overhead valve V-8. Cast-iron block and head. Displacement: 460 cid (7.5 liters). Bore & stroke: 4.36 x 3.85 in. Compression ratio: 8.0:1. Brake horsepower: 197 at 4000 rpm. Torque: 353 lbs.-ft. at 2000 rpm. Five main bearings. Hydraulic valve lifters. Carburetor: Motorcraft 4350 four-barrel. Serial number code: A.
Note: Horsepower and torque ratings of some Ford V-8s varied slightly, according to model.

1978 FORD

Base LTD II, Thunderbird, LTD V-8; Optional Granada V-8
90-degree, overhead valve V-8. Cast-iron block and head. Displacement: 302 cid (5.0 liters). Bore & stroke: 4.00 x 3.00 in. Compression ratio: 8.4:1. Brake horsepower: 134 at 3400 rpm. Torque: 248 lbs.-ft. at 1600 rpm. Five main bearings. Hydraulic valve lifters. Carburetor: Motorcraft 2150 two-barrel. Serial number code: F.

Optional Fairmont, Mustang V-8
90-degree, overhead valve V-8. Cast-iron block and head. Displacement: 302 cid (5.0 liters). Bore & stroke: 4.00 x 3.00 in. Compression ratio: 8.4:1. Brake horsepower: Fairmont, 139 at 3600 rpm. Torque: 250 lbs.-ft. at 1600 rpm. Five main bearings. Hydraulic valve lifters. Carburetor: Motorcraft 2150 two-barrel. Serial number code: F.

Base LTD wagon V-8; Optional LTD II V-8
90-degree, overhead valve V-8. Cast-iron block and head. Displacement: 351 cid (5.8 liters). Bore & stroke: 4.00 x 3.50 in. Compression ratio: 8.3:1. Brake horsepower: 144 at 3200 rpm. Torque: 277 lbs.-ft. at 1600 rpm. Five main bearings. Hydraulic valve lifters. Carburetor: Motorcraft 2150 two-barrel. Windsor engine. Serial number code: H.

Optional LTD V-8
90-degree, overhead valve V-8. Cast-iron block and head. Displacement: 351 cid (5.8 liters). Bore & stroke: 4.00 x 3.50 in. Compression ratio: 8.0:1. Brake horsepower: 145 at 3400 rpm. Torque: 273 lbs.-ft. at 1800 rpm. Five main bearings. Hydraulic valve lifters. Carburetor: Motorcraft 2150 two-barrel. Windsor engine. Serial number code: H.

Optional LTD II, Thunderbird V-8
90-degree, overhead valve V-8. Cast-iron block and head. Displacement: 351 cid (5.8 liters). Bore & stroke: 4.00 x 3.50 in. Compression ratio: 8.0:1. Brake horsepower: 152 at 3600 rpm. Torque: 278 lbs.-ft. at 1800 rpm. Five main bearings. Hydraulic valve lifters. Carburetor: Motorcraft 2150 two-barrel. Windsor engine. Serial number code: Q.

Optional LTD V-8
90-degree, overhead valve V-8. Cast-iron block and head. Displacement: 400 cid (6.6 liters). Bore & stroke: 4.00 x 4.00 in. Compression ratio: 8.0:1. Brake horsepower: 160 at 3800 rpm. Torque: 314 lbs.-ft. at 1800 rpm. Five main bearings. Hydraulic valve lifters. Carburetor: Motorcraft 2150 two-barrel. Serial number code: S.

Optional LTD II, Thunderbird V-8
90-degree, overhead valve V-8. Cast-iron block and head. Displacement: 400 cid (6.6 liters). Bore & stroke: 4.00 x 4.00 in. Compression ratio: 8.0:1. Brake horsepower: 166 at 3800 rpm. Torque: 319 lbs.-ft. at 1800 rpm. Five main bearings. Hydraulic valve lifters. Carburetor: Motorcraft 2150 two-barrel. Serial number code: S.

Optional LTD V-8
90-degree, overhead valve V-8. Cast-iron block and head. Displacement: 460 cid (7.5 liters). Bore & stroke: 4.36 x 3.85 in. Compression ratio: 8.0:1. Brake horsepower: 202 at 4000 rpm. Torque: 348 lbs.-ft. at 2000 rpm. Five main bearings. Hydraulic valve lifters. Carburetor: Motorcraft 4350 four-barrel. Serial number code: A.

Note: Horsepower and torque ratings of some Ford V-8s varied slightly, according to model.

1979 FORD

Base LTD V-8; Optional (Fairmont, Granada)
90-degree, overhead valve V-8. Cast-iron block and head. Displacement: 302 cid (5.0 liters). Bore & stroke: 4.00 x 3.00 in. Compression ratio: 8.4:1. Brake horsepower: 129 at 3600 rpm. Torque: 223 lbs.-ft. at 2600 rpm. Five main bearings. Hydraulic valve lifters. Carburetor: Motorcraft 2150 or 2700VV two-barrel. Serial number code: F.

Base LTD II, Thunderbird V-8
90-degree, overhead valve V-8. Cast-iron block and head. Displacement: 302 cid (5.0 liters). Bore & stroke: 4.00 x 3.00 in. Compression ratio: 8.4:1. Brake horsepower: 133 at 3400 rpm. Torque: LTD II/T-Bird 245 lbs.-ft. at 1600 rpm. Five main bearings. Hydraulic valve lifters. Carburetor: Motorcraft 2150 or 2700VV two-barrel. Serial number code: F.

Base Granada V-8
90-degree, overhead valve V-8. Cast-iron block and head. Displacement: 302 cid (5.0 liters). Bore & stroke: 4.00 x 3.00 in. Compression ratio: 8.4:1. Brake horsepower: 137 at 3600 rpm. Torque: 243 lbs.-ft. at 2000 rpm. Five main bearings. Hydraulic valve lifters. Carburetor: Motorcraft 2150 or 2700VV two-barrel. Serial number code: F.

Base Fairmont V-8; Optional Mustang V-8
90-degree, overhead valve V-8. Cast-iron block and head. Displacement: 302 cid (5.0 liters). Bore & stroke: 4.00 x 3.00 in. Compression ratio: 8.4:1. Brake horsepower: 140 at 3600 rpm. Torque: 250 lbs.-ft. at 1800 rpm. Five main bearings. Hydraulic valve lifters. Carburetor: Motorcraft 2150 or 2700VV two-barrel. Serial number code: F.

Optional LTD, Thunderbird V-8
90-degree, overhead valve V-8. Cast-iron block and head. Displacement: 351 cid (5.8 liters). Bore & stroke: 4.00 x 3.50 in. Compression ratio: 8.3:1. Brake horsepower: 135 or 142 at 3200 rpm. Torque: 286 lbs.-ft. at 1400 rpm. Five main bearings. Hydraulic valve lifters. Carburetor: Motorcraft 7200VV two-barrel. Windsor engine. Serial number code: H.

Optional LTD II, LTD, Thunderbird V-8
90-degree, modified overhead valve V-8. Cast-iron block and head. Displacement: 351 cid (5.8 liters). Bore & stroke: 4.00 x 3.50 in. Compression ratio: 8.0:1. Brake horsepower: 151 at 3600 rpm. Torque: 270 lbs.-ft. at 2200 rpm. Five main bearings. Hydraulic valve lifters. Carburetor: Motorcraft 2150 two-barrel. Windsor engine. Serial number code: H.
Note: Horsepower and torque ratings of some Ford V-8s varied slightly, according to model.

1980 FORD

Base Thunderbird V-8
90-degree, overhead valve V-8. Cast-iron block and head. Displacement: 255 cid (4.2 liters). Bore & stroke: 3.68 x 3.00 in. Compression ratio: 8.8:1. Brake horsepower: 115 at 3800 rpm. Torque: 194 lbs.-ft. at 2200 rpm. Five main bearings. Hydraulic valve lifters. Carburetor: Motorcraft 2150 two-barrel. Serial number code: D.

Optional Fairmont; Mustang V-8
90-degree, overhead valve V-8. Cast-iron block and head. Displacement: 255 cid (4.2 liters). Bore & stroke: 3.68 x 3.00 in. Compression ratio: 8.8:1. Brake horsepower: 119 at 3800 rpm. Torque: 194 lbs.-ft. at 2200 rpm. Five main bearings. Hydraulic valve lifters. Carburetor: Motorcraft 2150 two-barrel. Serial number code: D.

Base LTD, Thunderbird Silver Anniversary V-8
90-degree, overhead valve V-8. Cast-iron block and head. Displacement: 302 cid (5.0 liters). Bore & stroke: 4.00 x 3.00 in. Compression ratio: 8.4:1. Brake horsepower: 130 at 3600 rpm. Torque: 230 lbs.-ft. at 1600 rpm. Five main bearings. Hydraulic valve lifters. Carburetor: Motorcraft 2150 or 2700VV two-barrel. Serial number code: F.

Optional Thunderbird V-8
90-degree, overhead valve V-8. Cast-iron block and head. Displacement: 302 cid (5.0 liters). Bore & stroke: 4.00 x 3.00 in. Compression ratio: 8.4:1. Brake horsepower: 131 at 3600 rpm. Torque: 231 lbs.-ft. at 1600

rpm. Five main bearings. Hydraulic valve lifters. Carburetor: Motorcraft 2150 or 2700VV two-barrel. Serial number code: F.

Optional Granada V-8
90-degree, overhead valve V-8. Cast-iron block and head. Displacement: 302 cid (5.0 liters). Bore & stroke: 4.00 x 3.00 in. Compression ratio: 8.4:1. Brake horsepower: 134 at 3600 rpm. Torque: 232 lbs.-ft. at 1600 rpm. Five main bearings. Hydraulic valve lifters. Carburetor: Motorcraft 2150 or 2700VV two-barrel. Serial number code: F.

Optional LTD V-8
90-degree, overhead valve V-8. Cast-iron block and head. Displacement: 351 cid (5.8 liters). Bore & stroke: 4.00 x 3.50 in. Compression ratio: 8.3:1. Brake horsepower: 140 at 3400 rpm. Torque: 265 lbs.-ft. at 2000 rpm. Five main bearings. Hydraulic valve lifters. Carburetor: Motorcraft 7200VV two-barrel. Windsor engine. Serial number code: G.
Note: Horsepower and torque ratings of some Ford V-8s varied slightly, according to model.

1981 FORD

Base LTD V-8
90-degree, overhead valve V-8. Cast-iron block and head. Displacement: 255 cid (4.2 liters). Bore & stroke: 3.68 x 3.00 in. Compression ratio: 8.2:1. Brake horsepower: 120 at 3400 rpm. Torque: 205 lbs.-ft. at 2600 rpm. Five main bearings. Hydraulic valve lifters. Carburetor: Motorcraft 2150 or 7200VV two-barrel. Serial number code: D.

Optional Fairmont, Granada, Thunderbird, Mustang V-8
90-degree, overhead valve V-8. Cast-iron block and head. Displacement: 255 cid (4.2 liters). Bore & stroke: 3.68 x 3.00 in. Compression ratio: 8.2:1. Brake horsepower: 115 at 3400 rpm. Torque: 195 lbs.-ft. at 2200 rpm. Five main bearings. Hydraulic valve lifters. Carburetor: Motorcraft 2150 or 7200VV two-barrel. Serial number code: D.

Optional LTD V-8
90-degree, overhead valve V-8. Cast-iron block and head. Displacement: 302 cid (5.0 liters). Bore & stroke: 4.00 x 3.00 in. Compression ratio: 8.4:1. Brake horsepower: 130 at 3400 rpm. Torque: 235 lbs.-ft. at 1800 rpm. Five main bearings. Hydraulic valve lifters. Carburetor: Motorcraft 2150 or 7200VV two-barrel. Serial number code: F.

Optional Thunderbird V-8
90-degree, overhead valve V-8. Cast-iron block and head. Displacement: 302 cid (5.0 liters). Bore & stroke: 4.00 x 3.00 in. Compression ratio: 8.4:1. Brake horsepower: 130 at 3400 rpm. Torque: 235 lbs.-ft. at 1600 rpm. Five main bearings. Hydraulic valve lifters. Carburetor: Motorcraft 2150 or 7200VV two-barrel. Serial number code: F.

Optional LTD V-8
90-degree, overhead valve V-8. Cast-iron block and head. Displacement: 351 cid (5.8 liters). Bore & stroke: 4.00 x 3.50 in. Compression ratio: 8.3:1. Brake horsepower: 145 at 3200 rpm. Torque: 270 lbs.-ft. at 1800 rpm. Five main bearings. Hydraulic valve lifters. Carburetor: Motorcraft 7200VV two-barrel. Windsor engine. Serial number code: G.

1982 FORD

Base LTD V-8
90-degree, overhead valve V-8. Cast-iron block and head. Displacement: 255 cid (4.2 liters). Bore & stroke: 3.68 x 3.00 in. Compression ratio: 8.2:1. Brake horsepower: 122 at 3400 rpm. Torque: 209 lbs.-ft. at 2400 rpm. Five main bearings. Hydraulic valve lifters. Carburetor: Motorcraft 2150 or 7200VV two-barrel. Serial number code: D.
Note: The 255 cid V-8 was also offered in Fairmont police cars.

Optional Thunderbird, Mustang V-8
90-degree, overhead valve V-8. Cast-iron block and head. Displacement: 255 cid (4.2 liters). Bore & stroke: 3.68 x 3.00 in. Compression ratio: 8.2:1. Brake horsepower: 120 at 3400 rpm. Torque: 205 lbs.-ft. at 1600 rpm. Five main bearings. Hydraulic valve lifters. Carburetor: Motorcraft 2150 or 7200VV two-barrel. Serial number code: D.
Note: The 255 cid V-8 was also offered in Fairmont police cars.

Base LTD wagon V-8; Optional LTD sedan V-8
90-degree, overhead valve V-8. Cast-iron block and head. Displacement: 302 cid (5.0 liters). Bore & stroke: 4.00 x 3.00 in. Compression ratio: 8.4:1. Brake horsepower: 132 at 3400 rpm. Torque: 236 lbs.-ft. at 1800 rpm. Five main bearings. Hydraulic valve lifters. Carburetor: Motorcraft 7200VV two-barrel. Serial number code: F.

Optional 5.0-Liter High-Output Mustang V-8
90-degree, overhead valve V-8. Cast-iron block and head. Displacement: 302 cid (5.0 liters). Bore & stroke: 4.00 x 3.00 in. Compression ratio: 8.3:1. Brake horsepower: 157 at 4200 rpm. Torque: 240 lbs.-ft. at 2400 rpm. Five main bearings. Hydraulic valve lifters. Carburetor: Motorcraft 2150A two-barrel. Serial number code: F.

Police High-Output LTD V-8
90-degree, overhead valve V-8. Cast-iron block and head. Displacement: 351 cid (5.8 liters). Bore & stroke: 4.00 x 3.50 in. Compression ratio: 8.3:1. Brake horsepower: 165 at 3600 rpm. Torque: 285 lbs.-ft. at 2200 rpm. Five main bearings. Hydraulic valve lifters. Carburetor: Motorcraft 2150A or 7200VV two-barrel. Serial number code: G.

1983 FORD

Base Crown Victoria V-8; Optional LTD, Thunderbird V-8
90-degree, overhead valve V-8. Cast-iron block and head. Displacement: 302 cid (5.0 liters). Bore & stroke: 4.00 x 3.00 in. Compression ratio: 8.4:1. Brake horsepower: 130 at 3200 rpm. Torque: 240 lbs.-ft. at 2000 rpm. Five main bearings. Hydraulic valve lifters. Induction: Electronic fuel injection. Serial number code: F.

Optional Crown Victoria High-Output V-8
90-degree, overhead valve V-8. Cast-iron block and head. Displacement: 302 cid (5.0 liters). Bore & stroke: 4.00 x 3.00 in. Compression ratio: 8.4:1. Brake horsepower: 145 at 3600 rpm. Torque: 245 lbs.-ft. at 2200 rpm. Five main bearings. Hydraulic valve lifters. Induction: Electronic fuel injection. Serial number code: F.

Optional Mustang 5.0-Liter V-8
90-degree, overhead valve V-8. Cast-iron block and head. Displacement: 302 cid (5.0 liters). Bore & stroke: 4.00 x 3.00 in. Compression ratio: 8.3:1. Brake horsepower: 175 at 4000 rpm. Torque: 245 lbs.-ft. at 2400 rpm. Five main bearings. Hydraulic valve lifters. Carburetor: Holley four-barrel Model 4180. Serial number code: F.

Crown Victoria High-Output Police V-8
90-degree, overhead valve V-8. Cast-iron block and head. Displacement: 351 cid (5.8 liters). Bore & stroke: 4.00 x 3.50 in. Compression ratio: 8.3:1. Brake horsepower: 165 at 3600 rpm. Torque: 290 lbs.-ft. at 2200 rpm. Five main bearings. Hydraulic valve lifters. Carburetor: Two-barrel VV. Serial number code: G.

1984 FORD

Base Crown Victoria V-8; Optional Thunderbird V-8
90-degree, overhead valve V-8. Cast-iron block and head. Displacement: 302 cid (5.0 liters). Bore & stroke: 4.00 x 3.00 in. Compression ratio: 8.4:1. Brake horsepower: 140 at 3200 rpm. Torque: 250 lbs.-ft. at 1600 rpm. Five main bearings. Hydraulic valve lifters. Induction: Electronic fuel injection (TBI). Serial number code: F.

Base Crown Victoria Wagon High-Output V-8
90-degree, overhead valve V-8. Cast-iron block and head. Displacement: 302 cid (5.0 liters). Bore & stroke: 4.00 x 3.00 in. Compression ratio: 8.4:1. Brake horsepower: 155 at 3600 rpm. Torque: 265 lbs.-ft. at 2000 rpm. Five main bearings. Hydraulic valve lifters. Induction: Electronic fuel injection (TBI). Serial number code: F.

Optional Mustang 5.0-Liter V-8
90-degree, overhead valve V-8. Cast-iron block and head. Displacement: 302 cid (5.0 liters). Bore & stroke: 4.00 x 3.00 in. Compression ratio: 8.3:1. Brake horsepower: 175 at 4000 rpm. Torque: 245 lbs.-ft. at 2400 rpm. Five main bearings. Hydraulic valve lifters. Carburetor: Holley four-barrel Model 4180C. Serial number code: F.
Note: A high-output version of the four-barrel Mustang V-8 rated at 205 hp was announced, but delayed.

Optional Mustang 5.0-Liter V-8
90-degree, overhead valve V-8. Cast-iron block and head. Displacement: 302 cid (5.0 liters). Bore & stroke: 4.00 x 3.00 in. Compression ratio: 8.3:1. Brake horsepower: 165 at 3800 rpm. Torque: 245 lbs.-ft. at 2000

rpm. Five main bearings. Hydraulic valve lifters. Induction: Fuel-injected. Serial number code: F.

Crown Victoria High-Output Police V-8
90-degree, overhead valve V-8. Cast-iron block and head. Displacement: 351 cid (5.8 liters). Bore & stroke: 4.00 x 3.50 in. Compression ratio: 8.3:1. Brake horsepower: 180 at 3600 rpm. Torque: 285 lbs.-ft. at 2400 rpm. Five main bearings. Hydraulic valve lifters. Carburetor: Two-barrel VV. Serial number code: G.

1985 FORD

Base Crown Victoria V-8; Optional Thunderbird V-8
90-degree, overhead valve V-8. Cast-iron block and head. Displacement: 302 cid (5.0 liters). Bore & stroke: 4.00 x 3.00 in. Compression ratio: 8.4:1. Brake horsepower: 140 at 3200 rpm. Torque: 250 lbs.-ft. at 1600 rpm. Five main bearings. Hydraulic valve lifters. Induction: Electronic fuel injection (TBI). Serial number code: F.

Optional Crown Victoria High-Output V-8
90-degree, overhead valve V-8. Cast-iron block and head. Displacement: 302 cid (5.0 liters). Bore & stroke: 4.00 x 3.00 in. Compression ratio: 8.4:1. Brake horsepower: 155 at 3600 rpm. Torque: 265 lbs.-ft. at 2000 rpm. Five main bearings. Hydraulic valve lifters. Induction: Electronic fuel injection (TBI). Serial number code: F.

Base LTD LX V-8
90-degree, overhead valve V-8. Cast-iron block and head. Displacement: 302 cid (5.0 liters). Bore & stroke: 4.00 x 3.00 in. Compression ratio: 8.3:1. Brake horsepower: 165 at 3800 rpm. Torque: 245 lbs.-ft. at 2000 rpm. Five main bearings. Hydraulic valve lifters. Induction: Electronic fuel injection (TBI). Serial number code: F.

Optional Mustang 5.0-Liter V-8 (Automatic transmission only)
90-degree, overhead valve V-8. Cast-iron block and head. Displacement: 302 cid (5.0 liters). Bore & stroke: 4.00 x 3.00 in. Compression ratio: 8.3:1. Brake horsepower: 180 at 4200 rpm. Torque: 260 lbs.-ft. at 2600 rpm. Five main bearings. Hydraulic valve lifters. Induction: Electronic fuel injection. Serial number code: F.

Optional Mustang 5.0-Liter V-8 (Standard transmission only; first roller cam)
90-degree, overhead valve V-8. Cast-iron block and head. Displacement: 302 cid (5.0 liters). Bore & stroke: 4.00 x 3.00 in. Compression ratio: 8.3:1. Brake horsepower: 210 at 4400 rpm. Torque: 270 lbs.-ft. at 3200 rpm. Five main bearings. Roller cam. Carburetor: Holley four-barrel Model 4180C. Serial number code: M.

Optional Crown Victoria High-output Police V-8
90-degree, overhead valve V-8. Cast-iron block and head. Displacement: 351 cid (5.8 liters). Bore & stroke: 4.00 x 3.50 in. Compression ratio: 8.3:1. Brake horsepower: 180 at 3600 rpm. Torque: 285 lbs.-ft. at 2400 rpm. Five main bearings. Hydraulic valve lifters. Carburetor: Motorcraft 7200VV two-barrel. Serial number code: G.

1986 FORD

Base Crown Victoria V-8; Optional Thunderbird V-8
90-degree, overhead valve V-8. Cast-iron block and head. Displacement: 302 cid (5.0 liters). Bore & stroke: 4.00 x 3.00 in. Compression ratio: 8.9:1. Brake horsepower: 150 at 3200 rpm. Torque: 270 lbs.-ft. at 2000 rpm. Five main bearings. Roller cam. Induction: Sequential (port) fuel injection. Serial number code: F.

Optional Mustang 5.0-Liter V-8
90-degree, overhead valve V-8. Cast-iron block and head. Displacement: 302 cid (5.0 liters). Bore & stroke: 4.00 x 3.00 in. Compression ratio: 9.2:1. Brake horsepower: 200 at 4000 rpm. Torque: 285 lbs.-ft. at 3000 rpm. Five main bearings. Roller cam. Induction: Electronic fuel injection. Serial number code: M.

Optional Crown Victoria High-output Police V-8
90-degree, overhead valve V-8. Cast-iron block and head. Displacement: 351 cid (5.8 liters). Bore & stroke: 4.00 x 3.50 in. Compression ratio: 8.3:1. Brake horsepower: 180 at 3600 rpm. Torque: 285 lbs.-ft. at 2400 rpm. Five main bearings. Roller cam. Carburetor: Two-barrel. Serial number code: G.

1987 FORD

Base Mustang GT V-8: Optional Mustang LX V-8
90-degree, overhead valve V-8. Cast-iron block and head. Displacement: 302 cid (5.0 liters). Bore & stroke: 4.00 x 3.00 in. Compression ratio: 9.2:1. Brake horsepower: 225 at 4000 rpm. Torque: 300 lbs.-ft. at 3200 rpm. Five main bearings. Roller cam. Induction: Sequential (port) fuel injection. Serial number code: M.

Base LTD Crown Victoria V-8
90-degree, overhead valve V-8. Cast-iron block and head. Displacement: 302 cid (5.0 liters). Bore & stroke: 4.00 x 3.00 in. Compression ratio: 8.9:1. Brake horsepower: 150 at 3200 rpm. Torque: 270 lbs.-ft. at 2000 rpm. Five main bearings. Roller cam. Induction: Sequential (port) fuel injection. Serial number code: F.

Optional Thunderbird V-8
90-degree, overhead valve V-8. Cast-iron block and head. Displacement: 302 cid (5.0 liters). Bore & stroke: 4.00 x 3.00 in. Compression ratio: 8.9:1. Brake horsepower: 155 at 3400 rpm. Torque: 265 lbs.-ft. at 2200 rpm. Five main bearings. Roller cam. Induction: Sequential (port) fuel injection. Serial number code: F.

1988 FORD

Base Mustang GT V-8; Optional Mustang LX V-8
90-degree, overhead valve V-8. Cast-iron block and head. Displacement: 302 cid (5.0 liters). Bore & stroke: 4.00 x 3.00 in. Compression ratio: 9.2:1. Brake horsepower: 225 at 4200 rpm. Torque: 300 lbs.-ft. at 3200 rpm. Five main bearings. Roller cam. Induction: Sequential (port) fuel injection. Serial number code: M.

Base LTD Crown Victoria V-8
90-degree, overhead valve V-8. Cast-iron block and head. Displacement: 302 cid (5.0 liters). Bore & stroke: 4.00 x 3.00 in. Compression ratio: 8.9:1. Brake horsepower: 150 at 3200 rpm. Torque: 270 lbs.-ft. at 2000 rpm. Five main bearings. Roller cam. Induction: Sequential (port) fuel injection. Serial number code: F.

Optional Thunderbird V-8
90-degree, overhead valve V-8. Cast-iron block and head. Displacement: 302 cid (5.0 liters). Bore & stroke: 4.00 x 3.00 in. Compression ratio: 8.9:1. Brake horsepower: 155 at 3400 rpm. Torque: 265 lbs.-ft. at 2200 rpm. Five main bearings. Roller cam. Induction: Sequential (port) fuel injection. Serial number code: F.

1989 FORD

Base Mustang GT V-8; Optional Mustang LX V-8
90-degree, overhead valve V-8. Cast-iron block and head. Displacement: 302 cid (5.0 liters). Bore & stroke: 4.00 x 3.00 in. Compression ratio: 9.2:1. Brake horsepower: 225 at 4200 rpm. Torque: 300 lbs.-ft. at 3200 rpm. Five main bearings. Roller cam. Induction: Sequential (port) fuel injection. Serial number code: M.

Base LTD Crown Victoria V-8
90-degree, overhead valve V-8. Cast-iron block and head. Displacement: 302 cid (5.0 liters). Bore & stroke: 4.00 x 3.00 in. Compression ratio: 8.9:1. Brake horsepower: 150 at 3200 rpm. Torque: 270 lbs.-ft. at 2000 rpm. Five main bearings. Roller cam. Sequential (port) fuel injection.

1990 FORD

Base Mustang GT V-8; Optional Mustang LX V-8
90-degree, overhead valve V-8. Cast-iron block and head. Displacement:

302 cid (5.0 liters). Bore & stroke: 4.00 x 3.00 in. Compression ratio: 9.0:1. Brake horsepower: 225 at 4200 rpm. Torque: 300 lbs.-ft. at 3200 rpm. Five main bearings. Roller cam. Induction: Sequential (port) fuel injection. Serial number code: M.

Base LTD Crown Victoria V-8
90-degree, overhead valve V-8. Cast-iron block and head. Displacement: 302 cid (5.0 liters). Bore & stroke: 4.00 x 3.00 in. Compression ratio: 8.9:1. Brake horsepower: 150 at 3200 rpm. Torque: 270 lbs.-ft. at 2000 rpm. Five main bearings. Roller cam. Induction: Sequential (port) fuel injection.

1991 FORD

Base Mustang GT V-8; Optional Mustang LX V-8
90-degree, overhead valve V-8. Cast-iron block and head. Displacement: 302 cid (5.0 liters). Bore & stroke: 4.00 x 3.00 in. Compression ratio: 9.0:1. Brake horsepower: 225 at 4200 rpm. Torque: 300 lbs.-ft. at 3200 rpm. Five main bearings. Roller cam. Induction: Sequential (port) fuel injection. Serial number code: E.

Base LTD Crown Victoria V-8
90-degree, overhead valve V-8. Cast-iron block and head. Displacement: 302 cid (5.0 liters). Bore & stroke: 4.00 x 3.00 in. Compression ratio: 8.9:1. Brake horsepower: 150 at 3200 rpm. Torque: 270 lbs.-ft. at 2000 rpm. Five main bearings. Roller cam. Induction: Sequential fuel injection.

Optional Thunderbird, Thunderbird LX V-8
90-degree, overhead valve V-8. Cast-iron block and head. Displacement: 302 cid (5.0 liters). Bore & stroke: 4.00 x 3.00 in. Compression ratio: 9.0:1. Brake horsepower: 200 at 4000 rpm. Torque: 275 lbs.-ft. at 3000 rpm. Five main bearings. Roller cam. Induction: Sequential fuel injection.

1992 FORD

Base Mustang GT V-8; Optional Mustang LX V-8
90-degree, overhead valve V-8. Cast-iron block and head. Displacement: 302 cid (5.0 liters). Bore & stroke: 4.00 x 3.00 in. Compression ratio: 9.0:1. Brake horsepower: 225 at 4200 rpm. Torque: 300 lbs.-ft. at 3200 rpm. Five main bearings. Roller cam. Induction: Sequential (port) fuel injection. Serial number code: E.

Base Crown Victoria V-8
Modular, overhead valve V-8. Single overhead cam. Displacement: 281 cid (4.6 liters). Bore & stroke: 3.60 x 3.60 in. Compression ratio: 9.0:1. Brake horsepower: 190 at 4200 rpm. Torque: 260 lbs.-ft. at 3200 rpm. Roller cam. Induction: Sequential fuel injection.

Base Thunderbird Sport V-8; Optional Thunderbird, Thunderbird LX V-8
90-degree, overhead valve V-8. Cast-iron block and head. Displacement: 302 cid (5.0 liters). Bore & stroke: 4.00 x 3.00 in. Compression ratio: 9.0:1. Brake horsepower: 200 at 4000 rpm. Torque: 275 lbs.-ft. at 3000 rpm. Five main bearings. Roller cam. Induction: Sequential fuel injection.

1993 FORD

Base Mustang GT V-8; Optional Mustang LX V-8
90-degree, overhead valve V-8. Cast-iron block and head. Displacement: 302 cid (5.0 liters). Bore & stroke: 4.00 x 3.00 in. Compression ratio: 9.0:1. Brake horsepower: 205 at 4200 rpm. Torque: 275 lbs.-ft. at 3000 rpm. Five main bearings. Roller cam. Induction: Sequential (port) fuel injection. Serial number code: E.

Base Mustang Cobra V-8
90-degree, overhead valve V-8. Cast-iron block and head. Displacement: 302 cid (5.0 liters). Bore & stroke: 4.00 x 3.00 in. Compression ratio: 9.0:1. Brake horsepower: 230 at 4200 rpm. Torque: 280 lbs.-ft. at 3800 rpm. Five main bearings. Roller cam. Induction: Sequential (port) fuel injection. Serial number code: D.

Base Crown Victoria V-8
Modular, overhead cam V-8. Single overhead cam. Displacement: 281 cid (4.6 liters). Bore & stroke: 3.60 x 3.60 in. Compression ratio: 9.0:1. Brake horsepower: 190 at 4600 rpm. Torque: 260 lbs.-ft. at 3200 rpm. Roller cam. Induction: Sequential fuel injection.

Optional Crown Victoria V-8
Modular, overhead cam V-8. Single overhead cam. Displacement: 281 cid (4.6 liters). Bore & stroke: 3.60 x 3.60 in. Compression ratio: 9.0:1. Brake horsepower: 210 at 4600 rpm. Torque: 270 lbs.-ft. at 3400 rpm. Hydraulic valve lifters. Induction: Sequential fuel injection.

Optional Thunderbird LX, Thunderbird Super Coupe V-8
High output, 90-degree, overhead valve V-8. Cast-iron block and head. Displacement: 302 cid (5.0 liters). Bore & stroke: 4.00 x 3.00 in. Compression ratio: 9.0:1. Brake horsepower: 200 at 4000 rpm. Torque: 275 lbs.-ft. at 3000 rpm. Five main bearings. Roller cam. Induction: Sequential fuel injection.

1994 FORD

Base Mustang GT V-8: Optional Mustang LX V-8
90-degree, overhead valve V-8. Cast-iron block and head. Displacement: 302 cid (5.0 liters). Bore & stroke: 4.00 x 3.00 in. Compression ratio: 9.0:1. Brake horsepower: 215 at 4200 rpm. Torque: 285 lbs.-ft. at 3400 rpm. Five main bearings. Hydraulic valve lifters. Induction: Sequential (port) fuel injection. Serial number code: E.

Base Mustang Cobra V-8
90-degree, overhead valve V-8. Cast-iron block and head. Displacement: 302 cid (5.0 liters). Bore & stroke: 4.00 x 3.00 in. Compression ratio: 9.0:1. Brake horsepower: 240 at 4800 rpm. Torque: 285 lbs.-ft. at 4000 rpm. Five main bearings. Hydraulic valve lifters. Induction: Sequential (port) fuel injection. Serial number code: D.

Base Crown Victoria V-8
Modular, overhead cam V-8. Single overhead cam. Displacement: 281 cid (4.6 liters). Bore & stroke: 3.60 x 3.60 in. Compression ratio: 9.0:1. Brake horsepower: 190 at 4600 rpm. Torque: 260 lbs.-ft. at 3200 rpm. Hydraulic valve lifters. Induction: Sequential fuel injection.

Optional Thunderbird LX V-8
Modular, overhead cam V-8. Single overhead cam. Displacement: 281 cid (4.6 liters). Bore & stroke: 3.60 x 3.60 in. Compression ratio: 9.0:1. Brake horsepower: 205 at 4500 rpm. Torque: 265 lbs.-ft. at 3200 rpm. Hydraulic valve lifters. Induction: Sequential fuel injection.

1995 FORD

Base Mustang GT V-8; Optional Mustang LX V-8
90-degree, overhead cam. V-8. Cast-iron block and head. Displacement: 302 cid (5.0 liters). Bore & stroke: 4.00 x 3.00 in. Compression ratio: 9.0:1. Brake horsepower: 215 at 4200 rpm. Torque: 285 lbs.-ft. at 3400 rpm. Five main bearings. Hydraulic valve lifters. Induction: Sequential (port) fuel injection. Serial number code: E.

Base Mustang Cobra V-8
90-degree, overhead valve V-8. Cast-iron block and head. Displacement: 302 cid (5.0 liters). Bore & stroke: 4.00 x 3.00 in. Compression ratio: 9.0:1. Brake horsepower: 240 at 4800 rpm. Torque: 285 lbs.-ft. at 4000 rpm. Five main bearings. Roller lifters. Induction: Sequential (port) fuel injection. Serial number code: D.

Base Crown Victoria V-8
Modular, overhead cam V-8. Single overhead cam. Displacement: 281 cid (4.6 liters). Bore & stroke: 3.60 x 3.60 in. Compression ratio: 9.0:1. Brake horsepower: 190 at 4600 rpm. Torque: 260 lbs.-ft. at 3200 rpm. Hydraulic valve lifters. Induction: Sequential fuel injection.

Optional Thunderbird LX V-8
Modular, overhead cam V-8. Single overhead cam. Displacement: 281 cid (4.6 liters). Bore & stroke: 3.60 x 3.60 in. Compression ratio: 9.0:1. Brake horsepower: 215 at 4400 rpm. Torque: 285 lbs.-ft. at 4800 rpm. Hydraulic valve lifters. Induction: Sequential fuel injection.

1996 FORD

Base Mustang GT V-8
Modular, overhead cam V-8. Single overhead cam. Displacement: 281 cid (4.6 liters). Bore & stroke: 3.60 x 3.60 in. Compression ratio: 9.0:1. Brake horsepower: 190 at 4600 rpm. Torque: 260 lbs.-ft. at 3200 rpm. Hydraulic valve lifters. Induction: Sequential fuel injection. Serial number code: W.

Mustang Cobra V-8
Modular, overhead cam V-8. Double overhead cam. Aluminum block and heads. Displacement: 281 cid (4.6 liters). Bore & stroke: 3.60 x 3.60 in. Compression ratio: 9.0:1. Brake horsepower: 305 at 5800 rpm. Torque: 300 lbs.-ft. at 4800 rpm. Hydraulic valve lifters. Induction: Sequential fuel injection. Serial number code: V.

Base Taurus SHO V-8
Double overhead cam (32-valve) V-8. Aluminum block and heads. Displacement: 207 cid (3.4 liters). Bore & stroke: 3.50 x 3.15 in. Compression ratio: 9.8:1. Brake horsepower: 224 at 6500 rpm. Torque: 230 lbs.-ft. at 4800 rpm. Hydraulic valve lifters. Induction: Sequential fuel injection.

Base Crown Victoria V-8
Modular, overhead valve V-8. Single overhead cam. Displacement: 281 cid (4.6 liters). Bore & stroke: 3.60 x 3.60 in. Compression ratio: 9.0:1. Brake horsepower: 190 at 4250 rpm. Torque: 260 lbs.-ft. at 3200 rpm. Hydraulic valve lifters. Induction: Sequential fuel injection.

Optional Thunderbird LX V-8
Modular, overhead valve V-8. Double overhead cam. Displacement: 281 cid (4.6 liters). Bore & stroke: 3.60 x 3.60 in. Compression ratio: 9.0:1. Brake horsepower: 205 at 4500 rpm. Torque: 265 lbs.-ft. at 3200 rpm. Hydraulic valve lifters. Induction: Sequential fuel injection.

1997 FORD

Base Mustang GT V-8
Modular, overhead cam V-8. Single overhead cam. Displacement: 281 cid (4.6 liters). Bore & stroke: 3.60 x 3.60 in. Compression ratio: 9.0:1. Brake horsepower: 215 at 5000 rpm. Torque: 285 lbs.-ft. at 4800 rpm. Hydraulic valve lifters. Induction: Sequential fuel injection. Serial number code: W.

Mustang Cobra V-8
Modular, overhead cam V-8. Double overhead cam. Aluminum block and heads. Displacement: 281 cid (4.6 liters). Bore & stroke: 3.60 x 3.60 in. Compression ratio: 9.0:1. Brake horsepower: 305 at 5800 rpm. Torque: 300 lbs.-ft. at 4800 rpm. Hydraulic valve lifters. Induction: Sequential fuel injection. Serial number code: V.

Base Taurus SHO V-8
Double overhead cam (32-valve) V-8. Aluminum block and heads. Displacement: 207 cid (3.4 liters). Bore & stroke: 3.50 x 3.15 in. Compression ratio: 9.8:1. Brake horsepower: 235 at 6100 rpm. Torque: 230 lbs.-ft. at 4800 rpm. Hydraulic valve lifters. Induction: Sequential fuel injection.

Base Crown Victoria V-8
Modular, overhead cam V-8. Single overhead cam. Displacement: 281 cid (4.6 liters). Bore & stroke: 3.60 x 3.60 in. Compression ratio: 9.0:1. Brake horsepower: 190 at 4250 rpm. Torque: 265 lbs.-ft. at 3200 rpm. Hydraulic valve lifters. Induction: Sequential fuel injection.

Optional Thunderbird LX V-8
Modular, overhead cam V-8. Double overhead cam. Displacement: 281 cid (4.6 liters). Bore & stroke: 3.60 x 3.60 in. Compression ratio: 9.0:1. Brake horsepower: 205 at 4250 rpm. Torque: 280 lbs.-ft. at 3000 rpm. Hydraulic valve lifters. Induction: Sequential fuel injection.

1998 FORD

Base Mustang GT V-8
Modular, overhead cam V-8. Single overhead cam. Displacement: 281 cid (4.6 liters). Bore & stroke: 3.60 x 3.60 in. Compression ratio: 9.0:1. Brake horsepower: 225 at 4750 rpm. Torque: 290 lbs.-ft. at 3500 rpm. Hydraulic valve lifters. Induction: Sequential fuel injection. Serial number code: W.

Mustang Cobra V-8
Modular, overhead cam V-8. Double overhead cam. Aluminum block and heads. Displacement: 281 cid (4.6 liters). Bore & stroke: 3.60 x 3.60 in. Compression ratio: 9.0:1. Brake horsepower: 305 at 5800 rpm. Torque: 300 lbs.-ft. at 4800 rpm. Hydraulic valve lifters. Induction: Sequential fuel injection. Serial number code: V.

Base Taurus SHO V-8
Double overhead cam (32-valve) V-8. Aluminum block and heads. Displacement: 207 cid (3.4 liters). Bore & stroke: 3.50 x 3.15 in. Compression ratio: 9.8:1. Brake horsepower: 235 at 6100 rpm. Torque: 230 lbs.-ft. at 4800 rpm. Hydraulic valve lifters. Induction: Sequential fuel injection.

Base Crown Victoria V-8
Modular, overhead cam V-8. Single overhead cam. Displacement: 281 cid (4.6 liters). Bore & stroke: 3.60 x 3.60 in. Compression ratio: 9.0:1. Brake horsepower: 260 at 5250 rpm. Torque: 302 lbs.-ft. at 4000 rpm. Hydraulic valve lifters. Induction: Sequential fuel injection.

1999 FORD

Base Mustang GT V-8
Modular, 90-degree overhead cam V-8. Single overhead cam. Displacement: 281 cid (4.6 liters). Bore & stroke: 3.60 x 3.60 in. Compression ratio: 9.0:1. Brake horsepower: 260 at 5250 rpm. Torque: 302 lbs.-ft. at 4000 rpm. Hydraulic valve lifters. Induction: Sequential fuel injection. Serial number code X.
Note: Horsepower increase due to new higher lift, longer duration camshaft, bigger valves, and revised intake manifold runners that increased airflow above 2000 rpm. GT's also featured a new coil-on-plug ignition with increased reliability and higher energy spark.

Mustang Cobra V-8
Modular, 90-degree cam valve V-8. Double overhead cam. Aluminum block and heads. Displacement: 281 cid (4.6 liters). Bore & stroke: 3.60 x 3.60 in. Compression ratio: 9.0:1. Brake horsepower: 320 at 6000 rpm. Torque: 315 lbs.-ft. at 4000 rpm. Hydraulic valve lifters. Induction: Sequential fuel injection. Serial number code V.
Note: Horsepower increase due to new higher lift, longer duration camshaft, bigger valves, and revised intake manifold runners that increased airflow above 2000 rpm.

Base Taurus SHO V-8
Double overhead cam (32-valve) V-8. Aluminum block and heads. Displacement: 207 cid (3.4 liters). Bore & stroke: 3.50 x 3.15 in. Compression ratio: 9.8:1. Brake horsepower: 235 at 6100 rpm. Torque: 230 lbs.-ft. at 4800 rpm. Hydraulic valve lifters. Induction: Sequential fuel injection.

Base Crown Victoria V-8
Modular, overhead cam V-8. Displacement: 281 cid (4.6 liters). Bore & stroke: 3.60 x 3.60 in. Compression ratio: 9.0:1. Brake horsepower: 200 at 4250 rpm. Torque: 275 lbs.-ft. at 3000 rpm. Induction: Sequential fuel injection.

2000 FORD

Base Mustang GT V-8
Modular, 90-degree overhead cam V-8. Single overhead cam. Displacement: 281 cid (4.6 liters). Bore & stroke: 3.60 x 3.60 in. Compression ratio: 9.0:1. Brake horsepower: 260 at 5250 rpm. Torque: 302 lbs.-ft. at 4000 rpm. Hydraulic valve lifters. Induction: Sequential fuel injection. Serial number code: W.

Mustang Cobra "R" V-8
Double overhead camshaft V-8. Cast-iron block. Aluminum heads. Hydraulic valve lifters. Displacement: 330 cid (5.4 liters). Bore & stroke: 3.55 x 4.15 in. Compression ratio: 9.6:1. Brake horsepower: 385 at 6250 rpm. Torque: 385 lbs.-ft. at 4250 rpm. Induction: Sequential electronic fuel injection. VIN code: H.

Base Crown Victoria V-8
Modular, overhead cam V-8. Displacement: 281 cid (4.6 liters). Bore & stroke: 3.60 x 3.60 in. Compression ratio: 9.0:1. Brake horsepower: 200 at 4250 rpm. Torque: 275 lbs.-ft. at 3000 rpm. Induction: Sequential fuel injection.

Optional Natural Gas V-8
Modular, overhead cam V-8. Displacement: 281 cid (4.6 liters). Bore & stroke: 3.60 x 3.60 in. Compression ratio: 9.0:1. Brake horsepower: 175 at 4250 rpm. Torque: 235 lbs.-ft. at 3000 rpm. Natural gas engine.

2001 FORD

Base Mustang GT V-8
Modular, 90-degree overhead cam V-8. Single overhead cam. Displacement: 281 cid (4.6 liters). Bore & stroke: 3.60 x 3.60 in. Compression ratio: 9.0:1. Brake horsepower: 260 at 5250 rpm. Torque: 302 lbs.-ft. at 4000 rpm. Hydraulic valve lifters. Induction: Sequential fuel injection. Serial number code: X.

Base Bullit Mustang V-8
Modular, 90-degree overhead cam V-8. Single overhead cam. Displacement: 281 cid (4.6 liters). Bore & stroke: 3.60 x 3.60 in. Compression ratio: 9.0:1. Brake horsepower: 270 at unknown rpm. Torque: unknown at unknown rpm. Hydraulic valve lifters. Induction: Sequential fuel injection. Serial number code: X.

Mustang Cobra V-8
Modular, 90-degree overhead cam V-8. Double overhead cam. Aluminum block and heads. Displacement: 281 cid (4.6 liters). Bore & stroke: 3.60 x 3.60 in. Compression ratio: 9.0:1. Brake horsepower: 320 at 6000 rpm. Torque: 315 lbs.-ft. at 4000 rpm. Hydraulic valve lifters. Induction: Sequential fuel injection. Serial number code: V.

Base Crown Victoria V-8
Modular, overhead cam V-8. Displacement: 281 cid (4.6 liters). Bore & stroke: 3.60 x 3.60 in. Compression ratio: 9.0:1. Brake horsepower: 220 at 4750 rpm. Torque: 265 lbs.-ft. at 4000 rpm. Induction: Sequential fuel injection.

2002 FORD

Base Mustang GT V-8
Modular, 90-degree overhead cam V-8. Single overhead cam. Displacement: 281 cid (4.6 liters). Bore & stroke: 3.60 x 3.60 in. Compression ratio: 9.0:1. Brake horsepower: 260 at 5250 rpm. Torque: 302 lbs.-ft. at 4000 rpm. Hydraulic valve lifters. Induction: Sequential fuel injection. Serial number code: X.

Mustang Cobra V-8
Modular, 90-degree overhead cam V-8. Double overhead cam. Aluminum block and heads. Displacement: 281 cid (4.6 liters). Bore & stroke: 3.60 x 3.60 in. Compression ratio: 9.0:1. Brake horsepower: 320 at 6000 rpm. Torque: 315 lbs.-ft. at 4000 rpm. Hydraulic valve lifters. Induction: Sequential fuel injection. Serial number code: V.

Base Crown Victoria V-8
Modular, overhead cam V-8. Displacement: 281 cid (4.6 liters). Bore & stroke: 3.60 x 3.60 in. Compression ratio: 9.0:1. Brake horsepower: 220 at 4750 rpm. Torque: 265 lbs.-ft. at 4000 rpm. Induction: Sequential fuel injection.

Base Thunderbird V-8
Double overhead cam V-8. Aluminum block and heads. Displacement: 240 cid (3.9 liters). Bore & stroke: 3.39 x 3.35 in. Compression ratio: 10.55:1. Brake horsepower: 252 at 6100 rpm. Torque: 261 lbs.-ft. at 4300 rpm. Induction: Sequential multiport electronic fuel injection.

Ford V-8 Firing Order

1932-1953 (All Flathead V-8s)
1-5-4-8-6-3-7-2 with RH bank numbered
1-2-3-4 and LH bank numbered 5-6-7-8 front to rear

(Same as 1R-1L-4R-4L-2L-3R-3L-2R)
Note: Flathead V-8 plug wires go into the distributor cap in firing order in clockwise direction.

1954-1956 (Overhead valve V-8s)
1-5-4-8-6-3-7-2 with RH bank numbered
1-2-3-4 and LH bank numbered 5-6-7-8 front to rear

(Same as 1R-1L-4R-4L-2L-3R-3L-2R)
Note: Overhead valve V-8 plug wires go into the distributor cap in firing order in counter-clockwise direction.

1957-1962 (Overhead valve V-8s)
292-cid: 1-5-4-8-6-3-7-2 with RH bank numbered
1-2-3-4 and LH bank numbered 5-6-7-8 front to rear
Others: 1-5-4-2-6-3-7-8 with RH bank numbered 1-2-3-4 and LH bank numbered 5-6-7-8 front to rear

1963-1965 (Overhead valve V-8s)
Others: 1-5-4-2-6-3-7-8 with RH bank numbered 1-2-3-4 and LH bank numbered 5-6-7-8 front to rear

Selected Ford V-8 Tune-Up Specs 1932-1965

Year	Model	Spark Plugs Type	Spark Plugs Gap	Point Gap	Deg Cam Angle	Timing	Idle Speed	Compression Pressure	Battery (Delco)
1932	V-8-106	C-4	.025	.012-.014	35-37.5	4B	NA	100 lb.	18-10655
1933	V-8-112	C-7	.025	.012-.014	35-37.5	4B	NA	113 lb.	40-10655A
1934	V-8-112	C-7	.025	.012-.014	35-37.5	4B	NA	113 lb.	40-10655C
1935	V-8	C-7	.025	.012-.014	35-37.5	4B	NA	113 lb.	40-10655C
1936	V-8	C-7	.025	.012-.014	35-37.5	4B	5-7 mph	112 lb.	40-10655C
1937	V-8-60	H-10	.025	.014-.016	36	4B	5-7 mph	113 lb.	78-10655A
	V-8-85	C-7	.025	.014-.016	36	4B	5-7 mph	116 lb.	78-10655A
1938	V-8-60	H-10	.025	.014-.016	36	4B	5-7 mph	113 lb.	81A-10655A
	V-8-85	H-10	.025	.014-.016	36	4B	5-7 mph	116 lb.	81A-10655A
1939	Model 91A	H-10	.025	.014-.016	36	4B	5-7 mph	113 lb.	81A-10655A
	Model 922A	H-10	.025	.014-.016	36	4B	5-7 mph	116 lb.	81A-10655A
1940	Model 01A	H-10	.025	.014-.016	36	4B	485 rpm	113 lb.	81A-10655A
	Model 022A	H-10	.025	.014-.016	36	4B	485 rpm	116 lb.	81A-10655A
1941	Model 11A	H-10	.025	.014-.016	36	4B	485 rpm	113 lb.	81A-10655A
1942	Model 21A	H-10	.025	.014-.016	36	4B	485 rpm	113 lb.	01A-10655A
1946	Model 69A	H-10	.025	.014-.016	36	TDC	485 rpm	110 lb.	2F

1947	Model 69A	H-10	.025	.014-.016	36	TDC	485 rpm	110 lb.	2F
1948	Model 69A	H-10	.025	.014-.016	36	TDC	485 rpm	110 lb.	2F
1949	Model 98BA	H-10	.030	.014-.016	29	2B	485 rpm	120 lb.	2L
1950	Model 0BA	H-10	.032	.016	30	2B	425 rpm	110 lb.	2L
1951	Model 1BA	H-10	.032	.016	30	2B	425 rpm	110 lb.	2L
1952	Model B2	H-10	.032	.015	38.5	2B	485 rpm	110 lb.	2L
1953	V-8	H-10	.032	.015	38.5	2B	485 rpm	110 lb.	2L
1954	V-8 (OHV)	CH-870	.035	.015	27	6B	485 rpm	130 lb.	2HN/2N
1955	V-8	CH-870	.035	.015	27	6B	485 rpm	130 lb.	2HN/2N
	T-Bird	CH-870	.035	.015	27	6B	485 rpm	135 lb.	2HN/2N
1956	V-8	CH-870	.035	.015	27	6B	485 rpm	135 lb.	13HN/4HN

Note 1: Tune-up data for optional engines varies from above; changed to 12-volt neg. ground electrical system.

1957	V-8	CH-870	.035	.015	27	6B	485 rpm	160 lb.	13HN/4HN

Note 2: Tune-up data for optional engines varies from above.

1958	V-8 (Ford 292)	F-14-Y	.035	.014-.016	26-28.5	4B	475 rpm	160 lb.	55 amp-hr
	V-8 (Ford 332)	F-11-Y	.035	.014-.016	26-28.5	3B	475 rpm	160 lb.	55 amp-hr
	V-8 (Ford 352)	F-11-Y	.035	.014-.016	26-28.5	3B	600 rpm	180 lb.	55 amp-hr
	V-8 (T-bird 352)	F-11-Y	.035	.014-.016	26-28.5	3B	600 rpm	160 lb.	55 amp-hr

Note 3: Tune-up data for optional engines varies from above; idle speeds lower with automatic transmission; timing for 332- and 352-cid V-8s with automatic transmission is 6B.

1959	V-8 (Ford 292)	F-14-Y	.035	.014-.016	26-28.5	3/6B	475 rpm	160 lb.	12-66-55AH
	V-8 (Ford 332)	F-11-Y	.035	.014-.016	26-28.5	3/6B	475 rpm	160 lb.	12-66-55AH
	V-8 (Ford 352)	F-11-Y	.035	.014-.016	26-28.5	3/6B	475 rpm	180 lb.	12-78-65AH
	V-8 (T-bird 352)	F-11-Y	.035	.014-.016	26-28.5	3/6B	475 rpm	180 lb.	12-78-65AH
	V-8 (T-bird 430)	FEU12405A	.032	.014-.016	26-28.5	6B	475 rpm	200 lb.	12-78-65AH

Note 4: Tune-up data for optional engines varies from above; idle speeds lower with automatic transmission; timing for 332- and 352-cid V-8s with automatic transmission is 6B.

1960	V-8 (Ford 292)	F-14-Y	.035	.014-.016	26-28.5	3/6B	500 rpm	160 lb.	B9A-10654D-E-F
	V-8 (Ford 352)	F-11-Y	.035	.014-.016	26-28.5	3/6B	500 rpm	180 lb.	B9A-10654D-E-F
	V-8 (T-bird 352)	F-83-Y	.035	.014-.016	26-28.5	3/6B	500 rpm	180 lb.	B9A-10654D-E-F
	V-8 (T-bird 430)	F-11-Y	.032-.036	.014-.016	26-28.5	3/6B	500 rpm	200 lb.	B9A-10654D-E-F

Note 5: Tune-up data for optional engines may vary from above; idle speeds lower with automatic transmission; timing for 292-, 332- and 352-cid V-8s with automatic transmission is 6B.

1961	V-8 (Ford 292)	F-14-Y	.032-.036	.014-.016	26-28.5	2-10B	500 rpm	160 lb.	B9A-10654D-E-F
	V-8 (Ford 352)	F-11-Y	.032-.036	.014-.016	26-28.5	2-10B	500 rpm	180 lb.	B9A-10654D-E-F
	V-8 (Ford 390)	F-11-Y	.032-.036	.014-.016	26-28.5	2-10B	500 rpm	180 lb.	B9A-10654D-E-F
	V-8 (T-bird 430)	F-11-Y	.032-.036	.014-.016	26-28.5	2-10B	500 rpm	200 lb.	B9A-10654D-E-F

Note 6: Idle speeds lower with automatic transmission.
Note 7: General timing range for all engines is 2B to 10B.
Note 8: Recommended timing for 292-cid V-8 is 3B with standard transmission and 10B with automatic transmission.
Note 9: Recommended timing for 332- and 352-cid V-8s is 3B with standard transmission and 6B with automatic trans.

1962	V-8 (Ford 221)	BF-82	.035	.014-.016	26-28.5	6B	500 rpm	150 lb.	C2DF-10655A
	V-8 (Ford 260)	BF-82	.035	.014-.016	26-28.5	4B-6B	500 rpm	150 lb.	C2DF-10655A
	V-8 (Ford 292)	BF-82	.035	.014-.016	26-28.5	5B/12B	500 rpm	160 lb.	C2DF-10655A
	V-8 (Ford 352)	BF-42	.035	.014-.016	26-28.5	5B/8B	500 rpm	180 lb.	C2DF-10655A
	V-8 (Ford 390)	BF-42	.035	.014-.016	26-28.5	5B/8B	500 rpm	180 lb.	C2DF-10655A
	V-8 (T-bird 430)	BF-42	.035	.014-.016	26-28.5	8B	550 rpm	200 lb.	C2DF-10655A

Note 10: Idle speeds lower with automatic transmission.

1963	V-8 (Ford 221)	BF-82	.035	.014-.016	26-28.5	6B	500 rpm	150 lb.	C2DF-10655A
	V-8 (Ford 260)	BF-82	.035	.014-.016	26-28.5	4B-6B	500 rpm	150 lb	C2DF-10655A
	V-8 (Ford 289)	BF-82	.035	.014-.016	26-28.5	5B/12B	500 rpm	160 lb.	C2DF-10655A
	V-8 (Ford 352)	BF-42	.035	.014-.016	26-28.5	5B/8B	500 rpm	180 lb.	C2DF-10655A
	V-8 (Ford 390)	BF-42	.035	.014-.016	26-28.5	5B/8B	500 rpm	180 lb.	C2DF-10655A
	V-8 (T-bird 430)	BF-42	.035	.014-.016	26-28.5	8B	550 rpm	200 lb.	C2DF-10655A
1964	V-8 (Ford 260)	BF-82	.034	.015	27	6B	575 rpm	150 lb.	55-65-70 amp hr
	V-8 (Ford 289)	BF-42	.034	.015	27	6B/10B	575 rpm	150 lb.	55-65-70 amp hr
V-8 (Ford 352)		BF-42	.034	.015	27	6B/10B	500 rpm	180 lb.	55-65-70 amp hr
	V-8 (Ford 390)	BF-42	.034	.015	27	4B/6B	500 rpm	180 lb.	55-65-70 amp hr
	V-8 (Ford 427)	BF-42	.030	.020	35	8B	700 rpm	180 lb.	55-65-70 amp hr
1965	V-8 (Ford 289)	BF-42	.034	.015	27	6B/10B	750 rpm	150 lb.	55-65-70 amp hr
	V-8 (Ford 352)	BF-42	.034	.015	27	6B/10B	750 rpm	180 lb.	55-65-70 amp hr
	V-8 (Ford 390)	BF-42	.034	.015	27	4B/6B	750 rpm	180 lb.	55-65-70 amp hr
	V-8 (Ford 427)	BF-42	.030	.020	35	8B	750 rpm	180 lb.	55-65-70 amp hr

HUDSON
1955-1957

Founded by Roy D. Chapin and financed by J.L. Hudson, the Hudson Motor Car Co. pioneered modestly-priced closed cars from its creation in 1909. The company made a small profit during World War II building airplanes and landing craft engines and re-entered the automobile market in 1946 with face-lifted prewar models. Hudson retained both its L-head six and eight engines with the Super Six accounting for two-thirds of Hudson's 1946 production.

The famous Step Down Hudson bowed in 1948. It was offered in four models, the Commodore Six and Eight and the Super Six and Eight. This same year, Hudson introduced a new engine, the 262-cid Super Six, which developed 121 hp at 4000 rpm. In 1951 this same engine evolved into the "308" Hornet power plant which, from 1951 through 1954, was the king of stock car racing.

Although the Step Down proved to be one of America's most roadable cars from 1948-1954, Hudson lacked sufficient funds to add new models to the series and, combined with a lack of innovation (principally the lack of a V-8 engine) in subsequent years, found sales dropping through the early 1950s.

The 1950 Pacemaker model used a de-stroked version of the flathead Super Six and sold over 60,000 units. Carrying the same price tag as the Commodore Eight, the legendary Hudson Hornet was introduced in 1951. Its six-cylinder engine produced 145 hp at 3800 rpm in stock form. In the hands of skilled tuners, though, it was capable of considerably more and the most noted of these, Marshall Teague, achieved 112 mph from a NASCAR-certified stock Hornet. Hudson also offered factory severe-usage options that were designed for racing applications. Racing items were listed as "Export" options. They included "Twin H-Power" for improved breathing and a "7-X" racing engine which combined Twin H-power with other high-performance options to produce about 210 hp.

Tim Flock was the 1952 NASCAR champion in a Hudson Hornet. In 1953, Marshall Teague's Hornet won 12 of 13 AAA stock car events and drivers Herb Thomas, Dick Rathmann, Frank Mundy, and Al Keller drove Hornets to 65 NASCAR victories through 1954.

By late 1953, Hudson sales were slumping. The company merged with Nash and moved production to Kenosha, Wisconsin. The "all-new" 1955 Hudson was really a restyled Nash, but a new offering was a V-8 engine. The Hornet V-8 used a Packard 208-hp engine and a line of Ramblers and Metropolitans was offered.

In 1956, American Motors introduced its own 190-hp V-8 for the Hornet Special and a line of Hudson Ramblers. The 327-cid AMC V-8 was available with a four-barrel carburetor and dual exhaust. Cars so-equipped were excellent performers. Like other Hudsons, they also carried price reductions, but it was to be Hudson's last year. AMC decided to stop making Hudson and Nash cars in order to concentrate on building more economical Ramblers.

1955 HUDSON

Hornet base V-8
V-8. Overhead valves. Cast-iron block. Bore & stroke: 3-13/16 x 3-1/2 in. Displacement: 320 cid. Compression ratio: 7.8:1. Brake hp: 208 at 4200 rpm. Torque: 300 lbs.-ft. at 2300 rpm. Five main bearings. Hydraulic valve lifters (non-adjustable). Crankcase capacity less filter: 5 qt. Cooling system capacity: 20 qt. without heater; 21 qt. with heater. Carburetor: Carter WGD 2231-S two-barrel. Engine code: 3558. Engine numbers: P1001 and up. (Note: Packard V-8).

1956 HUDSON

Hornet Special base V-8
V-8. Overhead valves. Cast-iron block. Bore & stroke: 3-1/2 x 3-1/4 in. Displacement: 250 cid. Compression ratio: 8.0:1. Brake hp: 190 at 4900 rpm. Taxable hp: 39.2. Torque: 240 lbs.-ft. at 2500 rpm. Five main bearings. Hydraulic valve lifters (non-adjustable). Crankcase capacity less filter: 5 qt. Cooling system capacity: 20 qt. without heater; 21 qt. with heater. Carburetor: Carter WGD two-barrel model 2352SA. Engine code: 35650. Engine numbers: G-1001 and up.

Hornet V-8
V-8. Overhead valves. Cast-iron block. Bore & stroke: 4 x 3-1/2 in. Displacement: 352 cid. Compression ratio: 9.55:1. Brake hp: 220 at 4800 rpm. Taxable hp: 51.2. Torque: 320 lbs.-ft. at 2350 rpm. Five main bearings. Hydraulic valve lifters (non-adjustable). Crankcase capacity less filter: 5 qt. Cooling system capacity: 27 qt. without heater; 28 qt. with heater. Carburetor: Carter WGD 2231-S two-barrel. Engine code: 35680. Engine numbers: P-21001 and up. (Note: Packard V-8)

1957 HUDSON

Hornet Special V-8
V-8. Overhead valves. Cast-iron block. Bore & stroke: 4 x 3-1/4 in. Displacement: 327 cid. Compression ratio: 9.0:1. Brake hp: 255 at 4700 rpm. Taxable hp: 51.2. Torque: 345 lbs.-ft. at 2600 rpm. Five main bearings. Hydraulic valve lifters (non-adjustable). Crankcase capacity less filter: 5 qt. Cooling system capacity: 19 qt. without heater; 20 qt. with heater. Carburetor: Carter WCFB four-barrel model 2593SA. Equipped with dual exhaust system and 12-volt electrical system. Engine code: 35780. Engine numbers: N-1001 and up.

Selected Hudson V-8 Tune-Up Specs

Year	Model	Spark Plugs Type	Spark Plugs Gap	Point Gap	Deg Cam Angle	Timing	Idle Speed	Compression Pressure	Battery (Delco)
Hudson									
1955	Hornet V-8	AL-AG5	.035	.017	Note 1	5B	550	140 lb.	Group 1
1956	Hornet V-8	AL-AL7	.035	.015-.017	Note 1	5B	550	140 lb.	11HS/2SM
1957	Hornet V-8	AL-AL7	.035	.015-.017	30	5B	550	140 lb.	11HS/2SM

Note 1: Distributor has two sets of breaker points. Total dwell 34 degrees, single set 27 degrees.
Note 2: Idle speed 550 rpm with synchromesh, 450 rpm with Ultramatic, 425 rpm with Hydra-Matic.

Firing order

(1955-1957)
1-8-4-3-6-5-7-2

Note: *All Hudson V-8 engines have the cylinders numbered from front to back. On the left-hand bank the cylinders are numbered 1-3-5-7 and on the right-hand bank the cylinders are numbered 2-4-6-8.*

The 1957 Hudson engine offerings included the 327-cid Hornet Special V-8. This overhead-valve engine, which is sometimes confused with the better-known 327-cid Chevrolet V-8, had a 4 x 3-1/4-in. bore and stroke. With a 9.0:1 compression ratio, a single four-barrel carburetor and dual exhausts it generated 255 horsepower. Seen here and representing the marque's final-year swansong, is the V-8-powered Hornet four-door sedan.

LINCOLN
1921-2002

Henry M. Leland's firm, Leland & Falcouner, built a better engine for the Curved Dash Oldsmobile, but R.E. Olds wasn't interested. In 1902, Leland's small, efficient single-cylinder engine found its way into the first Cadillac. Leland next formed a company to build World War I aircraft engines. It sank into financial trouble during the postwar recession and was purchased by Ford, at the urging of Edsel Ford.

Edsel "souped up" the Lincoln's styling and engineering. By 1929, the mighty Model L had a 90-hp V-8 that was good for 90 mph. Cops loved it. The 1931 Model K was even faster, with its refined 120-hp V-8. Between 1932 and 1948, Lincolns came only with V-12 engines. These were actually of four different designs, but that's another story.

In 1949, the Lincoln got new styling and a new V-8 engine. The styling didn't last, but the V-8 did. It achieved motorsports fame in the Mexican Road Races. By 1953, stock 318-cid versions produced over 200 hp and by 1956, the 368-cid Lincoln V-8 was churning out 285 hp. Even more impressive was the 300-hp edition under the hood of the revived-for-1956 Lincoln Continental. The next year the 300-hp job was used in all Lincolns and from 1958 to 1965, the only Lincoln engine was a 430-cid V-8 tuned to generate well over 300 hp.

The 430 was the largest and heaviest (730 pounds) engine used in an American car in 1965. It was designed specifically as a luxury-car engine and built under strict quality-control conditions. It featured deep-skirted block construction, forged rods, and a precision alloy crankshaft. A two-stage four-barrel carburetor was fitted. In 1966, Lincoln retained its largest-engine status by boring and stroking the huge V-8 to 462 cid. At the same time, the block casting was redesigned, new cylinder heads were added, a new intake manifold was used, larger valves were fitted, the carburetion was upgraded and a new camshaft came out. The 462 was carried over for 1967 and was one of two Lincoln V-8s offered in 1968.

The new Lincoln Continental Mark III, which was based on the Thunderbird, used the other engine. It was a 460-cid V-8 based on the Thunderbird's 429-cid "thinwall" engine. It had "advanced wedge" combustion chambers, "porcupine" valves and a low-restriction "round-port" exhaust system. The first 460 produced 365 hp. From 1969-1976 it was the only Lincoln engine.

Starting in 1977, a variety of small- and big-block Ford V-8s found their way under the hoods of Lincolns. The 351 was used in the Versailles introduced that year, while the 400 was the base engine in Continentals and Mark Vs. Other Lincolns came standard with the 400 and the 460 was optional. In 1978, the Versailles got a fuel-injected version of the 302-cid V-8 used in the Ford Mustang. To increase fuel economy, Lincoln also offered the first automatic overdrive transmission in an American car.

Starting in 1980, the Mark VI became a variation of the Continental-Town Car platform. The next season the Versailles was dropped the name "Town Car" replaced the name "Continental" on the base Lincoln. Then, the Continental name was revived on an entirely new car in 1982. This V-6-powered model was the first Lincoln to offer an engine with less than eight cylinders. In 1984, a new Mark VII marked a big change in corporate philosophy. It was a real driver's car instead of a "land yacht." The Mark VII showcased an aerodynamic shape and used a 302-cid V-8 rated at 140 hp as standard equipment. (A BMW turbo-diesel was extra.)

The 302 remained the only Lincoln V-8 through 1991, when an all-new "modular" V-8 arrived. It was dropped in 1993 and the modular engine was the only type used for the next five years. In 1999, the Continental got the 4.6-liter DOHC engine as standard equipment. A 240-cid (3.9-liter) DOHC V-8 was base engine for the all-new 2000 Lincoln LS. This engine was based on a very sophisticated 4.0-liter DOHC V-8 used in the 2000 Jaguar S-Type. It featured four valves per cylinder, all-aluminum construction, coil-on-plug ignition, platinum-tipped spark plugs and a true dual exhaust system. Good off-the-line power and plenty of speed for passing were provided with this ultra-high-tech V-8.

SIDEBAR: MODULAR V-8

In the 1980s, Ford decided to design a new family of smoother, quieter engines with improved fuel economy, better sealing and better power-to-displacement ratios. The plan was to replace the 5.0-liter V-8s being used in the Ford Mustang and Crown Victoria, the Mercury Grand Marquis and the Lincoln Mark VII and Town Car. The FoMoCo engineers ultimately settled on a 4.6-liter overhead cam V-8 to be built at a factory in Romeo, Michigan.

It was called a "modular" V-8 for two reasons. First, many parts were designed to be interchangeable between single overhead cam (SOHC) and double overhead cam (DOHC) variations. In addition, Ford wanted to be able to machine and assemble the engines on the same lines, so that production scheduling for SOHC and DOHC engines could be adjusted to meet changes in customer demand.

The first car to get the 281-cid modular V-8 was the 1991 Lincoln Town Car. The Crown Victoria, Grand Marquis, Thunderbird, and Cougar got it later. A DOHC variant debuted in the 1993 Lincoln Mark VIII and was later added for the front-wheel-drive Continental. After 1996 the 4.6-liter SOHC V-8 wound up in the Mustang GT and the DOHC type was seen under the hood of the Cobra Mustang.

1921 LINCOLN

1921 LINCOLN
60 degree V-8. L-head. Two cast-iron cylinder blocks on top of an aluminum crankcase. Fork-and-blade connecting rods. Roller cam followers. Bore & stroke: 3-3/8 x 5 in. Displacement: 357.8 cid. Compression ratio: 4.8:1. Brake hp: 81 at 2600 rpm. Taxable hp: 36.45. Main bearings: five. Valve lifters: mechanical. Crankcase capacity: 10 qt. Cooling system capacity: 8 qt. Carburetor: Stromberg 1-1/2-in. updraft.
NOTE: Lincoln produced 834 cars in 1920. They had the same specifications as the 1921 model.

1922 LINCOLN

1922 LINCOLN
60 degree V-8. L-head. Two cast-iron cylinder blocks on top of an aluminum crankcase. Fork-and-blade connecting rods. Roller cam followers. Bore & stroke: 3-3/8 x 5 in. Displacement: 357.8 cid. Compression ratio: 4.8:1. Brake hp: 90 at 2800 rpm. Taxable hp: 36.45. Main bearings: five. Valve lifters: mechanical. Crankcase capacity: 10 qt. Cooling system capacity: 8 qt. Carburetor: Stromberg SF3 1-1/2-in. updraft.

1923 LINCOLN

1923 LINCOLN
60 degree V-8. L-head. Two cast-iron cylinder blocks on top of an aluminum crankcase. Fork-and-blade connecting rods. Roller cam followers. Bore & stroke: 3-3/8 x 5 in. Displacement: 357.8 cid. Compression ratio: 4.8:1. Brake hp: 90 at 2800 rpm. Taxable hp: 36.45. Main bearings: five. Valve lifters: mechanical. Crankcase capacity: 10 qt. Cooling system capacity: 7.5 qt. Carburetor: Stromberg 03 1-1/2-in. updraft.
Note: The 1923 Lincoln 8 averaged 13 mpg.

1924 LINCOLN

1924 LINCOLN
60 degree V-8. L-head. Two cast-iron cylinder blocks on top of an aluminum crankcase. Fork-and-blade connecting rods. Roller cam followers. Bore & stroke: 3-3/8 x 5 in. Displacement: 357.8 cid. Compression ratio: 4.8:1. Brake hp: 90 at 2800 rpm. Taxable hp: 36.45. Main bearings: five. Valve lifters: mechanical. Crankcase capacity: 10 qt. Cooling system capacity: 7.5 qt. Carburetor: Stromberg 03 1-1/2-in. updraft.

1925 LINCOLN

1925 LINCOLN
60 degree V-8. L-head. Two cast-iron cylinder blocks on top of an aluminum crankcase. Fork-and-blade connecting rods. Roller cam followers. Bore & stroke: 3-3/8 x 5 in. Displacement: 357.8 cid. Compression ratio: 4.8:1. Brake hp: 90 at 2800 rpm. Taxable hp: 36.45. Main bearings: five. Valve lifters: mechanical. Crankcase capacity: 10 qt. Cooling system capacity: 7.5 qt. Carburetor: Stromberg 03 1-1/2-in. updraft.

1926 LINCOLN

1926 LINCOLN
60 degree V-8. L-head. Two cast-iron cylinder blocks on top of an aluminum crankcase. Fork-and-blade connecting rods. Roller cam followers. Bore & stroke: 3-3/8 x 5 in. Displacement: 357.8 cid. Compression ratio: 4.8:1. Brake hp: 90 at 2800 rpm. Taxable hp: 36.45. Main bearings: five. Valve lifters: mechanical. Crankcase capacity: 10 qt. Cooling system capacity: 7.5 qt. Carburetor: Stromberg 03 1-1/2-in. updraft.

1927 LINCOLN

1927 LINCOLN
60 degree V-8. L-head. Two cast-iron cylinder blocks on top of an aluminum crankcase. Fork-and-blade connecting rods. Roller cam followers. Bore & stroke: 3-3/8 x 5 in. Displacement: 357.8 cid. Compression ratio: 4.8:1. Brake hp: 90 at 2800 rpm. Taxable hp: 36.45. Main bearings: five. Valve lifters: mechanical. Crankcase capacity: 10 qt. Cooling system capacity: 7.5 qt. Carburetor: Stromberg 03 1-1/2-in. updraft.

1928 LINCOLN

1928 LINCOLN
60 degree V-8. L-head. Two cast-iron cylinder blocks on top of an aluminum crankcase. Fork-and-blade connecting rods. Roller cam followers. Internal harmonic balancer. Bore & stroke: 3-1/2 x 5 in. Displacement: 384.8 cid. Compression ratio: 4.81:1. Brake hp: 90 at 2800 rpm. Taxable hp: 39.20. Main bearings: five. Valve lifters: mechanical. Crankcase capacity: 10 qt. Cooling system capacity: 7.5 qt. Carburetor: Stromberg 03 1-1/2-in. updraft.

1929 LINCOLN

1929 LINCOLN
60 degree V-8. L-head. Two cast-iron cylinder blocks on top of an aluminum crankcase. Fork-and-blade connecting rods. Roller cam followers. Internal harmonic balancer. Bore & stroke: 3-1/2 x 5 in. Displacement: 384.8 cid. Compression ratio: 4.81:1. Brake hp: 90 at 2800 rpm. Taxable hp: 39.20. Main bearings: five. Valve lifters: mechanical. Crankcase capacity: 10 qt. Cooling system capacity: 7.5 qt. Carburetor: Stromberg 03 1-1/2-in. updraft.

1930 LINCOLN

1930 LINCOLN
60 degree V-8. L-head. Two cast-iron cylinder blocks on top of an aluminum crankcase. Fork-and-blade connecting rods. Roller cam followers. Internal harmonic balancer. Bore & stroke: 3-1/2 x 5 in. Displacement: 384.8 cid. Compression ratio: 4.81:1. Brake hp: 90 at 2800 rpm. Taxable hp: 39.20. Main bearings: five. Valve lifters: mechanical. Crankcase capacity: 10 qt. Cooling system capacity: 7.5 qt. Carburetor: Stromberg 03 1-1/2-in. updraft.

1931 LINCOLN

1931 LINCOLN
60 degree V-8. L-head. Two cast-iron cylinder blocks on top of an aluminum crankcase. Fork-and-blade connecting rods. Roller cam followers. Internal harmonic balancer. Bore & stroke: 3-1/2 x 5 in. Displacement: 384.8 cid. Compression ratio: 4.95:1. Brake hp: 120 at 2900 rpm. Taxable hp: 39.20. Compression: 95 lbs. at 100 rpm. Main bearings: five. Valve lifters: mechanical. Crankcase capacity: 10 qt. Cooling system capacity: 8.5 qt. Carburetor: Stromberg DD 1-1/8-in. dual downdraft.

1932 LINCOLN

1932 LINCOLN
60 degree V-8. L-head. Two cast-iron cylinder blocks on top of an aluminum crankcase. Fork-and-blade connecting rods. Roller cam followers. Internal harmonic balancer. Bore & stroke: 3-1/2 x 5 in. Displacement: 384.8 cid. Compression ratio: 5.23:1. Brake hp: 125 at 2900 rpm. Main bearings: five. Valve lifters: mechanical. Crankcase capacity: 10 qt. Cooling system capacity: 8.5 qt. Carburetor: Stromberg DD3 1-1/8-in. dual downdraft.
Note: Lincoln provided only V-12 engines from 1933 through 1948.

1949 LINCOLN

1949 LINCOLN
V-8. L-head. Cast-iron block. Bore & stroke: 3.50 x 4.37 in. Displacement: 336.7 cid. Compression ratio: 7.0:1. Brake hp: 152 at 3600 rpm. Taxable hp: 39.2. Torque: 265 lbs.-ft. at 2000 rpm. Main bearings: three. Non-adjustable hydraulic valve lifters. Crankcase capacity: 6 qt. (add 1 qt. for new filter). Cooling system capacity: 34.5 qt. (add 1 qt. for heater). Carburetor: Holley 885-FFC two-barrel. Engine number same as serial number. Starting number: (Lincoln) 9EL-1; (Cosmopolitan) 9EH-1.

1950 LINCOLN

1950 LINCOLN
V-8. L-head. Cast-iron block. Bore & stroke: 3.50 x 4.37 in. Displacement: 336.7 cid. Compression ratio: 7.0:1. Brake hp: 152 at 3600 rpm. Taxable hp: 39.2. Torque: 265 lbs.-ft. at 2000 rpm. Main bearings: three. Non-adjustable hydraulic valve lifters. Crankcase capacity: 6.5 qt. (add 1 qt. for new filter). Cooling system capacity: 33 qt. (add 1 qt. for heater). Carburetor: Holley 885-FFC two-barrel. Engine number same as serial number. Starting number: (Lincoln) 0EL-1; (Cosmopolitan) 0EH-1.

1951 LINCOLN

1951 LINCOLN
V-8. L-head. Cast-iron block. Bore & stroke: 3.50 x 4.37 in. Displacement: 336.7 cid. Compression ratio: 7.0:1. Brake hp: 154 at 3600 rpm. Taxable hp: 39.2. Torque: 275 lbs.-ft. at 1800 rpm. Main bearings: three. Non-adjustable hydraulic valve lifters. Crankcase capacity: 6.5 qt. (add 1 qt. for new filter). Cooling system capacity: 34.5 qt. (add 1 qt. for heater). Carburetor: Holley 885-FFC two-barrel. Engine number same as serial number. Starting number: (Lincoln) 1EL-1; (Cosmopolitan) 1EH-1.

1952 LINCOLN

1952 LINCOLN
V-8. Overhead valve. Cast-iron block. Bore & stroke: 3.80 x 3.50 in. Displacement: 317.5 cid. Compression ratio: 7.5:1. Brake hp: 160 at 3900 rpm. Taxable hp: 46.2. Torque: 284 lbs.-ft. at 1800 rpm. Main bearings: five. Adjustable hydraulic valve lifters. Crankcase capacity: 5 qt. (add 1 qt. for new filter). Cooling system capacity: 22.5 qt. (add 2 qt. for heater). Carburetor: Holley two-barrel. Engine number same as serial number.

In 1952, Lincoln produced a new overhead valve V-8 engine. It achieved motorsports fame in the Mexican Road Races. By 1953, stock 318-cid versions were producing over 200 hp. In 1955, the Lincoln engine grew to 341-cid and developed 225 hp at 4400 rpm. It used a Holley four-barrel carburetor. Mobiloil featured this engine – with its handsome "Lincoln Eight" valve cover logo – in one of its advertisements.

Starting number: 52LP5001H.

1953 LINCOLN

1953 LINCOLN
V-8. Overhead valve. Cast-iron block. Bore & stroke: 3.80 x 3.50 in. Displacement: 317.5 cid. Compression ratio: 8.0:1. Brake hp: 205 at 4200 rpm. Taxable hp: 46.2. Torque: 305 lbs.-ft. at 2300 rpm. Main bearings: five. Hydraulic valve lifters. Crankcase capacity: 5 qt. (add 1 qt. for new filter). Cooling system capacity: 22.5 qt. (add 2 qt. for heater). Carburetor: Holley 2140 four-barrel. Engine number same as serial number. Starting number: 53WA10001.

1954 LINCOLN

1954 LINCOLN
V-8. Overhead valve. Cast-iron block. Bore & stroke: 3.80 x 3.50 in. Displacement: 317.5 cid. Compression ratio: 8.0:1. Brake hp: 205 at 4200 rpm. Taxable hp: 46.2. Torque: 305 lbs.-ft. at 2300 rpm. Main bearings: five. Hydraulic valve lifters. Crankcase capacity: 5 qt. (add 1 qt. for new filter). Cooling system capacity: 22.5 qt. (add 2 qt. for heater). Carburetor: Holley 2140 four-barrel. Engine number same as serial number. Starting number: 54WA5001H.

1955 LINCOLN

1955 LINCOLN
V-8. Overhead valve. Cast-iron block. Bore & stroke: 3.93 x 3.50 in. Displacement: 341 cid. Compression ratio: 8.5:1. Brake hp: 225 at 4400 rpm. Taxable hp: 49.6. Torque: 342 lbs.-ft. at 2500 rpm. Main bearings: five. Hydraulic valve lifters. Crankcase capacity: 5 qt. (add 1 qt. for new filter). Cooling system capacity: 23 qt. (add 2 qt. for heater). Carburetor: Holley four-barrel. Engine number same as serial number. Starting number: 55WA5001H.

1956 LINCOLN

Premiere, Capri base V-8
V-8. Overhead valve. Cast-iron block. Bore & stroke: 4.00 x 3.65 in. Displacement: 368 cid. Compression ratio: 9.0:1. Brake hp: 285 at 4600 rpm. Taxable hp: 51.2. Torque: 402 lbs.-ft. at 3000 rpm. Main bearings: five. Hydraulic valve lifters. Crankcase capacity: 5 qt. (add 1 qt. for new filter). Cooling system capacity: 22.5 qt. (add 2 qt. for heater). Carburetor: Lincoln four-barrel. Engine number same as serial number. Starting number: 56WA5001L.

Continental Mark II V-8
V-8. Overhead valve. Cast-iron block. Bore & stroke: 4.00 x 3.65 in. Displacement: 368 cid. Compression ratio: 10.0:1. Brake hp: 300 at 4800 rpm. Taxable hp: 51.2. Torque: 415 lbs.-ft. at 3000 rpm. Main bearings: five. Hydraulic valve lifters. Crankcase capacity: 5 qt. (add 1 qt. for new filter). Cooling system capacity: 22.5 qt. (add 2 qt. for heater). Carburetor: Carter four-barrel. Engine number same as serial number. Starting number: 56WA5001L.

1957 LINCOLN

Premiere, Capri base V-8
V-8. Overhead valve. Cast-iron block. Bore & stroke: 4.00 x 3.65 in. Displacement: 368 cid. Compression ratio: 10.0:1. Brake hp: 300 at 4800 rpm. Taxable hp: 51.2. Torque: 402 lbs.-ft. at 3000 rpm. Main bearings: five. Hydraulic valve lifters. Crankcase capacity: 5 qt. (add 1 qt. for new filter). Cooling system capacity: 22.5 qt. (add 2 qt. for heater). Carburetor: Carter four-barrel. Engine number same as serial number. Starting number: 57WA5001L.

Continental Mark II V-8
V-8. Overhead valve. Cast-iron block. Bore & stroke: 4.00 x 3.65 in. Displacement: 368 cid. Compression ratio: 10.0:1. Brake hp: 300 at 4800 rpm. Taxable hp: 51.2. Torque: 415 lbs.-ft. at 3000 rpm. Main bearings: five. Hydraulic valve lifters. Crankcase capacity: 5 qt. (add 1 qt. for new filter). Cooling system capacity: 22.5 qt. (add 2 qt. for heater). Carburetor: Carter four-barrel. Engine number same as serial number. Starting number: 57WA5001L.

1958 LINCOLN

Premiere, Capri, Continental base V-8
V-8. Overhead valve. Cast-iron block. Bore & stroke: 4.29 x 3.70 in. Displacement: 430 cid. Compression ratio: 10.5:1. Brake hp: 375 at 4800 rpm. Taxable hp: 59.17. Torque: 490 lbs.-ft. at 3100 rpm. Main bearings: five. Hydraulic valve lifters. Crankcase capacity: 5 qt. (add 1 qt. for new filter). Cooling system capacity: 23 qt. (add 3 qt. for heater). Carburetor: Holley 4150 four-barrel. Engine number same as serial number. Starting number: H8YG400004.

1959 LINCOLN

Premiere, Capri, Continental base V-8
V-8. Overhead valve. Cast-iron block. Bore & stroke: 4.29 x 3.70 in. Displacement: 430 cid. Compression ratio: 10.0:1. Brake hp: 350 at 4400 rpm. Taxable hp: 59.17. Torque: 490 lbs.-ft. at 2800 rpm. Main bearings: five. Hydraulic valve lifters. Crankcase capacity: 5 qt. (add 1 qt. for new filter). Cooling system capacity: 23 qt. (add 3 qt. for heater). Carburetor: Carter AFB-2853S four-barrel. Engine number same as serial number. Starting number: H9YA400004.

1960 LINCOLN

Lincoln, Premiere, Continental base V-8
V-8. Overhead valve. Cast-iron block. Bore & stroke: 4.29 x 3.70 in. Displacement: 430 cid. Compression ratio: 10.0:1. Brake hp: 315 at 4100 rpm. Taxable hp: 59.17. Torque: 465 lbs.-ft. at 2200 rpm. Main bearings: five. Hydraulic valve lifters. Crankcase capacity: 5 qt. (add 1 qt. for new filter). Cooling system capacity: 23 qt. (add 3 qt. for heater). Carburetor: Carter ABD-2965S two-barrel. Engine code: (U.S.) H; (export) K.

1961 LINCOLN

Lincoln Continental base V-8
V-8. Overhead valve. Cast-iron block. Bore & stroke: 4.29 x 3.70 in. Displacement: 430 cid. Compression ratio: 10.0:1. Brake hp: 300 at 4100 rpm. Taxable hp: 59.17. Torque: 465 lbs.-ft. at 2000 rpm. Main bearings: five. Hydraulic valve lifters. Crankcase capacity: 5 qt. (add 1 qt. for new filter). Cooling system capacity: 22 qt. (add 3 qt. for heater). Carburetor: Carter ABD two-barrel. Engine code: (U.S.) H; (export) K.

1962 LINCOLN

Lincoln Continental base V-8
V-8. Overhead valve. Cast-iron block. Bore & stroke: 4.29 x 3.70 in. Displacement: 430 cid. Compression ratio: 10.0:1. Brake hp: 300 at 4100 rpm. Taxable hp: 59.17. Torque: 465 lbs.-ft. at 2000 rpm. Main bearings: five. Hydraulic valve lifters. Crankcase capacity: 5 qt. (add 1 qt. for new filter). Cooling system capacity: 22 qt. (add 3 qt. for heater). Carburetor: Carter ABD two-barrel. Engine code: (U.S.) H; (export) K.

1963 LINCOLN

Lincoln Continental base V-8
V-8. Overhead valve. Cast-iron block. Bore & stroke: 4.29 x 3.70 in. Displacement: 430 cid. Compression ratio: 10.10:1. Brake hp: 320 at 4600 rpm. Taxable hp: 59.17. Torque: 465 lbs.-ft. at 2600 rpm. Main bearings: five. Hydraulic valve lifters. Crankcase capacity: 5 qt. (add 1 qt. for new filter). Cooling system capacity: 22 qt. (add 3 qt. for heater). Carburetor: Carter AFB four-barrel. Engine code: (U.S.) N; (export) 7.

1964 LINCOLN

Lincoln Continental base V-8
V-8. Overhead valve. Cast-iron block. Bore & stroke: 4.29 x 3.70 in. Displacement: 430 cid. Compression ratio: 10.10:1. Brake hp: 320 at 4600 rpm. Taxable hp: 59.17. Torque: 459 lbs.-ft. at 2600 rpm. Main bearings: five. Hydraulic valve lifters. Crankcase capacity: 5 qt. (add 1 qt. for new filter). Cooling system capacity: 22 qt. (add 3 qt. for heater). Carburetor: Carter C3VE-9510-B four-barrel. Engine code: (U.S.) N; (export) 7.

1965 LINCOLN

Lincoln Continental base V-8
V-8. Overhead valve. Cast-iron block. Bore & stroke: 4.29 x 3.70 in. Displacement: 430 cid. Compression ratio: 10.10:1. Brake hp: 320 at 4600 rpm. Taxable hp: 59.17. Torque: 459 lbs.-ft. at 2600 rpm. Main bearings: five. Hydraulic valve lifters. Crankcase capacity: 5 qt. (add 1 qt. for new filter). Cooling system capacity: 22 qt. (add 3 qt. for heater). Carburetor: Carter C3VE-9510-B four-barrel. Engine code: (U.S.) N; (export) 7.

The 1965 Lincoln V-8 was the largest (430 cid) and heaviest (735 pounds) engine in the American market. Its deep-skirted block construction made it extra strong. This engine was unusual in that the combustion chambers were in the block and the machined-flat cylinder heads served as a cover for the chambers. The timing chain ran on a sintered iron or steel crankshaft sprocket and a nylon-toothed aluminum camshaft sprocket.

1966 LINCOLN

Lincoln Continental base V-8
V-8. Overhead valve. Cast-iron block. Bore & stroke: 4.38 x 3.83 in. Displacement: 462 cid. Compression ratio: 10.25:1. Brake hp: 340 at 4600 rpm. Taxable hp: 61.4. Torque: 485 lbs.-ft. at 2800 rpm. Main bearings: five. Hydraulic valve lifters. Crankcase capacity: 5 qt. (add 1 qt. for new filter). Cooling system capacity: 23 qt. with heater. Carburetor: Carter C6VF-9510B four-barrel. Engine code: G.

1967 LINCOLN

Lincoln Continental base V-8
V-8. Overhead valve. Cast-iron block. Bore & stroke: 4.38 x 3.83 in. Displacement: 462 cid. Compression ratio: 10.25:1. Brake hp: 340 at 4600 rpm. Taxable hp: 61.4. Torque: 485 lbs.-ft. at 2800 rpm. Main bearings: five. Hydraulic valve lifters. Crankcase capacity: 5 qt. (add 1 qt. for new filter). Cooling system capacity: 23 qt. with heater. Carburetor: Carter 4362 four-barrel. Engine code: G.

In 1966, Lincoln retained its largest-engine status by boring and stroking the huge V-8 to 462 cid. At the same time, the block casting was redesigned, new cylinder heads were added, a new intake manifold was used, larger valves were fitted, the carburetion was upgraded, and a new camshaft came out. The 1967 V-8 seen here was unchanged. It weighed 750 pounds. This engine was based on every Lincoln V-8 that came before it since 1958, although it had undergone dramatic changes over that decade and its resemblance to the original was heavily masked. It was also largely unrelated to other Ford Motor Company engines.

1968 LINCOLN

Lincoln Continental base V-8
V-8. Overhead valve. Cast-iron block. Bore & stroke: 4.38 x 3.83 in. Displacement: 462 cid. Compression ratio: 10.25:1. Brake hp: 340 at 4600 rpm. Taxable hp: 61.4. Torque: 485 lbs.-ft. at 2800 rpm. Main bearings: five. Hydraulic valve lifters. Crankcase capacity: 5 qt. (add 1 qt. for new filter). Cooling system capacity: 23 qt. with heater. Carburetor: Carter 4362 four-barrel. Engine code: G.

Mark III base V-8
V-8. Overhead valve. Cast-iron block. Bore & stroke: 4.36 x 3.85 in. Displacement: 460 cid. Compression ratio: 10.50:1. Brake hp: 365 at 4600 rpm. Taxable hp: 60.83. Torque: 500 lbs.-ft. at 2800 rpm. Main bearings: five. Hydraulic valve lifters. Crankcase capacity: 4 qt. (add 1 qt. for new filter). Cooling system capacity: 19.6 qt. with heater. Carburetor: Autolite C8VF-9510J four-barrel. Engine code: A.

The 1968 Lincoln 462-cid V-8 had a new air cleaner and this one has lots of bolt-on accessories, too. A new 460-cid V-8 was introduced for the Lincoln Continental Mark III. This engine had a 10.5:1 compression ratio and developed 365 hp at 4600 rpm. It used an Autolite C8VF-9510J four-barrel carburetor.

1969 LINCOLN

Lincoln Continental, Mark III base V-8
V-8. Overhead valve. Cast-iron block. Bore & stroke: 4.36 x 3.85 in. Displacement: 460 cid. Compression ratio: 10.50:1. Brake hp: 365 at 4600 rpm. Taxable hp: 60.83. Torque: 500 lbs.-ft. at 2800 rpm. Main bearings: five. Hydraulic valve lifters. Crankcase capacity: 4 qt. (add 1 qt. for new filter). Cooling system capacity: 19.6 qt. with heater. Carburetor: Autolite C8VF-9510J four-barrel. Engine code: A.

1970 LINCOLN

Lincoln Continental, Mark III base V-8
V-8. Overhead valve. Cast-iron block. Bore & stroke: 4.36 x 3.85 in. Displacement: 460 cid. Compression ratio: 10.50:1. Brake hp: 365 at 4600 rpm. Taxable hp: 60.83. Torque: 500 lbs.-ft. at 2800 rpm. Main bearings: five. Hydraulic valve lifters. Crankcase capacity: 4 qt. (add 1 qt. for new filter). Cooling system capacity: 20.5 qt. with heater. Carburetor: Autolite D0VF-9510-AA four-barrel. Engine code: A.

1971 LINCOLN

Lincoln Continental, Mark III base V-8
V-8. Overhead valve. Cast-iron block. Bore & stroke: 4.36 x 3.85 in. Displacement: 460 cid. Compression ratio: 10.20:1. Brake hp: 365 at 4600 rpm. Taxable hp: 60.83. Torque: 500 lbs.-ft. at 2800 rpm. Main bearings: five. Hydraulic valve lifters. Crankcase capacity: 4 qt. (add 1 qt. for new filter). Cooling system capacity: 19.5 qt. with heater. Carburetor: Autolite D1VF-9510-AA four-barrel. Engine code: A.

1972 LINCOLN

Mark IV base V-8
V-8. Overhead valve. Cast-iron block. Bore & stroke: 4.36 x 3.85 in. Displacement: 460 cid. Compression ratio: 8.50:1. Net brake hp: 212 at 4400 rpm. Net torque: 342 lbs.-ft. at 2800 rpm. Main bearings: five. Hydraulic valve lifters. Crankcase capacity: 4 qt. (add 1 qt. for new filter). Cooling system capacity: 19.5 qt. with heater. Carburetor: Motorcraft D2VF-9510-AA four-barrel. Engine code: A.

Continental base V-8
V-8. Overhead valve. Cast-iron block. Bore & stroke: 4.36 x 3.85 in. Displacement: 460 cid. Compression ratio: 8.50:1. Net brake hp: 224 at 4400 rpm. Net torque: 357 lbs.-ft. at 2800 rpm. Main bearings: five. Hydraulic valve lifters. Crankcase capacity: 4 qt. (add 1 qt. for new filter). Cooling system capacity: 19.5 qt. with heater. Carburetor: Motorcraft D2VF-9510-AA four-barrel. Engine code: A.

1973 LINCOLN

Mark IV base V-8
V-8. Overhead valve. Cast-iron block. Bore & stroke: 4.36 x 3.85 in. Displacement: 460 cid. Compression ratio: 8.00:1. Net brake hp: 208 at 4400 rpm. Net torque: 338 lbs.-ft. at 2800 rpm. Main bearings: five. Hydraulic valve lifters. Crankcase capacity: 4 qt. (add 1 qt. for new filter). Cooling system capacity: 19.5 qt. with heater. Carburetor: four-barrel. Engine code: A.

Lincoln base V-8
V-8. Overhead valve. Cast-iron block. Bore & stroke: 4.36 x 3.85 in. Displacement: 460 cid. Compression ratio: 8.00:1. Net brake hp: 219 at 4400 rpm. Net torque: 360 lbs.-ft. at 2800 rpm. Main bearings: five. Hydraulic valve lifters. Crankcase capacity: 4 qt. (add 1 qt. for new filter). Cooling system capacity: 19.5 qt. with heater. Carburetor: four-barrel. Engine code: A.

1974 LINCOLN

Mark IV base V-8
V-8. Overhead valve. Cast-iron block. Bore & stroke: 4.36 x 3.85 in. Displacement: 460 cid. Compression ratio: 8.00:1. Net brake hp: 215 at 4000 rpm. Net torque: 350 lbs.-ft. at 2600 rpm. Main bearings: five. Hydraulic valve lifters. Crankcase capacity: 4 qt. (add 1 qt. for new filter). Cooling system capacity: 20.5 qt. with heater. Carburetor: four-barrel. Engine code: A.

Lincoln base V-8
V-8. Overhead valve. Cast-iron block. Bore & stroke: 4.36 x 3.85 in. Displacement: 460 cid. Compression ratio: 8.00:1. Net brake hp: 220 at 4000 rpm. Net torque: 355 lbs.-ft. at 2600 rpm. Main bearings: five. Hydraulic valve lifters. Crankcase capacity: 4 qt. (add 1 qt. for new filter). Cooling system capacity: 20.5 qt. with heater. Carburetor: four-barrel. Engine code: A.

1975 LINCOLN

All Lincolns base V-8
V-8. Overhead valve. Cast-iron block. Bore & stroke: 4.36 x 3.85 in. Displacement: 460 cid. Compression ratio: 8.00:1. Net brake hp: 253 at 4400 rpm. Net torque: 386 lbs.-ft. at 2600 rpm. Main bearings: five. Hydraulic valve lifters. Crankcase capacity: 4 qt. (add 1 qt. for new filter). Cooling system capacity with air conditioning: (Lincoln) 19.75 qt.; (Mark IV) 20 qt. Carburetor: four-barrel. Engine code: A.

1976 LINCOLN

All Lincolns base V-8
V-8: 90-degree, overhead valve. Cast-iron block and head. Bore and sroke: 4.36 x 3.85 in. Displacement: 460 cid (7.5 liters). Compression ratio: 8.0:1. Net brake hp: 202 at 3800 rpm. Net torque: 352 lbs.-ft. at 1600 rpm. Five main bearings. Hydraulic valve lifters. Carburetor: Motorcraft 4350 (9510) four-barrel. Engine code: A.

1977 LINCOLN

Base Versailles V-8
90-degree, overhead valve V-8. Cast-iron block and head. Bore & stroke: 4.00 x 3.50 in. Displacement: 351 cid (5.8 liters). Compression ratio: 8.1:1. Net brake hp: 135 at 3200 rpm. Net torque: 275 lbs.-ft. at 1600 rpm. Five main bearings. Hydraulic valve lifters. Carburetor: Motorcraft 2150 two-barrel. Engine code: H.

Base Mark V V-8; Base Continental V-8 (late production)
90-degree, overhead valve V-8. Cast-iron block and head. Bore & stroke: 4.00 x 4.00 in. Displacement: 400 cid (6.6 liters). Compression ratio: 8.0:1. Net brake hp: 179 at 4000 rpm. Net torque: 329 lbs.-ft. at 1600 rpm. Five main bearings. Hydraulic valve lifters. Carburetor: Motorcraft 2150 two-barrel. Engine code: S.

Base Continental V-8 (early production); Optional Mark V V-8;
90-degree, overhead valve V-8. Cast-iron block and head. Bore & stroke: 4.36 x 3.85 in. Displacement: 460 cid (7.5 liters). Compression ratio: 8.0:1. Net brake hp: 208 at 4000 rpm. Net torque: 356 lbs.-ft. at 2000

rpm. Five main bearings. Hydraulic valve lifters. Carburetor: Motorcraft 4350 four-barrel. Engine code: A.

1978 LINCOLN

Base Versailles V-8
90-degree, overhead valve V-8. Cast-iron block and head. Bore & stroke: 4.00 x 3.00 in. Displacement: 302 cid (5.0 liters). Compression ratio: 8.4:1. Net brake hp: 133 at 3600 rpm. Net torque: 243 lbs.-ft. at 1600 rpm. Five main bearings. Hydraulic valve lifters. Carburetor: Motorcraft variable-venturi 2150 two-barrel. Engine code: F.

Base Mark V, Continental V-8
90-degree, overhead valve V-8. Cast-iron block and head. Bore & stroke: 4.00 x 4.00 in. Displacement: 400 cid (6.6 liters). Compression ratio: 8.0:1. Net brake hp: 166 at 3800 rpm. Net torque: 319 lbs.-ft. at 1800 rpm. Five main bearings. Hydraulic valve lifters. Carburetor: Motorcraft 2150 two-barrel. Engine code: S.

Optional Mark V, Continental V-8
90-degree, overhead valve V-8. Cast-iron block and head. Bore & stroke: 4.36 x 3.85 in. Displacement: 460 cid (7.5 liters). Compression ratio: 8.0:1. Net brake hp: 210 at 4200 rpm. Net torque: 357 lbs.-ft. at 2200 rpm. Five main bearings. Hydraulic valve lifters. Carburetor: Motorcraft 4350 four-barrel. Engine code: A.

1979 LINCOLN

Base Versailles V-8
90-degree, overhead valve V-8. Cast-iron block and head. Bore & stroke: 4.00 x 3.00 in. Displacement: 302 cid (5.0 liters). Compression ratio: 8.4:1. Net brake hp: 130 at 3600 rpm. Net torque: 237 lbs.-ft. at 1600 rpm. Five main bearings. Hydraulic valve lifters. Carburetor: Motorcraft variable-venturi 2150 two-barrel. Engine code: F.

Base Mark V, Continental V-8
90-degree, overhead valve V-8. Cast-iron block and head. Bore & stroke: 4.00 x 4.00 in. Displacement: 400 cid (6.6 liters). Compression ratio: 8.0:1. Net brake hp: 159 at 3400 rpm. Net torque: 315 lbs.-ft. at 1800 rpm. Five main bearings. Hydraulic valve lifters. Carburetor: Motorcraft 2150 two-barrel. Engine code: S.

1980 LINCOLN

Base Versailles V-8
90-degree, overhead valve V-8. Cast-iron block and head. Displacement: 302 cid (5.0 liters). Bore & stroke: 4.00 x 3.00 in. Compression ratio: 8.4:1. Net brake hp: 132 at 3600 rpm. Net torque: 232 lbs.-ft. at 1400 rpm. Five main bearings. Hydraulic valve lifters. Carburetor: Motorcraft 2150 two-barrel. Engine code: F.

Base Mark V, Continental V-8
90-degree, overhead valve V-8. Cast-iron block and head. Displacement: 302 cid (5.0 liters). Bore & stroke: 4.00 x 3.00 in. Compression ratio: 8.4:1. Net brake hp: 129 at 3600 rpm. Net torque: 231 lbs.-ft. at 2000 rpm. Five main bearings. Hydraulic valve lifters. Induction system: Electronic fuel injection. Engine code: F.

Optional Mark V, Continental V-8
90-degree, overhead valve V-8. Cast-iron block and head. Displacement: 351 cid (5.8 liters). Bore & stroke: 4.00 x 3.50 in. Compression ratio: 8.3:1. Net brake hp: 140 at 3400 rpm. Net torque: 265 lbs.-ft. at 2000 rpm. Five main bearings. Hydraulic valve lifters. Carburetor: Motorcraft 7200VV two-barrel. Engine code: G.

1981 LINCOLN

Town Car, Mark VI base V-8
90-degree, overhead valve V-8. Cast-iron block and head. Bore & stroke: 4.00 x 3.00 in. Displacement: 302 cid (5.0 liters). Compression ratio: 8.4:1. Net brake hp: 130 at 3400 rpm. Net torque: 230 lbs.-ft. at 2200 rpm. Five main bearings. Hydraulic valve lifters. Induction system: Electronic fuel injection. Engine code: F.

1982 LINCOLN

Town Car, Mark VI base V-8
90-degree, overhead valve V-8. Cast-iron block and head. Bore & stroke: 4.00 x 3.00 in. Displacement: 302 cid (5.0 liters). Compression ratio: 8.4:1. Net brake hp: 134 at 3400 rpm. Net torque: 232 lbs.-ft. at 2200 rpm. Five main bearings. Hydraulic valve lifters. Induction system: Electronic fuel injection. Engine code: F.

Continental base V-8
90-degree, overhead valve V-8. Cast-iron block and head. Bore & stroke: 4.00 x 3.00 in. Displacement: 302 cid (5.0 liters). Compression ratio: 8.4:1. Net brake hp: 131 at 3400 rpm. Net torque: 229 lbs.-ft. at 1200 rpm. Five main bearings. Hydraulic valve lifters. Carburetor: Variable-venturi two-barrel. Engine code: F.

1983 LINCOLN

All models base V-8
90-degree, overhead valve V-8. Cast-iron block and head. Bore & stroke: 4.00 x 3.00 in. Displacement: 302 cid (5.0 liters). Compression ratio: 8.4:1. Net brake hp: 130 at 3200 rpm. Net torque: 240 lbs.-ft. at 2000 rpm. Five main bearings. Hydraulic valve lifters. Induction system: Throttle-body (electronic) fuel injection. Engine code: F.

Mark VI optional V-8
90-degree, overhead valve V-8. Cast-iron block and head. Bore & stroke: 4.00 x 3.00 in. Displacement: 302 cid (5.0 liters). Compression ratio: 8.4:1. Net brake hp: 145 at 3600 rpm. Net torque: 245 lbs.-ft. at 2200 rpm. Five main bearings. Hydraulic valve lifters. Induction system: Throttle-body (electronic) fuel injection. Engine code: F.

1984 LINCOLN

All models base V-8
90-degree, overhead valve V-8. Cast-iron block and head. Bore & stroke: 4.00 x 3.00 in. Displacement: 302 cid (5.0 liters). Compression ratio: 8.4:1. Net brake hp: 140 at 3200 rpm. Net torque: 250 lbs.-ft. at 1600 rpm. Five main bearings. Hydraulic valve lifters. Induction: Throttle-body (electronic) fuel injection. Engine code: F.

Town Car optional V-8
90-degree, overhead valve V-8. Cast-iron block and head. Bore & stroke: 4.00 x 3.00 in. Displacement: 302 cid (5.0 liters). Compression ratio: 8.4:1. Net brake hp: 155 at 3600 rpm. Net torque: 265 lbs.-ft. at 2000 rpm. Five main bearings. Hydraulic valve lifters. Induction: Throttle-body (electronic) fuel injection. Dual exhaust. Engine code: F.

1985 LINCOLN

All models base V-8
90-degree, overhead valve V-8. Cast-iron block and head. Bore & stroke: 4.00 x 3.00 in. Displacement: 302 cid (5.0 liters). Compression ratio: 8.4:1. Net brake hp: 140 at 3200 rpm. Net torque: 250 lbs.-ft. at 1600 rpm. Five main bearings. Hydraulic valve lifters. Induction: Throttle-body (electronic) fuel injection. Engine code: F.

Town Car optional V-8
90-degree, overhead valve V-8. Cast-iron block and head. Bore & stroke: 4.00 x 3.00 in. Displacement: 302 cid (5.0 liters). Compression ratio: 8.4:1. Net brake hp: 155 at 3600 rpm. Net torque: 265 lbs.-ft. at 2000 rpm. Five main bearings. Hydraulic valve lifters. Induction: Throttle-body (electronic) fuel injection. Dual exhaust. Engine code: F.

Base Mark VII LSC V-8
90-degree, overhead valve high-output V-8. Cast-iron block and head. Bore & stroke: 4.00 x 3.00 in. Displacement: 302 cid (5.0 liters). Compression ratio: 8.3:1. Net brake hp: 180 at 4200 rpm. Net torque: 260 lbs.-ft. at 2600 rpm. Five main bearings. Hydraulic valve lifters. Induction: Throttle-body (electronic) fuel injection. Dual exhaust. Engine code: M.

1986 LINCOLN

All models base V-8
90-degree, overhead valve V-8. Cast-iron block and head. Bore & stroke: 4.00 x 3.00 in. Displacement: 302 cid (5.0 liters). Compression ratio: 8.9:1. Net brake hp: 150 at 3200 rpm. Net torque: 270 lbs.-ft. at 2000 rpm. Five main bearings. Hydraulic valve lifters. Induction: Sequential port (electronic) fuel injection. Engine code: F.

Town Car optional V-8
90-degree, overhead valve V-8. Cast-iron block and head. Bore & stroke: 4.00 x 3.00 in. Displacement: 302 cid (5.0 liters). Compression ratio: 8.9:1. Net brake hp: 160 at 3400 rpm. Net torque: 280 lbs.-ft. at 2200 rpm. Five main bearings. Hydraulic valve lifters. Induction: Sequential port (electronic) fuel injection. Dual exhaust. Engine code: F.

Base Mark VII LSC V-8
90-degree, overhead valve high-output V-8. Cast-iron block and head. Bore & stroke: 4.00 x 3.00 in. Displacement: 302 cid (5.0 liters). Compression ratio: 9.2:1. Net brake hp: 200 at 4000 rpm. Net torque: 285 lbs.-ft. at 3000 rpm. Five main bearings. Hydraulic valve lifters. Induction: Throttle-body (electronic) fuel injection. Dual exhaust. Engine code: M.

1987 LINCOLN

All models base V-8
90-degree, overhead valve V-8. Cast-iron block and head. Bore & stroke: 4.00 x 3.00 in. Displacement: 302 cid (5.0 liters). Compression ratio: 8.9:1. Net brake hp: 150 at 3200 rpm. Net torque: 270 lbs.-ft. at 2000 rpm. Five main bearings. Hydraulic valve lifters. Induction: Sequential port (electronic) fuel injection. Engine code: F.

Base Mark VII LSC V-8
90-degree, overhead valve high-output V-8. Cast-iron block and head. Bore & stroke: 4.00 x 3.00 in. Displacement: 302 cid (5.0 liters). Compression ratio: 9.2:1. Net brake hp: 200 at 4000 rpm. Net torque: 285 lbs.-ft. at 3000 rpm. Five main bearings. Hydraulic valve lifters. Induction: Throttle-body (electronic) fuel injection. Dual exhaust. Engine code: M.

1988 LINCOLN

Town Car base V-8
90-degree, overhead valve V-8. Cast-iron block and head. Bore & stroke: 4.00 x 3.00 in. Displacement: 302 cid (5.0 liters). Compression ratio: 8.9:1. Net brake hp: 150 at 3200 rpm. Net torque: 270 lbs.-ft. at 2000 rpm. Five main bearings. Hydraulic valve lifters. Induction: Sequential port (electronic) fuel injection. Engine code: F.

Mark VII base V-8
90-degree, overhead valve high-output V-8. Cast-iron block and head. Bore & stroke: 4.00 x 3.00 in. Displacement: 302 cid (5.0 liters). Compression ratio: 9.2:1. Net brake hp: 225 at 4000 rpm. Net torque: 300 lbs.-ft. at 3200 rpm. Five main bearings. Hydraulic valve lifters. Induction: Sequential port (electronic) fuel injection. Engine code: M.

1989 LINCOLN

Town Car base V-8
90-degree, overhead valve V-8. Cast-iron block and head. Bore & stroke: 4.00 x 3.00 in. Displacement: 302 cid (5.0 liters). Compression ratio: 8.9:1. Net brake hp: 150 at 3200 rpm. Net torque: 270 lbs.-ft. at 2000 rpm. Five main bearings. Hydraulic valve lifters. Induction: Sequential port (electronic) fuel injection. Engine code: F.

Mark VII base V-8
90-degree, overhead valve high-output V-8. Cast-iron block and head. Bore & stroke: 4.00 x 3.00 in. Displacement: 302 cid (5.0 liters). Compression ratio: 9.2:1. Net brake hp: 225 at 4000 rpm. Net torque: 300 lbs.-ft. at 3200 rpm. Five main bearings. Hydraulic valve lifters. Induction: Sequential port (electronic) fuel injection. Engine code: M.

1990 LINCOLN

Town Car base V-8
90-degree, overhead valve V-8. Cast-iron block and head. Bore & stroke: 4.00 x 3.00 in. Displacement: 302 cid (5.0 liters). Compression ratio: 8.9:1. Net brake hp: 150 at 3200 rpm. Net torque: 270 lbs.-ft. at 2000 rpm. Five main bearings. Hydraulic valve lifters. Induction: Sequential port (electronic) fuel injection. Engine code: F.

Mark VII base V-8
90-degree, overhead valve high-output V-8. Cast-iron block and head. Bore & stroke: 4.00 x 3.00 in. Displacement: 302 cid (5.0 liters). Compression ratio: 9.2:1. Net brake hp: 225 at 4000 rpm. Net torque: 300 lbs.-ft. at 3200 rpm. Five main bearings. Hydraulic valve lifters. Induction: Sequential port (electronic) fuel injection. Engine code: T.

1991 LINCOLN

Town Car base V-8
Modular, overhead cam V-8. Cast-iron block and aluminum head. Bore & stroke: 3.60 x 3.60 in. Displacement: 281 cid (4.6 liters). Compression ratio: 9.0:1. Net brake hp: 190 at 4200 rpm (210 with dual exhaust). Net torque: 260 lbs.-ft. at 3600 rpm. Induction: Sequential fuel injection. Engine code: F.

Mark VII base V-8
90-degree, overhead valve high-output V-8. Cast-iron block and head. Bore & stroke: 4.00 x 3.00 in. Displacement: 302 cid (5.0 liters). Compression ratio: 9.2:1. Net brake hp: 225 at 4000 rpm. Net torque: 300 lbs.-ft. at 3200 rpm. Five main bearings. Hydraulic valve lifters. Induction: Sequential port (electronic) fuel injection. Engine code: T.

1992 LINCOLN

Town Car base V-8
Modular, overhead cam V-8. Cast-iron block and aluminum head. Bore & stroke: 3.60 x 3.60 in. Displacement: 281 cid (4.6 liters). Compression ratio: 9.0:1. Net brake hp: 190 at 4200 rpm (210 with dual exhaust). Net torque: 260 lbs.-ft. at 3600 rpm. Induction: Sequential fuel injection. Engine code: W.

Mark VII base V-8
90-degree, overhead valve high-output V-8. Cast-iron block and head. Bore & stroke: 4.00 x 3.00 in. Displacement: 302 cid (5.0 liters). Compression ratio: 9.2:1. Net brake hp: 225 at 4000 rpm. Net torque: 300 lbs.-ft. at 3200 rpm. Five main bearings. Hydraulic valve lifters. Induction: Sequential port (electronic) fuel injection. Engine code: T.

1993 LINCOLN

Town Car base V-8
Modular, overhead cam V-8. Aluminum block and head. Bore & stroke: 3.60 x 3.60 in. Displacement: 281 cid (4.6 liters). Compression ratio: 9.0:1. Net brake hp: 190 at 4600 rpm (210 at 4600 rpm with dual exhaust). Net torque: 260 lbs.-ft. at 3200 rpm (270 lbs.-ft. at 3400 rpm with dual exhaust). Induction: Sequential fuel injection. Engine code: W.

Mark VIII base V-8
Modular, dual-overhead cam V-8. Aluminum block and head. Bore & stroke: 3.60 x 3.60 in. Displacement: 281 cid (4.6 liters). Compression ratio: 9.8:1. Net brake hp: 280 at 5500 rpm. Net torque: 285 lbs.-ft. at 4500 rpm. Induction: Sequential fuel injection. Engine code: V.

1994 LINCOLN

Town Car base V-8
Modular, overhead cam V-8. Aluminum block and head. Bore & stroke: 3.60 x 3.60 in. Displacement: 281 cid (4.6 liters). Compression ratio: 9.0:1. Net brake hp: 210 at 4600 rpm. Net torque: 270 lbs.-ft. at 3400 rpm. Dual exhaust. Induction: Sequential fuel injection. Engine code: W.

Mark VIII base V-8
Modular, dual-overhead cam V-8. Aluminum block and head. Bore & stroke: 3.60 x 3.60 in. Displacement: 281 cid (4.6 liters). Compression ratio: 9.8:1. Net brake hp: 280 at 5500 rpm. Net torque: 285 lbs.-ft. at 4500 rpm. Induction: Sequential fuel injection. Engine code: V.

1995 LINCOLN

Town Car base V-8
Modular, overhead cam V-8. Aluminum block and head. Bore & stroke: 3.60 x 3.60 in. Displacement: 281 cid (4.6 liters). Compression ratio: 9.0:1. Net brake hp: 210 at 4600 rpm. Net torque: 270 lbs.-ft. at 3400 rpm. Dual exhaust. Induction: Sequential fuel injection. Engine code: W.

Continental base V-8
Modular, dual-overhead cam V-8. Aluminum block and head. Bore & stroke: 3.60 x 3.60 in. Displacement: 281 cid (4.6 liters). Compression ratio: 9.8:1. Net brake hp: 260 at 5750 rpm. Net torque: 265 lbs.-ft. at 4750 rpm. Induction: Sequential fuel injection. Engine code: V.

Mark VIII base V-8
Modular, dual-overhead cam V-8. Aluminum block and head. Bore & stroke: 3.60 x 3.60 in. Displacement: 281 cid (4.6 liters). Compression ratio: 9.8:1. Net brake hp: 280 at 5500 rpm. Net torque: 285 lbs.-ft. at 4500 rpm. Induction: Sequential fuel injection. Engine code: V.

1996 LINCOLN

Town Car base V-8
Modular, overhead cam V-8. Aluminum block and head. Bore & stroke: 3.60 x 3.60 in. Displacement: 281 cid (4.6 liters). Compression ratio: 9.0:1. Net brake hp: 210 at 4600 rpm. Net torque: 270 lbs.-ft. at 3400 rpm. Dual exhaust. Induction: Sequential fuel injection. Engine code: W (Romeo).

Continental base V-8
Modular, dual-overhead cam V-8. Aluminum block and head. Bore & stroke: 3.60 x 3.60 in. Displacement: 281 cid (4.6 liters). Compression ratio: 9.8:1. Net brake hp: 260 at 5750 rpm. Net torque: 265 lbs.-ft. at 4750 rpm. Sequential fuel injection. Engine code: V.

Mark VIII base V-8
Modular, dual-overhead cam V-8. Aluminum block and head. Bore & stroke: 3.60 x 3.60 in. Displacement: 281 cid (4.6 liters). Compression ratio: 9.8:1. Net brake hp: 280 at 5500 rpm. Net torque: 285 lbs.-ft. at 4500 rpm. Induction: Sequential fuel injection. Engine code: V.

1997 LINCOLN

Town Car base V-8
Modular, overhead cam V-8. Aluminum block and head. Bore & stroke: 3.60 x 3.60 in. Displacement: 281 cid (4.6 liters). Compression ratio: 9.0:1. Net brake hp: 210 at 4250 rpm. Net torque: 275 lbs.-ft. at 3250 rpm. Dual exhaust. Induction: Sequential fuel injection. Engine code: W (Romeo).

Continental base V-8
Modular, dual-overhead cam V-8. Aluminum block and head. Bore & stroke: 3.60 x 3.60 in. Displacement: 281 cid (4.6 liters). Compression ratio: 9.9:1. Net brake hp: 260 at 5750 rpm. Net torque: 265 lbs.-ft. at 4750 rpm. Induction: Sequential fuel injection. Engine code: V.

Mark VIII base V-8
Modular, dual-overhead cam V-8. Aluminum block and head. Bore & stroke: 3.60 x 3.60 in. Displacement: 281 cid (4.6 liters). Compression ratio: 9.9:1. Net brake hp: 280 at 5750 rpm. Net torque: 285 lbs.-ft. at 4500 rpm. Induction: Sequential fuel injection. Engine code: V.

Mark VIII LSC base V-8
Modular, dual-overhead cam V-8. Aluminum block and head. Bore & stroke: 3.60 x 3.60 in. Displacement: 281 cid (4.6 liters). Compression ratio: 9.9:1. Net brake hp: 290 at 5750 rpm. Net torque: 295 lbs.-ft. at 4500 rpm. Induction: Sequential fuel injection. Engine code: V.

1998 LINCOLN

Town Car Executive Edition, Town Car Signature Edition base V-8
Modular, overhead cam V-8. Aluminum block and head. Bore & stroke: 3.60 x 3.60 in. Displacement: 281 cid (4.6 liters). Compression ratio: 9.0:1. Net brake hp: 200 at 4500 rpm. Net torque: 265 lbs.-ft. at 3500 rpm. Induction: Sequential fuel injection. Engine code: W (Romeo).

Town Car Cartier Edition, Town Car Signature Touring Sedan Edition base V-8
Modular, overhead cam V-8. Aluminum block and head. Bore & stroke: 3.60 x 3.60 in. Displacement: 281 cid (4.6 liters). Compression ratio: 9.0:1. Net brake hp: 220 at 4500 rpm. Net torque: 275 lbs.-ft. at 3500 rpm. Dual exhaust. Induction: Sequential fuel injection. Engine code: W (Romeo).

Continental base V-8
Modular, dual-overhead cam V-8. Aluminum block and head. Bore & stroke: 3.60 x 3.60 in. Displacement: 281 cid (4.6 liters). Compression ratio: 9.9:1. Net brake hp: 260 at 5750 rpm. Net torque: 265 lbs.-ft. at 4750 rpm. Induction: Sequential fuel injection. Engine code: V.

Mark VIII base V-8
Modular, dual-overhead cam V-8. Aluminum block and head. Bore & stroke: 3.60 x 3.60 in. Displacement: 281 cid (4.6 liters). Compression ratio: 9.9:1. Net brake hp: 280 at 5750 rpm. Net torque: 285 lbs.-ft. at 4500 rpm. Induction: Sequential fuel injection. Engine code: V.

Mark VIII LSC base V-8
Modular, dual-overhead cam V-8. Aluminum block and head. Bore & stroke: 3.60 x 3.60 in. Displacement: 281 cid (4.6 liters). Compression ratio: 9.9:1. Net brake hp: 290 at 5750 rpm. Net torque: 295 lbs.-ft. at 4500 rpm. Induction: Sequential fuel injection. Engine code: V.

1999 LINCOLN

Town Car Executive Edition, Town Car Signature Edition base V-8
Modular, overhead cam V-8. Aluminum block and head. Bore & stroke: 3.60 x 3.60 in. Displacement: 281 cid (4.6 liters). Compression ratio: 9.0:1. Net brake hp: 200 at 4500 rpm. Net torque: 265 lbs.-ft. at 3500 rpm. Induction: Sequential fuel injection. Engine code: W (Romeo).

Town Car Cartier Edition, Town Car Signature Touring Sedan Edition base V-8
Modular, overhead cam V-8. Aluminum block and head. Bore & stroke: 3.60 x 3.60 in. Displacement: 281 cid (4.6 liters). Compression ratio: 9.0:1. Net brake hp: 220 at 4500 rpm. Net torque: 275 lbs.-ft. at 3500 rpm. Dual exhaust. Induction: Sequential fuel injection. Engine code: W (Romeo).

Continental base V-8
Double overhead cam V-8. Aluminum block and head. Bore & stroke: 3.60 x 3.60 in. Displacement: 281 cid (4.6 liters). Compression ratio: 9.89:1. Net brake hp: 260 at 5750 rpm. Net torque: 270 lbs.-ft. at 3000 rpm. Induction: Sequential multi-port electronic fuel injection. Engine code: V.

2000 LINCOLN

LS base V-8
Double overhead cam V-8. Aluminum block and head. Bore & stroke: 3.39 x 3.35 in. Displacement: 240 cid (3.9 liters). Compression ratio: 10.5:1. Net brake hp: 252 at 6100 rpm. Net torque: 267 lbs.-ft. at 4300 rpm. Induction: Sequential multi-port electronic fuel injection. Engine code: A.

Town Car Executive Edition, Town Car Signature Edition base V-8
Modular, overhead cam V-8. Aluminum block and head. Bore & stroke: 3.60 x 3.60 in. Displacement: 281 cid (4.6 liters). Compression ratio: 9.0:1. Net brake hp: 200 at 4500 rpm. Net torque: 265 lbs.-ft. at 3500 rpm. Sequential fuel injection. Engine code: W (Romeo).

Town Car Signature Edition with Touring Sedan optional V-8
Modular. Single overhead cam V-8. Aluminum block and head. Bore & stroke: 3.60 x 3.60 in. Displacement: 281 cid (4.6 liters). Compression ratio: 10.0:1. Net brake hp: 215 at 4500 rpm. Net torque: 285 lbs.-ft. at 3500 rpm. Sequential multi-port electronic fuel injection. Engine code: W (Romeo).

Town Car Cartier Edition optional V-8
Modular. Single overhead cam V-8. Aluminum block and head. Bore & stroke: 3.60 x 3.60 in. Displacement: 281 cid (4.6 liters). Compression ratio: 9.4:1. Net brake hp: 235 at 4650 rpm. Net torque: 275 lbs.-ft. at 4000 rpm. Dual exhaust. Sequential multi-port electronic fuel injection. Engine code: W (Romeo).

Continental base V-8
Double overhead cam V-8. Aluminum block and head. Bore & stroke: 3.55 x 3.54 in. Displacement: 281 cid (4.6 liters). Compression ratio: 9.89:1. Net brake hp: 275 at 5750 rpm. Net torque: 275 lbs.-ft. at 3000 rpm. Sequential multi-port electronic fuel injection. Engine code: V.

2001 LINCOLN

LS base V-8
Double overhead cam V-8. Aluminum block and head. Bore & stroke: 3.39 x 3.35 in. Displacement: 240 cid (3.9 liters). Compression ratio: 10.5:1. Net brake hp: 252 at 6100 rpm. Net torque: 267 lbs.-ft. at 4300 rpm. Induction: Sequential multi-port electronic fuel injection. Engine code: A.

Town Car base V-8
Modular. Single overhead cam V-8. Aluminum block and head. Bore & stroke: 3.60 x 3.60 in. Displacement: 281 cid (4.6 liters). Compression ratio: 9.4:1. Net brake hp: 220 at 4750 rpm. Net torque: 265 lbs.-ft. at 4000 rpm. Induction: Sequential multi-port electronic fuel injection. Engine code: W (Romeo).

Town Car Cartier Edition optional V-8
Modular. Single overhead cam V-8. Aluminum block and head. Bore & stroke: 3.60 x 3.60 in. Displacement: 281 cid (4.6 liters). Compression ratio: 9.4:1. Net brake hp: 235 at 4650 rpm. Net torque: 275 lbs.-ft. at 4000 rpm. Dual exhaust. Induction: Sequential multi-port electronic fuel injection. Engine code: W (Romeo).

Continental base V-8
Double overhead cam V-8. Aluminum block and head. Bore & stroke: 3.55 x 3.54 in. Displacement: 281 cid (4.6 liters). Compression ratio: 9.89:1. Net brake hp: 275 at 5750 rpm. Net torque: 275 lbs.-ft. at 4750 rpm. Induction: Sequential multi-port electronic fuel injection. Engine code: V.

2002 LINCOLN

LS base V-8
Double overhead cam V-8. Aluminum block and head. Bore & stroke: 3.39 x 3.35 in. Displacement: 240 cid (3.9 liters). Compression ratio: 10.5:1. Net brake hp: 252 at 6100 rpm. Net torque: 267 lbs.-ft. at 4300 rpm. Induction: Sequential multi-port electronic fuel injection. Engine code: A.

Town Car base V-8
Modular. Single overhead cam V-8. Aluminum block and head. Bore & stroke: 3.60 x 3.60 in. Displacement: 281 cid (4.6 liters). Compression ratio: 9.4:1. Net brake hp: 220 at 4750 rpm. Net torque: 265 lbs.-ft. at 4000 rpm. Sequential multi-port electronic fuel injection. Engine code: W (Romeo).

Town Car Cartier Edition optional V-8
Modular. Single overhead cam V-8. Aluminum block and head. Bore & stroke: 3.60 x 3.60 in. Displacement: 281 cid (4.6 liters). Compression ratio: 9.4:1. Net brake hp: 235 at 4650 rpm. Net torque: 275 lbs.-ft. at 4000 rpm. Dual exhaust. Sequential multi-port electronic fuel injection. Engine code: W (Romeo).

Continental base V-8
Double overhead cam V-8. Aluminum block and head. Bore & stroke: 3.55 x 3.54 in. Displacement: 281 cid (4.6 liters). Compression ratio: 9.89:1. Net brake hp: 275 at 5750 rpm. Net torque: 275 lbs.-ft. at 4750 rpm. Sequential multi-port electronic fuel injection. Engine code: V.

Selected Lincoln V-8 Tune-Up Specs 1921-1932 and 1949-1965

Year	Model	Spark Plugs Type	Spark Plugs Gap	Point Gap	Deg Cam Angle	Timing	Idle Speed	Compression Pressure	Battery (Delco)
1921-25	Lincoln 8	CH 7/8	NA	.020	20 deg.	5B	NA	NA	Exide 130
1926-27	Lincoln 8	CH Reg. S	NA	.020	20 deg.	5B	NA	NA	Exide 130
1928	Lincoln 8	CH 7/8	NA	.015	24 deg.	5B	NA	NA	Exide 135
1929	Lincoln 8	CH 7/8	.025	.018	NA	5B	NA	NA	Exide 135
1930	Lincoln 8	CH 7/8	.025	.018	NA	5B	NA	NA	Exide 135
1931-32	Lincoln 8	CH C4	NA	.018	NA	5B	NA	NA	Exide 135

1949	9EL, 9EH	CH-H10	.030	.015	27 deg.	4B	450	110 lb.	Group 2F
1950	All	CH-H10	.025	.015	27 deg.	4B	450	110 lb.	Group 2F
1951	All	CH-H10	.030	.015	27 deg.	3B	450	110 lb.	Group 2F
1952	All	CH-H10	.030	.015	27 deg.	3B	450	115 lb.	3HN/3NR
1953	All	CH-H10	.035	.015	27 deg.	3B	450	135 lb.	3HN/3NR
1954	All	CH-H10	.035	.015	27 deg.	3B	450	135 lb.	3HN/3NR
1955	All	CH-870	.035	.015	27 deg.	5B	450	135 lb.	3HN/3NR
1956	All	CH-860	.035	.015	27 deg.	5B	450	150 lb.	14HN/5HN//5SN
1957 14HN/5HN//5SN	All		CH-860	.035	.015	27 deg.	5B	450	150 lb.
1958 Ford 70 amp hr	All		CH-F11Y	.035	.015	27 deg.	6B	425-450	150 lb.
1959 FDU-10655-F	All	Ford FEU-12405-A		.032	.015	27 deg.	6B	450-475	150 lb.
1960 HJK-10655-F	All		CH-F11Y	.032-.034	.014-.016	27 deg.	8B	450-475	180 lb.
1961 CIVF-10655-F	All		CH-F11Y	.032-.036	.014-.016	27 deg.	8B	525	180 lb.
1962 CIVF-10655-A	All		BF42	.035	.017	27 deg.	8B	525	180 lb.
1963 CIVF-11001-D	All		BF42	.035	.017	27 deg.	8B	525	180 lb.
1964	Conv. Ign.	BF42	.034	.017	26-31 d	8B	525	180 lb.	C3VF-10655-C
	Trans. Ign.	BF42	.030	.020	22-24 d	8B	525	180 lb.	C3VF-10655-C
1965	Conv. Ign.	BF42	.034	.017	26-31 d	6B	525	180 lb.	C3VF-10655-C
	Trans. Ign.	BF42	.030	.020	22-24 d	6B	525	180 lb.	C3VF-10655-C

Lincoln V-8 Firing Order (1921-1976)

1921-1932
1R-4L-2R-3L-4R-1L-3R-2L

NOTE: *Some manuals express the firing order as 1-2-3-4-5-6-7-8 which can be confusing until you make reference to the accompanying diagram in which the cylinders on the left-hand bank are numbered 6-8-4-2 from front to rear and the cylinders on the right-hand bank are numbered 1-3-7-5 from front to rear.*

1949-1951
1-5-4-8-6-3-7-2

1952-1957
1-5-4-8-6-3-7-2

1958-1976
1-5-4-2-6-3-7-8

All postwar Lincoln V-8s are numbered from front to back. The cylinders on the right-hand bank are numbered 1-2-3-4 and the cylinders on the left-hand bank are numbered 5-6-7-8.

For the 1949-1951 V-8s the spark plug wires go into the distributor in firing order in a clockwise direction.
For 1952-1957 V-8s the spark plug wires go into the distributor in firing order in a counterclockwise direction.
For 1958-1976 V-8s the spark plug wires go into the distributor in firing order in a counterclockwise direction.

MERCURY
1939-2002

In 1939, Ford Motor Company released its third flathead V-8 with the addition of a 239-cid 95-hp version for the new Mercury. In addition to more cubic inches, this 24-stud flathead introduced heavier connecting rods, a heavier crankshaft, and other beefed-up internal components. There was little significant change in this engine until 1942, when the designs of the ignition coil and distributor were revised.

The postwar 239-cid "Mercury" flathead V-8 churned up 100 hp at 3600 rpm. Improvements included a pressurized cooling system, revised aluminum pistons, larger connecting rod journals, and a one-piece distributor cap. In 1949, the Mercury V-8 got a larger four-inch stroke that upped displacement to 255 cid and horsepower to 112. Other changes included a modern distributor, a new water pump, new valve guides, improved breathing, and locked-in-place connecting rod bearings. By 1952, the Mercury V-8 had a higher compression ratio and 125 hp.

Mercury's first overhead valve V-8 appeared in 1954 and had almost the same displacement—256 cid versus 255.4 cid—as the previous flathead. Its 161-hp rating was 36 hp higher than that of the old engine. Like Ford's new V-8, it was promoted as a "Y" block because the 90-degree cylinder banks appeared to form the upper arms of the letter "Y," while the deep skirt extending below the crankshaft centerline formed the vertical leg of a "Y."

Unlike Ford, Mercury offered just one V-8 in 1955. This 292-cid power plant produced 188 hp in stick-shift cars and 10 more when automatic transmission was ordered. By 1956, the base engine became a 312-cid V-8 that was joined, in 1957, by the 368-cid 290-hp Turnpike Cruiser engine. Even this was not "big enough" with the mid-'50s horsepower race rolling on at a fever pitch. By 1958, the Monterey and Montclair (along with their counterpart station wagons) were treated to a 383-cid V-8, while the top-of-the line Park Lane shared Lincoln's 430-cid V-8.

By the early 1960s, there was more direct sharing of V-8 engines between the various FoMoCo divisions. It's position between standard Fords and the upscale Thunderbird allowed Mercury to share engines with both models at various times. The 292-cid small-block V-8 that served as base engine in the 1961 Meteor and Monterey came from the Ford Fairlane and Galaxie range. The optional 352-cid V-8 had started out as a T-Bird engine. And the top-of-the-line 390-cid V-8 was standard in the then-current T-Bird. In 1962, Ford's new 406-cid V-8 was added to the options list. Big Mercurys also got the powerful 427 as a 1963 performance option.

Another famous FoMoCo V-8 of the '60s was the compact, lightweight 289-cid engine that made history in the Cougar and Mustang. It was originally a 221-cid V-8 designed for the compact 1963 Meteor, then grew to 260 cid in the 1964 Comet. Both of these configurations were dropped when the 289 appeared in 1965. It did not have the Y-block layout of earlier Mercury V-8s and featured a stronger "bottom end" that racing enthusiasts appreciated. The 289 was used as base engine in the first Cougar.

In 1968, only one 289 (200 hp) was offered, but the small V-8 was also stroked to 302 cid and offered in 210- and 230-hp formats. This made the engine competitive and "legal" for use in SCCA Trans-Am series sedan racing. The 302 was basically stroked again to achieve 351 cid in 1969. This provided a compact and lightweight V-8 that came in factory 250- and 290-hp versions and made a great platform for aftermarket performance hardware

The 428-cid big-block V-8 that bowed in the 1966 Thunderbird replaced the 427 option for big Mercurys. It was a bored-and-stroked 390 and was offered in 345- and 360-hp editions. Also available in this series of V-8s in 1968 was the 427-cid racing engine, which retained the 390's 3.784-inch stroke but had a larger 4.23-inch bore. It generated 390 hp.

The Thunderjet 429 V-8 that bowed as base engine in the 1969 Marquis was not a true performance engine. It was based on the 460-cid Lincoln engine. The base version of the 429 generated just 320 hp. However, the hotter versions with 360- and 375-hp were available, too. Ford brought out a 400-cid V-8 with 4.00 x 4.00-in. bore and stroke dimensions in 1971, but it was not a performance engine either.

Mercury listed 15 V-8 engines for 1973, but many were actually identical. The company gave each engine two or three different ratings, depending on which car it was put in. By 1975, the 302, the 351 Windsor, the 400, the regular 460 and the Police Interceptor 460 were the only V-8s left. By the end of the decade, Mercury offered only 302- and 351-cid V-8s.

With the trend towards fuel economy, a new 4.2-liter (255-cid) V-8 served as base engine in the 1980 Cougar and was optional in other Fords. The following season, a special "Police Package 5.0-liter H.O." 351-cid V-8 with 165 net hp arrived for the Grand Marquis By 1984, this limited-availability option was up to 180 hp in which form it survived three more years. From 1988 through 1991, only the 302 was offered and never with more than 160 hp in a Mercury.

The 1992 Grand Marquis was the first Mercury to use the 281-cid overhead cam "modular" V-8, while the Cougar retained the 302 that year. After 1983, the modular engine became the only V-8 offered in a Mercury.

1939 MERCURY

Mercury V-8
90 degree V-8. L-head. Cast-iron block. Bore & stroke: 3.187 x 3.75 in. Displacement: 239.4 cid. Compression ratio: 6.15:1. Brake hp: 95 at 3600 rpm. Taxable hp: 32.51. Torque: 170 lbs.-ft. at 2100 rpm. Three main bearings. Solid valve lifters. Crankcase capacity: 5 qt. Cooling system capacity: 21.25 qt. Carburetor: Holley (Chandler-Groves) Model AA-1 dual downdraft.

1940 MERCURY

Model 09A
90 degree V-8. L-head. Cast-iron block. Bore & stroke: 3.187 x 3.75 in. Displacement: 239.4 cid. Compression ratio: 6.15:1. Brake hp: 95 at 3600 rpm. Taxable hp: 32.51. Torque: 176 lbs.-ft. at 2100 rpm. Three main bearings. Solid valve lifters. Crankcase capacity: 5 qt. Cooling system capacity: 22 qt. Carburetor: Holley (Chandler-Groves) Model AA-1 dual downdraft.

1941 MERCURY

Model 09A
90 degree V-8. L-head. Cast-iron block. Bore & stroke: 3.187 x 3.75 in. Displacement: 239.4 cid. Compression ratio: 6.15:1. Brake hp: 95 at 3600 rpm. Taxable hp: 32.51. Torque: 170 lbs.-ft. at 2100 rpm. Three main bearings. Solid valve lifters. Crankcase capacity: 5 qt. Cooling system capacity: 23.75 qt. Carburetor: Holley (Chandler-Groves) Model AA-1 dual downdraft.

1942 MERCURY

Model 29A
90 degree V-8. L-head. Cast-iron block. Bore & stroke: 3.187 x 3.75 in. Displacement: 239.4 cid. Compression ratio: 6.40:1. Brake hp: 100 at 3800 rpm. Taxable hp: 32.51. Torque: 176 lbs.-ft. at 2100 rpm. Three main bearings. Solid valve lifters. Crankcase capacity: 5 qt. Cooling system capacity: 21.25 qt. Carburetor: Holley (Chandler-Groves) Model AA-1 dual downdraft.

The owners of this 1939 Mercury convertible and 1967 Mercury Cougar are likely to find many differences under the hoods of their cars, but the one thing that both cars share is a V-8 engine, which Mercurys used since the beginning of the brand, and still use in some models today. (LeRoy Byers photo)

This restored 1942 Mercury V-8 engine is attached to a rare Liquimatic transmission and this may be the only complete unit of its type still in existance. It was donated to, and restored by, the Early Ford V8 Foundation and was displayed at the Towe Ford Museum in Sacramento, California. Note the extended bellhousing, which was unique to the Liquimatic engine block, and the vacuum canister, which assisted in shifting between second and high gear.

1946 MERCURY

Model 69M
90 degree V-8. L-head. Cast-iron block. Bore & stroke: 3.187 x 3.75 in. Displacement: 239.4 cid. Compression ratio: 6.75:1. Brake hp: 100 at 3800 rpm. Taxable hp: 32.51. Torque: 180 lbs.-ft. at 2000 rpm. Three main bearings. Solid valve lifters. Crankcase capacity: 5 qt. Cooling system capacity: 22 qt. Carburetor: Holley 94 two-barrel. Serial numbers were the same as engine numbers and ranged from 99A50280 to 99A1412707.

1947 MERCURY

Model 79M
90 degree V-8. L-head. Cast-iron block. Bore & stroke: 3.187 x 3.75 in. Displacement: 239.4 cid. Compression ratio: 6.75:1. Brake hp: 100 at 3800 rpm. Taxable hp: 32.51. Torque: 180 lbs.-ft. at 2000 rpm. Three main bearings. Solid valve lifters. Crankcase capacity: 5 qt. Cooling system capacity: 22 qt. Carburetor: Holley 94 two-barrel. Serial numbers were the same as engine numbers and ranged from 799A1412708 to 799A2002282.

1948 MERCURY

Model 89M
90 degree V-8. L-head. Cast-iron block. Bore & stroke: 3.187 x 3.75 in. Displacement: 239.4 cid. Compression ratio: 6.75:1. Brake hp: 100 at 3800 rpm. Taxable hp: 32.51. Torque: 180 lbs.-ft. at 2000 rpm. Three main bearings. Solid valve lifters. Crankcase capacity: 5 qt. Cooling system capacity: 22 qt. Carburetor: Holley 94 two-barrel. Serial numbers were the same as engine numbers and ranged from 899A1990957 to 899A2374315.

1949 MERCURY

For day-in, day-out dependability, the 1950 Mercury featured improved "Split Second" starting plus "Hi-Power Compression." The Mercury's smoother, livelier V-8 engine was precision built to go farther with less maintenance.

Designed and built exclusively for the 1950 Mercury, this 100-hp eight-cylinder V-type power plant offered Mercury drivers real pickup and stamina. Its "Econ-O-Miser" carburetor made gas savings automatic. It was unusually compact in design and had a high power-to-weight ratio. The Mercury sales catalog described it as being a quiet engine that purred along the road.

90 degree V-8. L-head. Cast-iron block. Bore & stroke: 3.187 x 4.00 in. Displacement: 255.4 cid. Compression ratio: 6.80:1. Brake hp: 110 at 3600 rpm. Taxable hp: 32.51. Torque: 200 lbs.-ft. at 2000 rpm. Three main bearings. Solid valve lifters. Crankcase capacity: 5 qt. (add 1 qt. for new filter). Cooling system capacity: 22.25 qt. (add 1 qt. for heater). Carburetor: Holley 885FFC two-barrel. Serial numbers were the same as engine numbers and ranged from 9CM101 to 9CM302439.

1950 MERCURY

90 degree V-8. L-head. Cast-iron block. Bore & stroke: 3.187 x 4.00 in.

Displacement: 255.4 cid. Compression ratio: 6.80:1. Brake hp: 110 at 3600 rpm. Taxable hp: 32.51. Torque: 200 lbs.-ft. at 2000 rpm. Three main bearings. Solid valve lifters. Crankcase capacity: 5 qt. (add 1 qt. for new filter). Cooling system capacity: 22.25 qt. (add 1 qt. for heater). Carburetor: Holley 885FFC two-barrel. Serial numbers were the same as engine numbers. For 1950, they were: 50DA10001M to 50DA79027M; 50LA10001M to 50LA44958M; 50ME10001M to 50ME97749M; and 50SL10001M to 50SL110459M. The plant codes were: DA = Dearborn; LA = Los Angeles; ME = Metuchen; SL = St. Louis; and WA = Wayne.

1951 MERCURY

90 degree V-8. L-head. Cast-iron block. Bore & stroke: 3.187 x 4.00 in. Displacement: 255.4 cid. Compression ratio: 6.80:1. Brake hp: 112 at 3600 rpm. Taxable hp: 32.51. Torque: 206 lbs.-ft. at 2000 rpm. Three main bearings. Solid valve lifters. Crankcase capacity: 5 qt. (add 1 qt. for new filter). Cooling system capacity: 22.25 qt. (add 1 qt. for heater). Carburetor: Holley 885FFC two-barrel. Serial numbers were the same as engine numbers. Serial numbers for 1951 were: 51DA10001M to 51DA67910M; 51LA10001M to 51LA46772M; 51ME10001M to 51ME103515M; and 51SL10001M to 51SL127830M. The plant codes were: DA = Dearborn; LA = Los Angeles; ME = Metuchen; SL = St. Louis; and WA = Wayne.

1952 MERCURY

Custom, Monterey base V-8

90 degree V-8. L-head. Cast-iron block. Bore & stroke: 3.187 x 4.00 in. Displacement: 255.4 cid. Compression ratio: 7.20:1. Brake hp: 125 at 3700 rpm. Taxable hp: 32.51. Torque: 211 lbs.-ft. at 2100 rpm. Three main bearings. Solid valve lifters. Crankcase capacity: 4 qt. (add 1 qt. for new filter). Cooling system capacity: 21.50 qt. (add 1 qt. for heater). Carburetor: Holley 885FFC two-barrel. Serial numbers were the same as engine numbers. Serial numbers for 1952 were: 52SL10001M to 52SL86300M; 52WA10001M to 52WA19422M; 52LA10001M to 52LA38763M; and 52ME10001 to 52ME65500M. The plant codes were: DA = Dearborn; LA = Los Angeles; ME = Metuchen; SL = St Louis; and WA = Wayne.

1953 MERCURY

Custom, Monterey base V-8

90 degree V-8. L-head. Cast-iron block. Bore & stroke: 3.187 x 4.00 in. Displacement: 255.4 cid. Compression ratio: 7.20:1. Brake hp: 125 at 3800 rpm. Taxable hp: 32.51. Torque: 218 lbs.-ft. at 2100 rpm. Three main bearings. Solid valve lifters. Crankcase capacity: 4 qt. (add 1 qt. for new filter). Cooling system capacity: 21.50 qt. (add 1 qt. for heater). Carburetor: Holley 1901 FFC two-barrel. Serial numbers were the same as engine numbers. Serial numbers for 1953 were: 53LA10001M to 53LA50946M; 53WA10001M to 53WA45383M; 53ME10001M and up; and 53SL10001M to 53SL14285M. The plant codes were: DA = Dearborn; LA = Los Angeles; ME = Metuchen; SL = St. Louis; and WA = Wayne.

1954 MERCURY

Custom, Monterey base V-8

90 degree V-8. Overhead valve. Cast-iron block. Bore & stroke: 3.62 x 3.10 in. Displacement: 256 cid. Compression ratio: 7.50:1. Brake hp: 161 at 4400 rpm. Taxable hp: 42.05. Torque: 238 lbs.-ft. at 2500 rpm. Five main bearings. Solid valve lifters. Crankcase capacity: 5 qt. (add 1 qt. for new filter). Cooling system capacity: 19 qt. (add 1 qt. for heater). Carburetor: Holley 2140 four-barrel. Serial numbers were the same as engine numbers. Serial numbers ranged from 54WA10001M to 54WA75348M. The plant codes were: DA = Dearborn; LA = Los Angeles; ME = Metuchen; SL = St Louis; and WA = Wayne.

The 1954 Mercury V-8 had an unusual exhaust manifold system with a pipe that ran across the front of the engine connecting the exhaust manifolds. On the left-hand side of the engine, at the bottom, was a road draft tube that expelled spent gases to the atmosphere. This was before the days that a positive crankcase ventilation (PVC) system became mandatory.

Base engine in 1954 Mercury Custom and Monterey models was a brand new 256-cid overhead-valve V-8 that produced 161 hp at 4400 rpm. This engine featured a 7.5:1 compression ratio, which was considered to be "high-compression" in 1954. It had solid valve lifters and a Holley four-barrel carburetor.

1955 MERCURY

Custom, Monterey base V-8

90 degree V-8. Overhead valve. Cast-iron block. Bore & stroke: 3.75 x 3.30 in. Displacement: 292 cid. Compression ratio: 7.60:1. Brake hp: 188 at 4400 rpm. Taxable hp: 45.00. Torque: 274 lbs.-ft. at 2500 rpm. Five main bearings. Solid valve lifters. Crankcase capacity: 5 qt. (add 1 qt. for new filter). Cooling system capacity: 19 qt. (add 1 qt. for heater). Carburetor: Holley 2140 four-barrel. Serial numbers were the same as engine numbers. Custom serial numbers were: 55WA10001M to 55WA94613M and 55LA10001M to 55LA48892M. Monterey serial numbers were: 55WA10001M to 55WA94613M and 55LA10001M to 55LA46692. The plant codes were: DA = Dearborn; LA = Los Angeles;

New super-compression Super-Torque V-8 engines were used to power all three series of 1955 Mercurys. The low-rung Montclair came with a 198-hp V-8. The Mercury Monterey and Custom models utilized a 188-hp V-8.

ME = Metuchen; SL = St. Louis; and WA = Wayne.

Montclair base V-8; Custom, Monterey optional V-8 with automatic transmission
90 degree V-8. Overhead valve. Cast-iron block. Bore & stroke: 3.75 x 3.30 in. Displacement: 292 cid. Compression ratio: 8.50:1. Brake hp: 198 at 4400 rpm. Taxable hp: 45.00. Torque: 286 lbs.-ft. at 2500 rpm. Five main bearings. Solid valve lifters. Crankcase capacity: 5 qt. (add 1 qt. for new filter). Cooling system capacity: 19 qt. (add 1 qt. for heater). Carburetor: Holley 2140 four-barrel. Serial numbers were the same as engine numbers. Custom serial numbers were: 55WA10001M to 55WA94613M and 55LA10001M to 55LA48892M. Monterey serial numbers were: 55WA10001M to 55WA94613M and 55LA10001M to 55LA46692. Montclair serial numbers were: 55ME10001M to 55ME87345M and 55SL10001M to 55SL13753M. The plant codes were: DA = Dearborn; LA = Los Angeles; ME = Metuchen; SL = St. Louis; and WA = Wayne.

1956 MERCURY

Medalist, Custom, Monterey Montclair base V-8 (manual transmission)
90 degree V-8. Overhead valve. Cast-iron block. Bore & stroke: 3.80 x 3.44 in. Displacement: 312 cid. Compression ratio: 8.00:1. Brake hp: 210 at 4600 rpm. Taxable hp: 46.20. Torque: 312 lbs.-ft. at 2600 rpm. Five main bearings. Solid valve lifters. Crankcase capacity: 5 qt. (add 1 qt. for new filter). Cooling system capacity: 19 qt. (add 1 qt. for heater). Carburetor: Four-barrel. Serial numbers were the same as engine numbers. Medalist serial numbers were: 56WA10001M to 56WA69956M and 56LA10001M to 56LA51292M. Custom serial numbers were: 56WA10001M to 56WA89958M and 56LA10001M to 56LA51292M. Monterey serial numbers were: 56ME10001M to 56ME100055M; 56SL10001M to 56SL125006M. Montclair serial numbers were: 56ME10001M to 56ME100055M; 56SL10001M to 56SL125006M. The plant codes were: DA = Dearborn; LA = Los Angeles; ME = Metuchen; SL = St. Louis; and WA = Wayne.

Medalist, Custom, Monterey Montclair base V-8 (automatic transmission)
90 degree V-8. Overhead valve. Cast-iron block. Bore & stroke: 3.80 x 3.44 in. Displacement: 312 cid. Compression ratio: 8.40:1. Brake hp: 220 at 4600 rpm. Taxable hp: 46.20. Torque: 320 lbs.-ft. at 2600 rpm. Five main bearings. Solid valve lifters. Crankcase capacity: 5 qt. (add 1 qt. for new filter). Cooling system capacity: 19 qt. (add 1 qt. for heater). Carburetor: Four-barrel. Serial numbers were the same as engine numbers. Medalist serial numbers were: 56WA10001M to 56WA69956M and 56LA10001M to 56LA51292M. Custom serial numbers were: 56WA10001M to 56WA89958M and 56LA10001M to 56LA51292M. Monterey serial numbers were: 56ME10001M to 56ME100055M; 56SL10001M to 56SL125006M. Montclair serial numbers were: 56ME10001M to 56ME100055M; 56SL10001M to 56SL125006M. The plant codes were: DA = Dearborn; LA = Los Angeles, ME = Metuchen; SL = St. Louis; and WA = Wayne.

Powertrain Options
At midyear, a new camshaft raised the output in the standard and 225-hp V-8s by 10 hp. The M-260 package (two four-barrel carburetors, 260 hp) was offered in all series late in the year.

1957 MERCURY

For 1957, the big Mercury offered a choice of two V-8 engines and both were more powerful than anything available in the history of the company. Montclairs, Montereys, and Mercury station wagons came with this 255-hp "Safety-Surge" V-8 that featured a quiet-running "Power Booster" de-clutching fan.

Monterey, Montclair, Commuter, Voyager base V-8
90 degree V-8. Overhead valve. Cast-iron block. Bore & stroke: 3.80 x 3.44 in. Displacement: 312 cid. Compression ratio: 9.70:1. Brake hp: 255 at 4600 rpm. Taxable hp: 46.20. Torque: 340 lbs.-ft. at 2600 rpm. Five main bearings. Solid valve lifters. Crankcase capacity: 5 qt. (add 1 qt. for new filter). Cooling system capacity: 20 qt. (add 1 qt. for heater). Carburetor: Carter or Holley four-barrel. Serial numbers were the same as engine numbers. Monterey serial numbers were: 57WA10001M to 57WA90490M and 57LA10001M to 57LA40854M. Montclair serial numbers were: 57WA10001M to 57WA90490M and 57LA10001M to 57LA40854M. Turnpike Cruiser serial numbers were: 57ME10001M to 57ME65895M and 57SL10001M to 57SL98451M. Station wagon serial numbers were: 57WA10001M to 57WA90490M and 57LA10001M to 57LA40654M. The plant codes were: DA = Dearborn; LA = Los Angeles, ME = Metuchen; SL = St. Louis; and WA = Wayne.
NOTE: This engine was also used in Canadian Turnpike Cruisers.

Turnpike Cruiser, Colony Park base V-8; Monterey, Montclair, Commuter, Voyager optional V-8
90 degree V-8. Overhead valve. Cast-iron block. Bore & stroke: 4.00 x 3.65 in. Displacement: 368 cid. Compression ratio: 9.75:1. Brake hp: 290 at 4600 rpm. Taxable hp: 51.20. Torque: 405 lbs.-ft. at 2600 rpm. Five main bearings. Hydraulic valve lifters. Crankcase capacity: 5 qt. (add 1 qt. for new filter). Cooling system capacity: 23 qt. (add 1 qt. for heater). Carburetor: Carter or Holley four-barrel. Serial numbers were the same as engine numbers. Monterey serial numbers were: 57WA10001M to 57WA90490M and 57LA10001M to 57LA40854M. Montclair serial numbers were: 57WA10001M to 57WA90490M and 57LA10001M to 57LA40854M. Turnpike Cruiser serial numbers were: 57ME10001M to 57ME65895M and 57SL10001M to 57SL98451M. Station wagon serial numbers were: 57WA10001M to 57WA90490M and 57LA10001M to 57LA40654M. The plant codes were: DA = Dearborn; LA = Los Angeles, ME = Metuchen; SL = St. Louis; and WA = Wayne.

Powertrain Options
The M-335 power package (two four-barrel carburetors, 368 cid/335 hp V-8) was optional on Montereys.

1958 MERCURY

Medalist base V-8
90 degree V-8. Overhead valve. Cast-iron block. Bore & stroke: 3.80 x 3.44 in. Displacement: 312 cid. Compression ratio: 9.70:1. Brake hp: 235 at 4600 rpm. Taxable hp: 46.20. Torque: 340 lbs.-ft. at 2600 rpm. Five main bearings. Solid valve lifters. Crankcase capacity: 5 qt. (add 1 qt. for new filter). Cooling system capacity: 20.5 qt. (add 1 qt. for heater). Carburetor: Holley four-barrel. Serial numbers were the same as engine numbers. Medalist serial numbers were: W500001 to W547046. The plant codes were: DA = Dearborn; LA = Los Angeles, ME = Metuchen; SL = St. Louis; and WA = Wayne. VIN code: L.

Monterey, Commuter base V-8
90 degree V-8. Overhead valve. Cast-iron block. Bore & stroke: 4.30 x 3.30 in. Displacement: 383.4 cid. Compression ratio: 10.5:1. Brake hp: 312 at 4600 rpm. Taxable hp: 59.17. Torque: 405 lbs.-ft. at 2900 rpm. Five main bearings. Hydraulic valve lifters. Crankcase capacity: 5 qt. (add 1 qt. for new filter). Cooling system capacity: 20.5 qt. (add 1 qt. for heater). Carburetor: Holley four-barrel. Serial numbers were the same as engine numbers. Monterey serial numbers were: W500001 to W547046 and J500001 and up. Plant codes: W = Wayne; J = Los Angeles; T = Metuchen; Z = St. Louis. VIN code: M.

Montclair, Voyager, Colony Park base V-8
90 degree V-8. Overhead valve. Cast-iron block. Bore & stroke: 4.30 x 3.30 in. Displacement: 383.4 cid. Compression ratio: 10.5:1. Brake hp: 330 at 4800 rpm. Taxable hp: 59.17. Torque: 425 lbs.-ft. at 3000 rpm. Five main bearings. Hydraulic valve lifters. Crankcase capacity: 5 qt. (add 1 qt. for new filter). Cooling system capacity: 20.5 qt. (add 1 qt. for heater). Carburetor: Holley four-barrel. Serial numbers were the same as engine numbers. Montclair serial numbers were the same as for Montereys plus T500001 and up and, also, Z500001 and up. Plant codes: W = Wayne; J = Los Angeles; T = Metuchen; Z = St. Louis. VIN code: M.

Park Lane base V-8; Montclair optional V-8
90 degree V-8. Overhead valve. Cast-iron block. Bore & stroke: 4.30 x 3.70 in. Displacement: 430 cid. Compression ratio: 10.5:1. Brake hp: 360 at 4600 rpm. Taxable hp: 59.17. Torque: 480 lbs.-ft. at 3000 rpm. Five main bearings. Hydraulic valve lifters. Crankcase capacity: 5 qt. (add 1 qt. for new filter). Cooling system capacity: 20.5 qt. (add 1 qt. for heater). Carburetor: Holley four-barrel. Serial numbers were the same as engine numbers. Montclair serial numbers were the same as for Montereys plus T500001 and up and also Z500001 and up. Serial numbers were T500001 and up and, also, Z500001 and up. Plant codes: W = Wayne; J = Los Angeles; T = Metuchen; Z = St. Louis. VIN code: K.

Powertrain Options
A 430-cid/400-hp V-8 (VIN code: J) with three two-barrel carburetors was optional on all series. Dual exhaust was a $32.30 option.

1959 MERCURY

In 1959, each Mercury series had its own V-8 engine. The base Monterey model used a 312-cid 210-hp V-8. The Commuter station wagon used a 383-cid 280-hp as its base engine. Standard under the hood of the Montclair, Voyager, and Colony Park models was a 322-hp version of the 383 with a four-barrel carburetor. The Parklane listed a 430-cid 345-hp V-8 as standard equipment.

Monterey base V-8
90 degree V-8. Overhead valve. Cast-iron block. Bore & stroke: 3.80 x 3.44 in. Displacement: 312 cid. Compression ratio: 8.75:1. Brake hp: 210 at 4400 rpm. Taxable hp: 46.20. Torque: 325 lbs.-ft. at 2200 rpm. Five main bearings. Hydraulic valve lifters. Crankcase capacity: 5 qt. (add 1 qt. for new filter). Cooling system capacity: 20 qt. (add 1 qt. for heater). Carburetor: Holley 2300 two-barrel. VIN code: P.

Commuter base V-8; Monterey optional V-8
90 degree V-8. Overhead valve. Cast-iron block. Bore & stroke: 4.30 x 3.30 in. Displacement: 383.4 cid. Compression ratio: 10.0:1. Brake hp: 280 at 4400 rpm. Taxable hp: 46.20. Torque: 405 lbs.-ft. at 2200 rpm. Five main bearings. Hydraulic valve lifters. Crankcase capacity: 5 qt. (add 1 qt. for new filter). Cooling system capacity: 21 qt. (add 1 qt. for heater). Carburetor: Holley 2300 two-barrel. VIN code: N.

Montclair, Voyager, Colony Park base V-8
90 degree V-8. Overhead valve. Cast-iron block. Bore & stroke: 4.30 x 3.30 in. Displacement: 383.4 cid. Compression ratio: 10.0:1. Brake hp: 322 at 4600 rpm. Taxable hp: 59.17. Torque: 420 lbs.-ft. at 2800 rpm. Five main bearings. Hydraulic valve lifters. Crankcase capacity: 5 qt. (add 1 qt. for new filter). Cooling system capacity: 21 qt. (add 1 qt. for heater). Carburetor: Ford 5752425 four-barrel. VIN code: M.

Park Lane base V-8; Monterey, Montclair optional V-8
90 degree V-8. Overhead valve. Cast-iron block. Bore & stroke: 4.30 x 3.70 in. Displacement: 430 cid. Compression ratio: 10.5:1. Brake hp: 345 at 4400 rpm. Taxable hp: 59.17. Torque: 480 lbs.-ft. at 2800 rpm. Five main bearings. Hydraulic valve lifters. Crankcase capacity: 5 qt. (add 1 qt. for new filter). Cooling system capacity: 21 qt. (add 1 qt. for heater). Carburetor: AFB-2853-S four-barrel. VIN code: L.

Powertrain Options
A 430-cid/400-hp V-8 with three two-barrel carburetors was optional on all series. The VIN code for this engine was K.

1960 MERCURY

Monterey, Commuter base V-8
90 degree V-8. Overhead valve. Cast-iron block. Bore & stroke: 3.80 x 3.44 in. Displacement: 312 cid. Compression ratio: 8.90:1. Brake hp: 205 at 4000 rpm. Taxable hp: 46.21. Torque: 328 lbs.-ft. at 2100 rpm. Five main bearings. Solid valve lifters. Crankcase capacity: 5 qt. (add 1 qt. for new filter). Cooling system capacity: 20 qt. (add 1 qt. for heater). Carburetor: Holley 2300 two-barrel. VIN code: P.

Monterey, Commuter optional V-8
90 degree V-8. Overhead valve. Cast-iron block. Bore & stroke: 4.30 x 3.30 in. Displacement: 383.4 cid. Compression ratio: 10.0:1. Brake hp: 280 at 4200 rpm. Torque: 405 lbs.-ft. at 2200 rpm. Five main bearings. Hydraulic valve lifters. Crankcase capacity: 5 qt. (add 1 qt. for new filter). Cooling system capacity: 20 qt. (add 1 qt. for heater). Carburetor: Holley 2300 two-barrel. VIN code: N.

Montclair, Park Lane, Colony Park base V-8; Monterey, Commuter optional V-8
90 degree V-8. Overhead valve. Cast-iron block. Bore & stroke: 4.30 x 3.70 in. Displacement: 430 cid. Compression ratio: 10.0:1. Brake hp: 310 at 4100 rpm. Taxable hp: 59.17. Torque: 460 lbs.-ft. at 2200 rpm. Five main bearings. Hydraulic valve lifters. Crankcase capacity: 5 qt. (add 1 qt. for new filter). Cooling system capacity: 21 qt. (add 1 qt. for heater). Carburetor: Carter ABD-2965-S two-barrel. VIN code: M.

NOTE: Low-compression versions of the 383- and 430-cid V-8s were made for the export market. Both had two-barrel carburetors. They used VIN codes E and F, respectively.

1961 MERCURY

Meteor 600, Meteor 800, Monterey, Commuter base V-8
90 degree V-8. Overhead valve. Cast-iron block. Bore & stroke: 3.75 x 3.30 in. Displacement: 292 cid. Compression ratio: 8.80:1. Brake hp: 175 at 4200 rpm. Taxable hp: 45.00. Torque: 279 lbs.-ft. at 2200 rpm. Five main bearings. Solid valve lifters. Crankcase capacity: 5 qt. (add 1 qt. for new filter). Cooling system capacity: 19 qt. (add 1 qt. for heater). Carburetor: Ford two-barrel. VIN code: W.

Meteor 600, Meteor 800, Monterey optional Marauder V-8
90 degree V-8. Overhead valve. Cast-iron block. Bore & stroke: 4.00 x

The 1961 Mercury V-8 elminiated the exhaust system crossover pipe, which was famous for burning the elbows of mechanics who worked on these cars.

3.50 in. Displacement: 352 cid. Compression ratio: 8.90:1. Brake hp: 220 at 4400 rpm. Torque: 336 lbs.-ft. at 2400 rpm. Five main bearings. Hydraulic valve lifters. Crankcase capacity: 5 qt. (add 1 qt. for new filter). Cooling system capacity: 19 qt. (add 1 qt. for heater). Carburetor: Ford two-barrel. VIN code: X.

Meteor 800, Monterey optional Marauder V-8
90 degree V-8. Overhead valve. Cast-iron block. Bore & stroke: 4.047 x 3.781 in. Displacement: 390 cid. Compression ratio: 9.6:1. Brake hp: 300 at 4600 rpm. Taxable hp: 52.50. Torque: 427 lbs.-ft. at 2800 rpm. Five main bearings. Hydraulic valve lifters. Crankcase capacity: 5 qt. (add 1 qt. for new filter). Cooling system capacity: 19 qt. (add 1 qt. for heater). Carburetor: Ford four-barrel. VIN code: Z.

Meteor 600 optional Marauder V-8
90 degree V-8. Overhead valve. Cast-iron block. Bore & stroke: 4.047 x 3.781 in. Displacement: 390 cid. Compression ratio: 9.6:1. Brake hp: 330 at 5000 rpm. Taxable hp: 52.50. Torque: 427 lbs.-ft. at 3200 rpm. Five main bearings. Solid valve lifters. Crankcase capacity: 5 qt. (add 1 qt. for new filter). Cooling system capacity: 19 qt. (add 1 qt. for heater). Carburetor: Ford four-barrel. VIN code: P or Z.

1962 MERCURY

Meteor base V-8
90 degree V-8. Overhead valve. Cast-iron block. Bore & stroke: 3.50 x 2.875 in. Displacement: 221 cid. Compression ratio: 8.70:1. Brake hp: 145 at 4400 rpm. Taxable hp: 39.20. Torque: 216 lbs.-ft. at 2200 rpm. Five main bearings. Hydraulic valve lifters. Crankcase capacity: 4 qt. (add 1 qt. for new filter). Cooling system capacity: 13.5 qt. (add 1 qt. for heater). Carburetor: Ford two-barrel. VIN code: L.

Meteor optional
90 degree V-8. Overhead valve. Cast-iron block. Bore & stroke: 3.8 x 2.875 in. Displacement: 260 cid. Compression ratio: 8.70:1. Brake hp: 164 at 4400 rpm. Taxable hp: 46.20. Torque: 258 lbs.-ft. at 2200 rpm. Five main bearings. Hydraulic valve lifters. Crankcase capacity: 5 qt. (add 1 qt. for new filter). Cooling system capacity: 13.5 qt. (add 1 qt. for heater). Carburetor: Ford two-barrel. VIN code: F.

Monterey base V-8
90 degree V-8. Overhead valve. Cast-iron block. Bore & stroke: 3.75 x 3.30 in. Displacement: 292 cid. Compression ratio: 8.80:1. Brake hp: 170 at 4200 rpm. Taxable hp: 45.00. Torque: 279 lbs.-ft. at 2200 rpm. Five main bearings. Solid valve lifters. Crankcase capacity: 5 qt. (add 1 qt. for new filter). Cooling system capacity: 20 qt. (add 1 qt. for heater). Carburetor: Ford two-barrel. VIN code: W.

Monterey optional Marauder V-8
90 degree V-8. Overhead valve. Cast-iron block. Bore & stroke: 4.00 x 3.50 in. Displacement: 352 cid. Compression ratio: 8.90:1. Brake hp: 220 at 4300 rpm. Torque: 336 lbs.-ft. at 2600 rpm. Five main bearings. Hydraulic valve lifters. Crankcase capacity: 5 qt. (add 1 qt. for new filter). Cooling system capacity: 20.5 qt. (add 1 qt. for heater). Carburetor: Ford two-barrel. VIN code: X.

Monterey optional S-55 Marauder V-8
90 degree V-8. Overhead valve. Cast-iron block. Bore & stroke: 4.047 x 3.781 in. Displacement: 390 cid. Compression ratio: 9.6:1. Brake hp: 300 at 4600 rpm. Taxable hp: 52.50. Torque: 427 lbs.-ft. at 2800 rpm. Five main bearings. Hydraulic valve lifters. Crankcase capacity: 5 qt. (add 1 qt. for new filter). Cooling system capacity: 20.5 qt. (add 1 qt. for heater). Carburetor: Ford four-barrel. VIN code: Z.

Monterey optional Marauder V-8
90 degree V-8. Overhead valve. Cast-iron block. Bore & stroke: 4.047 x 3.781 in. Displacement: 390 cid. Compression ratio: 9.6:1. Brake hp: 330 at 5000 rpm. Taxable hp: 52.50. Torque: 427 lbs.-ft. at 3200 rpm. Five main bearings. Solid valve lifters. Crankcase capacity: 5 qt. (add 1 qt. for new filter). Cooling system capacity: 20.5 qt. (add 1 qt. for heater). Carburetor: Ford four-barrel. VIN code: P.

Monterey optional Marauder V-8
90 degree V-8. Overhead valve. Cast-iron block. Bore & stroke: 4.125 x 3.781 in. Displacement: 406 cid. Compression ratio: 9.6:1. Brake hp: 385 at 5800 rpm. Taxable hp: 54.58. Torque: 444 lbs.-ft. at 3400 rpm. Five main bearings. Solid valve lifters. Crankcase capacity: 5 qt. (add 1 qt. for new filter). Cooling system capacity: 20.5 qt. (add 1 qt. for heater). Carburetor: Holley four-barrel. VIN code: B.

Monterey optional Marauder V-8
90 degree V-8. Overhead valve. Cast-iron block. Bore & stroke: 4.125 x 3.781 in. Displacement: 406 cid. Compression ratio: 9.6:1. Brake hp: 405 at 5800 rpm. Taxable hp: 54.58. Torque: 448 lbs.-ft. at 3500 rpm. Five main bearings. Solid valve lifters. Crankcase capacity: 5 qt. (add 1 qt. for new filter). Cooling system capacity: 20.5 qt. (add 1 qt. for heater). Carburetor: Three Holley two-barrel carburetors. VIN code: G.

1963 MERCURY

Although they did not share the same bore and stroke, the 390- and 428-cid Mercury big-block V-8s looked alike and shared many internal parts. The sheet metal covering the left-hand exhaust manifold functioned as a duct that carried heat to the intake manifold pre-heater.

Meteor base V-8
90 degree V-8. Overhead valve. Cast-iron block. Bore & stroke: 3.50 x 2.875 in. Displacement: 221 cid. Compression ratio: 8.70:1. Brake hp: 145 at 4400 rpm. Taxable hp: 39.20. Torque: 216 lbs.-ft. at 2200 rpm. Five main bearings. Hydraulic valve lifters. Crankcase capacity: 3.5 qt. (add 1 qt. for new filter). Cooling system capacity: 13.5 qt. (add 1 qt. for heater). Carburetor: Ford two-barrel. VIN code: L.

Meteor, Comet optional V-8
90 degree V-8. Overhead valve. Cast-iron block. Bore & stroke: 3.8 x 2.875 in. Displacement: 260 cid. Compression ratio: 8.70:1. Brake hp: 164 at 4400 rpm. Taxable hp: 46.20. Torque: 258 lbs.-ft. at 2200 rpm. Five main bearings. Hydraulic valve lifters. Crankcase capacity: 3.5 qt. (add 1 qt. for new filter). Cooling system capacity: 13.5 qt. (add 1 qt. for heater). Carburetor: Ford two-barrel. VIN code: F.

Monterey base V-8
90 degree V-8. Overhead valve. Cast-iron block. Bore & stroke: 4.047 x 3.781 in. Displacement: 390 cid. Compression ratio: 8.90:1. Brake hp: 250 at 4400 rpm. Taxable hp: 52.49. Torque: 378 lbs.-ft. at 2400 rpm. Five main bearings. Hydraulic valve lifters. Crankcase capacity: 5 qt. (add 1 qt. for new filter). Cooling system capacity: 20.5 qt. (add 1 qt. for

heater). Carburetor: Ford two-barrel. VIN code: Y.

Monterey optional V-8
90 degree V-8. Overhead valve. Cast-iron block. Bore & stroke: 4.047 x 3.781 in. Displacement: 390 cid. Compression ratio: 9.6:1. Brake hp: 300 at 4600 rpm. Taxable hp: 52.49. Torque: 427 lbs.-ft. at 2800 rpm. Five main bearings. Hydraulic valve lifters. Crankcase capacity: 5 qt. (add 1 qt. for new filter). Cooling system capacity: 20.5 qt. (add 1 qt. for heater). Carburetor: Ford four-barrel. VIN code: Z.

Monterey optional V-8
90 degree V-8. Overhead valve. Cast-iron block. Bore & stroke: 4.047 x 3.781 in. Displacement: 390 cid. Compression ratio: 9.6:1. Brake hp: 330 at 5000 rpm. Taxable hp: 52.49. Torque: 427 lbs.-ft. at 3200 rpm. Five main bearings. Solid valve lifters. Crankcase capacity: 5 qt. (add 1 qt. for new filter). Cooling system capacity: 20.5 qt. (add 1 qt. for heater). Carburetor: Ford four-barrel. VIN code: P.

Monterey optional Marauder V-8
90 degree V-8. Overhead valve. Cast-iron block. Bore & stroke: 4.125 x 3.781 in. Displacement: 406 cid. Compression ratio: 11.4:1. Brake hp: 385 at 5800 rpm. Taxable hp: 54.58. Torque: 444 lbs.-ft. at 3400 rpm. Five main bearings. Solid valve lifters. Crankcase capacity: 5 qt. (add 1 qt. for new filter). Cooling system capacity: 20.5 qt. (add 1 qt. for heater). Carburetor: Holley four-barrel. VIN code: B.

Monterey optional Marauder V-8
90 degree V-8. Overhead valve. Cast-iron block. Bore & stroke: 4.125 x 3.781 in. Displacement: 406 cid. Compression ratio: 11.4:1. Brake hp: 405 at 5800 rpm. Taxable hp: 54.58. Torque: 448 lbs.-ft. at 3500 rpm. Five main bearings. Solid valve lifters. Crankcase capacity: 5 qt. (add 1 qt. for new filter). Cooling system capacity: 20.5 qt. (add 1 qt. for heater). Carburetor: Three Holley two-barrel carburetors. VIN code: G.

Monterey optional Marauder V-8
90 degree V-8. Overhead valve. Cast-iron block. Bore & stroke: 4.23 x 3.781 in. Displacement: 427 cid. Compression ratio: 11.5:1. Brake hp: 410 at 5600 rpm. Taxable hp: 54.58. Torque: 476 lbs.-ft. at 3400 rpm. Five main bearings. Hydraulic valve lifters. Crankcase capacity: 5 qt. (add 1 qt. for new filter). Cooling system capacity: 19.5 qt. (add 1 qt. for heater). Carburetor: Holley four-barrel. (Not available in station wagon). Engine code: Q.

Monterey optional Marauder V-8
90 degree V-8. Overhead valve. Cast-iron block. Bore & stroke: 4.23 x 3.781 in. Displacement: 427 cid. Compression ratio: 11.2:1. Brake hp: 425 at 6000 rpm. Taxable hp: 54.58. Torque: 480 lbs.-ft. at 3700 rpm. Five main bearings. Solid valve lifters. Crankcase capacity: 5 qt. (add 1 qt. for new filter). Cooling system capacity: 19.5 qt. (add 1 qt. for heater). Carburetor: Two Holley four-barrel carburetors. (Not available in station wagon.) Engine code: R.

1964 MERCURY

Comet 202, Comet 404, Comet Caliente base V-8
90 degree V-8. Overhead valve. Cast-iron block. Bore & stroke: 3.8 x 2.875 in. Displacement: 260 cid. Compression ratio: 8.80:1. Brake hp: 164 at 4400 rpm. Taxable hp: 46.20. Torque: 258 lbs.-ft. at 2200 rpm. Five main bearings. Hydraulic valve lifters. Crankcase capacity: 4 qt. (add 1 qt. for new filter). Cooling system capacity: 13.5 qt. (add 1 qt. for heater). Carburetor: Ford C4OF-9510-A two-barrel. Engine code: F.

Comet Cyclone V-8
90 degree V-8. Overhead valve. Cast-iron block. Bore & stroke: 4.00 x 2.875 in. Displacement: 289 cid. Compression ratio: 9.00:1. Brake hp: 210 at 4400 rpm. Taxable hp: 51.20. Torque: 300 lbs.-ft. at 2800 rpm. Five main bearings. Hydraulic valve lifters. Crankcase capacity: 4 qt. (add 1 qt. for new filter). Cooling system capacity: 14 qt. (add 1 qt. for heater). Carburetor: Ford C4GF-9510-D four-barrel. Engine code: K or D.

Monterey, Montclair base V-8 (manual transmission)
90 degree V-8. Overhead valve. Cast-iron block. Bore & stroke: 4.047 x 3.781 in. Displacement: 390 cid. Compression ratio: 9.40:1. Brake hp: 250 at 4400 rpm. Taxable hp: 52.49. Torque: 378 lbs.-ft. at 2400 rpm. Five main bearings. Hydraulic valve lifters. Crankcase capacity: 5 qt. (add 1 qt. for new filter). Cooling system capacity: 19.5 qt. (add 1 qt. for heater). Carburetor: Ford C4MF-9510-D two-barrel. Engine code: Y.

Monterey, Montclair base V-8 (automatic transmission)
90 degree V-8. Overhead valve. Cast-iron block. Bore & stroke: 4.047 x 3.781 in. Displacement: 390 cid. Compression ratio: 9.4:1. Brake hp: 266 at 4400 rpm. Taxable hp: 52.49. Torque: 378 lbs.-ft. at 2400 rpm. Five main bearings. Hydraulic valve lifters. Crankcase capacity: 5 qt. (add 1 qt. for new filter). Cooling system capacity: 19.5 qt. (add 1 qt. for heater). Carburetor: Ford C4MF-9510-D two-barrel. Engine code: H.

Meteor, Comet optional V-8
90 degree V-8. Overhead valve. Cast-iron block. Bore & stroke: 4.00 x 2.875 in. Displacement: 289 cid. Compression ratio: 10.50:1. Brake hp: 271 at 6000 rpm. Taxable hp: 46.20. Torque: 312 lbs.-ft. at 3400 rpm. Five main bearings. Solid valve lifters. Crankcase capacity: 4.0 qt. (add 1 qt. for new filter). Cooling system capacity: 14 qt. (add 1 qt. for heater). Carburetor: Ford C4GF-9510-D four-barrel. Engine code: K.

Parklane base V-8
90 degree V-8. Overhead valve. Cast-iron block. Bore & stroke: 4.047 x 3.781 in. Displacement: 390 cid. Compression ratio: 10.0:1. Brake hp: 300 at 4600 rpm. Taxable hp: 52.49. Torque: 427 lbs.-ft. at 2800 rpm. Five main bearings. Hydraulic valve lifters. Crankcase capacity: 5 qt. (add 1 qt. for new filter). Cooling system capacity: 19.5 qt. (add 1 qt. for heater). Carburetor: Ford four-barrel. Engine code: Z.

Mercury optional V-8
90 degree V-8. Overhead valve. Cast-iron block. Bore & stroke: 4.047 x 3.781 in. Displacement: 390 cid. Compression ratio: 10.0:1. Brake hp: 330 at 5000 rpm. Taxable hp: 52.50. Torque: 427 lbs.-ft. at 3200 rpm. Five main bearings. Solid valve lifters. Crankcase capacity: 5 qt. (add 1 qt. for new filter). Cooling system capacity: 19.5 qt. (add 1 qt. for heater). Carburetor: Ford four-barrel. Engine code: P.

Monterey optional Marauder V-8
90 degree V-8. Overhead valve. Cast-iron block. Bore & stroke: 4.23 x 3.781 in. Displacement: 427 cid. Compression ratio: 11.5:1. Brake hp: 410 at 5600 rpm. Taxable hp: 54.58. Torque: 476 lbs.-ft. at 3400 rpm. Five main bearings. Solid valve lifters. Crankcase capacity: 5 qt. (add 1 qt. for new filter). Cooling system capacity: 19.5 qt. (add 1 qt. for heater). Carburetor: Holley four-barrel. (Not available in station wagon.) Engine code: Q.

Monterey optional Marauder V-8
90 degree V-8. Overhead valve. Cast-iron block. Bore & stroke: 4.23 x 3.781 in. Displacement: 427 cid. Compression ratio: 11.2:1. Brake hp: 425 at 6000 rpm. Taxable hp: 54.58. Torque: 480 lbs.-ft. at 3700 rpm. Five main bearings. Solid valve lifters. Crankcase capacity: 5 qt. (add 1 qt. for new filter). Cooling system capacity: 19.5 qt. (add 1 qt. for heater). Carburetor: Two Holley four-barrel carburetors. (Not available in station wagon.) Engine code: R.
Note: A limited number of 1964 Mercury Comets with 427-cid, 425-hp engines were built as lightweight factory drag racing cars.

1965 MERCURY

Comet 202. Comet 404, Comet Caliente, base Cyclone V-8
90 degree V-8. Overhead valve. Cast-iron block. Bore & stroke: 4.00 x 2.875 in. Displacement: 289 cid. Compression ratio: 9.30:1. Brake hp: 200 at 4400 rpm. Taxable hp: 51.20. Torque: 282 lbs.-ft. at 2400 rpm. Five main bearings. Hydraulic valve lifters. Crankcase capacity: 4 qt. (add 1 qt. for new filter). Cooling system capacity: 14 qt. (add 1 qt. for heater). Carburetor: Ford C5MF-9510-A two-barrel. Engine code: C.

Meteor, Comet optional V-8
90 degree V-8. Overhead valve. Cast-iron block. Bore & stroke: 4.00 x 2.875 in. Displacement: 289 cid. Compression ratio: 10.9:1. Brake hp: 225 at 4800 rpm. Taxable hp: 46.20. Torque: 305 lbs.-ft. at 3200 rpm. Five main bearings. Hydraulic valve lifters. Crankcase capacity: 4 qt. (add 1 qt. for new filter). Cooling system capacity: 14 qt. (add 1 qt. for heater). Carburetor: Ford C4GF-9510-D four-barrel. Engine code: A.

Monterey, Montclair base V-8 (manual transmission)
90 degree V-8. Overhead valve. Cast-iron block. Bore & stroke: 4.047 x 3.781 in. Displacement: 390 cid. Compression ratio: 9.40:1. Brake hp: 250 at 4400 rpm. Taxable hp: 52.49. Torque: 378 lbs.-ft. at 2400 rpm. Five main bearings. Hydraulic valve lifters. Crankcase capacity: 5 qt. (add 1 qt. for new filter). Cooling system capacity: 19.5 qt. (add 1 qt. for heater). Carburetor: Ford C5MF-9510-A two-barrel. Engine code: Y.

Monterey, Montclair base V-8 (automatic transmission)
90 degree V-8. Overhead valve. Cast-iron block. Bore & stroke: 4.047 x 3.781 in. Displacement: 390 cid. Compression ratio: 9.40:1. Brake hp: 266 at 4400 rpm. Taxable hp: 52.49. Torque: 378 lbs.-ft. at 2400 rpm. Five main bearings. Hydraulic valve lifters. Crankcase capacity: 5 qt. (add 1 qt. for new filter). Cooling system capacity: 19.5 qt. (add 1 qt. for heater). Carburetor: Ford C4MF-9510-A two-barrel. Engine code: H.

Comet optional V-8
90 degree V-8. Overhead valve. Cast-iron block. Bore & stroke: 4.00 x

2.875 in. Displacement: 289 cid. Compression ratio: 10.50:1. Brake hp: 271 at 6000 rpm. Taxable hp: 46.20. Torque: 312 lbs.-ft. at 3400 rpm. Five main bearings. Solid valve lifters. Crankcase capacity: 4.0 qt. (add 1 qt. for new filter). Cooling system capacity: 14 qt. (add 1 qt. for heater). Carburetor: Ford C4GF-9510-D four-barrel. Engine code: K.

Parklane base V-8
90 degree V-8. Overhead valve. Cast-iron block. Bore & stroke: 4.047 x 3.781 in. Displacement: 390 cid. Compression ratio: 10.0:1. Brake hp: 300 at 4600 rpm. Taxable hp: 52.49. Torque: 427 lbs.-ft. at 2800 rpm. Five main bearings. Hydraulic valve lifters. Crankcase capacity: 5 qt. (add 1 qt. for new filter). Cooling system capacity: 19.5 qt. (add 1 qt. for heater). Carburetor: Ford C5AF-9510-E four-barrel. Engine code: Z.

Monterey, Montclair, Parklane optional V-8
90 degree V-8. Overhead valve. Cast-iron block. Bore & stroke: 4.047 x 3.781 in. Displacement: 390 cid. Compression ratio: 10.0:1. Brake hp: 330 at 5000 rpm. Taxable hp: 52.50. Torque: 427 lbs.-ft. at 3200 rpm. Five main bearings. Solid valve lifters. Crankcase capacity: 5 qt. (add 1 qt. for new filter). Cooling system capacity: 19.5 qt. (add 1 qt. for heater). Carburetor: Ford four-barrel. Engine code: P.

Monterey, Montclair, Parklane optional V-8
90 degree V-8. Overhead valve. Cast-iron block. Bore & stroke: 4.23 x 3.781 in. Displacement: 427 cid. Compression ratio: 11.2:1. Brake hp: 425 at 6000 rpm. Taxable hp: 54.58. Torque: 480 lbs.-ft. at 3700 rpm. Five main bearings. Solid valve lifters. Crankcase capacity: 5 qt. (add 1 qt. for new filter). Cooling system capacity: 19.5 qt. (add 1 qt. for heater). Carburetor: Two Holley four-barrel carburetors. Engine code: R.

1966 MERCURY

Comet 202, Capri, Caliente, Cyclone base V-8
90 degree V-8. Overhead valve. Cast-iron block. Bore & stroke: 4.00 x 2.875 in. Displacement: 289 cid. Compression ratio: 9.30:1. Brake hp: 200 at 4400 rpm. Taxable hp: 51.20. Torque: 282 lbs.-ft. at 2400 rpm. Five main bearings. Hydraulic valve lifters. Crankcase capacity: 4 qt. (add 1 qt. for new filter). Cooling system capacity: 14 qt. (add 1 qt. for heater). Carburetor: Ford 9510 two-barrel. Engine code: C.

Monterey, Montclair base V-8; Comet 202, 404, Caliente, Cyclone optional V-8
90 degree V-8. Overhead valve. Cast-iron block. Bore & stroke: 4.047 x 3.781 in. Displacement: 390 cid. Compression ratio: 9.50:1. Brake hp: 265 at 4400 rpm. Taxable hp: 52.49. Torque: 401 lbs.-ft. at 2600 rpm. Five main bearings. Hydraulic valve lifters. Crankcase capacity: 5 qt. (add 1 qt. for new filter). Cooling system capacity: 19.5 qt. (add 1 qt. for heater). Carburetor: Ford 9510 two-barrel. Engine code: Y.

Comet, Monterey, Montclair optional V-8
90 degree V-8. Overhead valve. Cast-iron block. Bore & stroke: 4.047 x 3.781 in. Displacement: 390 cid. Compression ratio: 9.50:1. Brake hp: 275 at 4400 rpm. Taxable hp: 52.49. Torque: 405 lbs.-ft. at 2600 rpm. Five main bearings. Hydraulic valve lifters. Crankcase capacity: (Comets) 4 qt.; (Mercurys) 5 qt. (add 1 qt. for new filter). Cooling system capacity: 19.5 qt. (add 1 qt. for heater). Carburetor: Ford C6AF-12127-B two-barrel. Engine code: H.

Although they did not share the same bore and stroke, the 390- and 428-cid Mercury big-block V-8s looked alike and shared many internal parts. The sheet metal covering the left-hand exhaust manifold functioned as a duct that carried heat to the intake manifold pre-heater.

Monterey, Montclair optional V-8
90 degree V-8. Overhead valve. Cast-iron block. Bore & stroke: 4.047 x 3.781 in. Displacement: 390 cid. Compression ratio: 10.50:1. Brake hp: 315 at 4600 rpm. Taxable hp: 52.49. Torque: 427 lbs.-ft. at 2800 rpm. Five main bearings. Hydraulic valve lifters. Crankcase capacity: (Comets) 4 qt.; (Mercurys) 5 qt. (add 1 qt. for new filter). Cooling system capacity: 19.5 qt. (add 1 qt. for heater). Carburetor: Four-barrel. Engine code: Z.

Parklane base V-8; Monterey, Montclair optional V-8
90 degree V-8. Overhead valve. Cast-iron block. Bore & stroke: 4.05 x 3.98 in. Displacement: 410 cid. Compression ratio: 10.5:1. Brake hp: 330 at 4600 rpm. Taxable hp: 52.49. Torque: 444 lbs.-ft. at 2800 rpm. Five main bearings. Hydraulic valve lifters. Crankcase capacity: 5 qt. (add 1 qt. for new filter). Cooling system capacity: 19.5 qt. (add 1 qt. for heater). Carburetor: Ford C6MF-12127-A four-barrel. Engine code: M.

As clearly stated atop its air cleaner, the new-for-1966 Mercury "Super Marauder" 410-cid V-8 carried a four-barrel carburetor on its intake manifold. Perhaps those crossed-flags decals were there to remind people that a Mercury Cyclone GT convertible was pacing the Indy 500 that year. Note the mechanical fuel pump and the screw-on oil filter on the lower left-hand side of the block.

A "Y" engine code identified the 1966 Mercury 390-cid two-barrel engine which was standard in Montereys and Montclairs and optional in Comets. This engine belonged to Ford's big-block family and developed 265 hp at 4400 rpm despite its small carburetor. As you can see, Autolite spark plugs were standard equipment.

Comet Cyclone GT, GTA base V-8
90 degree V-8. Overhead valve. Cast-iron block. Bore & stroke: 4.047 x 3.781 in. Displacement: 390 cid. Compression ratio: 10.50:1. Brake hp: 335 at 4800 rpm. Taxable hp: 52.49. Torque: 427 lbs.-ft. at 3200 rpm. Five main bearings. Hydraulic valve lifters. Crankcase capacity: 4 qt. (add 1 qt. for new filter). Cooling system capacity: 19.5 qt. (add 1 qt. for heater). Carburetor: Ford four-barrel. Engine code: S.

S-55 base V-8; Monterey, Montclair, Parklane optional V-8
90 degree V-8. Overhead valve. Cast-iron block. Bore & stroke: 4.13 x 3.98 in. Displacement: 428 cid. Compression ratio: 10.5:1. Brake hp: 345 at 4600 rpm. Taxable hp: 54.50. Torque: 462 lbs.-ft. at 2800 rpm. Five main bearings. Hydraulic valve lifters. Crankcase capacity: 5 qt. (add 1 qt. for new filter). Cooling system capacity: 19.5 qt. (add 1 qt. for heater). Carburetor: Ford C6AF-9510-AD four-barrel. Engine code: Q.

S-55 base V-8; Monterey, Montclair, Parklane optional V-8
90 degree V-8. Overhead valve. Cast-iron block. Bore & stroke: 4.13 x 3.98 in. Displacement: 428 cid. Compression ratio: 10.5:1. Brake hp: 360 at 5400 rpm. Taxable hp: 54.50. Torque: 459 lbs.-ft. at 3200 rpm. Five main bearings. Solid valve lifters. Crankcase capacity: 5 qt. (add 1 qt. for new filter). Cooling system capacity: 19.5 qt. (add 1 qt. for heater). Carburetor: Ford C6AF-9510-AD four-barrel. Engine code: Q.

New power combinations paced the 1966 Mercury to higher standards of smooth performance. Exclusive to Mercury for 1966 was this 410-cid V-8 with a 10.5:1 compression ratio. It could be teamed with heavy-duty Merc-O-Matic or three- or four-speed manual gearboxes. This engine was standard in Park Lane models and optional in all other models.

1967 MERCURY

Comet 202, Capri, Caliente, Cyclone, station wagon, Cougar base V-8
90 degree V-8. Overhead valve. Cast-iron block. Bore & stroke: 4.00 x 2.875 in. Displacement: 289 cid. Compression ratio: 9.30:1. Brake hp: 200 at 4400 rpm. Taxable hp: 51.20. Torque: 282 lbs.-ft. at 2400 rpm. Five main bearings. Hydraulic valve lifters. Crankcase capacity: 4 qt. (add 1 qt. for new filter). Cooling system capacity: 14 qt. (add 1 qt. for heater). Carburetor: Autolite C7DF-9510-Z two-barrel. Engine code: C.

Cougar optional V-8
90 degree V-8. Overhead valve. Cast-iron block. Bore & stroke: 4.00 x 2.875 in. Displacement: 289 cid. Compression ratio: 9.80:1. Brake hp: 225 at 4800 rpm. Taxable hp: 51.20. Torque: 305 lbs.-ft. at 3200 rpm. Five main bearings. Hydraulic valve lifters. Crankcase capacity: 4 qt. (add 1 qt. for new filter). Cooling system capacity: 14 qt. (add 1 qt. for heater). Carburetor: Autolite four-barrel. Engine code: A.

Monterey, Montclair, Commuter, Colony Park base V-8; Comet 202, Capri, Caliente, Cyclone, station wagon, Cougar optional V-8
90 degree V-8. Overhead valve. Cast-iron block. Bore & stroke: 4.047 x 3.781 in. Displacement: 390 cid. Compression ratio: 9.50:1. Brake hp: 270 at 4400 rpm. Taxable hp: 52.49. Torque: 403 lbs.-ft. at 2600 rpm. Five main bearings. Hydraulic valve lifters. Crankcase capacity: (Cougar/Comet) 4 qt.; (Full-size Mercury) 5 qt. (add 1 qt. for new filter). Cooling system capacity: 19.5 qt. (add 1 qt. for heater). Carburetor: Holley two-barrel. Engine code: H.
NOTE: The Cougar also had an optional 390-cid two-barrel V-8 that carried an "X" engine code.

Comet Cyclone GT base V-8; Cougar optional V-8
90 degree V-8. Overhead valve. Cast-iron block. Bore & stroke: 4.047 x 3.781 in. Displacement: 390 cid. Compression ratio: 10.50:1. Brake hp: 320 at 4800 rpm. Taxable hp: 52.49. Torque: 427 lbs.-ft. at 3200 rpm. Five main bearings. Hydraulic valve lifters. Crankcase capacity: 4 qt. Cooling system capacity: 19.5 qt. (add 1 qt. for heater). Carburetor: Holley C7AF-9510-AE four-barrel. Engine code: S.

Parklane, Brougham, Marquis base V-8; Monterey, Montclair, Commuter, Colony Park optional V-8
90 degree V-8. Overhead valve. Cast-iron block. Bore & stroke: 4.05 x 3.98 in. Displacement: 410 cid. Compression ratio: 10.5:1. Brake hp: 330 at 4600 rpm. Taxable hp: 52.49. Torque: 444 lbs.-ft. at 2800 rpm. Five main bearings. Hydraulic valve lifters. Crankcase capacity: 5 qt. (add 1 qt. for new filter). Cooling system capacity: 19.5 qt. (add 1 qt. for heater). Carburetor: Holley C7AF-9510-AE four-barrel. Engine code: M.

S-55 base V-8; Marquis, Commuter, Colony Park optional V-8
90 degree V-8. Overhead valve. Cast-iron block. Bore & stroke: 4.13 x 3.98 in. Displacement: 428 cid. Compression ratio: 10.5:1. Brake hp: 345 at 4600 rpm. Taxable hp: 54.50. Torque: 462 lbs.-ft. at 2800 rpm. Five main bearings. Hydraulic valve lifters. Crankcase capacity: 5 qt. (add 1 qt. for new filter). Cooling system capacity: 19.5 qt. (add 1 qt. for heater). Carburetor: Holley C7AF-9510-AE four-barrel. Engine code: Q.

Monterey, Montclair, Parklane, Brougham, Marquis, Commuter, Colony Park optional V-8
90 degree V-8. Overhead valve. Cast-iron block. Bore & stroke: 4.13 x 3.98 in. Displacement: 428 cid. Compression ratio: 10.5:1. Brake hp: 360 at 5400 rpm. Taxable hp: 54.50. Torque: 459 lbs.-ft. at 3200 rpm. Five main bearings. Solid valve lifters. Crankcase capacity: 5 qt. (add 1 qt. for new filter). Cooling system capacity: 19.5 qt. (add 1 qt. for heater). Carburetor: Holley C7AF-9510-AE four-barrel. Engine code: P (police).

1968 MERCURY

Cyclone, Montego Brougham, Cougar XR-7 base V-8 (after midyear)
90 degree V-8. Overhead valve. Cast-iron block. Bore & stroke: 4.00 x 2.875 in. Displacement: 289 cid. Compression ratio: 9.30:1. Brake hp: 200 at 4400 rpm. Taxable hp: 51.20. Torque: 282 lbs.-ft. at 2400 rpm. Five main bearings. Hydraulic valve lifters. Crankcase capacity: 4 qt. (add 1 qt. for new filter). Cooling system capacity: 14 qt. (add 1 qt. for heater). Carburetor: Autolite C7DF-9510-Z two-barrel. Engine code: C.
NOTE: The Code F engine was the base V-8 until midyear, when Mercury reverted to offering the Code C engine as the base V-8 in small and mid-size cars.

Cyclone, Montego Brougham, Cougar XR-7 base V-8; Comet, Montego, MX optional V-8
90 degree V-8. Overhead valve. Cast-iron block. Bore & stroke: 4.00 x 3.00 in. Displacement: 302 cid. Compression ratio: 9.00:1. Brake hp: 210 at 4600 rpm. Taxable hp: 51.20. Torque: 300 lbs.-ft. at 2600 rpm. Five main bearings. Hydraulic valve lifters. Crankcase capacity: 4 qt. (add 1 qt. for new filter). Cooling system capacity: 14 qt. (add 1 qt. for heater). Carburetor: Autolite C8AF-9510-AF two-barrel. Engine code F.

Comet, Montego, MX, Brougham, Cyclone, Cougar, XR-7 optional V-8
90 degree V-8. Overhead valve. Cast-iron block. Bore & stroke: 4.00 x 3.00 in. Displacement: 302 cid. Compression ratio: 10.0:1. Brake hp: 230 at 4800 rpm. Taxable hp: 51.20. Torque: 310 lbs.-ft. at 2800 rpm. Five main bearings. Hydraulic valve lifters. Crankcase capacity: 4 qt. (add 1 qt. for new filter). Cooling system capacity: 14 qt. (add 1 qt. for heater). Carburetor: Four-barrel. Engine code J.

Monterey, Montclair, Commuter, Colony Park base V-8; Comet, Montego, MX, Brougham, Cyclone optional V-8
90 degree V-8. Overhead valve. Cast-iron block. Bore & stroke: 4.047 x 3.781 in. Displacement: 390 cid. Compression ratio: 9.50:1. Brake hp: 265 at 4400 rpm. Taxable hp: 52.49. Torque: 390 lbs.-ft. at 2600 rpm. Five main bearings. Hydraulic valve lifters. Crankcase capacity: (Cougar/Comet) 4 qt.; (Full-size Mercury) 5 qt. (add 1 qt. for new filter). Cooling system capacity: 19.5 qt. (add 1 qt. for heater). Carburetor: Autolite C8AF-9510-M two-barrel. Engine code Y.

Cougar, XR-7, Monterey, Montclair, Commuter, Colony Park optional V-8
90 degree V-8. Overhead valve. Cast-iron block. Bore & stroke: 4.047 x 3.781 in. Displacement: 390 cid. Compression ratio: 10.50:1. Brake hp: 280 at 4400 rpm. Taxable hp: 52.49. Torque: 403 lbs.-ft. at 2600 rpm. Five main bearings. Hydraulic valve lifters. Crankcase capacity: (Cougar/Comet) 4 qt.; (Full-size Mercury) 5 qt. Cooling system capacity: 19.5 qt. (add 1 qt. for heater). Carburetor: Holley two-barrel.

Engine code X.

Parklane, Brougham, Marquis base V-8; Monterey, Montclair, Commuter, Colony Park optional V-8;
90 degree V-8. Overhead valve. Cast-iron block. Bore & stroke: 4.047 x 3.781 in. Displacement: 390 cid. Compression ratio: 10.50:1. Brake hp: 315 at 4600 rpm. Taxable hp: 52.49. Torque: 427 lbs.-ft. at 2800 rpm. Five main bearings. Hydraulic valve lifters. Crankcase capacity: 5 qt. (add 1 qt. for new filter). Cooling system capacity: 19.5 qt. (add 1 qt. for heater). Carburetor: Autolite C8AF-9510-R four-barrel. Engine code: Z.

Cougar, XR-7 optional V-8
90 degree V-8. Overhead valve. Cast-iron block. Bore & stroke: 4.047 x 3.781 in. Displacement: 390 cid. Compression ratio: 10.50:1. Brake hp: 325 at 4800 rpm. Taxable hp: 52.49. Torque: 427 lbs.-ft. at 3200 rpm. Five main bearings. Hydraulic valve lifters. Crankcase capacity: 4 qt. (add 1 qt. for new filter). Cooling system capacity: 19.5 qt. (add 1 qt. for heater). Carburetor: Holley four-barrel. Engine code: S.

Comet, Montego, MX, Brougham, Cyclone optional V-8
90 degree V-8. Overhead valve. Cast-iron block. Bore & stroke: 4.047 x 3.781 in. Displacement: 390 cid. Compression ratio: 10.50:1. Brake hp: 335 at 4800 rpm. Taxable hp: 52.49. Torque: 427 lbs.-ft. at 3200 rpm. Five main bearings. Hydraulic valve lifters. Crankcase capacity: 4 qt. (add 1 qt. for new filter). Cooling system capacity: 19.5 qt. (add 1 qt. for heater). Carburetor: Holley four-barrel. Engine code: Q.

Monterey, Montclair, Commuter, Colony Park, Parklane, Brougham, Marquis optional V-8
90 degree V-8. Overhead valve. Cast-iron block. Bore & stroke: 4.13 x 3.98 in. Displacement: 428 cid. Compression ratio: 10.5:1. Brake hp: 340 at 4600 rpm. Taxable hp: 54.50. Torque: 462 lbs.-ft. at 2800 rpm. Five main bearings. Hydraulic valve lifters. Crankcase capacity: 5 qt. (add 1 qt. for new filter). Cooling system capacity: 19.5 qt. (add 1 qt. for heater).

The 1968 Ford Cobra Jet version of the 428 was optional equipment in the Mercury Cougar and new Mercury Montego. The special through-the-hood air cleaner doubled as a functional ram-induction hood scoop. The 428-CJ did not share the 427's cross-bolted main bearing caps and could not be revved as high. It topped out at 5000-6000 rpm. It had a Holley 415OC four-barrel carburetor. Note how the alternator, under the air pump, is mounted. It was hard to reach when the engine was installed in a car.

Carburetor: Holley C7AF-9510-AE four-barrel. Engine code: R.

Mercury optional Police V-8
90 degree V-8. Overhead valve. Cast-iron block. Bore & stroke: 4.13 x 3.98 in. Displacement: 428 cid. Compression ratio: 10.5:1. Brake hp: 360 at 5400 rpm. Taxable hp: 54.50. Torque: 459 lbs.-ft. at 3200 rpm. Five main bearings. Solid valve lifters. Crankcase capacity: 5 qt. (add 1 qt. for new filter). Cooling system capacity: 19.5 qt. (add 1 qt. for heater). Carburetor: Holley C7AF-9510-AE four-barrel. Engine code: P (police).

Cougar GT-E base V-8; Cyclone GT optional V-8
90 degree V-8. Overhead valve. Cast-iron block. Bore & stroke: 4.23 x 3.78 in. Displacement: 427 cid. Compression ratio: 10.90:1. Brake hp: 390 at 5600 rpm. Torque: 460 lbs.-ft. at 3200 rpm. Five main bearings. Hydraulic valve lifters. Crankcase capacity: 4 qt. (add 1 qt. for new filter). Cooling system capacity: 19.5 qt. (add 1 qt. for heater). Carburetor: Holley C7AF-9510-AE four-barrel. Engine code: W.
NOTE: The 427-cid V-8 was only available for a short time before replacement by the 428-CJ V-8.

1969 MERCURY

The 351-cid V-8 was basically a version of the 302-cid Ford V-8 with the same bore size but a longer stroke. This motor was actually derived from the 260-cid V-8 of the early 1960s. It was compact, light, and sufficiently powerful to appeal to many car enthusiasts of the era.

Mercury's version of the CJ 428 four-barrel V-8 carried appropriate air cleaner decals and neat-looking finned rocker arm covers with "428 Cobra Jet" stamped right into them. This was the last year that the 428 was offered. The Cobra Jet edition was good for 335 hp at 5200 rpm while the Super Cobra Jet version produced 360 hp at 5400 rpm.

Cyclone base V-8
90 degree V-8. Overhead valve. Cast-iron block. Bore & stroke: 4.00 x 3.00 in. Displacement: 302 cid. Compression ratio: 9.50:1. Brake hp: 220 at 4600 rpm. Taxable hp: 51.20. Torque: 300 lbs.-ft. at 2600 rpm. Five main bearings. Hydraulic valve lifters. Crankcase capacity: 4 qt. (add 1 qt. for new filter). Cooling system capacity: 14 qt. (add 1 qt. for heater). Carburetor: Autolite C8AF-9510-BD two-barrel. Engine code: F.

Cougar, XR-7 base V-8
90 degree V-8. Overhead valve. Cast-iron block. Bore & stroke: 4.00 x 3.50 in. Displacement: 351 cid. Compression ratio: 9.5:1. Brake hp: 250 at 4600 rpm. Taxable hp: 51.20. Torque: 355 lbs.-ft. at 2600 rpm. Five main bearings. Hydraulic valve lifters. Crankcase capacity: 4 qt. (add 1 qt. for new filter). Cooling system capacity: 14 qt. (add 1 qt. for heater). Carburetor: Autolite C9ZF-9510-A two-barrel. Engine code: H.

Marauder, Monterey, Monterey Custom base V-8
90 degree V-8. Overhead valve. Cast-iron block. Bore & stroke: 4.047 x 3.781 in. Displacement: 390 cid. Compression ratio: 9.50:1. Brake hp: 265 at 4400 rpm. Taxable hp: 52.49. Torque: 390 lbs.-ft. at 2600 rpm. Five main bearings. Hydraulic valve lifters. Crankcase capacity: 4 qt. (add 1 qt. for new filter). Cooling system capacity: 19.5 qt. (add 1 qt. for heater). Carburetor: C9AF-9510-B two-barrel. Engine code: Y.

Marauder, Monterey optional V-8
90 degree V-8. Overhead valve. Cast-iron block. Bore & stroke: 4.047 x 3.781 in. Displacement: 390 cid. Compression ratio: 10.50:1. Brake hp: 280 at 4400 rpm. Taxable hp: 52.49. Torque: 403 lbs.-ft. at 2600 rpm. Five main bearings. Hydraulic valve lifters. Crankcase capacity: 4 qt. (add 1 qt. for new filter). Cooling system capacity: 19.5 qt. (add 1 qt. for heater). Carburetor: C9AF-9510-B two-barrel. Engine code: X.

Cougar Eliminator optional V-8
90 degree V-8. Overhead valve. Cast-iron block. Bore & stroke: 4.00 x 3.00 in. Displacement: 302 cid. Compression ratio: 10.50:1. Brake hp: 290 at 5800 rpm. Taxable hp: 51.20. Torque: 290 lbs.-ft. at 4300 rpm. Five main bearings. Solid valve lifters. Crankcase capacity: 4 qt. (add 1 qt. for new filter). Cooling system capacity: 14 qt. (add 1 qt. for heater). Carburetor: Holley four-barrel. Engine code: G.

Cougar, XR-7, Comet, Montego, MX, Cyclone optional V-8
90 degree V-8. Overhead valve. Cast-iron block. Bore & stroke: 4.00 x 3.50 in. Displacement: 351 cid. Compression ratio: 10.7:1. Brake hp: 290 at 4800 rpm. Taxable hp: 51.20. Torque: 385 lbs.-ft. at 3200 rpm. Five main bearings. Hydraulic valve lifters. Crankcase capacity: 4 qt. (add 1 qt. for new filter). Cooling system capacity: 14 qt. (add 1 qt. for heater). Carburetor: Four-barrel. Engine code: M.

Cougar, XR-7, Comet, Montego, MX, Cyclone optional V-8
90 degree V-8. Overhead valve. Cast-iron block. Bore & stroke: 4.047 x 3.781 in. Displacement: 390 cid. Compression ratio: 10.50:1. Brake hp: 320 at 4600 rpm. Taxable hp: 52.49. Torque: 427 lbs.-ft. at 3200 rpm. Five main bearings. Hydraulic valve lifters. Crankcase capacity: 4 qt. (add 1 qt. for new filter). Cooling system capacity: 14 qt. (add 1 qt. for heater). Carburetor: Four-barrel. Engine code: S.

Marquis base V-8
90 degree V-8. Overhead valve. Cast-iron block. Bore & stroke: 4.36 x 3.59 in. Displacement: 429 cid. Compression ratio: 10.5:1. Brake hp: 320 at 4400 rpm. Taxable hp: 60.83. Torque: 460 lbs.-ft. at 2200 rpm. Five main bearings. Hydraulic valve lifters. Crankcase capacity: 4 qt. (add 1 qt. for new filter). Cooling system capacity: 19.5 qt. (add 1 qt. for heater). Carburetor: C9AF-9510-J two-barrel. Engine code: K.

Cyclone CJ base V-8; Cougar optional V-8
90 degree V-8. Overhead valve. Cast-iron block. Bore & stroke: 4.13 x 3.98 in. Displacement: 428 cid. Compression ratio: 10.6:1. Brake hp: 335 at 5200 rpm. Taxable hp: 54.50. Torque: 440 lbs.-ft. at 3400 rpm. Five main bearings. Hydraulic valve lifters. Crankcase capacity. 4 qt. (add 1 qt. for new filter). Cooling system capacity: 19.5 qt. (add 1 qt. for heater). Carburetor: Holley C8AF-9510-B four-barrel. Engine code: Q.

Cyclone CJ, Cougar optional Ram Air V-8
90 degree V-8. Overhead valve. Cast-iron block. Bore & stroke: 4.13 x 3.98 in. Displacement: 428 cid. Compression ratio: 10.6:1. Brake hp: 335. Taxable hp: 54.50. Five main bearings. Hydraulic valve lifters. Crankcase capacity: 4 qt. (add 1 qt. for new filter). Cooling system capacity: 19.5 qt. (add 1 qt. for heater). Carburetor: Holley four-barrel. Engine code: R.

Mercury optional Police V-8
90 degree V-8. Overhead valve. Cast-iron block. Bore & stroke: 4.13 x 3.98 in. Displacement: 428 cid. Compression ratio: 10.5:1. Brake hp: 360 at 5400 rpm. Taxable hp: 54.50. Torque: 459 lbs.-ft. at 3200 rpm. Five main bearings. Solid valve lifters. Crankcase capacity: 4 qt. (add 1 qt. for new filter). Cooling system capacity: 19.5 qt. (add 1 qt. for heater). Carburetor: Autolite C8AF-9510-B four-barrel. Engine code: P.

Marauder X-100 base V-8
90 degree V-8. Overhead valve. Cast-iron block. Bore & stroke: 4.36 x 3.59 in. Displacement: 429 cid. Compression ratio: 10.5:1. Brake hp: 360 at 4600 rpm. Taxable hp: 60.83. Torque: 480 lbs.-ft. at 2800 rpm. Five main bearings. Hydraulic valve lifters. Crankcase capacity: 4 qt. (add 1 qt. for new filter). Cooling system capacity: 19.5 qt. (add 1 qt. for heater). Carburetor: Four-barrel. Engine code: N.

1970 MERCURY

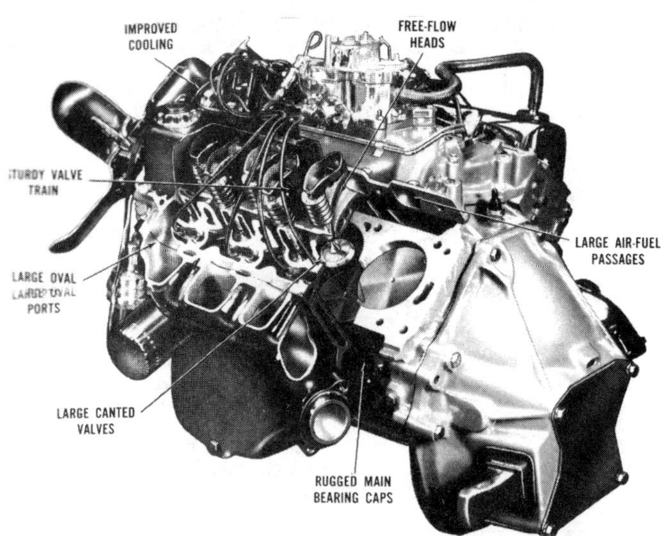

A factory cutaway illustration of the 1970 Mercury 351C V-8 illustrates its rugged main bearing caps, large canted valves, large oval dual ports, sturdy valve train, improved cooling, free-flowing head design, and large air-fuel passages.

Montego base V-8; Cougar optional V-8
90 degree V-8. Overhead valve. Cast-iron block. Bore & stroke: 4.00 x 3.00 in. Displacement: 302 cid. Compression ratio: 9.50:1. Brake hp: 210 at 4400 rpm. Taxable hp: 51.20. Torque: 295 lbs.-ft. at 2400 rpm. Five main bearings. Hydraulic valve lifters. Crankcase capacity: 4 qt. (add 1 qt. for new filter). Cooling system capacity: 14 qt. (add 1 qt. for heater). Carburetor: Two-barrel. Engine code: F.

Cougar, Cyclone base V-8
90 degree V-8. Overhead valve. Cast-iron block. Bore & stroke: 4.00 x 3.50 in. Displacement: 351 cid. Compression ratio: 9.5:1. Brake hp: 250 at 4600 rpm. Taxable hp: 51.20. Torque: 355 lbs.-ft. at 2600 rpm. Five main bearings. Hydraulic valve lifters. Crankcase capacity: 4 qt. (add 1 qt. for new filter). Cooling system capacity: 14 qt. (add 1 qt. for heater). Carburetor: Two-barrel. Engine code: H.

Monterey, Monterey Custom, Marauder, Marquis, Brougham base V-8
90 degree V-8. Overhead valve. Cast-iron block. Bore & stroke: 4.047 x 3.781 in. Displacement: 390 cid. Compression ratio: 9.50:1. Brake hp: 270 at 4400 rpm. Taxable hp: 52.54. Torque: 390 lbs.-ft. at 2600 rpm. Five main bearings. Hydraulic valve lifters. Crankcase capacity: (Cougar/Comet) 4 qt.; (Full-size Mercury) 5 qt. (add 1 qt. for new filter). Cooling system capacity: 19.5 qt. (add 1 qt. for heater). Carburetor: Two-barrel. Engine code: Y.

Cougar Eliminator optional 'BOSS' V-8
90 degree V-8. Overhead valve. Cast-iron block. Bore & stroke: 4.00 x 3.00 in. Displacement: 302 cid. Compression ratio: 10.50:1. Brake hp: 290 at 5800 rpm. Taxable hp: 51.20. Torque: 290 lbs.-ft. at 4300 rpm. Five main bearings. Solid valve lifters. Crankcase capacity: 4 qt. (add 1 qt. for new filter). Cooling system capacity: 14 qt. (add 1 qt. for heater). Carburetor: Holley four-barrel. Engine code: G.

Cougar Eliminator base V-8; Cougar optional V-8
90 degree V-8. Overhead valve. Cast-iron block. Bore & stroke: 4.00 x 3.50 in. Displacement: 351 cid. Compression ratio: 11.0:1. Brake hp: 300 at 5400 rpm. Taxable hp: 51.20. Torque: 380 lbs.-ft. at 3400 rpm. Five main bearings. Hydraulic valve lifters. Crankcase capacity: 4 qt. (add 1 qt. for new filter). Cooling system capacity: 14 qt. (add 1 qt. for heater). Carburetor: Four-barrel. Engine code: M.

Monterey, Monterey Custom, Marauder, Marquis, Brougham, Montego, MX, Brougham, optional V-8
90 degree V-8. Overhead valve. Cast-iron block. Bore & stroke: 4.36 x 3.59 in. Displacement: 429 cid. Compression ratio: 10.5:1. Brake hp: 320 at 4400 rpm. Taxable hp: 60.83. Torque: 460 lbs.-ft. at 2200 rpm. Five main bearings. Hydraulic valve lifters. Crankcase capacity: 5 qt. (add 1 qt. for new filter). Cooling system capacity: 19.5 qt. (add 1 qt. for heater). Carburetor: Two-barrel. Engine code: K.

Cougar, XR-7, Eliminator optional "428 CJ" V-8
90 degree V-8. Overhead valve. Cast-iron block. Bore & stroke: 4.13 x 3.98 in. Displacement: 428 cid. Compression ratio: 10.6:1. Brake hp: 335 at 5200 rpm. Taxable hp: 54.50. Torque: 440 lbs.-ft. at 3400 rpm. Five main bearings. Hydraulic valve lifters. Crankcase capacity: 5 qt. (add 1 qt. for new filter). Cooling system capacity: 19.5 qt. (add 1 qt. for heater). Carburetor: Four-barrel. Engine code: Q or R (Ram Air).

Compact and intermediate Mercury optional V-8
90 degree V-8. Overhead valve. Cast-iron block. Bore & stroke: 4.36 x 3.59 in. Displacement: 429 cid. Compression ratio: 11.0:1. Brake hp: 360 at 4600 rpm. Taxable hp: 60.83. Torque: 476 lbs.-ft. at 2800 rpm. Five main bearings. Hydraulic valve lifters. Crankcase capacity: 5 qt. (add 1 qt. for new filter). Cooling system capacity: 19.5 qt. (add 1 qt. for heater). Carburetor: Four-barrel. Engine code: N.

Cyclone Spoiler base V-8; Cyclone GT optional V-8
90 degree V-8. Overhead valve. Cast-iron block. Bore & stroke: 4.36 x 3.59 in. Displacement: 429 cid. Compression ratio: 11.3:1. Brake hp: 370 at 5400 rpm. Taxable hp: 60.83. Torque: 450 lbs.-ft. at 3400 rpm. Five main bearings. Hydraulic valve lifters. Crankcase capacity: 5 qt. (add 1 qt. for new filter). Cooling system capacity: 19.5 qt. (add 1 qt. for heater). Carburetor: Four-barrel. Engine code: C.

Cyclone, Cyclone Spoiler, Cyclone GT optional "BOSS" V-8
90 degree V-8. Overhead valve. Cast-iron block. Bore & stroke: 4.36 x 3.59 in. Displacement: 429 cid. Compression ratio: 10.5:1. Brake hp: 375 at 5200 rpm. Taxable hp: 60.83. Torque: 450 lbs.-ft. at 3400 rpm. Five main bearings. Solid valve lifters. Crankcase capacity: 5 qt. (add 1 qt. for new filter). Cooling system capacity: 19.5 qt. (add 1 qt. for heater). Carburetor: Four-barrel. Engine code: J.

1971 MERCURY

Comet, Montego optional V-8
90 degree V-8. Overhead valves. Cast-iron block. Displacement: 302 cid. Bore & stroke: 4.00 x 3.00 in. Compression ratio: 9.00:1. Brake hp: 210 at 4600 rpm. Torque: 296 lbs.-ft. at 2600 rpm. Five main bearings. Hydraulic valve lifters. Carburetor: Motorcraft two-barrel. Cooling system capacity: (Mid-size) 13.5 qt. with heater; (Full-size) 15.4 qt. with heater. Crankcase capacity: 4 qt. (add 1 qt. with new oil filter). Serial number code: F.

Cougar, Montego, Mercury base 351 "Windsor" V-8
90 degree V-8. Overhead valves. Cast-iron block. Displacement: 351 cid. Bore & stroke: 4.00 x 3.50 in. Compression ratio: 9.00:1. Brake hp: 240 at 4600 rpm. Torque: 350 lbs.-ft. at 2600 rpm. Hydraulic valve lifters. Carburetor: Motorcraft two-barrel. Cooling system capacity: (Mid-size) 14.6 qt. with heater; (Full-size) 16.5 qt. with heater. Crankcase capacity: 4 qt. (add 1 qt. with new oil filter). Serial number code: H.

Cougar, Montego, Mercury base 351 Cleveland V-8
90 degree V-8. Overhead valves. Cast-iron block. Displacement: 351 cid. Bore & stroke: 4.00 x 3.50 in. Compression ratio: 9.00:1. Brake hp: 240 at 4600 rpm. Torque: 350 lbs.-ft. at 2600 rpm. Hydraulic valve lifters. Carburetor: Motorcraft two-barrel. Cooling system capacity: (Mid-size) 14.6 qt. with heater; (Full-size) 16.5 qt. with heater. Crankcase capacity: 4 qt. (add 1 qt. with new oil filter). Serial number code: H.

Mercury base 400 Cleveland V-8
90 degree V-8. Overhead valves. Cast-iron block. Displacement: 400 cid. Bore & stroke: 4.00 x 4.00 in. Compression ratio: 9.5:1. Brake hp: 260 at 4400 rpm. Torque: 400 lbs.-ft. at 2200 rpm. Hydraulic valve lifters. Carburetor: Motorcraft two-barrel. Cooling system capacity: 20 qt. with heater. Crankcase capacity: 4 qt. (add 1 qt. with new oil filter). Code: S.

Cougar, Montego optional 351 "Cleveland" V-8 (Cobra Jet after May 1971)
90 degree V-8. Overhead valves. Cast-iron block. Displacement: 351 cid. Bore & stroke: 4.00 x 3.50 in. Compression ratio: 10.7:1. Brake hp: 285 at 5400 rpm. (280 hp after May 1971) Torque: 370 lbs.-ft. at 3400 rpm. Five main bearings. Hydraulic valve lifters. Carburetor: Holley four-barrel. Cooling system capacity: (Mid-size) 14.6 qt. with heater; (Full-size) 16.5 qt. with heater. Crankcase capacity: 4 qt. (add 1 qt. with new oil filter). Serial number code: M.

Mercury optional 429 V-8
90 degree V-8. Overhead valves. Cast-iron block. Displacement: 429 cid. Bore & stroke: 4.36 x 3.59 in. Compression ratio: 10.50:1. Brake hp: 320 at 4400 rpm. Torque: 460 lbs.-ft. at 2200 rpm. Five main bearings. Hydraulic valve lifters. Carburetor: Motorcraft two-barrel. Cooling system capacity: (midsize) 19 qt. with heater; (Full-size) 18.6 qt. with heater. Crankcase capacity: 4 qt. (add 1 qt. with new oil filter). Dual exhaust. Code: K.

Boss 351 V-8
90 degree V-8. Overhead valves. Cast-iron block. Displacement: 351 cid. Bore & stroke: 4.00 x 3.50 in. Compression ratio: 11.1:1. Brake hp: 330 at 5400 rpm. Torque: 380 lbs.-ft. at 3400 rpm. Five main bearings. Hydraulic valve lifters. Carburetor: Holley four-barrel. Cooling system capacity: 14.6 qt. with heater. Crankcase capacity: 4 qt. (add 1 qt. with new oil filter). Serial number code: Q.

Mercury optional 429 V-8
90 degree V-8. Overhead valves. Cast-iron block. Displacement: 429 cid. Bore & stroke: 4.36 x 3.59 in. Compression ratio: 10.50:1. Brake hp: 360 at 4600 rpm. Torque: 480 lbs.-ft. at 2800 rpm. Five main bearings. Hydraulic valve lifters. Carburetor: Holley four-barrel. Cooling system capacity: 18.6 qt. with heater. Crankcase capacity: 4 qt. (add 1 qt. with new oil filter). Dual exhaust. Code: N.

Cougar, Montego, Mercury optional 429 Cobra Jet V-8
90 degree V-8. Overhead valves. Cast-iron block. Displacement: 429 cid. Bore & stroke: 4.36 x 3.59 in. Compression ratio: 11.30:1. Brake hp: 370 at 5400 rpm. Torque: 450 lbs.-ft. at 3400 rpm. Five main bearings. Hydraulic valve lifters. Carburetor: Holley four-barrel. Cooling system capacity: (Mid-size) 19.6 qt. with heater; (Full-size) 18.6 qt. with heater. Crankcase capacity: 4 qt. (add 1 qt. with new oil filter). Dual exhaust. Code C.

Cougar, Montego, Mercury optional 429 Super Cobra Jet (Ram Air) V-8
90 degree V-8. Overhead valves. Cast-iron block. Displacement: 429 cid. Bore & stroke: 4.36 x 3.59 in. Compression ratio: 11.30:1. Brake hp: 375 at 5600 rpm. Torque: 450 lbs.-ft. at 3400 rpm. Five main bearings. Solid valve lifters. Carburetor: Holley four-barrel. Cooling system capacity: (Mid-size) 19.6 qt. with heater; (Full-size) 18.6 qt. with heater. Crankcase capacity: 6 qt. (includes 1 qt. for filter and 1 qt. for oil cooler). Dual exhaust. Code: J.

Starting in 1972, horsepower and torque ratings are net, as installed in the vehicle.

1972 MERCURY

Optional V-8
90 degree V-8. Overhead valves. Cast-iron block. Displacement: 302 cid. Bore & stroke: 4.00 x 3.00 in. Compression ratio: 8.50:1. Brake hp: 140 at 4000 rpm. Torque: 239 lbs.-ft. at 2000 rpm. Five main bearings. Hydraulic valve lifters. Carburetor: Motorcraft two-barrel. Cooling system capacity with air conditioning: (Maverick) 14.25 qt.; (Mustang) 15.5 qt.; (Torino) 16.25 qt. Crankcase capacity: 4 qt. (add 1 qt. with new oil filter). Serial number code: F.

Comet base V-8
90 degree V-8. Overhead valves. Cast-iron block. Displacement: 302 cid. Bore & stroke: 4.00 x 3.00 in. Compression ratio: 8.50:1. Brake hp: 143 at 4200 rpm. Torque: 242 lbs.-ft. at 2000 rpm. Five main bearings. Hydraulic valve lifters. Carburetor: Motorcraft two-barrel. Cooling system capacity with air conditioning: (Maverick) 14.25 qt.; (Mustang) 15.5 qt.; (Torino) 16.25 qt. Crankcase capacity: 4 qt. (add 1 qt. with new oil filter). Serial number code: F.

Optional Windsor V-8
90 degree V-8. Overhead valves. Cast-iron block. Displacement: 351 cid. Bore & stroke: 4.00 x 3.50 in. Compression ratio: 8.30:1. Brake hp: 153 at 3800 rpm. Torque: 266 lbs.-ft. at 2000 rpm. Five main bearings. Hydraulic valve lifters. Carburetor: Motorcraft two-barrel. Cooling system capacity with air conditioning: 16.75 qt. Crankcase capacity: 4 qt. (add 1 qt. with new oil filter). Serial number code H.

Montego optional V-8
90 degree V-8. Overhead valves. Cast-iron block. Displacement: 351 cid. Bore & stroke: 4.00 x 3.50 in. Compression ratio: 8.60:1. Brake hp: 161 at 4000 rpm. Torque: 276 lbs.-ft. at 2000 rpm. Five main bearings. Hydraulic valve lifters. Carburetor: Motorcraft two-barrel. Cooling system capacity with air conditioning: 15.75 qt. Crankcase capacity: 4 qt. (add 1 qt. with new oil filter). Serial number code: H.

Monterey base V-8
90 degree V-8. Overhead valves. Cast-iron block. Displacement: 351 cid. Bore & stroke: 4.00 x 3.50 in. Compression ratio: 8.60:1. Brake hp: 163 at 3800 rpm. Torque: 277 lbs.-ft. at 2000 rpm. Five main bearings. Hydraulic valve lifters. Carburetor: Motorcraft two-barrel. Cooling system capacity with air conditioning: 16.75 qt. Crankcase capacity: 4 qt. (add 1 qt. with new oil filter). Serial number code H. Serial number code: H.

Cougar base V-8
90 degree V-8. Overhead valves. Cast-iron block. Displacement: 351 cid. Bore & stroke: 4.00 x 3.50 in. Compression ratio: 8.60:1. Brake hp: 164 at 4000 rpm. Torque: 276 lbs.-ft. at 2000 rpm. Five main bearings. Hydraulic valve lifters. Carburetor: Motorcraft two-barrel. Cooling system capacity with air conditioning: 15.75 qt. Crankcase capacity: 4 qt. (add 1 qt. with new oil filter). Serial number code: H.

Montego optional "Cleveland" V-8
90 degree V-8. Overhead valves. Cast-iron block. Displacement: 400 cid. Bore & stroke: 4.00 x 4.00 in. Compression ratio: 8.40:1. Brake hp: 168 at 4200 rpm. Torque: 297 lbs.-ft. at 2200 rpm. Five main bearings. Hydraulic valve lifters. Carburetor: Motorcraft two-barrel. Cooling system capacity with air conditioning: 18.3 qt. Crankcase capacity: 4 qt. (add 1 qt. with new oil filter). Serial number code: S.

Monterey, Custom optional "Cleveland" V-8
90 degree V-8. Overhead valves. Cast-iron block. Displacement: 400 cid. Bore & stroke: 4.00 x 4.00 in. Compression ratio: 8.40:1. Brake hp: 172 at 4000 rpm. Torque: 298 lbs.-ft. at 2200 rpm. Five main bearings. Hydraulic valve lifters. Carburetor: Motorcraft two-barrel. Cooling system capacity with air conditioning: 17.75 qt. Crankcase capacity: 4 qt. (add 1 qt. with new oil filter). Serial number code: S.

Optional V-8
90 degree V-8. Overhead valves. Cast-iron block. Displacement: 351 cid. Bore & stroke: 4.00 x 3.50 in. Compression ratio: 8.60:1. Brake hp: 177 at 4000 rpm. Torque: 284 lbs.-ft. at 2000 rpm. Five main bearings. Hydraulic valve lifters. Carburetor: Motorcraft two-barrel. Cooling system capacity with air conditioning: 15.75 qt. Crankcase capacity: 4 qt. (add 1 qt. with new oil filter). Serial number code: H.

Montego optional V-8
90 degree V-8. Overhead valves. Cast-iron block. Displacement: 429 cid. Bore & stroke: 4.36 x 3.59 in. Compression ratio: 8.50:1. Brake hp: 205 at 4400 rpm. Torque: 322 lbs.-ft. at 2600 rpm. Five main bearings. Hydraulic valve lifters. Carburetor: Motorcraft two-barrel. Cooling system capacity with air conditioning: 20 qt. Crankcase capacity: 4 qt. (add 1 qt. with new oil filter). Serial number code: N.

Marquis base V-8
90 degree V-8. Overhead valves. Cast-iron block. Displacement: 429 cid. Bore & stroke: 4.36 x 3.59 in. Compression ratio: 8.50:1. Brake hp: 208 at 4400 rpm. Torque: 322 lbs.-ft. at 2800 rpm. Five main bearings. Hydraulic valve lifters. Carburetor: Motorcraft four-barrel. Cooling system capacity with air conditioning: 19.5 qt. Crankcase capacity: 4 qt. (add 1 qt. with new oil filter). Serial number code: N.

Optional V-8
90 degree V-8. Overhead valves. Cast-iron block. Displacement: 429 cid. Bore & stroke: 4.36 x 3.59 in. Compression ratio: 8.50:1. Brake hp: 212 at 4400 rpm. Torque: 327 lbs.-ft. at 2600 rpm. Five main bearings. Hydraulic valve lifters. Carburetor: Motorcraft four-barrel. Cooling system capacity with air conditioning: 20 qt. Crankcase capacity: 4 qt. (add 1 qt. with new oil filter). Serial number code: N.

Montego base V-8
90 degree V-8. Overhead valves. Cast-iron block. Displacement: 351 cid. Bore & stroke: 4.00 x 3.50 in. Compression ratio: 8.60:1. Brake hp: 248 at 5400 rpm. Torque: 299 lbs.-ft. at 3600 rpm. Five main bearings. Hydraulic valve lifters. Carburetor: Motorcraft four-barrel. Cooling system capacity with air conditioning: 15.75 qt. Crankcase capacity: 4 qt. (add 1 qt. with new oil filter). Serial number code: Q.

Cougar optional V-8
90 degree V-8. Overhead valves. Cast-iron block. Displacement: 351 cid. Bore & stroke: 4.00 x 3.50 in. Compression ratio: 8.60:1. Brake hp: 262 at 5400 rpm. Torque: 299 lbs.-ft. at 3600 rpm. Five main bearings. Hydraulic valve lifters. Carburetor: Motorcraft four-barrel. Carburetor: Motorcraft four-barrel. Cooling system capacity with air conditioning: 15.75 qt. Crankcase capacity: 4 qt. (add 1 qt. with new oil filter). Serial number code: Q.

Cougar optional 351 "Cleveland" optional HO V-8
90 degree V-8. Overhead valves. Cast-iron block. Displacement: 351 cid. Bore & stroke: 4.00 x 3.50 in. Compression ratio: 8.60:1. Brake hp: 266 at 5400 rpm. Torque: 301 lbs.-ft. at 3600 rpm. Five main bearings. Hydraulic valve lifters. Carburetor: Motorcraft four-barrel. Cooling system capacity with air conditioning: 15.75 qt. Crankcase capacity: 4 qt. (add 1 qt. with new oil filter). Serial number code: Q.

1973 MERCURY

302 Two-Barrel V-8
Overhead valve. Cast-iron block. Displacement: 302 cid. Bore & stroke: 4.00 x 3.00 in. Compression ratio: 8.0:1. Net hp: 135 at 4200 rpm. Net torque: 228 lbs.-ft. at 2200 rpm. Carburetor: Motorcraft two-barrel. Five main bearings. Cooling system capacity with air conditioning: 15.75 qt. Crankcase capacity: 4 qt (add 1 qt. with new oil filter). Serial number code: F.

302 Two-Barrel V-8
Overhead valve. Cast-iron block. Displacement: 302 cid. Bore & stroke: 4.00 x 3.00 in. Compression ratio: 8.0:1. Net hp: 137 at 4200 rpm. Net torque: 230 lbs.-ft. at 2200 rpm. Carburetor: Motorcraft two-barrel. Five main bearings. Cooling system capacity with air conditioning: 15.75 qt. Crankcase capacity: 4 qt (add 1 qt. with new oil filter). Serial number code: F.

302 Two-Barrel V-8
Overhead valve. Cast-iron block. Displacement: 302 cid. Bore & stroke: 4.00 x 3.00 in. Compression ratio: 8.0:1. Net hp: 138 at 4200 rpm. Net torque: 234 lbs.-ft. at 2200 rpm. Carburetor: Motorcraft two-barrel. Five main bearings. Cooling system capacity with air conditioning: 15.75 qt. Crankcase capacity: 4 qt (add 1 qt. with new oil filter). Serial number code: F.

351 "Cleveland" V-8
Overhead valve. Cast-iron block. Displacement: 351 cid. Bore & stroke: 4.00 x 3.50 in. Compression ratio: 8.0:1. Net hp: 154 at 4000 rpm. Net torque: 246 lbs.-ft. at 2400 rpm. Carburetor: Motorcraft two-barrel. Five main bearings. Cooling system capacity with air conditioning: 15.75 qt. (Full-size 16.25 qt.). Crankcase capacity: 4 qt (add 1 qt. with new oil filter). Serial number code: H.

351 "Windsor" V-8
Overhead valve. Cast-iron block. Displacement: 351 cid. Bore & stroke: 4.00 x 3.50 in. Compression ratio: 8.0:1. Net hp: 156 at 3800 rpm. Net torque: 260 lbs.-ft. at 2400 rpm. Carburetor: Motorcraft two-barrel. Five main bearings. Cooling system capacity with air conditioning: 15.75 qt. (Full-size 16.25 qt.). Crankcase capacity: 4 qt. (add 1 qt. with new oil filter). Serial number code: H.

351 "Cleveland" V-8
Overhead valve. Cast-iron block. Displacement: 351 cid. Bore & stroke: 4.00 x 3.50 in. Compression ratio: 8.0:1. Net hp: 159 at 4000 rpm. Net torque: 250 lbs.-ft. at 2400rpm. Carburetor: Motorcraft two-barrel. Five main bearings. Cooling system capacity with air conditioning: 15.75 qt. (Full-size 16.25 qt.). Crankcase capacity: 4 qt. (add 1 qt. with new oil filter). Serial number code: H.

351 "Cleveland" V-8
Overhead valve. Cast-iron block. Displacement: 351 cid. Bore & stroke: 4.00 x 3.50 in. Compression ratio: 8.6:1. Net hp: 161 at 4000 rpm. Net torque: 254 lbs.-ft. at 2400 rpm. Carburetor: Motorcraft two-barrel. Five main bearings. Cooling system capacity with air conditioning: 15.75 qt. (Full-size 16.25 qt.). Crankcase capacity: 4 qt. (add 1 qt. with new oil filter). Serial number code: H.

351 "Cobra Jet Cleveland" V-8
Overhead valve. Cast-iron block. Displacement: 351 cid. Bore & stroke: 4.00 x 3.50 in. Compression ratio: 8.0:1. Net hp: 246 at 5400 rpm. Net torque: 312 lbs.-ft. at 3600 rpm. Carburetor: Holley four-barrel. Five main bearings. Cooling system capacity with air conditioning: Cooling system capacity with air conditioning: 15.75 qt. Crankcase capacity: 4 qt. (add 1 qt. with new oil filter). Serial number code: Q.

351 "Cobra Jet Cleveland" V-8
Overhead valve. Cast-iron block. Displacement: 351 cid. Bore & stroke: 4.00 x 3.50 in. Compression ratio: 7.9:1. Net hp: 266 at 5400 rpm. Net torque: 301 lbs.-ft. at 3600 rpm. Carburetor: Holley four-barrel. Five main bearings. Cooling system capacity with air conditioning: 15.75 qt. Crankcase capacity: 4 qt. (add 1 qt. with new oil filter). Serial number code: Q.

400 "Cleveland" V-8
Overhead valve. Cast-iron block. Displacement: 400 cid. Bore & stroke: 4.00 x 4.00 in. Compression ratio: 8.0:1. Net hp: 163 at 3800 rpm. Net torque: 300 lbs.-ft. at 2000 rpm. Carburetor: Motorcraft two-barrel. Five main bearings. Cooling system capacity with air conditioning: 17.75 qt. (Full-size 18.00 qt.) Crankcase capacity: 4 qt. (add 1 qt. with new oil filter). Serial number code: S.

400 "Cleveland" V-8
Overhead valve. Cast-iron block. Displacement: 400 cid. Bore & stroke: 4.00 x 4.00 in. Compression ratio: 8.0:1. Net hp: 168 at 3800 rpm. Net torque: 310 lbs.-ft. at 2000 rpm. Carburetor: Motorcraft two-barrel. Five main bearings. Cooling system capacity with air conditioning: 17.75 qt. (Full-size 18.00 qt.) Crankcase capacity: 4 qt. (add 1 qt. with new oil filter). Serial number code: S.

400 "Cleveland" V-8
Overhead valve. Cast-iron block. Displacement: 400 cid. Bore & stroke: 4.00 x 4.00 in. Compression ratio: 8.4:1. Net hp: 171 at 3600 rpm. Net torque: 314 lbs.-ft. at 2000 rpm. Carburetor: Motorcraft two-barrel. Five main bearings. Cooling system capacity with air conditioning: 17.75 qt. (Full-size 18.00 qt.) Crankcase capacity: 4 qt (add 1 qt. with new oil filter). Serial number code: S.

429 V-8
Overhead valve. Cast-iron block. Displacement: 429 cid. Bore & stroke: 4.36 x 3.59 in. Compression ratio: 8.0:1. Net hp: 197 at 4400 rpm. Net torque: 320 lbs.-ft. at 2600 rpm. Carburetor: Motorcraft four-barrel. Five main bearings. Cooling system capacity with air conditioning: 18.75 qt. (Full-size 20.00 qt.) Crankcase capacity: 4 qt. (add 1 qt. with new oil filter). Serial number code: N.

429 V-8
Overhead valve. Cast-iron block. Displacement: 429 cid. Bore & stroke: 4.36 x 3.59 in. Compression ratio: 8.0:1. Net hp: 198 at 4400 rpm. Net torque: 320 lbs.-ft. at 2800 rpm. Carburetor: Motorcraft four-barrel. Five main bearings. Cooling system capacity with air conditioning: 18.75 qt. (Full-size 20.00 qt.) Crankcase capacity: 4 qt. (add 1 qt. with new oil filter). Serial number code: N.

429 V-8
Overhead valve. Cast-iron block. Displacement: 429 cid. Bore & stroke: 4.36 x 3.59 in. Compression ratio: 8.0:1. Net hp: 201 at 4400 rpm. Net torque: 322 lbs.-ft. at 2600 rpm. Carburetor: Motorcraft four-barrel. Five main bearings. Cooling system capacity with air conditioning: 18.75 qt. (Full-size 20.00 qt.) Crankcase capacity: 4 qt. (add 1 qt. with new oil filter). Serial number code: N.
Note: Mercury rated each 1973 engine with two or three different horsepower ratings, depending on the model each engine was installed in. As body size and weight increased between models, horsepower ratings increased correspondingly. Most engine ratings varied between one and five horsepower, with the 460-cid V-8 varying 17 hp.

1974 MERCURY

302 V-8
Overhead valve. Cast-iron block. Displacement: 302 cid. Bore & stroke: 4.00 x 3.00 in. Compression ratio: 8.0:1. Net hp: 140 at 3800 rpm. Net torque: 230 lbs.-ft. at 2600 rpm. Carburetor: Motorcraft two-barrel. Five main bearings. Cooling system capacity with air conditioning: (Comet) 14.25 qt.; (Montego) 15.5 qt. Crankcase capacity: (All) 4 qt. (add 1 qt. with new oil filter). Serial number code: F.

351 "Cleveland" V-8
Overhead valve. Cast-iron block. Displacement: 351 cid. Bore & stroke: 4.00 x 3.50 in. Compression ratio: 8.0:1. Net hp: 162 at 4000 rpm. Net torque: 278 lbs.-ft. at 2000 rpm. Carburetor: Motorcraft two-barrel. Five main bearings. Cooling system capacity with air conditioning: (Montego/Cougar) 16.5 qt.; (full-size) 17 qt. Crankcase capacity: (All) 4 qt. (add 1 qt. with new oil filter). Serial number code: Q.

351 "Windsor" V-8
Overhead valve. Cast-iron block. Displacement: 351 cid. Bore & stroke: 4.00 x 3.50 in. Compression ratio: 8.2:1. Net hp: 163 at 4200 rpm. Net torque: 278 lbs.-ft. at 2000 rpm. Carburetor: Motorcraft two-barrel. Five main bearings. Cooling system capacity with air conditioning: (Montego) 16.5 qt.; (full-size) 17 qt. Crankcase capacity: 4 qt. (add 1 qt. with new oil filter). Serial number code: H.

400 V-8
Overhead valve. Cast-iron block. Displacement: 400 cid. Bore & stroke: 4.00 x 4.00 in. Compression ratio: 8.0:1. Net hp: 170 at 3400 rpm. Net torque: 330 lbs.-ft. at 2000 rpm. Carburetor: Motorcraft two-barrel. Five main bearings. Cooling system capacity with air conditioning: (Cougar) 18.5 qt.; (Montego) 18.3; (full-size) 18.5 qt. Crankcase capacity: (All) 4 qt. (add 1 qt. with new oil filter). Serial number code: S.

460 V-8
Overhead valve. Cast-iron block. Displacement: 460 cid. Bore & stroke: 4.36 x 3.85 in. Compression ratio: 8.0:1. Net hp: 215 at 4000 rpm. Net torque: 350 lbs.-ft. at 2600 rpm. Carburetor: Carter four-barrel. Five main bearings. Cooling system capacity with air conditioning: (Torino) 19.5 qt.; (All) 19.5 qt. Crankcase capacity: (All) 4 qt. (add 1 qt. with new oil filter). Serial number code: A.

460 V-8
Overhead valve. Cast-iron block. Displacement: 460 cid. Bore & stroke: 4.36 x 3.85 in. Compression ratio: 8.0:1. Net hp: 220 at 4000 rpm. Net torque: 355 lbs.-ft. at 2600 rpm. Carburetor: Carter four-barrel. Five main bearings. Cooling system capacity with air conditioning: (Torino) 19.5 qt.; (All) 19.5 qt. Crankcase capacity: (All) 4 qt. (add 1 qt. with new oil filter). Serial number code: A.

351 "CJ" V-8
Overhead valve. Cast-iron block. Displacement: 351 cid. Bore & stroke: 4.00 x 3.50 in. Compression ratio: 7.9:1. Net hp: 255 at 5600 rpm. Net torque: 290 lbs.-ft. at 3400 rpm. Carburetor: Motorcraft four-barrel. Five main bearings. Cooling system capacity with air conditioning: (Cougar) 16.5 qt. Crankcase capacity: (All) 4 qt. (add 1 qt. with new oil filter). Serial number code: Q.

1975 MERCURY

302 V-8
Overhead valve. Cast-iron block. Displacement: 302 cid. Bore & stroke: 4.00 x 3.00 in. Compression ratio: 8.0:1. Net hp: 129 at 3800 rpm. Net torque: 220 lbs.-ft. at 1800 rpm. Carburetor: Motorcraft two-barrel. Five main bearings. Cooling system capacity with air conditioning: 14.1 qt. Crankcase capacity: 4 qt. (add 1 qt. with new oil filter). Serial number code: F.

351 "Windsor" V-8
Overhead valve. Cast-iron block. Displacement: 351 cid. Bore & stroke: 4.00 x 3.50 in. Compression ratio: 8.2:1. Net hp: 143 at 3600 rpm. Net torque: 255 lbs.-ft. at 2200 rpm. Carburetor: Motorcraft two-barrel. Five main bearings. Cooling system capacity with air conditioning: 16.2 qt. Crankcase capacity: 4 qt. (add 1 qt. with new oil filter). Serial number code: H.

351 "Modified" V-8
Overhead valve. Cast-iron block. Displacement: 351 cid. Bore & stroke: 4.00 x 3.50 in. Compression ratio: 8.0:1. Net hp: 148 at 3800 rpm. Net torque: 243 lbs.-ft. at 2400 rpm. Carburetor: Motorcraft two-barrel. Five main bearings. Cooling system capacity with air conditioning: 18.0 qt. Crankcase capacity: 4 qt. (add 1 qt. with new oil filter). Serial number code: H.

400 V-8
Overhead valve. Cast-iron block. Displacement: 400 cid. Bore & stroke: 4.00 x 4.00 in. Compression ratio: 8.0:1. Net hp: 158 at 3800 rpm. Net torque: 276 lbs.-ft. at 2000 rpm. Carburetor: Motorcraft two-barrel. Five main bearings. Cooling system capacity with air conditioning: (mid-size) 17.5 qt.; (full-size) 17.6 qt. Crankcase capacity: 4 qt. (add 1 qt. with new oil filter). Serial number code: S.

460 Police Interceptor V-8
Overhead valve. Cast-iron block. Displacement: 460 cid. Bore & stroke: 4.36 x 3.85 in. Compression ratio: 8.0:1. Net hp: 217 at 4000 rpm. Net torque: 365 lbs.-ft. at 2600 rpm. Carburetor: Carter four-barrel. Five main bearings. Cooling system capacity with air conditioning: (Mid-size) 19.7 qt.; (Full-size standard) 18.5 qt.; (Full-size medium-duty) 19 qt.; (Full-size heavy-duty and police) 20 qt. Crankcase capacity: (Mid-size) 6 qt.; (Police) 6.5 qt.; (Others) 4 qt. (add 1 qt. with new oil filter). Serial number code: A.

460 V-8
Overhead valve. Cast-iron block. Displacement: 460 cid. Bore & stroke: 4.36 x 3.85 in. Compression ratio: 8.0:1. Net hp: 218 at 4000 rpm. Net torque: 369 lbs.-ft. at 2600 rpm. Carburetor: Carter four-barrel. Five main bearings. Cooling system capacity with air conditioning: (Mid-size) 19.7 qt.; (Full-size standard) 18.5 qt.; (Full-size medium-duty) 19 qt.; (Full-size heavy-duty) 20 qt. Crankcase capacity: (Mid-size) 6 qt.; (Police) 6.5 qt.; (Others) 4 qt. (add 1 qt. with new oil filter). Serial number code: A.

1976 MERCURY

Comet base V-8
Overhead valve. Cast-iron block. Displacement: 302 cid (5.0 liters). Bore & stroke: 4.00 x 3.00 in. Compression ratio: 8.0:1. Net hp: 138 at 3600 rpm. Net torque: 245 lbs.-ft. at 2000 rpm. Carburetor: Ford 2150A 9510 (D5DE-AFA) two-barrel. Five main bearings. Cooling system capacity with air conditioning: 14.1 qt. Crankcase capacity: 4 qt. (add 1 qt. with new oil filter). Serial number code: F.

Monarch base V-8
Overhead valve. Cast-iron block. Displacement: 302 cid (5.0 liters). Bore

& stroke: 4.00 x 3.00 in. Compression ratio: 8.0:1. Net hp: 134 at 3600 rpm. Net torque: 242 lbs.-ft. at 2000 rpm. Carburetor: Ford 2150A 9510 (D5DE-AFA) two-barrel. Five main bearings. Cooling system capacity with air conditioning: 14.1 qt. Crankcase capacity: 4 qt. (add 1 qt. with new oil filter). Serial number code: F.

Cougar base V-8
90-degree, overhead valve V-8. Cast-iron block and head. Displacement: 351 cid (5.8 liters). Bore & stroke: 4.00 x 3.50 in. Compression ratio: 8.0:1. Brake hp: 152 at 3800 rpm. Torque: 274 lbs.-ft. at 1600 rpm. Five main bearings. Hydraulic valve lifters. Carburetor: Ford 2150A 9510 two-barrel. Cooling system capacity with air conditioning: 17.5 qt. Crankcase capacity: 4 qt. (add 1 qt. with new oil filter). Serial number code: H.

Montego base 351 Windsor V-8
90-degree, overhead valve V-8. Cast-iron block and head. Displacement: 351 cid (5.8 liters). Bore & stroke: 4.00 x 3.50 in. Compression ratio: 8.1:1. Brake hp: 154 at 3400 rpm. Torque: 286 lbs.-ft. at 1800 rpm. Five main bearings. Hydraulic valve lifters. Carburetor: Ford 2150A two-barrel. Serial number code: H.

Marquis base V-8; Cougar, Montego optional V-8
90-degree, overhead valve V-8. Cast-iron block and head. Displacement: 400 cid (6.6 liters). Bore & stroke: 4.00 x 4.00 in. Compression ratio: 8.0:1. Brake hp: 180 at 3800 rpm. Torque: 336 lbs.-ft. at 1800 rpm. Five main bearings. Hydraulic valve lifters. Carburetor: Ford 2150A 9510 two-barrel. Cooling system capacity with air conditioning: (Torino) 17.6 qt.; (Full-size Ford) 17.6 qt. Crankcase capacity: 4 qt. (add 1 qt. with new oil filter). Serial number code: S.

Cougar, Montego, Marquis optional V-8
90-degree, overhead valve V-8. Cast-iron block and head. Displacement: 460 cid (7.5 liters). Bore & stroke: 4.36 x 3.85 in. Compression ratio: 8.0:1. Brake hp: 202 at 3800 rpm. Torque: 352 lbs.-ft. at 1600 rpm. Five main bearings. Hydraulic valve lifters. Carburetor: Motorcraft 9510 or Ford 4350A9510-A four-barrel. Cooling system capacity with air conditioning: (Montego/Cougar) 19.2 qt.; (Mercury Marquis) 18.5 qt. Crankcase capacity: (Montego) 6 qt.; (Others) 4 qt. (add 1 qt. with new oil filter). Serial number code: A.

1977 MERCURY

Monarch base V-8
90-degree, overhead valve V-8. Cast-iron block and head. Displacement: 302 cid (5.0 liters). Bore & stroke: 4.00 x 3.00 in. Compression ratio: 8.4:1. Brake hp: 122 at 3200 rpm. Torque: 237 lbs.-ft. at 1600 rpm. Five main bearings. Hydraulic valve lifters. Carburetor: Motorcraft 2150 two-barrel. Serial number code: F.

Comet, Cougar base V-8
90-degree, overhead valve V-8. Cast-iron block and head. Displacement: 302 cid (5.0 liters). Bore & stroke: 4.00 x 3.00 in. Compression ratio: 8.4:1. Brake hp: 130-137 at 3400-3600 rpm. Torque: 243-245 lbs.-ft. at 1600-1800 rpm. Five main bearings. Hydraulic valve lifters. Carburetor: Motorcraft 2150 two-barrel. Serial number code: F.

Cougar, Monarch optional V-8
90-degree, overhead valve V-8, Cast-iron block and head. Displacement: 351 cid (5.8 liters). Bore & stroke: 4.00 x 3.50 in. Compression ratio: 8.3:1. Brake hp: 149 at 3200 rpm. Torque: 291 lbs.-ft. at 1600 rpm. Five main bearings. Hydraulic valve lifters. Carburetor: Motorcraft 2150 two-barrel. Windsor engine. Serial number code: H.

Monarch Ghia optional V-8
90-degree, overhead valve V-8, Cast-iron block and head. Displacement: 351 cid (5.8 liters). Bore & stroke: 4.00 x 3.50 in. Compression ratio: 8.3:1. Brake hp: 135 at 3200 rpm. Torque: 275 lbs.-ft. at 1600 rpm. Five main bearings. Hydraulic valve lifters. Carburetor: Motorcraft 2150 two-barrel. Windsor engine. Serial number code: Q.

Marquis base V-8; Cougar optional V-8
90-degree, overhead valve V-8. Cast-iron block and head. Displacement: 400 cid (6.6 liters). Bore & stroke: 4.00 x 4.00 in. Compression ratio: 8.0:1. Brake hp: 173 at 3800 rpm. Torque: 326 lbs.-ft. at 1600 rpm. Five main bearings. Hydraulic valve lifters. Carburetor: Motorcraft 2150 two-barrel. Serial number code: S.

Grand Marquis base V-8; Marquis optional V-8
90-degree, overhead valve V-8. Cast-iron block and head. Displacement: 460 cid (7.5 liters). Bore & stroke: 4.36 x 3.85 in. Compression ratio: 8.0:1. Brake hp: 197 at 4000 rpm. Torque: 353 lbs.-ft. at 2000 rpm. Five main bearings. Hydraulic valve lifters. Carburetor: Motorcraft 4350 four-barrel. Serial number code: A.

Note: Horsepower and torque ratings of some Mercury V-8s varied slightly, according to model.

1978 MERCURY

Zephyr, Monarch optional V-8
90-degree, overhead valve V-8. Cast-iron block and head. Displacement: 302 cid (5.0 liters). Bore & stroke: 4.00 x 3.00 in. Compression ratio: 8.4:1. Brake hp: 134 at 3400 rpm. Torque: 248 lbs.-ft. at 1600 rpm. Five main bearings. Hydraulic valve lifters. Carburetor: Motorcraft 2150 two-barrel. Serial number code: F.

Cougar base V-8
90-degree, overhead valve V-8. Cast-iron block and head. Displacement: 302 cid (5.0 liters). Bore & stroke: 4.00 x 3.00 in. Compression ratio: 8.4:1. Brake hp: Fairmont, 139 at 3600 rpm. Torque: 250 lbs.-ft. at 1600 rpm. Five main bearings. Hydraulic valve lifters. Carburetor: Motorcraft 2150 two-barrel. Serial number code: F.

Cougar optional V-8
90-degree, overhead valve V-8. Cast-iron block and head. Displacement: 351 cid (5.8 liters). Bore & stroke: 4.00 x 3.50 in. Compression ratio: 8.0:1. Brake hp: 145 at 3400 rpm. Torque: 273 lbs.-ft. at 1800 rpm. Five main bearings. Hydraulic valve lifters. Carburetor: Motorcraft 2150 two-barrel. Windsor engine. Serial number code: H.

Monarch Ghia optional V-8
90-degree, overhead valve V-8. Cast-iron block and head. Displacement: 351 cid (5.8 liters). Bore & stroke: 4.00 x 3.50 in. Compression ratio: 8.0:1. Brake hp: 152 at 3600 rpm. Torque: 278 lbs.-ft. at 1800 rpm. Five main bearings. Hydraulic valve lifters. Carburetor: Motorcraft 2150 two-barrel. Windsor engine. Serial number code: Q.

Cougar V-8
90-degree, overhead valve V-8. Cast-iron block and head. Displacement: 400 cid (6.6 liters). Bore & stroke: 4.00 x 4.00 in. Compression ratio: 8.0:1. Brake hp: 160 at 3800 rpm. Torque: 314 lbs.-ft. at 1800 rpm. Five main bearings. Hydraulic valve lifters. Carburetor: Motorcraft 2150 two-barrel. Serial number code: S.

Monarch V-8
90-degree, overhead valve V-8. Cast-iron block and head. Displacement: 400 cid (6.6 liters). Bore & stroke: 4.00 x 4.00 in. Compression ratio: 8.0:1. Brake hp: 166 at 3800 rpm. Torque: 319 lbs.-ft. at 1800 rpm. Five main bearings. Hydraulic valve lifters. Carburetor: Motorcraft 2150 two-barrel. Serial number code: S.

Optional V-8 (Cougar. Marquis)
90-degree, overhead valve V-8. Cast-iron block and head. Displacement: 400 cu. in. (6.6 liters). Bore & stroke: 4.00 x 4.00 in. Compression ratio: 8.0:1. Brake hp: 160-166 at 3800 rpm. Torque: 314-319 lbs.-ft. at 1800 rpm. Five main bearings. Hydraulic valve lifters. Carburetor: 2Bbl. Motorcraft 2150. VIN Code: S.

Grand Marquis base V-8; Marquis optional V-8
90-degree, overhead valve V-8. Cast-iron block and head. Displacement: 460 cid (7.5 liters). Bore & stroke: 4.36 x 3.85 in. Compression ratio: 8.0:1. Brake hp: 202 at 4000 rpm. Torque: 348 lbs.-ft. at 2000 rpm. Five main bearings. Hydraulic valve lifters. Carburetor: Motorcraft 4350 four-barrel. Serial number code: A.
Note: Horsepower and torque ratings of some Mercury V-8s varied slightly, according to model.

1979 MERCURY

Marquis base V-8
90-degree, overhead valve V-8. Cast-iron block and head. Displacement: 302 cid (5.0 liters). Bore & stroke: 4.00 x 3.00 in. Compression ratio: 8.4:1. Brake hp: 129 at 3600 rpm. Torque: 223 lbs.-ft. at 2600 rpm. Five main bearings. Hydraulic valve lifters. Carburetor: Motorcraft 2150 or 2700VV two-barrel. Serial number code: F.

Cougar XR-7 base V-8
90-degree, overhead valve V-8. Cast-iron block and head. Displacement: 302 cid (5.0 liters). Bore & stroke: 4.00 x 3.00 in. Compression ratio: 8.4:1. Brake hp: 133 at 3400 rpm. Torque: LTD II/TBird 245 lbs.-ft. at 1600 rpm. Five main bearings. Hydraulic valve lifters. Carburetor:

Motorcraft 2150 or 2700VV two-barrel. Serial number code: F.

Monarch base V-8
90-degree, overhead valve V-8. Cast-iron block and head. Displacement: 302 cid (5.0 liters). Bore & stroke: 4.00 x 3.00 in. Compression ratio: 8.4:1. Brake hp: 137 at 3600 rpm. Torque: 243 lbs.-ft. at 2000 rpm. Five main bearings. Hydraulic valve lifters. Carburetor: Motorcraft 2150 or 2700VV two-barrel. Serial number code: F.

Capri base V-8; Zephyr optional V-8
90-degree, overhead valve V-8. Cast-iron block and head. Displacement: 302 cid (5.0 liters). Bore & stroke: 4.00 x 3.00 in. Compression ratio: 8.4:1. Brake hp: 140 at 3600 rpm. Torque: 250 lbs.-ft. at 1800 rpm. Five main bearings. Hydraulic valve lifters. Carburetor: Motorcraft 2150 or 2700VV two-barel. Serial number code: F.

Cougar XR-7 optional V-8; Marquis optional V-8
90-degree, overhead valve V-8. Cast-iron block and head. Displacement: 351 cid (5.8 liters). Bore & stroke: 4.00 x 3.50 in. Compression ratio: 8.3:1. Brake hp: 135 or 142 at 3200 rpm. Torque: 286 lbs.-ft. at 1400 rpm. Five main bearings. Hydraulic valve lifters. Carburetor: Motorcraft 7200VV two-barrel. Windsor engine. Serial number code: H.

Cougar XR-7 optional V-8
90-degree, Modified overhead valve V-8. Cast-iron block and head. Displacement: 351 cid (5.8 liters). Bore & stroke: 4.00 x 3.50 in. Compression ratio: 8.0:1. Brake hp: 151 at 3600 rpm. Torque: 270 lbs.-ft. at 2200 rpm. Five main bearings. Hydraulic valve lifters. Carburetor: Motorcraft 2150 two-barrel. Windsor engine. Serial number code: H.
Note: Horsepower and torque ratings of some Mercury V-8s varied slightly, according to model.

1980 MERCURY

Cougar base V-8
90-degree, overhead valve V-8. Cast-iron block and head. Displacement: 255 cid (4.2 liters). Bore & stroke: 3.68 x 3.00 in. Compression ratio: 8.8:1. Brake hp: 115 at 3800 rpm. Torque: 194 lbs.-ft. at 2200 rpm. Five main bearings. Hydraulic valve lifters. Carburetor: Motorcraft 2150 two-barrel. Serial number code: D.

Monarch base V-8; Capri, Zephyr optional V-8
90-degree, overhead valve V-8. Cast-iron block and head. Displacement: 255 cid (4.2 liters). Bore & stroke: 3.68 x 3.00 in. Compression ratio: 8.8:1. Brake hp: 119 at 3800 rpm. Torque: 194 lbs.-ft. at 2200 rpm. Five main bearings. Hydraulic valve lifters. Carburetor: Motorcraft 2150 two-barrel. Serial number code: D.

Grand Marquis base V-8
90-degree, overhead valve V-8. Cast-iron block and head. Displacement: 302 cid (5.0 liters). Bore & stroke: 4.00 x 3.00 in. Compression ratio: 8.4:1. Brake hp: 130 at 3600 rpm. Torque: 230 lbs.-ft. at 1600 rpm. Five main bearings. Hydraulic valve lifters. Carburetor: Motorcraft 2150 or 2700VV two-barrel. Serial number code: F.

Cougar optional V-8
90-degree, overhead valve V-8. Cast-iron block and head. Displacement: 302 cid (5.0 liters). Bore & stroke: 4.00 x 3.00 in. Compression ratio: 8.4:1. Brake hp: 131 at 3600 rpm. Torque: 231 lbs.-ft. at 1600 rpm. Five main bearings. Hydraulic valve lifters. Carburetor: Motorcraft 2150 or 2700VV two-barrel. Serial number code: F.

Monarch optional V-8
90-degree, overhead valve V-8. Cast-iron block and head. Displacement: 302 cid (5.0 liters). Bore & stroke: 4.00 x 3.00 in. Compression ratio: 8.4:1. Brake hp: 134 at 3600 rpm. Torque: Granada 232 lbs.-ft. at 1600 rpm. Five main bearings. Hydraulic valve lifters. Carburetor: Motorcraft 2150 or 2700VV two-barrel. Serial number code: F.

Grand Marquis optional V-8
90-degree, overhead valve V-8. Cast-iron block and head. Displacement: 351 cid (5.8 liters). Bore & stroke: 4.00 x 3.50 in. Compression ratio: 8.3:1. Brake hp: 140 at 3400 rpm. Torque: 265 lbs.-ft. at 2000 rpm. Five main bearings. Hydraulic valve lifters. Carburetor: Motorcraft 7200VV two-barrel. Windsor engine. Serial number code: G.
Note: Horsepower and torque ratings of some Ford V-8s varied slightly, according to model.

1981 MERCURY

Capri, Cougar XR-7, Zephyr optional V-8
90-degree, overhead valve V-8. Cast-iron block and head. Displacement: 255 cid (4.2 liters). Bore & stroke: 3.68 x 3.00 in. Compression ratio: 8.8:1. Brake hp: 115 at 3800 rpm. Torque: 194 lbs.-ft. at 2200 rpm. Five main bearings. Hydraulic valve lifters. Carburetor: Motorcraft 2150 two-barrel. Serial number code: D.

Grand Marquis base V-8
90-degree, overhead valve V-8. Cast-iron block and head. Displacement: 255 cid (4.2 liters). Bore & stroke: 3.68 x 3.00 in. Compression ratio: 8.2:1. Brake hp: 120 at 3400 rpm. Torque: 205 lbs.-ft. at 2600 rpm. Five main bearings. Hydraulic valve lifters. Carburetor: Motorcraft 2150 or 7200VV two-barrel. Serial number code: D.

Grand Marquis optional V-8
90-degree, overhead valve V-8. Cast-iron block and head. Displacement: 302 cid (5.0 liters). Bore & stroke: 4.00 x 3.00 in. Compression ratio: 8.4:1. Brake hp: 130 at 3400 rpm. Torque: 235 lbs.-ft. at 1800 rpm. Five main bearings. Hydraulic valve lifters. Carburetor: Motorcraft 2150 or 7200VV two-barrel. Serial number code: F.

Grand Marquis optional V-8
90-degree, overhead valve V-8. Cast-iron block and head. Displacement: 351 cid (5.8 liters). Bore & stroke: 4.00 x 3.50 in. Compression ratio: 8.3:1. Brake hp: 145 at 3200 rpm. Torque: 270 lbs.-ft. at 1800 rpm. Five main bearings. Hydraulic valve lifters. Carburetor: Motorcraft 7200VV two-barrel. Windsor engine. Serial number code: H.

Grand Marquis optional police package H.O. V-8
90-degree, overhead valve V-8. Cast-iron block and head. Displacement: 351 cid (5.8 liters). Bore & stroke: 4.00 x 3.50 in. Compression ratio: 8.3:1. Brake hp: 165 at 3600 rpm. Torque: 285 lbs.-ft. at 2200 rpm. Five main bearings. Hydraulic valve lifters. Carburetor: Motorcraft 7200VV two-barrel. Windsor engine. Serial number code: G.

1982 MERCURY

Capri, Zephyr base V-8
90-degree, overhead valve V-8. Cast-iron block and head. Displacement: 255 cid (4.2 liters). Bore & stroke: 3.68 x 3.00 in. Compression ratio: 8.2:1. Brake hp: 120 at 3400 rpm. Torque: 205 lbs.-ft. at 1600 rpm. Five main bearings. Hydraulic valve lifters. Carburetor: Motorcraft 2150 or 7200VV two-barrel. Serial number code: D.

Grand Marquis base V-8
90-degree, overhead valve V-8. Cast-iron block and head. Displacement: 255 cid (4.2 liters). Bore & stroke: 3.68 x 3.00 in. Compression ratio: 8.2:1. Brake hp: 122 at 3400 rpm. Torque: 209 lbs.-ft. at 2400 rpm. Five main bearings. Hydraulic valve lifters. Carburetor: Motorcraft 2150 or 7200VV two-barrel. Serial number code: D.

Grand Marquis optional V-8
90-degree, overhead valve V-8. Cast-iron block and head. Displacement: 302 cid (5.0 liters). Bore & stroke: 4.00 x 3.00 in. Compression ratio: 8.4:1. Brake hp: 132 at 3400 rpm. Torque: 236 lbs.-ft. at 1800 rpm. Five main bearings. Hydraulic valve lifters. Carburetor: Motorcraft 7200VV two-barrel. Serial number code: F.

Capri optional V-8
90-degree, overhead valve V-8. Cast-iron block and head. Displacement: 302 cid (5.0 liters). Bore & stroke: 4.00 x 3.00 in. Compression ratio: 8.3:1. Brake hp: 157 at 4200 rpm. Torque: 240 lbs.-ft. at 2400 rpm. Five main bearings. Hydraulic valve lifters. Carburetor: Motorcraft 2150A two-barrel. Serial number code: F.

Grand Marquis police H.O. V-8
90-degree, overhead valve V-8. Cast-iron block and head. Displacement: 351 cid (5.8 liters). Bore & stroke: 4.00 x 3.50 in. Compression ratio: 8.3:1. Brake hp: 165 at 3600 rpm. Torque: 285 lbs.-ft. at 2200 rpm. Five main bearings. Hydraulic valve lifters. Carburetor: Motorcraft 2150A or 7200VV two-barrel. Serial number code: G.

1983 MERCURY

Cougar, Grand Marquis base V-8
90-degree, overhead valve V-8. Cast-iron block and head. Displacement: 302 cid (5.0 liters). Bore & stroke: 4.00 x 3.00 in. Compression ratio:

8.4:1. Brake hp: 130 at 3200 rpm. Torque: 240 lbs.-ft. at 2000 rpm. Five main bearings. Hydraulic valve lifters. Induction: Electronic fuel injection. Serial number code: F.

Grand Marquis optional H.O. V-8
90-degree, overhead valve V-8. Cast-iron block and head. Displacement: 302 cid (5.0 liters). Bore & stroke: 4.00 x 3.00 in. Compression ratio: 8.4:1. Brake hp: 145 at 3600 rpm. Torque: 245 lbs.-ft. at 2200 rpm. Five main bearings. Hydraulic valve lifters. Induction: Electronic fuel injection. Serial number code: F.

Capri base V-8
90-degree, overhead valve V-8. Cast-iron block and head. Displacement: 302 cid (5.0 liters). Bore & stroke: 4.00 x 3.00 in. Compression ratio: 8.3:1. Brake hp: 175 at 4000 rpm. Torque: 245 lbs.-ft. at 2400 rpm. Five main bearings. Hydraulic valve lifters. Carburetor: Holley four-barrel Model 4180. Serial number code: F.

Grand Marquis police H.O. V-8
90-degree, overhead valve V-8. Cast-iron block and head. Displacement: 351 cid (5.8 liters). Bore & stroke: 4.00 x 3.50 in. Compression ratio: 8.3:1. Brake hp: 165 at 3600 rpm. Torque: 290 lbs.-ft. at 2200 rpm. Five main bearings. Hydraulic valve lifters. Carburetor: Two-barrel VV. Serial number code: G.

1984 MERCURY

Cougar, Grand Marquis base V-8
90-degree, overhead valve V-8. Cast-iron block and head. Displacement: 302 cid (5.0 liters). Bore & stroke: 4.00 x 3.00 in. Compression ratio: 8.4:1. Brake hp: 140 at 3200 rpm. Torque: 250 lbs.-ft. at 1600 rpm. Five main bearings. Hydraulic valve lifters. Induction: Electronic fuel injection (TBI). Serial number code: F.

Grand Marquis optional V-8
90-degree, overhead valve V-8. Cast-iron block and head. Displacement: 302 cid (5.0 liters). Bore & stroke: 4.00 x 3.00 in. Compression ratio: 8.4:1. Brake hp: 155 at 3600 rpm. Torque: 265 lbs.-ft. at 2000 rpm. Five main bearings. Hydraulic valve lifters. Induction: Electronic fuel injection (TBI). Serial number code: F.

Capri base V-8
90-degree, overhead valve V-8. Cast-iron block and head. Displacement: 302 cid (5.0 liters). Bore & stroke: 4.00 x 3.00 in. Compression ratio: 8.3:1. Brake hp: 175 at 4000 rpm. Torque: 245 lbs.-ft. at 2400 rpm. Five main bearings. Hydraulic valve lifters. Carburetor: Holley four-barrel Model 4180C. Serial number code: M.
Note: A high-output version of the Capri V-8 rated at 205 hp was announced but delayed.

Grand Marquis optional police H.O. V-8
90-degree, overhead valve V-8. Cast-iron block and head. Displacement: 351 cid (5.8 liters). Bore & stroke: 4.00 x 3.50 in. Compression ratio: 8.3:1. Brake hp: 180 at 3600 rpm. Torque: 285 lbs.-ft. at 2400 rpm. Five main bearings. Hydraulic valve lifters. Carburetor: Two-barrel VV. Serial number code: G.

1985 MERCURY

Cougar, Grand Marquis base V-8
90-degree, overhead valve V-8. Cast-iron block and head. Displacement: 302 cid (5.0 liters). Bore & stroke: 4.00 x 3.00 in. Compression ratio: 8.4:1. Brake hp: 140 at 3200 rpm. Torque: 250 lbs.-ft. at 1600 rpm. Five main bearings. Hydraulic valve lifters. Induction: Electronic fuel injection (TBI). Serial number code: F.

Grand Marquis optional H.O. V-8
90-degree, overhead valve V-8. Cast-iron block and head. Displacement: 302 cid (5.0 liters). Bore & stroke: 4.00 x 3.00 in. Compression ratio: 8.4:1. Brake hp: 155 at 3600 rpm. Torque: 265 lbs.-ft. at 2000 rpm. Five main bearings. Hydraulic valve lifters. Induction: Electronic fuel injection (TBI). Serial number code: F.

Capri base V-8
90-degree, overhead valve V-8. Cast-iron block and head. Displacement: 302 cid (5.0 liters). Bore & stroke: 4.00 x 3.00 in. Compression ratio: 8.3:1. Brake hp: 180 at 4200 rpm. Torque: 260 lbs.-ft. at 2600 rpm. Five main bearings. Hydraulic valve lifters. Induction: Electronic fuel injection. Serial number code: F.

Capri optional V-8
90-degree, overhead valve V-8. Cast-iron block and head. Displacement: 302 cid (5.0 liters). Bore & stroke: 4.00 x 3.00 in. Compression ratio: 8.3:1. Brake hp: 210 at 4400 rpm. Torque: 270 lbs.-ft. at 3200 rpm. Five main bearings. Hydraulic valve lifters. Carburetor: Holley four-barrel Model 4180C. Serial number code: M.

Grand Marquis police H.O. V-8
90-degree, overhead valve V-8. Cast-iron block and head. Displacement: 351 cid (5.8 liters). Bore & stroke: 4.00 x 3.50 in. Compression ratio: 8.3:1. Brake hp: 180 at 3600 rpm. Torque: 285 lbs.-ft. at 2400 rpm. Five main bearings. Hydraulic valve lifters. Carburetor: Motorcraft 7200VV two-barrel. Serial number code: G.

1986 MERCURY

Cougar, Grand Marquis base V-8
90-degree, overhead valve V-8. Cast-iron block and head. Displacement: 302 cid (5.0 liters). Bore & stroke: 4.00 x 3.00 in. Compression ratio: 8.9:1. Brake hp: 150 at 3200 rpm. Torque: 270 lbs.-ft. at 2000 rpm. Five main bearings. Hydraulic valve lifters. Induction: Sequential (port) fuel injection. Serial number code: F.

Grand Marquis optional V-8
90-degree, overhead valve V-8. Cast-iron block and head. Displacement: 302 cid (5.0 liters). Bore & stroke: 4.00 x 3.00 in. Compression ratio: 8.9:1. Brake hp: 160. Five main bearings. Hydraulic valve lifters. Induction: Sequential (port) fuel injection. Serial number code: F.

Capri base V-8
90-degree, overhead valve V-8. Cast-iron block and head. Displacement: 302 cid (5.0 liters). Bore & stroke: 4.00 x 3.00 in. Compression ratio: 9.2:1. Brake hp: 200 at 4400 rpm. Torque: 285 lbs.-ft. at 3000 rpm. Five main bearings. Hydraulic valve lifters. Induction: Electronic fuel injection. Serial number code: M.

Grand Marquis police H.O. V-8
90-degree, overhead valve V-8. Cast-iron block and head. Displacement: 351 cid (5.8 liters). Bore & stroke: 4.00 x 3.50 in. Compression ratio: 8.3:1. Brake hp: 180 at 3600 rpm. Torque: 285 lbs.-ft. at 2400 rpm. Five main bearings. Hydraulic valve lifters. Carburetor: Two-barrel Serial number code: G.

1987 MERCURY

Cougar, Grand Marquis base V-8
90-degree, overhead valve V-8. Cast-iron block and head. Displacement: 302 cid (5.0 liters). Bore & stroke: 4.00 x 3.00 in. Compression ratio: 8.9:1. Brake hp: 150 at 3200 rpm. Torque: 270 lbs.-ft. at 2000 rpm. Five main bearings. Hydraulic valve lifters. Induction: Sequential (port) fuel injection. Serial number code: F.

Grand Marquis optional V-8
90-degree, overhead valve V-8. Cast-iron block and head. Displacement: 302 cid (5.0 liters). Bore & stroke: 4.00 x 3.00 in. Compression ratio: 8.9:1. Brake hp: 160 at 3400 rpm. Torque: 265 lbs.-ft. at 2200 rpm. Five main bearings. Hydraulic valve lifters. Induction: Sequential (port) fuel injection. Serial number code: F.

Grand Marquis police H.O. V-8
90-degree, overhead valve V-8. Cast-iron block and head. Displacement: 351 cid (5.8 liters). Bore & stroke: 4.00 x 3.50 in. Compression ratio: 8.3:1. Brake hp: 180 at 3600 rpm. Torque: 285 lbs.-ft. at 2400 rpm. Five main bearings. Hydraulic valve lifters. Carburetor: Two-barrel Serial number code: G.

1988 MERCURY

Grand Marquis base V-8
90-degree, overhead valve V-8. Cast-iron block and head. Displacement: 302 cid (5.0 liters). Bore & stroke: 4.00 x 3.00 in. Compression ratio: 8.9:1. Brake hp: 150 at 3200 rpm. Torque: 270 lbs.-ft. at 2000 rpm. Five main bearings. Hydraulic valve lifters. Induction: Sequential (port) fuel injection. Serial number code: F.

Cougar base V-8
90-degree, overhead valve V-8. Cast-iron block and head. Displacement: 302 cid (5.0 liters). Bore & stroke: 4.00 x 3.00 in. Compression ratio: 8.9:1. Brake hp: 155 at 3400 rpm. Torque: 265 lbs.-ft. at 2200 rpm. Five main bearings. Hydraulic valve lifters. Induction: Sequential (port) fuel injection. Serial number code: F.

Grand Marquis optional V-8
90-degree, overhead valve V-8. Cast-iron block and head. Displacement: 302 cid (5.0 liters). Bore & stroke: 4.00 x 3.00 in. Compression ratio: 8.9:1. Brake hp: 160 at 3400 rpm. Torque: 265 lbs.-ft. at 2200 rpm. Five main bearings. Hydraulic valve lifters. Induction: Sequential (port) fuel injection. Serial number code: F.

1989 MERCURY

Grand Marquis base V-8
90-degree, overhead valve V-8. Cast-iron block and head. Displacement: 302 cid (5.0 liters). Bore & stroke: 4.00 x 3.00 in. Compression ratio: 8.9:1. Brake hp: 150 at 3200 rpm. Torque: 270 lbs.-ft. at 2000 rpm. Five main bearings. Hydraulic valve lifters. Induction: Sequential (port) fuel injection. Serial number code: F.

Grand Marquis optional V-8
90-degree, overhead valve V-8. Cast-iron block and head. Displacement: 302 cid (5.0 liters). Bore & stroke: 4.00 x 3.00 in. Compression ratio: 8.9:1. Brake hp: 160 at 3400 rpm. Torque: 265 lbs.-ft. at 2200 rpm. Five main bearings. Hydraulic valve lifters. Induction: Sequential (port) fuel injection. Serial number code: F.

1990 MERCURY

Grand Marquis base V-8
90-degree, overhead valve V-8. Cast-iron block and head. Displacement: 302 cid (5.0 liters). Bore & stroke: 4.00 x 3.00 in. Compression ratio: 8.9:1. Brake hp: 150 at 3200 rpm. Torque: 270 lbs.-ft. at 2000 rpm. Five main bearings. Hydraulic valve lifters. Induction: Sequential (port) fuel injection. Serial number code: F.

Grand Marquis optional V-8
90-degree, overhead valve V-8. Cast-iron block and head. Displacement: 302 cid (5.0 liters). Bore & stroke: 4.00 x 3.00 in. Compression ratio: 8.9:1. Brake hp: 160 at 3400 rpm. Torque: 265 lbs.-ft. at 2200 rpm. Five main bearings. Hydraulic valve lifters. Induction: Sequential (port) fuel injection. Serial number code: F.

1991 MERCURY

Grand Marquis base V-8
90-degree, overhead valve V-8. Cast-iron block and head. Displacement: 302 cid (5.0 liters). Bore & stroke: 4.00 x 3.00 in. Compression ratio: 8.9:1. Brake hp: 150 at 3200 rpm. Torque: 270 lbs.-ft. at 2000 rpm. Five main bearings. Hydraulic valve lifters. Induction: Sequential (port) fuel injection. Serial number code: F.

Grand Marquis optional V-8
90-degree, overhead valve V-8. Cast-iron block and head. Displacement: 302 cid (5.0 liters). Bore & stroke: 4.00 x 3.00 in. Compression ratio: 8.9:1. Brake hp: 160 at 3400 rpm. Torque: 265 lbs.-ft. at 2200 rpm. Five main bearings. Hydraulic valve lifters. Induction: Sequential (port) fuel injection. Serial number code: F.

Cougar LS base V-8
90-degree, overhead valve V-8. Cast-iron block and head. Displacement: 302 cid (5.0 liters). Bore & stroke: 4.00 x 3.00 in. Compression ratio: 9.0:1. Brake hp: 200 at 4000 rpm. Torque: 275 lbs.-ft. at 3000 rpm. Five main bearings. Hydraulic valve lifters. Induction: Sequential fuel injection. Serial number code: E.

1992 MERCURY

Grand Marquis base V-8
Modular, overhead valve V-8. Single overhead cam. Displacement: 281 cid (4.6 liters). Bore & stroke: 3.60 x 3.60 in. Compression ratio: 9.0:1. Brake hp: 190 at 4200 rpm. Torque: 260 lbs.-ft. at 3200 rpm. Hydraulic valve lifters. Induction: Sequential fuel injection. Serial number code: 6.

Cougar XR-7 base V-8; Cougar LS and 25th Anniversary Edition optional V-8
90-degree, overhead valve V-8. Cast-iron block and head. Displacement: 302 cid (5.0 liters). Bore & stroke: 4.00 x 3.00 in. Compression ratio: 9.0:1. Brake hp: 200 at 4000 rpm. Torque: 275 lbs.-ft. at 3000 rpm. Five main bearings. Hydraulic valve lifters. Induction: Sequential fuel injection. Serial number code: E.

Grand Marquis optional V-8
Modular, overhead valve V-8. Single overhead cam. Displacement: 281 cid (4.6 liters). Bore & stroke: 3.60 x 3.60 in. Compression ratio: 9.0:1. Brake hp: 210 at 4600 rpm. Torque: 270 lbs.-ft. at 3400 rpm. Hydraulic valve lifters. Dual exhaust. Induction: Sequential fuel injection. Serial number code: 6.

1993 MERCURY

Grand Marquis base V-8
Modular, overhead valve V-8. Single overhead cam. Displacement: 281 cid (4.6 liters). Bore & stroke: 3.60 x 3.60 in. Compression ratio: 9.0:1. Brake hp: 190 at 4200 rpm. Torque: 260 lbs.-ft. at 3200 rpm. Hydraulic valve lifters. Induction: Sequential fuel injection. Serial number code: 6.

Cougar XR-7 optional V-8
90-degree, overhead valve V-8. Cast-iron block and head. Displacement: 302 cid (5.0 liters). Bore & stroke: 4.00 x 3.00 in. Compression ratio: 9.0:1. Brake hp: 200 at 4000 rpm. Torque: 275 lbs.-ft. at 3000 rpm. Five main bearings. Hydraulic valve lifters. Induction: Sequential fuel injection. Serial number code: E.

Grand Marquis optional V-8
Modular, overhead valve V-8. Single overhead cam. Displacement: 281 cid (4.6 liters). Bore & stroke: 3.60 x 3.60 in. Compression ratio: 9.0:1. Brake hp: 210 at 4600 rpm. Torque: 270 lbs.-ft. at 3400 rpm. Hydraulic valve lifters. Dual exhaust. Induction: Sequential fuel injection. Serial number code: 6.

1994 MERCURY

Grand Marquis base V-8
Modular, overhead valve V-8. Single overhead cam. Displacement: 281 cid (4.6 liters). Bore & stroke: 3.60 x 3.60 in. Compression ratio: 9.0:1. Brake hp: 190 at 4200 rpm. Torque: 260 lbs.-ft. at 3200 rpm. Hydraulic valve lifters. Induction: Sequential fuel injection. Serial number code: 6.

Cougar XR7 optional V-8
Modular, overhead valve V-8. Single overhead cam. Displacement: 281 cid (4.6 liters). Bore & stroke: 3.60 x 3.60 in. Compression ratio: 9.0:1. Brake hp: 205 at 4500 rpm. Torque: 265 lbs.-ft. at 3200 rpm. Hydraulic

valve lifters. Induction: Sequential fuel injection. Serial number code: W.

Grand Marquis optional V-8
Modular, overhead valve V-8. Single overhead cam. Displacement: 281 cid (4.6 liters). Bore & stroke: 3.60 x 3.60 in. Compression ratio: 9.0:1. Brake hp: 210 at 4600 rpm. Torque: 270 lbs.-ft. at 3400 rpm. Hydraulic valve lifters. Dual exhaust. Induction: Sequential fuel injection. Serial number code: 6.

1995 MERCURY

Grand Marquis base V-8
Modular, overhead valve V-8. Single overhead cam. Displacement: 281 cid (4.6 liters). Bore & stroke: 3.60 x 3.60 in. Compression ratio: 9.0:1. Brake hp: 190 at 4200 rpm. Torque: 260 lbs.-ft. at 3200 rpm. Hydraulic valve lifters. Induction: Sequential fuel injection. Serial number code: 6.

Cougar XR7 optional V-8
Modular, overhead valve V-8. Single overhead cam. Displacement: 281 cid (4.6 liters). Bore & stroke: 3.60 x 3.60 in. Compression ratio: 9.0:1. Brake hp: 205 at 4500 rpm. Torque: 265 lbs.-ft. at 3200 rpm. Hydraulic valve lifters. Induction: Sequential fuel injection. Serial number code: W.

Grand Marquis optional V-8
Modular, overhead valve V-8. Single overhead cam. Displacement: 281 cid (4.6 liters). Bore & stroke: 3.60 x 3.60 in. Compression ratio: 9.0:1. Brake hp: 210 at 4600 rpm. Torque: 270 lbs.-ft. at 3400 rpm. Hydraulic valve lifters. Dual exhaust. Induction: Sequential fuel injection. Serial number code: 6.

1996 MERCURY

Grand Marquis base V-8
Modular, overhead valve V-8. Single overhead cam. Displacement: 281 cid (4.6 liters). Bore & stroke: 3.60 x 3.60 in. Compression ratio: 9.0:1. Brake hp: 190 at 4200 rpm. Torque: 260 lbs.-ft. at 3200 rpm. Hydraulic valve lifters. Induction: Sequential fuel injection. Serial number code: 6.

Cougar XR7 optional V-8
Modular, overhead valve V-8. Single overhead cam. Displacement: 281 cid (4.6 liters). Bore & stroke: 3.60 x 3.60 in. Compression ratio: 9.0:1. Brake hp: 205 at 4500 rpm. Torque: 265 lbs.-ft. at 3200 rpm. Hydraulic valve lifters. Induction: Sequential fuel injection. Serial number code: W.

Grand Marquis optional V-8
Modular, overhead valve V-8. Single overhead cam. Displacement: 281 cid (4.6 liters). Bore & stroke: 3.60 x 3.60 in. Compression ratio: 9.0:1. Brake hp: 210 at 4600 rpm. Torque: 270 lbs.-ft. at 3400 rpm. Hydraulic valve lifters. Dual exhaust. Induction: Sequential fuel injection. Serial number code: 6.

1997 MERCURY

Grand Marquis base V-8
Modular, overhead valve V-8. Single overhead cam. Displacement: 281 cid (4.6 liters). Bore & stroke: 3.60 x 3.60 in. Compression ratio: 9.0:1. Brake hp: 190 at 4200 rpm. Torque: 260 lbs.-ft. at 3200 rpm. Hydraulic valve lifters. Induction: Sequential fuel injection. Serial number code: 6.

Cougar XR7 optional V-8
Modular, overhead valve V-8. Single overhead cam. Displacement: 281 cid (4.6 liters). Bore & stroke: 3.60 x 3.60 in. Compression ratio: 9.0:1. Brake hp: 205 at 4500 rpm. Torque: 265 lbs.-ft. at 3200 rpm. Hydraulic valve lifters. Induction: Sequential fuel injection. Serial number code: W.

Grand Marquis optional V-8
Modular, overhead valve V-8. Single overhead cam. Displacement: 281 cid (4.6 liters). Bore & stroke: 3.60 x 3.60 in. Compression ratio: 9.0:1. Brake hp: 210 at 4600 rpm. Torque: 270 lbs.-ft. at 3400 rpm. Hydraulic valve lifters. Dual exhaust. Induction: Sequential fuel injection. Serial number code: 6.

1998 MERCURY

Grand Marquis base V-8
Modular, overhead valve V-8. Single overhead cam. Displacement: 281 cid (4.6 liters). Bore & stroke: 3.60 x 3.60 in. Compression ratio: 9.0:1. Brake hp: 200 at 4250 rpm. Torque: 265 lbs.-ft. at 3000 rpm. Hydraulic valve lifters. Induction: Sequential fuel injection. Serial number code: W.

Grand Marquis optional "Handling Package" V-8
Modular, overhead valve V-8. Single overhead cam. Displacement: 281 cid (4.6 liters). Bore & stroke: 3.60 x 3.60 in. Compression ratio: 9.0:1. Brake hp: 215 at 4500 rpm. Torque: 275 lbs.-ft. at 3000 rpm. Hydraulic valve lifters. Quasi-dual exhaust. Induction: Sequential fuel injection. Serial number code: W.

1999 MERCURY

Grand Marquis base V-8
Modular, overhead valve V-8. Single overhead cam. Displacement: 281 cid (4.6 liters). Bore & stroke: 3.60 x 3.60 in. Compression ratio: 9.0:1. Brake hp: 200 at 4250 rpm. Torque: 275 lbs.-ft. at 3000 rpm. Hydraulic valve lifters. Induction: Sequential fuel injection. Serial number code: W.

Grand Marquis optional "Handling Package" V-8
Modular, overhead valve V-8. Single overhead cam. Displacement: 281 cid (4.6 liters). Bore & stroke: 3.60 x 3.60 in. Compression ratio: 9.0:1. Brake hp: 215 at 4500 rpm. Torque: 285 lbs.-ft. at 3000 rpm. Hydraulic valve lifters. Quasi-dual exhaust. Induction: Sequential fuel injection. Serial number code: W.

2000 MERCURY

Grand Marquis GS/LS base V-8
Modular, overhead valve V-8. Single overhead cam. Displacement: 281 cid (4.6 liters). Bore & stroke: 3.60 x 3.60 in. Compression ratio: 9.0:1. Brake hp: 200 at 4250 rpm. Torque: 275 lbs.-ft. at 3000 rpm. Hydraulic valve lifters. Induction: Sequential fuel injection. Serial number code: W.

Grand Marquis optional "Handling Package" V-8
Modular, overhead valve V-8. Single overhead cam. Displacement: 281 cid (4.6 liters). Bore & stroke: 3.60 x 3.60 in. Compression ratio: 9.0:1. Brake hp: 215 at 4500 rpm. Torque: 285 lbs.-ft. at 3000 rpm. Hydraulic valve lifters. Quasi-dual exhaust. Induction: Sequential fuel injection. Serial number code: W.

2001 MERCURY

Grand Marquis GS/LS base V-8
Modular, overhead valve V-8. Single overhead cam. Displacement: 281 cid (4.6 liters). Bore & stroke: 3.60 x 3.60 in. Compression ratio: 9.0:1. Brake hp: 220 at 4750 rpm. Torque: 265 lbs.-ft. at 4000 rpm. Hydraulic valve lifters. Induction: Sequential fuel injection. Serial number code: W.

Grand Marquis LSE or Grand Marquis LS/GS with optional "Handling Package" V-8
Modular, overhead valve V-8. Single overhead cam. Displacement: 281 cid (4.6 liters). Bore & stroke: 3.60 x 3.60 in. Compression ratio: 9.0:1. Brake hp: 235 at 4750 rpm. Torque: 275 lbs.-ft. at 4000 rpm. Hydraulic valve lifters. Quasi-dual exhaust. Induction: Sequential fuel injection. Serial number code: W.

2002 MERCURY

Grand Marquis GS/LS base V-8
Modular, overhead valve V-8. Single overhead cam. Displacement: 281 cid (4.6 liters). Bore & stroke: 3.60 x 3.60 in. Compression ratio: 9.0:1. Brake hp: 220 at 4750 rpm. Torque: 265 lbs.-ft. at 4000 rpm. Hydraulic valve lifters. Induction: Sequential fuel injection. Serial number code: W.

Grand Marquis LSE or Grand Marquis LS/GS with optional "Handling Package" V-8
Modular, overhead valve V-8. Single overhead cam. Displacement: 281 cid (4.6 liters). Bore & stroke: 3.60 x 3.60 in. Compression ratio: 9.0:1. Brake hp: 235 at 4750 rpm. Torque: 275 lbs.-ft. at 4000 rpm. Hydraulic valve lifters. Quasi-dual exhaust. Induction: Sequential fuel injection. Serial number code: W.

Selected Mercury V-8 Tune-Up Specs 1939-1965

Year	Model	Spark Plugs Type	Spark Plugs Gap	Point Gap	Deg Cam Angle	Timing	Idle Speed	Compression Pressure	Battery (Delco)
1939	Mercury V-8	H-10	.025	.016	36	4B	425 rpm	105-110 lb.	2F
1940	Model 09A	CH-H10	.025	.016	36	4B	425 rpm	105-110 lb.	2F
1941	Model 19A	CH-H10	.025	.015	36	4B	425 rpm	105-110 lb.	2F
1942	Model 29A	CH-H10	.025	.015	36	4B	425 rpm	105-110 lb.	2F
1946	Model 69M	CH-H10	.025	.015	36	4B	425 rpm	105-125 lb.	2F
1947	Model 79M	CH-H10	.025	.015	36	4B	425 rpm	105-125 lb.	2L
1948	Model 89M	CH-H10	.025	.015	36	4B	425 rpm	105-125 lb.	2F
1949	Mercury V-8	CH-H10	.030	.014-.016	28	2B	450 rpm	115 lb.	2L
1950	Mercury V-8	CH-H10	.030	.014-.016	27	2B	450 rpm	115 lb.	2L
1951	Mercury V-8	CH-H10	.030	.014-.016	27	2B	450 rpm	115 lb.	2L
1952	Mercury V-8	CH-H10	.030	.014-.016	27	2B	450 rpm	115 lb.	2L
1953	Mercury V-8	CH-H10	.030	.014-.016	27	2B	450 rpm	115 lb.	2L
1954	Mercury V-8	CH-H10	.030	.014-.016	27	3B	485 rpm	115 lb.	2HN/2N
1955	Mercury V-8	CH-870	.034	.014-.016	27	3B	485 rpm	135 lb.	2HN/2N
1956	Mercury V-8	CH-870	.034	.014-.016	27	5B	485 rpm	155 lb.	2HN/2N
1957	312-cid V-8	CH-870	.034	.014-.016	27	3B/6B	485 rpm	165 lb.	13HN/4HN/4SN
	368-cid V-8	CH-870	.034	.014-.016	27	5B	485 rpm	160 lb.	13HN/4HN/4SN

Note 1: Timing for 312-cid V-8 with green valve covers is 3 degrees before top dead center (3B) with manual transmission and 6 degrees before top dead center (6B) with automatic transmission.

Year	Model	Spark Plugs Type	Spark Plugs Gap	Point Gap	Deg Cam Angle	Timing	Idle Speed	Compression Pressure	Battery (Delco)
1958	312-cid V-8	CH-F11Y	.032-.036	.014-.016	26-28.5	3B/6B	Note 2	160 lb.	55 amp. hr.
	383-cid V-8	CH-F11Y	.032-.036	.014-.016	26-28.5	3B/6B	Note 2	160 lb.	65 amp. hr.
	430-cid V-8	CH-F11Y	.032-.036	.014-.016	26-28.5	3B/6B	Note 2	180 lb.	65 amp. hr.

Note 1: Timing is 3 degrees before top dead center (3B) with manual transmission and 6 degrees before top dead center (6B) with automatic transmission.
Note 2: Recommended idle setting is 475-500 rpm with standard transmission or 425-450 rpm with automatic transmission in drive.

Year	Model	Spark Plugs Type	Spark Plugs Gap	Point Gap	Deg Cam Angle	Timing	Idle Speed	Compression Pressure	Battery (Delco)
1959	312-cid V-8	FEH-12405A	.032	.014-.016	26-28.5	3B/6B	Note 2	160 lb.	55 amp. hr.
	383-cid V-8	FEV-12405A	.032	.014-.016	26-28.5	3B/6B	Note 2	160 lb.	65 amp. hr.
	430-cid V-8	FEV-12405A	.032	.014-.016	26-28.5	3B/6B	Note 2	180 lb.	65 amp. hr.

Note 1: Timing is 3 degrees before top dead center (3B) with manual transmission and 6 degrees before top dead center (6B) with automatic transmission.
Note 2: Recommended idle setting is 475-500 rpm with standard transmission or 425-450 rpm with automatic transmission in drive.

Year	Model	Spark Plugs Type	Spark Plugs Gap	Point Gap	Deg Cam Angle	Timing	Idle Speed	Compression Pressure	Battery (Delco)
1960	312-cid V-8	CH-F-14Y	.032-.036	.014-.016	26-28.5	3B/6B	Note 2	160 lb.	B9A-10654-D
	383-cid V-8	CH-F-11Y	.032-.036	.014-.016	26-28.5	8B	Note 2	160 lb.	B9A-10654-F
	430-cid V-8	CH-F-11Y	.032-.036	.014-.016	26-28.5	8B	Note 2	180 lb.	B9A-10654-F

Note 1: Timing is 3 degrees before top dead center (3B) with manual transmission and 6 degrees before top dead center (6B) with automatic transmission.
Note 2: Recommended idle setting is 500-525 rpm with standard transmission in neutral or 470-475 rpm with automatic transmission in drive.

Year	Model	Spark Plugs Type	Spark Plugs Gap	Point Gap	Deg Cam Angle	Timing	Idle Speed	Compression Pressure	Battery (Delco)
1961	292-cid V-8	CH-F-14Y	.032-.036	.014-.016	26-28.5	5B/12B	Note 2	160 lb.	B9A-10654-D
	352-cid V-8	CH-F-11Y	.032-.036	.014-.016	26-28.5	5B/8B	Note 2	180 lb.	B9A-10654-F
	390-cid V-8	CH-F-11Y	.032-.036	.014-.016	26-28.5	5B/8B	Note 2	180 lb.	B9A-10654-F

Note 1: Timing specification before slash is for manual transmission/specification after slash is for automatic transmission. Set timing to 14 degrees before top dead center (14B) for 390-cid V-8 with 375 or 401 hp.
Note 2: Recommended idle setting is 500-525 rpm with standard transmission in neutral or 470-475 rpm with automatic transmission in drive.

Year	Model	Spark Plugs Type	Spark Plugs Gap	Point Gap	Deg Cam Angle	Timing	Idle Speed	Compression Pressure	Battery (Delco)
1962	221-cid V-8	AL BF-82	.035	.014-.016	26-28.5	4B/12B	Note 2	150 lb.	C2DF-10655A
	260-cid V-8	AL BF-82	.035	.014-.016	26-28.5	4B/12B	Note 2	150 lb.	C2DF-10655A
	292-cid V-8	AL BF-82	.035	.014-.016	26-28.5	5B/12B	Note 2	160 lb.	C2AF-10655A
	352-cid V-8	AL BF-42	.035	.014-.016	26-28.5	5B/8B	Note 2	180 lb.	C25F-10655A
	390-cid V-8	AL BF-42	.035	.014-.016	26-28.5	8B	Note 2	180 lb.	C25F-10655A

	406-cid V-8 AL BF-42	.035	.014-.016	26-28.5	8B	Note 2	180 lb.	C25F-10655A

Note 1: Timing specification before slash is for manual transmission/specification after slash is for automatic transmission. Set timing to 5B for 390-cid V-8 with 375 hp or 401 hp. Set timing to 8B for 406-cid V-8 with 385 hp or 405 hp.
Note 2: Recommended idle setting with manual transmission is 400-525 rpm for 221-cid V-8; 500-525 rpm for 260-, 292-, and 352-cid V-8; and 575-600 rpm for 390-cid V-8. Recommended idle setting with automatic transmission is 400-500 rpm for 221-cid V-8; 450-475 rpm for 292- and 352-cid V-8; 475-500 rpm for 260- or 390-cid V-8; and 550-575 rpm for "police" V-8.

1963	221-cid V-8 AL BF-82	.034	.017	26-31	4B/12B	Note 2	150 lb.	C2DF-10655A
	260-cid V-8 AL BF-82	.034	.017	26-31	6B/10B	Note 2	150 lb.	C2DF-10655A
	289-cid V-8 AL BF-82	.034	.017	26-31	6B/10B	Note 2	150 lb.	C2DF-10655A
	352-cid V-8 AL BF-42	.034	.017	26-31	5B/8B	Note 2	180 lb.	C25F-10655A
	390-cid V-8 AL BF-42	.034	.017	26-31	5B/8B	Note 2	180 lb.	C25F-10655A
	406-cid V-8 AL BF-42	.034	.017	26-31	5B	Note 2	180 lb.	C25F-10655A
	427-cid V-8 AL BF-42	.034	.017	26-31	5B	Note 2	180 lb.	C25F-10655A

Note 1: Timing specification before slash is for manual transmission/specification after slash is for automatic transmission. Set timing to 5B for 390-cid V-8 with two-barrel carburetor. Set timing to 10B for 390-cid V-8 with 330 hp. Set timing to 8B for 406- or 427-cid V-8 with multiple carburetors.
Note 2: Recommended idle setting with manual transmission is 400-525 rpm for 221-cid V-8; 500-525 rpm for 260-, 289-, and 352-cid V-8; and 575-600 rpm for 390-cid V-8. Recommended idle setting with automatic transmission is 400-500 rpm for 221-cid V-8; 450-475 rpm for 289- and 352-cid V-8; 475-500 rpm for 260- or 390-cid V-8; and 550-575 rpm for "police" V-8.

1964	260-cid V-8 AL BF-42	.034	.015	27	6B/10B	Note 2	150 lb.	C2DF-10655A
	289-cid V-8 AL BF-42 (3)	.034	.015	27	6B/10B	Note 2	150 lb.	C2DF-10655A
	390-cid V-8 AL BF-42	.035	.015	27	6B	Note 2	180 lb.	C25F-10655A
	390-cid V-8 Hi Po AL BF-32	.035	.020	27	10B	Note 2	180 lb.	C25F-10655A
	406-cid V-8 AL BF-32	.035	.020	27	8B	Note 2	180 lb.	C25F-10655A
	427-cid V-8 AL BF-32	.035	.020	35	8B	Note 2	180 lb.	C25F-10655A

Note 1: Timing specification before slash is for manual transmission/specification after slash is for automatic transmission.
Note 2: Recommended idle setting with manual transmission is 575 rpm for 260- and 289-cid V-8s, 500 rpm for 390-cid V-8, and 675 rpm for 406- and 427-cid V-8s. Recommended idle setting with automatic transmission is 500 rpm for 260- and 289-cid V-8s; 500 for 390-cid V-8 and 675 rpm for 406- and 427-cid V-8s.
Note 3: High-performance 289-cid V-8 uses Autolite BF-32 spark plugs.

1965	260-cid V-8 AL BF-42	.034	.015	27	6B/10B	Note 2	150 lb.	C2DF-10655A
	289-cid V-8 AL BF-42 (3)	.034	.015	27	6B/10B	Note 2	150 lb.	C2DF-10655A
	390-cid 2V V-8 AL BF-42	.035	.015	27	4B	Note 2	180 lb.	C25F-10655A
	390-cid 4V V-8 AL BF-32	.035	.020	27	6B	Note 2	180 lb.	C25F-10655A
	427-cid V-8 AL BF-32	.035	.020	35	8B	Note 2	180 lb.	C25F-10655A

Note 1: Timing specification before slash is for manual transmission/specification after slash is for automatic transmission.
Note 2: Recommended idle setting with manual transmission is 575 rpm fcr 260- and 289-cid V-8s, 575 rpm for 390-cid V-8, and 800 rpm for 427-cid V-8. Recommended idle setting with automatic transmission is 500 rpm for 260-, 289-cid V-8s; 500 rpm for 390-cid V-8, 575 rpm for high-performance 390-cid V-8, and 800 rpm for 427-cid V-8.
Note 3: High-performance 289-cid V-8 uses Autolite BF-32 spark plugs.

Mercury V-8 Firing Order

1940-1948 (All Flathead V-8s)
1-5-4-8-6-3-7-2 with RH bank numbered 1-2-3-4 and LH bank numbered 5-6-7-8 front to rear

(Same as 1R-1L-4R-4L-2L-3R-3L-2R)

Note: For the double-ended distributor spark plug wires go into the distributor cap in firing order in clockwise direction 1-2-3-4 on right side starting in the 11 o'clock position and 8-7-6-5 on the left side starting in the 11 o'clock position. For the "crab" type distributor (all wires on front) the holes are numbered to correspond to the appropriate cylinder.

1949-1953 (All Flathead V-8s)
1-5-4-8-6-3-7-2 with RH bank numbered 1-2-3-4 and LH bank numbered 5-6-7-8 front to rear

(Same as 1R-1L-4R-4L-2L-3R-3L-2R)

Note: The spark plug wires go in the distributor cab of flathead V-8s in the firing order in a clockwise direction.

1954-1957 (Overhead valve V-8s)
1-5-4-8-6-3-7-2 with RH bank numbered 1-2-3-4 and LH bank numbered 5-6-7-8 front to rear

(Same as 1R-1L-4R-4L-2L-3R-3L-2R)

Note: Overhead valve V-8 plug wires go into the distributor cap in firing order in counter-clockwise direction.

1958-1960 (Overhead valve V-8s)
312-cid: 1-5-4-8-6-3-7-2 with RH bank numbered 1-2-3-4 and LH bank numbered 5-6-7-8 front to rear

Others: 1-5-4-2-6-3-7-8 with RH bank numbered 1-2-3-4 and LH bank numbered 5-6-7-8 front to rear

1961-1962 (Overhead valve V-8s)
292-cid: 1-5-4-8-6-3-7-2 with RH bank numbered 1-2-3-4 and LH bank numbered 5-6-7-8 front to rear

Others: 1-5-4-2-6-3-7-8 with RH bank numbered 1-2-3-4 and LH bank numbered 5-6-7-8 front to rear

1963-1969 (Overhead valve V-8s)
All V-8s: 1-5-4-2-6-3-7-8 with RH bank numbered 1-2-3-4 and LH bank numbered 5-6-7-8 front to rear

1970-1976 (Overhead valve V-8s)
351-cid: 1-3-7-2-6-5-4-8 with RH bank numbered 1-2-3-4 and LH bank numbered 5-6-7-8 front to rear

Other V-8s: 1-5-4-2-6-3-7-8 with RH bank numbered 1-2-3-4 and LH bank numbered 5-6-7-8 front to rear

OLDSMOBILE
1916-2002

Oldsmobile V-8s can be grouped into five different "generations." The company produced its first V-8 engine—a L-head design—in the years 1916-1923. The year 1949 marked the debut of the famous "Rocket V-8." A major change in the design and engineering of Oldsmobile V-8s took place in 1964. The last two-valve-per-cylinder overhead valve Olds V-8 was seen in 1990. In 1995, the fifth generation of Oldsmobile V-8s began with the double-overhead-cam Aurora engine. Oldsmobile also built the 1929-1930 Viking V-8, which is covered separately in this catalog.

The term "Rocket V-8" can be applied to all 1949-1990 Olds V-8s, even though Olds officially stopped using it in the late '70s. When it was installed in the small Olds 88, this engine turned that model into the first "factory hot rod." Some even consider it the first muscle car, though others give this credit to the Chrysler 300 or Pontiac GTO.

Whatever you call it, the Rocket V-8 was hot. Bowing with 303 cid, the engine was based on the engineering principles of GM genius Charles Kettering. It was even built in a factory known as the "Kettering Engine Plant." Over the years, it was called by a variety of names like Rocket, Rocket 88, Golden Rocket, Sky Rocket, Super Rocket, 4-4-2 Rocket, Jetfire Rocket, and Starfire Rocket. It was the first real mass-produced overhead-valve V-8 and it started the long-lasting trend towards the use of such engines in American cars.

The Olds V-8 grew to 324 cid in 1954-1956, and 371 cid in 1957. The J-2 "Golden Rocket" option of 1957-1958 carried three two-barrel carburetors and generated 312 hp, but was not recommended for street use. In 1961, the 371 was bored out to 394 cid. Until 1964, all of the Rocket V-8 engines had a strong "family resemblance." The main differences were in compression ratios, carburetion, ignition, air cleaners, intake and exhaust manifolds, cylinder heads, and piston design.

Introduced for the F-85 and Jetstar models in 1964, a new 330-cid Olds V-8 was a blend of old and new. It traced its ancestry to the first Rocket V-8 in terms of dimensions like the distance between cylinder bores and the crankshaft-to-camshaft span, so that the old tooling could be re-used to save money. At the same time, the 330 was a more compact engine with entirely new cylinder head, block and manifold castings, new internal parts, and new valve gears. The options list was also expanded. For the big cars, the 394 remained one last year.

In 1965, the 330 was bored and stroked to create another brand-new engine—the "high-deck" 425. There was also a new 400-cid 345-hp Olds V-8 for the year-old 4-4-2 high-performance model. With muscle car fever in full swing, there were many different compression ratios, valve setups, and carburetion options available. The biggest ever Rocket V-8 was a bored-out 400 that arrived in 1968 with 455 cubic inches of displacement. It offered five horsepower ratings from 310 to 400 that year. The 330 was also bored out to create the Olds 350-cid V-8, which powered F-85s, Cutlasses, and Delmont 88s. It put up to 310 hp on tap.

The 400-cid V-8 disappeared in 1970, leaving behind only the 350 and the 455. An EGR (exhaust gas re-circulation) system was added to Oldsmobile engines in 1973 and the following season a High-Energy Ignition system was introduced as an option, only to become standard the following year. All Oldsmobile V-8s got a catalytic converter for 1975. Also introduced that year was an Olds-built 260-cid V-8 for the new Omega compact. A 400-cid V-8 returned as an option for the big Oldsmobiles, but was not the same one dropped in 1970. It was a Pontiac engine used only this one year. Similarly, a new 305-cid V-8 for the 1977 Omega was a Pontiac power plant.

Oldsmobile did build the 403-cid V-8 first used in 1977 Cutlass, Delta 88, and 98 models. The Toronado version of the same year was a bit special, with 200 net hp. "Corporate" engines, like the 301-cid Pontiac V-8 in the 1979 Delta 88 and the 265-cid Pontiac V-8 in the 1980 Delta 88, were an outgrowth of the '80s, but Oldsmobile Division did produce the 305-cid V-8 that appeared in many 1981 models. By 1986, this 307 was the only V-8 that Oldsmobile offered. The Hurst/Olds version put out 170 net hp. As mentioned earlier, the Olds-built 307 disappeared after 1990. After that, a Chevy 305-cid V-8 was available, but only in Custom Cruiser station wagons.

No Olds V-8s were marketed in 1993 or 1994. In 1995, the double overhead cam Aurora V-8 made its debut. This all-aluminum V-8 with sequential fuel injection produced 250 hp at 5600 rpm. It was the base V-8 in the Aurora until the end of the model's production history, which was also the end of Oldsmobile.

This publicity photo indicated that Oldsmobile engineers used 12 separate tests and more than 3,000,000 miles of actual driving to assure that the diesel 350 started each time it was used. The tests were said to simulate actual operational conditions from freezing winter days, to sizzling summer afternoons, to driving in rain or heavy snow.

Noting the runaway success of the Pontiac GTO with its 389-cid engine, Oldsmobile engineers saw the need to cram more cubes into their mid-size Cutlass and created the 4-4-2 option. Reducing the bore of the new 425-cid Oldsmobile engine from 4.125 inches to an even 4.0 produced a 400-cid V-8. Power rose to 345 hp at 4800 rpm and torque increased to 440 at 3200 rpm. The new engine had a 10.25:1 compression ratio and a Rochester four-barrel carburetor.

1916 OLDSMOBILE

Series 44 V-type
V-8. Cast-iron block. Cylinders cast in fours. Valves in V. Bore & stroke: 2-7/8 x 4-3/4 in. Displacement: 246.7 cid. Brake hp: 40 at 2000 rpm. NACC taxable hp: 26.5. Valve lifters: mechanical. Two crankshaft bearings. Pump cooling. Force feed lubrication. Helical camshaft drive. Delco single ignition. Carburetor: Tillotson gravity feed.
Note: Actually started production in 1915. In 1916 Miss Amada Preuss drove an Oldsmobile Model 44 V-8 roadster from San Francisco to New York via the Lincoln Highway in 11 days, five hours and 45 minutes to establish a new woman's transcontinental driving record.

1917 OLDSMOBILE

Model 45
V-8. Cast-iron block. Cylinders cast in fours. Valves in V. Bore & stroke: 2-7/8 x 4-3/4 in. Brake hp: 58 at 2510 rpm. Displacement: 246.7 cid. NACC taxable hp: 26.55. Valve lifters: mechanical. Two crankshaft bearings. Pump cooling. Force feed lubrication. Helical camshaft drive. Delco single ignition. Carburetor: Johnson vertical carburetor.

1918 OLDSMOBILE

Model 45-A
V-8. Cast-iron block. Cylinders cast in fours. Valves in V. Bore & stroke: 2-7/8 x 4-3/4 in. Displacement: 246.7 cid. Brake hp: 58 at 2510 rpm. NACC taxable hp: 26.55. Valve lifters: mechanical. Two crankshaft bearings. Crankshaft drilled for oil. Pump cooling. Force feed lubrication. Helical camshaft drive. Delco single ignition. Carburetor: Ball & Ball automatic compensating vertical carburetor.

1919 OLDSMOBILE

Model 45-A
V-8. Cast-iron block. Cylinders cast in fours. Valves in V. Bore & stroke: 2-7/8 x 4-3/4 in. Displacement: 246.7 cid. Brake hp: 58 at 2510 rpm. NACC taxable hp: 26.55. Valve lifters: mechanical. Two crankshaft bearings. Crankshaft drilled for oil. Pump cooling. Force feed lubrication. Helical camshaft drive. Valve lifters: mechanical. Delco single ignition. Carburetor: Ball & Ball automatic compensating vertical carburetor.

1920 OLDSMOBILE

Model 45-B
V-8. Cast-iron block. Detachable cylinder head. Cylinders cast in fours. Valve location: side. Bore & stroke: 2-7/8 x 4-3/4 in. Displacement: 246.7 cid. Brake hp: 58 at 2510 rpm. NACC taxable hp: 26.55 at 2600 rpm. Valve lifters: mechanical. Two crankshaft bearings. Crankshaft drilled for oil. Pump cooling. Force feed lubrication. Helical camshaft drive. Valve lifters: mechanical. Delco single ignition. Carburetor: Ball & Ball automatic compensating vertical carburetor.

1921 OLDSMOBILE

Model 46
V-8. Cast-iron block. Detachable cylinder head. Cylinders cast in fours. Valve location: side. Bore & stroke: 2-7/8 x 4-3/4 in. Displacement: 246.7 cid. Brake hp: 58 at 2510 rpm. NACC taxable hp: 26.55 at 2600 rpm. Valve lifters: mechanical. Two crankshaft bearings. Crankshaft drilled for oil. Pump cooling. Force feed lubrication. Helical camshaft drive. Delco single ignition. Crankcase capacity: 4 qt. Cooling system capacity: 5-1/4 gal. Carburetor: Ball & Ball 1-1/4-in. two-stage type.
Note: The 1921 Oldsmobile Model 46 V-8 averaged 8 miles per gallon of gasoline.

1922 OLDSMOBILE

Model 46
V-8. Cast-iron block. Detachable cylinder head. Cylinders cast in fours. Valve location: side. Bore & stroke: 2-7/8 x 4-3/4 in. Displacement: 246.7 cid. Brake hp: 58 at 2510 rpm. NACC taxable hp: 26.55 at 2600 rpm. Valve lifters: mechanical. Two crankshaft bearings. Crankshaft drilled for oil. Pump cooling. Force feed lubrication. Helical camshaft drive. Delco single ignition. Crankcase capacity: 4 qt. Cooling system capacity: 5-1/4 gal. Carburetor: Ball & Ball 1-1/4-in. two-stage type.

Model 47
V-8. Cast-iron block. Detachable cylinder head. Cylinders cast in fours. Valve location: side. Bore & stroke: 2-7/8 x 4-1/2 in. Displacement: 233.7 cid. Compression ratio: 4.25:1. Brake hp: 60 at 2710 rpm. NACC taxable hp: 26.55 at 2600 rpm. Valve lifters: mechanical. Two crankshaft bearings. Crankshaft drilled for oil. Pump cooling. Force feed lubrication. Helical camshaft drive. Delco single ignition. Crankcase capacity: 6 qt. Cooling system capacity: 3-1/2 gal. Carburetor: Johnson 1-1/4.
Note: The 1921 Oldsmobile Model 46 V-8 averaged 12 miles per gallon of gasoline.

1923 OLDSMOBILE

Model 47
V-8. Cast-iron block. Detachable cylinder head. Cylinders cast in fours. Valve location: side. Bore & stroke: 2-7/8 x 4-1/2 in. Displacement: 233.7 cid. Compression ratio: 4.25:1. Brake hp: 63.5 at 2710 rpm. NACC taxable hp: 26.55 at 2600 rpm. Valve lifters: mechanical. Two crankshaft bearings. Crankshaft drilled for oil. Pump cooling. Force feed lubrication. Helical camshaft drive. Delco single ignition. Crankcase capacity: 6 qt. Cooling system capacity: 3-1/2 gal. Carburetor: Johnson 1-1/8.
Note: The 1923 Oldsmobile Model 47 V-8 averaged 17 miles per gallon of gasoline.

The Oldsmobile V-8 was dropped in 1924. The Oldsmobile-built Viking of 1930 had a V-8 that is covered separately in this catalog. An inline Oldsmobile 8 was introduced in 1932. Oldsmobile then offered a straight 8 through 1942 and again in 1946-1948. There was no Oldsmobile V-8 during this period. Oldsmobile then introduced its famous overhead valve "Rocket V-8" engine in 1949 models.

1949 OLDSMOBILE

Series 88, 98 V-8
V-8. Overhead valves. Cast-iron block. Bore & stroke: 3-3/4 x 3-7/16 in. Displacement: 303.7 cid. Compression ratio: 7.25:1. Brake hp: 135 at 3600 rpm. AMA taxable hp: 45. Torque: 263 lbs.-ft. at 1800 rpm. Non-adjustable hydraulic valve lifters. Five main bearings. Crankcase capacity: 5 qt. (add 1 qt. if new filter installed). Cooling system capacity: 21-1/2 qt. (add 1 qt. for heater). Carburetor: Carter WGD-714S dual downdraft.

"The revolutionary new "ROCKET" engine!" heralded a 1949 Oldsmobile advertisement in *The Saturday Evening Post.* "This is the engine you've heard so much about – a completely new, high-compression, valve-in-head eight that actually gives you more power on less gasoline!"

1950 OLDSMOBILE

Series 88, 98 V-8
V-8. Overhead valves. Cast-iron block. Bore & stroke: 3-3/4 x 3-7/16 in. Displacement: 303.7 cid. Compression ratio: 7.50:1. Brake hp: 135 at 3600 rpm. AMA taxable hp: 45. Torque: 263 lbs.-ft. at 1800 rpm. Non-adjustable hydraulic valve lifters. Five main bearings. Crankcase capacity: 5 qt. (add 1 qt. if new filter installed). Cooling system capacity: 21-1/2 qt. (add 1 qt. for heater). Carburetor: Carter WGD-714S dual downdraft. Starting engine number: 8A or 194001 and up.

1951 OLDSMOBILE

All Series V-8
V-8. Overhead valves. Cast-iron block. Bore & stroke: 3-3/4 x 3-7/16 in. Displacement: 303.7 cid. Compression ratio: 7.50:1. Brake hp: 135 at 3600 rpm. AMA taxable hp: 45. Torque: 263 lbs.-ft. at 1800 rpm. Non-adjustable hydraulic valve lifters. Five main bearings. Crankcase capacity: 5 qt. (add 1 qt. if new filter installed). Cooling system capacity: 21-1/2 qt. (add 1 qt. for heater). Carburetor: Rochester BB two-barrel. Starting engine number: 8C1001 and up.

1952 OLDSMOBILE

Deluxe 88 Series V-8
V-8. Overhead valves. Cast-iron block. Bore & stroke: 3-3/4 x 3-7/16 in. Displacement: 303.7 cid. Compression ratio: 7.50:1. Brake hp: 145 at 3600 rpm. AMA taxable hp: 45. Torque: 280 lbs.-ft. at 1800 rpm. Non-adjustable hydraulic valve lifters. Five main bearings. Crankcase capacity: 5 qt. (add 1 qt. if new filter installed). Cooling system capacity: 21-1/2 qt. (add 1 qt. for heater). Carburetor: Rochester BB two-barrel or Carter dual downdraft. Starting engine number: R1001 and up.

Super 88, 98 Series V-8
V-8. Overhead valves. Cast-iron block. Bore & stroke: 3-3/4 x 3-7/16 in. Displacement: 303.7 cid. Compression ratio: 7.50:1. Brake hp: 160 at 3600 rpm. AMA taxable hp: 45. Torque: 283 lbs.-ft. at 1800 rpm. Non-adjustable hydraulic valve lifters. Five main bearings. Crankcase capacity: 5 qt. (add 1 qt. if new filter installed). Cooling system capacity:

The Oldsmobile Rocket V-8 engine gained 25 horsepower in 1952. Among changes made was the switch from a two-barrel carburetor to a Rochester 4GC "Quadri-Jet" four-barrel and the addition of a new high-lift valve mechanism that boosted horsepower to 160 at 3600 rpm. The new high-lift cam increased the engine's volumetric efficiency by allowing the pistons to compress more gas and air into the same space.

21-1/2 qt. (add 1 qt. for heater). Carburetor: Either Rochester 4GC or Carter WGD four-barrel. Starting engine number: R1001 and up.

1953 OLDSMOBILE

Deluxe 88 series V-8
V-8. Overhead valves. Cast-iron block. Bore & stroke: 3-3/4 x 3-7/16 in. Displacement: 303.7 cid. Compression ratio: 8.00:1. Brake hp: 150 at 3600 rpm. AMA taxable hp: 45. Torque: 280 lbs.-ft. at 1800 rpm. Non-adjustable hydraulic valve lifters. Five main bearings. Crankcase capacity: 5 qt. (add 1 qt. if new filter installed). Cooling system capacity: 21-1/2 qt. (add 1 qt. for heater). Carburetor: Carter WGD dual downdraft. Starting engine number: R-215001 and up.

Super 88, 98 Series V-8
V-8. Overhead valves. Cast-iron block. Bore & stroke: 3-3/4 x 3-7/16 in. Displacement: 303.7 cid. Compression ratio: 8.00:1. Brake hp: 165 at 3600 rpm. AMA taxable hp: 45. Torque: 284 lbs.-ft. at 1800 rpm. Non-adjustable hydraulic valve lifters. Five main bearings. Crankcase capacity: 5 qt. (add 1 qt. if new filter installed). Cooling system capacity: 21-1/2 qt. (add 1 qt. for heater). Carburetor: Either Rochester 4GC or Carter WCFB four-barrel. Starting engine number: R-215001 and up.

Fiesta model V-8
V-8. Overhead valves. Cast-iron block. Bore & stroke: 3-3/4 x 3-7/16 in. Displacement: 303.7 cid. Compression ratio: 8.30:1. Brake hp: 170 at 4000 rpm. AMA taxable hp: 48. Torque: 295 lbs.-ft. at 2000 rpm. Non-adjustable hydraulic valve lifters. Five main bearings. Crankcase capacity: 5 qt. (add 1 qt. if new filter installed). Cooling system capacity: 21-1/2 qt. (add 1 qt. for heater). Carburetor: Rochester 4GC four-barrel. Starting engine number: R-215001 and up.

1954 OLDSMOBILE

88 Series V-8
V-8. Overhead valves. Cast-iron block. Bore & stroke: 3-7/8 x 3-7/16 in. Displacement: 324.3 cid. Compression ratio: 8.25:1. Brake hp: 170 at 4000 rpm. AMA taxable hp: 48. Torque: 295 lbs.-ft. at 2000 rpm. Non-adjustable hydraulic valve lifters. Five main bearings. Crankcase capacity: 5 qt. (add 1 qt. if new filter installed). Cooling system capacity: 20-1/2 qt. (add 1 qt. for heater). Carburetor: Carter dual downdraft WGD. Starting engine number: V-1001 and up.

Super 88, 98 Series V-8
V-8. Overhead valves. Cast-iron block. Bore & stroke: 3-7/8 x 3-7/16 in. Compression ratio: 8.25:1. Displacement: 324.3 cid. Brake hp: 185 at 4000 rpm. AMA taxable hp: 48. Torque: 300 lbs.-ft. at 2000 rpm. Non-adjustable hydraulic valve lifters. Five main bearings. Crankcase capacity: 5 qt. (add 1 qt. if new filter installed). Cooling system capacity: 20-1/2 qt. (add 1 qt. for heater). Carburetor: Either Carter WCFB or Rochester 4GC four-barrel. Starting engine number: V-1001 and up.

1955 OLDSMOBILE

88 Series V-8
V-8. Overhead valves. Cast-iron block. Bore & stroke: 3-7/8 x 3-7/16 in. Displacement: 324.3 cid. Compression ratio: 8.50:1. Brake hp: 185 at 4000 rpm. AMA taxable hp: 48. Torque: 320 lbs.-ft. at 2000 rpm. Non-adjustable hydraulic valve lifters. Five main bearings. Crankcase capacity: 5 qt. (add 1 qt. if new filter installed). Cooling system capacity: 20-1/2 qt. (add 1 qt. for heater). Carburetor: Rochester 2GC dual downdraft. Starting engine number: V-400001 and up.

Super 88, 98 Series V-8
V-8. Overhead valves. Cast-iron block. Bore & stroke: 3-7/8 x 3-7/16 in. Displacement: 324.3 cid. Compression ratio: 8.50:1. Brake hp: 202 at 4000 rpm. AMA taxable hp: 48. Torque: 332 lbs.-ft. at 2400 rpm. Non-adjustable hydraulic valve lifters. Five main bearings. Crankcase capacity: 5 qt. (add 1 qt. if new filter installed). Cooling system capacity: 20-1/2 qt. (add 1 qt. for heater). Carburetor: Rochester 4GC Quadra-jet four-barrel. Starting engine number: V-400001 and up.

Oldsmobile dubbed the V-8 engine used in its 1955 Super 88 and 98 models the "Rocket 202." This description was based on its 202-hp rating. This 324-cid four-barrel engine had an 8.5:1 compression ratio compared to 8.25:1 in 1954. Also available in 1955 was a two-barrel version of the 324 that put out 185 hp.

1956 OLDSMOBILE

88 Series V-8
V-8. Overhead valves. Cast-iron block. Bore & stroke: 3-7/8 x 3-7/16 in. Displacement: 324.3 cid. Compression ratio: 9.25:1. Brake hp: 230 at 4400 rpm. AMA taxable hp: 48. Torque: 340 lbs.-ft. at 2400 rpm. Non-adjustable hydraulic valve lifters. Five main bearings. Crankcase capacity: 5 qt. (add 1 qt. if new filter installed). Cooling system capacity: 20-1/2 qt. (add 1 qt. for heater). Carburetor: Rochester 2GC dual downdraft. Starting engine number: V-1000001 and up.

Super 88, 98 Series V-8; Optional 88 Series V-8
V-8. Overhead valves. Cast-iron block. Bore & stroke: 3-7/8 x 3-7/16 in. Displacement: 324.3 cid. Compression ratio: 9.25:1. Brake hp: 240 at 4400 rpm. AMA taxable hp: 48. Torque: 350 lbs.-ft. at 2800 rpm. Non-adjustable hydraulic valve lifters. Five main bearings. Crankcase capacity: 5 qt. (add 1 qt. if new filter installed). Cooling system capacity: 20-1/2 qt. (add 1 qt. for heater). Carburetor: Rochester 4GC Quadra-jet four-barrel. Starting engine number: V-1000001 and up.

1957 OLDSMOBILE

88, Super 88, 98 Series V-8
V-8. Overhead valves. Cast-iron block. Bore & stroke: 4 x 3-11/16 in. Displacement: 370.7 cid. Compression ratio: 9.25:1. Brake hp: 277 at 4400 rpm. AMA taxable hp: 51. Torque: 400 lbs.-ft. at 2800 rpm. Non-adjustable hydraulic valve lifters. Five main bearings. Crankcase capacity: 5 qt. (add 1 qt. if new filter installed). Cooling system capacity: 20 qt. (add 1 qt. for heater). Carburetor: Rochester 4GC Quadra-jet four-barrel. Engine code: T-400.
Notes: J-2 induction system (300 hp) with three two-barrel carburetors ($83). Special J-2 induction system (312 hp) not recommended for street use and offered to drag racers and stock car racers for off-road use ($395). Engine code: J-2.

1958 OLDSMOBILE

Dynamic 88 Series V-8
V-8. Overhead valves. Cast-iron block. Bore & stroke: 4 x 3-11/16 in. Displacement: 370.7 cid. Compression ratio: 10.0:1. Brake hp: 265 at 4400 rpm. AMA taxable hp: 51. Torque: 390 lbs.-ft. at 2400 rpm. Hydraulic valve lifters. Five main bearings. Crankcase capacity: 4 qt. (add 1 qt. if new filter installed). Cooling system capacity: 20 qt. (add 1 qt. for heater). Carburetor: Rochester 2GC "Econ-o-way" two-barrel. Starting

engine number: V-400001 and up.

Super 88, 98 Series V-8
V-8. Overhead valves. Cast-iron block. Bore & stroke: 4 x 3-11/16 in. Displacement: 370.7 cid. Compression ratio: 10.0:1. Brake hp: 305 at 4600 rpm. AMA taxable hp: 51. Torque: 410 lbs.-ft. at 2800 rpm. Hydraulic valve lifters. Five main bearings. Crankcase capacity: 4 qt. (add 1 qt. if new filter installed). Cooling system capacity: 20 qt. (add 1 qt. for heater). Carburetor: Rochester 4GC Quadra-jet four-barrel. Starting engine number: V-400001 and up.
Note: A triple carbureted J-2 engine option could be ordered on any 1958 Olds. It boosted power to 312 hp. Other engine options were a heavy-duty air cleaner and a heavy-duty crankcase ventilation system.

1959 OLDSMOBILE

Dynamic 88 Series V-8
V-8. Overhead valves. Cast-iron block. Bore & stroke: 4 x 3-11/16 in. Displacement: 370.7 cid. Compression ratio: 9.75:1. Brake hp: 270 at 4600 rpm. AMA taxable hp: 51. Torque: 390 lbs.-ft. at 2400 rpm. Hydraulic valve lifters. Five main bearings. Crankcase capacity: 4 qt. (add 1 qt. if new filter installed). Cooling system capacity: 20 qt. (add 1 qt. for heater). Carburetor: Rochester 2GC "Econ-o-way" two-barrel. Engine code: C.

Super 88, 98 Series V-8
V-8. Overhead valves. Cast-iron block. Bore & stroke: 4-1/8 x 3-11/16 in. Displacement: 394.0 cid. Compression ratio: 9.75:1. Brake hp: 315 at 4600 rpm. AMA taxable hp: 54. Torque: 435 lbs.-ft. at 2800 rpm. Hydraulic valve lifters. Five main bearings. Crankcase capacity: 4 qt. (add 1 qt. if new filter installed). Cooling system capacity: 20 qt. (add 1 qt. for heater). Carburetor: Rochester 4GC Quadra-jet four-barrel. Engine code: D.

1960 OLDSMOBILE

Dynamic 88 Series V-8
V-8. Overhead valves. Cast-iron block. Bore & stroke: 4 x 3-11/16 in. Displacement: 370.7 cid. Compression ratio: 8.75:1. Brake hp: 240 at 4400 rpm. AMA taxable hp: 51. Torque: 375 lbs.-ft. at 2400 rpm. Hydraulic valve lifters. Five main bearings. Crankcase capacity: 4 qt. (add 1 qt. if new filter installed). Cooling system capacity: 20 qt. (add 1 qt. for heater). Carburetor: Rochester 2GC "Econ-o-way" two-barrel. Engine code: C.

Dynamic 88 Series optional V-8
V-8. Overhead valves. Cast-iron block. Bore & stroke: 4 x 3-11/16 in. Displacement: 370.7 cid. Compression ratio: 9.75:1. Brake hp: 260 at 4400 rpm. AMA taxable hp: 51. Torque: 390 lbs.-ft. at 2400 rpm. Hydraulic valve lifters. Five main bearings. Crankcase capacity: 4 qt. (add 1 qt. if new filter installed). Cooling system capacity: 20 qt. (add 1 qt. for heater). Carburetor: Rochester 2GC "Econ-o-way' two-barrel. Engine code: CH.

Super 88, 98 Series V-8
V-8. Overhead valves. Cast-iron block. Bore & stroke: 4-1/8 x 3-11/16 in. Displacement: 394.0 cid. Compression ratio: 9.75:1. Brake hp: 315 at 4600 rpm. AMA taxable hp: 54. Torque: 435 lbs.-ft. at 2800 rpm. Hydraulic valve lifters. Five main bearings. Crankcase capacity: 4 qt. (add 1 qt. if new filter installed). Cooling system capacity: 20 qt. (add 1 qt. for heater). Carburetor: Rochester 4GC Multi-jet four-barrel. Engine code: D.

1961 OLDSMOBILE

F-85 Series V-8
V-8. Overhead valves. Aluminum block. Liquid cooled. Bore & stroke: 3.5 x 2.8 in. Displacement: 215.0 cid. Compression ratio: 8.75:1. Brake hp: 155 at 4800 rpm. AMA taxable hp: 39.20. Torque: 210 lbs.-ft. at 3200 rpm. Hydraulic valve lifters. Five main bearings. Crankcase capacity: 4 qt. (add 1 qt. if new filter installed). Cooling system capacity with heater: 12.5 qt. Carburetor: Rochester 2GC two-barrel. Engine code: S.

Sports-car dash was promised in the new high-performance, aluminum Rockette V-8 engine that Olds introduced in the F-85 during 1961. It had a high 10.25-to-1 compression ratio and a new high-capacity manifold to improve engine breathing. Other features included an exceptionally-responsive four-barrel carburetor and dual outlet exhausts to increase top-range responsiveness.

Dynamic 88 Series V-8
V-8. Overhead valves. Cast-iron block. Bore & stroke: 4-1/8 x 3-11/16 in. Displacement: 394.0 cid. Compression ratio: 8.75:1. Brake hp: 250 at 4200 rpm. AMA taxable hp: 54. Torque: 405 lbs.-ft. at 2400 rpm. Hydraulic valve lifters. Five main bearings. Crankcase capacity: 4 qt. (add 1 qt. if new filter installed). Cooling system capacity: 19.3 qt. (add 1 qt. for heater). Carburetor: Rochester 2GC Econ-o-way two-barrel. Engine code: F.

Super 88, 98 Series V-8
V-8. Overhead valves. Cast-iron block. Bore & stroke: 4-1/8 x 3-11/16 in. Displacement: 394.0 cid. Compression ratio: 10.0:1. Brake hp: 325 at 4600 rpm. AMA taxable hp: 54. Torque: 435 lbs.-ft. at 2800 rpm. Hydraulic valve lifters. Five main bearings. Crankcase capacity: 4 qt. (add 1 qt. if new filter installed). Cooling system capacity: 19.3 qt. (add 1 qt. for heater). Carburetor: Rochester 4GC four-barrel. Engine code: G.

Starfire model V-8
V-8. Overhead valves. Cast-iron block. Bore & stroke: 4-1/8 x 3-11/16 in. Displacement: 394.0 cid. Compression ratio: 10.25:1. Brake hp: 330 at 4600 rpm. AMA taxable hp: 54. Torque: 440 lbs.-ft. at 2800 rpm. Hydraulic valve lifters. Five main bearings. Crankcase capacity: 4 qt. (add 1 qt. if new filter installed). Cooling system capacity: 19.3 qt. (add 1 qt. for heater). Carburetor: Rochester 4GC four-barrel. Engine code: G.

New-for-1961 was the 155-hp Rockette V-8 engine that was standard equipment on the new F-85 compact model. This regular-fuel aluminum engine was actually built by Buick. It displaced 215 cubic inches. Advertisements for this power plant promised "remarkable performance, plus outstanding fuel economy."

1962 OLDSMOBILE

F-85 Series base V-8
V-8. Overhead valves. Aluminum block. Liquid cooled. Bore & stroke: 3.5 x 2.8 in. Displacement: 215.0 cid. Compression ratio: 8.75:1. Brake hp: 155 at 4800 rpm. AMA taxable hp: 39.20. Torque: 210 lbs.-ft. at 3200 rpm. Hydraulic valve lifters. Five main bearings. Crankcase capacity: 4 qt. (add 1 qt. if new filter installed). Cooling system capacity with heater: 12.5 qt. Carburetor: Rochester 2GC two-barrel. Engine code: S.

F-85 Cutlass Series base V-8
V-8. Overhead valves. Aluminum block. Liquid cooled. Bore & stroke: 3.5 x 2.8 in. Displacement: 215.0 cid. Compression ratio: 10.25:1. Brake hp: 185 at 4800 rpm. Torque: 230 lbs.-ft. at 3200 rpm. Hydraulic valve lifters. Five main bearings. Crankcase capacity: 4 qt. (add 1 qt. if new filter installed). Cooling system capacity with heater: 12.5 qt. Carburetor: Rochester 4GC four-barrel. Engine code: SG.

F-85 Cutlass Series base V-8
V-8. Overhead valves. Aluminum block. Liquid cooled. Bore & stroke: 3.5 x 2.8 in. Displacement: 215.0 cid. Compression ratio: 10.25:1. Brake hp: 215 at 4800 rpm. Torque: 300 lbs.-ft. at 3200 rpm. Hydraulic valve lifters. Five main bearings. Crankcase capacity: 4 qt. (add 1 qt. if new filter installed). Cooling system capacity with heater: 12.5 qt. Induction system: Turbocharged and fluid-injected with Rochester Model RC carburetor. Engine code: ST.

Dynamic 88 Series optional Economy V-8
V-8. Overhead valves. Cast-iron block. Bore & stroke: 4-1/8 x 3-11/16 in. Displacement: 394.0 cid. Compression ratio: 8.75:1. Brake hp: 260 at 4400 rpm. Torque: 410 lbs.-ft. at 2400 rpm. Hydraulic valve lifters. Five main bearings. Crankcase capacity: 4 qt. (add 1 qt. if new filter installed). Cooling system capacity with heater: 20.3 qt. Carburetor: Rochester 2GC Econ-o-way two-barrel. Engine code: FL.

Dynamic 88 Series base V-8
V-8. Overhead valves. Cast-iron block. Bore & stroke: 4-1/8 x 3-11/16 in. Displacement: 394.0 cid. Compression ratio: 10.25:1. Brake hp: 280 at 4400 rpm. AMA taxable hp: 54. Torque: 430 lbs.-ft. at 2400 rpm. Hydraulic valve lifters. Five main bearings. Crankcase capacity: 4 qt. (add 1 qt. if new filter installed). Cooling system capacity with heater: 20.3 qt. Carburetor: Rochester 2GC Econ-o-way two-barrel. Engine code: F.

Super 88, 98 Series base V-8; Optional Dynamic 88 Series V-8
V-8. Overhead valves. Cast-iron block. Bore & stroke: 4-1/8 x 3-11/16 in. Displacement: 394.0 cid. Compression ratio: 10.25:1. Brake hp: 330 at 4600 rpm. AMA taxable hp: 54. Torque: 440 lbs.-ft. at 2800 rpm. Hydraulic valve lifters. Five main bearings. Crankcase capacity: 4 qt. (add 1 qt. if new filter installed). Cooling system capacity with heater: 20.3 qt. Carburetor: Rochester 4GC four-barrel. Engine code: G.

Starfire Series base V-8; Optional Super 88, 98 Series V-8
V-8. Overhead valves. Cast-iron block. Bore & stroke: 4-1/8 x 3-11/16 in. Displacement: 394.0 cid. Compression ratio: 10.50:1. Brake hp: 345 at 4800 rpm. AMA taxable hp: 54. Torque: 440 lbs.-ft. at 3200 rpm. Hydraulic valve lifters. Five main bearings. Crankcase capacity: 4 qt. (add 1 qt. if new filter installed). Cooling system capacity with heater: 20.3 qt. Carburetor: Rochester 4GC four-barrel. Engine code: GS.

1963 OLDSMOBILE

F-85 Series base V-8
V-8. Overhead valves. Aluminum block. Liquid cooled. Bore & stroke: 3.5 x 2.8 in. Displacement: 215.0 cid. Compression ratio: 8.75:1. Brake hp: 155 at 4800 rpm. AMA taxable hp: 39.20. Torque: 210 lbs.-ft. at 3200 rpm. Hydraulic valve lifters. Five main bearings. Crankcase capacity: 4 qt. (add 1 qt. if new filter installed). Cooling system capacity with heater: 12.25 qt. Carburetor: Rochester 2GC two-barrel. Engine code: S.

F-85 Cutlass Series base V-8
V-8. Overhead valves. Aluminum block. Liquid cooled. Bore & stroke: 3.5 x 2.8 in. Displacement: 215.0 cid. Compression ratio: 10.25:1. Brake hp: 195 at 4800 rpm. Torque: 235 lbs.-ft. at 3200 rpm. Hydraulic valve lifters. Five main bearings. Crankcase capacity: 4 qt. (add 1 qt. if new filter installed). Cooling system capacity with heater: 12.25 qt. Carburetor: Rochester 4GC four-barrel. Engine code: SG.

F-85 Cutlass Series base V-8
V-8. Overhead valves. Aluminum block. Liquid cooled. Bore & stroke: 3.5 x 2.8 in. Displacement: 215.0 cid. Compression ratio: 10.75:1. Brake hp: 185 at 4800 rpm. Torque: 230 lbs.-ft. at 3200 rpm. Hydraulic valve lifters. Five main bearings. Crankcase capacity: 4 qt. (add 1 qt. if new filter installed). Cooling system capacity with heater: 12.25 qt. Carburetor: Rochester 4GC four-barrel. Engine code: SG.

F-85 Cutlass Series base V-8
V-8. Overhead valves. Aluminum block. Liquid cooled. Bore & stroke: 3.5 x 2.8 in. Displacement: 215.0 cid. Compression ratio: 10.25:1. Brake hp: 215 at 4800 rpm. Torque: 300 lbs.-ft. at 3200 rpm. Hydraulic valve lifters. Five main bearings. Crankcase capacity: 4 qt. (add 1 qt. if new filter installed). Cooling system capacity with heater: 12.25 qt. Induction system: Turbocharged and fluid-injected with Rochester Model RC carburetor. Engine code: ST.

Dynamic 88 Series optional Economy V-8
V-8. Overhead valves. Cast-iron block. Bore & stroke: 4-1/8 x 3-11/16 in. Displacement: 394.0 cid. Compression ratio: 8.75:1. Brake hp: 260 at 4400 rpm. Torque: 410 lbs.-ft. at 2400 rpm. Hydraulic valve lifters. Five main bearings. Crankcase capacity: 4 qt. (add 1 qt. if new filter installed). Cooling system capacity with heater: 20.25 qt. Carburetor: Rochester 2GC "Econ-o-way" two-barrel. Engine code: HL.

Dynamic 88 Series base V-8
V-8. Overhead valves. Cast-iron block. Bore & stroke: 4-1/8 x 3-11/16 in. Displacement: 394.0 cid. Compression ratio: 10.25:1. Brake hp: 280 at 4400 rpm. AMA taxable hp: 54. Torque: 430 lbs.-ft. at 2400 rpm. Hydraulic valve lifters. Five main bearings. Crankcase capacity: 4 qt. (add 1 qt. if new filter installed). Cooling system capacity with heater: 20.25 Carburetor: Rochester 2GC two-barrel. Engine code: H.

Super 88, 98 Series base V-8; Optional Dynamic 88 Series V-8
V-8. Overhead valves. Cast-iron block. Bore & stroke: 4-1/8 x 3-11/16 in. Displacement: 394.0 cid. Compression ratio: 10.25:1. Brake hp: 330 at 4600 rpm. AMA taxable hp: 54. Torque: 440 lbs.-ft. at 2800 rpm. Hydraulic valve lifters. Five main bearings. Crankcase capacity: 4 qt. (add 1 qt. if new filter installed). Cooling system capacity with heater: 20.25 qt. Carburetor: Rochester 4GC four-barrel. Engine code: J.

Starfire Series base V-8; Optional Super 88, 98 Series V-8
V-8. Overhead valves. Cast-iron block. Bore & stroke: 4-1/8 x 3-11/16 in. Displacement: 394.0 cid. Compression ratio: 10.50:1. Brake hp: 345 at 4800 rpm. AMA taxable hp: 54. Torque: 440 lbs.-ft. at 3200 rpm. Hydraulic valve lifters. Five main bearings. Crankcase capacity: 4 qt. (add 1 qt. if new filter installed). Cooling system capacity with heater: 20.25 qt. Carburetor: Rochester 4GC four-barrel. Engine code: JS.

1964 OLDSMOBILE

F-85, Vista Cruiser model V-8
V-8. Overhead valves. Cast-iron block. Bore & stroke: 3.938 x 3.385 in. Displacement: 330.0 cid. Compression ratio: 9.00:1. Brake hp: 230 at 4400 rpm. Taxable hp: 49.60. Torque: 325 lbs.-ft. at 2400 rpm. Hydraulic valve lifters. Five main bearings. Crankcase capacity: 4 qt. (add 1 qt. if new filter installed). Cooling system capacity with heater: 17 qt. Carburetor: Rochester 2GC two-barrel. Engine code: T.

Jetstar model V-8
V-8. Overhead valves. Cast-iron block. Bore & stroke: 3.938 x 3.385 in. Displacement: 330.0 cid. Compression ratio: 10.25:1. Brake hp: 245 at 4600 rpm. Taxable hp: 49.60. Torque: 345 lbs.-ft. at 2400 rpm. Hydraulic valve lifters. Five main bearings. Crankcase capacity: 4 qt. (add 1 qt. if new filter installed). Cooling system capacity with heater: 17 qt. Carburetor: Rochester 2GC two-barrel. Engine code: TK.

F-85 Cutlass Series base V-8; F-85 Series optional V-8; Jetstar 88 Series optional V-8
V-8. Overhead valves. Cast-iron block. Bore & stroke: 3.938 x 3.385 in. Displacement: 330.0 cid. Compression ratio: 10.25:1. Brake hp: 290 at 4800 rpm. Taxable hp: 49.60. Torque: 355 lbs.-ft. at 2800 rpm. Hydraulic valve lifters. Five main bearings. Crankcase capacity: 4 qt. (add 1 qt. if new filter installed). Cooling system capacity with heater: 17 qt. Carburetor: Rochester 2GC two-barrel. Engine code: TG.

4-4-2 model V-8
V-8. Overhead valves. Cast-iron block. Bore & stroke: 3.938 x 3.385 in. Displacement: 330.0 cid. Compression ratio: 10.25:1. Brake hp: 310 at 5200 rpm. Taxable hp: 49.60. Torque: 355 lbs.-ft. at 3600 rpm. Hydraulic valve lifters. Five main bearings. Crankcase capacity: 4 qt. (add 1 qt. if new filter installed). Cooling system capacity with heater: 17 qt. Carburetor: Rochester 2GC two-barrel. Engine code: * (4-4-2 option)

Note: RPO B09 "police apprehender-pursuit (4-4-2) package" was a midyear option for the F-85 priced at $285.14. The 4-4-2 engine can be identified by special air cleaner with two intakes.

Dynamic 88 Series optional Economy V-8
V-8. Overhead valves. Cast-iron block. Bore & stroke: 4-1/8 x 3-11/16 in. Displacement: 394.0 cid. Compression ratio: 8.75:1. Brake hp: 260 at 4400 rpm. Torque: 410 lbs.-ft. at 2400 rpm. Hydraulic valve lifters. Five main bearings. Crankcase capacity: 4 qt. (add 1 qt. if new filter installed). Cooling system capacity with heater: 20.25 qt. Carburetor: Rochester 2GC "Econ-o-way" two-barrel. Engine code: HL.

Dynamic 88 Series base V-8
V-8. Overhead valves. Cast-iron block. Bore & stroke: 4-1/8 x 3-11/16 in. Displacement: 394.0 cid. Compression ratio: 10.25:1. Brake hp: 280 at 4400 rpm. AMA taxable hp: 54. Torque: 430 lbs.-ft. at 2400 rpm. Hydraulic valve lifters. Five main bearings. Crankcase capacity: 4 qt. (add 1 qt. if new filter installed). Cooling system capacity with heater: 20.25 qt. Carburetor: Rochester 2GC two-barrel. Engine code: H.

Super 88, 98 Series base V-8; Optional Dynamic 88 Series V-8
V-8. Overhead valves. Cast-iron block. Bore & stroke: 4-1/8 x 3-11/16 in. Displacement: 394.0 cid. Compression ratio: 10.25:1. Brake hp: 330 at 4600 rpm. AMA taxable hp: 54. Torque: 440 lbs.-ft. at 2800 rpm. Hydraulic valve lifters. Five main bearings. Crankcase capacity: 4 qt. (add 1 qt. if new filter installed). Cooling system capacity with heater: 20.25 qt. Carburetor: Rochester 4GC four-barrel. Engine code: J.

Jetstar I, Starfire Series base V-8; Optional Super 88, 98 Series V-8
V-8. Overhead valves. Cast-iron block. Bore & stroke: 4-1/8 x 3-11/16 in. Displacement: 394.0 cid. Compression ratio: 10.50:1. Brake hp: 345 at 4800 rpm. AMA taxable hp: 54. Torque: 440 lbs.-ft. at 3200 rpm. Hydraulic valve lifters. Five main bearings. Crankcase capacity: 4 qt. (add 1 qt. if new filter installed). Cooling system capacity with heater: 20.25 qt. Carburetor: Rochester 4GC four-barrel. Engine code: JS.

1965 OLDSMOBILE

F-85, Vista Cruiser model base V-8
V-8. Overhead valves. Cast-iron block. Bore & stroke: 3.938 x 3.385 in. Displacement: 330.0 cid. Compression ratio: 9.00:1. Brake hp: 250 at 4800 rpm. Taxable hp: 49.60. Torque: 335 lbs.-ft. at 2800 rpm. Hydraulic valve lifters. Five main bearings. Crankcase capacity: 4 qt. (add 1 qt. if new filter installed). Cooling system capacity with heater: 17 qt. Carburetor: Rochester 2GC two-barrel. Engine code: T.

Jetstar 88 Series base V-8
V-8. Overhead valves. Cast-iron block. Bore & stroke: 3.938 x 3.385 in. Displacement: 330.0 cid. Compression ratio: 10.25:1. Brake hp: 260 at 4800 rpm. Taxable hp: 49.60. Torque: 355 lbs.-ft. at 2800 rpm. Hydraulic valve lifters. Five main bearings. Crankcase capacity: 4 qt. (add 1 qt. if new filter installed). Cooling system capacity with heater: 17 qt. Carburetor: Rochester 4GC four-barrel. Engine code: U.

Dynamic 88, Delta 88 Series base V-8
V-8. Overhead valves. Cast-iron block. Bore & stroke: 4.125 x 3.975 in. Displacement: 425.0 cid. Compression ratio: 9.00:1. Brake hp: 300 at 4400 rpm. AMA taxable hp: 54. Torque: 430 lbs.-ft. at 2400 rpm. Hydraulic valve lifters. Five main bearings. Crankcase capacity: 4 qt. (add 1 qt. if new filter installed). Cooling system capacity with heater: 18 qt. Carburetor: Rochester 2GC two-barrel. Engine code: ML.

Dynamic 88, Delta 88 Series base V-8
V-8. Overhead valves. Cast-iron block. Bore & stroke: 4.125 x 3.975 in. Displacement: 425.0 cid. Compression ratio: 10.50:1. Brake hp: 310 at 4400 rpm. AMA taxable hp: 54. Torque: 450 lbs.-ft. at 2400 rpm. Hydraulic valve lifters. Five main bearings. Crankcase capacity: 4 qt. (add 1 qt. if new filter installed). Cooling system capacity with heater: 18 qt. Carburetor: Rochester 2GC two-barrel.

F-85 Cutlass Series base V-8; Jetstar 88 Series optional V-8
V-8. Overhead valves. Cast-iron block. Bore & stroke: 3.938 x 3.385 in. Displacement: 330.0 cid. Compression ratio: 10.25:1. Brake hp: 315 at 5200 rpm. Taxable hp: 49.60. Torque: 360 lbs.-ft. at 3600 rpm. Hydraulic valve lifters. Five main bearings. Crankcase capacity: 4 qt. (add 1 qt. if new filter installed). Cooling system capacity with heater: 17 qt. Carburetor: Rochester 4GC four-barrel. Engine code: UG.

4-4-2 model base V-8
V-8. Overhead valves. Cast-iron block. Bore & stroke: 4.00 x 3.975 in. Displacement: 400.0 cid. Compression ratio: 10.25:1. Brake hp: 345 at 4800 rpm. Taxable hp: 51.20. Torque: 400 lbs.-ft. at 3200 rpm. Hydraulic valve lifters. Five main bearings. Crankcase capacity: 4 qt. (add 1 qt. if new filter installed). Cooling system capacity with heater: 17 qt. Carburetor: Rochester 4GC four-barrel. Engine code: V (4-4-2 option).

Dynamic 88, Delta 88 Series base V-8
V-8. Overhead valves. Cast-iron block. Bore & stroke: 4.125 x 3.975 in. Displacement: 425.0 cid. Compression ratio: 10.25:1. Brake hp: 360 at 4800 rpm. AMA taxable hp: 54. Torque: 470 lbs.-ft. at 2800 rpm. Hydraulic valve lifters. Five main bearings. Crankcase capacity: 4 qt. (add 1 qt. if new filter installed). Cooling system capacity with heater: 18 qt. Carburetor: Rochester 4GC four-barrel. Engine code: N.

Starfire model base V-8; Delta 88, 98 Series optional V-8
V-8. Overhead valves. Cast-iron block. Bore & stroke: 4.125 x 3.975 in. Displacement: 425.0 cid. Compression ratio: 10.50:1. Brake hp: 370 at 4800 rpm. AMA taxable hp: 54. Torque: 470 lbs.-ft. at 3200 rpm. Hydraulic valve lifters. Five main bearings. Crankcase capacity: 4 qt. (add 1 qt. if new filter installed). Cooling system capacity with heater: 18 qt. Carburetor: Rochester 4GC four-barrel. Engine code: NS.

F-85 Cutlass Series base V-8; Jetstar 88 Series optional V-8
V-8. Overhead valves. Cast-iron block. Bore & stroke: 3.938 x 3.385 in. Displacement: 330.0 cid. Compression ratio: 10.25:1. Brake hp: 315 at 5200 rpm. Taxable hp: 49.60. Torque: 360 lbs.-ft. at 3600 rpm. Hydraulic valve lifters. Five main bearings. Crankcase capacity: 4 qt. (add 1 qt. if new filter installed). Cooling system capacity with heater: 17 qt. Carburetor: Rochester 4GC four-barrel. Engine code: UG.

1966 OLDSMOBILE

F-85, Vista Cruiser Series V-8
V-8. Overhead valves. Cast-iron block. Bore & stroke: 3.938 x 3.385 in. Displacement: 330.0 cid. Compression ratio: 9.00:1. Brake hp: 250 at 4800 rpm. Taxable hp: 49.60. Torque: 335 lbs.-ft. at 2800 rpm. Hydraulic valve lifters. Five main bearings. Crankcase capacity: 4 qt. (add 1 qt. if new filter installed). Cooling system capacity with heater: 17 qt. Carburetor: Rochester 2GC two-barrel. Engine code: W or XL.

Jetstar 88 Series base V-8
V-8. Overhead valves. Cast-iron block. Bore & stroke: 3.938 x 3.385 in. Displacement: 330.0 cid. Compression ratio: 10.25:1. Brake hp: 260 at 4800 rpm. Taxable hp: 49.60. Torque: 355 lbs.-ft. at 2800 rpm. Hydraulic valve lifters. Five main bearings. Crankcase capacity: 4 qt. (add 1 qt. if new filter installed). Cooling system capacity with heater: 17 qt. Carburetor: Rochester 4GC four-barrel. Engine code: X.

Jetstar 88 Series optional V-8
V-8. Overhead valves. Cast-iron block. Bore & stroke: 3.938 x 3.385 in. Displacement: 330.0 cid. Compression ratio: 10.25:1. Brake hp: 320 at 5200 rpm. Taxable hp: 54. Torque: 360 lbs.-ft. at 3600 rpm. Hydraulic valve lifters. Five main bearings. Crankcase capacity: 4 qt. (add 1 qt. if new filter installed). Cooling system capacity with heater: 16.5 qt. Carburetor: Rochester 4MV four-barrel. Engine code: WG or XG.

An interesting view of the 1966 Oldsmobile Toronado engine, power train, and wheels showing the heart of the front-wheel-drive design in which all drive train components were located below the hood, well forward of the passenger compartment. Note the 425-cid V-8's unique twin-snorkel air cleaner.

Dynamic 88, Delta 88 Series base V-8
V-8. Overhead valves. Cast-iron block. Bore & stroke: 4.125 x 3.975 in. Displacement: 425.0 cid. Compression ratio: 10.25:1. Brake hp: 300 at 4400 rpm. AMA taxable hp: 54. Torque: 450 lbs.-ft. at 2400 rpm. Hydraulic valve lifters. Five main bearings. Crankcase capacity: 4 qt. (add 1 qt. if new filter installed). Cooling system capacity with heater: 17.5 qt. Carburetor: Rochester 2GC two-barrel. Engine code: ML.

F-85 Cutlass Series base V-8
V-8. Overhead valves. Cast-iron block. Bore & stroke: 3.938 x 3.385 in. Displacement: 330.0 cid. Compression ratio: 9.00:1. Brake hp: 310 at 5200 rpm. AMA taxable hp: 49.6. Torque: 360 lbs.-ft. at 3600 rpm. Hydraulic valve lifters. Five main bearings. Crankcase capacity: 4 qt. (add 1 qt. if new filter installed). Cooling system capacity with heater: 17 qt. Carburetor: Rochester 2GC two-barrel. Engine code: WL.

Dynamic 88, Delta 88 Series base V-8
V-8. Overhead valves. Cast-iron block. Bore & stroke: 4.125 x 3.975 in. Displacement: 425.0 cid. Compression ratio: 10.25:1. Brake hp: 310 at 4400 rpm. AMA taxable hp: 54. Torque: 450 lbs.-ft. at 2400 rpm. Hydraulic valve lifters. Five main bearings. Crankcase capacity: 4 qt. (add 1 qt. if new filter installed). Cooling system capacity with heater: 17.5 qt. Carburetor: Rochester 2GC two-barrel. Engine code: M.

4-4-2 Series base V-8
V-8. Overhead valves. Cast-iron block. Bore & stroke: 4.00 x 3.975 in. Displacement: 400.0 cid. Compression ratio: 10.50:1. Brake hp: 350 at 4400 rpm. Taxable hp: 51.20. Torque: 440 lbs.-ft. at 3600 rpm. Hydraulic valve lifters. Five main bearings. Crankcase capacity: 4 qt. (add 1 qt. if new filter installed). Cooling system capacity with heater: 17 qt. Carburetor: Rochester 4MV four-barrel. Engine code: V.

98 Series base V-8
V-8. Overhead valves. Cast-iron block. Bore & stroke: 4.125 x 3.975 in. Displacement: 425.0 cid. Compression ratio: 10.25:1. Brake hp: 365 at 4800 rpm. AMA taxable hp: 54.45. Torque: 470 lbs.-ft. at 3200 rpm. Hydraulic valve lifters. Five main bearings. Crankcase capacity: 4 qt. (add 1 qt. if new filter installed). Cooling system capacity with heater: 17.5 qt. Carburetor: Rochester 4MV four-barrel. Engine code: N.

Starfire Series base V-8
V-8. Overhead valves. Cast-iron block. Bore & stroke: 4.125 x 3.975 in. Displacement: 425.0 cid. Compression ratio: 10.50:1. Brake hp: 375 at 4800 rpm. AMA taxable hp: 54.45. Torque: 470 lbs.-ft. at 3200 rpm. Hydraulic valve lifters. Five main bearings. Crankcase capacity: 4 qt. (add 1 qt. if new filter installed). Cooling system capacity with heater: 17.5 qt. Carburetor: Rochester 4MV four-barrel. Engine code: NS.

This elongated view of the 1966 Oldsmobile Toronado chassis clearly shows the relocation of all power train components ahead of the passenger compartment. You can also see the car's advanced torsion-bar suspension system and the conspicuous absence of rear-end power train components.

F-85 Cutlass Series base V-8
V-8. Overhead valves. Cast-iron block. Bore & stroke: 3.938 x 3.385 in. Displacement: 330.0 cid. Compression ratio: 9.00:1. Brake hp: 310 at 5200 rpm. AMA taxable hp: 49.6. Torque: 360 lbs.-ft. at 3600 rpm. Hydraulic valve lifters. Five main bearings. Crankcase capacity: 4 qt. (add 1 qt. if new filter installed). Cooling system capacity with heater: 17 qt. Carburetor: Rochester 2GC two-barrel. Engine code: WL.

Toronado Series base V-8
V-8. Overhead valves. Cast-iron block. Bore & stroke: 4.125 x 3.975 in. Displacement: 425.0 cid. Compression ratio: 10.50:1. Brake hp: 385 at 4800 rpm. AMA taxable hp: 54.45. Torque: 475 lbs.-ft. at 3200 rpm. Hydraulic valve lifters. Five main bearings. Crankcase capacity: 5 qt. (add 1 qt. if new filter installed). Cooling system capacity with heater: 17.5 qt. Carburetor: Rochester 4MV four-barrel. Engine code: NT.

1967 OLDSMOBILE

F-85, Cutlass, Vista Cruiser, Delmont 88 Series base V-8
V-8. Overhead valves. Cast-iron block. Bore & stroke: 3.938 x 3.385 in. Displacement: 330.0 cid. Compression ratio: 9.00:1. Brake hp: 250 at 4800 rpm. Taxable hp: 49.60. Torque: 335 lbs.-ft. at 2800 rpm. Hydraulic valve lifters. Five main bearings. Crankcase capacity: 4 qt. (add 1 qt. if new filter installed). Cooling system capacity with heater: 17 qt. Carburetor: Rochester 2GC two-barrel. Engine code: WE.

The "small" Oldsmobile V-8 used in 1967 was a 330-cid job that came in 250-, 260-, 310-, and 320-hp versions. It was basically the same F-85, Cutlass, and Cutlass Supreme base V-8 introduced in 1965 and carried over in 1966. This engine weighed 560-565 pounds. The 250- and 310-hp versions could be safely operated on regular gasoline, while the 260- and 320-hp version required premium fuel.

The 400-cid Oldsmobile V-8 was again offered for exclusive use in the 4-4-2 model in 1967. This engine was basically similar to the small 330-cid V-8 in its basic design, but used larger and heavier components, such as three-inch diameter main bearings (instead of 2-1/2-inch diameter bearings). The larger 425-cid V-8 had a larger bore, but the same stroke as the 400.

F-85, Cutlass, Vista Cruiser, Delmont 88 Series optional V-8
V-8. Overhead valves. Cast-iron block. Bore & stroke: 3.938 x 3.385 in. Displacement: 330.0 cid. Compression ratio: 10.25:1. Brake hp: 260 at 4800 rpm. Taxable hp: 49.60. Torque: 355 lbs.-ft. at 2800 rpm. Hydraulic valve lifters. Five main bearings. Crankcase capacity: 4 qt. (add 1 qt. if new filter installed). Cooling system capacity with heater: 17 qt. Carburetor: Rochester 4GC four-barrel. Engine code: W or X.

Cutlass Supreme Series optional V-8
V-8. Overhead valves. Cast-iron block. Bore & stroke: 4.00 x 3.975 in. Displacement: 400.0 cid. Compression ratio: 10.50:1. Brake hp: 300 at 4600 rpm. Taxable hp: 51.20. Torque: 425 lbs.-ft. at 3000 rpm. Hydraulic valve lifters. Five main bearings. Crankcase capacity: 4 qt. (add 1 qt. if new filter installed). Cooling system capacity with heater: 17 qt. Carburetor: Rochester 4MV four-barrel. Engine code: V.

Delmont 88, Delta 88 Custom Series base V-8
V-8. Overhead valves. Cast-iron block. Bore & stroke: 4.125 x 3.975 in. Displacement: 425.0 cid. Compression ratio: 10.25:1. Brake hp: 300 at 4400 rpm. AMA taxable hp: 54.45. Torque: 430 lbs.-ft. at 2400 rpm. Hydraulic valve lifters. Five main bearings. Crankcase capacity: 4 qt. (add 1 qt. if new filter installed). Cooling system capacity with heater: 17.5 qt. Carburetor: Rochester 2GC two-barrel. Engine code: PL.

F-85 standard, Cutlass, Cutlass Supreme Series optional V-8
V-8. Overhead valves. Cast-iron block. Bore & stroke: 3.938 x 3.385 in. Displacement: 330.0 cid. Compression ratio: 9.00:1. Brake hp: 310 at 5200 rpm. AMA taxable hp: 49.6. Torque: 360 lbs.-ft. at 3600 rpm. Hydraulic valve lifters. Five main bearings. Crankcase capacity: 4 qt. (add 1 qt. if new filter installed). Cooling system capacity with heater: 17 qt. Carburetor: Rochester 2GC two-barrel.

Delmont 88, Delta 88 and Delta 88 Custom Series optional V-8
V-8. Overhead valves. Cast-iron block. Bore & stroke: 4.125 x 3.975 in. Displacement: 425.0 cid. Compression ratio: 10.25:1. Brake hp: 310 at 4400 rpm. AMA taxable hp: 54. Torque: 450 lbs.-ft. at 2400 rpm. Hydraulic valve lifters. Five main bearings. Crankcase capacity: 4 qt. (add 1 qt. if new filter installed). Cooling system capacity with heater: 17.5 qt. Carburetor: Rochester 2GC two-barrel. Engine code: P.

Cutlass Supreme Series base V-8; Delmont 88 Standard and Custom Series optional V-8
V-8. Overhead valves. Cast-iron block. Bore & stroke: 3.938 x 3.385 in. Displacement: 330.0 cid. Compression ratio: 10.25:1. Brake hp: 320 at 5200 rpm. Taxable hp: 49.60. Torque: 360 lbs.-ft. at 3600 rpm. Hydraulic valve lifters. Five main bearings. Crankcase capacity: 4 qt. (add 1 qt. if new filter installed). Cooling system capacity with heater: 16.50 qt. Carburetor: Rochester 4MV four-barrel. Engine code: XG.

4-4-2 Series base V-8
V-8. Overhead valves. Cast-iron block. Bore & stroke: 4.00 x 3.975 in. Displacement: 400.0 cid. Compression ratio: 10.50:1. Brake hp: 350 at 5000 rpm. Taxable hp: 51.20. Torque: 440 lbs.-ft. at 3600 rpm. Hydraulic valve lifters. Five main bearings. Crankcase capacity: 4 qt. (add 1 qt. if new filter installed). Cooling system capacity with heater: 17 qt. Carburetor: Rochester 4MV four-barrel. Engine code: VG.

98 Series base V-8; Delmont 88, Delta 88 and Delta 88 Custom Series optional V-8
V-8. Overhead valves. Cast-iron block. Bore & stroke: 4.125 x 3.975 in. Displacement: 425.0 cid. Compression ratio: 10.25:1. Brake hp: 365 at 4800 rpm. AMA taxable hp: 54.45. Torque: 470 lbs.-ft. at 3200 rpm. Hydraulic valve lifters. Five main bearings. Crankcase capacity: 4 qt. (add 1 qt. if new filter installed). Cooling system capacity with heater: 17.5 qt. Carburetor: Rochester 4MV four-barrel. Engine code: R.

Delmont 88, Delta 88, Delta 88 Custom and 98 Series optional V-8
V-8. Overhead valves. Cast-iron block. Bore & stroke: 4.125 x 3.975 in. Displacement: 425.0 cid. Compression ratio: 10.50:1. Brake hp: 375 at 4800 rpm. AMA taxable hp: 54.45. Torque: 470 lbs.-ft. at 3200 rpm. Hydraulic valve lifters. Five main bearings. Crankcase capacity: 4 qt. (add 1 qt. if new filter installed). Cooling system capacity with heater: 17.5 qt. Carburetor: Rochester 4MV four-barrel. Engine code: RS.

Toronado Series base V-8
V-8. Overhead valves. Cast-iron block. Bore & stroke: 4.125 x 3.975 in. Displacement: 425.0 cid. Compression ratio: 10.50:1. Brake hp: 385 at 4800 rpm. AMA taxable hp: 54.45. Torque: 475 lbs.-ft. at 3200 rpm. Hydraulic valve lifters. Five main bearings. Crankcase capacity: 5 qt. (add 1 qt. if new filter installed). Cooling system capacity with heater: 17.5 qt. Carburetor: Rochester 4MV four-barrel. Engine code: RT.

1968 OLDSMOBILE

For 1968, what had been the largest Olds V-8 (425 cid) was punched out to a full 455 cid. GM brass would not OK it for the 4-4-2, but this didn't stop George Hurst, of Hurst Performance Products, from doing so. In the Hurst/Olds, the 4-4-2's standard 400-cid engine was replaced by a Force-Air 455 with a 10.5:1 compression ratio. It was beefed-up internally to develop 390 hp at 5000 rpm and 500 foot-pounds of torque at 3600 rpm. The big W45 motor was based on the Toronado V-8, but used a special crankshaft, a custom-curved distributor, special carburetor jets, a 308-degree camshaft with a .474-inch lift, and hand-assembled Ram-Air cylinder heads.

As in 1967, a 400-cid V-8 was standard, but it was a totally new one with 3.87 x 4.25-inch bore and stroke (4 x 3.975-inch bore and stroke in 1967). There were three basic four-barrel versions of this 10.5:1 compression ratio engine, the hottest with the W-30 Force Air package added. They gave 325 hp (automatic), 350 hp (stick), and 360 hp, respectively.

F-85, Cutlass, Cutlass Supreme, Vista Cruiser, Delmont 88 Series base V-8
V-8. Overhead valves. Cast-iron block. Bore & stroke: 4.06 x 3.39 in. Displacement: 350.0 cid. Compression ratio: 9.00:1. Brake hp: 250 at 4800 rpm. Torque: 335 lbs.-ft. at 2600 rpm. Hydraulic valve lifters. Five main bearings. Crankcase capacity: 4 qt. (add 1 qt. if new filter installed). Cooling system capacity with heater: 15.25 qt. Carburetor: Rochester two-barrel. Engine code: QI, QJ, T, TB, or TD.

4-4-2, Vista Cruiser Series optional V-8
V-8. Overhead valves. Cast-iron block. Bore & stroke: 4.00 x 3.975 in. Displacement: 400.0 cid. Compression ratio: 9.00:1. Brake hp: 290 at 4600 rpm. Torque: 425 lbs.-ft. at 2400 rpm. Hydraulic valve lifters. Five main bearings. Crankcase capacity: 4 qt. (add 1 qt. if new filter installed). Cooling system capacity with heater: (Vista Cruiser) 15.25 qt.; (4-4-2) 16.25 qt. Carburetor: Rochester two-barrel model 7028251. Engine code: QL.

F-85, Cutlass, Cutlass Supreme, Vista Cruiser, Delmont 88 Series optional V-8
V-8. Overhead valves. Cast-iron block. Bore & stroke: 4.06 x 3.39 in.

In 1968, the Oldsmobile 330 V-8 was also bored out to create the Olds 350-cid V-8. It was used to power F-85s, Cutlasses, and Delmont 88s. There were two-barrel (250-hp) and four-barrel (310-hp) versions. This engine's evolution could be traced all the way back to 1949, when Olds brought out its first modern, overhead-valve V-8.

In 1968, the stroke of the biggest Oldsmobile V-8 was increased to match that of the 400-cid V-8. This increased its displacement to 455 cubic inches. This engine came in five variations with 310, 320, 365, 375, and 400 hp. It looked very large with the air conditioning pump mounted on the right and the alternator mounted on the left.

Displacement: 350.0 cid. Compression ratio: 10.25:1. Brake hp: 310 at 4800 rpm. Torque: 390 lbs.-ft. at 3200 rpm. Hydraulic valve lifters. Five main bearings. Crankcase capacity: 4 qt. (add 1 qt. if new filter installed). Cooling system capacity with heater: (Delmont) 17.50 qt.; (Others) 15.25 qt. Carburetor: Rochester 4MV four-barrel. Engine code: QV. QX, QN, QP, or TN.

Delta 88, Delta 88 Custom Series base V-8
V-8. Overhead valves. Cast-iron block. Bore & stroke: 4.126 x 4.250 in. Displacement: 455 cid. Compression ratio: 10.25:1. Brake hp: 310 at 4200 rpm. Torque: 490 lbs.-ft. at 2400 rpm. Five main bearings. Hydraulic valve lifters. Crankcase capacity: 4 qt. (add 1 qt. if new filter installed). Cooling system capacity with heater: 17.50 qt. Carburetor: Rochester 2GC two-barrel model 709250. Engine code: UJ, UC, or UD.

Delta 88, Delta 88 Custom Series optional V-8
V-8. Overhead valves. Cast-iron block. Bore & stroke: 4.126 x 4.250 in. Displacement: 455 cid. Compression ratio: 10.75:1. Brake hp: 320 at 4200 rpm. Torque: 500 lbs.-ft. at 2400 rpm. Five main bearings. Hydraulic valve lifters. Crankcase capacity: 4 qt. (add 1 qt. if new filter installed). Cooling system capacity with heater: 17.50 qt. Carburetor: Rochester 4MV four-barrel. Engine code: UA or UB.

4-4-2, Vista Cruiser Series V-8 (automatic transmission)
V-8. Overhead valves. Cast-iron block. Bore & stroke: 4.00 x 3.975 in. Displacement: 400.0 cid. Compression ratio: 10.50:1. Brake hp: 325 at 4600 rpm. Torque: (Vista Cruiser) 440 lbs.-ft. at 3200 rpm; (4-4-2) 440 at 3000 rpm. Hydraulic valve lifters. Five main bearings. Crankcase capacity: 4 qt. (add 1 qt. if new filter installed). Cooling system capacity with heater: (Vista Cruiser) 15.25 qt.; (4-4-2) 16.25 qt. Carburetor: Rochester 4MV four-barrel. Engine code: QR, QS, or QT.

4-4-2 Series base V-8 (manual transmission)
V-8. Overhead valves. Cast-iron block. Bore & stroke: 4.00 x 3.975 in. Displacement: 400.0 cid. Compression ratio: 10.50:1. Brake hp: 350 at 4800 rpm. Torque: 440 lbs.-ft. at 3200 rpm. Hydraulic valve lifters. Five main bearings. Crankcase capacity: 4 qt. (add 1 qt. if new filter installed). Cooling system capacity with heater: 16.25 qt. Carburetor: Rochester 4MV four-barrel. Engine code: QW or QU.

4-4-2 Series optional V-8
V-8. Overhead valves. Cast-iron block. Bore & stroke: 4.00 x 3.975 in. Displacement: 400.0 cid. Compression ratio: 10.50:1. Brake hp: 360 at 5400 rpm. Torque: 440 lbs.-ft. at 3600 rpm. Hydraulic valve lifters. Five main bearings. Crankcase capacity: 4 qt. (add 1 qt. if new filter installed). Cooling system capacity with heater: 16.25 qt. Carburetor: Rochester 4MV four-barrel. Engine code: (W-30 package).

98 Series base V-8; Delta 88, Delta 88 Custom Series optional V-8
V-8. Overhead valves. Cast-iron block. Bore & stroke: 4.126 x 4.250 in. Displacement: 455 cid. Compression ratio: 10.25:1. Brake hp: 365 at 4600 rpm. Torque: 510 lbs.-ft. at 3000 rpm. Five main bearings. Hydraulic valve lifters. Crankcase capacity: 4 qt. (add 1 qt. if new filter installed). Cooling system capacity with heater: 17.50 qt. Carburetor: Rochester 4MV four-barrel. Engine code: UN or UO.

Toronado Series base V-8
V-8. Overhead valves. Cast-iron block. Bore & stroke: 4.126 x 4.250 in. Displacement: 455 cid. Compression ratio: 10.25:1. Brake hp: 375 at 4600 rpm. Torque: 510 lbs.-ft. at 3000 rpm. Five main bearings. Hydraulic valve lifters. Crankcase capacity: 5 qt. (add 1 qt. if new filter installed). Cooling system capacity with heater: 20.5 qt. Carburetor: Rochester 4MV four-barrel. Engine code: US, UT, UV, and UW.

Toronado Series optional V-8
V-8. Overhead valves. Cast-iron block. Bore & stroke: 4.126 x 4.250 in. Displacement: 455 cid. Compression ratio: 10.25:1. Brake hp: 400 at 4800 rpm. Torque: 500 lbs.-ft. at 3200 rpm. Five main bearings. Hydraulic valve lifters. Crankcase capacity: 5 qt. (add 1 qt. if new filter installed). Cooling system capacity with heater: 20.5 qt. Carburetor: Rochester 4MV four-barrel.

1969 OLDSMOBILE

Delta 88 Series base V-8
V-8. Overhead valves. Cast-iron block. Bore & stroke: 4.06 x 3.39 in. Displacement: 350.0 cid. Compression ratio: 9.00:1. Brake hp: 250 at 4400 rpm. Torque: 355 lbs.-ft. at 2600 rpm. Hydraulic valve lifters. Five main bearings. Crankcase capacity: 4 qt. (add 1 qt. if new filter installed). Cooling system capacity with heater: 17.50 qt. Carburetor: Rochester 2GC two-barrel. Engine code: QI, QA. QB, TL, and (Cutlass only) QJ.

Cutlass Supreme base V-8
V-8. Overhead valves. Cast-iron block. Bore & stroke: 4.06 x 3.39 in. Displacement: 350.0 cid. Compression ratio: 10.25:1. Brake hp: 310 at 4800 rpm. Torque: 390 lbs.-ft. at 3200 rpm. Hydraulic valve lifters. Five main bearings. Crankcase capacity: 4 qt. (add 1 qt. if new filter installed). Cooling system capacity with heater: 15.25 qt. Carburetor: Rochester 4MV four-barrel. Engine code: QV, QN, and QP.

Delta 88 Royale, Delta 88 Custom Series base V-8
V-8. Overhead valves. Cast-iron block. Bore & stroke: 4.126 x 4.250 in. Displacement: 455 cid. Compression ratio: 10.25:1. Brake hp: 310 at 4200 rpm. Torque: 490 lbs.-ft. at 2400 rpm. Five main bearings. Hydraulic valve lifters. Crankcase capacity: 4 qt. (add 1 qt. if new filter installed). Cooling system capacity with heater: 17.50 qt. Carburetor: Rochester 2GC two-barrel model 709250. Engine code: UJ, UC, and UD.

F-85, Cutlass, Cutlass Supreme optional V-8
V-8. Overhead valves. Cast-iron block. Bore & stroke: 4.06 x 3.39 in. Displacement: 350.0 cid. Compression ratio: 10.50:1. Brake hp: 325 at

For 1969, the big-block Olds V-8s, including the 455-cid version shown here, were essentially the same as they were the year before. This was an almost square, long-stroke V-8 with its 4.125-inch bore and 4.250-inch stroke. The air cleaner decal read "Oldsmobile Rocket 455."

5400 rpm. Torque: 360 lbs.-ft. at 3600 rpm. Hydraulic valve lifters. Five main bearings. Crankcase capacity: 4 qt. (add 1 qt. if new filter installed). Cooling system capacity with heater: 15.25 qt. Carburetor: Rochester 4MV four-barrel. Engine code: QX (W-31).

4-4-2 base W-31 base V-8 (automatic transmission)
V-8. Overhead valves. Cast-iron block. Bore & stroke: 4.00 x 3.975 in. Displacement: 400.0 cid. Compression ratio: 10.50:1. Brake hp: 325 at 4600 rpm. Torque: 440 lbs.-ft. at 3000 rpm. Hydraulic valve lifters. Five main bearings. Crankcase capacity: 4 qt. (add 1 qt. if new filter installed). Cooling system capacity with heater: (4-4-2) 16.25 qt. Carburetor: Rochester 4MV four-barrel. Engine code: QR and QS.

4-4-2 Series base V-8 (manual transmission)
V-8. Overhead valves. Cast-iron block. Bore & stroke: 4.00 x 3.975 in. Displacement: 400.0 cid. Compression ratio: 10.50:1. Brake hp: 350 at 4800 rpm. Torque: 440 lbs.-ft. at 3200 rpm. Hydraulic valve lifters. Five main bearings. Crankcase capacity: 4 qt. (add 1 qt. if new filter installed). Cooling system capacity with heater: 16.25 qt. Carburetor: Rochester 4MV four-barrel. Engine code: QW.

4-4-2 Series optional V-8
V-8. Overhead valves. Cast-iron block. Bore & stroke: 4.00 x 3.975 in. Displacement: 400.0 cid. Compression ratio: 10.50:1. Brake hp: 360 at 5400 rpm. Torque: 400 lbs.-ft. at 3600 rpm. Hydraulic valve lifters. Five main bearings. Crankcase capacity: 4 qt. (add 1 qt. if new filter installed). Cooling system capacity with heater: 16.25 qt. Carburetor: Rochester 4MV four-barrel. Engine code: QU and QT.

98 Series base V-8
V-8. Overhead valves. Cast-iron block. Bore & stroke: 4.126 x 4.250 in. Displacement: 455 cid. Compression ratio: 10.25:1. Brake hp: 365 at 4600 rpm. Torque: 510 lbs.-ft. at 3000 rpm. Five main bearings. Hydraulic valve lifters. Crankcase capacity: 4 qt. (add 1 qt. if new filter installed). Cooling system capacity with heater: 17.50 qt. Carburetor: Rochester 4MV four-barrel. Engine code: QN and QO.

Toronado Series base V-8
V-8. Overhead valves. Cast-iron block. Bore & stroke: 4.126 x 4.250 in. Displacement: 455 cid. Compression ratio: 10.25:1. Brake hp: 375 at 4600 rpm. Torque: 510 lbs.-ft. at 3000 rpm. Five main bearings. Hydraulic valve lifters. Crankcase capacity: 5 qt. (add 1 qt. if new filter installed). Cooling system capacity with heater: 18 qt. Carburetor: Rochester 4MV four-barrel. Engine code: US, UT, and UV.

Delta 88, Delta 88 Custom, Delta 88 Royale optional V-8
V-8. Overhead valves. Cast-iron block. Bore & stroke: 4.126 x 4.250 in. Displacement: 455 cid. Compression ratio: 10.25:1. Brake hp: 390 at 5000 rpm. Torque: 500 lbs.-ft. at 3200 rpm. Five main bearings. Hydraulic valve lifters. Crankcase capacity: 4 qt. (add 1 qt. if new filter installed). Cooling system capacity with heater: 17.5 qt. Carburetor: Rochester 4MV four-barrel. Engine code: UL.

Toronado Series optional V-8
V-8. Overhead valves. Cast-iron block. Bore & stroke: 4.126 x 4.250 in. Displacement: 455 cid. Compression ratio: 10.25:1. Brake hp: 400 at 4800 rpm. Torque: 500 lbs.-ft. at 3200 rpm. Five main bearings. Hydraulic valve lifters. Crankcase capacity: 5 qt. (add 1 qt. if new filter installed). Cooling system capacity with heater: 18 qt. Carburetor: Rochester 4MV four-barrel. Engine code: UW.
Optional: W-31 package including special Force-Air 350-cid engine option for Cutlass ($310). W-30 special 400-cid Force-Air engine package for 4-4-2 ($264).

1970 OLDSMOBILE

F-85, Cutlass Series base V-8
V-8. Overhead valves. Cast-iron block. Bore & stroke: 4.06 x 3.39 in. Displacement: 350.0 cid. Compression ratio: 9.00:1. Brake hp: 250 at 4400 rpm. Torque: 355 lbs.-ft. at 2600 rpm. Hydraulic valve lifters. Five main bearings. Crankcase capacity: 4 qt. (add 1 qt. if new filter installed). Cooling system capacity with heater: 15.50 qt. Carburetor: Rochester 2GC two-barrel. Engine code: QI, QA, QJ, TC, TD, and TL.

Cutlass Supreme base V-8
V-8. Overhead valves. Cast-iron block. Bore & stroke: 4.06 x 3.39 in. Displacement: 350.0 cid. Compression ratio: 10.25:1. Brake hp: 310 at 4800 rpm. Torque: 390 lbs.-ft. at 3200 rpm. Hydraulic valve lifters. Five main bearings. Crankcase capacity: 4 qt. (add 1 qt. if new filter installed). Cooling system capacity with heater: 15.50 qt. Carburetor: Rochester 4MV four-barrel. Engine code: QN, QP, and QV.

Delta 88 Royale, Delta 88 Custom Series base V-8
V-8. Overhead valves. Cast-iron block. Bore & stroke: 4.126 x 4.250 in. Displacement: 455 cid. Compression ratio: 10.25:1. Brake hp: 310 at 4200 rpm. Torque: 510 lbs.-ft. at 3000 rpm. Five main bearings. Hydraulic valve lifters. Crankcase capacity: 4 qt. (add 1 qt. if new filter installed). Cooling system capacity with heater: 17.50 qt. Carburetor: Rochester 2GC two-barrel model 709250. Engine code: UC, UD, and UJ.

Delta Series optional V-8
V-8. Overhead valves. Cast-iron block. Bore & stroke: 4.126 x 4.250 in. Displacement: 455 cid. Compression ratio: 8.50:1. Brake hp: 320 at 4400 rpm. Net hp: 225 at 3600 rpm. Torque: 460 lbs.-ft. at 2800 rpm. Net torque: 360 at 2600 rpm. Hydraulic valve lifters. Crankcase capacity: 4 qt. (add 1 qt. if new filter installed). Cooling system capacity with air conditioning: 18 qt. Carburetor: Rochester 4MV four-barrel. Engine code: TX and TY.

F-85, Cutlass optional V-8
V-8. Overhead valves. Cast-iron block. Bore & stroke: 4.06 x 3.39 in. Displacement: 350.0 cid. Compression ratio: 10.50:1. Brake hp: 325 at 5400 rpm. Torque: 360 lbs.-ft. at 3600 rpm. Hydraulic valve lifters. Five main bearings. Crankcase capacity: 4 qt. (add 1 qt. if new filter installed). Cooling system capacity with heater: 15.25 qt. Carburetor: Rochester 4MV four-barrel. Engine code: QD and QX.

4-4-2, 98 Series base V-8;
V-8. Overhead valves. Cast-iron block. Bore & stroke: 4.126 x 4.250 in. Displacement: 455 cid. Compression ratio: 10.25:1. Brake hp: 365 at 4600 rpm. Torque: 510 lbs.-ft. at 3000 rpm. Five main bearings. Hydraulic valve lifters. Crankcase capacity: 4 qt. (add 1 qt. if new filter installed). Cooling system capacity with heater: (4-4-2) 16.3 qt.; (98) 17.50. Carburetor: Rochester 4MV four-barrel. Engine code: TP, TQ, TU, TV, TW, UN, and UO.

4-4-2 Series optional V-8
V-8. Overhead valves. Cast-iron block. Bore & stroke: 4.126 x 4.250 in. Displacement: 455 cid. Compression ratio: 10.50:1. Brake hp: 370 at 5200 rpm. Torque: 500 lbs.-ft. at 3600 rpm. Five main bearings. Hydraulic valve lifters. Crankcase capacity: 5 qt. (add 1 qt. if new filter installed). Cooling system capacity with heater: 16.3 qt. Carburetor: Rochester 4MV four-barrel. Engine code: TS and TT.

Toronado Series base V-8
V-8. Overhead valves. Cast-iron block. Bore & stroke: 4.126 x 4.250 in. Displacement: 455 cid. Compression ratio: 10.25:1. Brake hp: 375 at 4600 rpm. Torque: 510 lbs.-ft. at 3000 rpm. Five main bearings. Hydraulic valve lifters. Crankcase capacity: 5 qt. (add 1 qt. if new filter installed). Cooling system capacity with heater: 18 qt. Carburetor: Rochester 4MV four-barrel. Engine code: US and UT.

Delta 88, Delta 88 Custom, Delta 88 Royale optional V-8
V-8. Overhead valves. Cast-iron block. Bore & stroke: 4.126 x 4.250 in. Displacement: 455 cid. Compression ratio: 10.25:1. Brake hp: 390 at 5000 rpm. Torque: 500 lbs.-ft. at 3200 rpm. Five main bearings. Hydraulic valve lifters. Crankcase capacity: 4 qt. (add 1 qt. if new filter installed). Cooling system capacity with heater: 17.5 qt. Carburetor: Rochester 4MV four-barrel. Engine code: UL.

Toronado Series optional V-8
V-8. Overhead valves. Cast-iron block. Bore & stroke: 4.126 x 4.250 in. Displacement: 455 cid. Compression ratio: 10.25:1. Brake hp: 400 at 4800 rpm. Torque: 500 lbs.-ft. at 3200 rpm. Five main bearings. Hydraulic valve lifters. Crankcase capacity: 5 qt. (add 1 qt. if new filter installed). Cooling system capacity with heater: 18 qt. Carburetor: Rochester 4MV four-barrel. Engine code: UV and UW.
Optional: Air-induction package with W-30 or W-31.

1971 OLDSMOBILE

Cutlass, Delta Series base V-8
V-8. Overhead valves. Cast-iron block. Bore & stroke: 4.06 x 3.39 in. Displacement: 350.0 cid. Compression ratio: 8.50:1. Brake hp: 240 at 4200 rpm. Net hp: 160 at 4000 rpm. Torque: 350 lbs.-ft. at 2400 rpm. Net torque: 275 lbs.-ft. at 2400 rpm. Hydraulic valve lifters. Five main bearings. Crankcase capacity: 4 qt. (add 1 qt. if new filter installed). Cooling system capacity with air conditioning: 17 qt. Carburetor: Rochester 2GC two-barrel. Engine code: QA, QJ, QI, TC, TD, and TE.

F-85, Cutlass optional V-8
V-8. Overhead valves. Cast-iron block. Bore & stroke: 4.06 x 3.39 in. Displacement: 350.0 cid. Compression ratio: 8.50:1. Brake hp: 260 at 4600 rpm. Net hp: 180 at 4000 rpm. Torque: 360 lbs.-ft. at 3200 rpm. Net torque: 275 lbs.-ft. at 2800 rpm. Hydraulic valve lifters. Five main bearings. Crankcase capacity: 4 qt. (add 1 qt. if new filter installed). Cooling system capacity with air conditioning: 17 qt. Carburetor: Rochester 4MV four-barrel. Engine code: QN, QP, QB, and QO.

Delta Series base V-8
V-8. Overhead valves. Cast-iron block. Bore & stroke: 4.126 x 4.250 in. Displacement: 455 cid. Compression ratio: 8.50:1. Brake hp: 280 at 4000 rpm. Net hp: 185 at 3600 rpm. Torque: 445 lbs.-ft. at 2000 rpm. Net torque: 355 lbs.-ft. at 2000 rpm. Hydraulic valve lifters. Crankcase capacity: 4 qt. (add 1 qt. if new filter installed). Cooling system capacity with air conditioning: 18 qt. Carburetor: Rochester 2GC two-barrel. Engine code: UC, UD, UE, TY, and TX.

Delta Series base V-8
V-8. Overhead valves. Cast-iron block. Bore & stroke: 4.126 x 4.250 in. Displacement: 455 cid. Compression ratio: 8.50:1. Brake hp: 320 at 4400 rpm. Net hp: 225 at 3600 rpm. Torque: 460 lbs.-ft. at 2800 rpm. Net torque: 360 lbs.-ft. at 2600 rpm. Hydraulic valve lifters. Crankcase capacity: 4 qt. (add 1 qt. if new filter installed). Cooling system capacity with air conditioning: 18 qt. Carburetor: Rochester 4MV four-barrel. Engine code: TQ, TP, TW, TV, TA, TU, TN, UN, and UO.

4-4-2 Series base V-8
V-8. Overhead valves. Cast-iron block. Bore & stroke: 4.126 x 4.250 in. Displacement: 455 cid. Compression ratio: 8.50:1. Brake hp: 340 at 4600 rpm. Net hp: 270 at 4400 rpm. Torque: 460 lbs.-ft. at 3200 rpm. Net torque: 370 lbs.-ft. at 3200 rpm. Hydraulic valve lifters. Crankcase capacity: 4 qt. (add 1 qt. if new filter installed). Cooling system capacity with air conditioning: 18 qt. Carburetor: Rochester 4MV four-barrel. Engine code: TT, TL, TS, and TB.

Toronado base V-8
V-8. Overhead valves. Cast-iron block. Bore & stroke: 4.126 x 4.250 in. Displacement: 455 cid. Compression ratio: 8.50:1. Brake hp: 350 at 4400 rpm. Net hp: 275 at 4200 rpm. Torque: 465 lbs.-ft. at 2800 rpm. Net torque: 375 lbs.-ft. at 2800 rpm. Hydraulic valve lifters. Crankcase capacity: 4 qt. (add 1 qt. if new filter installed). Cooling system capacity with air conditioning: 19 qt. Carburetor: Rochester 4MV four-barrel. Engine code: US and UT.

1972 OLDSMOBILE

F-85, Cutlass, Cutlass 'S,' Cutlass Cruiser, Delta 88, Delta 88 Royale Series base V-8
V-8. Overhead valves. Cast-iron block. Bore & stroke: 4.06 x 3.39 in. Displacement: 350.0 cid. Compression ratio: 8.50:1. Net hp: 160 at 4000 rpm. Net torque: 275 lbs.-ft. at 2400 rpm. Hydraulic valve lifters. Five main bearings. Crankcase capacity: 4 qt. (add 1 qt. if new filter installed). Cooling system capacity with air conditioning: (midsize) 15.7 qt.; (full-size) 16.7 qt. Carburetor: Rochester 2GC two-barrel. VIN code: H.

F-85, Cutlass Series optional V-8
V-8. Overhead valves. Cast-iron block. Bore & stroke: 4.06 x 3.39 in. Displacement: 350.0 cid. Compression ratio: 8.50:1. Net hp: 175 at 4000 rpm. Net torque: 295 lbs.-ft. at 2600 rpm. Hydraulic valve lifters. Five main bearings. Crankcase capacity: 4 qt. (add 1 qt. if new filter installed). Cooling system capacity with air conditioning: 15.7 qt. Carburetor: Rochester 4MV four-barrel. VIN code: J.

F-85, Cutlass Series optional V-8
V-8. Overhead valves. Cast-iron block. Bore & stroke: 4.06 x 3.39 in. Displacement: 350.0 cid. Compression ratio: 8.50:1. Net hp: 180 at 4000 rpm. Net torque: 275 lbs.-ft. at 2800 rpm. Hydraulic valve lifters. Five main bearings. Crankcase capacity: 4 qt. (add 1 qt. if new filter installed). Cooling system capacity with air conditioning: 15.7 qt. Carburetor: Rochester 4MV four-barrel. VIN code: K.

Cutlass Supreme Series optional V-8
V-8. Overhead valves. Cast-iron block. Bore & stroke: 4.06 x 3.39 in. Displacement: 350.0 cid. Compression ratio: 8.50:1. Net hp: 200 at 4400 rpm. Net torque: 300 lbs.-ft. at 3200 rpm. Hydraulic valve lifters. Five main bearings. Crankcase capacity: 4 qt. (add 1 qt. if new filter installed). Cooling system capacity with air conditioning: 15.7 qt. Carburetor: Rochester 4MC four-barrel. VIN code: M.

Custom Cruiser, 98 Series base V-8
V-8. Overhead valves. Cast-iron block. Bore & stroke: 4.126 x 4.250 in. Displacement: 455 cid. Compression ratio: 8.50:1. Net hp: 225 at 3600 rpm. Net torque: 360 lbs.-ft. at 2600 rpm. Hydraulic valve lifters. Crankcase capacity: 4 qt. (add 1 qt. if new filter installed). Cooling system capacity with air conditioning: 17.5 qt. Carburetor: Rochester 4MC four-barrel. VIN code: T.

F-85, Cutlass, Delta, Royale, 98 Series optional V-8
V-8. Overhead valves. Cast-iron block. Bore & stroke: 4.126 x 4.250 in. Displacement: 455 cid. Compression ratio: 8.50:1. Net hp: 250 at 4200 rpm. Net torque: 370 lbs.-ft. at 2800 rpm. Hydraulic valve lifters. Crankcase capacity: 4 qt. (add 1 qt. if new filter installed). Cooling system capacity with air conditioning: (mid-size) 15.7 qt.; (full-size) 18 qt. Carburetor: Rochester 4MC four-barrel. VIN code: U.

Toronado base V-8
V-8. Overhead valves. Cast-iron block. Bore & stroke: 4.126 x 4.250 in. Displacement: 455 cid. Compression ratio: 8.50:1. Net hp: 265 at 4200 rpm. Net torque: 375 lbs.-ft. at 2800 rpm. Hydraulic valve lifters. Crankcase capacity: 4 qt. (add 1 qt. if new filter installed). Cooling system capacity with air conditioning: 20 qt. Carburetor: Rochester 4MV four-barrel. VIN code: W.

F-85, Cutlass optional V-8
V-8. Overhead valves. Cast-iron block. Bore & stroke: 4.126 x 4.250 in. Displacement: 455 cid. Compression ratio: 8.50:1. Net hp: 270 at 4400 rpm. Net torque: 370 lbs.-ft. at 3200 rpm. Hydraulic valve lifters. Crankcase capacity: 4 qt. (add 1 qt. if new filter installed). Cooling system capacity with air conditioning: 15.7 qt. Carburetor: Rochester 4MC four-barrel. VIN code: V.

F-85, Cutlass optional V-8
V-8. Overhead valves. Cast-iron block. Bore & stroke: 4.126 x 4.250 in. Displacement: 455 cid. Compression ratio: 8.50:1. Net hp: 300 at 4700 rpm. Net torque: 410 lbs.-ft. at 3200 rpm. Hydraulic valve lifters. Crankcase capacity: 4 qt. (add 1 qt. if new filter installed). Cooling system capacity with air conditioning: 15.7 qt. Carburetor: Rochester 4MC four-barrel. VIN code: X.

1973 OLDSMOBILE

Delta 88, Delta 88 Royale Series base V-8
V-8. Overhead valves. Cast-iron block. Bore & stroke: 4.06 x 3.39 in. Displacement: 350.0 cid. Compression ratio: 8.50:1. Net hp: 160 at 4000 rpm. Net torque: 275 lbs.-ft. at 2400 rpm. Hydraulic valve lifters. Five main bearings. Crankcase capacity: 4 qt. (add 1 qt. if new filter installed). Cooling system capacity with air conditioning: 16.25 qt. Carburetor: Rochester 2GC two-barrel. VIN code: H.

Omega, Cutlass, Cutlass 'S,' Cutlass Supreme, Vista Cruiser Series base V-8
V-8. Overhead valves. Cast-iron block. Bore & stroke: 4.06 x 3.39 in. Displacement: 350.0 cid. Compression ratio: 8.50:1. Net hp: 180 at 3800 rpm. Net torque: 275 lbs.-ft. at 2800 rpm. Hydraulic valve lifters. Five main bearings. Crankcase capacity: 4 qt. (add 1 qt. if new filter installed). Cooling system capacity with air conditioning: 16 qt. Carburetor: Rochester 4MV four-barrel. VIN code: K.

Custom Cruiser, 98 Series base V-8
V-8. Overhead valves. Cast-iron block. Bore & stroke: 4.126 x 4.250 in. Displacement: 455 cid. Compression ratio: 8.50:1. Net hp: 225 at 3600 rpm. Net torque: 360 lbs.-ft. at 2600 rpm. Hydraulic valve lifters. Crankcase capacity: 4 qt. (add 1 qt. if new filter installed). Cooling system capacity with air conditioning: 17.5 qt. Carburetor: Rochester 4MC four-barrel. VIN code: T.

Toronado Series base V-8
V-8. Overhead valves. Cast-iron block. Bore & stroke: 4.126 x 4.250 in. Displacement: 455 cid. Compression ratio: 8.50:1. Net hp: 250 at 4000 rpm. Net torque: 375 lbs.-ft. at 2800 rpm. Hydraulic valve lifters. Crankcase capacity: 4 qt. (add 1 qt. if new filter installed). Cooling system capacity with air conditioning: 20 qt. Carburetor: Rochester 4MC four-barrel. VIN code: U.

Custom Cruiser, Ninety-Eight Series optional V-8
V-8. Overhead valves. Cast-iron block. Bore & stroke: 4.126 x 4.250 in. Displacement: 455 cid. Compression ratio: 8.50:1. Net hp: 270 at 4200 rpm. Net torque: 370 lbs.-ft. at 3200 rpm. Hydraulic valve lifters. Crankcase capacity: 4 qt. (add 1 qt. if new filter installed). Cooling system capacity with air conditioning: 17.5 qt. Carburetor: Rochester 4MC four-barrel. VIN code: V.

1974 OLDSMOBILE

Omega, Cutlass, Cutlass 'S,' Cutlass Supreme, Vista Cruiser, Delta 88, Royale base V-8
V-8. Overhead valves. Cast-iron block. Bore & stroke: 4.06 x 3.39 in. Displacement: 350.0 cid. Compression ratio: 8.50:1. Net hp: 180 at 3800 rpm. Net torque: 275 lbs.-ft. at 2800 rpm. Hydraulic valve lifters. Five main bearings. Crankcase capacity: 4 qt. (add 1 qt. if new filter installed). Cooling system capacity with air conditioning: 20 qt. Carburetor: Rochester 4MV four-barrel. Serial number code: K.

Delta 88, Delta 88 Royale optional V-8
V-8. Overhead valves. Cast-iron block. Bore & stroke: 4.06 x 3.39 in. Displacement: 350.0 cid. Compression ratio: 8.50:1. Net hp: 200 at 4200 rpm. Net torque: 300 lbs.-ft. at 3200 rpm. Hydraulic valve lifters. Five main bearings. Crankcase capacity: 4 qt. (add 1 qt. if new filter installed). Cooling system capacity with air conditioning: 20/21 qt. Carburetor: Rochester 4MV four-barrel. Dual exhaust. Serial number code: M.

Custom Cruiser, Ninety-Eight Series base V-8; optional Delta 88, Delta 88 Royale, 98 V-8
V-8. Overhead valves. Cast-iron block. Bore & stroke: 4.126 x 4.250 in. Displacement: 455 cid. Compression ratio: 8.50:1. Net hp: 210 at 3600 rpm. Net torque: 350 lbs.-ft. at 2400 rpm. Hydraulic valve lifters. Crankcase capacity: 4 qt. (add 1 qt. if new filter installed). Cooling system capacity with air conditioning: 21.5 qt. Carburetor: Rochester 4MC four-barrel. Serial number code: T.

Toronado Series base V-8; optional Cutlass V-8; optional Delta 88, Delta 88 Royale, 98 V-8
V-8. Overhead valves. Cast-iron block. Bore & stroke: 4.126 x 4.250 in. Displacement: 455 cid. Compression ratio: 8.50:1. Net hp: 230 at 3800 rpm. Net torque: 370 lbs.-ft. at 2800 rpm. Hydraulic valve lifters. Crankcase capacity: 4 qt. (add 1 qt. if new filter installed). Cooling system capacity with air conditioning: 21.5 qt. Carburetor: Rochester 4MC four-barrel. Dual exhaust. Serial number code: U.

Cutlass, Custom Cruiser, Ninety-Eight Series optional V-8
V-8. Overhead valves. Cast-iron block. Bore & stroke: 4.126 x 4.250 in. Displacement: 455 cid. Compression ratio: 8.50:1. Net hp: 275 at 4200 rpm. Net torque: 395 lbs.-ft. at 3200 rpm. Hydraulic valve lifters. Crankcase capacity: 4 qt. (add 1 qt. if new filter installed). Cooling system capacity with air conditioning: 21.5 qt. Carburetor: Rochester 4MC four-barrel. Serial number code: V.

1975 OLDSMOBILE

Cutlass Series base V-8; Omega Series base V-8
V-8. Overhead valves. Cast-iron block. Bore & stroke: 3.50 x 3.385 in. Displacement: 260 cid. Compression ratio: 8.0:1. Net hp: 110 at 3400 rpm. Net torque: 205 lbs.-ft. at 1600 rpm. Five main bearings. Hydraulic valve lifters. Crankcase capacity: 4 qt. (add 1 qt. if new filter installed). Cooling system capacity with air conditioning: (Cutlass) 23-1/2 qt.; (Omega) 19.5 qt. Carburetor: Rochester Model 2GC, two-barrel. Serial number code: F.

Omega Series optional V-8 (Used only in Oldsmobile Omega)
V-8. Overhead valves. Cast-iron block. Bore & stroke: 3.80 x 3.85 in. Displacement: 350.0 cid. Compression ratio: 8.50:1. Net hp: 145 at 3600 rpm. Net torque: 250 lbs.-ft. at 2200 rpm. Hydraulic valve lifters. Five main bearings. Crankcase capacity: 4 qt. (add 1 qt. if new filter installed). Cooling system capacity with air conditioning: 19.5 qt. Carburetor: Rochester 4MV four-barrel. Built by Buick. Serial number code: J.

Omega Series optional V-8 (Used only in Oldsmobile Omega)
V-8. Overhead valves. Cast-iron block. Bore & stroke: 3.80 x 3.85 in. Displacement: 350.0 cid. Compression ratio: 8.50:1. Net hp: 165 at 3800 rpm. Net torque: 260 lbs.-ft. at 2200 rpm. Hydraulic valve lifters. Five main bearings. Crankcase capacity: 4 qt. (add 1 qt. if new filter installed). Cooling system capacity with air conditioning: 19.5 qt. Carburetor: Rochester 4MV four-barrel. Built by Buick. Serial number code: J.

Delta 88 Series base V-8; Cutlass Series optional V-8
V-8. Overhead valves. Cast-iron block. Bore & stroke: 4.06 x 3.39 in. Displacement: 350.0 cid. Compression ratio: 8.50:1. Net hp: 170 at 3800 rpm. Net torque: 275 lbs.-ft. at 2400 rpm. Hydraulic valve lifters. Five main bearings. Crankcase capacity: 4 qt. (add 1 qt. if new filter installed). Cooling system capacity with air conditioning: 20 qt. Carburetor: Rochester 4MV four-barrel. Serial number code: R.

Custom Cruiser Series base V-8; Delta 88 Series optional V-8; 98 Series optional V-8
V-8. Overhead valves. Cast-iron block. Bore & stroke: 4.121 x 3.75 in. Displacement: 400 cid. Compression ratio: 7.6:1. Net hp: 190 at 3400 rpm. Net torque: 350 lbs.-ft. at 2000 rpm. Five main bearings. Hydraulic valve lifters. Crankcase capacity: 4 qt. (add 1 qt. if new filter installed). Cooling system capacity with air conditioning: 22 qt. Carburetor: M4MC, four-barrel. Built by Pontiac. Serial number code: S.

98 Series base V-8; Cutlass Series optional V-8; Delta 88 optional V-8
V-8. Overhead valves. Cast-iron block. Bore & stroke: 4.125 x 4.250 in. Displacement: 455 cid. Compression ratio: 8.5:1. Net hp: 190 at 3600 rpm. Net torque: 350 lbs.-ft. at 2400 rpm. Five main bearings. Hydraulic valve lifters. Crankcase capacity: 4 qt. (add 1 qt. if new filter installed). Cooling system capacity with air conditioning: 21.5 qt. Carburetor: M4MC, four-barrel. Serial number code: W.

Toronado Series base V-8
V-8. Overhead valves. Cast-iron block. Bore & stroke: 4.125 x 4.250 in. Displacement: 455 cid. Compression ratio: 8.5:1. Net hp: 215 at 3600 rpm. Net torque: 370 lbs.-ft. at 2400 rpm. Five main bearings. Hydraulic valve lifters. Crankcase capacity: 4 qt. (add 1 qt. if new filter installed). Cooling system capacity with air conditioning: 21.5 qt. Carburetor: M4MC, four-barrel. Serial number code: T.

1976 OLDSMOBILE

Omega, Cutlass base V-8
90-degree, overhead valve V-8. Cast-iron block and head. Bore & stroke: 3.50 x 3.39 in. Displacement: 260 cid (4.3 liters). Compression ratio: 8.0:1. Brake horsepower: 110 at 3400 rpm. Torque: 205 lbs.-ft. at 1600 rpm. Five main bearings. Hydraulic valve lifters. Crankcase capacity: 4 qt. (add 1 qt. if new filter installed). Cooling system capacity with air conditioning: 23.5 qt. Carburetor: Rochester two-barrel. VIN code: F.

Omega optional V-8
90-degree, overhead valve V-8. Cast-iron block and head. Bore & stroke: 3.80 x 3.85 in. Displacement: 350 cid (5.7 liters). Compression ratio: 8.0:1. Brake horsepower: 140 at 3200 rpm. Torque: 280 lbs.-ft. at 1800 rpm. Five main bearings. Hydraulic valve lifters. Crankcase capacity: 4 qt. (add 1 qt. if new filter installed). Cooling system capacity with air conditioning: 21.5 qt. Carburetor: Rochester 2GC two-barrel. Built by Buick. Serial number code: J.

Delta 88, Cutlass Supreme, Vista Cruiser base V-8; Optional Omega V-8
90-degree, overhead valve V-8. Cast-iron block and head. Bore & stroke: 3.80 x 3.85 in. Displacement: 350 cid (5.7 liters). Compression ratio: 8.0:1. Brake horsepower: 155 at 3400 rpm. Torque: 280 lbs.-ft. at 1800 rpm. Five main bearings. Hydraulic valve lifters. Crankcase capacity: 4 qt. (add 1 qt. if new filter installed). Cooling system capacity with air conditioning: (Omega) 22 qt.; (Others) 20 qt. Carburetor: Rochester 4MV four-barrel. Built by Buick. Serial number code: H.

Optional Cutlass V-8
90-degree, overhead valve V-8. Cast-iron block and head. Bore & stroke: 4.06 x 3.39 in. Displacement: 350 cid (5.7 liters). Compression ratio: 8.5:1. Brake horsepower: 170 at 3800 rpm. Torque: 275 lbs.-ft. at 2400 rpm. Five main bearings. Hydraulic valve lifters. Crankcase capacity: 4 qt. (add 1 qt. if new filter installed). Cooling system capacity with air conditioning: 20 qt. Carburetor: Rochester four barrel. VIN code: R.

Base V-8 Ninety-Eight, Custom Cruiser; Optional Cutlass, Delta 88
90-degree, overhead valve V-8. Cast-iron block and head. Bore & stroke: 4.13 x 4.25 in. Displacement: 455 cid (7.5 liters). Compression ratio: 8.5:1. Brake horsepower: 190 at 3400 rpm. Torque: 350 lbs.-ft. at 2000 rpm. Five main bearings. Hydraulic valve lifters. Crankcase capacity: 4 qt. (add 1 qt. if new filter installed). Cooling system capacity with air conditioning: 21.5 qt. Carburetor: Rochester four-barrel. VIN code: T.

Toronado base V-8
90-degree, overhead valve V-8. Cast-iron block and head. Bore & stroke: 4.13 x 4.25 in. Displacement: 455 cid (7.5 liters). Compression ratio: 8.5:1. Brake horsepower: 215 at 3600 rpm. Torque: 370 lbs.-ft. at 2400 rpm. Five main bearings. Hydraulic valve lifters. Crankcase capacity: 4 qt. (add 1 qt. if new filter installed). Cooling system capacity with air conditioning: 21.5 qt. Carburetor: Rochester four-barrel. VIN code: S.

1977 OLDSMOBILE

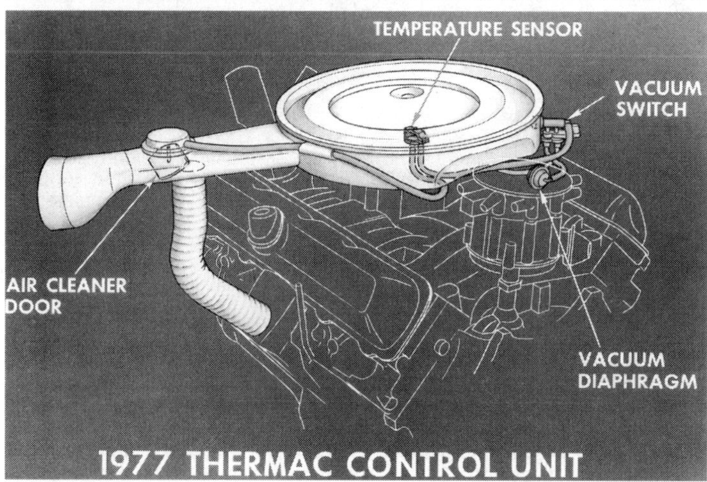

New for 1977 Oldsmobile V-8 engines was a Thermac Control unit that consisted of a temperature sensor in the air cleaner, a vacuum switch on the outside of the air cleaner, a vacuum diaphragm, and an air cleaner door. It was designed to make Oldsmobile engines conform to ever-tightening emissions requirements.

Cutlass Salon Series base V-8; Omega, Cutlass, Delta 88 Series optional V-8
90-degree, overhead valve V-8. Cast-iron block and head. Bore & stroke: 3.50 x 3.39 in. Displacement: 260 cid (4.3 liters). Bore & stroke: 3.50 x 3.39 in. Compression ratio: 7.5:1. Brake horsepower: 110 at 3400 rpm. Torque: 205 lbs.-ft. at 1800 rpm. Five main bearings. Hydraulic valve lifters. Carburetor: Rochester two-barrel. VIN code: F.

Omega Series optional V-8
90-degree, overhead valve V-8. Cast-iron block and head. Bore & stroke: 3.74 x 3.48 in. Displacement: 305 cid (5.0 liters). Compression ratio: 8.5:1. Brake horsepower: 145 at 3800 rpm. Torque: 245 lbs.-ft. at 2400 rpm. Five main bearings. Hydraulic valve lifters. Carburetor: Two-barrel. Built by Chevrolet. VIN code: U.

Omega Series optional V-8
90-degree, overhead valve V-8. Cast-iron block and head. Bore & stroke: 4.00 x 3.48 in. Displacement: 350 cid (5.7 liters). Compression ratio: 8.5:1. Brake horsepower: 170 at 3800 rpm. Torque: 270 lbs.-ft. at 2400 rpm. Five main bearings. Hydraulic valve lifters. Carburetor: Rochester four-barrel M4MC. VIN code: L.

98, Vista Cruiser, Custom Cruiser wagon base V-8; Cutlass, Delta 88 Series optional V-8
90-degree, overhead valve V-8. Cast-iron block and head. Bore & stroke: 4.06 x 3.39 in. Displacement: 350 cid (5.7 liters). Compression ratio: 8.0:1. Brake horsepower: 170 at 3800 rpm. Torque: 275 lbs.-ft. at 2000 rpm. Five main bearings. Hydraulic valve lifters. Carburetor: Rochester four-barrel. VIN code: R.

Cutlass, Delta 88, 98 optional V-8
90-degree, overhead valve V-8. Cast-iron block and head. Bore & stroke: 4.35 x 3.39 in. Displacement: 403 cid (6.6 liters). Compression ratio: 8.0:1. Brake horsepower: 185 at 3600 rpm. Torque: 320 lbs.-ft. at 2200 rpm. Five main bearings. Hydraulic valve lifters. Carburetor: Rochester four-barrel. VIN code: K.

Toronado base V-8
90-degree, overhead valve V-8. Cast-iron block and head. Bore & stroke: 4.35 x 3.39 in. Displacement: 403 cid (6.6 liters). Compression ratio: 8.0:1. Brake horsepower: 200 at 3600 rpm. Torque: 330 lbs.-ft. at 2400 rpm. Five main bearings. Hydraulic valve lifters. Carburetor: Rochester four-barrel. VIN code: K.

Oldsmobile built the 403-cid V-8 first used in 1977 Cutlass, Delta 88, and 98 models. It had a 4.35 x 3.39-inch bore and stroke and an 8.0:1 compression ratio. With a four-barrel carburetor, it developed 185 hp at 2200 rpm. A second version used in the 1977 Toronado had dual exhausts and cranked up an additional 15 horsepower.

1978 OLDSMOBILE

Cutlass, Cutlass wagon, Delta 88 optional V-8
90-degree, overhead valve V-8. Cast-iron block and head. Bore & stroke: 3.50 x 3.39 in. Displacement: 260 cid (4.3 liters). Compression ratio: 7.5:1. Brake horsepower: 110 at 3400 rpm. Torque: 205 lbs.-ft. at 1800 rpm. Five main bearings. Hydraulic valve lifters. Carburetor: Rochester two-barrel 2GC. VIN code: F.

Starfire, Omega, Cutlass optional V-8
90-degree, overhead valve V-8. Cast-iron block and head. Bore & stroke: 3.74 x 3.48 in. Displacement: 305 cid (5.0 liters). Compression ratio: 8.4:1. Brake horsepower: 145 at 3800 rpm. Torque: 245 lbs.-ft. at 2400

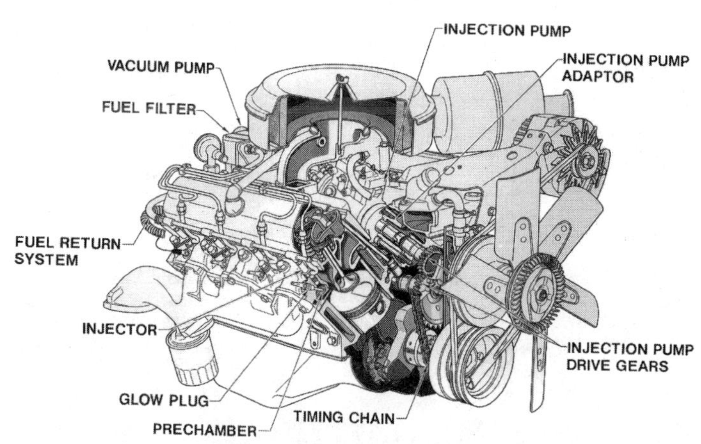

This factory cutaway illustration details the important features of the 1978 General Motors diesel engine produced by Oldsmobile. Used as optional equipment in Delta 88, 98, and Custom Cruiser models (as well as other GM cars), it was a 350-cid 90-degree, overhead valve V-8 with a 4.06 x 3.39 inch bore and stroke.

This special cutaway engine was made to promote the features of the General Motors diesel V-8 produced by Oldsmobile starting in 1978. It had a 22.5:1 compression ratio and developed 120 hp at 3600 rpm. Torque was 220 foot-pounds at 1600 rpm. The diesel V-8 had five main bearings, hydraulic valve lifters and fuel injection. Its VIN engine code was "N."

Here's a look at the left-hand side of the 1978 GM diesel V-8 that Oldsmobile Motor Division produced for the corporation. The release of a diesel was a reaction to gas shortages, but this first-generation version proved to lack the necessary stamina to give it a high reliability rating. Olds salesmen did a good job of selling the diesel, however, due to the gas crunch.

rpm. Five main bearings. Hydraulic valve lifters. Carburetor: Rochester two-barrel 2GC. Built by Chevrolet or GM of Canada. VIN code: U.

Cutlass optional V-8
90-degree, overhead valve V-8. Cast-iron block and head. Bore & stroke: 3.74 x 3.48 in. Displacement: 305 cid (5.0 liters). Compression ratio: 8.4:1. Brake horsepower: 160 at 4000 rpm. Torque: unknown. Five main bearings. Hydraulic valve lifters. Carburetor: Rochester four-barrel. Built by Chevrolet or GM of Canada. VIN code: H.

California/high-altitude Omega, Cutlass Cruiser optional V-8
90-degree, overhead valve V-8. Cast-iron block and head. Bore & stroke: 4.00 x 3.48 in. Displacement: 350 cid (5.7 liters). Compression ratio: 8.2:1. Brake horsepower: 160 at 3800 rpm. Torque: 260 lbs.-ft. at 2400 rpm. Five main bearings. Hydraulic valve lifters. Carburetor: Rochester four-barrel M4MC. Built by Chevrolet or GM of Canada. VIN code: L.

98, Custom Cruiser wagon base V-8; Delta 88 optional V-8
90-degree, overhead valve V-8. Cast-iron block and head. Bore & stroke: 4.06 x 3.39 in. Displacement: 350 cid (5.7 liters). Compression ratio: 7.9:1. Brake horsepower: 170 at 3800 rpm. Torque: 275 lbs.-ft. at 2000 rpm. Five main bearings. Hydraulic valve lifters. Carburetor: Rochester four-barrel M4MC. VIN code: R.

Delta 88, 98, Custom Cruiser Series optional V-8
90-degree, overhead valve V-8. Cast-iron block and head. Bore & stroke: 4.35 x 3.39 in. Displacement: 403 cid (6.6 liters). Compression ratio: 7.9:1. Brake horsepower: 185 at 3600 rpm. Torque: 320 lbs.-ft. at 2000 rpm. Five main bearings. Hydraulic valve lifters. Carburetor: Rochester four-barrel M4MC. VIN code: K.

Toronado Series optional V-8
90-degree, overhead valve V-8. Cast-iron block and head. Bore & stroke: 4.35 x 3.39 in. Displacement: 403 cid (6.6 liters). Compression ratio: 7.9:1. Brake horsepower: 190 at 3600 rpm. Torque: 325 lbs.-ft. at 2000 rpm. Five main bearings. Hydraulic valve lifters. Carburetor: Rochester four-barrel M4MC. VIN code: K.

Delta 88, 98, Custom Cruiser optional diesel V-8
90-degree, overhead valve V-8. Cast-iron block and head. Bore & stroke: 4.06 x 3.39 in. Displacement: 350 cid (5.7 liters). Compression ratio: 22.5:1. Brake horsepower: 120 at 3600 rpm. Torque: 220 lbs.-ft. at 1600 rpm. Five main bearings. Hydraulic valve lifters. Fuel injection. VIN code: N.

1979 OLDSMOBILE

Cutlass, Cutlass wagon, Delta 88 Series optional V-8
90-degree, overhead valve V-8. Cast-iron block and head. Bore & stroke: 3.50 x 3.39 in. Displacement: 260 cid (4.3 liters). Compression ratio: 7.5:1. Brake horsepower: 105 at 3600 rpm. Torque: 205 lbs.-ft. at 1800 rpm. Five main bearings. Hydraulic valve lifters. Carburetor: Rochester two-barrel M2MC. VIN code: F.

Cutlass Series optional diesel V-8
90-degree, overhead valve V-8. Cast-iron block and head. Bore & stroke: 3.50 x 3.39 in. Displacement: 260 cid (4.3 liters). Compression ratio: 22.5:1. Brake horsepower: 90 at 3600 rpm. Torque: 160 lbs.-ft. at 1600 rpm. Five main bearings. Hydraulic valve lifters. Induction: Fuel injection. VIN code: P.

Delta 88 Series optional V-8
90-degree, overhead valve V-8. Cast-iron block and head. Bore & stroke: 4.00 x 3.00 in. Displacement: 301 cid (4.9 liters). Compression ratio: 8.2:1. Brake horsepower: 135 at 3800 rpm. Torque: 240 lbs.-ft. at 1600 rpm. Five main bearings. Hydraulic valve lifters. Carburetor: Rochester two-barrel M2MC. Pontiac-built. VIN code: Y.

Starfire, Omega Series optional V-8
90-degree, overhead valve V-8. Cast-iron block and head. Bore & stroke: 3.74 x 3.48 in. Displacement: 305 cid (5.0 liters). Compression ratio: 8.4:1. Brake horsepower: 130 at 3200 rpm. Torque: 245 lbs.-ft. at 2000 rpm. Five main bearings. Hydraulic valve lifters. Carburetor: Rochester two-barrel M2MC. Chevrolet-built. VIN code: G.

Cutlass Series optional V-8
90-degree, overhead valve V-8. Cast-iron block and head. Bore & stroke: 3.74 x 3.48 in. Displacement: 305 cid (5.0 liters). Compression ratio: 8.4:1. Brake horsepower: 160 at 4000 rpm. Torque: 235 lbs.-ft. at 2400 rpm. Five main bearings. Hydraulic valve lifters. Carburetor: Rochester four-barrel M4MC. Chevrolet-built. VIN code: H.

California/high-altitude Omega, Cutlass Cruiser Series optional V-8
90-degree, overhead valve V-8. Cast-iron block and head. Bore & stroke: 4.00 x 3.48 in. Displacement: 350 cid (5.7 liters). Compression ratio: 8.5:1. Brake horsepower: 160 at 3800 rpm. Torque: 260 lbs.-ft. at 2400 rpm. Five main bearings. Hydraulic valve lifters. Carburetor: Rochester four-barrel M4MC. Chevrolet-built. VIN code: L.

Custom Cruiser wagon, Toronado, 98 Series base V-8; Delta 88 Series optional V-8
90-degree, overhead valve V-8. Cast-iron block and head. Bore & stroke: 4.06 x 3.39 in. Displacement: 350 cid (5.7 liters). Compression ratio: 8.0:1. Brake horsepower: 160-165 at 3800 rpm. Torque: 270-275 lbs.-ft. at 2000 rpm. Five main bearings. Hydraulic valve lifters. Carburetor: Rochester four-barrel M4MC. VIN code: R.

98, Custom Cruiser Series optional V-8
90-degree, overhead valve V-8. Cast-iron block and head. Bore & stroke: 4.35 x 3.39 in. Displacement: 403 cid (6.6 liters). Compression ratio: 7.8:1. Brake horsepower: 175 at 3600 rpm. Torque: 310 lbs.-ft. at 2000 rpm. Five main bearings. Hydraulic valve lifters. Carburetor: Rochester four-barrel M4MC. VIN code: K.

Cutlass Cruiser, Delta 88, 98, Custom Cruiser, Toronado optional diesel V-8
90-degree, overhead valve V-8. Cast-iron block and head. Bore & stroke: 4.06 x 3.39 in. Displacement: 350 cid (5.7 liters). Compression ratio: 22.5:1. Brake horsepower: 125 at 3600 rpm. Torque: 225 lbs.-ft. at 1600 rpm. Five main bearings. Hydraulic valve lifters. Induction: Fuel injection. VIN code: N.

1980 OLDSMOBILE

Optional V-8 Cutlass
90-degree, overhead valve V-8. Cast-iron block and head. Bore & stroke: 3.50 x 3.39 in. Displacement: 260 cid (4.3 liters). Compression ratio: 7.5:1. Brake horsepower: 105 at 3400 rpm. Torque: 195 lbs.-ft. at 1600 rpm. Five main bearings. Hydraulic valve lifters. Carburetor: Rochester two-barrel M2MC. VIN code: F.

Optional V-8 Delta 88
90-degree, overhead valve V-8. Cast-iron block and head. Bore & stroke: 3.75 x 3.00 in. Displacement: 265 cid (4.3 liters). Compression ratio: 8.3:1. Brake horsepower: 120 at 3600 rpm. Torque: 210 lbs.-ft. at 1800 rpm. Five main bearings. Hydraulic valve lifters. Carburetor: Two-barrel VIN code: S.

Base V-8 98, Toronado; Optional Delta 88
90-degree, overhead valve V-8. Cast-iron block and head. Bore & stroke: 3.80 x 3.39 in. Displacement: 307 cid (5.0 liters). Compression ratio: 8.0:1. Brake horsepower: 150 at 3600 rpm. Torque: 245 lbs.-ft. at 1600 rpm. Five main bearings. Hydraulic valve lifters. Carburetor: Rochester four-barrel M4MC. VIN code: Y.

Optional V-8 Cutlass
90-degree, overhead valve V-8. Cast-iron block and head. Displacement: 305 cid (5.0 liters). Bore & stroke: 3.74 x 3.48 in. Compression ratio: 8.6:1. Brake horsepower: 155 at 4000 rpm. Torque: 240 lbs.-ft. at 1600 rpm. Five main bearings. Hydraulic valve lifters. Carburetor: Rochester four-barrel M4MC. Chevrolet-built. VIN code: H.

Optional V-8 Delta 88, 98, Toronado
90-degree, overhead valve V-8. Cast-iron block and head. Displacement: 350 cid (5.7 liters). Bore & stroke: 4.06 x 3.39 in. Compression ratio: 8.0:1. Brake horsepower: 160 at 3600 rpm. Torque: 270 lbs.-ft. at 1600 rpm. Five main bearings. Hydraulic valve lifters. Carburetor: Rochester four-barrel E4MC. VIN code: R.

Diesel V-8 Cutlass, Delta 88, 98, Custom Cruiser, Toronado
90-degree, overhead valve V-8. Cast-iron block and head. Displacement: 350 cid (5.7 liters). Bore & stroke: 4.06 x 3.39 in. Compression ratio: 22.5:1. Brake horsepower: 105 at 3200 rpm. Torque: 205 lbs.-ft. at 1600 rpm. Five main bearings. Hydraulic valve lifters. Induction: Fuel injection. VIN code: N.

1981 OLDSMOBILE

Cutlass Supreme, Delta 88 Series base V-8
90-degree, overhead valve V-8. Cast-iron block and head. Bore & stroke: 3.50 x 3.39 in. Displacement: 260 cid (4.3 liters). Compression ratio: 7.5:1. Brake horsepower: 105 at 3600 rpm. Torque: 190 lbs.-ft. at 1600 rpm. Five main bearings. Hydraulic valve lifters. Carburetor. Rochester 2MC two-barrel. VIN code: F.

"Diesel is Popular" stated the caption that came with this photo in 1981. It suggested that Oldsmobile diesel-powered cars had increased in popularity since the 350-cid diesel V-8's introduction in 1978. A total of 31,610 diesel-powered models were delivered in 1978 and the following year the number jumped to 111,000. In 1980, the number of diesel-equipped units jumped to 115,000.

Regency 98, Toronado base V-8; Cutlass Supreme wagon, Delta 88 Series optional V-8
90-degree, overhead valve V-8. Cast-iron block and head. Bore & stroke: 3.80 x 3.38 in. Displacement: 307 cid (5.0 liters). Compression ratio: 8.0:1. Brake horsepower: 140 at 3800 rpm. Torque: 240 lbs.-ft. at 1600 rpm. Five main bearings. Hydraulic valve lifters. Carburetor: Rochester M4ME four-barrel. VIN code: Y.

Cutlass, Delta 88, Regency 98, Custom Cruiser, Toronado Series optional diesel V-8
90-degree, overhead valve V-8. Cast-iron block and head. Bore & stroke: 4.06 x 3.39 in. Displacement: 350 cid (5.7 liters). Compression ratio: 22.5:1. Brake horsepower: 105 at 3200 rpm. Torque: 200 lbs.-ft. at 1600 rpm. Five main bearings. Hydraulic valve lifters. Induction: Fuel injection. VIN code: N.
Note: A V-6 was the lowest-priced engine in all lines, but in the Cutlass Supreme, Delta 88, Regency 98, and Toronado series the smallest available V-8 was considered the "base" V-8. Also, the Delta 88 station wagon came with the 307-cid V-8 as standard equipment.

Oldsmobile promoted the diesel V-8 as a fuel efficient power plant. In 1981, it was offered as an option on 19 of the 23 models marketed by Oldsmobile. Several engineering improvements were carried out in 1981 to make diesel ownership more convenient. They included an increase in the recommended oil and filter change intervals and a light to signal if excessive amounts of water were in the fuel tank. The diesel engine's fuel economy was said to be 25 percent better than that of a gas engine.

1982 OLDSMOBILE

Cutlass Supreme Series base V-8; Delta 88 Series (except station wagon) base V-8
90-degree, overhead valve V-8. Cast-iron block and head. Bore & stroke: 3.50 x 3.38 in. Displacement: 260 cid (4.3 liters). Compression ratio: 7.5:1. Brake horsepower: 100 at 3600 rpm. Torque: 190 lbs.-ft. at 1600 rpm. Five main bearings. Hydraulic valve lifters. Carburetor: Rochester two-barrel E2ME. VIN code: 8.

Regency 98, Custom Cruiser, Toronado Series base V-8; Cutlass Supreme station wagon optional V-8; Delta 88 Series optional V-8
90-degree, overhead valve V-8. Cast-iron block and head. Bore & stroke: 3.80 x 3.38 in. Displacement: 307 cid (5.0 liters). Compression ratio: 8.0:1. Brake horsepower: 140 at 3600 rpm. Torque: 240 lbs. ft. at 1600 rpm. Five main bearings. Hydraulic valve lifters. Carburetor: Rochester four-barrel M4MC. VIN code: Y.

Cutlass Supreme, Delta 88, Regency 98, Custom Cruiser, Toronado optional diesel V-8
90-degree, overhead valve V-8. Cast-iron block and head. Bore & stroke: 4.06 x 3.39 in. Displacement: 350 cid (5.7 liters). Compression ratio: 21.6:1. Brake horsepower: 105 at 3200 rpm. Torque: 200 lbs.-ft. at 1600 rpm. Five main bearings. Hydraulic valve lifters. Induction: Fuel injection. VIN code: N.

1983 OLDSMOBILE

Custom Cruiser Series base V-8; Cutlass Supreme, Delta 88, 98, Toronado Series base V-8
90-degree, overhead valve V-8. Cast-iron block and head. Bore & stroke: 3.80 x 3.38 in. Displacement: 307 cid (5.0 liters). Compression ratio: 8.0:1. Brake horsepower: 140 at 3600 rpm. Torque: 240 lbs.-ft. at 1600 rpm. Five main bearings. Hydraulic valve lifters. Carburetor: Rochester four-barrel. VIN code: Y.

Hurst-Olds base V-8; Cutlass Calais series optional V-8
90-degree, overhead valve V-8. Cast-iron block and head. Bore & stroke: 3.80 x 3.38 in. Displacement: 307 cid (5.0 liters). Compression ratio: 9.0:1. Brake horsepower: 180 at 4000 rpm. Torque: 245 lbs.-ft. at 3200 rpm. Five main bearings. Hydraulic valve lifters. Carburetor: Rochester four-barrel. VIN code: 9.

Cutlass Supreme, Delta 88, 98, Custom Cruiser, Toronado Series optional diesel V-8
90-degree, overhead valve V-8. Bore & stroke: 4.06 x 3.39 in. Cast-iron block and head. Displacement: 350 cid (5.7 liters). Compression ratio: 22.5:1. Brake horsepower: 105 at 3200 rpm. Torque: 200 lbs.-ft. at 1600 rpm. Five main bearings. Hydraulic valve lifters. Induction: Fuel injection. VIN code: N.

1984 OLDSMOBILE

Cutlass Supreme, Delta 88, Custom cruiser, Regency 98, Toronado Series base V-8
90-degree, overhead valve V-8. Cast-iron block and head. Bore & stroke: 3.80 x 3.38 in. Displacement: 307 cid (5.0 liters). Compression ratio: 8.0:1. Brake horsepower: 140 at 3600 rpm. Torque: 240 lbs.-ft. at 1600 rpm. Five main bearings. Hydraulic valve lifters. Carburetor: Rochester four-barrel. VIN code: Y.

Hurst-Olds base V-8; Cutlass Calais Series optional V-8
90-degree, overhead valve V-8. Cast-iron block and head. Bore & stroke: 3.80 x 3.38 in. Displacement: 307 cid (5.0 liters). Compression ratio: 9.0:1. Brake horsepower: 180 at 4000 rpm. Torque: 245 lbs.-ft. at 3200 rpm. Five main bearings. Hydraulic valve lifters. Carburetor: Rochester four-barrel. VIN code: 9.

Cutlass Supreme, Delta 88, 98, Custom Cruiser, Toronado Series optional diesel V-8
90-degree, overhead valve V-8. Cast-iron block and head. Bore & stroke: 4.06 x 3.39 in. Displacement: 350 cid (5.7 liters). Compression ratio: 22.7:1. Brake horsepower: 105 at 3200 rpm. Torque: 200 lbs.-ft. at 1600 rpm. Five main bearings. Hydraulic valve lifters. Fuel injection. VIN code: N.

1985 OLDSMOBILE

Delta 88, Delta 88 LS, Custom Cruiser, Toronado Series base V-8
90-degree, overhead valve V-8. Cast-iron block and head. Bore & stroke: 3.80 x 3.38 in. Displacement: 307 cid (5.0 liters). Compression ratio: 8.0:1. Brake horsepower: 140 at 3600 rpm. (Cutlass, 140 at 3200 rpm). Torque: 240 lbs.-ft. at 1600 rpm. (Cutlass, 255 lbs.-ft. at 2000 rpm). Five main bearings. Hydraulic valve lifters. Carburetor: Rochester four-barrel 4MC. VIN code: Y.

Cutlass Supreme Series base V-8
90-degree, overhead valve V-8. Cast-iron block and head. Bore & stroke: 3.80 x 3.38 in. Displacement: 307 cid (5.0 liters). Compression ratio: 8.0:1. Brake horsepower: 140 at 3200 rpm. Torque: 255 lbs.-ft. at 2000 rpm. Five main bearings. Hydraulic valve lifters. Carburetor: Rochester four-barrel 4MC. VIN code: Y.

Hurst-Olds base V-8; Cutlass Calais, Cutlass Supreme Series optional V-8
90-degree, overhead valve V-8. Cast-iron block and head. Bore & stroke: 3.80 x 3.38 in. Displacement: 307 cid (5.0 liters). Compression ratio: 9.0:1. Brake horsepower: 180 at 4000 rpm. Torque: 245 lbs.-ft. at 3200 rpm. Five main bearings. Hydraulic valve lifters. Carburetor: Rochester four-barrel. VIN code: 9.

Cutlass Supreme, Delta 88, Custom Cruiser, Toronado optional Diesel V-8
90-degree, overhead valve V-8. Cast-iron block and head. Bore & stroke: 4.06 x 3.39 in. Displacement: 350 cid (5.7 liters). Compression ratio: 22.7:1. Brake horsepower: 105 at 3200 rpm. Torque: 200 lbs.-ft. at 1600 rpm. Five main bearings. Hydraulic valve lifters. Induction: Fuel injection. VIN code: N.

1986 OLDSMOBILE

Custom Cruiser, Cutlass Supreme Series base V-8
90-degree, overhead valve V-8. Cast-iron block and head. Bore & stroke: 3.80 x 3.38 in. Displacement: 307 cid (5.0 liters). Compression ratio: 8.0:1. Brake horsepower: 140 at 3200 rpm. Torque: 255 lbs.-ft. at 2000 rpm. Five main bearings. Hydraulic valve lifters. Carburetor: Rochester four-barrel E4MC. VIN code: Y.

Hurst-Olds, Cutlass Supreme Series optional V-8
90-degree, overhead valve V-8. Cast-iron block and head. Bore & stroke: 3.80 x 3.38 in. Displacement: 307 cid (5.0 liters). Compression ratio: 8.0:1. Brake horsepower: 170 at 4000 rpm. Torque: 250 lbs.-ft. at 2600 rpm. Five main bearings. Hydraulic valve lifters. Carburetor: Rochester four-barrel E4MC. VIN code: 9.
Note: The Custom Cruiser station wagon was available only with the 307-cid V-8.

1987 OLDSMOBILE

Custom Cruiser, Cutlass Supreme Series base V-8
90-degree, overhead valve V-8. Cast-iron block and head. Bore & stroke: 3.80 x 3.38 in. Displacement: 307 cid (5.0 liters). Compression ratio: 8.0:1. Brake horsepower: 140 at 3200 rpm. Torque: 255 lbs.-ft. at 2000 rpm. Five main bearings. Hydraulic valve lifters. Carburetor: Rochester four-barrel E4MC. VIN code: Y.

Cutlass Supreme Series optional V-8
90-degree, overhead valve V-8. Cast-iron block and head. Bore & stroke: 3.80 x 3.38 in. Displacement: 307 cid (5.0 liters). Compression ratio: 8.0:1. Brake horsepower: 170 at 4000 rpm. Torque: 250 lbs.-ft. at 2600 rpm. Five main bearings. Hydraulic valve lifters. Carburetor: Rochester four-barrel E4MC. VIN code: 9.
Note: The Custom Cruiser station wagon was available only with the 307-cid V-8.

1988 OLDSMOBILE

Cutlass Supreme Classic, Custom Cruiser Series base V-8
Overhead valve V-8. Cast-iron block and head. Bore & stroke: 3.80 x 3.40 in. Displacement: 307 cid (5.0 liters). Compression ratio: 7.9:1. Brake horsepower: 140 at 3200 rpm. Torque: 255 lbs.-ft. at 2000 rpm. Carburetor: Rochester four-barrel. VIN code: Y.

1989 OLDSMOBILE

Custom Cruiser base V-8
90-degree overhead valve V-8. Cast-iron block and head. Bore & stroke: 3.80 x 3.39 in. Displacement: 307 cid (5.0 liters). Compression ratio: 8.0:1. Brake horsepower: 140 at 3200 rpm. Torque: 255 lbs.-ft. at 2000 rpm. Carburetor: Four-barrel. VIN code: Y.

1990 OLDSMOBILE

Custom Cruiser base V-8
90-degree overhead valve V-8. Cast-iron block and head. Bore & stroke: 3.80 x 3.39 in. Displacement: 307 cid (5.0 liters). Compression ratio: 8.0:1. Brake horsepower: 140 at 3200 rpm. Torque: 255 lbs.-ft. at 2000 rpm. Carburetor: Four-barrel. VIN code: Y.

1991 OLDSMOBILE

Custom Cruiser base V-8
90-degree overhead valve V-8. Cast-iron block and head. Bore & stroke: 3.74 x 3.48 in. Displacement: 305 cid (5.0 liters). Compression ratio: 9.3:1. Brake horsepower: 170 at 4200 rpm. Torque: 255 lbs.-ft. at 2500 rpm. Induction: Fuel injected (EFI). VIN code: E.

1992 OLDSMOBILE

Custom Cruiser base V-8
90-degree overhead valve V-8. Cast-iron block and head. Bore & stroke: 3.74 x 3.48 in. Displacement: 305 cid (5.0 liters). Compression ratio: 9.3:1. Brake horsepower: 170 at 4200 rpm. Torque: 255 lbs.-ft. at 2500 rpm. Induction: Fuel injected (EFI). VIN code: E.

Custom Cruiser optional V-8
90-degree overhead valve V-8. Cast-iron block and head. Bore & stroke: 4.00 x 3.48 in. Displacement: 350 cid (5.7 liters). Compression ratio: 9.8:1. Brake horsepower: 180 at 4000 rpm. Torque: 300 lbs.-ft. at 2400 rpm. Induction: Fuel injected (EFI). VIN code: E.

1993 OLDSMOBILE

No Oldsmobile V-8s.

1994 OLDSMOBILE

No Oldsmobile V-8s.

1995 OLDSMOBILE

Base V-8
Aurora 90-degree dual overhead-cam V-8. Aluminum block and heads. Displacement: 244 cid (4.0 liters). Bore & stroke: 3.43 x 3.31 in. Compression ratio: 10.3:1. Brake horsepower: 250 at 5600 rpm. Torque: 260 lbs.-ft. at 4400 rpm. Crankcase capacity: 5.5 qt. Cooling system capacity: 9.6 qt. Induction: Fuel injected (SFI). Aurora V-8 indicated by C as the eighth symbol in VIN.

1996 OLDSMOBILE

Base V-8
Aurora 90-degree dual overhead-cam V-8. Aluminum block and heads. Displacement: 244 cid (4.0 liters). Bore & stroke: 3.43 x 3.31 in. Compression ratio: 10.3:1. Brake horsepower: 250 at 5600 rpm. Torque: 260 lbs.-ft. at 4400 rpm. Crankcase capacity: 5.5 qt. Cooling system capacity: 9.6 qt. Induction: Fuel injected (SFI). Aurora V-8 indicated by C as the eighth symbol in VIN.

1997 OLDSMOBILE

Base V-8
Aurora 90-degree dual overhead-cam V-8. Aluminum block and heads. Displacement: 244 cid (4.0 liters). Bore & stroke: 3.43 x 3.31 in. Compression ratio: 10.3:1. Brake horsepower: 250 at 5600 rpm. Torque: 260 lbs.-ft. at 4400 rpm. Crankcase capacity: 5.5 qt. Cooling system capacity: 9.6 qt. Induction: Fuel injected (SFI). Aurora V-8 indicated by C as the eighth symbol in VIN.

1998 OLDSMOBILE

Base V-8
Aurora 90-degree dual overhead-cam V-8. Aluminum block and heads. Displacement: 244 cid (4.0 liters). Bore & stroke: 3.43 x 3.31 in. Compression ratio: 10.3:1. Brake horsepower: 250 at 5600 rpm. Torque: 260 lbs.-ft. at 4400 rpm. Crankcase capacity: 5.5 qt. Cooling system capacity: 9.6 qt. Induction: Fuel injected (SFI). Aurora V-8 indicated by C as the eighth symbol in VIN.

1999 OLDSMOBILE

Base V-8
Aurora 90-degree dual overhead-cam V-8. Aluminum block and heads. Displacement: 244 cid (4.0 liters). Bore & stroke: 3.43 x 3.31 in. Compression ratio: 10.3:1. Brake horsepower: 250 at 5600 rpm. Torque: 260 lbs.-ft. at 4400 rpm. Crankcase capacity: 5.5 qt. Cooling system capacity: 9.6 qt. Induction: Fuel injected (SFI). Aurora V-8 indicated by C as the eighth symbol in VIN.

2000 OLDSMOBILE

Base V-8
Aurora 90-degree dual overhead-cam V-8. Aluminum block and heads. Displacement: 244 cid (4.0 liters). Bore & stroke: 3.43 x 3.31 in. Compression ratio: 10.3:1. Brake horsepower: 250 at 5600 rpm. Torque: 260 lbs.-ft. at 4400 rpm. Crankcase capacity: 5.5 qt. Cooling system capacity: 9.6 qt. Induction: Fuel injected (SFI). Aurora V-8 indicated by C as the eighth symbol in VIN.

2001 OLDSMOBILE

Base V-8
Aurora 90-degree dual overhead-cam V-8. Aluminum block and heads. Displacement: 244 cid (4.0 liters). Bore & stroke: 3.43 x 3.31 in. Compression ratio: 10.3:1. Brake horsepower: 250 at 5600 rpm. Torque: 260 lbs.-ft. at 4400 rpm. Crankcase capacity: 5.5 qt. Cooling system capacity: 9.6 qt. Induction: Fuel injected (SFI). Aurora V-8 indicated by C as the eighth symbol in VIN.

2002 OLDSMOBILE

Base V-8
Aurora 90-degree dual overhead-cam V-8. Aluminum block and aluminum heads. Displacement: 244 cid (4.0 liters). Bore & stroke: 3.43 x 3.31 in. Compression ratio: 10.3:1. Brake horsepower: 250 at 5600 rpm. Torque: 260 lbs.-ft. at 4400 rpm. Crankcase capacity: 5.5 qt. Cooling system capacity: 9.6 qt. Induction: Fuel injected (SFI). Aurora V-8 indicated by C as the eighth symbol in VIN

Selected Oldsmobile V-8 Tune-Up Specs 1949-1965

Year	Model	Spark Plugs Type	Gap	Point Gap	Deg Cam Angle	Timing	Idle Speed	Compression Pressure	Battery (Delco)
1939	Mercury V-8	H-10	.025	.016	36	4B	425 rpm	105-110 lb.	2F
1949	88	AC-46.5	.030	.015	22	2.5B	425 rpm	136	Group 2
	98	AC-46.5	.030	.015	22	2.5B	425 rpm	136	Group 2
1950	88	AC-46.5	.030	.015	22	2.5B	425 rpm	136	Group 2
	98	AC-46.5	.030	.015	22	2.5B	425 rpm	136	Group 2
1951	All V-8	AC-46.5	.030	.016	30	2.5B	425 rpm	136	Group 2
1952	All V-8	AC-46.5	.030	.016	30	2.5B	425 rpm	136	Group 2
1953	All V-8	AC-46.5	.030	.016	30	2.5B	425 rpm	136	3EM/3EE
1954	All V-8	AC-46.5	.030	.016	30	5B	425 rpm	136	3KM/3EA
1955	All V-8	AC-44	.030	.016	30	5B	425 rpm	140	3KM/3EA
1956	All V-8	AC-44	.030	.016	30	5B	425 rpm	150	3KM/3EA
1957	All V-8	AC-44	.030	.016	30	5B	450 rpm	150	3KM/3EA
1958	All V-8	AC-44	.028-.030	.016	28-32	5B	460 rpm	NA	Delc 562
1959	All V-8		AC-44	.028-.030	.01628-32	5B	460 rpm	NA	Delc 562

Note: Recommended idle speed with air conditioning is 520 rpm.

Year	Model	Spark Plugs Type	Gap	Point Gap	Deg Cam Angle	Timing	Idle Speed	Compression Pressure	Battery (Delco)
1960	88	AC-45	.030	.016	28-32	5B	460 rpm	NA	Delc 462
	Super 88, 98	AC-44	.030	.016	28-32	7.5B	460 rpm	NA	Delc 562

Note: Recommended idle speed with air conditioning is 520 rpm.

Year	Model	Spark Plugs Type	Gap	Point Gap	Deg Cam Angle	Timing	Idle Speed	Compression Pressure	Battery (Delco)
1961	F-85	AC-46FF	.035	.016	28-32	5B	500 rpm	NA	1980455
	88	AC-44	.030	.016	28-32	5B	550 rpm	NA	462 - 463
	Super 88, 98	AC-45	.030	.016	28-32	5B	550 rpm	NA	562 – 563

Note: Idle speeds vary from 500-550 rpm depending on transmission.

Year	Model	Spark Plugs Type	Gap	Point Gap	Deg Cam Angle	Timing	Idle Speed	Compression Pressure	Battery (Delco)
1962	F-85	AC46FFX	.030	.016	28-32	5B	550 rpm	NA	1980454
	88, Super 88	AC-44	.030	.016	28-32	5B	550 rpm	NA	562
	98	AC-44	.030	.016	28-32	5B	500 rpm	NA	562
1962	F-85	AC46FFX	.030	.016	28-32	5B	550 rpm	NA	1980544
	F-85 (optional V-8)	AC45FF	.030	.016	28-32	7.5B	550 rpm	NA	1980544
	Jetfire	AC45FF	.030	.016	28-32	10B	550 rpm	NA	1980544
	88, Super 88	AC-44	.030	.016	28-32	2.5B	550 rpm	NA	562
	98	AC-44	.030	.016	28-32	5B	550 rpm	NA	562
1964	F-85 (2 barrel)	AC-44S	.035	.016	30	7.5B	550 rpm	180	61amp hr
	F-85 (4 barrel)	AC-45S	.035	.016	30	7.5B	550 rpm	180	61amp hr
	Jetstar (2 barrel)	AC-45	.030	.016	30	7.5B	550 rpm	180	61amp hr
	Jetstar (4 barrel)	AC-44	.030	.016	30	7.5B	550 rpm	180	61amp hr
	Dynamic 88 (man)	AC-45	.030	.016	30	2.5B	550 rpm	185	62amp hr
	Dynamic 88 (auto)	AC-45	.030	.016	30	5B	550 rpm	185	62amp hr
	Others (man)	AC-44	.030	.016	30	2.5B	550 rpm	185	62amp hr
	Others (auto)	AC-44	.030	.016	30	5B	550 rpm	185	62amp hr

Note: 70 amp-hr battery with high-compression engines

Year	Model	Spark Plugs Type	Gap	Point Gap	Deg Cam Angle	Timing	Idle Speed	Compression Pressure	Battery (Delco)
1965	F-85 (2 barrel)	AC-45S	.035	.016	30	5B	550 rpm	180	44amp hr
	F-85 (4 barrel)	AC-44S	.035	.016	30	7.5B	550 rpm	180	44amp hr
	Jetstar (2 barrel)	AC-45S	.030	.016	30	7.5B	550 rpm	180	44amp hr
	Jetstar (4 barrel)	AC-44S	.030	.016	30	7.5B	550 rpm	180	44amp hr
	Dynamic 88 (2 bar)	AC-45S	.030	.016	30	7.5B	850 rpm	170	61amp hr
	98	AC-44S	.030	.016	30	5B	850 rpm	180	61amp hr
	Jetstar 1	AC-44S	.030	.016	30	5B	850 rpm	180	61amp hr

Starfire	AC-44S	.030	.016	30	5B	850 rpm	180	61amp hr

Note: 70 amp-hr battery with high-compression engines

Oldsmobile V-8 Firing Order (1949-1976)

1949-1960
(All V-8) 1-8-7-3-6-5-4-2
Note: Cylinders are numbered front to back (left) 1-3-5-7; (right) 2-4-6-8

1961-1964
(F-85 V-8) 1-8-4-3-6-5-7-2
(Olds V-8) 1-8-7-3-6-5-4-2
Note: Cylinders are numbered front to back (left) 1-3-5-7; (right) 2-4-6-8

1965-1976
(All V-8) 1-8-4-3-6-5-7-2
Note: Cylinders are numbered front to back (left) 1-3-5-7; (right) 2-4-6-8

PACKARD
1955-1958

One of America's oldest carmakers, Packard continued building outstanding automobiles after World War II. Unfortunately, its cars failed to find popularity with a new breed of buyers who wanted a radically different type of product. The marque lasted only until 1958.

For years, Packard's business philosophy avoided annual model-year changes. The directors of the company felt that such customers preferred a quality product that could retain its style value one season after the next. After 1949, the growing emphasis on styling, sportiness, and high-performance teamed up to reek havoc with Packard sales.

The first postwar Packards saw daylight on a chilly, gray October morning in 1945 and looked much like 1942 models. The year 1948 brought a new look. It was one that seemed to have mass appeal and 146,441 were sold, but in 1949, Packard's weaknesses began to show. In contrast to the models from other firms, buyers began viewing Packard as old-fashioned. One thing that did little to combat the image was Packard's lack of a modern, overhead valve V-8. Packard made the largest straight eight available at the time, but it wasn't enough.

While neglecting the need for a V-8, Packard made a valiant attempt to get back on track in 1951 with an almost completely new car. The addition of a handsome Mayfair hardtop was also a step in the right direction. Unfortunately, it wasn't enough and the indicators were still being ignored. Despite flashier and fancier models in 1953 and 1954, the new Packards remained virtually unchanged in basic terms. By the end of the Korean War in 1954, Packard assemblies were down to 27,593 cars per year.

Forced into a corner, Packard finally went to V-8 power for 1955. There were also new three-tone color combinations, hooded headlights, torsion bars, wraparound windshields, a hint of tailfins, and a medium-priced Clipper line. These revisions came with bills for new tooling, technology, and factory expansion; costs that Packard could no longer bear alone. Company president James Nance went looking for a financial partner and purchased Studebaker. This move seemed logical on the surface, but brought more problems because Studebaker was also overextended.

When sales dropped to 13,000 cars in 1956, Nance bailed out. Curtiss-Wright Corp. gained management control of Studebaker-Packard, primarily as a tax write-off, and quickly decided to consolidate its holdings at Studebaker's South Bend, Ind., facilities. Many profitable aspects of the auto company (defense contracts, etc.) were acquired by Curtiss-Wright. What remained of Studebaker-Packard was eventually released under its own management.

For the next two years, the Packard became a badge-engineered car built off a Studebaker platform. Though technically a fine automobile, and certainly rare and unique, this new type of Packard was unable to attract repeat customers due to its lack of identity and some controversial styling. After 1958, the name Packard disappeared from the automotive scene.

Arranged in order by horsepower ratings.

1955 PACKARD

Clipper Deluxe, Super 5540 base V-8
Overhead valve V-8. Cast-iron block. Bore & stroke: 3.813 x 3.50 in. Displacement: 320 cid. Compression ratio: 8.5:1. Brake hp: 225 at 4600 rpm. Taxable hp: 46.5. Torque: 325 lbs.-ft. at 2600 rpm. Five main bearings. Hydraulic valve lifters. Crankcase capacity: 5 qt. (add 1 qt. if new filter installed). Cooling system capacity: 26 qt. (add 1 qt. for heater). Carburetor: Carter WCFB-2232-S four-barrel. Engine code: 5540.

Clipper Custom 5560 base V-8
Overhead valve V-8. Cast-iron block. Bore & stroke: 4.00 x 3.50 in. Displacement: 352 cid. Compression ratio: 8.5:1. Brake hp: 245 at 4600 rpm. Taxable hp: 51.2. Torque: 355 lbs.-ft. at 2600 rpm. Five main bearings. Hydraulic valve lifters. Crankcase capacity: 5 qt. (add 1 qt. if new filter installed). Cooling system capacity: 26 qt. (add 1 qt. for heater). Carburetor: Rochester 4GC four-barrel. Engine code: 5560.

Packard 5580 base V-8
Overhead valve V-8. Cast-iron block. Bore & stroke: 4.00 x 3.50 in. Displacement: 352 cid. Compression ratio: 8.5:1. Brake hp: 260 at 4600 rpm. Taxable hp: 51.2. Torque: 355 lbs.-ft. at 2600 rpm. Five main bearings. Hydraulic valve lifters. Crankcase capacity: 5 qt. (add 1 qt. if new filter installed). Cooling system capacity: 26 qt. (add 1 qt. for heater). Carburetor: Rochester 4GC four-barrel. Engine code: 5580.

Packard Caribbean base V-8
Overhead valve V-8. Cast-iron block. Bore & stroke: 4.00 x 3.50 in. Displacement: 352 cid. Compression ratio: 8.5:1. Brake hp: 275 at 4800 rpm. Taxable hp: 51.2. Torque: 355 lbs.-ft. at 2800 rpm. Five main bearings. Hydraulic valve lifters. Crankcase capacity: 5 qt. (add 1 qt. if new filter installed). Cooling system capacity: 26 qt. (add 1 qt. for heater). Carburetors: Two Rochester 4GC four-barrels. Engine code: 5588.

1956 PACKARD

Clipper Deluxe, Super 5640 base V-8
Overhead valve V-8. Cast-iron block. Bore & stroke: 4.00 x 3.50 in. Displacement: 352 cid. Compression ratio: 9.5:1. Brake hp: 240 at 4600 rpm. Taxable hp: 51.2. Torque: 350 lbs.-ft. at 2800 rpm. Five main bearings. Hydraulic valve lifters. Crankcase capacity: 5 qt. (add 1 qt. if new filter installed). Cooling system capacity: 26 qt. (add 1 qt. for heater). Carburetor: Carter two-barrel. Engine code: 5540.

Clipper Custom, Executive 5660 base V-8
Overhead valve V-8. Cast-iron block. Bore & stroke: 4.00 x 3.50 in. Displacement: 352 cid. Compression ratio: 9.5:1. Brake hp: 275 at 4600 rpm. Taxable hp: 51.2. Torque: 380 lbs.-ft. at 2800 rpm. Five main bearings. Hydraulic valve lifters. Crankcase capacity: 5 qt. (add 1 qt. if new filter installed). Cooling system capacity: 26 qt. (add 1 qt. for heater). Carburetor: Carter four-barrel. Engine code: (Custom) 5560 or (Executive) 5570.

Packard sedan, hardtop 5680 base V-8
Overhead valve V-8. Cast-iron block. Bore & stroke: 4.00 x 3.50 in. Displacement: 352 cid. Compression ratio: 10.00:1. Brake hp: 290 at 4600 rpm. Taxable hp: 51.2. Torque: 405 lbs.-ft. at 2800 rpm. Five main bearings. Hydraulic valve lifters. Crankcase capacity: 5 qt. (add 1 qt. if new filter installed). Cooling system capacity: 26 qt. (add 1 qt. for heater). Carburetor: Rochester four-barrel. Engine code: 5580.

Packard Caribbean convertible 5688 base V-8
Overhead valve V-8. Cast-iron block. Bore & stroke: 4.00 x 3.50 in. Displacement: 352 cid. Compression ratio: 10.00:1. Brake hp: 310 at 4600 rpm. Taxable hp: 51.2. Torque: 405 lbs.-ft. at 2800 rpm. Five main bearings. Hydraulic valve lifters. Crankcase capacity: 5 qt. (add 1 qt. if new filter installed). Cooling system capacity: 26 qt. (add 1 qt. for heater). Carburetors: Two Rochester four-barrels. Engine code: 5588.

1957 PACKARD

Clipper 57L base Supercharged V-8
Overhead valve supercharged V-8. Cast-iron block. Bore & stroke: 3.562 x 3.625 in. Displacement: 289.1 cid. Compression ratio: 7.80:1. Brake hp: 275 at 4800 rpm. Taxable hp: 40.6. Torque: 333 lbs.-ft. at 3200 rpm. Five main bearings. Hydraulic valve lifters. Crankcase capacity: 5 qt. (add 1 qt. if new filter installed). Cooling system capacity: 17 qt. (add 1 qt. for heater). Carburetor: Stromberg WW-6-121 two-barrel. Included as standard equipment on this engine was a McCulloch Model 57S supercharger designed to cut in at 3000 rpm. Engine code: (Clipper Sedan) 57LY or (Clipper Station Wagon) 57LP.

1958 PACKARD

Packard 58L base V-8
Overhead valve V-8. Cast-iron block. Bore & stroke: 3.562 x 3.625 in. Displacement: 289.1 cid. Compression ratio: 8.30:1. Brake hp: 225 at 4500 rpm. Taxable hp: 40.6. Torque: 305 lbs.-ft. at 3000 rpm. Five main bearings. Hydraulic valve lifters. Crankcase capacity: 5 qt. (add 1 qt. if new filter installed). Cooling system capacity: 17 qt. (add 1 qt. for heater). Carburetor: Carter 2575-S four-barrel. (A 7.50:1 compression head was optional). Engine code: 58L.

Packard Hawk 58LS base V-8
Overhead valve supercharged V-8. Cast-iron block. Bore & stroke: 3.562 x 3.625 in. Displacement: 289.1 cid. Compression ratio: 7.80:1. Brake hp: 275 at 4800 rpm. Taxable hp: 40.6. Torque: 333 lbs.-ft. at 3200 rpm. Five main bearings. Hydraulic valve lifters. Crankcase capacity: 5 qt. (add 1 qt. if new filter installed). Cooling system capacity: 17 qt. (add 1 qt. for heater). Carburetor: Stromberg WW-6-122-A two-barrel. Included as standard equipment on this engine was a McCulloch Model VS-57S supercharger designed to cut in at 3000 rpm. (A 7.50:1 compression head was optional.) Engine code: 58LS.

Selected Packard V-8 Tune-Up Specs 1955-1958

Year	Model	Spark Plugs Type	Spark Plugs Gap	Point Gap	Deg Cam Angle	Timing	Idle Speed	Compression Pressure	Battery (Delco)
1955	Clipper Deluxe	CH-H10	.035	.015-.017	27	6B	425 rpm	150	3EA/3KM
	Clipper Deluxe	CH-H10	.035	.015-.017	27	6B	425 rpm	150	3EA/3KM
	Clipper Custom	CH-H10	.035	.015-.017	29.5	6B	425 rpm	150	3EA/3KM
	Packard	CH-H10	.035	.015-.017	29.5	6B	425 rpm	150	3EA/3KM
	Caribbean	CH-H10	.035	.015-.017	29.5	6B	425 rpm	150	3EA/3KM
1956	Clipper Deluxe	CH-N18	.035	.015-.017	27	5B	475 rpm	160	3EA/3KM
	Clipper Super	CH-N18	.035	.015-.017	27	5B	475 rpm	160	3EA/3KM
	Clipper Custom	CH-N18	.035	.015-.017	27	5B	475 rpm	160	3EA/3KM
	Packard	CH-N18	.035	.015-.017	29.5	10B	475 rpm	170	3EA/3KM
	Caribbean	CH-N18	.035	.015-.017	29.5	10B	475 rpm	170	3EA/3KM
1957	Clipper	CH-H10	.035	.013-.018	26-33	4B	550 rpm	135	3EA/3KM
1958	Packard	CH-H18Y	.035	.013-.018	26-33	4B	550 rpm	135	HO-11-50

Packard V-8 Firing Order

1952-1961
1-8-4-3-6-5-7-2

All Packard V-8s have the cylinders numbered from front to back. The cylinders on the left-hand bank are numbered 1-3-5-7 and the cylinders on the right-hand bank are numbered 2-4-6-8.

PLYMOUTH
1955-1989

In 1955, Plymouth was the last Chrysler family brand to receive a V-8. It was basically the '54 Dodge Hemi fitted with cheaper-to-make polyspherical cylinder heads for use on the economical Plymouth. This 241-cid 157-hp V-8 was available early in the 1955 model year. Later, a 259-cid V-8 was offered in standard and "power pack" formats.

There were three new V-8s in 1956 and the hottest was the 303-cid 240-hp version placed under the new Fury's hood. By 1957, the base Fury V-8 was up to 300 cid and 235 hp with a four-barrel carburetor, but the hot ticket was a 290-hp 318-cid V-8 with dual four-barrel carburetors. This remained a 1958 offering, but Plymouth upped the ante again with a 350-cid 305-hp "Golden Commando" V-8.

The 361-cid Plymouth engine bowed in 1960. This "big" Mopar mill was also the basis of the 383- and 426- "wedge" V-8s (a.k.a. "B" block engines). In the same family were the 413-cid V-8 and the 440-cid V-8 that was later derived from it. All were 90-degree V-8s with Cast-iron blocks and heads and generous bore spacing that made them very suitable for high-performance modifications. The 413- and 426-cid engines were dropped in 1966 and the 361 disappeared in 1967, but their 383- and 440-cid cousins stuck around.

Also available starting in 1964 was the "small" Mopar V-8 that started at 273-cid and grew to 318- and 340-cid. It was created to give the Valiant compact a high-performance option and came in handy when the Barracuda was launched to compete with the Mustang. The 318- and 340-cid versions were also popular in mid-sized cars like the Belvedere.

As the muscle car era exploded, the compact and intermediate Mopars were also fitted with big V-8s and during the 1966 model year, a "street" version of the second-generation Chrysler Hemi became a Belvedere option. It had the same 426-cid displacement as the "wedge" version and the same 425 maximum horsepower as Plymouth's "Super Stock Max Wedge Stage II" V-8. However, the official rating for the Hemi was extremely conservative and Richard Petty's NASCAR version put out 550 hp. Other engines received hot dual four-barrel or "Six Pack" carburetor setups with three two-barrel carburetors and a crowning touch on some street racers was an "Air Grabber" hood.

Amazingly, the street Hemi lasted until 1971, during which year a 360-cid V-8 arrived as a Fury option. It was basically the 383 with a reduced bore size and produced 255 hp. The 400-cid V-8 introduced in 1972 was the opposite—a bored out 383 producing 190 net hp (as a new rating system was adopted that year). That change was a hint of things to come as government smog regulations and rising insurance rates teamed up to kill the high-performance car hobby from 1973 on. By 1979, the only Plymouth V-8s remaining were the 318 and the 360 and the top output was 195 net hp.

Even the 360 was dropped past 1980, leaving the 318 to "hold the fort" as the sole Plymouth V-8. It survived only until 1989, when the last V-8-powered Plymouth was built. If you ask us, 25 years of Plymouth V-8s just wasn't enough!

Arranged in horsepower rating order.

1955 PLYMOUTH

Plymouth P27 base "Hy-fire" V-8
Overhead valve V-8. Cast-iron block. Polyspherical cylinder heads. Bore & stroke: 3.44 x 3.25 in. Displacement: 241 cid. Compression ratio: 7.6:1. Brake hp: 157 at 4400 rpm. Taxable hp: 37.9. Torque: 217 lbs.-ft. at 2400 rpm. Five main bearings. Non-adjustable hydraulic valve lifters. Crankcase capacity: 5 qt. (add 1 qt. if new filter installed). Cooling system capacity: 19 qt. (add 1 qt. for heater). Carburetor: Carter BBD-2262-S two-barrel. Engine number: P27-1001 and up.
Note: This was essentially a 1954 Dodge Hemi V-8 equipped with Plymouth polyspherical cylinder heads and was available early in the 1955 model year only. The polyspherical head was less expensive to produce than the hemispherical cylinder head, although it did position the spark plug between the intake and exhaust valves to provide an open combustion chamber with good volumetric efficiency.

Plymouth P27 optional V-8
Overhead valve V-8. Cast-iron block. Polyspherical cylinder heads. Bore & stroke: 3.563 x 3.25 in. Displacement: 259.2 cid. Compression ratio: 7.6:1. Brake hp: 167 at 4400 rpm. Taxable hp: 40.6. Torque: 231 lbs.-ft. at 2400 rpm. Five main bearings. Non-adjustable hydraulic valve lifters. Crankcase capacity: 5 qt. (add 1 qt. if new filter installed). Cooling system capacity: 19 qt. (add 1 qt. for heater). Carburetor: Carter BBD-2155-S two-barrel. Engine number: P27-60201 and up.

Plymouth P27 optional "Power Pak" V-8
Overhead valve V-8. Cast-iron block. Polyspherical cylinder heads. Bore & stroke: 3.563 x 3.25 in. Displacement: 259.2 cid. Compression ratio: 7.6:1. Brake hp: 177 at 4400 rpm. Taxable hp: 40.6. Torque: 231 lbs.-ft. at 2800 rpm. Five main bearings. Non-adjustable hydraulic valve lifters. Crankcase capacity: 5 qt. (add 1 qt. if new filter installed). Cooling system capacity: 19 qt. (add 1 qt. for heater). Carburetor: Carter WCFB-2253-S four-barrel. Engine number: P27-60201 and up.

In 1955, Plymouth advertisements claimed that the company had the top two engines in the low-priced field. The "Hy-Fire" V-8 shown here was the top engine option, while a flathead six was standard equipment in most models. This 167-hp overhead-valve V-8 could also be equipped with a four-barrel carburetor which produced an additional 10 horsepower.

1956 PLYMOUTH

Plymouth P29 Plaza, Savoy base V-8
Overhead valve V-8. Cast-iron block. Polyspherical cylinder heads. Bore & stroke: 3.63 x 3.25 in. Displacement: 270.2 cid. Compression ratio: 8.0:1. Brake hp: 180 at 4400 rpm. Taxable hp: 42.2. Torque: 260 lbs.-ft. at 2400 rpm. Five main bearings. Non-adjustable hydraulic valve lifters. Crankcase capacity: 5 qt. (add 1 qt. if new filter installed). Cooling

Plymouth's still-new V-8 was big for a Plymouth engine, but not as big as this 1956 advertisement might lead you to believe. Chrysler claimed that it had spent $50,000,000 developing this power plant in Plymouth's new Qualimatic engine plant. This year, the top power option climbed to 200 hp. A special V-8 was standard in the Fury model and the V-8 shown here was available, at extra cost, in any other Plymouth model.

system capacity: 19 qt. (add 1 qt. for heater). Carburetor: Carter BBD-2259-SB two-barrel. Engine number: P29-41001 or P29-250001 and up.

Plymouth P29 Plaza, Savoy, Belvedere optional V-8
Overhead valve V-8. Cast-iron block. Polyspherical cylinder heads. Bore & stroke: 3.75 x 3.125 in. Displacement: 276.1 cid. Compression ratio: 8.0:1. Brake hp: 187 at 4400 rpm. Taxable hp: 45. Torque: 265 lbs.-ft. at 2400 rpm. Five main bearings. Non-adjustable hydraulic valve lifters. Crankcase capacity: 5 qt. (add 1 qt. if new filter installed). Cooling system capacity: 19 qt. (add 1 qt. for heater). Carburetor: Carter BBD-2407-S two-barrel. Engine number: P29-80001 and up or P29-274101 and up.

Plymouth P29 Plaza, Savoy, Belvedere optional V-8
Overhead valve V-8. Cast-iron block. Polyspherical cylinder heads. Bore & stroke: 3.75 x 3.125 in. Displacement: 276.1 cid. Compression ratio: 8.0:1. Brake hp: 200 at 4400 rpm. Taxable hp: 45. Torque: 272 lbs.-ft. at 2400 rpm. Five main bearings. Solid valve lifters. Crankcase capacity: 5 qt. (add 1 qt. if new filter installed). Cooling system capacity: 19 qt. (add 1 qt. for heater). Carburetor: Carter WCFB-2442-S four-barrel. P29-161829 and up.

Plymouth Fury base V-8
Overhead valve V-8. Cast-iron block. Polyspherical cylinder heads. Bore & stroke: 3.81 x 3.31nches. Displacement: 303 cid. Compression ratio: 9.25:1. Brake hp: 240 at 4800 rpm. Taxable hp: 46.5. Torque: 310 lbs.-ft. at 2800 rpm. Five main bearings. Solid valve lifters. Crankcase capacity: 5 qt. (add 1 qt. if new filter installed). Cooling system capacity: 19 qt. (add 1 qt. for heater). Dual exhaust. Carburetor: Carter WCFB-2442-S four-barrel. P29-1630429 and up. (The prefix FP was used with the Fury engine serial number.)
Note: The Fury V-8 was built in Windsor, Canada. It featured balanced rods, a high-performance cam, adjustable rocker arms, a dual-breaker ignition system, and reinforced dome-type pistons.

1957 PLYMOUTH

Plymouth P31 passenger car base V-8
Overhead valve V-8. Cast-iron block. Bore & stroke: 3.75 x 3.125 in. Displacement: 276.1 cid. Compression ratio: 8.0:1. Brake hp: 197 at 4400 rpm. Taxable hp: 45. Torque: 270 lbs.-ft. at 2400 rpm. Five main bearings. Solid valve lifters. Crankcase capacity: 5 qt. (add 1 qt. if new filter installed). Cooling system capacity: 19 qt. (add 1 qt. for heater). Carburetor: Carter BBD-2407-S two-barrel. Engine number: P31-1001 and up.

Plymouth P-31 station wagon base V-8; other models optional V-8
Overhead valve V-8. Cast-iron block. Bore & stroke: 3.906 x 3.125 in. Displacement: 299.6 cid. Compression ratio: 8.5:1. Brake hp: 215 at 9700 rpm. Taxable hp: 48.9. Torque: 285 lbs.-ft. at 2800 rpm. Five main bearings. Solid valve lifters. Crankcase capacity: 5 qt. (add 1 qt. if new filter installed). Cooling system capacity: 19 qt. (add 1 qt. for heater). Carburetor: Carter BBD-2512-S two-barrel. Engine number: P31-1001

This underhood shot shows the 1957 Fury 318-cid V-8. The photo was taken at the Golden Fin Society's second national gathering at the Walter P. Chrysler Museum in Auburn Hills, Michigan. This 1957 Plymouth Fury was on display there. This engine is also done in silver with gold valve covers and air cleaners. The oil filler cap is painted black. The 1958 Fury 318 motor is identical, with the exception of the air cleaners. (Courtesy Jack Lewis of the Golden Fin Society)

A 1958 Plymouth Fury with Golden Commando 350 cubic inch motor. This fine looking motor resides in GFSer, Paul Schmalz's car. Mr. Schmalz lives in Fair Oaks, California. (Courtesy Jack Lewis of the Golden Fin Society)

and up.
Note: This engine is normally referred to as the "301-cid V-98."

Plymouth optional "Super-Pak" V-8
Overhead valve V-8. Cast-iron block. Bore & stroke: 3.906 x 3.125 in. Displacement: 299.6 cid. Compression ratio: 8.5:1. Brake hp: 235 at 4400 rpm. Taxable hp: 48.9. Torque: 305 lbs.-ft. at 2800 rpm. Five main bearings. Solid valve lifters. Crankcase capacity: 5 qt. (add 1 qt. if new filter installed). Cooling system capacity: 19 qt. (add 1 qt. for heater). Carburetor: Carter Type WCFB-2631-S four-barrel. Engine number: P31-1001 and up.

Plymouth Fury base V-8
Overhead valve. Cast-iron block. Bore & stroke: 3.906 x 3.31 in. Displacement: 318 cid. Compression ratio: 9.25:1. Brake hp: 290 at 5400 rpm. Taxable hp: 48.90. Torque: 325 lbs.-ft. at 4000 rpm. Five main bearings. Hydraulic valve lifters. Crankcase capacity: 5 qt. (add 1 qt. if new filter installed). Cooling system capacity: 16 qt. (add 1 qt. for heater). Carburetion: Dual Carter WCFB-2631-S four-barrel. Engine number: FP31-1001 and up.

1958 PLYMOUTH

Plymouth Plaza, Savoy, Belvedere base V-8
Overhead valve. Cast-iron block. Bore & stroke: 3.906 x 3.312 in. Displacement: 318 cid. Compression ratio: 9.00:1. Brake hp: 225 at 4400 rpm. Taxable hp: 48.90. Torque: 330 lbs.-ft. at 2800 rpm. Five main bearings. Hydraulic valve lifters. Crankcase capacity: 5 qt. (add 1 qt. if new filter installed). Cooling system capacity: 16 qt. (add 1 qt. for heater). Carter BBD-2644-S two-barrel. Engine number: LP8-1001 and up.

Plymouth Plaza, Savoy, Belvedere optional "V-800 Fury Super-Pak" V-8
Overhead valve. Cast-iron block. Bore & stroke: 3.906 x 3.312 in. Displacement: 318 cid. Compression ratio: 9.00:1. Brake hp: 250 at 4400 rpm. Taxable hp: 48.90. Torque: unknown. Five main bearings. Hydraulic valve lifters. Crankcase capacity: 5 qt. (add 1 qt. if new filter installed). Cooling system capacity: 16 qt. (add 1 qt. for heater). Carter BBD-2644-S two-barrel. Engine number: LP8-1001 and up.
Note: The "V-800 Fury Super-Pak" V-8 was not available in the Fury.

Plymouth Fury base "V-800 Dual Fury" V-8
Overhead valve. Cast-iron block. Bore & stroke: 3.906 x 3.31 in. Displacement: 318 cid. Compression ratio: 9.25:1. Brake hp: 290 at 5400 rpm. Taxable hp: 48.90. Torque: 325 lbs.-ft. at 4000 rpm. Five main bearings. Hydraulic valve lifters. Crankcase capacity: 5 qt. (add 1 qt. if new filter installed). Cooling system capacity: 16 qt. (add 1 qt. for heater). Carburetion: Dual Carter WCFB-2631-S four-barrel. Engine number: LP8-1001 and up.

Plymouth "Golden Commando" optional V-8
Overhead valve. Cast-iron block. Bore & stroke: 4.06 x 3.39 in. Displacement: 350 cid. Compression ratio: 10.0:1. Brake hp: 305 at 5000 rpm. Taxable hp: 52.7. Torque: 385 lbs.-ft. at 2800 rpm. Five main bearings. Hydraulic valve lifters. Crankcase capacity: 4 qt. (add 1 qt. if new filter installed). Cooling system capacity: 16 qt. (add 1 qt. for heater). Carburetor: Dual Carter AFB four-barrel. Engine number: LP8-1001 and up.

Plymouth "Golden Commando" optional EFI V-8
Overhead valve. Cast-iron block. Bore & stroke: 4.06 x 3.39 in. Displacement: 350 cid. Compression ratio: 10.0:1. Brake hp: 315. Five main bearings. Hydraulic valve lifters. Crankcase capacity: 4 qt. (add 1 qt. if new filter installed). Cooling system capacity: 16 qt. (add 1 qt. for heater). Induction: Bendix electronic fuel injection. Engine number: LP8-1001 and up.
Note: A very limited number of EFI-equipped Golden Commando V-8s were made before this option was recalled and cars so-equipped were retro-fitted with twin four-barrel carburetors.

1959 PLYMOUTH

Plymouth Savoy, Belvedere and Fury base V-8
Overhead valve. Cast-iron block. Bore & stroke: 3.91 x 3.31 in. Displacement: 318 cid. Bore & stroke: 3.91 x 3.31 in. Compression ratio: 9.0:1. Brake hp: 230 at 4400 rpm. Taxable hp: 48.90. Torque: 340 lbs.-ft. at 2400 rpm. Five main bearings. Hydraulic valve lifters. Crankcase capacity: 5 qt. (add 1 qt. if new filter installed). Cooling system capacity: 20 qt. (add 1 qt. for heater). Carburetor: Carter BBD-2824-S two-barrel.

Plymouth Sport Fury base V-8; Savoy, Belvedere, Fury optional "Super-Pak" V-8
Overhead valve. Cast-iron block. Bore & stroke: 3.91 x 3.31 in. Displacement: 318 cid. Bore & stroke: 3.91 x 3.31 in. Compression ratio: 9.0:1. Brake hp: 260 at 4400 rpm. Taxable hp: 48.90. Torque: 345 lbs.-ft. at 2800 rpm. Five main bearings. Hydraulic valve lifters. Crankcase capacity: 5 qt. (add 1 qt. if new filter installed). Cooling system capacity: 20 qt. (add 1 qt. for heater). Carburetor: Carter AFB-2813-S four-barrel.

Plymouth "Golden Commando 395" optional V-8
Overhead valve. Cast-iron block. Bore & stroke: 4.125 x 3.375 in. Displacement: 361 cid. Compression ratio: 10.00:1. Brake hp: 305 at 4600 rpm. Taxable hp: 54.30. Torque: 395 lbs.-ft. at 3000 rpm. Five main bearings. Hydraulic valve lifters. Crankcase capacity: 5 qt. (add 1 qt. if new filter installed). Cooling system capacity: 20 qt. (add 1 qt. for heater). Carburetor: Carter AFB-2813-S four-barrel.

1960 PLYMOUTH

Plymouth Savoy, Belvedere and Fury base V-8
Overhead valve. Cast-iron block. Bore & stroke: 3.91 x 3.31 in. Displacement: 318 cid. Bore & stroke: 3.91 x 3.31 in. Compression ratio: 9.0:1. Brake hp: 230 at 4400 rpm. Taxable hp: 48.90. Torque: 340 lbs.-ft. at 2400 rpm. Five main bearings. Hydraulic valve lifters. Crankcase capacity: 5 qt. (add 1 qt. if new filter installed). Cooling system capacity: 16 qt. (add 1 qt. for heater). Carburetor: Carter BBD-2921-S two-barrel. Engine code: P-318.

Plymouth Sport Fury base V-8; Savoy, Belvedere, Fury optional "Super-Pak" V-8
Overhead valve. Cast-iron block. Bore & stroke: 3.91 x 3.31 in. Displacement: 318 cid. Bore & stroke: 3.91 x 3.31 in. Compression ratio: 9.0:1. Brake hp: 260 at 4400 rpm. Taxable hp: 48.90. Torque: 345 lbs.-ft. at 2800 rpm. Five main bearings. Hydraulic valve lifters. Crankcase capacity: 5 qt. (add 1 qt. if new filter installed). Cooling system capacity: 20 qt. (add 1 qt. for heater). Carburetor: Carter Type AFB-2925-S four-barrel. Engine code: P-318.

Plymouth Savoy, Belvedere, Fury optional "Golden Commando" V-8
Overhead valve. Cast-iron block. Bore & stroke: 4.125 x 3.375 in. Displacement: 361 cid. Compression ratio: 10.00:1. Brake hp: 305 at 4800 rpm. Taxable hp: 54.30. Torque: 395 lbs.-ft. at 3000 rpm. Five main bearings. Hydraulic valve lifters. Crankcase capacity: 5 qt. (add 1 qt. if new filter installed). Cooling system capacity: 20 qt. (add 1 qt. for heater). Carburetor: Carter Type AFB-2969-S four-barrel. Engine code: P-36.

Plymouth Savoy, Belvedere, Fury optional "Sonoramic Commando" V-8
Overhead valve. Cast-iron block. Bore & stroke: 4.125 x 3.375 in. Displacement: 361 cid. Compression ratio: 10.00:1. Brake hp: 310 at 4800 rpm. Taxable hp: 54.30. Torque: 435 lbs.-ft. at 2800 rpm. Five main bearings. Hydraulic valve lifters. Crankcase capacity: 5 qt. (add 1 qt. if new filter installed). Cooling system capacity: 20 qt. (add 1 qt. for heater). Carburetor: Dual Carter AFB-2903-S four-barrel. Engine code: P-36.

Plymouth Savoy, Belvedere, Fury optional "Golden Commando" V-8
Overhead valve. Cast-iron block. Bore & stroke: 4.25 x 3.375 in. Displacement: 383 cid. Compression ratio: 10.00:1. Brake hp: 325 at 4600 rpm. Taxable hp: 57.80. Torque: 435 lbs.-ft. at 2800 rpm. Five main bearings. Hydraulic valve lifters. Crankcase capacity: 5 qt. (add 1 qt. if new filter installed). Cooling system capacity: 20 qt. (add 1 qt. for heater). Carburetor: Carter AFB-2968-S four-barrel. Engine code: P-38.

Plymouth Savoy, Belvedere, Fury optional "Sonoramic Commando" V-8
Overhead valve. Cast-iron block. Bore & stroke: 4.25 x 3.375 in. Displacement: 383 cid. Compression ratio: 10.00:1. Brake hp: 330 at 5200 rpm. Taxable hp: 57.80. Torque: 425 lbs.-ft. at 3600 rpm. Five main bearings. Hydraulic valve lifters. Crankcase capacity: 5 qt. (add 1 qt. if new filter installed). Cooling system capacity: 20 qt. (add 1 qt. for heater). Carburetor: Dual Carter AFB (Front) 2970-S; (Rear) 2971-S four-barrel. Engine code: P-38.

There were five Plymouth engines to select from in 1960 and two were entirely new SonoRamic Commando V-8s like the one seen here. This engine featured high-performance goodies like cross-ram intake manifolding. A four-barrel carburetor sat above each valve cover, atop criss-crossng ram tubes that ran to the intake on the top of the engine block. This was a wild-looking induction set up.

Plymouth Savoy, Belvedere, Fury optional "Sonoramic Commando" V-8 (Ram Tuned)
Overhead valve. Cast-iron block. Bore & stroke: 4.25 x 3.375 in. Displacement: 383 cid. Compression ratio: 10.00:1. Brake hp: 330 at 4800 rpm. Taxable hp: 57.80. Torque: 460 lbs.-ft. at 2800 rpm. Five main bearings. Hydraulic valve lifters. Crankcase capacity: 5 qt. (add 1 qt. if new filter installed). Cooling system capacity: 20 qt. (add 1 qt. for heater). Carburetor: Dual Carter AFB four-barrel. Engine code: P-38.

1961 PLYMOUTH

Plymouth Savoy, Belvedere and Fury base V-8
Overhead valve. Cast-iron block. Bore & stroke: 3.91 x 3.31 in. Displacement: 318 cid. Bore & stroke: 3.91 x 3.31 in. Compression ratio: 9.0:1. Brake hp: 230 at 4400 rpm. Taxable hp: 48.90. Torque: 340 lbs.-ft. at 2400 rpm. Five main bearings. Hydraulic valve lifters. Crankcase capacity: 5 qt. (add 1 qt. if new filter installed). Cooling system capacity: 20 qt. (add 1 qt. for heater). Carburetor: Carter BBD-2921-S two-barrel. Engine code: R318.

Plymouth Savoy, Belvedere, Fury optional "Golden Commando-361" V-8
Overhead valve. Cast-iron block. Bore & stroke: 4.125 x 3.375 in. Displacement: 361 cid. Compression ratio: 10.00:1. Brake hp: 305 at 4800 rpm. Taxable hp: 54.30. Torque: 395 lbs.-ft. at 3000 rpm. Five main bearings. Hydraulic valve lifters. Crankcase capacity: 5 qt. (add 1 qt. if new filter installed). Cooling system capacity: 20 qt. (add 1 qt. for heater). Carburetor: Carter Type AFB-3105-S four-barrel. Engine code: R-36.

Plymouth Savoy, Belvedere, Fury optional "Golden Commando-383" V-8
Overhead valve. Cast-iron block. Bore & stroke: 4.25 x 3.375 in. Displacement: 383 cid. Compression ratio: 10.00:1. Brake hp: 325 at 4600 rpm. Taxable hp: 57.80. Torque: 435 lbs.-ft. at 2800 rpm. Five main bearings. Hydraulic valve lifters. Crankcase capacity: 5 qt. (add 1 qt. if new filter installed). Cooling system capacity: 20 qt. (add 1 qt. for heater). Carburetor: Carter AFB-2968-S four-barrel. Engine code: R-38.

Plymouth Savoy, Belvedere, Fury optional "Sonoramic Commando" V-8
Overhead valve. Cast-iron block. Bore & stroke: 4.25 x 3.375 in. Displacement: 383 cid. Compression ratio: 10.00:1. Brake hp: 330 at 4800 rpm. Taxable hp: 57.80. Torque: 460 lbs.-ft. at 2800 rpm. Five main bearings. Hydraulic valve lifters. Crankcase capacity: 5 qt. (add 1 qt. if new filter installed). Cooling system capacity: 20 qt. (add 1 qt. for heater). Carburetor: Dual Carter AFB-2903-S. Engine code: R-38.

Plymouth Savoy, Belvedere, Fury optional "Sonoramic Commando" V-8
Overhead valve. Cast-iron block. Bore & stroke: 4.25 x 3.375 in. Displacement: 383 cid. Compression ratio: 10.00:1. Brake hp: 330 at 5200 rpm. Taxable hp: 57.80. Torque: 425 lbs.-ft. at 3600 rpm. Five main bearings. Hydraulic valve lifters. Crankcase capacity: 5 qt. (add 1 qt. if new filter installed). Cooling system capacity: 20 qt. (add 1 qt. for heater). Carburetor: Dual Carter AFB (Front) 2970-S; (Rear) 2971-S four-barrel. Engine code: R-38.

Plymouth Savoy, Belvedere, Fury "Sonoramic Ram Induction" optional V-8
Overhead valve. Cast-iron block. Bore & stroke: 4.25 x 3.375 in. Displacement: 383 cid. Compression ratio: 10.00:1. Brake hp: 340 at 5000 rpm. Taxable hp: 57.80. Torque: 440 lbs.-ft. at 2800 rpm. Five main bearings. Hydraulic valve lifters. Crankcase capacity: 5 qt. (add 1 qt. if new filter installed). Cooling system capacity: 20 qt. (add 1 qt. for heater). Carburetor: Dual Carter AFB-3084-S four-barrel. Engine code: R-38.

Plymouth Savoy, Belvedere, Fury optional V-8
Overhead valve. Cast-iron block. Bore & stroke: 4.19 x 3.75 in. Displacement: 413 cid. Compression ratio: 10.00:1. Brake hp: 350 at 4600 rpm. Torque: 470 lbs.-ft. at 2800 rpm. Five main bearings. Hydraulic valve lifters. Crankcase capacity: 5 qt. (add 1 qt. if new filter installed). Cooling system capacity: 20 qt. (add 1 qt. for heater). Carburetor: Carter AFB-3108-S four-barrel. Engine code: R-41.

Plymouth Savoy, Belvedere, Fury "Sonoramic Ram Induction" optional V-8
Overhead valve. Cast-iron block. Bore & stroke: 4.19 x 3.75 in. Displacement: 413 cid. Compression ratio: 10.00:1. Brake hp: 375 at 5000 rpm. Torque: 465 lbs.-ft. at 2800 rpm. Five main bearings. Solid valve lifters. Crankcase capacity: 5 qt. (add 1 qt. if new filter installed). Cooling system capacity: 20 qt. (add 1 qt. for heater). Carburetor: Dual Carter AFB-3084-S four-barrel. (Option added at midyear; includes "long" ram-tuned induction system.) Engine code: R-41.

Plymouth Savoy, Belvedere, Fury "Sonoramic Ram Induction" optional V-8
Overhead valve. Cast-iron block. Bore & stroke: 4.19 x 3.75 in. Displacement: 413 cid. Compression ratio: 10.00:1. Brake hp: 375 at 5200 rpm. Torque: 450 lbs.-ft. at 2800 rpm. Five main bearings. Solid valve lifters. Crankcase capacity: 5 qt. (add 1 qt. if new filter installed). Cooling system capacity: 20 qt. (add 1 qt. for heater). Carburetor: Dual Carter AFB four-barrels: (front) 2790-S; (rear) 2791-S. (Option added at midyear; includes "short" ram-tuned induction system.) Engine code: R-41.

1962 PLYMOUTH

Plymouth Savoy, Belvedere and Fury base V-8
Overhead valve. Cast-iron block. Bore & stroke: 3.91 x 3.31 in. Displacement: 318 cid. Bore & stroke: 3.91 x 3.31 in. Compression ratio: 9.0:1. Brake hp: 230 at 4400 rpm. Taxable hp: 48.90. Torque: 340 lbs.-ft. at 2400 rpm. Five main bearings. Hydraulic valve lifters. Crankcase capacity: 5 qt. (add 1 qt. if new filter installed). Cooling system capacity: 20 qt. (add 1 qt. for heater). Carburetor: Carter AFB-3240-S or Bendix WW-3-198 two-barrel. Engine code: S-318.

Plymouth Sport Fury base V-8; Savoy, Belvedere, Fury optional "Super-Pak" V-8
Overhead valve. Cast-iron block. Bore & stroke: 3.91 x 3.31 in. Displacement: 318 cid. Bore & stroke: 3.91 x 3.31 in. Compression ratio: 9.0:1. Brake hp: 260 at 4400 rpm. Taxable hp: 48.90. Torque: 345 lbs.-ft. at 2800 rpm. Five main bearings. Hydraulic valve lifters. Crankcase capacity: 5 qt. (add 1 qt. if new filter installed). Cooling system capacity: 20 qt. (add 1 qt. for heater). Carburetor: Carter AFB-3247-S four-barrel. Engine code: S-318.

Plymouth Savoy, Belvedere, Fury optional "Golden Commando-361" V-8
Overhead valve. Cast-iron block. Bore & stroke: 4.125 x 3.375 in. Displacement: 361 cid. Compression ratio: 10.00:1. Brake hp: 305 at 4800 rpm. Taxable hp: 54.30. Torque: 395 lbs.-ft. at 3000 rpm. Five main bearings. Hydraulic valve lifters. Crankcase capacity: 5 qt. (add 1 qt. if new filter installed). Cooling system capacity: 20 qt. (add 1 qt. for heater). Carburetor: Carter AFB-3252-S four-barrel. Engine code: S-36.

This is the special 303-cid V-8 engine used in the high-performance Plymouth Fury hardtop. This particular engine is a freshly-rebuilt unit owned by Larry Gammon of the Golden Fin Society, who lives in Calagary, Alberta, Canada. This engine is correctly finished in silver with gold valve covers and a gold air cleaner. The radiator fan is painted black as are some other accessories. (Courtesy Jack Lewis of the Golden Fin Society)

Plymouth Savoy, Belvedere, Fury optional Cross Ram V-8
Overhead valve. Cast-iron block. Bore & stroke: 4.125 x 3.375 in. Displacement: 361 cid. Compression ratio: 9.00:1. Brake hp: 310 at 5200 rpm. Taxable hp: 54.30. Torque: 390 lbs.-ft. at 3400 rpm. Five main bearings. Hydraulic valve lifters. Crankcase capacity: 4 qt. (add 1 qt. if new filter installed). Cooling system capacity: 16 qt. (add 1 qt. for heater). Carburetor: Two Carter AFB four-barrels: (front) 2970-S; (rear) 2971-S. Engine code: S-36.

Plymouth Savoy, Belvedere, Fury optional optional V-8
Overhead valve. Cast-iron block. Bore & stroke: 4.25 x 3.375 in. Displacement: 383 cid. Compression ratio: 10.00:1. Brake hp: 330 at 4600 rpm. Taxable hp: 57.80. Torque: 452 lbs.-ft. at 2800 rpm. Five main bearings. Hydraulic valve lifters. Crankcase capacity: 4 qt. (add 1 qt. if new filter installed). Cooling system capacity: 16 qt. (add 1 qt. for heater). Carburetor: Carter AFB-3438-S four-barrel. Engine code: S-38.

Plymouth Savoy, Belvedere, Fury optional Cross Ram V-8
Overhead valve. Cast-iron block. Bore & stroke: 4.25 x 3.375 in. Displacement: 383 cid. Compression ratio: 10.00:1. Brake hp: 335 at 5200 rpm. Taxable hp: 57.80. Torque: 420 lbs.-ft. at 3600 rpm. Five main bearings. Hydraulic valve lifters. Crankcase capacity: 4 qt. (add 1 qt. if new filter installed). Cooling system capacity: 16 qt. (add 1 qt. for heater). Carburetor: Two Carter AFB four-barrels: (front) 2970-S; (rear) 2971-S. Engine code: S-38.

Plymouth Savoy, Belvedere, Fury optional V-8
Overhead valve. Cast-iron block. Bore & stroke: 4.19 x 3.75 in. Displacement: 413 cid. Compression ratio: 11.00:1. Brake hp: 365 at 4600 rpm. Taxable hp: 55.90. Torque: 460 lbs.-ft. at 2800 rpm. Five main bearings. Hydraulic valve lifters. Crankcase capacity: 4 qt. (add 1 qt. if new filter installed). Cooling system capacity: 16 qt. (add 1 qt. for heater). Carburetor: Carter AFB 3251-S four-barrel. Engine code: S-41.

Plymouth Savoy, Belvedere, Fury "Sonoramic Ram Induction" optional long-ram V-8
Overhead valve. Cast-iron block. Bore & stroke: 4.19 x 3.75 in. Displacement: 413 cid. Compression ratio: 11.0:1. Brake hp: 410 at 5400 rpm. Taxable hp: unknown. Torque: 460 lbs.-ft. at 4400 rpm. Five main bearings. Hydraulic valve lifters. Crankcase capacity: 5 qt. qt. (add 1 qt. for new filter). Cooling system capacity: 20 qt. (add 1 qt. for heater). Carburetor: Dual Carter four-barrel. Engine code: S-41.

Plymouth Savoy, Belvedere, Fury "Sonoramic Ram Induction" optional short-ram V-8
Overhead valve. Cast-iron block. Bore & stroke: 4.19 x 3.75 in. Displacement: 413 cid. Compression ratio: 13.5:1. Brake hp: 420 at 5400 rpm. Taxable hp: unknown. Torque: 470 lbs.-ft. at 4400 rpm. Five main bearings. Solid valve lifters. Crankcase capacity: 5 qt. qt. (add 1 qt. for new filter). Cooling system capacity: 20 qt. (add 1 qt. for heater). Carburetor: Dual Carter four-barrel. Engine code: S-41.

1963 PLYMOUTH

Plymouth base V-8
Overhead valve. Cast-iron block. Bore & stroke: 3.91 x 3.31 in. Displacement: 318 cid. Compression ratio: 9.0:1. Brake hp: 230 at 4400 rpm. Taxable hp: 48.90. Torque: 340 lbs.-ft. at 2400 rpm. Five main bearings. Hydraulic valve lifters. Crankcase capacity: 4 qt. (add 1 qt. for new filter). Cooling system capacity: 20 qt. (add 1 qt. for heater). Carburetor: Carter BBD-3472-S or Stromberg WW-3-222 two-barrel. Engine code: T-318.

Plymouth optional "Commando" V-8
Overhead valve. Cast-iron block. Bore & stroke: 4.13 x 3.38 in. Displacement: 361 cid. Compression ratio: 9.0:1. Brake hp: 265 at 4400 rpm. Taxable hp: 54.30. Torque: 380 lbs.-ft. at 2400 rpm. Five main bearings. Hydraulic valve lifters. Crankcase capacity: 4 qt. (add 1 qt. for new filter). Cooling system capacity: 16 qt. (add 1 qt. for heater). Carburetor: Carter BBD-3475-S two-barrel. Engine code: T-36.

Plymouth optional "Golden Commando" V-8
Overhead valve. Cast-iron block. Bore & stroke: 4.25 x 3.38 in. Displacement: 383 cid. Compression ratio: 10.0:1. Brake hp: 330 at 4600 rpm. Taxable hp: 57.80. Torque: 425 lbs.-ft. at 2800 rpm. Five main bearings. Hydraulic valve lifters. Crankcase capacity: 4 qt. (add 1 qt. for new filter). Cooling system capacity: 16 qt. (add 1 qt. for heater). Carburetor: Carter AFB-3437-S four-barrel. Engine code: T-38.

Plymouth optional "Max Wedge Stage II" V-8
Overhead valve. Cast-iron block. Bore & stroke: 4.25 x 3.75 in. Displacement: 426 cid. Compression ratio: 11.0:1. Brake hp: 415 at 5800 rpm. Taxable hp: unknown. Torque: unknown at unknown rpm. Five main bearings. Solid valve lifters. Crankcase capacity: 4 qt. (add 1 qt. for new filter). Cooling system capacity: 16 qt. (add 1 qt. for heater). Carburetor: Dual Carter four-barrel. Engine code: T-42.

Plymouth optional "Max Wedge Stage II" V-8
Overhead valve. Cast-iron block. Bore & stroke: 4.25 x 3.75 in. Displacement: 426 cid. Compression ratio: 13.5:1. Brake hp: 425 at 5600 rpm. Taxable hp: unknown. Torque: unknown at unknown rpm. Five main bearings. Solid valve lifters. Crankcase capacity: 4 qt. (add 1 qt. for new filter). Cooling system capacity: 16 qt. (add 1 qt. for heater). Carburetor: Dual Carter four-barrel. Engine code: T-42.

1964 PLYMOUTH

Plymouth Valiant V-100, V-200, Signet optional V-8
Overhead valve. Cast-iron block. Bore & stroke: 3.63 x 3.31 in. Displacement: 273 cid. Compression ratio: 8.8:1. Brake hp: 180 at 4200 rpm. Taxable hp: 42.20. Torque: 260 lbs.-ft. at 1600 rpm. Five main bearings. Hydraulic valve lifters. Crankcase capacity: 4 qt. (add 1 qt. for new filter). Cooling system capacity: 16 qt. (add 1 qt. for heater). Carburetor: Carter BBD-3767-S two-barrel. Engine code: V-273.

Plymouth Savoy, Belvedere, Fury, Sport Fury base V-8
Overhead valve. Cast-iron block. Bore & stroke: 3.91 x 3.31 in. Displacement: 318 cid. Compression ratio: 9.0:1. Brake hp: 230 at 4400 rpm. Taxable hp: 48.90. Torque: 340 lbs.-ft. at 2400 rpm. Five main bearings. Hydraulic valve lifters. Crankcase capacity: 4 qt. (add 1 qt. for new filter). Cooling system capacity: 20 qt. (add 1 qt. for heater). Carburetor: Carter BBD-3682-S two-barrel. Engine code: V-318.

Plymouth Savoy, Belvedere, Fury, Sport Fury optional V-8
Overhead valve. Cast-iron block. Bore & stroke: 4.13 x 3.38 in. Displacement: 361 cid. Compression ratio: 9.0:1. Brake hp: 265 at 4400 rpm. Taxable hp: 54.30. Torque: 380 lbs.-ft. at 2400 rpm. Five main bearings. Hydraulic valve lifters. Crankcase capacity: 4 qt. (add 1 qt. for new filter). Cooling system capacity: 16 qt. (add 1 qt. for heater). Carburetor: Stromberg WWC-3-239 two-barrel. Engine code: V-36.

Plymouth Savoy, Belvedere, Fury, Sport Fury optional V-8
Overhead valve. Cast-iron block. Displacement: 383 cid. Bore & stroke: 4.25 x 3.38 in. Compression ratio: 10.00:1. Brake hp: 305 at 4600 rpm. Taxable hp: 57.80. Torque: 410 lbs.-ft. at 2400 rpm. Five main bearings. Hydraulic valve lifters. Crankcase capacity: 4 qt. (add 1 qt. for new filter). Cooling system capacity: 16 qt. (add 1 qt. for heater). Carburetor: Carter BBD 3682-S two-barrel. Engine code: V-38.

Plymouth Savoy, Belvedere, Fury, Sport Fury optional V-8
Overhead valve. Cast-iron block. Displacement: 383 cid. Bore & stroke: 4.25 x 3.38 in. Compression ratio: 10.0:1. Brake hp: 330 at 4600 rpm. Taxable hp: 57.80. Torque: 425 lbs.-ft. at 2800 rpm. Five main bearings. Hydraulic valve lifters. Crankcase capacity: 4 qt. (add 1 qt. for new filter). Cooling system capacity: 16 qt. (add 1 qt. for heater). Carburetor: Carter AFB-3611-S four-barrel. Engine code: V-38.

Plymouth Savoy, Belvedere, Fury, Sport Fury optional V-8
Overhead valve. Cast-iron block. Displacement: 426 cid. Bore & stroke: 4.25 x 3.75 in. Compression ratio: 10.30:1. Brake hp: 365 at 4800 rpm. Taxable hp: 57.80. Torque: 470 lbs.-ft. at 4800 rpm. Five main bearings. Hydraulic valve lifters. Crankcase capacity: 5 qt. (add 1 qt. for new filter). Cooling system capacity: 16 qt. (add 1 qt. for heater). Carburetor: Carter AFB-3704-S four-barrel. Engine code: V-42.

Plymouth optional " NASCAR" Hemi V-8 (Race Hemi)
Overhead valve with hemispherical segment combustion chambers. Cast-iron block. Bore & stroke: 4.25 x 3.75 in. Displacement: 426 cid. Compression ratio: 12.5:1. Brake hp: 400 at 5600 rpm. Taxable hp: unknown. Torque: unknown at unknown rpm. Five main bearings. Solid valve lifters. Crankcase capacity: 5 qt. (add 1 qt. for new filter). Cooling system capacity: 16 qt. (add 1 qt. for heater). Carburetor: Holley four-barrel. Engine code: V-426.

Plymouth optional "Super Stock Max Wedge Stage II" V-8
Overhead valve. Cast-iron block. Bore & stroke: 4.25 x 3.75 in. Displacement: 426 cid. Compression ratio: 11.0:1. Brake hp: 415 at 5800 rpm. Taxable hp: unknown. Torque: unknown at unknown rpm. Five main bearings. Solid valve lifters. Crankcase capacity: 5 qt. (add 1 qt. for new filter). Cooling system capacity: 16 qt. (add 1 qt. for heater). Carburetor: Dual Carter four-barrel. Engine code: V-426.

Plymouth optional " Super Stock Max Wedge Stage II" V-8
Overhead valve. Cast-iron block. Bore & stroke: 4.25 x 3.75 in. Displacement: 426 cid. Compression ratio: 13.5:1. Brake hp: 425 at 5600 rpm. Taxable hp: unknown. Torque: unknown at unknown rpm. Five main

bearings. Solid valve lifters. Crankcase capacity: 5 qt. (add 1 qt. for new filter). Cooling system capacity: 16 qt. (add 1 qt. for heater). Carburetor: Dual Carter four-barrel. Engine code: V-426.

Plymouth optional " Super Commando" Hemi V-8 (Race Hemi)
Overhead valve with hemispherical segment combustion chambers. Cast-iron block. Displacement: 426 cid. Bore & stroke: 4.25 x 3.75 in. Compression ratio: 11.0:1. Brake hp: 415 at 5600 rpm. Taxable hp: 57.80. Torque: 470 lbs.-ft. at 4400 rpm. Five main bearings. Solid valve lifters. Crankcase capacity: 5 qt. (add 1 qt. for new filter). Cooling system capacity: 16 qt. (add 1 qt. for heater). Carburetor: Dual Carter AFB-3859-S four-barrel. Engine code: V-426.

Plymouth optional "Super Commando" Hemi V-8 (Race Hemi)
Overhead valve with hemispherical segment combustion chambers. Cast-iron block. Displacement: 426 cid. Bore & stroke: 4.25 x 3.75 in. Compression ratio: 12.5:1. Brake hp: 425 at 5600 rpm. Taxable hp: 57.80. Torque: 480 lbs.-ft. at 4600 rpm. Five main bearings. Solid valve lifters. Crankcase capacity: 5 qt. (add 1 qt. for new filter). Cooling system capacity: 16 qt. (add 1 qt. for heater). Carburetor: Dual Carter AFB-3084-S four-barrel. Engine code: V-426.

1965 PLYMOUTH

Plymouth Valiant, Signet, Barracuda, Belvedere, Satellite base V-8
Overhead valve. Cast-iron block. Bore & stroke: 3.63 x 3.31 in. Displacement: 273 cid. Compression ratio: 8.80:1. Brake hp: 180 at 4200 rpm. Taxable hp: 42.20. Torque: 260 lbs.-ft. at 1600 rpm. Five main bearings. Solid valve lifters. Crankcase capacity: 4 qt. (add 1 qt. for new filter). Cooling system capacity: 17 qt. (add 1 qt. for heater). Carburetor: Carter BBD-3843-S or Stromberg WW3-248 two-barrel. Engine code: A-273.

Plymouth Sport Fury base V-8; Belvedere, Satellite, Fury optional "Comando" V-8
Overhead valve. Cast-iron block. Bore & stroke: 3.91 x 3.31 in. Displacement: 318 cid. Compression ratio: 9.0:1. Brake hp: 230 at 4400 rpm. Taxable hp: 48.90. Torque: 340 lbs.-ft. at 2400 rpm. Five main bearings. Solid valve lifters. Crankcase capacity: 4 qt. (add 1 qt. for new filter). Cooling system capacity: 20 qt. (add 1 qt. for heater). Carburetor: Carter BBD-3847-S or Stromberg WW3-250 two-barrel. Engine code: A-318.

Plymouth Formula 'S' Barracuda base V-8; Valiant, Signet, Barracuda optional V-8
Overhead valve. Cast-iron block. Bore & stroke: 3.63 x 3.31 in. Displacement: 273 cid. Compression ratio: 10.50:1. Brake hp: 235 at 5200 rpm. Taxable hp: 42.20. Torque: 280 lbs.-ft. at 4000 rpm. Five main bearings. Solid valve lifters. Crankcase capacity: 4 qt. (add 1 qt. for new filter). Cooling system capacity: 17 qt. (add 1 qt. for heater). Carburetor: Carter AFB-3853-S four-barrel. Engine code: A-273.

Plymouth Belvedere, Satellite optional "Commando" V-8
Overhead valve. Cast-iron block. Bore & stroke: 4.13 x 3.38 in. Displacement: 361 cid. Compression ratio: 9.0:1. Brake hp: 265 at 4400 rpm. Taxable hp: 54.30. Torque: 380 lbs.-ft. at 2400 rpm. Five main bearings. Hydraulic valve lifters. Crankcase capacity: 4 qt. (add 1 qt. for new filter). Cooling system capacity: 16 qt. (add 1 qt. for heater). Carburetor: Carter BBD-4849-S or Stromberg WWC3-254 two-barrel. Engine code: A-361.

Plymouth Fury, Sport Fury optional "Commando 383" V-8
Overhead valve. Cast-iron block. Displacement: 383 cid. Bore & stroke: 4.25 x 3.38 in. Compression ratio: 9.20:1. Brake hp: 270 at 4400 rpm. Taxable hp: 57.80. Torque: 390 lbs.-ft. at 2800 rpm. Five main bearings. Hydraulic valve lifters. Crankcase capacity: 4 qt. (add 1 qt. for new filter). Cooling system capacity: 16 qt. (add 1 qt. for heater). Carburetor: Carter BBD-4125-S two-barrel. Engine code: A-383.

Plymouth Belvedere, Satellite, Fury optional "Commando 383" V-8
Overhead valve. Cast-iron block. Displacement: 383 cid. Bore & stroke: 4.25 x 3.38 in. Compression ratio: 10.0:1. Brake hp: 330 at 4600 rpm. Taxable hp: 57.80. Torque: 425 lbs.-ft. at 2800 rpm. Five main bearings. Hydraulic valve lifters. Crankcase capacity: 4 qt. (add 1 qt. for new filter). Cooling system capacity: 16 qt. (add 1 qt. for heater). Carburetor: Carter AFB-3855-S four-barrel. Engine code: A-383.

Plymouth Belvedere, Satellite, Fury, Sport Fury optional "426 S" V-8
Overhead valve. Cast-iron block. Displacement: 426 cid. Bore & stroke: 4.25 x 3.75 in. Compression ratio: 10.10:1. Brake hp: 365 at 4800 rpm. Taxable hp: 57.80. Torque: 470 lbs.-ft. at 3200 rpm. Five main bearings. Hydraulic valve lifters. Crankcase capacity: 4 qt. (add 1 qt. for new filter). Cooling system capacity: 16 qt. (add 1 qt. for heater). Carburetor: Carter AFB-3859-S four-barrel. Engine code: A-426.

Plymouth optional "426 R" V-8
Overhead valve. Cast-iron block. Displacement: 426 cid. Bore & stroke: 4.25 x 3.75 in. Compression ratio: 11.00:1. Brake hp: 415 at 5600 rpm. Taxable hp: 57.80. Torque: 470 lbs.-ft. at 4400 rpm. Five main bearings. Solid valve lifters. Crankcase capacity: 4 qt. (add 1 qt. for new filter). Cooling system capacity: 16 qt. (add 1 qt. for heater). Carburetor: Dual Carter AFB 3859-S four-barrel. Engine code: A-426.

Plymouth optional "426 R" V-8
Overhead valve with hemispherical segment combustion chambers. Cast-iron block. Displacement: 426 cid. Bore & stroke: 4.25 x 3.75 in. Compression ratio: 12.0:1. Brake hp: 425 at 5600 rpm. Taxable hp: 57.80. Torque: 480 lbs.-ft. at 4600 rpm. Five main bearings. Solid valve lifters. Crankcase capacity: 4 qt. (add 1 qt. for new filter). Cooling system capacity: 16 qt. (add 1 qt. for heater). Carburetor: Dual Carter AFB 3084-S four-barrel. Engine code: A-426.

1966 PLYMOUTH

Note: Beginning in 1966, the fifth symbol in the Plymouth Vehicle Identification Number identified the engine.

Plymouth Valiant, Signet, Barracuda, Belvedere, Satellite base V-8
Overhead valve. Cast-iron block. Bore & stroke: 3.63 x 3.31 in. Displacement: 273 cid. Compression ratio: 8.80:1. Brake hp: 180 at 4200 rpm. Taxable hp: 42.20. Torque: 260 lbs.-ft. at 1600 rpm. Five main bearings. Solid valve lifters. Crankcase capacity: 4 qt. (add 1 qt. for new filter). Cooling system capacity: 18 qt. (With heater). Carburetor: Carter BBD-4113-S or Stromberg WW3-248 two-barrel. Engine code: D.

Plymouth Formula 'S' Barracuda base V-8; Valiant, Signet, Barracuda optional V-8
Overhead valve. Cast-iron block. Bore & stroke: 3.63 x 3.31 in. Displacement: 273 cid. Compression ratio: 10.50:1. Brake hp: 235 at 5200 rpm. Taxable hp: 42.20. Torque: 280 lbs.-ft. at 4000 rpm. Five main bearings. Hydraulic valve lifters. Crankcase capacity: 4 qt. (add 1 qt. for new filter). Cooling system capacity: 18 qt. (With heater standard). Carburetor: Carter AFB-4119-S four-barrel. Engine code: D.

Plymouth Sport Fury, VIP base V-8; Belvedere, Satellite, Fury optional "Commando" V-8
Overhead valve. Cast-iron block. Bore & stroke: 3.91 x 3.31 in. Displacement: 318 cid. Compression ratio: 9.0:1. Brake hp: 230 at 4400 rpm. Taxable hp: 48.90. Torque: 340 lbs.-ft. at 2400 rpm. Five main bearings. Solid valve lifters. Crankcase capacity: 4 qt. (add 1 qt. for new filter). Cooling system capacity: 21 qt. (With heater). Carburetor: Stromberg WW3-258 two-barrel. Engine code: E.

Plymouth Belvedere, Satellite optional "Commando" V-8
Overhead valve. Cast-iron block. Bore & stroke: 4.13 x 3.38 in. Displacement: 361 cid. Compression ratio: 9.0:1. Brake hp: 265 at 4400 rpm. Taxable hp: 54.30. Torque: 380 lbs.-ft. at 2400 rpm. Five main bearings. Hydraulic valve lifters. Crankcase capacity: 4 qt. (add 1 qt. for new filter). Cooling system capacity: 17 qt. (With heater). Carburetor: Carter BBD-4125-S two-barrel. Engine code: F.

Plymouth Fury, Sport Fury, VIP optional "Commando 383" V-8
Overhead valve. Cast-iron block. Displacement: 383 cid. Bore & stroke: 4.25 x 3.38 in. Compression ratio: 9.20:1. Brake hp: 270 at 4400 rpm. Taxable hp: 57.80. Torque: 390 lbs.-ft. at 2800 rpm. Five main bearings. Hydraulic valve lifters. Crankcase capacity: 4 qt. (add 1 qt. for new filter). Cooling system capacity: 17 qt. (With heater). Carburetor: Carter BBD-4125-S two-barrel. Engine code: G.

Plymouth Belvedere, Satellite, Fury, VIP optional "Commando" V-8
Overhead valve. Cast-iron block. Displacement: 383 cid. Bore & stroke: 4.25 x 3.38 in. Compression ratio: 10.0:1. Brake hp: 325 at 4800 rpm. Taxable hp: 57.80. Torque: 425 lbs.-ft. at 2800 rpm. Five main bearings. Hydraulic valve lifters. Crankcase capacity: 4 qt. (add 1 qt. for new filter). Cooling system capacity: 17 qt. (With heater standard). Carburetor: Carter AFB-4130-S four-barrel. Engine code: G.

Plymouth Fury, Sport Fury, VIP optional "440 TNT" (Wedge) V-8
Overhead valve. Cast-iron block. Bore & stroke: 4.32 x 3.75 in. Displacement: 440 cid. Compression ratio: 10.10:1. Brake hp: 365 at 4600 rpm. Taxable hp: 59.70. Torque: 480 lbs.-ft. at 3200 rpm. Five main bearings. Hydraulic valve lifters. Crankcase capacity: 4 qt. (add 1 qt. for new filter). Cooling system capacity: 17 qt. (With heater standard). Carburetor: Carter AFB 4130-S four-barrel. Engine code: J.

Plymouth Belvedere, Satellite optional "Street Hemi" Eight-Barrel V-8
Overhead valve with hemispherical segment combustion chambers. Cast-iron block. Displacement: 426 cid. Bore & stroke: 4.25 x 3.75 in. Compression ratio: 10.25:1. Brake hp: 425 at 5000 rpm. Taxable hp: 57.80. Torque: 490 lbs.-ft. at 4000 rpm. Five main bearings. Hydraulic valve lifters. Crankcase capacity: 4 qt. (add 1 qt. for new filter). Cooling system capacity: 17 qt. (With heater standard). Carburetor: Dual Carter (Front) AFB-4139-S; (Rear) AFB-4140-S four-barrel. Engine code: H.

Plymouth Belvedere, Satellite optional "Street Hemi" Eight-Barrel V-8
Overhead valve with hemispherical segment combustion chambers. Cast-iron block. Displacement: 426 cid. Bore & stroke: 4.25 x 3.75 in. Compression ratio: 10.25:1. Compression ratio: 12.0:1. Brake hp: 425 at 5600 rpm. Taxable hp: 57.80. Torque: 480 lbs.-ft. at 4600 rpm. Five main bearings. Solid valve lifters. Crankcase capacity: 6 qt. (add 1 qt. for new filter). Cooling system capacity: 17 qt. (With heater standard). Carburetor: Dual Carter (Front) AFB-4139-S; (Rear) AFB-4140-S four-barrel. Engine code: H.

1967 PLYMOUTH

Plymouth Valiant 100, Signet, Barracuda, Belvedere, Satellite base V-8
Overhead valve. Cast-iron block. Bore & stroke: 3.63 x 3.31 in. Displacement: 273 cid. Compression ratio: 8.80:1. Brake hp: 180 at 4200 rpm. Taxable hp: 42.20. Torque: 260 lbs.-ft. at 1600 rpm. Five main bearings. Solid valve lifters. Crankcase capacity: 4 qt. (add 1 qt. for new filter). Cooling system capacity: 18 qt. (With heater). Carburetor: Carter BBD-4113-SA two-barrel. Engine code: D.

Plymouth Fury, Sport Fury, VIP base V-8; Belvedere, Satellite optional "Commando" V-8
Overhead valve. Cast-iron block. Bore & stroke: 3.91 x 3.31 in. Displacement: 318 cid. Compression ratio: 9.0:1. Brake hp: 230 at 4400 rpm. Taxable hp: 48.90. Torque: 340 lbs.-ft. at 2400 rpm. Five main bearings. Solid valve lifters. Crankcase capacity: 4 qt. (add 1 qt. for new filter). Cooling system capacity: 21 qt. (With heater). Carburetor: Stromberg WW3-272 two-barrel. Engine code: F.

Valiant 100, Signet, Barracuda optional V-8
Overhead valve. Cast-iron block. Bore & stroke: 3.63 x 3.31 in. Displacement: 273 cid. Compression ratio: 10.50:1. Brake hp: 235 at 5200 rpm. Taxable hp: 42.20. Torque: 280 lbs.-ft. at 4000 rpm. Five main bearings. Solid valve lifters. Crankcase capacity: 4 qt. (add 1 qt. for new filter). Cooling system capacity: 18 qt. (With heater standard). Carburetor: Carter AFB 4326-S four-barrel. Engine code: E.

Plymouth Belvedere, Satellite, Fury, Sport Fury, VIP optional "Commando 383" V-8
Overhead valve. Cast-iron block. Displacement: 383 cid. Bore & stroke: 4.25 x 3.38 in. Compression ratio: 9.20:1. Brake hp: 270 at 4400 rpm. Taxable hp: 57.80. Torque: 390 lbs.-ft. at 2800 rpm. Five main bearings. Hydraulic valve lifters. Crankcase capacity: 4 qt. (add 1 qt. for new filter). Cooling system capacity: 17 qt. (With heater). Carburetor: Carter BBD-4296-S two-barrel. Engine code: G.

Barracuda optional V-8
Overhead valve. Cast-iron block. Bore & stroke: 4.32 x 3.75 in. Displacement: 440 cid. Compression ratio: 10.00:1. Brake hp: 280 at 4200 rpm. Taxable hp: 57.80. Torque: 400 lbs.-ft. at 2400 rpm. Five main bearings. Hydraulic valve lifters. Crankcase capacity: 4 qt. (add 1 qt. for new filter). Cooling system capacity: 17 qt. (With heater standard). Carburetor: Carter AFB 4326-S four-barrel. Engine code: K.

Plymouth Belvedere, Satellite, Fury, Sport Fury, VIP optional "Commando 383" V-8
Overhead valve. Cast-iron block. Displacement: 383 cid. Bore & stroke: 4.25 x 3.38 in. Compression ratio: 10.0:1. Brake hp: 325 at 4800 rpm. Taxable hp: 57.80. Torque: 425 lbs.-ft. at 2800 rpm. Five main bearings. Hydraulic valve lifters. Crankcase capacity: 4 qt. (add 1 qt. for new filter). Cooling system capacity: 17 qt. (With heater). Carburetor: Carter AFB 4326-S four-barrel. Engine code: H.

Fury station wagon optional 440 "Wedge" V-8
Overhead valve. Cast-iron block. Bore & stroke: 4.32 x 3.75 in. Displacement: 440 cid. Compression ratio: 10.10:1. Brake hp: 350 at 4400 rpm. Taxable hp: 59.70. Torque: 480 lbs.-ft. at 2800 rpm. Five main bearings. Hydraulic valve lifters. Crankcase capacity: 4 qt. (add 1 qt. for new filter). Cooling system capacity: 18 qt. (With heater). Carburetor: Carter AFB 4326-S four-barrel.

Plymouth Belvedere GTX base V-8; Fury, Sport Fury, VIP (cars) optional 440 V-8
Overhead valve. Cast-iron block. Bore & stroke: 4.32 x 3.75 in. Displacement: 440 cid. Compression ratio: 10.10:1. Brake hp: 375 at 4600 rpm. Taxable hp: 59.70. Torque: 480 lbs.-ft. at 3200 rpm. Five main bearings. Hydraulic valve lifters. Crankcase capacity: 4 qt. (add 1 qt. for new filter). Cooling system capacity: 17 qt. (With heater standard). Carburetor: Carter AFB 4326-S four-barrel. Engine code: L.

Plymouth Belvedere GTX optional "Street Hemi" Eight-Barrel V-8
Overhead valve with hemispherical segment combustion chambers. Cast-iron block. Displacement: 426 cid. Bore & stroke: 4.25 x 3.75 in. Compression ratio: 10.25:1. Brake hp: 425 at 5000 rpm. Taxable hp: 57.80. Torque: 490 lbs.-ft. at 4000 rpm. Five main bearings. Hydraulic valve lifters. Crankcase capacity: 6 qt. (add 1 qt. for new filter). Cooling system capacity: 17 qt. (With heater). Carburetor: Dual Carter AFB 3084-S four-barrel. Engine code: J.

1968 PLYMOUTH

Plymouth Valiant 100, Signet, Belvedere, Satellite base V-8
Overhead valve. Cast-iron block. Bore & stroke: 3.63 x 3.31 in. Displacement: 273 cid. Compression ratio: 9.00:1. Brake hp: 190 at 4400 rpm. Taxable hp: 42.20. Torque: 260 lbs.-ft. at 2000 rpm. Five main bearings. Hydraulic valve lifters. Crankcase capacity: 4 qt. (add 1 qt. for new filter). Cooling system capacity: 19 qt. (with heater). Carburetor: Carter BBD-4116-S two-barrel. Engine code: D.

Plymouth Barracuda, Sport Satellite, Fury, Sport Fury, VIP, suburban base V-8; Valiant 100, Signet, Belvedere, Satellite optional V-8
Overhead valve. Cast-iron block. Bore & stroke: 3.91 x 3.31 in. Displacement: 318 cid. Compression ratio: 9.0:1. Brake hp: 230 at 4400 rpm. Taxable hp: 48.90. Torque: 340 lbs.-ft. at 2400 rpm. Five main bearings. Hydraulic valve lifters. Crankcase capacity: 4 qt. (add 1 qt. for new filter). Cooling system capacity: 18 qt. (with heater). Carburetor: Carter BBD-4420-S two-barrel. Engine code: F.

Plymouth Barracuda Formula 'S' base V-8
Overhead valve. Cast-iron block. Bore & stroke: 4.04 x 3.31 in. Displacement: 340 cid. Compression ratio: 10.5:1. Brake hp: 275 at 5000 rpm. Taxable hp: 52.20. Torque: 340 lbs.-ft. at 3200 rpm. Five main bearings. Hydraulic valve lifters. Crankcase capacity: 4 qt. (add 1 qt. for new filter). Cooling system capacity: 17 qt. (with heater). Carburetor: Carter Thermo-Quad AVS-4424-S four-barrel. Engine code: P.

Plymouth Sport Fury, VIP, suburban optional V-8
Overhead valve. Cast-iron block. Displacement: 383 cid. Bore & stroke: 4.25 x 3.38 in. Compression ratio: 9.20:1. Brake hp: 290 at 4400 rpm. Taxable hp: 57.80. Torque: 390 lbs.-ft. at 2800 rpm (except Polara 500 is 380 lbs.-ft. at 2400 rpm). Five main bearings. Hydraulic valve lifters. Crankcase capacity: 4 qt. (add 1 qt. for new filter). Cooling system capacity: 17 qt. (with heater). Carburetor: Carter BBD-4422-S two-barrel. Engine code: G.

Although the 360-cid Plymouth V-8 and this 383-cid V-8 were technically similar, they did not look identical by any means. Many of the accessories are of totally different designs on each of these engines. Note the heftier power steering pump and starter used with the 383. Also, the 383's air cleaner has the Plymouth name at the front of the engine.

Plymouth Barracuda Formula 'S' optional V-8
Overhead valve. Cast-iron block. Displacement: 383 cid. Bore & stroke: 4.25 x 3.38 in. Compression ratio: 10.00:1. Brake hp: 300 at 4400 rpm. Taxable hp: 57.80. Torque: 400 lbs.-ft. at 2400 rpm. Five main bearings. Hydraulic valve lifters. Crankcase capacity: 4 qt. (add 1 qt. for new filter). Cooling system capacity: 17 qt. (with heater). Carburetor: Carter BBD-4422-S two-barrel. Engine code: G.

Plymouth Belvedere, Road Runner, Satellite, Fury, Sport Fury, VIP, suburban optional V-8
Overhead valve. Cast-iron block. Displacement: 383 cid. Bore & stroke: 4.25 x 3.38 in. Compression ratio: 10.10:1. Brake hp: 330 at 5000 rpm. Taxable hp: 57.80. Torque: 425 lbs.-ft. at 3200 rpm. Five main bearings. Hydraulic valve lifters. Crankcase capacity: 4 qt. (add 1 qt. for new filter). Cooling system capacity: 17 qt. (with heater). Carburetor: Carter four-barrel. Engine code: H.

Plymouth Road Runner base V-8
Overhead valve. Cast-iron block. Displacement: 383 cid. Bore & stroke: 4.25 x 3.38 in. Compression ratio: 10.00:1. Brake hp: 335 at 5200 rpm. Taxable hp: 57.80. Torque: 425 lbs.-ft. at 3400 rpm. Five main bearings. Hydraulic valve lifters. Crankcase capacity: 4 qt. (add 1 qt. for new filter). Cooling system capacity: 17 qt. (with heater). Carburetor: Carter four-barrel. Engine code: H.

Plymouth Suburban optional 440 "Wedge" V-8
Overhead valve. Cast-iron block. Bore & stroke: 4.32 x 3.75 in. Displacement: 440 cid. Compression ratio: 10.10:1. Brake hp: 350 at 4400 rpm. Taxable hp: 59.70. Torque: 480 lbs.-ft. at 2800 rpm. Five main bearings. Hydraulic valve lifters. Crankcase capacity: 4 qt. (add 1 qt. for new filter). Cooling system capacity: 18 qt. (with heater). Carter AVS-4428-S four-barrel. Engine code: K.

Plymouth GTX base V-8
Overhead valve. Cast-iron block. Bore & stroke: 4.32 x 3.75 in. Displacement: 440 cid. Compression ratio: 10.10:1. Brake hp: 375 at 4600 rpm. Taxable hp: 59.70. Torque: 480 lbs.-ft. at 3200 rpm. Five main bearings. Hydraulic valve lifters. Crankcase capacity: 4 qt. (add 1 qt. for new filter). Cooling system capacity: 17 qt. (with heater). Carburetor: Carter AVS-4428-S four-barrel. Engine code: L.

Plyouth Belvedere, Road Runner, Satellite, GTX optional "Street Hemi" Eight-Barrel V-8
Overhead valve with hemispherical segment combustion chambers. Cast-iron block. Displacement: 426 cid. Bore & stroke: 4.25 x 3.75 in. Compression ratio: 10.25:1. Brake hp: 425 at 5000 rpm. Taxable hp: 57.80. Torque: 490 lbs.-ft. at 4000 rpm. Five main bearings. Hydraulic valve lifters. Crankcase capacity: 6 qt. (add 1 qt. for new filter). Cooling system capacity: 18 qt. (with heater). Carburetor: Dual Carter AFB 3084-S four-barrel. Engine code: J.

1969 PLYMOUTH

Plymouth Valiant 100, Signet optional V-8
Overhead valve. Cast-iron block. Bore & stroke: 3.63 x 3.31 in. Displacement: 273 cid. Compression ratio: 9.00:1. Brake hp: 190 at 4400 rpm. Taxable hp: 42.20. Torque: 260 lbs.-ft. at 2000 rpm. Five main bearings. Hydraulic valve lifters. Crankcase capacity: 4 qt. (add 1 qt. for new filter). Cooling system capacity: 19 qt. (with heater). Carburetor: Carter BBD-4605-S two-barrel. Engine code: D.

Plymouth Sport Satellite, Sport Fury, VIP base V-8
Overhead valve. Cast-iron block. Bore & stroke: 3.91 x 3.31 in. Displacement: 318 cid. Compression ratio: 9.0:1. Brake hp: 230 at 4400 rpm. Taxable hp: 48.90. Torque: 340 lbs.-ft. at 2400 rpm. Five main bearings. Hydraulic valve lifters. Crankcase capacity: 4 qt. (add 1 qt. for new filter). Cooling system capacity: 18 qt. (with heater). Carburetor: Carter BBD-4607-S two-barrel. Engine code: F.

Plymouth Barracuda optional V-8
Overhead valve. Cast-iron block. Bore & stroke: 4.04 x 3.31 in. Displacement: 340 cid. Compression ratio: 10.5:1. Brake hp: 275 at 5000 rpm. Taxable hp: 52.20. Torque: 340 lbs.-ft. at 3200 rpm. Five main bearings. Hydraulic valve lifters. Crankcase capacity: 4 qt. (add 1 qt. for new filter). Cooling system capacity: 17 qt. (with heater). Carburetor: Carter Thermo-Quad AVS-4611-S four-barrel. Engine code: P.

Belvedere, Satellite, Sport Satellite, Fury, Sport Fury, VIP optional V-8
Overhead valve. Cast-iron block. Displacement: 383 cid. Bore & stroke: 4.25 x 3.38 in. Compression ratio: 9.20:1. Brake hp: 290 at 4400 rpm. Taxable hp: 57.80. Torque: 390 lbs.-ft. at 2800 rpm (except Polara 500 is 380 lbs.-ft. at 2400 rpm). Five main bearings. Hydraulic valve lifters.

Base engine in the 1969 Plymouth Road Runner was a five-main-bearings 383-cid V-8 that had a 4.25 x 3.38-inch bore and stroke, 10.00:1 compression ratio, and 335 hp at 5000 rpm. It developed 425 foot-pounds of torque at 3400 rpm. The engine had hydraulic valve lifters and a Carter Thermo-Quad AVS-4615-S four-barrel carburetor. The block and heads were painted orange and the flat-black air cleaner had an orange-and-white Road Runner decal.

Crankcase capacity: 4 qt. (add 1 qt. for new filter). Cooling system capacity: 17 qt. (with heater). Carburetor: Carter BBD-4613-S two-barrel. Engine code: G.

Belvedere, Satellite, Sport Satellite, Fury, Sport Fury, VIP optional V-8
Overhead valve. Cast-iron block. Displacement: 383 cid. Bore & stroke: 4.25 x 3.38 in. Compression ratio: 10.00:1. Brake hp: 330 at 5200 rpm. Taxable hp: 57.80. Torque: 425 lbs.-ft. at 3200 rpm. Five main bearings. Hydraulic valve lifters. Crankcase capacity: 4 qt. (add 1 qt. for new filter). Cooling system capacity: 17 qt. (with heater). Carburetor: Carter four-barrel. Engine code: H.

Plymouth Road Runner base V-8
Overhead valve. Cast-iron block. Displacement: 383 cid. Bore & stroke: 4.25 x 3.38 in. Compression ratio: 10.00:1. Brake hp: 335 at 5000 rpm. Taxable hp: 57.80. Torque: 425 lbs.-ft. at 3400 rpm. Five main bearings. Hydraulic valve lifters. Crankcase capacity: 4 qt. (add 1 qt. for new filter). Cooling system capacity: 17 qt. (with heater). Carburetor: Carter Thermo-Quad AVS-4615-S four-barrel. Engine code: H.

Plymouth GTX base V-8
Overhead valve. Cast-iron block. Bore & stroke: 4.32 x 3.75 in. Displacement: 440 cid. Compression ratio: 10.10:1. Brake hp: 375 at 4600 rpm. Taxable hp: 59.70. Torque: 480 lbs.-ft. at 3200 rpm. Five main bearings. Hydraulic valve lifters. Crankcase capacity: 4 qt. (add 1 qt. for new filter). Cooling system capacity: 17 qt. (with heater). Carburetor: Carter AVS-4617-S four-barrel. Engine code: K.

Plymouth 'Cuda, Road Runner, GTX, Sport Fury, Sport Fury GT optional "6-Pak" V-8
Overhead valve. Cast-iron block. Bore & stroke: 4.32 x 3.75 in. Displacement: 440 cid. Compression ratio: 10.50:1. Brake hp: 390 at 4700 rpm. Taxable hp: 59.70. Torque: 490 lbs.-ft. at 3200 rpm. Five main bearings. Hydraulic valve lifters. Crankcase capacity: 4 qt. (add 1 qt. for new filter). Cooling system capacity: 15.5 qt. (with heater). Carburetor: three Holley two-barrel. Engine code: L.

Belvedere, Road Runner, Satellite, Sport Satellite, GTX optional V-8
Overhead valve with hemispherical segment combustion chambers. Cast-iron block. Displacement: 426 cid. Bore & stroke: 4.25 x 3.75 in. Compression ratio: 10.25:1. Brake hp: 425 at 5000 rpm. Taxable hp: 57.80. Torque: 490 lbs.-ft. at 4000 rpm. Five main bearings. Hydraulic valve lifters. Crankcase capacity: 6 qt. (add 1 qt. for new filter). Cooling system capacity: 18 qt. (with heater). Carburetor: Dual Carter AFB 3084-S four-barrel. Engine code: J.

1970 PLYMOUTH

Plymouth Barracuda, Gran Coupe, Belvedere, Satellite, Fury base V-8
Overhead valve. Cast-iron block. Bore & stroke: 3.91 x 3.31 in. Displacement: 318 cid. Compression ratio: 8.80:1. Brake hp: 230 at 4400

rpm. Taxable hp: 48.90. Torque: 320 lbs.-ft. at 2000 rpm. Five main bearings. Hydraulic valve lifters. Crankcase capacity: 4 qt. (add 1 qt. for new filter). Cooling system capacity: 16 qt. (with heater). Carburetor: Carter BBD-4721-S two-barrel. Engine code: G.

Plymouth Duster "340" base V-8
Overhead valve. Cast-iron block. Bore & stroke: 4.04 x 3.31 in. Displacement: 340 cid. Compression ratio: 10.5:1. Brake hp: 275 at 5000 rpm. Taxable hp: 52.20. Torque: 340 lbs.-ft. at 3200 rpm. Five main bearings. Hydraulic valve lifters. Crankcase capacity: 4 qt. (add 1 qt. for new filter). Cooling system capacity: 16 qt. (with heater). Carburetor: Carter Thermo-Quad AVS-4933-S four-barrel. Engine code: H.

Plymouth 'Cuda, Road Runner base V-8
Plymouth Barracuda, Gran Coupe, Belvedere, Satellite, Fury (except Sport GT) optional V-8
Overhead valve. Cast-iron block. Displacement: 383 cid. Bore & stroke: 4.25 x 3.38 in. Compression ratio: 8.70:1. Brake hp: 290 at 4400 rpm. Taxable hp: 57.80. Torque: 390 lbs.-ft. at 2800 rpm (except Fury is 380 ft.-lbs. at 2400 rpm). Five main bearings. Hydraulic valve lifters. Crankcase capacity: 4 qt. (add 1 qt. for new filter). Cooling system capacity: 14.5 qt. (with heater). Carburetor: Carter BBD-4613-S two-barrel (Federal). Engine code: L.

Plymouth 'Cuda, Road Runner base V-8
Plymouth Barracuda, Gran Coupe, Belvedere, Satellite, Fury (except Sport GT) optional V-8
Overhead valve. Cast-iron block. Displacement: 383 cid. Bore & stroke: 4.25 x 3.38 in. Compression ratio: 9.50:1. Brake hp: 330 at 5000 rpm. Taxable hp: 57.80. Torque: 425 lbs.-ft. at 3200 rpm. Five main bearings. Hydraulic valve lifters. Crankcase capacity: 4 qt. (add 1 qt. for new filter). Cooling system capacity: 14.5 qt. (with heater). Carburetor: Carter AVS-4376-S four-barrel (Federal). Engine code: N.

Plymouth 'Cuda optional high-performance V-8
Overhead valve. Cast-iron block. Displacement: 383 cid. Bore & stroke: 4.25 x 3.38 in. Compression ratio: 9.50:1. Brake hp: 335 at 5200 rpm. Taxable hp: 57.80. Torque: 425 lbs.-ft. at 3400 rpm. Five main bearings. Hydraulic valve lifters. Crankcase capacity: 4 qt. (add 1 qt. for new filter). Cooling system capacity: 14.5 qt. (with heater). Carburetor: Carter AVS-4376-S four-barrel (Federal). Engine code: N.

Plymouth Sport Fury GT base 440 "Wedge" V-8
Overhead valve. Cast-iron block. Bore & stroke: 4.32 x 3.75 in. Displacement: 440 cid. Compression ratio: 9.70:1. Brake hp: 350 at 4400 rpm. Taxable hp: 59.70. Torque: 480 lbs.-ft. at 2800 rpm. Five main bearings. Hydraulic valve lifters. Crankcase capacity: 4 qt. (add 1 qt. for new filter). Cooling system capacity: 15.5 qt. (with heater). Carburetor: Carter AVS-4366-A four-barrel (Federal). Engine code: T.

Plymouth GTX base V-8
Overhead valve. Cast-iron block. Bore & stroke: 4.32 x 3.75 in. Displacement: 440 cid. Compression ratio: 9.70:1. Brake hp: 375 at 4600 rpm. Taxable hp: 59.70. Torque: 480 lbs.-ft. at 2800 rpm. Five main bearings. Hydraulic valve lifters. Crankcase capacity: 4 qt. (add 1 qt. for new filter). Cooling system capacity: 15.5 qt. (with heater). Carburetor: Carter AVS-4737-S four-barrel (Federal). Engine code: U.

Plymouth 'Cuda, Road Runner, GTX, Sport Fury, Sport Fury GT optional "6-Pak" V-8
Overhead valve. Cast-iron block. Bore & stroke: 4.32 x 3.75 in.

The owner of a 1970 Plymouth Road Runner shows off the special, large air cleaner used with the 440 Magnum Six Pack V-8. It carried the wording "440 Magnum Six Pack" in a band across the center flanked by black-and-white "checkered flag" graphics. The air cleaner is designed to fit into the air-induction opening in the car's hood.

Displacement: 440 cid. Compression ratio: 10.50:1. Brake hp: 390 at 4700 rpm. Taxable hp: 59.70. Torque: 490 lbs.-ft. at 3200 rpm. Five main bearings. Hydraulic valve lifters. Crankcase capacity: 4 qt. (add 1 qt. for new filter). Cooling system capacity: 15.5 qt. (with heater). Carburetor: three Holley two-barrel. Engine code: V.

Plymouth 'Cuda, Road Runner, GTX optional "Street Hemi" V-8
Overhead valve with hemispherical segment combustion chambers. Cast-iron block. Displacement: 426 cid. Bore & stroke: 4.25 x 3.75 in. Compression ratio: 10.20:1. Brake hp: 425 at 5000 rpm. Taxable hp: 57.80. Torque: 490 lbs.-ft. at 4000 rpm. Five main bearings. Hydraulic valve lifters. Crankcase capacity: 6 qt. (add 1 qt. for new filter). Cooling system capacity: 17 qt. (with heater). Carburetor: Dual Carter AFB 3084-S four-barrel. Engine code: R.

This 1970 Road Runner 440 Magnum V-8 carried the hot-performing "Six Pack" carburetor setup with three Holley two-barrel carburetors. The crowning touch on such street racers was the "Air Grabber" hood. This particular engine was optional for the Plymouth 'Cuda, Road Runner, GTX, Sport Fury, and Sport Fury GT models. It produced 390 hp at 4700 rpm.

1971 PLYMOUTH

Plymouth Valiant base V-8
Overhead valve. Cast-iron block. Bore & stroke: 3.91 x 3.31 in. Displacement: 318 cid. Compression ratio: 8.60:1. Brake hp: 230 at 4400 rpm. Taxable hp: 48.90. Torque: 320 lbs.-ft. at 2000 rpm. Five main bearings. Hydraulic valve lifters. Crankcase capacity: 4 qt. (add 1 qt. for new filter). Cooling system capacity: 16 qt. (with heater). Carburetor: Carter BBD-4957-S two-barrel. Engine code: G.

Plymouth Fury optional V-8
Overhead valve. Cast-iron block. Bore & stroke: 4.00 x 3.38 in.

Plymouth's street Hemi lasted until 1971, the same year that this 360-cid V-8 arrived as a Fury option. It was basically the 383 with a reduced bore size and produced 255 hp. Note the Plymouth name on the left-hand side of the black-finished air cleaner.

An underdressed engine expert checks out the high-performance "mill" in a 1971 Plymouth 'Cuda 340 coupe. The 340-cid V-8 was a small-block V-8 that originated in the early '60s as the 273 and evolved into the 318, before going up to 340 cid. This was a strong little engine generating 275 hp in base four-barrel form and 290 hp with the Six-Pack option. The only '71 Plymouth it came in was the Barracuda.

Displacement: 360 cid. Compression ratio: 8.70:1. Brake hp: 255 at 4400 rpm. Taxable hp: 51.0. Torque: 360 lbs.-ft. at 2400 rpm. Five main bearings. Hydraulic valve lifters. Crankcase capacity: 6 qt. (add 1 qt. for new filter). Cooling system capacity: 15 qt. (with heater). Carburetor: two-barrel. Engine code: K.

Plymouth Barracuda base V-8
Overhead valve. Cast-iron block. Bore & stroke: 4.04 x 3.31 in. Displacement: 340 cid. Compression ratio: 10.50:1. Brake hp: 275 at 5000 rpm. Taxable hp: 52.20. Torque: 340 lbs.-ft. at 3200 rpm. Five main bearings. Hydraulic valve lifters. Crankcase capacity: 4 qt. (add 1 qt. for new filter). Cooling system capacity: 16 qt. (with heater). Carburetor: Carter Thermo-Quad TQ-4972-S four-barrel. Engine code: H.

Plymouth Barracuda, Satellite, Fury optional V-8
Overhead valve. Cast-iron block. Displacement: 383 cid. Bore & stroke: 4.25 x 3.38 in. Compression ratio: 8.50:1. Brake hp: 275 at 4400 rpm. Taxable hp: 57.80. Torque: 375 lbs.-ft. at 2800 rpm. Five main bearings. Hydraulic valve lifters. Crankcase capacity: 4 qt. (add 1 qt. for new filter). Cooling system capacity: 14.5 qt. (with heater). Carburetor: Carter BBD-4961-S two-barrel. Engine code: L.

Plymouth Barracuda optional "340 6-Pak" V-8
Overhead valve. Cast-iron block. Bore & stroke: 4.04 x 3.31 in. Displacement: 340 cid. Compression ratio: 10.20:1. Brake hp: 290 at 5000 rpm. Taxable hp: 52.20. Torque: 340 lbs.-ft. at 3200 rpm (except Fury is 380 lbs.-ft. at 2400 rpm). Five main bearings. Hydraulic valve lifters. Crankcase capacity: 4 qt. (add 1 qt. for new filter). Cooling system capacity: 16 qt. (with heater). Carburetor: three Holley two-barrel. Engine code: J.

Plymouth 'Cuda base V-8
Overhead valve. Cast-iron block. Displacement: 383 cid. Bore & stroke: 4.25 x 3.38 in. Compression ratio: 8.50:1. Brake hp: 300 at 4800 rpm. Taxable hp: 57.80. Torque: 410 lbs.-ft. at 3400 rpm. Five main bearings. Hydraulic valve lifters. Crankcase capacity: 4 qt. (add 1 qt. for new filter). Cooling system capacity: 14.5 qt. Carburetor: Holley R-4668A four-barrel. Engine code: N.

Plymouth Fury base V-8
Overhead valve. Cast-iron block. Bore & stroke: 4.32 x 3.75 in. Displacement: 440 cid. Compression ratio: 8.80:1. Brake hp: 335 at 4600 rpm. Taxable hp: 59.70. Torque: 460 lbs.-ft. at 3200 rpm. Five main bearings. Hydraulic valve lifters. Crankcase capacity: 4 qt. (add 1 qt. for new filter). Cooling system capacity: 15.5 qt. (with heater). Carburetor: Carter AVS-4967-S four-barrel. Engine code: T.

Plymouth Satellite base V-8
Overhead valve. Cast-iron block. Bore & stroke: 4.32 x 3.75 in. Displacement: 440 cid. Compression ratio: 9.50:1. Brake hp: 370 at 4600 rpm. Taxable hp: 59.70. Torque: 480 lbs.-ft. at 3200 rpm. Five main bearings. Hydraulic valve lifters. Crankcase capacity: 4 qt. (add 1 qt. for

The 1971 Plymouth 440-cid Six-Pack V-8 was offered in the Plymouth Barracuda, Satellite, and Fury (as well as the Satellite GTX). It had a 4.32 x 3.75 inch bore and stroke and a 10.3:1 compression ratio. Generating 385 hp at 4700 rpm and 490 foot-pounds of torque at 3200 rpm, this V-code engine carried three Holley two-barrel carburetors under that big air cleaner with the Plymouth name on it.

new filter). Cooling system capacity 15.5 qt. (with heater). Carburetor: Carter AVS-4967-S four-barrel. Engine code: U.

Plymouth Barracuda, satellite, Fury "440 6-Pak" V-8
Overhead valve. Cast-iron block. Bore & stroke: 4.32 x 3.75 in. Displacement: 440 cid. Compression ratio: 10.30:1. Brake hp: 385 at 4700 rpm. Taxable hp: 59.70. Torque: 490 lbs.-ft. at 3200 rpm. Five main bearings. Hydraulic valve lifters. Crankcase capacity: 6 qt. (add 1 qt. for new filter). Cooling system capacity: 15.5 qt. (with heater). Carburetor: three Holley two-barrel. Engine code: V.

Plymouth Barracuda, Satellite optional "Street Hemi" V-8
Overhead valve with hemispherical segment combustion chambers. Cast-iron block. Displacement: 426 cid. Bore & stroke: 4.25 x 3.75 in. Compression ratio: 10.20:1. Brake hp: 425 at 5000 rpm. Taxable hp: 57.80. Torque: 490 lbs.-ft. at 4000 rpm. Five main bearings. Hydraulic valve lifters. Crankcase capacity: 6 qt. (add 1 qt. for new filter). Cooling system capacity: 15.5 qt. (with heater). Carburetor: Dual Carter AFB 3084-S four-barrel. Engine code: R.

1972 PLYMOUTH

Plymouth Valiant, Barracuda, Satellite, Fury base V-8
Overhead valve. Cast-iron block. Bore & stroke: 3.91 x 3.31 in. Displacement: 318 cid. Compression ratio: 8.60:1. Net brake hp: 150 at 4000 rpm. Net torque: 260 lbs.-ft. at 1600 rpm. Four main bearings. Hydraulic valve lifters. Crankcase capacity: 4 qt. (add 1 qt. for new filter). Cooling system capacity: 16.5 qt. (with air conditioning). Carburetor: Carter BBD-6150-S two-barrel. Engine code: G.

Plymouth Fury optional V-8
Overhead valve. Cast-iron block. Bore & stroke: 4.00 x 3.58 in. Displacement: 360 cid. Compression ratio: 8.80:1. Net brake hp: 175 at 4000 rpm. Net torque: 285 lbs.-ft. at 2400 rpm. Five main bearings. Hydraulic valve lifters. Crankcase capacity: 4 qt. (add 1 qt. for new filter). Cooling system capacity: 15 qt. (with air conditioning). Carburetor: Holley R-6161A two-barrel. Engine code: K.

Plymouth Satellite, Fury optional V-8
Overhead valve. Cast-iron block. Bore & stroke: 4.34 x 3.38 in. Displacement: 400 cid. Compression ratio: 8.20:1. Net brake hp: 190 at 4400 rpm. Net torque: 310 lbs.-ft. at 2400 rpm. Five main bearings. Hydraulic valve lifters. Crankcase capacity: 4 qt. (add 1 qt. for new filter). Cooling system capacity: (Satellite) 15 qt.; (Fury) 14 qt. (with air conditioning). Carburetor: Holley ZR-6368A two-barrel. Engine code: M.

Plymouth Fury optional V-8
Overhead valve. Cast-iron block. Bore & stroke: 4.32 x 3.75 in. Displacement: 440 cid. Compression ratio: 8.20:1. Net brake hp: 225 at 4400 rpm. Net torque: 345 lbs.-ft. at 3200 rpm. Five main bearings. Hydraulic valve lifters. Crankcase capacity: 4 qt. (add 1 qt. for new filter). Cooling system capacity: 16 qt. (with air conditioning). Carburetor: Holley R-6160A four-barrel. Engine code: T.

Plymouth Duster "340," 'Cuda, Road Runner base V-8
Overhead valve. Cast-iron block. Bore & stroke: 4.04 x 3.31 in. Displacement: 340 cid. Compression ratio: 8.50:1. Net brake hp: 240 at 4800 rpm. Net torque: 290 lbs.-ft. at 3600 rpm. Five main bearings. Hydraulic valve lifters. Crankcase capacity: 4 qt. (add 1 qt. for new filter). Cooling system capacity: 14.5 qt. (with air conditioning). Carburetor: Carter Thermo-Quad TQ-6139-S four-barrel. Engine code: H.

Plymouth optional V-8
Overhead valve. Cast-iron block. Bore & stroke: 4.32 x 3.75 in. Displacement: 440 cid. Compression ratio: 8.20:1. Net brake hp: 245 at 4400 rpm. Net torque: 360 lbs.-ft. at 3200 rpm. Five main bearings. Hydraulic valve lifters. Crankcase capacity: 4 qt. (add 1 qt. for new filter). Cooling system capacity: 15 qt. (with air conditioning). Carburetor: Holley R-6160A four-barrel. Engine code: U.

Plymouth Satellite base V-8; Optional Fury V-8
Overhead valve. Cast-iron block. Bore & stroke: 4.34 x 3.38 in. Displacement: 400 cid. Compression ratio: 8.2:1. Net brake hp: 255 at 4800 rpm (250 hp in Fury). Net torque: 340 lbs.-ft. at 3200 rpm. Five main bearings. Hydraulic valve lifters. Crankcase capacity: 4 qt. (add 1 qt. for new filter). Cooling system capacity: (Satellite) 15 qt.; (Fury) 14 qt. (with air conditioning). Carburetor: Carter Thermo-Quad TQ-6090-S four-barrel. Engine code: P.

Plymouth Satellite base V-8
Overhead valve. Cast-iron block. Bore & stroke: 4.34 x 3.38 in. Displacement: 400 cid. Compression ratio: 8.2:1. Net brake hp: 265 at 4800 rpm. Net torque: 345 lbs.-ft. at 3200 rpm. Five main bearings. Hydraulic valve lifters. Crankcase capacity: 4 qt. (add 1 qt. for new filter). Cooling system capacity: 15 qt. (with air conditioning). Carburetor: Carter Thermo-Quad TQ-6090-S four-barrel. Engine code: P.

Plymouth Satellite optional V-8
Overhead valve. Cast-iron block. Bore & stroke: 4.32 x 3.75 in. Displacement: 440 cid. Compression ratio: 8.20:1. Net brake hp: 280 at 4800 rpm. Net torque: 375 lbs.-ft. at 3200 rpm. Five main bearings. Hydraulic valve lifters. Crankcase capacity: 4 qt. (add 1 qt. for new filter). Cooling system capacity: 15 qt. (with air conditioning). Carburetor: Holley R-6160A four-barrel. Engine code: U.

Plymouth Satellite, Fury optional V-8
Overhead valve. Cast-iron block. Bore & stroke: 4.32 x 3.75 in. Displacement: 440 cid. Compression ratio: 8.20:1. Net brake hp: 290 at 4800 rpm. Net torque: 380 lbs.-ft. at 3200 rpm. Five main bearings. Hydraulic valve lifters. Crankcase capacity: 4 qt. (add 1 qt. for new filter). Cooling system capacity: 15 qt. (with air conditioning). Carburetor: Holley R-6160A four-barrel. Engine code: U.

Plymouth Satellite optional "440 6-Pak" V-8
Overhead valve. Cast-iron block. Displacement: 440 cid. Bore & stroke: 4.32 x 3.75 in. Compression ratio: 10.30:1. Net brake hp: 330 at 4800 rpm. Net torque: 410 lbs.-ft. at 3600 rpm. Five main bearings. Hydraulic valve lifters. Crankcase capacity: 4 qt. (add 1 qt. for new filter). Cooling system capacity: 15 qt. (with air conditioning). Carburetor: Three Holley two-barrel. Engine code: V.

1973 PLYMOUTH

Plymouth's 1973 version of the small-block V-8 was a 318-cid job that featured improvements to Chrysler's much-heralded Cleaner Air System. These included an electric-assist choke and an exhaust gas recirculation (EGR) system. This engine was designed to operate on lead-free fuel.

Plymouth Valiant, Barracuda, Satellite (except Road Runner), Fury base V-8
Overhead valve. Cast-iron block. Bore & stroke: 3.91 x 3.31 in. Displacement: 318 cid. Compression ratio: 8.60:1. Net brake hp: 150 at 4000 rpm. Net torque: 265 lbs.-ft. at 2000 rpm. Four main bearings. Hydraulic valve lifters. Crankcase capacity: 4 qt. (add 1 qt. for new filter). Cooling system capacity: (Fury) 19 qt.; (Other models) 16 qt. Carburetor: Carter Two-barrel. (Note: Not available in Valiant, Duster or Scamp with 4-speed manual transmission). Engine code: G.

Road Runner base V-8
Overhead valve. Cast-iron block. Bore & stroke: 3.91 x 3.31 in. Displacement: 318 cid. Compression ratio: 8.60:1. Net brake hp: 170 at 4000 rpm. Net torque: 270 lbs.-ft. at 2000 rpm. Four main bearings. Hydraulic valve lifters. Crankcase capacity: 4 qt. (add 1 qt. for new filter). Cooling system capacity: (Fury) 19 qt.; (Other models) 16 qt. Dual exhaust. Carburetor: Carter Two-barrel. Engine code: G.

Plymouth Fury Station wagon base V-8; Fury (cars) optional V-8
Overhead valve. Cast-iron block. Bore & stroke: 4.00 x 3.58 in. Displacement: 360 cid. Compression ratio: 8.40:1. Net brake hp: 170 at 4000 rpm. Net torque: 285 lbs.-ft. at 2400 rpm. (When equipped with emission package N95 subtract 7 hp and 5 lbs.-ft. of torque.) Five main bearings. Hydraulic valve lifters. Crankcase capacity: 4 qt. (add 1 qt. for new filter). Cooling system capacity: 16 qt. (with air conditioning). Carburetor: Two-barrel. Engine code: K.

The 1973 Plymouth 340-cid V-8 was completely different from the 318- and 360-cid V-8s and served a different purpose in the engine lineup. It was a high-performance engine for compact and intermediate models. The compact block featured very large cylinder bores teamed with large intake manifold passages, large intake and exhaust valve ports, and a free-breathing dual exhaust system. Plymouth said that the aluminum-alloy-on-steel engine bearings used on four of the 340 engine's five main bearings were of the "type used on the Plymouth high-performance cars raced by Richard Petty."

The 1973 Plymouth 440-cid V-8 was the largest engine offered that year. Its size, four-barrel induction, and electronic ignition system made it one of the most versatile and dependable V-8s in the auto industry. It was optional for all Fury and Road Runner models. For 1973, the carburetor had a number of improvements including better low-speed driveability and more precise control of fuel metering. In addition, good cold-weather starting was aided by the Thermo-Quad design, which permitted the engine to operate well on a lean fuel mixture. The 440 featured durable tri-metal connecting rod bearings and a heavy-duty 78-plate battery. On Road Runner models, the 440 four-barrel engine included a high-performance high-lift camshaft and valve springs with surge dampeners to control valve bounce at high speeds.

Plymouth Satellite (except Road Runner) optional V-8
Overhead valve. Cast-iron block. Bore & stroke: 4.34 x 3.38 in. Displacement: 400 cid. Compression ratio: 8.20:1. Net brake hp: 175 at 3600 rpm. Net torque: 305 lbs.-ft. at 2400 rpm. Five main bearings. Hydraulic valve lifters. Crankcase capacity: 4 qt. (add 1 qt. for new filter). Cooling system capacity: 17 qt. (with air conditioning). Carburetor: Two-barrel. (Note: Not available in Satellite station wagon with manual transmission). Engine code: M.

Plymouth Fury (all) optional V-8
Overhead valve. Cast-iron block. Bore & stroke: 4.34 x 3.38 in. Displacement: 400 cid. Compression ratio: 8.20:1. Net brake hp: 185 at 3600 rpm. Net torque: 310 lbs.-ft. at 2400 rpm. Five main bearings. Hydraulic valve lifters. Crankcase capacity: 4 qt. (add 1 qt. for new filter). Cooling system capacity: 17 qt. (with air conditioning). Carburetor: Two-barrel. Engine code: P.

Plymouth Fury optional "Magnum" V-8
Overhead valve. Cast-iron block. Bore & stroke: 4.32 x 3.75 in. Displacement: 440 cid. Compression ratio: 8.20:1. Net brake hp: 220 at 3600 rpm. Net torque: 350 lbs.-ft. at 2400 rpm. (When equipped with

The 1973 Plymouth 318-cid V-8 had the Plymouth name decaled on both sides of the air cleaner's single snorkel. The air cleaner was finished in black and the "Plymouth" name was spelled out in white letters. New this year was an electronic ignition system that eliminated breaker points and provided longer spark plug life. It was standard equipment.

emission package N95 subtract 7 hp and 5 lbs.-ft. of torque.) Five main bearings. Hydraulic valve lifters. Crankcase capacity: 4 qt. (add 1 qt. for new filter). Cooling system capacity: 18 qt. (with air conditioning). Carburetor: Four-barrel. Engine code: T.

Plymouth Duster "340" base V-8; Barracuda, Satellite, Road Runner optional V-8
Overhead valve. Cast-iron block. Bore & stroke: 4.04 x 3.31 in. Displacement: 340 cid. Compression ratio: 8.50:1. Net brake hp: 240 at 4800 rpm. Net torque: 295 lbs.-ft. at 3600 rpm. Five main bearings. Hydraulic valve lifters. Crankcase capacity: 4 qt. (add 1 qt. for new filter). Cooling system capacity: 16 qt. (with air conditioning). Carburetor: Four-barrel. (Note: Not available with 3-speed manual transmission.) Engine code: H.

Plymouth Satellite, Road Runner, Satellite station wagons optional V-8
Overhead valve. Cast-iron block. Bore & stroke: 4.34 x 3.38 in. Displacement: 400 cid. Compression ratio: 8.20:1. Net brake hp: 260 at 4800 rpm. Net torque: 335 lbs.-ft. at 3600 rpm. Five main bearings. Hydraulic valve lifters. Crankcase capacity: 4 qt. (add 1 qt. for new filter). Cooling system capacity: 17 qt. (with air conditioning). Low-restriction dual exhaust. Carburetor: Four-barrel. (Note: Not available with 3-speed manual transmission.) Engine code: P.

Plymouth Road Runner optional V-8
Overhead valve. Cast-iron block. Bore & stroke: 4.32 x 3.75 in. Displacement: 440 cid. Compression ratio: 8.20:1. Net brake hp: 280 at 4800 rpm. Net torque: 380 lbs.-ft. at 3200 rpm. Five main bearings. Hydraulic valve lifters. Crankcase capacity: 4 qt. (add 1 qt. for new filter). Cooling system capacity: 17 qt. (with air conditioning). Dual exhaust. Carburetor: Four-barrel. (Note: Not available with manual transmission.) Engine code: U.

1974 PLYMOUTH

Plymouth Valiant, Scamp, Duster, Barracuda, Satellite, Satellite wagon, Sebring base V-8
Overhead valve. Cast-iron block. Bore & stroke: 3.91 x 3.31 in. Displacement: 318 cid. Compression ratio: 8.60:1. Net brake hp: 150 at 4000 rpm. Net torque: 255 lbs.-ft. at 2200 rpm. Four main bearings. Hydraulic valve lifters. Crankcase capacity: 4 qt. (add 1 qt. for new filter). Cooling system capacity: 16 qt. Carburetor: Carter Two-barrel. Engine code: G.

Road Runner base V-8
Overhead valve. Cast-iron block. Bore & stroke: 3.91 x 3.31 in. Displacement: 318 cid. Compression ratio: 8.60:1. Net brake hp: 170 at 4000 rpm. Net torque: 265 lbs.-ft. at 2600 rpm. Four main bearings. Hydraulic valve lifters. Crankcase capacity: 4 qt. (add 1 qt. for new filter). Cooling system capacity: 16 qt. Dual exhaust. Carburetor: Carter Two-barrel. Engine code: G.

Plymouth Fury I, II, III base V-8
Overhead valve. Cast-iron block. Bore & stroke: 4.00 x 3.58 in. Displacement: 360 cid. Compression ratio: 8.40:1. Net brake hp: 180 at 4000 rpm. Net torque: 290 lbs.-ft. at 2400 rpm. Five main bearings. Hydraulic valve lifters. Crankcase capacity: 4 qt. (add 1 qt. for new filter). Cooling system capacity: (Fury) 16 qt.; (Other models) 16.5 qt. (with air conditioning). Carburetor: Two-barrel. Engine code: K.

Plymouth Fury Gran Coupe and Grand Sedan base V-8
Plymouth Satellite (except Road Runner), Sebring, Satellite station wagon optional V-8
Overhead valve. Cast-iron block. Bore & stroke: 4.34 x 3.38 in. Displacement: 400 cid. Compression ratio: 8.20:1. Net brake hp: 185 at 4000 rpm. Net torque: 315 lbs.-ft. at 2400 rpm. Five main bearings. Hydraulic valve lifters. Crankcase capacity: 4 qt. (add 1 qt. for new filter). Cooling system capacity: 16.5 qt. (with air conditioning). Carburetor: Two-barrel.Engine code: M.

Plymouth Satellite (except Road Runner), Satellite station wagon, Fury I, II, III optional V-8
Overhead valve. Cast-iron block. Bore & stroke: 4.00 x 3.58 in. Displacement: 360 cid. Compression ratio: 8.40:1. Net brake hp: 200 at 4000 rpm. Net torque: 290 lbs.-ft. at 3200 rpm. Five main bearings. Hydraulic valve lifters. Crankcase capacity: 4 qt. (add 1 qt. for new filter). Cooling system capacity: (Fury) 16 qt.; (Other models) 16.5 qt. (with air conditioning). Carburetor: Four-barrel. (Note: This version is equipped with the N95 exhaust emission package required in California.) Engine code: J.

Plymouth Fury, Satellite (exc. Road Runner), Sebring, Satellite station wagon optional V-8
Overhead valve. Cast-iron block. Bore & stroke: 4.34 x 3.38 in. Displacement: 400 cid. Compression ratio: 8.20:1. Net brake hp: 200 at 4400 rpm. Net torque: 310 lbs.-ft. at 2400 rpm. Five main bearings. Hydraulic valve lifters. Crankcase capacity: 4 qt. (add 1 qt. for new filter). Cooling system capacity: 16.5 qt. (with air conditioning). Carburetor: Four-barrel. (Note: This version is equipped with the N95 exhaust emission package required in California.) Engine code: N.

Plymouth Fury, Satellite (exc. Road Runner), Sebring, Satellite station wagon optional V-8
Overhead valve. Cast-iron block. Bore & stroke: 4.34 x 3.38 in. Displacement: 400 cid. Compression ratio: 8.20:1. Net brake hp: 205 at 4400 rpm. Net torque: 315 lbs.-ft. at 2400 rpm. Five main bearings. Hydraulic valve lifters. Crankcase capacity: 4 qt. (add 1 qt. for new filter). Cooling system capacity: 16.5 qt. (with air conditioning). Carburetor: Four-barrel. Engine code: T.

Plymouth Fury I, II, III, Fury Suburban optional V-8
Overhead valve. Cast-iron block. Bore & stroke: 4.32 x 3.75 in. Displacement: 440 cid. Compression ratio: 8.20:1. Net brake hp: 220 at 4000 rpm. Net torque: 345 lbs.-ft. at 3200 rpm. Five main bearings. Hydraulic valve lifters. Crankcase capacity: 4 qt. (add 1 qt. for new filter). Cooling system capacity: 16 qt. (with air conditioning). Carburetor: Four-barrel. (Note: This version is equipped with the N95 exhaust emission package required in California.) Engine code: T.

The 360-cid Plymouth V-8 seen here was a larger-displacement version of the compact 318-cid V-8. This made it the perfect engine for fully-accessorized cars, which were becoming more common in 1975. Valve timing, valve lift and duration were carefully engineered for low emissions, economy, power and smooth operation at all speeds. The 318's hydraulic valve lifters required no periodic adjustments. The 1975 camshaft was redesigned to seat the valves more smoothly and to decrease the possibilty of valve bounce and the stress in caused in valve stems.

Plymouth Fury I, II, III, Fury Suburban optional V-8
Overhead valve. Cast-iron block. Bore & stroke: 4.32 x 3.75 in. Displacement: 440 cid. Compression ratio: 8.20:1. Net brake hp: 230 at 4000 rpm. Net torque: 350 lbs.-ft. at 3200 rpm. Five main bearings. Hydraulic valve lifters. Crankcase capacity: 4 qt. (add 1 qt. for new filter). Cooling system capacity: 16 qt. (with air conditioning). Carburetor: Four-barrel. Engine code: U.

Plymouth Duster "360" base V-8; 'Cuda, Road Runner optional V-8
Overhead valve. Cast-iron block. Bore & stroke: 4.00 x 3.58 in. Displacement: 360 cid. Compression ratio: 8.40:1. Net brake hp: 245 at 4800 rpm. Net torque: 320 lbs.-ft. at 3600 rpm. Five main bearings. Hydraulic valve lifters. Crankcase capacity: 4 qt. (add 1 qt. for new filter). Cooling system capacity: 16.5 qt. (with air conditioning). Dual exhaust. Carburetor: Four-barrel. Engine code: H.

Plymouth Satellite Road Runner optional V-8
Overhead valve. Cast-iron block. Bore & stroke: 4.34 x 3.38 in. Displacement: 400 cid. Compression ratio: 8.20:1. Net brake hp: 250 at 4800 rpm. Net torque: 330 lbs.-ft. at 3400 rpm. Five main bearings. Hydraulic valve lifters. Crankcase capacity: 4 qt. (add 1 qt. for new filter). Cooling system capacity: 16.5 qt. (with air conditioning). Dual exhaust. Dual snorkel air cleaner. Carburetor: Four-barrel. Engine code: P.

Plymouth Road Runner optional V-8
Overhead valve. Cast-iron block. Bore & stroke: 4.32 x 3.75 in. Displacement: 440 cid. Compression ratio: 8.20:1. Net brake hp: 275 at 4400 rpm. Net torque: 375 lbs.-ft. at 3200 rpm. Five main bearings. Hydraulic valve lifters. Crankcase capacity: 4 qt. (add 1 qt. for new filter). Cooling system capacity: 16 qt. (with air conditioning). Dual exhaust. Carburetor: Four-barrel. Engine code: U.

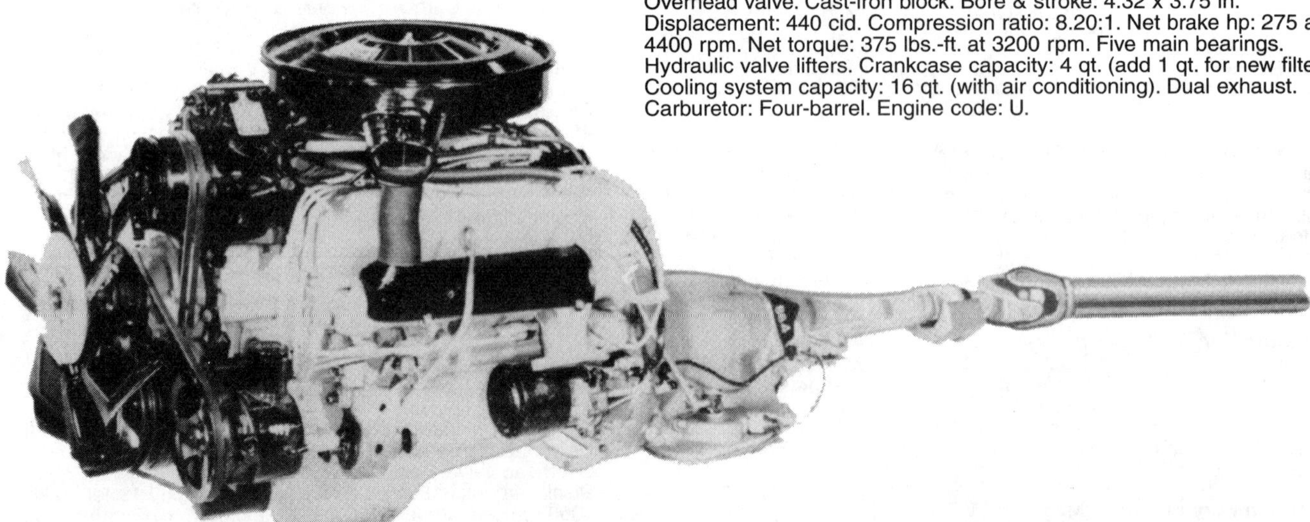

Plymouth promoted a number of engine and driveline features in its 1974 models including an oval snorkel on the carburetor air cleaner to reduce air intake noise, a vibration damper to smooth engine vibrations, spool-type engine mounts with bonded rubber cushions. A power plant bending damper, mounted at the rear of the transmission, absorbed vibration and a propeller shaft vibration absorber reduced rear axle noise levels.

Optional at extra cost on the 1974 Road Runner, the four-barrel 440 high-performance V-8 included a high-lift cam, dual exhausts, and valve-spring surge dampeners. A heavy-duty suspension was required in cars getting this heavy power plant. The 275-nhp engine featured a sturdy crankshaft made of forged steel.

1975 PLYMOUTH

Plymouth Fury, Road Runner, Fury station wagons base V-8 (California)
Overhead valve. Cast-iron block. Bore & stroke: 3.91 x 3.31 in. Displacement: 318 cid. Compression ratio: 8.5:1. Net brake hp: 135 at 3600 rpm. Net torque: 245 lbs.-ft. at 1600 rpm. Five main bearings. Hydraulic valve lifters. Crankcase capacity: 4 qt. (add 1 qt. for new filter). Cooling system capacity: 16 qt. (with air conditioning). Carburetor: Two-barrel. Sales code: E44.

Plymouth Valiant, Duster (except 360) base V-8 (California)
Overhead valve. Cast-iron block. Bore & stroke: 3.91 x 3.31 in. Displacement: 318 cid. Compression ratio: 8.5:1. Net brake hp: 140 at 3600 rpm. Net torque: 255 lbs.-ft. at 1600 rpm. Five main bearings. Hydraulic valve lifters. Crankcase capacity: 4 qt. (add 1 qt. for new filter). Cooling system capacity: 16 qt. (with air conditioning). Carburetor: Two-barrel. Sales code: E44.

Plymouth Gran Fury, Gran Fury Custom optional V-8 (California)
Overhead valve. Cast-iron block. Bore & stroke: 3.91 x 3.31 in. Displacement: 318 cid. Compression ratio: 8.5:1. Net brake hp: 145 at 3600 rpm. Net torque: 250 lbs.-ft. at 2000 rpm. Five main bearings. Hydraulic valve lifters. Crankcase capacity: 4 qt. (add 1 qt. for new filter). Cooling system capacity: 16 qt. (with air conditioning). Carburetor: Two-barrel. Sales code: E44.

Plymouth Valiant, Duster (except 360) base V-8 (Federal)
Overhead valve. Cast-iron block. Bore & stroke: 3.91 x 3.31 in. Displacement: 318 cid. Compression ratio: 8.5:1. Net brake hp: 145 at 4000 rpm. Net torque: 255 lbs.-ft. at 1600 rpm. Five main bearings. Hydraulic valve lifters. Crankcase capacity: 4 qt. (add 1 qt. for new filter). Cooling system capacity: 16 qt. (with air conditioning). Carburetor: Two-barrel. Sales code: E44.

Plymouth Fury, Road Runner, Fury station wagons base V-8 (Federal)
Overhead valve. Cast-iron block. Bore & stroke: 3.91 x 3.31 in. Displacement: 318 cid. Compression ratio: 8.5:1. Net brake hp: 150 at 4000 rpm. Net torque: 255 lbs.-ft. at 1600 rpm. Five main bearings. Hydraulic valve lifters. Crankcase capacity: 4 qt. (add 1 qt. for new filter).

The 360-cid Plymouth V-8 seen here was a larger-displacement version of the compact 318-cid V-8. This made it the perfect engine for fully-accessorized cars, which were becoming more common in 1975. Valve timing, valve lift, and duration were carefully engineered for low emissions, economy, power, and smooth operation at all speeds. The 318's hydraulic valve lifters required no periodic adjustments. The 1975 camshaft was redesigned to seat the valves more smoothly and to decrease the possibilty of valve bounce and the stress it caused in valve stems.

Cooling system capacity: 16 qt. (with air conditioning). Carburetor: Two-barrel. Sales code: E44.

Plymouth Gran Fury, Gran Fury Custom optional V-8 (Federal)
Overhead valve. Cast-iron block. Bore & stroke: 3.91 x 3.31 in. Displacement: 318 cid. Compression ratio: 8.5:1. Net brake hp: 150 at 4000 rpm. Net torque: 260 lbs.-ft. at 1600 rpm. Five main bearings. Hydraulic valve lifters. Crankcase capacity: 4 qt. (add 1 qt. for new filter). Cooling system capacity: 16 qt. (with air conditioning). Carburetor: Two-barrel. Sales code: E44.

Plymouth Fury, Road Runner optional V-8 (Federal)
Overhead valve. Cast-iron block. Bore & stroke: 4.34 x 3.38 in. Displacement: 400 cid. Compression ratio: 8.2:1. Net brake hp: 165 at

Plymouth's 400-cid four-barrel V-8 provided a moderate horsepower increase over the two-barrel version. This was accomplished primarily through its increased fuel-air induction capacity. The increased engine breathing gave stronger performance at cruising and passing speeds. This 1975 version had a single exhaust system. It was available, at extra cost only, with the N95 California emission control system on all Fury models including the Road Runner. A different high-performance version of the 400-cid four-barrel engine was introduced late in 1975. It featured a free-breathing four-barrel carburetor and a dual low-restriction exhaust system. A dual-snorkel air cleaner was a police car option.The high-performance version generated 235 nhp, compared to the regular engine's 190. The high-performance V-8's carburetor had a plastic main body that did not conduct heat as readily as metal. The valves had hot-pressed valve springs to reduce load loss after extended operation. Tri-metal connecting rods were designed to accommodate this engine's higher horsepower. The high-performance 400-cid four-barrel V-8 was available in Fury hardtop models, including Road Runners, but not in California.

4000 rpm. Net torque: 295 lbs.-ft. at 3200 rpm. Five main bearings. Hydraulic valve lifters. Crankcase capacity: 4 qt. (add 1 qt. for new filter). Cooling system capacity: 16 qt. (with air conditioning). Carburetor: Two-barrel. Sales code: E63.

Plymouth Gran Fury, Gran Fury Custom, Wagons, Gran Fury Brougham base V-8 (Federal)
Overhead valve. Cast-iron block. Bore & stroke: 4.34 x 3.38 in. Displacement: 400 cid. Compression ratio: 8.2:1. Net brake hp: 175 at 4000 rpm. Net torque: 300 lbs.-ft. at 2400 rpm. Five main bearings. Hydraulic valve lifters. Crankcase capacity: 4 qt. (add 1 qt. for new filter). Cooling system capacity: 16 qt. (with air conditioning). Carburetor: Two-barrel. Sales code: E63.

Plymouth Gran Fury Sedan and Hardtop base V-8 (Federal only) Fury and Gran Fury Brougham optional V-8 (not available on Gran Fury station wagons)
Overhead valve. Cast-iron block. Bore & stroke: 4.00 x 3.58 in. Displacement: 360 cid. Compression ratio: 8.4:1. Net brake hp: 180 at 4000 rpm. Net torque: 290 lbs.-ft. at 2400 rpm. Five main bearings. Hydraulic valve lifters. Crankcase capacity: 4 qt. (add 1 qt. for new filter). Cooling system capacity: 16 qt. (with air conditioning). Carburetor: Two-barrel. Sales code: E57.

Plymouth Fury, Sport Fury, Fury Custom, Road Runner optional V-8 (California)
Overhead valve. Cast-iron block. Bore & stroke: 4.34 x 3.38 in. Displacement: 400 cid. Compression ratio: 8.2:1. Net brake hp: 185 at 4000 rpm. Net torque: 285 lbs.-ft. at 3200 rpm. Five main bearings. Hydraulic valve lifters. Crankcase capacity: 4 qt. (add 1 qt. for new filter). Cooling system capacity: 16 qt. (with air conditioning). Carburetor: Four-barrel. Sales code: E64.

Fury, Fury Sport, Fury Custom, Gran Fury, Gran Fury Custom optional V-8 (California only)
Overhead valve. Cast-iron block. Bore & stroke: 4.00 x 3.58 in. Displacement: 360 cid. Compression ratio: 8.4:1. Net brake hp: 190 at 4000 rpm. Net torque: 270 lbs.-ft. at 3200 rpm. Five main bearings. Hydraulic valve lifters. Crankcase capacity: 4 qt. (add 1 qt. for new filter). Cooling system capacity: 16 qt. (with air conditioning). Carburetor: Four-barrel. Sales code: E56.

Plymouth Duster 360 base V-8; Road Runner optional V-8 (California only)
Overhead valve. Cast-iron block. Bore & stroke: 4.00 x 3.58 in. Displacement: 360 cid. Compression ratio: 8.4:1. Net brake hp: 190 at 4000 rpm. Net torque: 270 lbs.-ft. at 3200 rpm. Five main bearings. Hydraulic valve lifters. Crankcase capacity: 4 qt. (add 1 qt. for new filter). Cooling system capacity: 16 qt. (with air conditioning). Carburetor: Four-barrel. Sales code: E58.

Plymouth Fury, Sport Fury, Fury Custom, Road Runner optional V-8 (Federal)
Overhead valve. Cast-iron block. Bore & stroke: 4.34 x 3.38 in. Displacement: 400 cid. Compression ratio: 8.2:1. Net brake hp: 190 at 4000 rpm. Net torque: 290 lbs.-ft. at 3200 rpm. Five main bearings. Hydraulic valve lifters. Crankcase capacity: 4 qt. (add 1 qt. for new filter). Cooling system capacity: 16 qt. (with air conditioning). Carburetor: Four-barrel. Sales code: E64.

Plymouth Gran Fury, Gran Fury Custom, Grand Fury Brougham optional V-8 (California)
Overhead valve. Cast-iron block. Bore & stroke: 4.34 x 3.38 in. Displacement: 400 cid. Compression ratio: 8.2:1. Net brake hp: 195 at 4000 rpm. Net torque: 285 lbs.-ft. at 3200 rpm. Five main bearings. Hydraulic valve lifters. Crankcase capacity: 4 qt. (add 1 qt. for new filter). Cooling system capacity: 16 qt. (with air conditioning). Carburetor: Four-barrel. Sales code: E64.

Plymouth Gran Fury, Gran Fury Custom, Grand Fury Brougham optional V-8 (California)
Overhead valve. Cast-iron block. Bore & stroke: 4.32 x 3.75 in. Displacement: 440 cid. Compression ratio: 8.20:1. Net brake hp: 210 at 4000 rpm. Net torque: 320 lbs.-ft. at 3200 rpm. Five main bearings. Hydraulic valve lifters. Crankcase capacity: 4 qt. (add 1 qt. for new filter). Cooling system capacity: 16 qt. (with air conditioning). Carburetor: Four-barrel. Sales code: E85.

Plymouth Gran Fury, Gran Fury Custom, Grand Fury Brougham optional V-8 (Federal)
Overhead valve. Cast-iron block. Bore & stroke: 4.32 x 3.75 in. Displacement: 440 cid. Compression ratio: 8.20:1. Net brake hp: 215 at 4000 rpm. Net torque: 330 lbs.-ft. at 3200 rpm. Five main bearings. Hydraulic valve lifters. Crankcase capacity: 4 qt. (add 1 qt. for new filter). Cooling system capacity: 16 qt. (with air conditioning). Carburetor: Four-barrel. Sales code: E85.

Plymouth Duster "360" base V-8 (Federal)
Overhead valve. Cast-iron block. Bore & stroke: 4.00 x 3.58 in. Displacement: 360 cid. Compression ratio: 8.4:1. Net brake hp: 230 at 4200 rpm. Net torque: 320 lbs.-ft. at 3200 rpm. Five main bearings. Hydraulic valve lifters. Crankcase capacity: 4 qt. (add 1 qt. for new filter). Cooling system capacity: 16 qt. (with air conditioning). Dual exhaust. High-performance camshaft. Carburetor: Four-barrel. Sales code: E58.

Plymouth Fury hardtop, Road Runner hardtop optional heavy-duty V-8 (Federal)
Overhead valve. Cast-iron block. Bore & stroke: 4.34 x 3.38 in. Displacement: 400 cid. Compression ratio: 8.2:1. Net brake hp: 235 at 4200 rpm. Net torque: 320 lbs.-ft. at 3200 rpm. Five main bearings. Hydraulic valve lifters. Crankcase capacity: 4 qt. (add 1 qt. for new filter). Cooling system capacity: 16 qt. (with air conditioning). Dual exhaust. Dual snorkel air cleaner. Carburetor: Four-barrel. Sales code: E68.

Plymouth Gran Fury optional heavy-duty V-8 (Federal)
Overhead valve. Cast-iron block. Bore & stroke: 4.34 x 3.38 in. Displacement: 400 cid. Compression ratio: 8.2:1. Net brake hp: 240 at 4400 rpm. Net torque: 325 lbs.-ft. at 3200 rpm. Five main bearings. Hydraulic valve lifters. Crankcase capacity: 4 qt. (add 1 qt. for new filter). Cooling system capacity: 16 qt. (with air conditioning). Dual exhaust. Carburetor: Four-barrel. Sales code: E68.

1976 PLYMOUTH

Plymouth Volare, Valiant, Duster, Fury (all models), Gran Fury base V-8 (California)
Overhead valve. Cast-iron block and head. Bore & stroke: 3.91 x 3.31 in. Displacement: 318 cid (5.2 liters). Compression ratio: 8.5:1. Net brake hp: 140 at 3600 rpm. Net torque: 250 lbs.-ft. at 2000 rpm. Five main bearings. Hydraulic valve lifters. Crankcase capacity: 4 qt. (add 1 qt. for new filter). Cooling system capacity: 17.5-18 qt. (with air conditioning). Carburetor: Carter BBD 8068S or 8069S two-barrel. VIN code: G. Sales code: E44. (Note: Available on the 318 only was package N91 that deleted the standard catalytic converter and used an air pump only instead. With N91 the 318 could be operated on regular, low-lead or lead-free gasoline.)

Plymouth Volare, Valiant, Duster, Fury (All models), Gran Fury base V-8 (Federal)
Overhead valve. Cast-iron block and head. Bore & stroke: 3.91 x 3.31 in. Displacement: 318 cid (5.2 liters). Compression ratio: 8.5:1. Net brake hp: 150 at 4000 rpm. Net torque: 255 lbs.-ft. at 1600 rpm. Five main

The 440-cid four-barrel V-8 remained Plymouth's largest engine in 1976. It was now rated for 205 nhp at 3600 rpm. Features of the big-block V-8 were the same as in past years and there were no changes to the engine's appearance either.

In 1976, Plymouth again offered standard and high-performance versions of its 400-cid four-barrel V-8. The hotter engine gained five horsepower and now carried a 240 nhp rating. That was 55 nhp higher than the standard type engine. All other features of the high-performance V-8 were the same as in 1975.

bearings. Hydraulic valve lifters. Crankcase capacity: 4 qt. (add 1 qt. for new filter). Cooling system capacity: 17.5-18 qt. (with air conditioning). Carburetor: Carter BBD 8068S or 8069S two-barrel. VIN code: G. Sales code: E44. (Note: Available on the 318 only was package N91 that deleted the standard catalytic converter and used an air pump only instead. With N91 the 318 could be operated on regular, low-lead or lead-free gasoline.)

Plymouth Gran Fury Custom, Gran Fury Brougham, Fury wagon base V-8 (Federal)
Plymouth Volare, Fury, Fury Sport, Fury Salon optional V-8 (Federal)
Overhead valve. Cast-iron block and head. Bore & stroke: 4.00 x 3.58 in. Displacement: 360 cid (5.9 liters). Compression ratio: 8.4:1. Net brake hp: 170 at 4000 rpm. Net torque: 280 lbs.-ft. at 2400 rpm. Five main bearings. Hydraulic valve lifters. Crankcase capacity: 4 qt. (add 1 qt. for new filter). Cooling system capacity: 16 qt. (with air conditioning). Carburetor: Holley 2245 (R7364A) two-barrel. VIN code: K. Sales code: E57.

Plymouth Fury, Fury Sport, Fury Salon optional V-8 (California)
Plymouth Gran Fury Custom, Gran Fury Brougham optional V-8 (California)
Overhead valve. Cast-iron block and head. Bore & stroke: 4.00 x 3.58 in. Displacement: 360 cid (5.9 liters). Compression ratio: 8.4:1. Net brake hp: 175 at 4000 rpm. Net torque: 270 lbs.-ft. at 1600 rpm. Five main bearings. Hydraulic valve lifters. Crankcase capacity: 4 qt. (add 1 qt. for new filter). Cooling system capacity: 16 qt. (with air conditioning). Carburetor: Carter TQ four-barrel carburetor. VIN code: J. Sales code: E56.

Plymouth Gran Fury wagon base V-8 (Federal)
Plymouth Fury, Fury Sport, Fury Salon, Gran Fury Custom, Gran Fury Brougham optional V-8 (Federal)
Overhead valve. Cast-iron block and head. Bore & stroke: 4.34 x 3.38 in. Displacement: 400 cid (6.6 liters). Compression ratio: 8.2:1. Net brake hp: 175 at 4000 rpm. Net torque: 300 lbs.-ft. at 2400 rpm. Five main

A number of modifications were also made to Plymouth's 360-cid two-barrel V-8 for 1976. The vacuum spark advance time delay was shortened by 10 seconds in the OSAC system to improve engine response and fuel economy, the temperature sensing valve for the EGR system was relocated to the radiator header tank (it was formerly in the engine) to provide more accurate EGR control, and idle speed was reduced from 750 to 700 rpm to improve fuel economy.

As with Plymouth's other 1976 V-8s, the 400-cid two-barrel V-8 had three changes related to improved fuel economy and drivability. The vacuum spark advance time delay was shortened from 27 to 17 seconds in the OSAC system to improve engine response and fuel economy, engine idle speed was reduced from 750 to 700 rpm to save gas, and the carburetor power mixture was set to operate at higher road speed to improve low-speed and suburban fuel economy.

A number of modifications were made to Plymouth's 318-cid small-block V-8 for 1976. The Orifice Spark Advance Control (OSAC) system was eliminated for better response and fuel economy; the EGR system was recalibrated for improved fuel economy, a new single-element catalytic converter was fitted (except to California cars) and the low-rpm spark advance was improved by the utilization of a manifold vacuum instead of carburetor port vacuum (except in California).

bearings. Hydraulic valve lifters. Crankcase capacity: 4 qt. (add 1 qt. for new filter). Cooling system capacity: 16.5 qt. (with air conditioning). Carburetor: Holley 2245 (R7366A) two-barrel. VIN code: M. Sales code: E63.

Plymouth Fury, Fury Sport, Fury Salon optional V-8 (California) Plymouth Gran Fury Custom, Gran Fury Brougham optional V-8 (California)
Overhead valve. Cast-iron block and head. Bore & stroke: 4.34 x 3.38 in. Displacement: 400 cid (6.6 liters). Compression ratio: 8.2:1. Net brake hp: 185 at 3600 rpm. Net torque: 285 lbs.-ft. at 3200 rpm. Five main bearings. Hydraulic valve lifters. Crankcase capacity: 4 qt. (add 1 qt. for new filter). Cooling system capacity: 16.5 qt. (with air conditioning). Carburetor: Carter TQ-9056-S four-barrel. VIN code: N.

Plymouth Gran Fury Custom, Gran Fury Brougham optional V-8 (California)
Overhead valve. Cast-iron block and heads. Bore & stroke: 4.32 x 3.75 in. Displacement: 440 cid (7.2 liters). Compression ratio: 8.2:1. Net brake hp: 200 at 3600 rpm. Net torque: 310 lbs.-ft. at 2400 rpm. Five main bearings. Hydraulic valve lifters. Crankcase capacity: 5 qt. (add 1 qt. for new filter). Cooling system capacity: 16 qt. (with air conditioning). Carburetor: Carter TQ-9058-S four-barrel. VIN code: T. Sales code: E85.

Plymouth Gran Fury Custom, Gran Fury Brougham optional V-8 (Federal)
Overhead valve. Cast-iron block and heads. Bore & stroke: 4.32 x 3.75 in. Displacement: 440 cid (7.2 liters). Compression ratio: 8.2:1. Net brake hp: 205 at 3600 rpm. Net torque: 320 lbs.-ft. at 2000 rpm. Five main bearings. Hydraulic valve lifters. Crankcase capacity: 5 qt. (add 1 qt. for new filter). Cooling system capacity: 16 qt. (with air conditioning). Carburetor: Carter TQ-9058-S four-barrel. VIN code: T. Sales code: E85.

Duster "360" base V-8 (Federal)
Overhead valve. Cast-iron block and head. Bore & stroke: 4.00 x 3.58 in. Displacement: 360 cid (5.9 liters). Compression ratio: 8.4:1. Net brake hp: 220 at 4400 rpm. Net torque: 280 lbs.-ft. at 3200 rpm. Five main bearings. Hydraulic valve lifters. Crankcase capacity: 4 qt. (add 1 qt. for new filter). Cooling system capacity: 16 qt. (with air conditioning). Dual exhaust. High-performance cam. Single snorkel air cleaner. Carburetor: Carter TQ-9002-S four-barrel carburetor. VIN code: L. Sales code: E58. (Note: Automatic transmission only).

Plymouth Fury, Fury Sport, Fury Salon optional V-8 (Federal)
Overhead valve. Cast-iron block and head. Bore & stroke: 4.34 x 3.38 in. Displacement: 400 cid (6.6 liters). Compression ratio: 8.2:1. Net brake hp: 240 at 4400 rpm. Net torque: 325 lbs.-ft. at 3200 rpm. Five main bearings. Hydraulic valve lifters. Crankcase capacity: 5 qt. (add 1 qt. for new filter). Cooling system capacity: 16 qt. (with air conditioning). Dual exhaust. Dual snorkel air cleaner. High-performance cam. Carburetor: Carter TQ-9054-S four-barrel. VIN code: P. Sales code: E68.

1977 PLYMOUTH

Plymouth Volare, Valiant, Duster, Fury (All models), Gran Fury base V-8
Overhead valve. Cast-iron block and head. Displacement: 318 cid (5.2 liters). Bore & stroke: 3.91 x 3.31 in. Compression ratio: 8.5:1. Net brake hp: 145 at 4000 rpm. Net torque: 245 lbs.-ft. at 1600 rpm. Five main bearings. Hydraulic valve lifters. Carburetor: Carter BBD 8093S or 8094S two-barrel. VIN code: G.

Plymouth Gran Fury Custom, Gran Fury Brougham, Fury wagon base V-8Plymouth Volare, Fury, Fury Sport, Fury Salon optional V-8
Overhead valve. Cast-iron block and head. Bore & stroke: 4.00 x 3.58 in. Displacement: 360 cid (5.9 liters). Compression ratio: 8.4:1. Net brake hp: 155 at 3600 rpm. Net torque: 275 lbs.-ft. at 2000 rpm. Five main bearings. Hydraulic valve lifters. Carburetor: Holley 2245 (R7671A) two-barrel. VIN code: K.

Plymouth Volare optional V-8
Overhead valve. Cast-iron block and head. Bore & stroke: 4.00 x 3.58 in. Displacement: 360 cid (5.9 liters). Compression ratio: 8.4:1. Net brake hp: 175 at 4000 rpm. Net torque: 275 lbs.-ft. at 2000 rpm. Five main bearings. Hydraulic valve lifters. Carburetor: Carter TQ four-barrel and Lean Burn system. VIN code: L.

Plymouth Fury optional V-8
Overhead valve. Cast-iron block and head. Bore & stroke: 4.34 x 3.38 in. Displacement: 400 cid (6.6 liters). Compression ratio: 8.2:1. Net brake hp: 190 at 3600 rpm. Net torque: 305 lbs.-ft. at 3200 rpm. Five main bearings. Hydraulic valve lifters. Carburetor: Carter TQ9102S four-barrel. VIN code: N.

Plymouth Fury optional V-8
Overhead valve. Cast-iron block and head. Bore & stroke: 4.32 x 3.75 in.

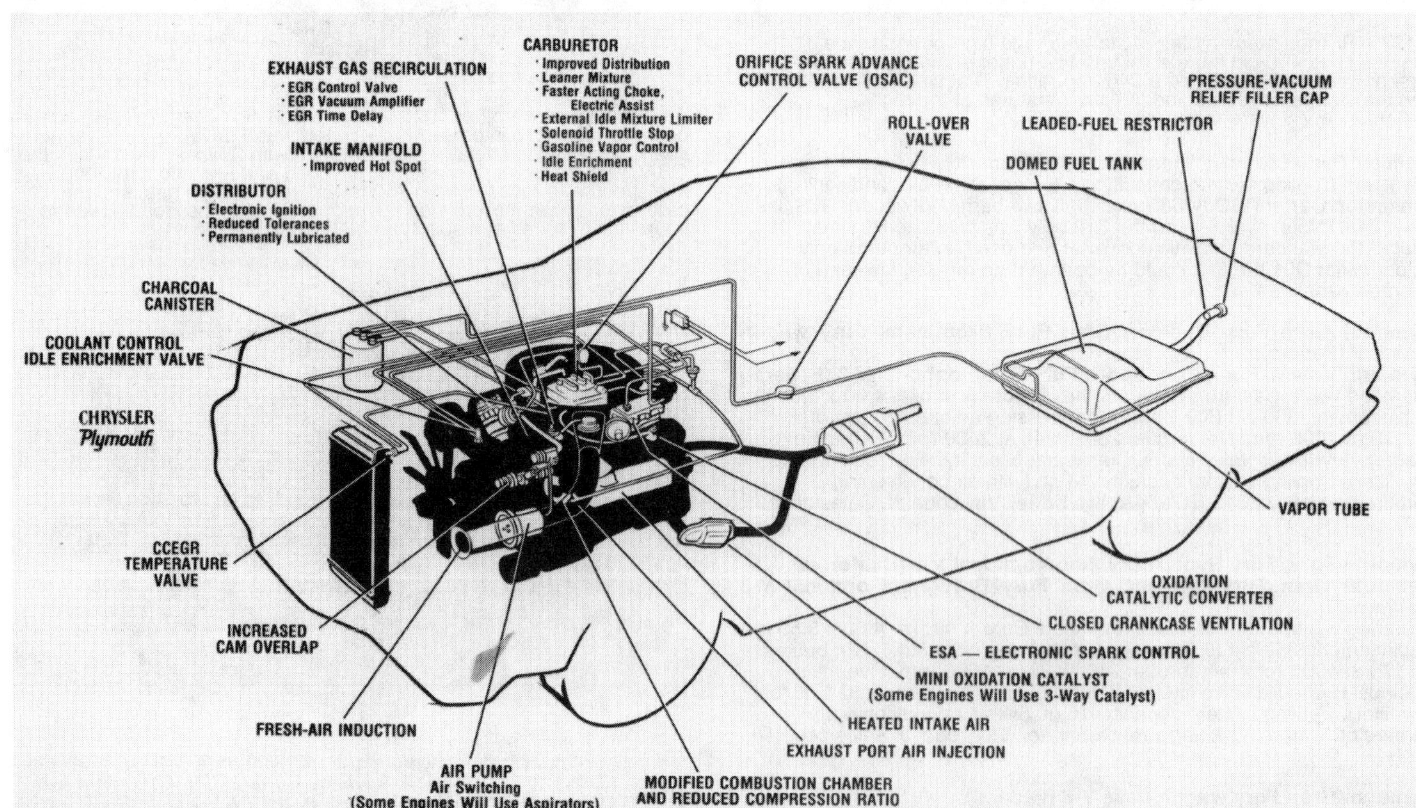

All of Plymouth's 1977 models were equipped with an improved Cleaner Air System. It was revised to meet both new Environmental Protection Agency (EPA) standards for oxides of nitrogen and the new California Air Resources Board standards for hydrocarbons and oxides of nitrogen. Improvements included new mini-oxidation (or "3-way") catalysts, fresh-air induction, air aspirators, high-altitude calibration, power-heat control valves, air injection, and air switching.

Displacement: 440 cid (7.2 liters). Compression ratio: 8.2:1. Net brake hp: 195 at 3600 rpm. Net torque: 320 lbs.-ft. at 2000 rpm. Five main bearings. Hydraulic valve lifters. Carburetor: Carter TQ9078S four-barrel. VIN code: T.

1978 PLYMOUTH

A 360-cid two-barrel lean-burn V-8 (code E57) was available on all 1978 Plymouth Fury and Volare models sold in federal areas. This was a bigger version of the 318 with a valve train that was carefully engineered for low emissions, power and smooth operation at all speed ranges. Also available for California Fury models and some Volares was the 360-cid four-barrel heavy-duty V-8 (code E58). When installed in cars sold in California or high-altitude counties, this engine did not carry the electronic lean-burn system.

The 1978 Plymouth V-8 seen in this photo featured an electronic lean-burn system that the automaker described as "one of the most important developments in engine control since the 1920s and early '30s, when manual spark control was replaced by the automatic spark advance." The electronic spark-control computer responded to data from all sensors and instantly adjusted the timing of sprak plug firings. In promoting the system, Plymouth promised smoother engine warm ups, improved engine performance, improved smoothness of engine operation, controlled emissions, and less emission-control hardware, but the lean-burn system ultimately proved troublesome and hard to maintain.

Plymouth Volare Coupe, Sedan, Station Wagon, Fury (cars) optional V-8 (Federal)
Overhead valve. Cast-iron block and head. Bore & stroke: 3.91 x 3.31 in. Displacement: 318 cid (5.2 liters). Compression ratio: 8.5:1. Net brake hp: 140 at 4000 rpm. Net torque: 245 lbs.-ft. at 1600 rpm. Five main bearings. Hydraulic valve lifters. Electronic Lean burn system. Carburetor: Carter BBD two-barrel. VIN code: G. Sales code: E44.

Plymouth Volare Coupe, Sedan, Station Wagon, Fury (cars) optional V-8 (High-Altitude)
Overhead valve. Cast-iron block and head. Bore & stroke: 3.91 x 3.31 in. Displacement: 318 cid (5.2 liters). Compression ratio: 8.5:1. Net brake

For 1978, Plymouth's rugged 400-cid four-barrel V-8 with electronic lean-burn system (code E64) was available on Furys only. It had replaced the 440-cid V-8 as the company's largest power plant. The heavy-duty version of this motor (code E68) was also available on Fury models, but only those equipped with the A36 heavy-duty package and not on cars sold in California or high-altitude counties.

hp: 140 at 4000 rpm. Net torque: 245 lbs.-ft. at 1600 rpm. Five main bearings. Hydraulic valve lifters. Electronic Lean burn system. Carburetor: Carter BBD two-barrel. VIN code: G. Sales code: E44.

Plymouth Volare Coupe, Sedan, Station Wagon, Fury (cars) optional V-8 (California)
Overhead valve. Cast-iron block and head. Bore & stroke: 3.91 x 3.31 in. Displacement: 318 cid (5.2 liters). Compression ratio: 8.5:1. Net brake hp: 155 at 4000 rpm. Net torque: 245 lbs.-ft. at 1600 rpm. Five main bearings. Hydraulic valve lifters. Electronic Lean burn system. Carburetor: Carter four-barrel. VIN code: G. Sales code: E46.

Plymouth Fury Suburban, Fury Sport Suburban base V-8 (California) Plymouth Volare Coupe, Sedan, Station Wagon; Fury cars optional V-8 (California)
Overhead valve. Cast-iron block and head. Bore & stroke: 4.00 x 3.58 in. Displacement: 360 cid (5.9 liters). Compression ratio: 8.4:1. Net brake hp: 155 at 3600 rpm. Net torque: 270 lbs.-ft. at 2000 rpm. Five main bearings. Hydraulic valve lifters. Electronic Lean burn system. Carburetor: Holley 2245 two-barrel. VIN code: K. Sales code: E57.

Plymouth Volare Sedan, Station Wagon optional "Heavy-Duty" V-8 (Federal)
Overhead valve. Cast-iron block and head. Bore & stroke: 4.00 x 3.58 in. Displacement: 360 cid (5.9 liters). Compression ratio: 8.4:1. Net brake hp: 155 at 3600 rpm. Net torque: 270 lbs.-ft. at 2400 rpm. Five main bearings. Hydraulic valve lifters. Electronic Lean burn system. Carburetor: Holley 2245 two-barrel. VIN code: K. Sales code: E55.

Plymouth Volare Coupe, Sedan, Station Wagon optional "Heavy-Duty" V-8 (California)
Overhead valve. Cast-iron block and head. Bore & stroke: 4.00 x 3.58 in. Displacement: 360 cid (5.9 liters). Compression ratio: 8.4:1. Net brake hp: 160 at 3600 rpm. Net torque: 265 lbs.-ft. at 1600 rpm. Five main bearings. Hydraulic valve lifters. Electronic Lean Burn system. Carburetor: Carter TQ four-barrel. VIN code: J. Sales code: E58. Requires package N95.

Plymouth Volare, Fury, Fury Salon Sport, Fury Suburban and Sport Suburban optional V-8 (High-Altitude)
Overhead valve. Cast-iron block and head. Bore & stroke: 4.00 x 3.58 in. Displacement: 360 cid (5.9 liters). Compression ratio: 8.4:1. Net brake hp: 160 at 3600 rpm. Net torque: 265 lbs.-ft. at 1600 rpm. Five main bearings. Hydraulic valve lifters. Electronic Lean Burn system. Carburetor: Carter TQ four-barrel. VIN code: J. Sales code: E58.

Plymouth Fury, Fury Salon Sport, Fury Suburban and Sport Suburban base V-8 (California)
Overhead valve. Cast-iron block and head. Bore & stroke: 4.00 x 3.58 in. Displacement: 360 cid (5.9 liters). Compression ratio: 8.4:1. Net brake hp: 170 at 4000 rpm. Net torque: 270 lbs.-ft. at 1600 rpm. Five main bearings. Hydraulic valve lifters. Carburetor: Carter TQ four-barrel. VIN code: J. Sales code: E56. Requires package N95.

Plymouth Fury, Fury Salon Sport, Fury Suburban and Sport Suburban base V-8 (High-Altitude)
Overhead valve. Cast-iron block and head. Bore & stroke: 4.00 x 3.58 in. Displacement: 360 cid (5.9 liters). Compression ratio: 8.4:1. Net brake hp: 170 at 4000 rpm. Net torque: 270 lbs.-ft. at 1600 rpm. Five main bearings. Hydraulic valve lifters. Carburetor: Carter TQ four-barrel. VIN

code: J. Sales code: E56. Requires package N96.

Plymouth Fury, Fury Salon Sport, Fury Suburban and Sport Suburban base V-8 (Federal)
Overhead valve. Cast-iron block and head. Bore & stroke: 4.00 x 3.58 in. Displacement: 360 cid (5.9 liters). Compression ratio: 8.4:1. Net brake hp: 175 at 4000 rpm. Net torque: 260 lbs.-ft. at 1600 rpm. Five main bearings. Hydraulic valve lifters. Electronic Lean Burn system. Carburetor: Carter TQ four-barrel. VIN code: J. Sales code: E58.

Plymouth Fury, Fury Salon Sport, Fury Suburban and Sport Suburban base V-8 (Federal)
Overhead valve. Cast-iron block and head. Bore & stroke: 4.34 x 3.38 in. Displacement: 400 cid (6.6 liters). Compression ratio: 8.2:1. Net brake hp: 190 at 3600 rpm. Net torque: 305 lbs.-ft. at 3200 rpm. Five main bearings. Hydraulic valve lifters. Electronic Lean Burn system. Carburetor: Carter TQ four-barrel. VIN code: N. Sales code: E64.

Plymouth Fury, Fury Salon Sport, Fury Suburban and Sport Suburban base V-8 (Federal)
Overhead valve. Cast-iron block and head. Bore & stroke: 4.34 x 3.38 in. Displacement: 400 cid (6.6 liters). Compression ratio: 8.2:1. Net brake hp: 190 at 3600 rpm. Net torque: 305 lbs.-ft. at 3200 rpm. Five main bearings. Hydraulic valve lifters. Electronic Lean Burn system. Carburetor: Carter TQ four-barrel. VIN code: N. Sales code: E68.

1979 PLYMOUTH

In 1979, Plymouth used the term Electronic Spark Control (ESC), instead of Lean-Burn, to identify its engine management system. The system used sensors to feed information to an electronic computer on this 1979 Plymouth V-8 engine. The computer controlled engine rpms, engine load conditions, throttle position, coolant temperatures, the opening and closing time of the throttle, and engine starting functions.

Plymouth Volare optional V-8 (Federal)
Overhead valve. Cast-iron block and head. Bore & stroke: 3.91 x 3.31 in. Displacement: 318 cid (5.2 liters). Compression ratio: 8.5:1. Net brake hp: 135 at 4000 rpm. Net torque: 250 lbs.-ft. at 1600 rpm. Five main bearings. Hydraulic valve lifters. Carburetor: Holley 2280 (R8448A) two-barrel. VIN code: G. Sales code: E44

Plymouth Volare optional V-8 (California)
Overhead valve. Cast-iron block and head. Bore & stroke: 3.91 x 3.31 in. Displacement: 318 cid (5.2 liters). Compression ratio: 8.5:1. Net brake hp: 155 at 4000 rpm. Net torque: 245 lbs.-ft. at 1600 rpm. Five main bearings. Hydraulic valve lifters. Electronic Spark Control. Carburetor: Four-barrel. VIN code: G. Sales code: E46.

Plymouth Volare optional V-8 (California)
Overhead valve. Cast-iron block and head. Bore & stroke: 4.00 x 3.58 in. Displacement: 360 cid (5.9 liters). Compression ratio: 8.0:1. Net brake hp: 190 at 4000 rpm. Net torque: 275 lbs.-ft. at 2000 rpm. Five main bearings. Hydraulic valve lifters. Dual exhaust on Magnum XE and St. Regis models; other models utilize dual mufflers and catalysts converging into a single tailpipe. Heavy-duty Electronic Spark Control. Carburetor: Carter TQ four-barrel. VIN code: L. Sales code E58.

Plymouth Volare optional V-8 (Federal)
Overhead valve. Cast-iron block and head. Bore & stroke: 4.00 x 3.58 in.

For 1979, the 318-cid two-barrel V-8 (code E44) seen here was not available in California. Cars sold there could be had with a 318-cid four-barrel V-8 with the ESC system (code E46) which was equipped to meet the state's more stringent emission requirements. Also available on selected models was the 360-cid four-barrel heavy-duty V-8 with ESC and dual exhausts (code E58).

Displacement: 360 cid (5.9 liters). Compression ratio: 8.0:1. Net brake hp: 195 at 4000 rpm. Net torque: 280 lbs.-ft. at 2400 rpm. Five main bearings. Hydraulic valve lifters. Dual exhaust on Magnum XE and St. Regis models; other models utilize dual mufflers and catalysts converging into a single tailpipe. Heavy-duty Electronic Spark Control. Carburetor: Carter TQ four-barrel. VIN code: L. Sales code E58.

1980 PLYMOUTH

Plymouth Volare, Gran Fury base V-8 (Federal)
Overhead valve. Cast-iron block and heads. Bore & stroke: 3.91 x 3.31 in. Displacement: 318 cid (5.2 liters). Compression ratio: 8.5:1. Net brake hp: 120 at 3600 rpm. Net torque: 245 lbs.-ft. at 1600 rpm. Five main bearings. Hydraulic valve lifters. Carburetor: Carter BBD two-barrel. VIN code: G or H.

Plymouth Volare, Gran Fury base V-8 (High-Altitude)
Overhead valve. Cast-iron block and heads. Bore & stroke: 3.91 x 3.31 in. Displacement: 318 cid (5.2 liters). Compression ratio: 8.5:1. Net brake hp: 155 at 4000 rpm. Net torque: 245 lbs.-ft. at 1600 rpm. Five main bearings. Hydraulic valve lifters. Carburetor: Carter four-barrel. VIN code: G or H.

Gran Fury optional V-8
Overhead valve. Cast-iron block and head. Bore & stroke: 4.00 x 3.58 in. Displacement: 360 cid (5.9 liters). Compression ratio: 8.4:1. Net brake hp: 130 at unknown rpm. Net torque: unknown at unknown rpm. Five main bearings. Hydraulic valve lifters. Carburetor: Carter BBD two-barrel. VIN code: K.

1981 PLYMOUTH

Plymouth Gran Fury base V-8
Overhead valve. Cast-iron block and heads. Bore & stroke: 3.91 x 3.31 in. Displacement: 318 cid (5.2 liters). Compression ratio: 8.5:1. Net brake hp: 130 at 4000 rpm. Net torque: 230 lbs.-ft. at 2000 rpm. Five main bearings. Hydraulic valve lifters. Carburetor: Carter BBD two-barrel. VIN code: K.

Plymouth Gran Fury optional V-8
Overhead valve. Cast-iron block and heads. Bore & stroke: 3.91 x 3.31 in. Displacement: 318 cid (5.2 liters). Compression ratio: 8.5:1. Net brake hp: 165 at 4000 rpm. Net torque: 240 lbs.-ft. at 2000 rpm. Five main bearings. Hydraulic valve lifters. Carburetor: Carter TQ-9283-S four-barrel. VIN code: M.

1982 PLYMOUTH

Plymouth Gran Fury base V-8
Overhead valve. Cast-iron block and heads. Bore & stroke: 3.91 x 3.31 in. Displacement: 318 cid (5.2 liters). Compression ratio: 8.5:1. Net brake hp: 130 at 4000 rpm. Net torque: 230 lbs.-ft. at 2000 rpm. Five main bearings. Hydraulic valve lifters. Carburetor: Carter BBD two-barrel. VIN code: K.

Plymouth Gran Fury optional V-8
Overhead valve. Cast-iron block and heads. Bore & stroke: 3.91 x 3.31 in. Displacement: 318 cid (5.2 liters). Compression ratio: 8.5:1. Net brake hp: 165 at 4000 rpm. Net torque: 240 lbs.-ft. at 2000 rpm. Five main bearings. Hydraulic valve lifters. Carburetor: Carter TQ-9283-S four-barrel. VIN code: M.

1983 PLYMOUTH

Plymouth Gran Fury base V-8
Overhead valve. Cast-iron block and heads. Bore & stroke: 3.91 x 3.31 in. Displacement: 318 cid (5.2 liters). Compression ratio: 8.5:1. Net brake hp: 130 at 4000 rpm. Net torque: 230 lbs.-ft. at 1600 rpm. Five main bearings. Hydraulic valve lifters. Carburetor: Carter BBD two-barrel. VIN code: P.

1984 PLYMOUTH

Plymouth Gran Fury base V-8
Overhead valve. Cast-iron block and heads. Bore & stroke: 3.91 x 3.31 in. Displacement: 318 cid (5.2 liters). Compression ratio: 8.5:1. Net brake hp: 130 at 4000 rpm. Net torque: 235 lbs.-ft. at 1600 rpm. Five main bearings. Hydraulic valve lifters. Carburetor: Carter BBD two-barrel. VIN code: P.

Plymouth Gran Fury police V-8
Overhead valve. Cast-iron block and heads. Bore & stroke: 3.91 x 3.31 in. Displacement: 318 cid (5.2 liters). Compression ratio: 8.5:1. Net brake hp: 165 at 4000 rpm. Net torque: 240 lbs.-ft. at 2000 rpm. Five main bearings. Hydraulic valve lifters. Carburetor: Carter TQ 9295 four-barrel. VIN code: P.

1985 PLYMOUTH

Plymouth Gran Fury base V-8
Overhead valve. Cast-iron block and heads. Bore & stroke: 3.91 x 3.31 in. Displacement: 318 cid (5.2 liters). Compression ratio: 9.1:1. Net brake hp: 140 at 3600 rpm. Net torque: 265 lbs.-ft. at 1600 rpm. Five main bearings. Hydraulic valve lifters. Carburetor: Holley 6280 two-barrel.

Plymouth Gran Fury police V-8
Overhead valve. Cast-iron block and heads. Bore & stroke: 3.91 x 3.31 in. Displacement: 318 cid (5.2 liters). Compression ratio: 8.5:1. Net brake hp: 175 at unknown rpm. Net torque: unknown at unknown rpm. Five main bearings. Hydraulic valve lifters. Carburetor: Four-barrel. VIN code: P.

A 360-cid four-barrel V-8 (code E56) was specifically designed to meet California and high-altitude emission-control regulations. This engine did not carry the electronic lean-burn system. Instead, it came equipped with an air pump, as well as a catalytic converter. It could be ordered for use in federal areas, but only with the N95 emission-control package installed. All versions of this engine used a single exhaust system.

"The 318-cid has a long and enviable record for performance and fuel economy," said Plymouth in its 1974 salesmans guide. This was the number one seller in Plymouth's lineup of V-8 engines because of its reliability and low-maintenance requirements. The 318 was the standard V-8 in Valiant and Duster (except Duster 340) models, plus Barracudas, 'Cudas, and all Satellites. When used in the Road Runner, it came equipped with dual exhausts, which added 20 net horsepower.

1986 PLYMOUTH

Plymouth Gran Fury base V-8
Overhead valve. Cast-iron block and heads. Bore & stroke: 3.91 x 3.31 in. Displacement: 318 cid (5.2 liters). Compression ratio: 9.1:1. Net brake hp: 140 at 3600 rpm. Net torque: 265 lbs.-ft. at 1600 rpm. Five main bearings. Hydraulic valve lifters. Carburetor: Holley 6280 two-barrel.

Plymouth Gran Fury police V-8
Overhead valve. Cast-iron block and heads. Bore & stroke: 3.91 x 3.31 in. Displacement: 318 cid (5.2 liters). Compression ratio: 8.5:1. Net brake hp: 175 at unknown rpm. Net torque: unknown at unknown rpm. Five main bearings. Hydraulic valve lifters. Carburetor: Rochester four-barrel. VIN code: P.

1987 PLYMOUTH

Plymouth Gran Fury base V-8
Overhead valve. Cast-iron block and heads. Bore & stroke: 3.91 x 3.31 in. Displacement: 318 cid (5.2 liters). Compression ratio: 9.1:1. Net brake hp: 140 at 3600 rpm. Net torque: 265 lbs.-ft. at 1600 rpm. Five main bearings. Hydraulic valve lifters. Carburetor: Two-barrel.

Plymouth Gran Fury police V-8
Overhead valve. Cast-iron block and heads. Bore & stroke: 3.91 x 3.31 in. Displacement: 318 cid (5.2 liters). Compression ratio: 8.5:1. Net brake hp: 175 at unknown rpm. Net torque: unknown at unknown rpm. Five main bearings. Hydraulic valve lifters. Carburetor: Four-barrel. VIN code: P.

1988 PLYMOUTH

Plymouth Gran Fury base V-8
Overhead valve. Cast-iron block and heads. Bore & stroke: 3.91 x 3.31 in. Displacement: 318 cid. (5.2 liters). Compression ratio: 9.1:1. Net brake hp: 140 at 3600 rpm. Net torque: 265 lbs.-ft. at 1600 rpm. Five main bearings. Hydraulic valve lifters. Carburetor: Two-barrel.

Plymouth Gran Fury police V-8
Overhead valve. Cast-iron block and heads. Bore & stroke: 3.91 x 3.31 in. Displacement: 318 cid. (5.2 liters). Compression ratio: 8.5:1. Net brake hp: 175 at unknown rpm. Net torque: unknown at unknown rpm. Five main bearings. Hydraulic valve lifters. Carburetor: Four-barrel. VIN code: P.

1989 PLYMOUTH

Plymouth Gran Fury Salon base V-8
Overhead valve. Cast-iron block and heads. Bore & stroke: 3.91 x 3.31 in. Displacement: 318 cid. (5.2 liters). Compression ratio: 9.1:1. Net brake hp: 140 at 3600 rpm. Net torque: 265 lbs.-ft. at 1600 rpm. Five main bearings. Hydraulic valve lifters. Carburetor: Two-barrel.

Plymouth V-8 Firing Order

__1955-1976__
1-8-4-3-6-5-7-2

All Plymouth V-8s have the cylinders numbered from front to back. The cylinders on the left-hand bank are numbered 1-3-5-7 and the cylinders on the right-hand bank are numbered 2-4-6-8.

Standard on the 1974 Plymouth Fury Gran coupe, Grand sedan and Fury wagons sold in Federal areas was the 400-cid two-barrel V-8. A four-barrel version was the "minimum-size" engine required for the same models sold in California. This was a different power plant than the smaller Plymouth V-8s. The two-barrel version was great at low- and middle-speed ranges because of its displacement, large intake manifold, large valves and easy-breathing characteristics. Four of its five main bearings were made of steel-backed Babbitt, while the thrust bearing was made of high-strength aluminum-alloy-on-steel material. This engine had its distributor located at the front for easy accessibility.

Selected Plymoth V-8 Tune-Up Specs 1955-1965

Year	Model	Spark Plugs Type	Spark Plugs Gap	Point Gap	Deg Cam Angle	Timing	Idle Speed	Compression Pressure	Battery (Delco)
1955	240-cid	AL-AR51	.035	.017-.019	Note 1	4B	475 rpm	130	Group 1
	260-cid	AL-AR51	.035	.017-.019	Note 1	4B	475 rpm	135	Group 1
Note 1: These V-8s have two sets of distributor points. The total cam angle is 32-36 degrees and for each point set the cam angle should be set at 26-28 degrees.									
1956	270-cid	AL-AR52	.035	.016-.018	30.5	4B	475 rpm	130	11HS/2SM
	277-cid	AL-AR52	.035	.016-.018	30.5	4B	475 rpm	135	11HS/2SM
	303-cid	AL-4S-250	.035	.016-.018	37.5	4B	475 rpm	145	11HS/2SM
1957	277-cid	AL-AR52	.035	.016-.018	31.5	4B	475 rpm	135	11HS/2SM
	301-cid	AL-AR52	.035	.016-.018	31.5	8B	475 rpm	140	1HS/2SM
	318-cid	AL-AR32	.035	.016-.018	Note 2	8B	475 rpm	145	11HS/2SM
Note 2: The 318-cid Fury V-8 had a double breaker point distributor and total dwell angle was 36-39 degrees.									
1958	318 4003F	AL-AR42	.035	.015-.018	27-32	10B	500 rpm	145	HO-11-50
	318 4003D	AL-AR42	.035	.015-.018	27-32	10B	500 rpm	145	HO-11-50
	318 4003	AL-AR42	.035	.015-.018	Note 3	8B	500 rpm	145	HO-11-50
	350	AL-AR42	.035	.015-.018	Note 3	8B	500 rpm	150	HO-11-50
Note 3: These V-8s have two sets of distributor points. The total cam angle is 36-39 degrees and for each point set the cam angle should be set at 29-32 degrees.									
1959	318	AL-AR42	.035	.015-.018	27-32	10B	500 rpm	145	11-HS-50
	361	AL-AR42	.035	.015-.018	Note 4	7.5B	500 rpm	150	11-HS-60
Note 4: These V-8s have two sets of distributor points. The total cam angle is 34-40 degrees and for each point set the cam angle should be set at 27-32 degrees.									
1960	318 (man.)	AL-A42	.035	.014-.019	27-32	5B	500 rpm	125-165	11-HS-50
	318 (auto.)	AL-A42	.035	.014-.019	27-32	10B	500 rpm	125-165	11-HS-50
	361, 383	AL-A42	.035	.014-.019	27-32	10B	500 rpm	150-180	11-HS-60
	383 Ram-Tuned	AL-A42	.035	.014-.019	27-32	7.5B	500 rpm	150-180	11-HS-60
1961	318 (man.)	AL-A42	.035	.014-.019	27-32	5B	500 rpm	135-165	11-HS-50
	318 (auto.)	AL-A42	.035	.014-.019	27-32	10B	500 rpm	135-165	11-HS-50
	361	AL-A42	.035	.014-.019	Note 5	10B	500 rpm	150-180	11-HS-59B
	383	AL-A32	.035	.014-.019	Note 5	7.5B	500 rpm	150-180	11-HS-59B
Note 5: These V-8s have two sets of distributor points. The total cam angle is 34-40 degrees and for each point set the cam angle should be set at 27-32 degrees.									
1962	318 (man.)	AL-A42	.035	.014-.019	27-32	5B	500 rpm	120-150	11HS-488
	318 (auto.)	AL-A42	.035	.014-.019	27-32	10B	500 rpm	120-150	11HS-488
	361	AL-A42	.035	.014-.019	Note 6	10B	500 rpm	125-155	11HS-50
Note 6: These V-8s have two sets of distributor points. The total cam angle is 34-40 degrees and for each point set the cam angle should be set at 27-32 degrees.									
1963	318 (man.)	CH-J12Y	.035	.017	28-33	5B	500 rpm	135	48 amp hr
	318 (auto.)	CH-J12Y	.035	.017	28-33	10B	500 rpm	135	48 amp hr
	361, 383 2-barrel	AL-A42	.035	.017	28-33	10B	500 rpm	140	59 amp hr
	383 4-barrel	CH-J12Y	.035	.017	Note 7	10B	500 rpm	145	59 amp hr
	426	CH-J12Y	.035	.017	Note 7	10B	500 rpm	145	70 amp hr
Note 7: These V-8s have two sets of distributor points. The total cam angle is 34-40 degrees and for each point set the cam angle should be set at 27-32 degrees.									
1964	318 (man.)	CH-J12Y	.035	.017	30	5B	550 rpm	135	48 amp hr
	318 (auto.)	CH-J12Y	.035	.017	30	10B	475 rpm	135	48 amp hr
	361	CH-J12Y	.035	.017	30	10B	500 rpm	140	59 amp hr
	383 2-barrel	CH-J12Y	.035	.017	30	10B	500 rpm	145	59 amp hr
	383 4-barrel	CH-J10Y	.035	.017	Note 8	10B	500 rpm	145	59 amp hr
	426	CH-N14Y	.035	.017	Note 8	10B	500 rpm	145	70 amp hr
	426 Hemi	CH-N61Y	.020	.012	Note 8	31B	1400 rpm	145	70 amp hr
Note 8: These V-8s have two sets of distributor points. The total cam angle is 37-42 degrees and for each point set the cam angle should be set at 27-32 degrees.									
1965	318 (man.)	CH-J14Y	.035	.017	30	5B	550 rpm	140	48 amp hr
	318 (auto.)	CH-J14Y	.035	.017	30	105B	475 rpm	140	48 amp hr
	361	CH-J14Y	.035	.017	30	10B	500 rpm	140	59 amp hr
	383 2-barrel	CH-J14Y	.035	.017	Note 9	10B	550 rpm	145	59 amp hr
	383 4-barrel	CH-J10Y	.035	.017	Note 9	10B	500 rpm	145	59 amp hr
	426 Hemi	CH-J10Y	.035	.017	Note 9	10B	500 rpm	145	70 amp hr
Note 9: These V-8s have two sets of distributor points. The total cam angle is 37-42 degrees and for each point set the cam angle should be set at 27-32 degrees.									

PONTIAC

1954-2002

The first Pontiac V-8 lasted only one year, 1932. It was the same 251-cid V-8 used in Pontiac's almost-dead parent make, Oakland, during 1930 and 1931. The 85-hp flathead was basically unchanged in the Pontiac, except that the mounting system was slightly changed. Pontiac had really just renamed the Oakland V-8 as an upper-series Pontiac. It was a good way to clear out the parts bin. In 1933 and 1934, Pontiac built only eight-cylinder cars, but they used a straight eight rather than a V-8.

In 1955, Pontiac started towards a youthful new, high-performance image that it would nurture during the last half of the '50s. The first important change came under the hood, where a 287-cid overhead valve V-8 was nestled. When linked to a manual transmission, this motor had 7.4:1 compression and 173 brake horsepower. Adding Hydra-Matic Drive upped compression to 8.0:1 and horsepower to 180.

Engineers Clayton B. Leach and Ed Windeler developed this cast-iron engine with ball-and-stud rocker arms, gusher type cooling, pressure-suction crankcase ventilation, and easy-to-cast block construction to keep costs low and make it efficient. It had a very short, stiff crankcase, a lightweight valve train, five main bearings, hydraulic valve lifters, and a Carter WGD or Rochester 2GC two-barrel carburetor. By mid-1955, an optional "power-pack" engine with a four-barrel carburetor and dual exhausts was made available.

In 1956, the 287 was bored to 317 cid. Pontiac really jumped in the horsepower race with built-for-racing NASCAR options carrying dual-quad carburetion and delivering up to 285 hp. This was just the beginning. In 1957, the stroke was increased and displacement rose to 347 cid. The limited-edition Bonneville got a 315-hp fuel-injected version of this engine. There was also a 290-hp "Tri-Power" (three two-barrel carbs) option, plus a pair of 317-hp NASCAR Tri-Power engines. Pontiac engineers got the boring bar out again, in 1958, to reach 370 cid. Fuel-injection was made an option and again there were Tri-Power and NASCAR (a.k.a. Tempest 395-A) options delivering horsepower numbers as high as 330. In 1959, the 370 was stroked out to become the famous 389-cid Pontiac V-8. Fuel-injection was gone, but the highest output level was now up to 345 hp thanks to a Tempest 395-A option with a 10.5:1 compression ratio and Tri Power carburetion.

The 389-cid V-8 was used in all Pontiacs in 1960. In 1961, the new compact Tempest offered a different type of V-8. This was a 215-cid 155-hp job built by Buick. Also showing up very late in the calendar year was a 421-cid V-8 that became an RPO in 1962. It produced 405 hp. The 421 did not differ greatly in basic design from the 389-cid V-8, but featured lighter connecting rods and much larger crankshaft journals. The 421-cid 405-hp designation became well known after it appeared on the hoods of many NASCAR stock cars and Super-Duty drag cars.

More directly derived from the 389, a new 326-cid V-8 arrived in the 1963 Tempest. It was a de-stroked version of the bigger engine with a four-barrel carburetor and 260 hp. Pontiac continued with Tri-Power options for both the 389 and the 421. Two 421 H.O. V-8s produced 353 and 370 hp, respectively, but the hottest options of all were the 390-, 405-, and 410-hp 421 Super-Duty choices. These featured up to 13.0:1 compression and came with a single four-barrel, three two-barrel, or dual four-barrel carburetors. They were intended strictly for racing and to make Pontiac the competition king.

In 1964, GM management banned corporate racing involvement and Pontiac was forced to dump the Super-Duty program. The 421 H.O. options were allowed to remain and 370 hp was tops. The 389 was still around and the Tempest retained the 326, until the GTO arrived at midyear. The first true "muscle car," this so-called "Goat" came with a base 389-cid 325-hp V-8, plus an optional Tri-Power version of the same engine with 348 hp. Pontiac's intention was to use the GTO to maintain its high-performance image after the racing ban. A good job was done of this. By 1965, the Tri-Power GTO was up to 360 hp.

The 400-cid V-8 that bowed in 1967 Pontiacs was a bored-out version of the 389 and a new 428 was a bored-out 421. The extra cubes were added to mask the fact that Tri-Power options were no longer offered. The top 400-cid V-8 was a four-barrel version for the GTO that had the same 360-hp rating as the previous 389 Tri-Power option. Likewise, the top 428-cid engine carried the same 376-hp rating as last year's 421 Tri-Power job. The new Firebird used the 326 as its base V-8 and offered 400 and 400 Ram Air options. For 1968, the Tempest and Firebird received a new Pontiac-built 350 that was created by boring the 326.

Never seeing the light of day was Pontiac's only really new 1969 engine, a 303-cid small-block V-8 that never got too far beyond the prototype stage. It was developed to make the Firebird competitive in Sports Car Club of America Trans Am racing, but only about 25 racing cars were fitted with these short-stroke, tunnel-port V-8s. For 1970, a bored-and-stroked version of the big-block Pontiac V-8 arrived. This 455-cid engine was not really one of the year's high-performance options, but it had huge amounts of torque and potential for future development. Steps in this direction were made in 1971, when the Trans Am got a 335-hp Ram Air version (also optional in the LeMans/GTO).

When a Nova-based compact car revived the Ventura nameplate in 1972, a Chevy-built 307-cid V-8 was under its hood. Otherwise Pontiac stuck with the 350-, 400-, and 455-cid V-8s with up to 300 net horsepower (a new rating system). In 1973, the Ventura got the Pontiac 350 instead of the Chevy 307; in 1975 it got an Olds-built 260. By that time Pontiac V-8s had catalytic converters and other power-robbing economy and emissions changes. The last year for the big 455 was 1976. In 1977, a 301-cid engine was the base V-8 in many models, while the Ventura used a 305-cid V-8 made by Chevrolet or GM of Canada. There was also an Oldsmobile 403-cid "alternate" V-8 for Pontiacs sold in California.

It was the 301-, 305-, 350-, 400-, and 403-cid V-8s for the next several years, but in 1980 all engines over the 350 were dropped and a new 265-cid V-8 was optional in Firebird, LeMans, Grand Prix, Catalina, and Bonneville models. This 4.3-liter V-8 was built by Pontiac and also supplied to Oldsmobile. Also new in 1980 was a 350-cid Olds-built diesel.

By 1982, V-8 options were down to a pair of 305-cid V-8s for Firebirds and a single 350-cid Olds-built diesel for Grand Prixs and Bonnevilles. By 1986, the V-8 lineup consisted of five 305s with 150 to 205 net hp. Then a 307-cid V-8 became the standard power plant for Safari station wagons. An H.O. version of the Trans Am 305 was added in 1988 along with an even hotter 350-cid 225-hp V-8 for the GTA (optional in Formula and Trans Am). From 1990 on, the only Pontiac offered with a V-8 was the Firebird. These got pretty hot with up to 325 hp available in some models in some years. Only 5.7-liter V-8s were offered from 1993 on, but in 1998 the motor was changed from the 350-cid LT1 to the 346-cid LS1.

1954 PONTIAC

Pontiac V-8
L-head. V-block. Eight. Cast-iron block. Bore & stroke: 3-7/16 x 3-3/8 in. Displacement: 251 cid. Compression ratio: 5.2:1. Brake hp: 85 at 3200 rpm. NACC hp: 37.80. Main bearings: three. Valve lifters: Solid. Carburetor: Marvel one-barrel. Starting engine number: 310001. Engine number: 316282.

1955 PONTIAC

In 1955, Pontiac started towards a youthful new, high-performance image that it would nurture during the last half of the '50s. The first important change came under the hood, where a 287-cid overhead valve V-8 was nestled. When linked to a manual transmission, this motor had 7.4:1 compression ratio and 173 brake horsepower. Adding Hydra-Matic Drive upped the compression ratio to 8.0:1 and horsepower to 180.

Pontiac base V-8 (synchromesh)
Overhead valve. Cast-iron block. Bore & stroke: 3.75 x 3.25 in. Displacement: 287.2 cid. Compression ratio: 7.4:1. Brake hp: 173 at 4400 rpm. Taxable hp: 45.00. Torque: 256 lbs.-ft. at 2400 rpm. Five main bearings. Hydraulic valve lifters. Crankcase capacity: 5 qt. (add 1 qt. for new filter). Cooling system capacity: 24 qt. (add 1 quart with heater). Carburetor: Carter WGD two-barrel or Rochester 2GC two-barrel. The engine number on a pad on the front of the right-hand cylinder bank should match the vehicle's serial number (VIN).

Pontiac base V-8 (Hydra-Matic)
Overhead valve. Cast-iron block. Bore & stroke: 3.75 x 3.25 in. Displacement: 287.2 cid. Compression ratio: 8.0:1. Brake hp: 180 at 4600 rpm. Taxable hp: 45.00. Torque: 264 lbs.-ft. at 2400 rpm. Five main bearings. Hydraulic valve lifters. Crankcase capacity: 5 qt. (add 1 qt. for new filter). Cooling system capacity: 24 qt. (add 1 quart with heater). Carburetor: Carter WGD two-barrel or Rochester 2GC two-barrel. The engine number on a pad on the front of the right-hand cylinder bank should match the vehicle's serial number (VIN).

Pontiac base V-8 (Hydra-Matic)
Overhead valve. Cast-iron block. Bore & stroke: 3.75 x 3.25 in. Displacement: 287.2 cid. Compression ratio: 8.0:1. Brake hp: 200 at 4600 rpm. Taxable hp: 45.00. Torque: 264 lbs.-ft. at 2400 rpm. Five main bearings. Hydraulic valve lifters. Crankcase capacity: 5 qt. (add 1 qt. for new filter). Cooling system capacity: 24 qt. (add 1 quart with heater). Carburetor: Rochester 4GC four-barrel or Carter WCFB model 2268S or 2283S four-barrel. The engine number on a pad on the front of the right-hand cylinder bank should match the vehicle's serial number (VIN).
NOTE: Optional 200-hp "power pack" available for Pontiac V-8 after March 1, 1955 for $35.

1956 PONTIAC

Pontiac 860, 870 base V-8 (synchromesh)
Overhead valve. Cast-iron block. Bore & stroke: 3.94 x 3.25 in. Displacement: 316.6 cid. Compression ratio: 7.9:1. Brake hp: 192 at 4400 rpm. Taxable hp: 49.60. Torque: 294 lbs.-ft. at 2600 rpm. Five main bearings. Hydraulic valve lifters. Crankcase capacity: 5 qt. (add 1 qt. for new filter). Cooling system capacity: 22.7 qt. (add 1 quart with heater). Carburetor: Rochester 2GC two-barrel with black tag number 8696. The engine number on a pad on the front of the right-hand cylinder bank should match the vehicle's serial number (VIN).

Pontiac Chieftain base V-8 (Hydra-Matic)
Overhead valve. Cast-iron block. Bore & stroke: 3.94 x 3.25 in. Displacement: 316.6 cid. Compression ratio: 8.9:1. Brake hp: 205 at 4800 rpm. Taxable hp: 49.60. Torque: 294 lbs.-ft. at 2600 rpm. Five main bearings. Hydraulic valve lifters. Crankcase capacity: 5 qt. (add 1 qt. for new filter). Cooling system capacity: 22.7 qt. (add 1 quart with heater). Carburetor: Rochester 2GC two-barrel with brass tag number 8695. The engine number on a pad on the front of the right-hand cylinder bank should match the vehicle's serial number (VIN).

Pontiac Star Chief base V-8 (synchromesh); optional Pontiac 860, 870
Overhead valve. Cast-iron block. Bore & stroke: 3.94 x 3.25 in. Displacement: 316.6 cid. Compression ratio: 7.9:1. Brake hp: 216 at 4800 rpm. Taxable hp: 49.60. Five main bearings. Hydraulic valve lifters. Crankcase capacity: 5 qt. (add 1 qt. for new filter). Cooling system capacity: 22.7 qt. (add 1 quart with heater). Carburetor: Rochester 4GC four-barrel with black tag number 7900. The engine number on a pad on the front of the right-hand cylinder bank should match the vehicle's serial number (VIN).

Pontiac Star Chief base V-8 (Hydra-Matic); optional Pontiac 860, 870
Overhead valve. Cast-iron block. Bore & stroke: 3.94 x 3.25 in. Displacement: 316.6 cid. Compression ratio: 8.9:1. Brake hp: 227 at 4800 rpm. Taxable hp: 49.60. Torque: 312 lbs.-ft. at 3000 rpm. Five main bearings. Hydraulic valve lifters. Crankcase capacity: 5 qt. (add 1 qt. for new filter). Cooling system capacity: 22.7 qt. (add 1 quart with heater). Carburetor: Rochester 4GC four-barrel with brass tag number 8697. Some cars were also equipped with Carter WCFB model 2364S four-barrel carburetors. The engine number on a pad on the front of the right-hand cylinder bank should match the vehicle's serial number (VIN).

Pontiac 860, 870, Star Chief optional extra-horsepower "NASCAR" V-8
Overhead valve. Cast-iron block. Bore & stroke: 3.94 x 3.25 in. Displacement: 316.6 cid. Compression ratio: 10.0:1. Brake hp: 285 at 5100 rpm. Taxable hp: 49.60. Five main bearings. Anti pump-up hydraulic valve lifters. Crankcase capacity: 5 qt. (add 1 qt. for new filter). Cooling system capacity: 22.7 qt. (add 1 quart with heater). Special intake manifold. Special fuel pump with larger valves. Dual element delta-wing air cleaner. Greater overlap high-lift camshaft. Revised valve timing. Long-stem valves. High-tension valve springs. Modified electrics. Carburetor: Two (2) Rochester four-barrel. The engine number on a pad on the front of the right-hand cylinder bank should match the vehicle's serial number (VIN).
NOTE: Released March 1956; installed in approximately 200 cars.

1957 PONTIAC

Pontiac Chieftain base V-8 (synchromesh)
Overhead valve. Cast-iron block. Bore & stroke: 3.94 x 3.56 in. Displacement: 347 cid. Compression ratio: 8.5:1. Brake hp: 227 at 4600 rpm. Taxable hp: 49.6. Torque: 333 lbs.-ft. at 2300 rpm. Five main bearings. Hydraulic valve lifters. Crankcase capacity: 5 qt. (add 1 qt. for new filter). Cooling system capacity: 21.7 qt. (add 1 quart with heater). Carburetor: Rochester 2GC two-barrel. The engine number on a pad on the front of the right-hand cylinder bank should match the vehicle's serial number (VIN).

Pontiac Star Chief base V-8 (synchromesh); Optional Chieftain V-8
Overhead valve. Cast-iron block. Bore & stroke: 3.94 x 3.56 in. Displacement: 347 cid. Compression ratio: 8.5:1. Brake hp: 244 at 4800 rpm. Taxable hp: 49.6. Torque: 338 lbs.-ft. at 2700 rpm. Five main bearings. Hydraulic valve lifters. Crankcase capacity: 5 qt. (add 1 qt. for new filter). Cooling system capacity: 21.7 qt. (add 1 quart with heater).

In 1957, the stroke of the Pontiac V-8 was increased and displacement rose to 347 cid. The limited-edition Bonneville got a 315-hp fuel-injected version of this engine. There was also a 290-hp "Tri-Power" (three two-barrel carbs) option, which is shown here. A pair of 317-hp NASCAR Tri-Power engines were available for racing.

The unique, limited-edition Custom Bonneville Convertible was released on Jan. 11, 1957 on a one-to-a-dealer basis. Pontiac made 630 production examples and two prototypes. A 347-cid fuel-injected V-8 producing about 315 hp was used in all of these cars.

Carburetor: (Early production) Rochester 4GC four-barrel; (Late production) Carter AFB used in mixed production with Rochester 4GC. The engine number on a pad on the front of the right-hand cylinder bank should match the vehicle's serial number (VIN).

Pontiac Chieftain base V-8 (Hydra-Matic)
Overhead valve. Cast-iron block. Bore & stroke: 3.94 x 3.56 in. Displacement: 347 cid. Compression ratio: 10.0:1. Brake hp: 252 at 4600 rpm. Taxable hp: 49.6. Torque: 354 lbs.-ft. at 2400 rpm. Five main bearings. Hydraulic valve lifters. Crankcase capacity: 5 qt. (add 1 qt. for new filter). Cooling system capacity: 21.7 qt. (add 1 quart with heater). Carburetor: Rochester 2GC two-barrel. The engine number on a pad on the front of the right-hand cylinder bank should match the vehicle's serial number (VIN).

Pontiac Star Chief base V-8 (Hydra-Matic); Optional Chieftain V-8
Overhead valve. Cast-iron block. Bore & stroke: 3.94 x 3.56 in. Displacement: 347 cid. Compression ratio: 10.0:1. Brake hp: 270 at 4800 rpm. Taxable hp: 49.6. Torque: 359 lbs.-ft. at 2800 rpm. Five main bearings. Hydraulic valve lifters. Crankcase capacity: 5 qt. (add 1 qt. for new filter). Cooling system capacity: 21.7 qt. (add 1 quart with heater). Carburetor: (Early production) Rochester 4GC four-barrel with brass tag number 8697. Some cars were also equipped with Carter WCFB model 2364S four-barrel carburetors. Carter AFB used in mixed production with Rochester 4GC. The engine number on a pad on the front of the right-hand cylinder bank should match the vehicle's serial number (VIN).

Pontiac optional Tri-Power V-6 (Hydra-Matic)
Overhead valve. Cast-iron block. Bore & stroke: 3.94 x 3.56 in. Displacement: 347 cid. Compression ratio: 10.0:1. Brake hp: 290 at 5000 rpm. Taxable hp: 49.6. Five main bearings. Hydraulic valve lifters. Crankcase capacity: 5 qt. (add 1 qt. for new filter). Cooling system capacity: 21.7 qt. (add 1 quart with heater). Carburetion: Three (3) two-barrel Rochester carburetors. Ignition: Single breaker point. The engine number on a pad on the front of the right-hand cylinder bank should match the vehicle's serial number (VIN).

Bonneville base V-8 (Hydra-Matic only)
Overhead valve. Cast-iron block. Bore & stroke: 3.94 x 3.56 in. Displacement: 347 cid. Compression ratio: 10.0:1. Brake hp: 315 at 4800 rpm. Taxable hp: 49.6. Five main bearings. Hydraulic valve lifters. Crankcase capacity: 5 qt. (add 1 qt. for new filter). Cooling system capacity: 21.7 qt. (add 1 quart with heater). Induction: Rochester mechanical fuel injection. The engine number on a pad on the front of the right-hand cylinder bank should match the vehicle's serial number (VIN).

Pontiac optional NASCAR Tri-Power Extra Horsepower V-8 (synchromesh)
Overhead valve. Cast-iron block. Bore & stroke: 3.94 x 3.56 in. Displacement: 347 cid. Compression ratio: 10.0:1. Brake hp: 317 at 5200 rpm. Taxable hp: 49.6. Five main bearings. Hydraulic valve lifters. Crankcase capacity: 5 qt. (add 1 qt. for new filter). Cooling system capacity: 21.7 qt. (add 1 quart with heater). Carburetion: Three (3) two-barrel Rochester carburetors. Ignition: Dual breaker point. The engine number on a pad on the front of the right-hand cylinder bank should match the vehicle's serial number (VIN).

Pontiac optional NASCAR Tri-Power Extra Horsepower V-8 (Hydra-Matic)
Overhead valve. Cast-iron block. Bore & stroke: 3.94 x 3.56 in. Displacement: 347 cid. Compression ratio: 10.0:1. Brake hp: 317 at 5200 rpm. Taxable hp: 49.6. Five main bearings. Hydraulic valve lifters. Crankcase capacity: 5 qt. (add 1 qt. for new filter). Cooling system capacity: 21.7 qt. (add 1 quart with heater). Carburetion: Three (3) two-barrel Rochester carburetors. Ignition: Single breaker point. The engine number on a pad on the front of the right-hand cylinder bank should match the vehicle's serial number (VIN).

1958 PONTIAC

Chieftain, Super Chief base V-8 (synchromesh)
Overhead valve. Cast-iron block. Bore & stroke: 4.06 x 3.56 in. Displacement: 370 cid. Compression ratio: 8.6:1. Brake hp: 240 at 4500 rpm. Taxable hp: 52.80. Torque: 354 lbs.-ft. at 2600 rpm. Five main bearings. Hydraulic valve lifters. Crankcase capacity: 5 qt. (add 1 qt. for new filter). Cooling system capacity: 21.2 qt. (add 1 quart with heater). Carburetor: (Early production) Rochester 2GC two-barrel. (Late Production) Rochester 2GC two-barrel with air bypass idle. The engine number on a pad on the front of the right-hand cylinder bank should match the vehicle's serial number (VIN).

Star Chief, Bonneville base V-8 (synchromesh); Chieftain, Super Chief optional V-8
Overhead valve. Cast-iron block. Bore & stroke: 4.06 x 3.56 in. Displacement: 370 cid. Compression ratio: 8.6:1. Brake hp: 255 at 4500 rpm. Taxable hp: 52.80. Torque: 360 lbs.-ft. at 2600 rpm. Five main bearings. Hydraulic valve lifters. Crankcase capacity: 5 qt. (add 1 qt. for new filter). Cooling system capacity: 21.2 qt. (add 1 quart with heater). Carburetor: (Regular production) Carter AFB four-barrel. (Special Production) 500 cars built with Rochester 4GC four-barrel late in the year. These were available in only four sales zones and were likely "PM" option cars assembled for NASCAR certification. The engine number on a pad on the front of the right-hand cylinder bank should match the vehicle's serial number (VIN).

Chieftain, Super Chief base V-8 (Hydra-Matic)
Overhead valve. Cast-iron block. Bore & stroke: 4.06 x 3.56 in. Displacement: 370 cid. Compression ratio: 10.0:1. Brake hp: 270 at 4600 rpm. Taxable hp: 52.80. Torque: 388 lbs.-ft. at 2800 rpm. Five main bearings. Hydraulic valve lifters. Crankcase capacity: 5 qt. (add 1 qt. for new filter). Cooling system capacity: 21.2 qt. (add 1 quart with heater). Carburetor: (Early production) Rochester 2GC two-barrel. (Late Production) Rochester 2GC two-barrel with air bypass idle. The engine number on a pad on the front of the right-hand cylinder bank should match the vehicle's serial number (VIN).

Star Chief, Bonneville base V-8 (Hydra-Matic); Chieftain, Super Chief optional V-8
Overhead valve. Cast-iron block. Bore & stroke: 4.06 x 3.56 in. Displacement: 370 cid. Compression ratio: 10.0:1. Brake hp: 285 at 4600 rpm. Taxable hp: 52.80. Torque: 395 lbs.-ft. at 2800 rpm. Five main bearings. Hydraulic valve lifters. Crankcase capacity: 5 qt. (add 1 qt. for new filter). Cooling system capacity: 21.2 qt. (add 1 quart with heater). Carburetor: (Regular production) Carter AFB four-barrel. (Special Production) 500 cars built with Rochester 4GC four-barrel late in the year. These were available in only four sales zones and were likely "PM" option cars assembled for NASCAR certification. The engine number on a pad on the front of the right-hand cylinder bank should match the vehicle's serial number (VIN).

Pontiac optional Tri-Power V-8
Overhead valve. Cast-iron block. Bore & stroke: 4.06 x 3.56 in. Displacement: 370 cid. Compression ratio: 10.5:1. Brake hp: 300 at 4600 rpm. Taxable hp: 52.80. Torque: 400 lbs.-ft. at 3000 rpm. Five main bearings. Hydraulic valve lifters. Crankcase capacity: 5 qt. (add 1 qt. for new filter). Cooling system capacity: 21.2 qt. (add 1 quart with heater). Carburetion: Three (3) two-barrel Rochester carburetors. The engine number on a pad on the front of the right-hand cylinder bank should match the vehicle's serial number (VIN).

Pontiac Bonneville optional fuel-injection V-8
Overhead valve. Cast-iron block. Bore & stroke: 4.06 x 3.56 in. Displacement: 370 cid. Compression ratio: 10.5:1. Brake hp: 310 at 4800 rpm. Taxable hp: 52.80. Torque: 400 lbs.-ft. at 3400 rpm. Five main bearings. Hydraulic valve lifters. Crankcase capacity: 5 qt. (add 1 qt. for new filter). Cooling system capacity: 21.2 qt. (add 1 quart with heater). Induction: Rochester mechanical fuel injection. The engine number on a pad on the front of the right-hand cylinder bank should match the vehicle's serial number (VIN).

Pontiac optional "Tempest 395-A" NASCAR extra-horsepower V-8 (four-barrel)
Overhead valve. Cast-iron block. Bore & stroke: 4.06 x 3.56 in. Displacement: 370 cid. Compression ratio: 10.5:1. Brake hp: 315 at 5000 rpm. Taxable hp: 52.80. Five main bearings. High-lift camshaft. Low-restriction dual exhausts. Crankcase capacity: 5 qt. (add 1 qt. for new filter). Cooling system capacity: 21.2 qt. (add 1 quart with heater). Carburetor: Rochester 4GC four-barrel. The engine number on a pad on the front of the right-hand cylinder bank should match the vehicle's serial number (VIN). Engine code: PK.

Pontiac optional "Tempest 395-A" NASCAR extra-horsepower V-8 (Tri-Power)
Overhead valve. Cast-iron block. Bore & stroke: 4.06 x 3.56 in. Displacement: 370 cid. Compression ratio: 10.5:1. Brake hp: 330 at 5200 rpm. Taxable hp: 52.80. Five main bearings. High-lift camshaft. Low-restriction dual exhausts. Crankcase capacity: 5 qt. (add 1 qt. for new filter). Cooling system capacity: 21.2 qt. (add 1 quart with heater). Carburetion: Three (3) two-barrel Rochester carburetors. The engine number on a pad on the front of the right-hand cylinder bank should match the vehicle's serial number (VIN). Engine code: PM.

1959 PONTIAC

Bonneville Custom, Star Chief, Catalina optional V-8 (Hydra-Matic)
Overhead valve. Cast-iron block. Bore & stroke: 4.06 x 3.75 in. Displacement: 389 cid. Compression ratio: 8.6:1. Brake hp: 215 at 3600 rpm. Taxable hp: 52.80. Torque: 290 lbs.-ft. at 2000 rpm. Five main bearings. Hydraulic valve lifters. Crankcase capacity: 5 qt. (add 1 qt. for new filter). Cooling system capacity: 21.4 qt. (add 1 quart with heater). Carburetor: Rochester two-barrel. Includes special fuel-efficient camshaft no. 7015958. The engine number on a pad on the front of the right-hand cylinder bank should match the vehicle's serial number (VIN).

Catalina, Custom Star Chief base V-8 (synchromesh)
Overhead valve. Cast-iron block. Bore & stroke: 4.06 x 3.75 in. Displacement: 389 cid. Compression ratio: 8.6:1. Brake hp: 245 at 4200 rpm. Taxable hp: 52.80. Torque: 392 lbs.-ft. at 2000 rpm. Five main bearings. Hydraulic valve lifters. Crankcase capacity: 5 qt. (add 1 qt. for new filter). Cooling system capacity: 21.4 qt. (add 1 quart with heater). Carburetor: Rochester model no. 7015909 two-barrel. The engine number on a pad on the front of the right-hand cylinder bank should match the vehicle's serial number (VIN).

Bonneville Custom base V-8 (synchromesh); Optional Catalina, Star Chief (synchromesh)
Overhead valve. Cast-iron block. Bore & stroke: 4.06 x 3.75 in. Displacement: 389 cid. Compression ratio: 8.6:1. Brake hp: 260 at 4200 rpm. Taxable hp: 52.80. Torque: 400 lbs.-ft. at 2800 rpm. Five main bearings. Hydraulic valve lifters. Crankcase capacity: 5 qt. (add 1 qt. for new filter). Cooling system capacity: 21.4 qt. (add 1 quart with heater). Carburetor: Carter AFB-2820S four-barrel (no. 532301). The engine number on a pad on the front of the right-hand cylinder bank should match the vehicle's serial number (VIN).

Catalina, Custom Star Chief optional V-8 (Hydra-Matic)
Overhead valve. Cast-iron block. Bore & stroke: 4.06 x 3.75 in. Compression ratio: 10.0:1. Displacement: 389 cid. Brake hp: 280 at 4400 rpm. Taxable hp: 52.80. Torque: 408 lbs.-ft. at 2800 rpm. Five main bearings. Hydraulic valve lifters. Crankcase capacity: 5 qt. (add 1 qt. for new filter). Cooling system capacity: 21.4 qt. (add 1 quart with heater). Carburetor: Rochester model no. 7015910 two-barrel. The engine number on a pad on the front of the right-hand cylinder bank should match the vehicle's serial number (VIN).

Bonneville Custom, Star Chief, Catalina optional V-8 (Hydra-Matic)
Overhead valve. Cast-iron block. Bore & stroke: 4.06 x 3.75 in. Displacement: 389 cid. Compression ratio: 10.0:1. Brake hp: 300 at 4600 rpm. Taxable hp: 52.80. Torque: 420 lbs.-ft. at 2800 rpm. Five main bearings. Hydraulic valve lifters. Crankcase capacity: 5 qt. (add 1 qt. for new filter). Cooling system capacity: 21.4 qt. (add 1 quart with heater). Carburetor: Carter AFB-2820S four-barrel (no. 532302). The engine number on a pad on the front of the right-hand cylinder bank should match the vehicle's serial number (VIN).

Catalina, Star Chief, Bonneville Custom optional Tri-Power V-8 (all transmissions)
Overhead valve. Cast-iron block. Bore & stroke: 4.06 x 3.75 in. Displacement: 389 cid. Compression ratio: 10.0:1. Brake hp: 315 at 4600 rpm. Taxable hp: 52.80. Torque: 425 lbs.-ft. at 3200 rpm. Five main bearings. Crankcase capacity: 5 qt. (add 1 qt. for new filter). Cooling system capacity: 21.4 qt. (add 1 quart with heater). Carburetor: Rochester four-barrel. The engine number on a pad on the front of the right-hand cylinder bank should match the vehicle's serial number (VIN).

Pontiac optional "Tempest 395-A NASCAR Extra-Horsepower" V-8 (synchromesh)
Overhead valve. Cast-iron block. Bore & stroke: 4.06 x 3.75 in. Displacement: 389 cid. Compression ratio: 10.5:1. Brake hp: 330 at 4800 rpm. Taxable hp: 52.80. Torque: 420 lbs.-ft. at 2800 rpm. Five main bearings. Performance camshaft. Dual exhausts. Crankcase capacity: 5 qt. (add 1 qt. for new filter). Cooling system capacity: 21.4 qt. (add 1 quart with heater). Carburetor: Four-barrel. The engine number on a pad on the front of the right-hand cylinder bank should match the vehicle's serial number (VIN). Engine code: PK.

Pontiac optional "Tempest 395-A NASCAR Extra-Horsepower" V-8 (Hydra-Matic)
Overhead valve. Cast-iron block. Bore & stroke: 4.06 x 3.75 in. Displacement: 389 cid. Compression ratio: 10.5:1. Brake hp: 345 at 4800 rpm. Taxable hp: 52.80. Torque: 425 lbs.-ft. at 3200 rpm. Five main bearings. Performance camshaft. Dual exhausts. Crankcase capacity: 5 qt. (add 1 qt. for new filter). Cooling system capacity: 21.4 qt. (add 1 quart with heater). Carburetion: Three (3) two-barrel Rochester carburetors. The engine number on a pad on the front of the right-hand cylinder bank should match the vehicle's serial number (VIN). Engine code: PM.

1960 PONTIAC

Catalina, Ventura, Star Chief base V-8 (synchromesh); All Pontiacs base V-8 (Hydra-Matic)
Overhead valve. Cast-iron block. Bore & stroke: 4.06 x 3.75 in. Displacement: 389 cid. Bore & stroke: 4.06 x 3.75. Compression ratio: 8.6:1. Brake hp: 215 at 3600 rpm. Taxable hp: 52.80. Torque: 290 lbs.-ft. at 2000 rpm. Five main bearings. Hydraulic valve lifters. Crankcase capacity: 5 qt. (add 1 qt. for new filter). Cooling system capacity: 21.2 qt. (add 1 quart with heater). Carburetor: Rochester 7015073 two-barrel and special fuel-efficient camshaft. Engine code: (Catalina, Ventura and Star Chief) A-2; (Bonneville) E-3.

Bonneville base V-8 (synchromesh)
Overhead valve. Cast-iron block. Bore & stroke: 4.06 x 3.75 in. Displacement: 389 cid. Compression ratio: 8.6:1. Brake hp: 281 at 4400 rpm. Taxable hp: 52.80. Torque: 407 lbs.-ft. at 2800 rpm. Five main bearings. Hydraulic valve lifters. Crankcase capacity: 5 qt. (add 1 qt. for new filter). Cooling system capacity: 21.2 qt. (add 1 quart with heater). Carburetor: Carter AFB four-barrel. Engine code: B-2.

Catalina, Ventura, Star Chief optional V-8 (Hydra-Matic)
Overhead valve. Cast-iron block. Bore & stroke: 4.06 x 3.75 in. Displacement: 389 cid. Compression ratio: 10.00:1. Brake hp: 283 at 4400 rpm. Taxable hp: 52.80. Torque: 413 lbs.-ft. at 2800 rpm. Five main bearings. Hydraulic valve lifters. Crankcase capacity: 5 qt. (add 1 qt. for new filter). Cooling system capacity: 21.2 qt. (add 1 quart with heater). Carburetor: Rochester 2GC 7015073 two-barrel. Engine code: A-1.

Catalina, Ventura, Star Chief, Bonneville optional V-8 (Hydra-Matic)
Overhead valve. Cast-iron block. Bore & stroke: 4.06 x 3.75 in. Displacement: 389 cid. Compression ratio: 10.00:1. Brake hp: 303 at 4600 rpm. Taxable hp: 52.80. Torque: 425 lbs.-ft. at 2800 rpm. Five main bearings. Hydraulic valve lifters. Crankcase capacity: 5 qt. (add 1 qt. for new filter). Cooling system capacity: 21.2 qt. (add 1 quart with heater). Carburetor: Carter AFB-2975-S four-barrel. Engine code: (All) B-1.

Pontiac optional Tri-Power V-8 (synchromesh)
Overhead valve. Cast-iron block. Bore & stroke: 4.06 x 3.75 in. Displacement: 389 cid. Compression ratio: 10.0:1. Brake hp: 315 at 4600 rpm. Taxable hp: 52.80. Five main bearings. Crankcase capacity: 5 qt. (add 1 qt. for new filter). Cooling system capacity: 21.2 qt. (add 1 quart with heater). Carburetor: Rochester four-barrel.

Catalina, Ventura, Star Chief, Bonneville optional Tri-Power V-8 (Hydra-Matic)
Overhead valve. Cast-iron block. Bore & stroke: 4.06 x 3.75 in. Displacement: 389 cid. Compression ratio: 10.0:1. Brake hp: 318 at 4600 rpm. Torque: 430 lbs.-ft. at 3200 rpm. Taxable hp: 52.80. Five main bearings. Crankcase capacity: 5 qt. (add 1 qt. for new filter). Cooling system capacity: 21.2 qt. (add 1 quart with heater). Carburetion: Three Rochester two-barrel carburetors: (front) 7013063; (center) 7015068; (rear) 7013065. Engine code: C-1 or C-4.

Pontiac optional V-8
Overhead valve. Cast-iron block. Bore & stroke: 4.06 x 3.75 in. Displacement: 389 cid. Compression ratio: 10.75:1. Brake hp: 333 at 4600 rpm. Taxable hp: 52.80. Torque: 420 lbs.-ft. at 2800 rpm. Five main bearings. Performance camshaft. Dual exhausts. Crankcase capacity: 5 qt. (add 1 qt. for new filter). Cooling system capacity: 21.2 qt. (add 1 quart with heater). Carburetor: Carter AFB-2975-S four-barrel. Engine code: F-1 or F-4.

Pontiac optional "Tempest 425 NASCAR Extra-Horsepower" V-8
Overhead valve. Cast-iron block. Bore & stroke: 4.06 x 3.75 in. Displacement: 389 cid. Compression ratio: 10.75:1. Brake hp: 345 at 4800 rpm. Taxable hp: 52.80. Torque: 425 lbs.-ft. at 3200 rpm. Five main bearings. Performance camshaft. Dual exhausts. Crankcase capacity: 5 qt. (add 1 qt. for new filter). Cooling system capacity: 21.2 qt. (add 1 quart with heater). Carburetor: Four-barrel. Engine code: M-1 or M-4.

Pontiac optional "Tempest 425 NASCAR Extra-Horsepower" Tri-Power V-8 (Hydra-Matic)
Overhead valve. Cast-iron block. Bore & stroke: 4.06 x 3.75 in. Displacement: 389 cid. Compression ratio: 10.75:1. Brake hp: 348 at 4600 rpm. Taxable hp: 52.80. Torque: 425 lbs.-ft. at 3200 rpm. Five main bearings. Performance camshaft. Dual exhausts. Crankcase capacity: 5 qt. (add 1 qt. for new filter). Cooling system capacity: 21.2 qt. (add 1 quart with heater). Carburetion: Three (3) two-barrel Rochester carburetors; (front) 7013063; (center) 7013974; (rear) 7013975. Engine code: K-1 or K-4.

1961 PONTIAC

Base V-8 for the 1961 Pontiac Bonneville with synchromesh transmission was a 389-cid V-8 with 8.6:1 compression, a Carter AFB four-barrel carburetor, and 235 hp at 3600 rpm. Hydra-Matic transmission was a comon option on these cars and the V-8 used in this case had a 10.25:1 compression ratio and 303 hp. The Hydra-Matic engine is shown here.

Pontiac Tempest optional V-8 (Built by Buick)
Overhead valve. Bore & stroke: 3.50 x 2.80 in. Displacement: 215 cid. Compression ratio: 8.8:1. Brake hp: 155 at 4600 rpm. Taxable hp: 39.20. Torque: 220 lbs.-ft. at 2400 rpm. Five main bearings. Hydraulic valve lifters. Crankcase capacity: 4 qt. (add 1 qt. for new filter). Cooling system capacity: 11.6 qt. (add 1 quart with heater). Carburetor: Rochester 7019060 two-barrel. Engine code: XS, YS, or XA.

Catalina, Ventura, Star Chief base V-8 (synchromesh)
Overhead valve. Cast-iron block. Bore & stroke: 4.06 x 3.75 in.

Standard in the 1961 Ventura was the 389-cid, 215-hp V-8 with 8.6:1 compression ratio and a two-barrel carburetor. But options ranged to a Tri-Power version of the 389 with 10.75:1 compression heads and 348 hp at 4800 rpm. (John Lee photo)

Another view of the 1961 Bonnville Hydra-Matic engine. The engine block, cylinder heads, and valve covers were finished in light Pontiac engine blue. The air cleaner was painted black and decorated with an orange decal. This car has the heavy-duty version of the AC air cleaner. Accessories like the power steering pump are also done in black.

Displacement: 389 cid. Compression ratio: 8.6:1. Brake hp: 215 at 3600 rpm. Taxable hp: 52.80. Torque: 290 lbs.-ft. at 2000 rpm. Five main bearings. Hydraulic valve lifters. Crankcase capacity: 4 qt. (add 1 qt. for new filter). Cooling system capacity: 18.5 qt. (add 1 quart with heater). Carburetor: Rochester 2GC two-barrel no. 7019060. Engine code: A2 or G4.

Pontiac optional (Hydra-Matic)
Overhead valve. Cast-iron block. Bore & stroke: 4.06 x 3.75 in. Displacement: 389 cid. Compression ratio: 8.6:1. Brake hp: 230 at 4000 rpm. Torque: 380 lbs.-ft. at 2000 rpm. Five main bearings. Hydraulic valve lifters. Crankcase capacity: 4 qt. (add 1 qt. for new filter). Cooling system capacity: 18.5 qt. (add 1 quart with heater). Carburetor: Rochester 2GC two-barrel no. 7020075. Ebgine code: E3, W3, E7, or W7.

Bonneville base V-8 (synchromesh); Other full-size Pontiacs optional V-8 (synchromesh)
Overhead valve. Cast-iron block. Bore & stroke: 4.06 x 3.75 in. Displacement: 389 cid. Compression ratio: 8.6:1. Brake hp: 235 at 3600 rpm. Taxable hp: 52.80. Torque: 402 lbs.-ft. at 2000 rpm. Five main bearings. Hydraulic valve lifters. Crankcase capacity: 4 qt. (add 1 qt. for new filter). Cooling system capacity: 18.5 qt. (add 1 quart with heater). Carburetor: Carter AFB-3123-S four-barrel. Engine code: P4, B4, or H4.

Catalina and Ventura optional V-8 (Hydra-Matic)
Overhead valve. Cast-iron block. Bore & stroke: 4.06 x 3.75 in. Displacement: 389 cid. Compression ratio: 10.25:1. Brake hp: 267 at 4200 rpm. Taxable hp: 52.80. Torque: 405 lbs.-ft. at 2400 rpm. Five main bearings. Hydraulic valve lifters. Crankcase capacity: 4 qt. (add 1 qt. for new filter). Cooling system capacity: 18.5 qt. (add 1 quart with heater). Carburetor: Rochester 7020071 two-barrel. Engine code: S1 or S5.

Star Chief optional V-8 (Hydra-Matic)
Overhead valve. Cast-iron block. Bore & stroke: 4.06 x 3.75 in. Displacement: 389 cid. Compression ratio: 10.25:1. Brake hp: 283 at 4400 rpm. Taxable hp: 52.80. Torque: 413 lbs.-ft. at 2800 rpm. Five main bearings. Hydraulic valve lifters. Crankcase capacity: 4 qt. (add 1 qt. for new filter). Cooling system capacity: 18.5 qt. (add 1 quart with heater). Carburetor: Rochester 7020070 two-barrel.

Catalina optional V-8 (Hydra-Matic)
Overhead valve. Cast-iron block. Bore & stroke: 4.06 x 3.75 in. Displacement: 389 cid. Compression ratio: 10.25:1. Brake hp: 287 at 4400 rpm. Taxable hp: 52.80. Torque: 416 lbs.-ft. at 2400 rpm. Five main bearings. Hydraulic valve lifters. Crankcase capacity: 4 qt. (add 1 qt. for new filter). Cooling system capacity: 18.5 qt. (add 1 quart with heater). Carburetor: Four-barrel. Engine code: A1 or A5.

Bonneville optional V-8 (Hydra-Matic)
Overhead valve. Cast-iron block. Bore & stroke: 4.06 x 3.75 in. Displacement: 389 cid. Compression ratio: 10.25:1. Brake hp: 303 at 4600 rpm. Taxable hp: 52.80. Torque: 425 lbs.-ft. at 2800 rpm. Five main bearings. Hydraulic valve lifters. Crankcase capacity: 4 qt. (add 1 qt. for new filter). Cooling system capacity: 18.5 qt. (add 1 quart with heater). Carburetor: Carter AFB- 3123-S four-barrel. Engine code: B1, T1, P0, B5, or T5.

All Pontiacs optional Super-Duty Tri-Power V-8 (any transmission)
Overhead valve. Cast-iron block. Bore & stroke: 4.06 x 3.75 in. Displacement: 389 cid. Compression ratio: 10.75:1. Brake hp: 318 at 4600 rpm. Taxable hp: 52.80. Torque: 430 lbs.-ft. at 3200 rpm. Five main bearings. Hydraulic valve lifters. Crankcase capacity: 4 qt. (add 1 qt. for new filter). Cooling system capacity: 18.5 qt. (add 1 quart with heater). Carburetion: Three (3) two-barrel Rochester carburetors; (front) 7013063; (center) 7013974; (rear) 7013975. Engine code: C4, RC4, CO, RC0, 19, or R19.

All Pontiacs optional Super-Duty four-barrel V-8 (any transmission)
Overhead valve. Cast-iron block. Bore & stroke: 4.06 x 3.75 in. Displacement: 389 cid. Compression ratio: 10.75:1. Brake hp: 333 at 4800 rpm. Taxable hp: 52.80. Torque: 425 lbs.-ft. at 2800 rpm. Five main bearings. Hydraulic valve lifters. Crankcase capacity: 4 qt. (add 1 qt. for new filter). Cooling system capacity: 18.5 qt. (add 1 quart with heater). Carburetor: Carter AFB-3123-S four-barrel. Engine code: F4, FO, RMO, or U9.

All Pontiacs optional Super-Duty Tri-Power V-8 (any transmission)
Overhead valve. Cast-iron block. Bore & stroke: 4.06 x 3.75 in. Displacement: 389 cid. Compression ratio: 10.75:1. Brake hp: 348 at 4800 rpm. Taxable hp: 52.80. Torque: 430 lbs.-ft. at 3200 rpm. Five main bearings. Hydraulic valve lifters. Crankcase capacity: 4 qt. (add 1 qt. for new filter). Cooling system capacity: 18.5 qt. (add 1 quart with heater). Carburetion: Three (3) two-barrel Rochester carburetors; (front) 7013063; (center) 7013974; (rear) 7013975. Engine code: M4, RM4, MO, or V9.

Catalina optional Super-Duty V-8 (Catalina two-door models only)
Overhead valve. Cast-iron block. Bore & stroke: 4.09 x 4.00 in. Displacement: 421 cid. Compression ratio: 11.0:1. Brake hp: 405 at 5600 rpm. Taxable hp: 53.60. Torque: 425 lbs.-ft. at 4400 rpm. Five main bearings. Hydraulic valve lifters. Crankcase capacity: 4 qt. (add 1 qt. for new filter). Cooling system capacity: 18.5 qt. (add 1 quart with heater). Carburetor: Two (2) Rochester four-barrels.

1962 PONTIAC

Pontiac Tempest optional V-8 (Built by Buick)
Overhead valve. Bore & stroke: 3.50 x 2.80 in. Displacement: 215 cid. Compression ratio: 8.8:1. Brake hp: 155 at 4600 rpm. Taxable hp: 39.20. Torque: 220 lbs.-ft. at 2400 rpm. Five main bearings. Hydraulic valve lifters. Crankcase capacity: 4 qt. (add 1 qt. for new filter). Cooling system capacity: 11.6 qt. (add 1 quart with heater). Carburetor: Rochester two-barrel. Engine code: 91Z.

Pontiac Tempest optional V-8 (Built by Buick)
Overhead valve. Bore & stroke: 3.50 x 2.80 in. Displacement: 215 cid. Compression ratio: 10.25:1. Brake hp: 185 at 4800 rpm. Taxable hp: 39.20. Torque: 230 lbs.-ft. at 2800 rpm. Five main bearings. Hydraulic valve lifters. Crankcase capacity: 4 qt. (add 1 qt. for new filter). Cooling system capacity: 11.6 qt. (add 1 quart with heater). Carburetor: Rochester 4GC model 7020078 four-barrel. Engine code: 97Z.

Catalina, Star Chief base V-8 (synchromesh)
Overhead valve. Cast-iron block. Bore & stroke: 4.06 x 3.75 in. Displacement: 389 cid. Compression ratio: 8.60:1. Brake hp: 215 at 3600 rpm. Taxable hp: 52.80. Torque: 390 lbs.-ft. at 2000 rpm. Five main bearings. Hydraulic valve lifters. Crankcase capacity: 4 qt. (add 1 qt. for new filter). Cooling system capacity: 18.5 qt. (add 1 quart with heater). Carburetor: Rochester 2GC model 7020060 two-barrel. Engine code 0A1 or 03B.

Pontiac optional V-8 (Hydra-Matic)
Overhead valve. Cast-iron block. Bore & stroke: 4.06 x 3.75 in. Displacement: 389 cid. Compression ratio: 8.6:1. Brake hp: 230 at 4000 rpm. Taxable hp: 52.80. Torque: 380 lbs.-ft. at 2000 rpm. Five main bearings. Hydraulic valve lifters. Crankcase capacity: 4 qt. (add 1 qt. for

The 1962 Grand Prix was identified by clean side styling with a checkered flag badge in the concave section of side spears, rocker panel molding, an anodized grille insert and nosepiece, and special rear end styling. The from-the-factory equipment list included all Bonneville features starting with a 389-cid four-barrel V-8. (John Lee photo)

new filter). Cooling system capacity: 18.5 qt. (add 1 quart with heater). Carburetor: Rochester two-barrel. Engine code: 20L, 21L, 40R, or 41R.

Bonneville base V-8 (synchromesh); optional all other Pontiacs (synchromesh)
Overhead valve. Cast-iron block. Bore & stroke: 4.06 x 3.75 in. Displacement: 389 cid. Compression ratio: 8.6:1. Brake hp: 235 at 3600 rpm. Taxable hp: 52.80. Torque: 402 lbs.-ft. at 2000 rpm. Five main bearings. Hydraulic valve lifters. Crankcase capacity: 4 qt. (add 1 qt. for new filter). Cooling system capacity: 18.5 qt. (add 1 quart with heater). Carburetor: Carter AFB-3123-S four-barrel. Engine code 02B.

Catalina optional V-8 (Hydra-Matic)
Overhead valve. Cast-iron block. Bore & stroke: 4.06 x 3.75 in. Displacement: 389 cid. Compression ratio: 10.25:1. Brake hp: 267 at 4200 rpm. Taxable hp: 52.80. Torque: 405 lbs.-ft. at 2400 rpm. Five main bearings. Hydraulic valve lifters. Crankcase capacity: 4 qt. (add 1 qt. for new filter). Cooling system capacity: 18.5 qt. (add 1 quart with heater). Carburetor: Rochester 2GC two-barrel. Engine code: 15H or 17H.

Star Chief optional V-8 (Hydra-Matic)
Overhead valve. Cast-iron block. Bore & stroke: 4.06 x 3.75 in. Displacement: 389 cid. Compression ratio: 10.25:1. Brake hp: 283 at 4400 rpm. Taxable hp: 52.80. Torque: 413 lbs.-ft. at 2800. Five main bearings. Hydraulic valve lifters. Crankcase capacity: 4 qt. (add 1 qt. for new filter). Cooling system capacity: 18.5 qt. (add 1 quart with heater). Carburetor: Rochester 2GC two-barrel. Engine code: 35M or 37M.

Bonneville, Bonneville Custom, Grand Prix optional V-8 (Hydra-Matic)
Overhead valve. Cast-iron block. Bore & stroke: 4.06 x 3.75 in. Displacement: 389 cid. Compression ratio: 10.25:1. Brake hp: 303 at 4600 rpm. Taxable hp: 52.80. Torque: 402 lbs.-ft. at 2000 rpm. Five main bearings. Hydraulic valve lifters. Crankcase capacity: 4 qt. (add 1 qt. for new filter). Cooling system capacity: 18.5 qt. (add 1 quart with heater). Carburetor: Carter AFB-3123-S four-barrel. Engine code: 16J, 16K, 18K, 39N, 36P, or 38P.

Pontiac optional Tri-Power V-8 (all transmissions)
Overhead valve. Cast-iron block. Bore & stroke: 4.06 x 3.75 in. Displacement: 389 cid. Compression ratio: 10.75:1. Brake hp: 318 at 4600 rpm. Taxable hp: 52.80. Torque: 430 lbs.-ft. at 3200 rpm. Five main bearings. Hydraulic valve lifters. Crankcase capacity: 4 qt. (add 1 qt. for new filter). Cooling system capacity: 18.5 qt. (add 1 quart with heater). Carburetion: Three Rochester two-barrel carburetors. Engine code: 10B, 27J, or 49N.

Pontiac optional V-8 (all transmissions)
Overhead valve. Cast-iron block. Displacement: 389 cid. Bore & stroke: 4.06 x 3.75 in. Compression ratio: 10.75:1. Brake hp: 333 at 4800 rpm. Taxable hp: 52.80. Torque: 425 lbs.-ft. at 2800 rpm. Five main bearings. Hydraulic valve lifters. Crankcase capacity: 4 qt. (add 1 qt. for new filter). Cooling system capacity: 18.5 qt. (add 1 quart with heater). Carburetor: Four-barrel. Engine code: 08B, 25, or 47NJ.

Pontiac optional Tri-Power V-8 (all transmissions)
Overhead valve. Cast-iron block. Displacement: 389 cid. Bore & stroke: 4.06 x 3.75 in. Compression ratio: 10.75:1. Brake hp: 348 at 4800 rpm. Taxable hp: 52.80. Torque: 430 lbs.-ft. at 3200 rpm. Five main bearings. Hydraulic valve lifters. Crankcase capacity: 4 qt. (add 1 qt. for new filter). Cooling system capacity: 18.5 qt. (add 1 quart with heater). Carburetion: Three Rochester two-barrel carburetors. Engine code: 11B, 28J or 50N.

Catalina optional Super-Duty V-8 (Catalina two-door models only)
Overhead valve. Cast-iron block. Bore & stroke: 4.06 x 3.75 in. Displacement: 389 cid. Compression ratio: 10.75:1. Brake hp: 385 at 5200 rpm. Taxable hp: 52.80. Five main bearings. Hydraulic valve lifters. Crankcase capacity: 4 qt. (add 1 qt. for new filter). Cooling system capacity: 18.5 qt. (add 1 quart with heater). Carburetor: Rochester four-barrel.

Catalina optional Super-Duty V-8 (Catalina two-door models only)
Overhead valve. Cast-iron block. Bore & stroke: 4.09 x 4.00 in. Displacement: 421 cid. Compression ratio: 11.0:1. Brake hp: 405 at 5600 rpm. Taxable hp: 53.60. Torque: 425 lbs.-ft. at 4400 rpm. Five main bearings. Hydraulic valve lifters. Crankcase capacity: 4 qt. (add 1 qt. for new filter). Cooling system capacity: 18.5 qt. (add 1 quart with heater). Carburetion: Two (2) Rochester four-barrels.

1963 PONTIAC

Catalina, Star Chief base V-8 (synchromesh)
Overhead valve. Cast-iron block. Bore & stroke: 4.06 x 3.75 in.

Pontiac was famous for its Tri-Power option. In this case you're looking at the Tri-Power HO V-8 in a 1964 Pontiac Catalina 2 + 2. The base engine in this baby was a big-block 421 with a 10.75:1 compression ratio, a four-barrel carb, and 320 hp on tap at 4400 rpm. There were two Tri-Power options, both on the HO block with three two-barrel Rochester carbs. One was good for 350 hp and the other churned up 370.

Displacement: 389 cid. Compression ratio: 8.6:1. Brake hp: 215 at 3600 rpm. Taxable hp: 52.80. Torque: 390 lbs.-ft. at 2000 rpm. Five main bearings. Hydraulic valve lifters. Crankcase capacity: 4 qt. (add 1 qt. for new filter). Cooling system capacity: 18.5 qt. (add 1 quart with heater). Carburetor: Rochester 2GC two-barrel. Engine code: 01A or 03B.

Pontiac optional V-8 (Hydra-Matic)
Overhead valve. Cast-iron block. Bore & stroke: 4.06 x 3.75 in. Displacement: 389 cid. Compression ratio: 8.6:1. Brake hp: 230 at 4000 rpm. Taxable hp: 52.80. Torque: 380 lbs.-ft. at 2000 rpm. Five main bearings. Hydraulic valve lifters. Crankcase capacity: 4 qt. (add 1 qt. for new filter). Cooling system capacity: 18.5 qt. (add 1 quart with heater). Carburetor: Rochester two-barrel. Engine code: 20L, 21L, 40R, or 41R.

Bonneville base V-8 (synchromesh); optional all other Pontiacs (synchromesh)
Overhead valve. Cast-iron block. Bore & stroke: 4.06 x 3.75 in. Displacement: 389 cid. Compression ratio: 8.6:1. Brake hp: 235 at 3600 rpm. Taxable hp: 52.80. Torque: 402 lbs.-ft. at 2000 rpm. Five main bearings. Hydraulic valve lifters. Crankcase capacity: 4 qt. (add 1 qt. for new filter). Cooling system capacity: 18.5 qt. (add 1 quart with heater). Carburetor: Carter AFB four-barrel. Engine code: 02B or 04B.

Tempest, Lemans optional V-8
Overhead valve. Bore & stroke: 3.718 x 3.75 in. Displacement: 326 cid. Compression ratio: 10.25:1. Brake hp: 260 at 4800 rpm. Taxable hp: 44.4. Torque: 352 lbs.-ft. at 2800 rpm. Five main bearings. Hydraulic valve lifters. Crankcase capacity: 4 qt. (add 1 qt. for new filter). Cooling system capacity: 18.5 qt. (add 1 quart with heater). Carburetor: Four-barrel. Engine code: 68X, 71X, 60O, or 69O.

Catalina optional V-8 (Hydra-Matic)
Overhead valve. Cast-iron block. Bore & stroke: 4.06 x 3.75 in. Displacement: 389 cid. Compression ratio: 10.25:1. Brake hp: 267 at 4200 rpm. Taxable hp: 52.80. Torque: 405 lbs.-ft. at 2400 rpm. Five main bearings. Hydraulic valve lifters. Crankcase capacity: 4 qt. (add 1 qt. for new filter). Cooling system capacity: 18.5 qt. (add 1 quart with heater). Carburetor: Rochester 2GC two-barrel. Engine code: 15H or 17H.

Star Chief optional V-8 (Hydra-Matic)
Overhead valve. Cast-iron block. Bore & stroke: 4.06 x 3.75 in. Displacement: 389 cid. Compression ratio: 10.25:1. Brake hp: 283 at 4400 rpm. Taxable hp: 52.80. Torque: 413 lbs.-ft. at 2800 rpm. Five main bearings. Hydraulic valve lifters. Crankcase capacity: 4 qt. (add 1 qt. for new filter). Cooling system capacity: 18.5 qt. (add 1 quart with heater). Carburetor: Rochester 2GC two-barrel. Engine code: 35M or 37M.

Bonneville, Bonneville Custom, Grand Prix optional V-8 (Hydra-Matic)
Overhead valve. Cast-iron block. Bore & stroke: 4.06 x 3.75 in. Displacement: 389 cid. Compression ratio: 10.25:1. Brake hp: 303 at 4600 rpm. Taxable hp: 52.80. Torque: 425 lbs.-ft. at 2800 rpm. Five main bearings. Hydraulic valve lifters. Crankcase capacity: 4 qt. (add 1 qt. for new filter). Cooling system capacity: 18.5 qt. (add 1 quart with heater). Carburetor: Carter AFB four-barrel. Engine code: O6B, 16K, 18K, 36P, or 38P.

Pontiac optional Tri-Power V-8 (all transmissions)
Overhead valve. Cast-iron block. Bore & stroke: 4.06 x 3.75 in. Displacement: 389 cid. Compression ratio: 10.25:1. Brake hp: 313 at

4600 rpm. Taxable hp: 52.80. Torque: 430 lbs.-ft. at 3200 rpm. Five main bearings. Hydraulic valve lifters. Crankcase capacity: 4 qt. (add 1 qt. for new filter). Cooling system capacity: 18.5 qt. (add 1 quart with heater). Carburetion: Three Rochester two-barrel carburetors. Engine code: O7B, 48N, or 26G.

Pontiac optional 421 HO V-8 (all transmissions)
Overhead valve. Cast-iron block. Bore & stroke: 4.09 x 4.00 in. Displacement: 421 cid. Compression ratio: 10.75:1. Brake hp: 353 at 5000 rpm. Taxable hp: 53.60. Torque: 455 lbs.-ft. at 3400 rpm. Five main bearings. Hydraulic valve lifters. Crankcase capacity: 4 qt. (add 1 qt. for new filter). Cooling system capacity: 18.5 qt. (add 1 quart with heater). Carburetor: Four-barrel. Engine code: O8B, 47Q, or 25G.

Pontiac optional 421 HO Tri-Power V-8 (all transmissions)
Overhead valve. Cast-iron block. Displacement: 421 cid. Bore & stroke: 4.09 x 4.00 in. Compression ratio: 10.75:1. Brake hp: 370 at 5200 rpm. Taxable hp: 53.60. Torque: 460 lbs.-ft. at 3800 rpm. Five main bearings. Hydraulic valve lifters. Crankcase capacity: 4 qt. (add 1 qt. for new filter). Cooling system capacity: 18.5 qt. (add 1 quart with heater). Carburetion: Three Rochester two-barrel carburetors. Engine code: 11B, 28G, or 50Q.

Catalina optional Super-Duty V-8 (Catalina two-door models only)
Overhead valve. Cast-iron block. Displacement: 421 cid. Bore & stroke: 4.09 x 4.00 in. Compression ratio: 12.0:1. Brake hp: 390 at 5800 rpm. Taxable hp: 53.60. Torque: 425 lbs.-ft. at 3600 rpm. Five main bearings. Hydraulic valve lifters. Crankcase capacity: 4 qt. (add 1 qt. for new filter). Cooling system capacity: 18.5 qt. (add 1 quart with heater). Carburetor: Four-barrel.

Catalina optional Super-Duty V-8 (Catalina two-door models only)
Overhead valve. Cast-iron block. Bore & stroke: 4.09 x 4.00 in. Displacement: 421 cid. Compression ratio: 11.0:1. Brake hp: 405 at 5600 rpm. Taxable hp: 53.60. Torque: 425 lbs.-ft. at 4400 rpm. Five main bearings. Hydraulic valve lifters. Crankcase capacity: 4 qt. (add 1 qt. for new filter). Cooling system capacity: 18.5 qt. (add 1 quart with heater). Carburetor: Two (2) Rochester four-barrels. Engine code: 12-5.

Catalina optional Super-Duty V-8 (Catalina two-door models only)
Overhead valve. Cast-iron block. Displacement: 421 cid. Bore & stroke: 4.09 x 4.00 in. Compression ratio: 13.0:1. Brake hp: 410 at 5600 rpm. Taxable hp: 53.60. Torque: 435 lbs.-ft. at 4400 rpm. Five main bearings. Hydraulic valve lifters. Crankcase capacity: 4 qt. (add 1 qt. for new filter). Cooling system capacity: 18.5 qt. (add 1 quart with heater). Carburetor: Two (2) Rochester four-barrels. Engine code: 13-5.

1964 PONTIAC

Full-size Pontiac optional Economy V-8 (Hydra-Matic)
Overhead valve. Cast-iron block. Bore & stroke: 4.06 x 3.75 in. Displacement: 389 cid. Compression ratio: 8.6:1. Brake hp: 230 at 4000 rpm. Taxable hp: 52.80. Torque: 386 lbs.-ft. at 2000 rpm. Five main bearings. Hydraulic valve lifters. Crankcase capacity: 4 qt. (add 1 qt. for new filter). Cooling system capacity: 19.5 qt. (add 1 quart with heater). Carburetor: Rochester 2GC two-barrel. Engine code: O4I, O5L, O8R, or O9R.

Two V-8s were available in the 1964 Tempest LeMans. Both were based on a 326-cid block with 10.25:1 compression heads. The two-barrel engine produced 250 hp and a 280-hp four-barrel option was available at additional cost. (John Lee photo)

Catalina, Star Chief base V-8 (synchromesh)
Overhead valve. Cast-iron block. Bore & stroke: 4.06 x 3.75 in. Displacement: 389 cid. Compression ratio: 8.6:1. Brake hp: 235 at 4000 rpm. Taxable hp: 52.80. Torque: 386 lbs.-ft. at 2000 rpm. Five main bearings. Hydraulic valve lifters. Crankcase capacity: 4 qt. (add 1 qt. for new filter). Cooling system capacity: 19.5 qt. (add 1 quart with heater). Carburetor: Rochester 2GC two-barrel. Engine code: 22B.

Tempest, LeMans optional V-8
Overhead valve. Displacement: 326 cid. Bore & stroke: 3.718 x 3.75 in. Compression ratio: 8.6:1. Brake hp: 250 at 4600 rpm. Taxable hp: 44.4. Torque: 333 lbs.-ft. at 2800 rpm. Five main bearings. Hydraulic valve lifters. Crankcase capacity: 4 qt. (add 1 qt. for new filter). Cooling system capacity: 19.5 qt. (add 1 quart with heater). Carburetor: Two-barrel. Engine code: 92X or 96O.

Bonneville base V-8 (synchromesh)
Overhead valve. Cast-iron block. Bore & stroke: 4.06 x 3.75 in. Displacement: 389 cid. Compression ratio: 8.6:1. Brake hp: 255 at 4000 rpm. Taxable hp: 52.80. Torque: 407 lbs.-ft. at 2000 rpm. Five main bearings. Hydraulic valve lifters. Crankcase capacity: 4 qt. (add 1 qt. for new filter). Cooling system capacity: 19.5 qt. (add 1 quart with heater). Carburetor: Four-barrel. Engine code: 19M.

Catalina base V-8 (Hydra-Matic)
Overhead valve. Cast-iron block. Bore & stroke: 4.06 x 3.75 in. Displacement: 389 cid. Compression ratio: 10.50:1. Brake hp: 267 at 4200 rpm. Taxable hp: 52.80. Torque: 410 lbs.-ft. at 2400 rpm. Five main bearings. Hydraulic valve lifters. Crankcase capacity: 4 qt. (add 1 qt. for new filter). Cooling system capacity: 19.5 qt. (add 1 quart with heater). Carburetor: Rochester 2GC two-barrel. Engine code: 11H or 12H.

Star Chief base V-8 (Hydra-Matic)
Overhead valve. Cast-iron block. Bore & stroke: 4.06 x 3.75 in. Displacement: 389 cid. Compression ratio: 10.50:1. Brake hp: 283 at 4400 rpm. Taxable hp: 52.80. Torque: 418 lbs.-ft. at 2800 rpm. Five main bearings. Hydraulic valve lifters. Crankcase capacity: 4 qt. (add 1 qt. for new filter). Cooling system capacity: 19.5 qt. (add 1 quart with heater). Carburetor: Rochester 2GC two-barrel. Engine code: 10A, 17M, or 18M.

Tempest, LeMans optional V-8
Overhead valve. Bore & stroke: 3.718 x 3.75 in. Displacement: 326 cid. Compression ratio: 10.5:1. Brake hp: 280 at 4800 rpm. Taxable hp: 44.4. Torque: 355 lbs.-ft. at 3200 rpm. Five main bearings. Hydraulic valve lifters. Crankcase capacity: 4 qt. (add 1 qt. for new filter). Cooling system capacity: 19.5 qt. (add 1 quart with heater). Carburetor: Carter AFB four-barrel. Engine code: 94X or 97O.

Bonneville, Grand Prix base V-8 (Hydra-Matic)
Overhead valve. Cast-iron block. Bore & stroke: 4.06 x 3.75 in. Displacement: 389 cid. Compression ratio: 10.50:1. Brake hp: 303 at 4600 rpm. Taxable hp: 52.80. Torque: 430 lbs.-ft. at 2800 rpm. Five main bearings. Hydraulic valve lifters. Crankcase capacity: 4 qt. (add 1 qt. for new filter). Cooling system capacity: 19.5 qt. (add 1 quart with heater). Carburetor: Carter AFB four-barrel. Engine code: 25K, 26K, 27P, and 28P.

Full-size Pontiac optional V-8
Overhead valve. Cast-iron block. Bore & stroke: 4.06 x 3.75 in. Displacement: 389 cid. Compression ratio: 10.50:1. Brake hp: 306 at 4800 rpm. Taxable hp: 52.80. Torque: 420 lbs.-ft. at 2800 rpm. Five main bearings. Hydraulic valve lifters. Crankcase capacity: 4 qt. (add 1 qt. for new filter). Cooling system capacity: 19.5 qt. (add 1 quart with heater). Carburetor: Carter AFB four-barrel. Engine code: 23B and 29N.

Catalina 2+2 base; optional V-8 in other full-size Pontiacs
Overhead valve. Cast-iron block. Bore & stroke: 4.09 x 4.00 in. Displacement: 421 cid. Compression ratio: 10.75:1. Brake hp: 320 at 4400 rpm. Taxable hp: 53.60. Torque: 455 lbs.-ft. at 2800 rpm. Five main bearings. Hydraulic valve lifters. Crankcase capacity: 4 qt. (add 1 qt. for new filter). Cooling system capacity with heater: 19.5 qt. Carburetor: Four-barrel. Engine code: 35B, 38S, or 43N.

GTO base V-8
Overhead valve. Cast-iron block. Bore & stroke: 4.06 x 3.75 in. Displacement: 389 cid. Compression ratio: 10.75:1. Brake hp: 325 at 4800 rpm. Taxable hp: 52.80. Torque: 428 lbs.-ft. at 3200 rpm. Five main bearings. Hydraulic valve lifters. Crankcase capacity: 5 qt. (add 1 qt. for new filter). Cooling system capacity with heater: 19.5 qt. Carburetor: Four-barrel. Engine code: 78X or 79J.

Full-size Pontiac optional tri-Power V-8
Overhead valve. Cast-iron block. Bore & stroke: 4.06 x 3.75 in. Displacement: 389 cid. Compression ratio: 10.75:1. Brake hp: 330 at 4600 rpm. Taxable hp: 52.80. Torque: 430 lbs.-ft. at 3200 rpm. Five main bearings. Hydraulic valve lifters. Crankcase capacity: 5 qt. (add 1 qt. for new filter). Cooling system capacity with heater: 19.5 qt. Carburetion: Three Rochester two-barrel carburetors. Engine code: 32B, 33G, or 34N.

GTO optional V-8
Overhead valve. Cast-iron block. Bore & stroke: 4.06 x 3.75 in. Displacement: 389 cid. Compression ratio: 10.75:1. Brake hp: 348 at 4900 rpm. Taxable hp: 52.80. Torque: 428 lbs.-ft. at 3600 rpm. Five main bearings. Hydraulic valve lifters. Crankcase capacity: 5 qt. (add 1 qt. for new filter). Cooling system capacity with heater: 19.5 qt. Carburetion: Three Rochester two-barrel carburetors. Engine code: 76X or 77J.

Full-size Pontiac optional 421 HO V-8
Overhead valve. Cast-iron block. Bore & stroke: 4.09 x 4.00 in. Displacement: 421 cid. Compression ratio: 10.75:1. Brake hp: 350 at 4600 rpm. Taxable hp: 53.60. Torque: 454 lbs.-ft. at 3200 rpm. Five main bearings. Hydraulic valve lifters. Crankcase capacity: 4 qt. (add 1 qt. for new filter). Cooling system capacity with heater: 19.5 qt. Carburetion: Three Rochester two-barrel carburetors. Engine code: 44B, 47S, or 49N.

Full-size Pontiac optional 421 HO V-8
Overhead valve. Cast-iron block. Bore & stroke: 4.09 x 4.00 in. Displacement: 421 cid. Compression ratio: 10.75:1. Brake hp: 370 at 5200 rpm. Taxable hp: 53.60. Torque: 460 lbs.-ft. at 3800 rpm. Five main bearings. Hydraulic valve lifters. Crankcase capacity: 4 qt. (add 1 qt. for new filter). Cooling system capacity with heater: 19.5 qt. Carburetion: Three Rochester two-barrel carburetors. Engine code: 45B, 46G, or 50Q.

The well-known 389-cid Pontiac V-8 was the common engine in full-size cars, but was also offered in the 1965 LeMans GTO in special 335- and 360-hp versions. Pontiac offered numerous carburetors, compression ratios, and camshafts and a wide range of horsepower options for each of its displacement groups.

1965 PONTIAC

Tempest base V-8
Overhead valve. Bore & stroke: 3.718 x 3.75 in. Displacement: 326 cid. Compression ratio: 9.2:1. Brake hp: 250 at 4600 rpm. Taxable hp: 44.4. Torque: 333 lbs.-ft. at 2800 rpm. Five main bearings. Hydraulic valve lifters. Crankcase capacity: 4 qt. Cooling system capacity with heater: 20.5 qt. (add 1 qt. for new filter). Carburetor: Two-barrel. Engine code: WP, WX, YN, or XF.

Full-size Pontiac base V-8
Overhead valve. Cast-iron block. Bore & stroke: 4.06 x 3.75 in. Displacement: 389 cid. Compression ratio: 8.6:1. Brake hp: 256 at 4600 rpm. Taxable hp: 52.80. Torque: 388 lbs.-ft. at 2400 rpm. Five main bearings. Hydraulic valve lifters. Crankcase capacity: 5 qt. (add 1 qt. for new filter). Cooling system capacity with heater: 20 qt. Carburetor: Rochester 2GC two-barrel. Engine code: WA, WB, or YA.

Tempest optional V-8
Overhead valve. Displacement: 326 cid. Bore & stroke: 3.718 x 3.75 in. Compression ratio: 10.5:1. Brake hp: 285 at 5000 rpm. Taxable hp: 44.4. Torque: 359 lbs.-ft. at 3200 rpm. Five main bearings. Hydraulic valve lifters. Crankcase capacity: 4 qt. Cooling system capacity with heater: 20.5 qt. (add 1 qt. for new filter). Carburetor: Carter AFB four-barrel Model 3899-S. Engine code: WR, YP, or XG.

In 1965, Catalina coupes and convertibles with the optional 2 + 2 Sports package were trimmed with "421" engine badges on front fenders, 2 + 2 numbering on rear fenders and deck lid, and simulated louvers behind the front wheel cutouts. The 338-hp version was standard in the high-performance Pontiac. (John Lee photo)

Full-size Pontiac optional V-8
Overhead valve. Cast-iron block. Bore & stroke: 4.06 x 3.75 in. Displacement: 389 cid. Compression ratio: 10.5:1. Brake hp: 290 at 4600 rpm. Taxable hp: 52.80. Torque: 418 lbs.-ft. at 2400 rpm. Five main bearings. Hydraulic valve lifters. Crankcase capacity: 5 qt. (add 1 qt. for new filter). Cooling system capacity with heater: 20 qt. Carburetor: Rochester 2GC two-barrel. Engine code: WC, YC, YD, YU, or YV.

Full-size Pontiac optional V-8
Overhead valve. Cast-iron block. Bore & stroke: 4.06 x 3.75 in. Displacement: 389 cid. Compression ratio: 10.5:1. Brake hp: 325 at 4800 rpm. Torque: 429 lbs.-ft. at 2800 rpm. Five main bearings. Hydraulic valve lifters. Crankcase capacity: 5 qt. (add 1 qt. for new filter). Cooling system capacity with heater: 20 qt. Carburetor: Carter AFB four-barrel Model 3895-S. Engine code: YE, YF, YL, YW, or YX.

Full-size Pontiac optional V-8
Overhead valve. Cast-iron block. Bore & stroke: 4.06 x 3.75 in. Displacement: 389 cid. Compression ratio: 10.5:1. Brake hp: 333 at 5000 rpm. Taxable hp: 52.80. Torque: 429 lbs.-ft. at 3200 rpm. Five main bearings. Hydraulic valve lifters. Crankcase capacity: 5 qt. (add 1 qt. for new filter). Cooling system capacity with heater: 20 qt. Carburetor: Carter AFB four-barrel Model 3896-S. Engine code: WE.

GTO base V-8
Overhead valve. Cast-iron block. Bore & stroke: 4.06 x 3.75 in. Displacement: 389 cid. Compression ratio: 10.75:1. Brake hp: 335 at 5000 rpm. Taxable hp: 52.80. Torque: 431 lbs.-ft. at 3200 rpm. Five main bearings. Hydraulic valve lifters. Crankcase capacity: 5 qt. (add 1 qt. for new filter). Cooling system capacity with heater: 20.5 qt. Carburetor: Four-barrel. Engine code: WT, WW, YS, or XE.

Full-size Pontiac optional V-8
Overhead valve. Cast-iron block. Bore & stroke: 4.09 x 4.00 in. Displacement: 421 cid. Compression ratio: 10.50:1. Brake hp: 338 at 4600 rpm. Taxable hp: 53.60. Torque: 459 lbs.-ft. at 2800 rpm. Five main bearings. Hydraulic valve lifters. Crankcase capacity: 5 qt. (add 1 qt. for new filter). Cooling system capacity with heater: 20 qt. Carburetor: Carter AFB Four-barrel. Engine code: WK, YZ, YT, WG, or YH.

Full-size Pontiac optional V-8
Overhead valve. Cast-iron block. Bore & stroke: 4.09 x 4.00 in. Displacement: 421 cid. Compression ratio: 10.75:1. Brake hp: 356 at 4800 rpm. Taxable hp: 53.60. Torque: 459 lbs.-ft. at 3200 rpm. Five main bearings. Hydraulic valve lifters. Crankcase capacity: 5 qt. (add 1 qt. for new filter). Cooling system capacity with heater: 20 qt. Carburetion: Three Rochester two-barrel carburetors. Engine code: YJ, YM, or WH.

GTO optional V-8
Overhead valve. Cast-iron block. Bore & stroke: 4.06 x 3.75 in. Displacement: 389 cid. Compression ratio: 10.75:1. Brake hp: 360 at 5200 rpm. Taxable hp: 52.80. Torque: 424 lbs.-ft. at 3600 rpm. Five main bearings. Hydraulic valve lifters. Crankcase capacity: 5 qt. (add 1 qt. for new filter). Cooling system capacity with heater: 20.5 qt. Carburetion: Three Rochester two-barrel carburetors. Engine code: WS, WV. XS, or YR.

Full-size Pontiac optional V-8
Overhead valve. Cast-iron block. Bore & stroke: 4.09 x 4.00 in. Displacement: 421 cid. Compression ratio: 10.75:1. Brake hp: 376 at 5000 rpm. Taxable hp: 53.60. Torque: 461 lbs.-ft. at 3600 rpm. Five main bearings. Hydraulic valve lifters. Crankcase capacity: 5 qt. (add 1 qt. for new filter). Cooling system capacity with heater: 20 qt. Carburetion: Three Rochester two-barrel carburetors. Engine code: WJ or YK.

1966 PONTIAC

Tempest base V-8
Overhead valve. Bore & stroke: 3.718 x 3.75 in. Displacement: 326 cid. Compression ratio: 9.2:1. Brake hp: 250 at 4600 rpm. Taxable hp: 44.4. Torque: 333 lbs.-ft. at 2800 rpm. Five main bearings. Hydraulic valve lifters. Crankcase capacity: 5 qt. (add 1 qt. for new filter). Cooling system capacity with heater: 20.5 qt. Carburetor: Two-barrel. Engine code: WP, WX, YN, or XF.

This car has one of the two 421-cid Tri-Power H.O. V-8s offered in 1965 Catalina 2 + 2s. The most powerful version had a 10.75:1 compression ratio and generated 376 hp. (John Lee photo)

Catalina base V-8 (synchromesh); All models optional V-8 (Hydra-Matic)
Overhead valve. Cast-iron block. Bore & stroke: 4.06 x 3.75 in. Displacement: 389 cid. Compression ratio: 8.6:1. Brake hp: 256 at 4600 rpm. Taxable hp: 52.80. Torque: 388 lbs.-ft. at 2400 rpm. Five main bearings. Hydraulic valve lifters. Crankcase capacity: 5 qt. (add 1 qt. for new filter).Cooling system capacity with heater: 20 qt. Carburetor: Rochester 2GC two-barrel Model 7026066. Engine code: WA, WB, or YA.

Tempest optional V-8
Overhead valve. Bore & stroke: 3.718 x 3.75 in. Displacement: 326 cid. Compression ratio: 10.5:1. Brake hp: 285 at 5000 rpm. Taxable hp: 44.4. Torque: 359 lbs.-ft. at 3200 rpm. Five main bearings. Hydraulic valve lifters. Crankcase capacity: 5 qt. (add 1 qt. for new filter). Cooling system capacity with heater: 20.5 qt. Carburetor: Four-barrel. Engine code: WR, YP, or XG.

Catalina, Star Chief, Executive base V-8 (Hydra-Matic)
Overhead valve. Cast-iron block. Bore & stroke: 4.06 x 3.75 in. Displacement: 389 cid. Compression ratio: 10.5:1. Brake hp: 290 at 4600 rpm. Taxable hp: 52.80. Torque: 418 lbs.-ft. at 2400 rpm. Five main bearings. Hydraulic valve lifters. Crankcase capacity: 5 qt. (add 1 qt. for new filter). Cooling system capacity with heater: 20 qt. Carburetor: Rochester 2GC two-barrel Model 7026066. Engine code: WC, YC, YD, YU, or YV.

Bonneville, Grand Prix base V-8 (Hydra-Matic); Optional all full-Size Pontiacs except 2+2
Overhead valve. Cast-iron block. Bore & stroke: 4.06 x 3.75 in. Displacement: 389 cid. Compression ratio: 10.5:1. Brake hp: 325 at 4800 rpm. Taxable hp: 52.80. Torque: 429 lbs.-ft. at 2800 rpm. Five main bearings. Hydraulic valve lifters. Crankcase capacity: 5 qt. (add 1 qt. for new filter). Cooling system capacity with heater: 20 qt. Carburetor: Carter AFB four-barrel Model 4033S. Engine code: YE, YF, YL, YW, or YX.

Bonneville, Grand Prix base V-8 (synchromesh); Optional all full-Size Pontiacs except 2+2
Overhead valve. Cast-iron block. Bore & stroke: 4.06 x 3.75 in. Displacement: 389 cid. Compression ratio: 10.5:1. Brake hp: 333 at 5000 rpm. Taxable hp: 52.80. Torque: 429 lbs.-ft. at 3200 rpm. Five main bearings. Hydraulic valve lifters. Crankcase capacity: 5 qt. (add 1 qt. for new filter). Cooling system capacity with heater: 20 qt. Carburetor: Carter AFB four-barrel Model 4033S. Engine code: WE.

GTO base V-8
Overhead valve. Cast-iron block. Bore & stroke: 4.06 x 3.75 in. Displacement: 389 cid. Compression ratio: 10.75:1. Brake hp: 335 at 5000 rpm. Taxable hp: 52.80. Torque: 431 lbs.-ft. at 3200 rpm. Five main bearings. Hydraulic valve lifters. Crankcase capacity: 5 qt. (add 1 qt. for new filter). Cooling system capacity with heater: 20 qt. Carburetor: Four-barrel. Engine code: WT, WW, YS, or XE.

Catalina 2+2 base V-8; Optional in other full-size Pontiacs
Overhead valve. Cast-iron block. Bore & stroke: 4.09 x 4.00 in. Displacement: 421 cid. Compression ratio: 10.5:1. Brake hp: 338 at 4600 rpm. Taxable hp: 53.60. Torque: 459 lbs.-ft. at 2800 rpm. Five main bearings. Hydraulic valve lifters. Cooling system capacity with heater: 19.5 qt. Crankcase capacity: 5 qt. (add 1 qt. for new filter). Carburetor: Carter AFB four-barrel Model 4033S. Engine code: WK, YZ, YT, WG, or YH.

Full-size Pontiac optional Tri-Power V-8
Overhead valve. Cast-iron block. Bore & stroke: 4.09 x 4.00 in. Displacement: 421 cid. Compression ratio: 10.75:1. Brake hp: 356 at 4800 rpm. Taxable hp: 53.60. Torque: 459 lbs.-ft. at 3200 rpm. Five main bearings. Hydraulic valve lifters. Crankcase capacity: 5 qt. (add 1 qt. for new filter). Cooling system capacity with heater: 19.5 qt. Carburetion: Three Rochester two-barrel carburetors. Engine code: YJ, YM, or WH.

GTO optional V-8
Overhead valve. Cast-iron block. Bore & stroke: 4.06 x 3.75 in. Displacement: 389 cid. Compression ratio: 10.75:1. Brake hp: 360 at 5200 rpm. Taxable hp: 52.80. Torque: 424 lbs.-ft. at 3600 rpm. Five main bearings. Hydraulic valve lifters. Crankcase capacity: 5 qt. (add 1 qt. for new filter). Cooling system capacity with heater: 20 qt. Carburetion: Three Rochester two-barrel carburetors. Engine code: WS, WV, XS, or YR.

Full-size Pontiac optional Tri-Power V-8
Overhead valve. Cast-iron block. Bore & stroke: 4.09 x 4.00 in. Displacement: 421 cid. Compression ratio: 10.75:1. Brake hp: 376 at 5000 rpm. Taxable hp: 53.60. Torque: 461 lbs.-ft. at 3600 rpm. Five main bearings. Hydraulic valve lifters. Crankcase capacity: 5 qt. (add 1 qt. for new filter). Cooling system capacity with heater: 19.5 qt. Carburetion: Three Rochester two-barrel carburetors. Engine code: WJ or YK.

The 1966 Catalina 2 + 2 included a four-barrel 421-cid V-8 with 338 hp, low-restriction exhausts, chrome air cleaner, chrome valve covers, three-speed manual transmission with Hurst shifter, and heavy-duty suspension. Available axle ratios included 3.08, 3.23, and 3.42. (John Lee Photo)

1967 PONTIAC

Tempest, Firebird base V-8
Overhead valve. Bore & stroke: 3.718 x 3.75 in. Displacement: 326 cid. Compression ratio: 9.2:1. Brake hp: 250 at 4600 rpm. Taxable hp: 44.40. Torque: 333 lbs.-ft. at 2800 rpm. Five main bearings. Hydraulic valve lifters. Crankcase capacity: 6 qt. (add 1 qt. for new filter). Cooling system capacity with heater: 18.5 qt. Carburetor: Two-barrel. Engine code: WP, WX, YN, XF, WH, WC, XJ, and XL.

GTO optional "Economy" V-8 (automatic)
Overhead valve. Cast-iron block. Bore & stroke: 4.12 x 3.75 in. Displacement: 400 cid. Compression ratio: 8.6:1. Brake hp: 255 at 4400 rpm. Taxable hp: 54.3. Torque: 397 lbs.-ft. at 2400 rpm. Five main bearings. Hydraulic valve lifters. Crankcase capacity: 6 qt. (add 1 qt. for new filter). Cooling system capacity with heater: 17.75 qt. Carburetor: Two-barrel. Engine code: YB.

Catalina, Executive base V-8 (synchromesh); "Economy" option for all full-size Pontiacs
Overhead valve. Cast-iron block. Bore & stroke: 4.12 x 3.75 in. Displacement: 400 cid. Compression ratio: 8.6:1. Brake hp: 265 at 4600 rpm. Taxable hp: 54.3. Torque: 397 lbs.-ft. at 2400 rpm. Five main bearings. Hydraulic valve lifters. Crankcase capacity: 6 qt. (add 1 qt. for new filter). Cooling system capacity with heater: 18 qt. Carburetor: Two-barrel. Engine code: WA, YA, YB, and WB.

Tempest, Firebird optional V-8
Overhead valve. Bore & stroke: 3.718 x 3.75 in. Displacement: 326 cid. Compression ratio: 10.5:1. Brake hp: 285 at 5000 rpm. Taxable hp: 44.40. Torque: 359 lbs.-ft. at 3200 rpm. Five main bearings. Hydraulic valve lifters. Crankcase capacity: 6 qt. (add 1 qt. for new filter). Cooling system capacity with heater: 18.5 qt. Carburetor: Four-barrel. Engine code: WK, WO, WR, YM, YP, XG, XO, and XR.

Catalina, Executive base V-8 (automatic); optional in other full-size Pontiacs
Overhead valve. Cast-iron block. Bore & stroke: 4.12 x 3.75 in. Displacement: 400 cid. Compression ratio: 10.5:1. Brake hp: 290 at 4600 rpm. Taxable hp: 54.3. Torque: 428 lbs.-ft. at 2500 rpm. Five main bearings. Hydraulic valve lifters. Crankcase capacity: 6 qt. (add 1 qt. for new filter). Cooling system capacity with heater: 18 qt. Carburetor: Two-barrel. Engine code: YC and YD.

Firebird 400 base V-8
Overhead valve. Cast-iron block. Bore & stroke: 4.12 x 3.75 in. Displacement: 400 cid. Compression ratio: 10.75:1. Brake hp: 325 at 4800 rpm. Taxable hp: 54.3. Torque: 445 lbs.-ft. at 2900 rpm. Five main bearings. Hydraulic valve lifters. Crankcase capacity: 6 qt. (add 1 qt. for new filter). Cooling system capacity with heater: 17.75 qt. Carburetor: Four-barrel. Engine code: YE, YF, YT, WI, WQ, WZ, WU, and XN.

Firebird 400 optional "Ram Air"
Overhead valve. Cast-iron block. Bore & stroke: 4.12 x 3.75 in. Displacement: 400 cid. Compression ratio: 10.75:1. Brake hp: 325 at 5200 rpm. Taxable hp: 54.3. Torque: 410 lbs.-ft. at 3600 rpm. Five main bearings. Hydraulic valve lifters. Crankcase capacity: 6 qt. (add 1 qt. for new filter). Cooling system capacity with heater: 17.75 qt. Carburetor: Four-barrel. Engine code: YE, YF, YT, WI, WQ, WZ, WU, and XN.

Bonneville base V-8 (automatic); Optional other full-size Pontiacs (automatic)
Overhead valve. Cast-iron block. Bore & stroke: 4.12 x 3.75 in. Displacement: 400 cid. Compression ratio: 10.5:1. Brake hp: 325 at 4800 rpm. Taxable hp: 54.3. Torque: 445 lbs.-ft. at 2900 rpm. Five main bearings. Hydraulic valve lifters. Crankcase capacity: 6 qt. (add 1 qt. for new filter). Cooling system capacity with heater: 18 qt. Carburetor: Four-barrel.

Bonneville base V-8 (synchromesh); Catalina, Executive optional V-8 (synchromesh)
Overhead valve. Cast-iron block. Bore & stroke: 4.12 x 3.75 in. Displacement: 400 cid. Compression ratio: 10.5:1. Brake hp: 333 at 5000 rpm. Taxable hp: 54.3. Torque: 445 lbs.-ft. at 3000 rpm. Five main bearings. Hydraulic valve lifters. Crankcase capacity: 6 qt. (add 1 qt. for new filter). Cooling system capacity with heater: 18 qt. Carburetor: Four-barrel. Engine code: WE, WD, XZ, XY, and XH.

GTO base V-8
Overhead valve. Cast-iron block. Bore & stroke: 4.12 x 3.75 in. Displacement: 400 cid. Compression ratio: 10.75:1. Brake hp: 335 at 5000 rpm. Taxable hp: 54.3. Torque: 441 lbs.-ft. at 3400 rpm. Five main bearings. Hydraulic valve lifters. Crankcase capacity: 6 qt. (add 1 qt. for new filter). Cooling system capacity with heater: 17.75 qt. Carburetor: Four-barrel. Engine code: WT, WW, and YS.

Grand Prix base V-8
Overhead valve. Cast-iron block. Bore & stroke: 4.12 x 3.75 in. Displacement: 400 cid. Compression ratio: 10.5:1. Brake hp: 350 at 5000 rpm. Taxable hp: 54.3. Torque: 440 lbs.-ft. at 3200 rpm. Five main bearings. Hydraulic valve lifters. Crankcase capacity: 6 qt. (add 1 qt. for new filter). Cooling system capacity with heater: 18 qt. Carburetor: Four-barrel. Engine code: XZ, XY, and XJ.

GTO optional V-8
Overhead valve. Cast-iron block. Bore & stroke: 4.12 x 3.75 in. Displacement: 400 cid. Compression ratio: 10.75:1. Brake hp: 360 at 5100 rpm. Taxable hp: 54.3. Torque: 438 lbs.-ft. at 3600 rpm. Five main bearings. Hydraulic valve lifters. Crankcase capacity: 6 qt. (add 1 qt. for new filter). Cooling system capacity with heater: 17.75 qt. Carburetor: Four-barrel. Engine code: WS, WV, XP, XS, YR, and YZ.

Full-size Pontiac optional V-8
Overhead valve. Cast-iron block. Bore & stroke: 4.12 x 4.00 in. Displacement: 428 cid. Compression ratio: 10.5:1. Brake hp: 360 at 4600 rpm. Taxable hp: 54.3. Torque: 472 lbs.-ft. at 3200 rpm. Five main bearings. Hydraulic valve lifters. Crankcase capacity: 6 qt. (add 1 qt. for new filter). Cooling system capacity with heater: (2+2) 17-1/4 qt.; (Grand Prix) 18.50 qt. Carburetor: Four-barrel. Engine code: WG, WK, XD, Y2, YH, YY, YZ, and YT.

Full-size Pontiac optional V-8
Overhead valve. Cast-iron block. Bore & stroke: 4.12 x 4.000 in. Displacement: 428 cid. Compression ratio: 10.75:1. Brake hp: 376 at 5100 rpm. Taxable hp: 54.3. Torque: 462 lbs.-ft. at 3400 rpm. Five main bearings. Hydraulic valve lifters. Crankcase capacity: 6 qt. (add 1 qt. for new filter). Cooling system capacity with heater: (2+2) 17-1/4 qt.; (Grand Prix) 18.50 qt. Carburetor: Four-barrel. Engine code: WJ, WL, XK, and YK.

For 1967, Pontiac motor Division blew the 389-cid V-8 out to 400 cubes and increased the 421 to 428 cid. This big-block engine tipped the scales at about 600 pounds and offered performance buffs a 2.40 power-to-weight ratio. Options up to 376 hp were available on these big-block V-8s.

1968 PONTIAC

GTO optional "Economy" V-8
Overhead valve. Cast-iron block. Bore & stroke: 4.12 x 3.75 in. Displacement: 400 cid. Compression ratio: 8.6:1. Brake hp: 265 at 4600 rpm. Taxable hp: 54.3. Torque: 397 lbs.-ft. at 2400 rpm. Five main bearings. Hydraulic valve lifters. Crankcase capacity: 5 qt. (add 1 qt. for new filter). Cooling system capacity with heater: 17.75 qt. Carburetor: Two-barrel.

Tempest, Firebird V-8 base V-8
Overhead valve. Bore & stroke: 3.88 x 3.75 in. Displacement: 350 cid. Compression ratio: 9.2:1. Brake hp: 265 at 4600 rpm. Taxable hp: 48. Torque: 255 lbs.-ft. at 2800. Five main bearings. Hydraulic valve lifters.

Crankcase capacity: 5 qt. (add 1 qt. for new filter). Cooling system capacity with heater: 19.5 qt. Carburetor: Two-barrel. Engine code: WP, YN, WD, WC, and YJ.

Full-size Pontiac optional V-8 (automatic); GTO optional V-8 (automatic)
Overhead valve. Cast-iron block. Bore & stroke: 4.12 x 3.75 in. Displacement: 400 cid. Compression ratio: 8.6:1. Brake hp: 265 at 4600 rpm. Taxable hp: 54.3. Torque: 397 lbs.-ft. at 2400 rpm. Five main bearings. Hydraulic valve lifters. Crankcase capacity: 5 qt. (add 1 qt. for new filter). Cooling system capacity with heater: 18 qt. (18.6 qt. Grand Prix). Carburetor: Two-barrel. Engine code: XM and YA.

Catalina, Executive base V-8
Overhead valve. Cast-iron block. Bore & stroke: 4.12 x 3.75 in. Displacement: 400 cid. Compression ratio: 10.5:1. Brake hp: 290 at 4600 rpm. Taxable hp: 54.3. Torque: 428 lbs.-ft. at 2500 rpm. Five main bearings. Hydraulic valve lifters. Crankcase capacity: 5 qt. (add 1 qt. for new filter). Cooling system capacity with heater: 18 qt. Carburetor: Two-barrel. Engine code: WA, WB, and YC.

Firebird 400 base V-8
Overhead valve. Cast-iron block. Bore & stroke: 4.12 x 3.75 in. Displacement: 400 cid. Compression ratio: 10.75:1. Brake hp: 330 at 4800 rpm. Taxable hp: 54.3. Torque: 430 lbs.-ft. at 3300 rpm. Five main bearings. Hydraulic valve lifters. Crankcase capacity: 5 qt. (add 1 qt. for new filter). Cooling system capacity with heater: 17.75 qt. Carburetor: Four-barrel. Engine code: WZ and YT.

Firebird 400 optional V-8
Overhead valve. Cast-iron block. Bore & stroke: 4.12 x 3.75 in. Displacement: 400 cid. Compression ratio: 10.75:1. Brake hp: 335 at 5000 rpm. Taxable hp: 54.3. Torque: 430 lbs.-ft. at 3400 rpm. Five main bearings. Hydraulic valve lifters. Crankcase capacity: 5 qt. (add 1 qt. for new filter). Cooling system capacity with heater: 17.75 qt. Carburetor: Four-barrel. Engine code: WQ, WI, YW, and XN.

Firebird 400 optional "Ram Air" V-8
Overhead valve. Cast-iron block. Bore & stroke: 4.12 x 3.75 in. Displacement: 400 cid. Compression ratio: 10.75:1. Brake hp: 335 at 5300 rpm. Taxable hp: 54.3. Torque: 430 lbs.-ft. at 3600 rpm. Five main bearings. Hydraulic valve lifters. Crankcase capacity: 5 qt. (add 1 qt. for new filter). Cooling system capacity with heater: 17.75 qt. Carburetor: Four-barrel. Engine code: WQ, WI, YW, and XN.

Bonneville base V-8; Other full-size Pontiacs optional V-8
Overhead valve. Cast-iron block. Bore & stroke: 4.12 x 3.75 in. Displacement: 400 cid. Compression ratio: 10.5:1. Brake hp: 340 at 4800 rpm. Taxable hp: 54.3. Torque: 445 lbs.-ft. at 2900 rpm. Five main bearings. Hydraulic valve lifters. Crankcase capacity: 5 qt. (add 1 qt. for new filter). Cooling system capacity with heater: 18 qt. Carburetor: Four-barrel. Engine code: WE and YE.

Grand Prix base V-8
Overhead valve. Cast-iron block. Bore & stroke: 4.12 x 3.75 in. Displacement: 400 cid. Compression ratio: 10.5:1. Brake hp: 350 at 5000 rpm. Taxable hp: 54.3. Torque: 445 lbs.-ft. at 3000 rpm. Five main bearings. Hydraulic valve lifters. Crankcase capacity: 5 qt. (add 1 qt. for new filter). Cooling system capacity with heater: 18 qt. (18.6 qt. Grand Prix). Carburetor: Four-barrel. Engine code: XZ and XH.

GTO base V-8
Overhead valve. Cast-iron block. Bore & stroke: 4.12 x 3.75 in. Displacement: 400 cid. Compression ratio: 10.5:1. Brake hp: 350 at 5000 rpm. Taxable hp: 54.3. Torque: 445 lbs.-ft. at 3000 rpm. Five main bearings. Hydraulic valve lifters. Crankcase capacity: 5 qt. (add 1 qt. for new filter). Cooling system capacity with heater: 18 qt. (18.6 qt. Grand Prix). Carburetor: Four-barrel. Engine code: WT.

GTO optional "Ram Air" V-8
Overhead valve. Cast-iron block. Bore & stroke: 4.12 x 3.75 in. Displacement: 400 cid. Compression ratio: 10.75:1. Brake hp: 360 at 5100 rpm. Taxable hp: 54.3. Torque: 445 lbs.-ft. at 3600 rpm. Five main bearings. Hydraulic valve lifters. Crankcase capacity: 5 qt. (add 1 qt. for new filter). Cooling system capacity with heater: 17.75 qt. Carburetor: Four-barrel. Engine code: WS and XS.

GTO optional "Ram Air II" V-8
Overhead valve. Cast-iron block. Bore & stroke: 4.12 x 3.75 in. Displacement: 400 cid. Compression ratio: 10.75:1. Brake hp: 366 at 5400 rpm. Taxable hp: 54.3. Torque: 445 lbs.-ft. at 3800 rpm. Five main bearings. Hydraulic valve lifters. Crankcase capacity: 5 qt. (add 1 qt. for new filter). Cooling system capacity with heater: 17.75 qt. Carburetor: Four-barrel.

Catalina, Executive, Bonneville, Grand Prix optional V-8
Overhead valve. Cast-iron block. Bore & stroke: 4.12 x 4.00 in. Displacement: 428 cid. Compression ratio: 10.5:1. Brake hp: 375 at 4800 rpm. Taxable hp: 54.3. Torque: 472 lbs.-ft. at 3200 rpm. Five main bearings. Hydraulic valve lifters. Crankcase capacity: 5 qt. (add 1 qt. for new filter). Cooling system capacity with heater: 17.2 qt.; (Grand Prix) 18.6 qt. Carburetor: Four-barrel. Engine code: WG and YH.

Catalina, Executive, Bonneville, Grand Prix optional V-8
Overhead valve. Cast-iron block. Bore & stroke: 4.12 x 4.00 in. Displacement: 428 cid. Compression ratio: 10.75:1. Brake hp: 390 at 5200 rpm. Taxable hp: 54.3. Torque: 465 lbs.-ft. at 3400 rpm. Five main bearings. Hydraulic valve lifters. Crankcase capacity: 5 qt. (add 1 qt. for new filter). Cooling system capacity with heater: 17.2 qt.; (Grand Prix) 18.6 qt. Carburetor: Four-barrel. Engine code: WJ and YK.

1969 PONTIAC

Firebird 350 baseV-8; Tempest optional V-8
Overhead valve. Bore & stroke: 3.88 x 3.75 in. Displacement: 350 cid. Compression ratio: 9.2:1. Brake hp: 265 at 4600 rpm. Taxable hp: 48.0. Torque: 355 lbs.-ft. at 2800 rpm. Five main bearings. Hydraulic valve lifters. Crankcase capacity: 5 qt. (add 1 qt. for new filter). Cooling system capacity with heater: 19.5 qt. Carburetor: Four-barrel. Engine code: WU, YU, XS, XR, YN, WP, WM, YE, XB, WC, XL, and YJ.

Catalina, Executive, GTO optional "Economy" V-8
Overhead valve. Cast-iron block. Bore & stroke: 4.12 x 3.75 in. Displacement: 400 cid. Compression ratio: 8.6:1. Brake hp: 265 at 4600 rpm. Taxable hp: 54.3. Torque: 397 lbs.-ft. at 2400 rpm. Five main bearings. Hydraulic valve lifters. Crankcase capacity: 5 qt. (add 1 qt. for new filter). Cooling system capacity with heater: 18 qt. (18.6 qt. Grand Prix). Carburetor: Two-barrel. Engine code: XM, XX, YA, YB, and YF.

Catalina, Executive base V-8
Overhead valve. Cast-iron block. Bore & stroke: 4.12 x 3.75 in. Displacement: 400 cid. Compression ratio: 10.5:1. Brake hp: 290 at 4600 rpm. Taxable hp: 54.3. Torque: 428 lbs.-ft. at 2500 rpm. Five main bearings. Hydraulic valve lifters. Crankcase capacity: 5 qt. (add 1 qt. for new filter). Cooling system capacity with heater: 18 qt. Carburetor: Rochester two-barrel Model 7028066. Engine code: WD, WE, YD, WA, and WB.

Firebird 350 HO base V-8
Overhead valve. Bore & stroke: 3.88 x 3.75 in. Displacement: 350 cid. Compression ratio: 10.5:1. Brake hp: 325 at 5100 rpm. Taxable hp: 48.0. Torque: 380 lbs.-ft. at 3200 rpm. Five main bearings. Hydraulic valve lifters. Cooling system capacity with heater: 19.5 qt. Crankcase capacity: 5 qt. (add 1 qt. for new filter). Carburetor: Four-barrel. Engine code: WN and XG.

Tempest optional V-8
Overhead valve. Bore & stroke: 3.88 x 3.75 in. Displacement: 350 cid. Compression ratio: 10.5:1. Brake hp: 330 at 5100 rpm. Taxable hp: 48.0. Torque: 380 lbs.-ft. at 3200 rpm. Five main bearings. Hydraulic valve lifters. Crankcase capacity: 5 qt. (add 1 qt. for new filter). Cooling system

The family tree of the 1969 Pontiac V-8 could be traced all the way back to 1958. Three displacements were available: 350-cid (as shown here); 400-cid, and 428-cid. All were closely related. The 400 was basically a bored 350 with no change in stroke and the 428 was a stroked 400, with no change in bore. A 390-hp version of the 428 was the year's most powerful engine.

capacity with heater: 19.5 qt. Carburetor: Four-barrel. Engine code: WV and XU.

Firebird 400 base V-8
Overhead valve. Cast-iron block. Bore & stroke: 4.12 x 3.75 in. Displacement: 400 cid. Compression ratio: 10.75:1. Brake hp: 330 at 4800 rpm. Taxable hp: 54.3. Torque: 430 lbs.-ft. at 3300 rpm. Five main bearings. Hydraulic valve lifters. Crankcase capacity: 5 qt. (add 1 qt. for new filter). Cooling system capacity with heater: 17.75 qt. Carburetor: Four-barrel. Engine code: WZ and YT.

Firebird 400 optional HO V-8
Overhead valve. Cast-iron block. Bore & stroke: 4.12 x 3.75 in. Displacement: 400 cid. Compression ratio: 10.75:1. Brake hp: 335 at 5000 rpm. Taxable hp: 54.3. Torque: 430 lbs.-ft. at 3400 rpm. Five main bearings. Hydraulic valve lifters. Crankcase capacity: 5 qt. (add 1 qt. for new filter). Cooling system capacity with heater: 17.75 qt. Carburetor: Four-barrel. Engine code: WQ, WD, WE, YD, WA and WB.

Firebird 400, Trans optional Ram Air IV V-8
Overhead valve. Cast-iron block. Bore & stroke: 4.12 x 3.75 in. Displacement: 400 cid. Compression ratio: 10.75:1. Brake hp: 345 at 5400 rpm. Taxable hp: 54.3. Torque: 430 lbs.-ft. at 3700 rpm. Five main bearings. Hydraulic valve lifters. Crankcase capacity: 5 qt. (add 1 qt. for new filter). Cooling system capacity with heater: 17.75 qt. Carburetor: Four-barrel. Engine code: WH and XN.

Grand Prix base V-8
Overhead valve. Cast-iron block. Bore & stroke: 4.12 x 3.75 in. Displacement: 400 cid. Compression ratio: 10.5:1. Brake hp: 350 at 5000 rpm. Taxable hp: 54.3. Torque: 445 lbs.-ft. at 3000 rpm. Five main bearings. Hydraulic valve lifters. Crankcase capacity: 5 qt. (add 1 qt. for new filter). Cooling system capacity with heater: 18.6 qt. Carburetor: Rochester four-barrel Model 7029263. Engine code: WX, XH, and WT.

GTO base V-8
Overhead valve. Cast-iron block. Bore & stroke: 4.12 x 3.75 in. Displacement: 400 cid. Compression ratio: 10.75:1. Brake hp: 350 at 5000 rpm. Taxable hp: 54.3. Torque: 445 lbs.-ft. at 3000 rpm. Five main bearings. Hydraulic valve lifters. Crankcase capacity: 5 qt. (add 1 qt. for new filter). Cooling system capacity with heater: 17.75 qt. Carburetor: Four-barrel. Engine code: WX, XH, and WT.

Bonneville base V-8
Overhead valve. Cast-iron block. Bore & stroke: 4.12 x 4.00 in. Displacement: 428 cid. Compression ratio: 10.5:1. Brake hp: 360 at 4600 rpm. Taxable hp: 54.3. Torque: 472 lbs.-ft. at 3200 rpm. Five main bearings. Hydraulic valve lifters. Crankcase capacity: 5 qt. (add 1 qt. for new filter). Cooling system capacity with heater: 17.2 qt. Carburetor: Rochester four-barrel Model 7029262. Engine code: WG, YH, XJ, YL, and XE.

GTO optional V-8
Overhead valve. Cast-iron block. Bore & stroke: 4.12 x 3.75 in. Displacement: 400 cid. Compression ratio: 10.75:1. Brake hp: 366 at 5100 rpm. Taxable hp: 54.3. Torque: 445 lbs.-ft. at 3600 rpm. Five main bearings. Hydraulic valve lifters. Crankcase capacity: 5 qt. (add 1 qt. for new filter). Cooling system capacity with heater: 17.75 qt. Carburetor: Four-barrel. Engine code: WS, WW, and YZ.

GTO optional "Ram Air" V-8
Overhead valve. Cast-iron block. Bore & stroke: 4.12 x 3.75 in. Displacement: 400 cid. Compression ratio: 10.75:1. Brake hp: 370 at 5500 rpm. Taxable hp: 54.3. Torque: 445 lbs.-ft. at 3900 rpm. Five main bearings. Hydraulic valve lifters. Crankcase capacity: 5 qt. (add 1 qt. for new filter). Cooling system capacity with heater: 17.75 qt. Carburetor: Four-barrel. Engine code: XK, WF, and XF.

Grand Prix optional V-8
Overhead valve. Cast-iron block. Bore & stroke: 4.12 x 4.00 in. Displacement: 428 cid. Compression ratio: 10.5:1. Brake hp: 370 at 4800 rpm. Taxable hp: 54.3. Torque: 472 lbs.-ft. at 3200 rpm. Five main bearings. Hydraulic valve lifters. Crankcase capacity: 5 qt. (add 1 qt. for new filter). Cooling system capacity with heater: 17.2 qt. Carburetor: Four-barrel. Engine code: XK, WF, and XF.

Catalina, Executive, Bonneville, Grand Prix optional V-8
Overhead valve. Cast-iron block. Bore & stroke: 4.12 x 4.00 in. Displacement: 428 cid. Compression ratio: 10.75:1. Brake hp: 390 at 5200 rpm. Taxable hp: 54.3. Torque: 465 lbs.-ft. at 3400 rpm. Five main bearings. Hydraulic valve lifters. Crankcase capacity: 5 qt. (add 1 qt. for new filter). Cooling system capacity with heater: 17.2 qt. Carburetor: Four-barrel. Engine code: WJ, YK, WL, and XG.

1970 PONTIAC

Catalina, Firebird Esprit base V-8; LeMans, Firebird optional V-8
Overhead valve. Cast-iron block. Bore & stroke: 3.88 x 3.53 in. Displacement: 350 cid. Compression ratio: 8.80:1. Brake hp: 255 at 4600 rpm. Taxable hp: 48.0. Torque: 355 lbs.-ft. at 2800 rpm. Five main bearings. Hydraulic valve lifters. Crankcase capacity: 5 qt. (add 1 qt. for new filter). Cooling system capacity with heater: (Firebird) 19.5 qt.; (Others) 18 qt. Carburetor: Rochester two-barrel Model 7040071. Engine code: WU, YU, W7, and X7.

Full-size Pontiac, Grand Prix, LeMans, Firebird Esprit optional V-8
Overhead valve. Cast-iron block. Bore & stroke: 4.125 x 3.75 in. Displacement: 400 cid. Compression ratio: 8.80:1. Brake horsepower: 265 at 4600 rpm. Taxable hp: 54.3. Torque: 397 lbs.-ft. at 2400 rpm. Five main bearings. Hydraulic valve lifters. Crankcase capacity: 5 qt. (add 1 qt. for new filter). Cooling system capacity with heater: 18.5 qt. Carburetor: Two-barrel. Engine code: XX and YB.

Executive, Safari base V-8; Catalina optional V-8
Overhead valve. Cast-iron block. Bore & stroke: 4.12 x 3.75 in. Displacement: 400 cid. Compression ratio: 10.0:1. Brake hp: 290 at 4600 rpm. Taxable hp: 54.3. Torque: 428 lbs.-ft. at 2500 rpm. Five main bearings. Hydraulic valve lifters. Crankcase capacity: 5 qt. (add 1 qt. for new filter). Cooling system capacity with heater: 18 qt. Carburetor: Rochester four-barrel Model 7040263. Engine code: WE and YD.

Full-size Pontiac, LeMans optional V-8
Overhead valve. Cast-iron block. Bore & stroke: 4.125 x 3.75 in. Displacement: 400 cid. Compression ratio: 10.0:1. Brake horsepower: 330 at 4800 rpm. Taxable hp: 54.3. Torque: 445 lbs.-ft. at 2900 rpm. Five main bearings. Hydraulic valve lifters. Crankcase capacity: 5 qt. (add 1 qt. for new filter). Cooling system capacity with heater: (Pontiac) 18.0 qt.; (LeMans) 18.5 qt. Carburetor: Four-barrel. Engine code: YS, XV, and XZ.

Firebird Formula 400 base V-8
Overhead valve. Cast-iron block. Bore & stroke: 4.125 x 3.75 in. Displacement: 400 cid. Compression ratio: 10.25:1. Brake horsepower: 330 at 4800 rpm. Taxable hp: 54.3. Torque: 430 lbs.-ft. at 3000 rpm. Five main bearings. Hydraulic valve lifters. Crankcase capacity: 5 qt. (add 1 qt. for new filter). Cooling system capacity with heater: 18.5 qt. Carburetor: Four-barrel. Engine code: WT.

Firebird 400, Trans Am optional Ram Air III V-8
Overhead valve. Cast-iron block. Bore & stroke: 4.12 x 3.75 in. Displacement: 400 cid. Compression ratio: 10.75:1. Brake hp: 345 at 5000 rpm. Taxable hp: 54.3. Torque: 430 lbs.-ft. at 3400 rpm. Five main bearings. Hydraulic valve lifters. Crankcase capacity: 5 qt. (add 1 qt. for new filter). Cooling system capacity with heater: 18.5 qt. Carburetor: Four-barrel. Engine code: WS.

Firebird 400, Trans Am optional Ram Air IV V-8
Overhead valve. Cast-iron block. Bore & stroke: 4.12 x 3.75 in.

In 1970, Pontiac offered V-8s with displacements of 350, 400, and 455 cid. All of these engines were conventional 90-degree V-8s with cast iron blocks and heads, overhead valves, and 4.62-inch center-to-center cylinder bore spacing, which resulted in many components being interchangeable. The top two engines of the year were the 400-cid Ram Air engines. The Ram Air II version generated 366 hp and the hotter Ram Air IV edition was good for 370 hp.

Displacement: 400 cid. Compression ratio: 10.75:1. Brake hp: 345 at 5400 rpm. Taxable hp: 54.3. Torque: 430 lbs.-ft. at 3700 rpm. Five main bearings. Hydraulic valve lifters. Crankcase capacity: 5 qt. (add 1 qt. for new filter). Cooling system capacity with heater: 18.5 qt. Carburetor: Four-barrel. Engine code: YZ.

Grand Prix, GTO base V-8
Overhead valve. Cast-iron block. Bore & stroke: 4.12 x 3.75 in. Displacement: 400 cid. Compression ratio: 10.25:1. Brake hp: 350 at 5000 rpm. Taxable hp: 54.3. Torque: 445 lbs.-ft. at 3000 rpm. Five main bearings. Hydraulic valve lifters. Crankcase capacity: 5 qt. (add 1 qt. for new filter). Cooling system capacity with heater: (GTO) 18.5 qt.; (Grand Prix) 18.6 qt. Carburetor: Rochester four-barrel Model 7040263. Engine code: WT, WX, YS, and XH.

Bonneville base V-8
Overhead valve. Cast-iron block. Bore & stroke: 4.15 x 4.21 in. Displacement: 455 cid. Compression ratio: 10.0:1. Brake hp: 360 at 4300 rpm. Taxable hp: 54.5. Torque: 500 lbs.-ft. at 2700 rpm. Five main bearings. Hydraulic valve lifters. Crankcase capacity: 5 qt. (add 1 qt. for new filter). Cooling system capacity with heater: 18.0 qt. Carburetor: Rochester four-barrel Model 7040071. Engine code: YH.

GTO optional V-8
Overhead valve. Cast-iron block. Bore & stroke: 4.12 x 3.75 in. Displacement: 400 cid. Compression ratio: 10.50:1. Brake hp: 366 at 5100 rpm. Taxable hp: 54.3. Torque: 445 lbs.-ft. at 3600 rpm. Five main bearings. Hydraulic valve lifters. Crankcase capacity: 5 qt. (add 1 qt. for new filter). Cooling system capacity with heater: 18.5 qt. Carburetor: Four-barrel. Engine code: WS and YZ.

GTO, Trans Am, Formula Firebird optional V-8
Overhead valve. Cast-iron block. Bore & stroke: 4.12 x 3.75 in. Displacement: 400 cid. Compression ratio: 10.50:1. Brake hp: 370 at 5500 rpm. Taxable hp: 54.3. Torque: 445 lbs.-ft. at 3900 rpm. Five main bearings. Hydraulic valve lifters. Crankcase capacity: 5 qt. (add 1 qt. for new filter). Cooling system capacity with heater: 18.5 qt. Carburetor: Four-barrel. Engine code: WW, WH, XP, and XN.

Grand Prix optional V-8
Overhead valve. Cast-iron block. Bore & stroke: 4.15 x 4.21 in. Displacement: 455 cid. Compression ratio: 10.25:1. Brake hp: 370 at 4600 rpm. Taxable hp: 54.5. Torque: 500 lbs.-ft. at 3100 rpm. Five main bearings. Hydraulic valve lifters. Crankcase capacity: 5 qt. (add 1 qt. for new filter). Cooling system capacity with heater: 18.0 qt. Carburetor: Four-barrel. Engine code: WA, WG, YC, YA, and XF.

1971 PONTIAC

Catalina, Firebird Esprit base V-8; LeMans optional V-8
Overhead valve. Cast-iron block. Bore & stroke: 3.88 x 3.75 in. Displacement: 350 cid. Compression ratio: 8.00:1. Brake hp: 250 at 4400 rpm. Net hp: 165 at 4200 rpm. Taxable hp: 48.0. Torque: 350 lbs.-ft. at 2400 rpm. Net Torque: 275 lbs.-ft. at 2200 rpm. Five main bearings. Hydraulic valve lifters. Crankcase capacity: 5 qt. (add 1 qt. for new filter). Cooling system capacity with heater: 18 qt. Carburetor: Rochester two-barrel. Engine code: WR, WU, YU, XR, WN, WP, YN, and YP.

Firebird base V-8
Overhead valve. Cast-iron block. Bore & stroke: 3.88 x 3.75 in. Displacement: 350 cid. Compression ratio: 8.20:1. Brake hp: 250 at 4400 rpm. Taxable hp: 48.0. Torque: 350 lbs.-ft. at 2400 rpm. Five main bearings. Hydraulic valve lifters. Crankcase capacity: 5 qt. (add 1 qt. for new filter). Cooling system capacity with heater: 19.50 qt. Carburetor: Rochester two-barrel. Engine code: WR, WU, YU, XR, WN, WP, YN, and YP.

Catalina Brougham base V-8; Catalina, LeMans, Firebird Esprit optional V-8
Overhead valve. Cast-iron block. Bore & stroke: 4.125 x 3.75 in. Displacement: 400 cid. Compression ratio: 8.2:1. Brake horsepower: 265 at 4400 rpm. Net hp: 180 at 3800 rpm. Taxable hp: 54.3. Torque: 400 lbs.-ft. at 2400 rpm. Net torque: 320 lbs.-ft. at 2200 rpm. Five main bearings. Hydraulic valve lifters. Crankcase capacity: 5 qt. Cooling system capacity with heater: (Firebird and LeMans) 18.5 qt.; (Catalina) 18 qt. (add 1 qt. for new filter). Carburetor: Four-barrel. Engine code: WS, WX, XX, and XY.

Bonneville base V-8; Catalina optional V-8
Overhead valve. Cast-iron block. Bore & stroke: 4.15 x 4.21 in. Displacement: 455 cid. Compression ratio: 8.2:1. Brake hp: 280 at 4400 rpm. Net hp: 190 at 3200 rpm. Taxable hp: 54.5. Torque: 455 lbs.-ft. at 2000 rpm. Net torque: 360 lbs.-ft. at 2000 rpm. Five main bearings. Hydraulic valve lifters. Crankcase capacity: 5 qt. (add 1 qt. for new filter). Cooling system capacity with heater: 17.25 qt. Carburetor: Rochester four-barrel. Engine code: WG and YG.

GTO, Grand Prix, Firebird Formula base V-8; Catalina, LeMans optional V-8
Overhead valve. Cast-iron block. Bore & stroke: 4.12 x 3.75 in. Displacement: 400 cid. Compression ratio: 8.2:1. Brake hp: 300 at 4800 rpm. Net hp: 255 at 4400 rpm. Taxable hp: 54.3. Torque: 400 lbs.-ft. at 3600 rpm. Net torque: 340 lbs.-ft. at 3200 rpm. Five main bearings. Hydraulic valve lifters. Crankcase capacity: 5 qt. (add 1 qt. for new filter). Cooling system capacity with heater: (Grand Prix) 18.6 qt.; (Firebird) 18.50 qt. Carburetor: Rochester four-barrel. Engine code: WT, WK, and YS.

Grand Ville base V-8; Firebird Formula optional V-8
Overhead valve. Cast-iron block. Bore & stroke: 4.15 x 4.21 in. Displacement: 455 cid. Compression ratio: 8.2:1. Brake hp: 325 at 4400 rpm. Net hp: 255 at 4000 rpm. Taxable hp: 54.5. Torque: 455 lbs.-ft. at 3200 rpm. Net torque: 360 lbs.-ft. at 2800 rpm. Five main bearings. Hydraulic valve lifters. Crankcase capacity: 5 qt. (add 1 qt. for new filter). Cooling system capacity with heater: 17.25 qt. Carburetor: Rochester four-barrel. Engine code: WJ and YC.

Trans Am base V-8; GTO, LeMans optional V-8
Overhead valve. Cast-iron block. Bore & stroke: 4.15 x 4.21 in. Displacement: 455 cid. Compression ratio: 8.4:1. Brake hp: 335 at 4800 rpm. Taxable hp: 54.5. Torque: 480 lbs.-ft. at 360 rpm. Five main bearings. Hydraulic valve lifters. Crankcase capacity: 5 qt. (add 1 qt. for new filter). Cooling system capacity with heater: 17.25 qt. Carburetor: Rochester four-barrel. Engine code: WL, WC, YE, and YA.

1972 PONTIAC

NOTE: Beginning in 1972 the fifth symbol in the Pontiac Vehicle Identification Number identified the engine.

Ventura II base V-8 (Made by Chevrolet)
Overhead valve. Cast-iron block. Bore & stroke: 3.88 x 3.25 in. Displacement: 307 cid. Compression ratio: 8.50:1. Net hp: 130 at 4400 rpm. Taxable hp: 48.00. Net Torque: 230 lbs.-ft. at 2400 rpm. Five main bearings. Hydraulic valve lifters. Crankcase capacity: 4 qt. (add 1 qt. for new filter). Cooling system capacity with heater: 15 qt. Carburetor: Rochester two-barrel. VIN code: F.

Firebird Esprit, Luxury Lemans base V-8; Ventura, Firebird, LeMans optional V-8
Overhead valve. Cast-iron block. Bore & stroke: 3.88 x 3.75 in. Displacement: 350 cid. Compression ratio: 8.50:1. Net hp: 160 at 4400 rpm. Taxable hp: 48.00. Net Torque: 270 lbs.-ft. at 2000 rpm. Five main bearings. Hydraulic valve lifters. Crankcase capacity: 5 qt. (add 1 qt. for new filter). Cooling system capacity with heater: (Firebird) 19.4 qt.; (LeMans) 20.2 qt. Carburetor: Rochester two-barrel. Single exhaust. VIN code: M.

Firebird 350 base V-8; Firebird, LeMans optional V-8
Overhead valve. Cast-iron block. Bore & stroke: 3.88 x 3.53 in. Displacement: 350 cid. Compression ratio: 8.0:1. Net hp: 175 at 4400 rpm. Taxable hp: 48.00. Torque: 275 lbs.-ft. at 2000 rpm. Five main bearings. Hydraulic valve lifters. Carburetor: Rochester two-barrel. Crankcase capacity: 5 qt. (add 1 qt. for new filter). Cooling system capacity with heater: 20.2 qt. Carburetor: Rochester four-barrel. Dual exhaust. VIN code: N.

Catalina base V-8; LeMans optional V-8
Overhead valve. Cast-iron block. Bore & stroke: 4.12 x 3.75 in. Displacement: 400 cid. Compression ratio: 8.2:1. Taxable hp: 54.30. Net hp: 175 at 4000 rpm. Net torque: 310 lbs.-ft. at 2400 rpm. Five main bearings. Hydraulic valve lifters. Crankcase capacity: 5 qt. (add 1 qt. for new filter). Cooling system capacity with heater: 18.6 qt. Carburetor: Rochester two-barrel. Single exhaust. VIN code: R.

Bonneville base V-8; Catalina optional V-8
Overhead valve. Cast-iron block. Bore & stroke: 4.15 x 4.21 in. Displacement: 455 cid. Compression ratio: 8.2:1. Net hp: 185 at 4000 rpm. Taxable hp: 55.20. Net Torque: 350 lbs.-ft. at 2000 rpm. Five main bearings. Hydraulic valve lifters. Crankcase capacity: 5 qt. (add 1 qt. for new filter). Cooling system capacity with heater: 17.9 qt. Carburetor: Rochester four-barrel. Single exhaust. VIN code: V.

Optional LeMans, Catalina V-8
Overhead valve. Cast-iron block. Bore & stroke: 4.12 x 3.75 in. Displacement: 400 cid. Compression ratio: 8.2:1. Net hp: 200 at 4000

rpm. Taxable hp: 54.30. Net torque: 325 lbs.-ft. at 2400 rpm. Five main bearings. Hydraulic valve lifters. Crankcase capacity: 5 qt. (add 1 qt. for new filter). Cooling system capacity with heater: 18.6 qt. Carburetor: Rochester two-barrel. Dual exhaust. VIN code: P.

Optional LeMans, Catalina V-8
Overhead valve. Cast-iron block. Bore & stroke: 4.12 x 3.75 in. Displacement: 400 cid. Compression ratio: 8.2:1. Net hp: 200 at 4000 rpm. Taxable hp: 54.30. Net torque: 295 lbs.-ft. at 2800 rpm. Five main bearings. Hydraulic valve lifters. Crankcase capacity: 5 qt. (add 1 qt. for new filter). Cooling system capacity with heater: 18.6 qt. Carburetor: Rochester four-barrel. Single exhaust. VIN code: S.

Catalina, Bonneville optional V-8
Overhead valve. Cast-iron block. Bore & stroke: 4.15 x 4.21 in. Displacement: 455 cid. Compression ratio: 8.2:1. Net hp: 200 at 4000 rpm. Taxable hp: 55.20. Net Torque: 370 lbs.-ft. at 2000 rpm. Five main bearings. Hydraulic valve lifters. Crankcase capacity: 5 qt. (add 1 qt. for new filter). Cooling system capacity with heater: 17.9 qt. Carburetor: Rochester two-barrel. Single exhaust. VIN code: U.

Grand Ville base V-8; Catalina, Bonneville optional V-8
Overhead valve. Cast-iron block. Bore & stroke: 4.15 x 4.21 in. Displacement: 455 cid. Compression ratio: 8.2:1. Net hp: 220 at 3600 rpm. Taxable hp: 55.20. Net Torque: 350 lbs.-ft. at 2400 rpm. Five main bearings. Hydraulic valve lifters. Crankcase capacity: 5 qt. (add 1 qt. for new filter). Cooling system capacity with heater: 17.9 qt. Carburetor: Rochester four-barrel. Dual exhaust. VIN code: W.

LeMans, GTO optional V-8
Overhead valve. Cast-iron block. Bore & stroke: 4.15 x 4.21 in. Displacement: 455 cid. Compression ratio: 8.2:1. Net hp: 230 at 4400 rpm. Taxable hp: 55.20. Net Torque: 360 lbs.-ft. at 2800 rpm. Five main bearings. Hydraulic valve lifters. Crankcase capacity: 5 qt. (add 1 qt. for new filter). Cooling system capacity with heater: 17.9 qt. Carburetor: Rochester four-barrel. Dual exhaust. VIN code: X.

Firebird Formula 400, GTO, Grand Prix base V-8; LeMans, Catalina optional V-8
Overhead valve. Cast-iron block. Bore & stroke: 4.12 x 3.75 in. Displacement: 400 cid. Compression ratio: 8.2:1. Net hp: 250 at 4400 rpm. Taxable hp: 54.30. Net torque: 325 lbs.-ft. at 3200 rpm. Five main bearings. Hydraulic valve lifters. Crankcase capacity: 5 qt. (add 1 qt. for new filter). Cooling system capacity with heater: 18.6 qt. Carburetor: Rochester four-barrel. Dual exhaust. VIN code: T.

Grand Prix, Catalina, Bonneville, Grand Ville optional V-8
Overhead valve. Cast-iron block. Bore & stroke: 4.15 x 4.21 in. Displacement: 455 cid. Compression ratio: 8.2:1. Net hp: 250 at 3600 rpm. Taxable hp: 55.20. Net Torque: 375 lbs.-ft. at 2400 rpm. Five main bearings. Hydraulic valve lifters. Crankcase capacity: 5 qt. (add 1 qt. for new filter). Cooling system capacity with heater: 17.9 qt. Carburetor: Rochester four-barrel. VIN code: Y.

LeMans, GTO optional V-8
Overhead valve. Cast-iron block. Bore & stroke: 4.15 x 4.21 in. Displacement: 455 cid. Compression ratio: 8.2:1. Net hp: 250 at 3600 rpm. Taxable hp: 55.20. Net Torque: 370 lbs.-ft. at 2400 rpm. Five main bearings. Hydraulic valve lifters. Crankcase capacity: 5 qt. (add 1 qt. for new filter). Cooling system capacity with heater: 17.9 qt. Carburetor: Rochester four-barrel. VIN code: Y.

Trans Am, Firebird Formula 455 base V-8; LeMans, GTO optional V-8
Overhead valve. Cast-iron block. Bore & stroke: 4.15 x 4.21 in. Displacement: 455 cid. Compression ratio: 8.4:1. Net hp: 300 at 4000 rpm. Taxable hp: 55.20. Net torque: 415 lbs.-ft. at 3200 rpm. Five main bearings. Hydraulic valve lifters. Crankcase capacity: 5 qt. (add 1 qt. for new filter). Cooling system capacity with heater: 17.9 qt. Carburetor: Rochester four-barrel. VIN code: X.

1973 PONTIAC

Base Catalina, Luxury LeMans, Firebird, Esprit, Formula V-8
Overhead valve. Cast-iron block. Bore & stroke: 3.88 x 3.53 in. Displacement: 350 cid. Compression ratio: 7.6:1. Net hp: 150 at 4000 rpm. Taxable hp: 48.00. Net torque: 270 lbs.-ft. at 2000 rpm. Five main bearings. Hydraulic valve lifters. Crankcase capacity: (Ventura) 4 qt.; (Others) 5 qt. (add 1 qt. for new filter). Cooling system capacity with heater: (Ventura, Firebird) 19.5 qt.; (Others) 20.2 qt. Carburetor: Rochester two-barrel. Engine code: M.

Bonneville, Safari, Grand Am base V-8; LeMans, Catalina Firebird Esprit optional V-8
Overhead valve. Cast-iron block. Bore & stroke: 4.12 x 3.75 in. Displacement: 400 cid. Compression ratio: 8.00:1. Net hp: 170 at 3600 rpm. Taxable hp: 54.30. Net torque: 320 lbs.-ft. at 2000 rpm. Five main bearings. Hydraulic valve lifters. Crankcase capacity: 5 qt. (add 1 qt. for new filter). Cooling system capacity with heater: 18.6 qt. Carburetor: Rochester two-barrel. Engine code: R.

Firebird Formula base 350 V-8; Ventura, LeMans, Catalina optional V-8
Overhead valve. Cast-iron block. Bore & stroke: 3.88 x 3.53 in. Displacement: 350 cid. Compression ratio: 7.6:1. Net hp: 175 at 4400 rpm. Taxable hp: 48.00. Net torque: 280 lbs.-ft. at 2400 rpm. Five main bearings. Hydraulic valve lifters. Crankcase capacity: (Ventura) 4 qt.; (Others) 5 qt. (add 1 qt. for new filter). Cooling system capacity with heater: (Ventura, Firebird) 19.5 qt.; (Others) 20.2 qt. Dual exhausts. Carburetor: Rochester two-barrel. Engine code: N.

Catalina, LeMans, Grand Am optional V-8
Overhead valve. Cast-iron block. Bore & stroke: 4.12 x 3.75 in. Displacement: 400 cid. Compression ratio: 8.00:1. Net hp: 185 at 4000 rpm. Taxable hp: 54.30. Net torque: 320 lbs.-ft. at 2400 rpm. Five main bearings. Hydraulic valve lifters. Crankcase capacity: 5 qt. (add 1 qt. for new filter). Cooling system capacity with heater: 18.6 qt. Dual exhausts. Carburetor: Rochester two-barrel. Engine code: P.

Grand Safari base V-8; Catalina, Bonneville optional V-8
Overhead valve. Cast-iron block. Bore & stroke: 4.12 x 3.75 in. Displacement: 400 cid. Compression ratio: 8.00:1. Net hp: 200 at 4000 rpm. Taxable hp: 54.30. Net torque: 310 lbs.-ft. at 2400 rpm. Five main bearings. Hydraulic valve lifters. Crankcase capacity: 5 qt. (add 1 qt. for new filter). Cooling system capacity with heater: 18.6 qt. Dual exhausts. Carburetor: Rochester two-barrel. Engine code: S.

Grand Ville base V-8; Catalina, Bonneville optional V-8
Overhead valve. Cast-iron block. Bore & stroke: 4.15 x 4.21 in. Displacement: 455 cid. Compression ratio: 8.00:1. Net hp: 215 at 3600 rpm. Taxable hp: 55.20. Net torque: 350 lbs.-ft. at 2400 rpm. Five main bearings. Hydraulic valve lifters. Crankcase capacity: 5 qt. (add 1 qt. for new filter). Cooling system capacity with heater: 18.0 qt. Carburetor: Rochester four-barrel. Engine code: W.

GTO, Grand Prix base V-8; LeMans, Grand Am, Formula 400, Catalina Bonneville optional V-8
Overhead valve. Cast-iron block. Bore & stroke: 4.12 x 3.75 in. Displacement: 400 cid. Compression ratio: 8.00:1. Net hp: 230 at 4400 rpm. Taxable hp: 54.30. Net torque: 325 lbs.-ft. at 3200 rpm. Five main bearings. Hydraulic valve lifters. Crankcase capacity: 5 qt. (add 1 qt. for new filter). Cooling system capacity with heater: 18.6 qt. Carburetor: Rochester two-barrel. Engine code: T.

Trans Am base V-8; Catalina, Bonneville, GTO, Grand Ville, Grand Prix, LeMans, Formula optional-8
Overhead valve. Cast-iron block. Bore & stroke: 4.15 x 4.21 in. Displacement: 455 cid. Compression ratio: 8.00:1. Net hp: 250 at 4000 rpm. Taxable hp: 55.20. Net torque: 370 lbs.-ft. at 2800 rpm. Five main bearings. Hydraulic valve lifters. Crankcase capacity: 5 qt. (add 1 qt. for new filter). Cooling system capacity with heater: 18.0 qt. Dual exhaust. Carburetor: Rochester four-barrel. Engine code: Y.

Trans Am, Formula optional V-8
Overhead valve. Cast-iron block. Bore & stroke: 4.15 x 4.21 in. Displacement: 455 cid. Compression ratio: 8.4:1. Net hp: 290 at 4000 rpm. Taxable hp: 55.20. Net torque: 395 lbs.-ft. at 3200 rpm. Five main bearings. Hydraulic valve lifters. Crankcase capacity: 5 qt. (add 1 qt. for new filter). Cooling system capacity with heater: 18.0 qt. Dual exhaust. Carburetor: Rochester four-barrel. Engine code: X.

Trans Am, Formula optional "Super-Duty SD-455" V-8
Overhead valve. Cast-iron block. Bore & stroke: 4.15 x 4.21 in. Displacement: 455 cid. Compression ratio: 8.4:1. Net hp: 310 at 4000 rpm. Taxable hp: 55.20. Net torque: 390 lbs.-ft. at 3600 rpm. Five main bearings. Hydraulic valve lifters. Crankcase capacity: 5 qt. (add 1 qt. for new filter). Cooling system capacity with heater: 18.0 qt. Dual exhaust. Carburetor: Rochester four-barrel. Engine code: X.

1974 PONTIAC

Ventura, Ventura Custom, Federal LeMans Safari, Esprit, Formula base V-8
Overhead valve. Cast-iron block. Bore & stroke: 3.88 x 3.53 in. Displacement: 350 cid. Compression ratio: 7.6:1. Net hp: 155 at 3600 rpm. Taxable hp: 48.00. Net torque: 275 lbs.-ft. at 2400 rpm. Five main

bearings. Hydraulic valve lifters. Crankcase capacity: 5 qt. Cooling system capacity with heater: (Ventura) 20 qt.; (Firebird) 22 qt.; (LeMans) 21.3 qt. Carburetor: Two-barrel. Not available in LeMans Safaris sold in California. Engine code: M.

LeMans, Ventura optional V-8
Overhead valve. Cast-iron block. Bore & stroke: 3.88 x 3.53 in. Displacement: 350 cid. Compression ratio: 7.6:1. Net hp: 170 at 4000 rpm. Taxable hp: 48.00. Net torque: 290 lbs.-ft. at 2400 rpm. Five main bearings. Hydraulic valve lifters. Crankcase capacity: 5 qt. Cooling system capacity with heater: (Ventura) 20 qt.; (LeMans) 21.3 qt. Dual exhausts. Carburetor: Two-barrel. Engine code: N.

Firebird Formula 350 base V-8; LeMans, Firebird optional V-8
Overhead valve. Cast-iron block. Bore & stroke: 3.88 x 3.53 in. Displacement: 350 cid. Compression ratio: 7.6:1. Net hp: 170 at 4000 rpm. Taxable hp: 48.00. Net torque: 280 lbs.-ft. at 2000 rpm. Five main bearings. Hydraulic valve lifters. Crankcase capacity: 5 qt. Cooling system capacity with heater: (Ventura) 20 qt.; (Firebird) 22 qt.; (LeMans) 21.3 qt. Dual exhausts. Carburetor: Two-barrel. Engine code: N.

Grand Am, Catalina, Bonneville, Safari base V-8; LeMans, Firebird Esprit optional V-8
Overhead valve. Cast-iron block. Bore & stroke: 4.12 x 3.75 in. Displacement: 400 cid. Compression ratio: 8.0:1. Net hp: 175 at 3600 rpm. Taxable hp: 54.30. Net torque: 315 lbs.-ft. at 2000 rpm. Five main bearings. Hydraulic valve lifters. Crankcase capacity: 5 qt. Cooling system capacity with heater: (Firebird) 21. 9 qt.; (LeMans) 21.3 qt.; (Other) 21.6 qt. Carburetor: Two-barrel. Engine code: R.

LeMans, LeMans Sport optional V-8 (California)
Overhead valve. Cast-iron block. Bore & stroke: 3.88 x 3.53 in. Displacement: 350 cid. Compression ratio: 7.6:1. Net hp: 185 at 4000 rpm. Taxable hp: 48.00. Net torque: 320 lbs.-ft. at 4400 rpm. Five main bearings. Hydraulic valve lifters. Crankcase capacity: 5 qt. Cooling system capacity with heater: 21.3 qt. Carburetor: Four-barrel. Engine code: J.

LeMans, Grand Am, Firebird Esprit, Firebird Formula 400 optional V-8
Overhead valve. Cast-iron block. Bore & stroke: 4.12 x 3.75 in. Displacement: 400 cid. Compression ratio: 8.00:1. Net hp: 190 at 4000 rpm. Taxable hp: 54.30. Net torque: 330 lbs.-ft. at 2400 rpm. Five main bearings. Hydraulic valve lifters. Crankcase capacity: 5 qt. Cooling system capacity with heater: (Firebird) 21. 9 qt.; (LeMans) 21.3 qt. Dual exhausts. Carburetor: Two-barrel. Engine code: P.

GTO base V-8; Ventura, LeMans optional V-8
Overhead valve. Cast-iron block. Bore & stroke: 3.88 x 3.53 in. Displacement: 350 cid. Compression ratio: 7.6:1. Net hp: 200 at 4400 rpm. Taxable hp: 48.00. Net torque: 295 lbs.-ft. at 2800 rpm. Five main bearings. Hydraulic valve lifters. Crankcase capacity: 5 qt. Cooling system capacity with heater: (Ventura) 20 qt; (LeMans) 21.3 qt. Dual exhaust. Carburetor: Four-barrel. Engine code: K.

Optional Catalina, Bonneville, LeMans, LeMans Sport, Luxury LeMans V-8
Overhead valve. Cast-iron block. Bore & stroke: 4.12 x 3.75 in. Displacement: 400 cid. Compression ratio: 8.00:1. Net hp: 200 at 4000 rpm. Taxable hp: 54.30. Net torque: 320 lbs.-ft. at 2400 rpm. Five main bearings. Hydraulic valve lifters. Crankcase capacity: 5. Cooling system capacity with heater: (LeMans) 21.3 qt.; (Pontiac, Grand Prix) 21.6 qt. Not available in LeMans Safaris, Catalinas and Bonnevilles sold in California. Carburetor: Rochester two-barrel. Engine code: S.

Catalina, Bonneville optional V-8
Overhead valve. Cast-iron block. Bore & stroke: 4.15 x 4.21 in. Displacement: 455 cid. Compression ratio: 8.00:1. Net hp: 215 at 3600 rpm. Taxable hp: 55.20. Net torque: 355 lbs.-ft. at 2400 rpm. Five main bearings. Hydraulic valve lifters. Crankcase capacity: 5 qt. Cooling system capacity with heater: 19.8 qt. Carburetor: Four-barrel. Engine code: W.

Grand Prix, Trans Am base V-8; Formula 400, LeMans, Grand Am optional V-8
Overhead valve. Cast-iron block. Bore & stroke: 4.12 x 3.75 in. Displacement: 400 cid. Compression ratio: 8.00:1. Net hp: 225 at 4000 rpm. Taxable hp: 54.30. Net torque: 330 lbs.-ft. at 2800 rpm. Five main bearings. Hydraulic valve lifters. Crankcase capacity: 5 qt. Cooling system capacity with heater: (Firebird) 21. 9 qt.; (LeMans) 21.3 qt.; (Grand Prix) 21.6 qt. Not available in LeMans Safaris, Catalinas and Bonnevilles sold in California. Carburetor: Rochester two-barrel. Engine code: S.

Grand Prix SJ base V-8; Catalina, Bonneville, Grand Ville, Grand Prix, Federal LeMans, Federal LeMans Sport, Federal Luxury LeMans, Grand Am, Formula optional V-8
Overhead valve. Cast-iron block. Bore & stroke: 4.15 x 4.21 in. Displacement: 455 cid. Compression ratio: 8.00:1. Net hp: 250 at 4000 rpm. Taxable hp: 55.20. Net torque: 380 lbs.-ft. at 2800 rpm. Five main bearings. Hydraulic valve lifters. Crankcase capacity: 5 qt. Cooling system capacity with heater: (Firebird) 20.3 qt.; (LeMans) 21.1 qt.; (Grand Prix) 20.2 qt.); (Pontiac) 19.8 qt. Unknown. Dual exhaust. Carburetor: Four-barrel. Engine code: Y.

Trans Am, Formula optional "Super-Duty SD-455" V-8
Overhead valve. Cast-iron block. Bore & stroke: 4.15 x 4.21 in. Displacement: 455 cid. Compression ratio: 8.4:1. Net hp: 290 at 4000 rpm. Taxable hp: 55.20. Net torque: 395 lbs.-ft. at 3200 rpm. Five main bearings. Hydraulic valve lifters. Crankcase capacity: 5 qt. Cooling system capacity with heater: 20.3 qt. Dual exhaust. Carburetor: Four-barrel. Engine code: X.

1975 PONTIAC

Ventura base V-8 (Made by Oldsmobile)
Overhead valve. Cast-iron block. Bore & stroke: 3.50 x 3.385 in. Displacement: 260 cid. Compression ratio: 8.5:1. Net hp: 110 at 3400 rpm. Taxable hp: 39.20. Net torque: 205 lbs.-ft. at 1600 rpm. Five main bearings. Hydraulic valve lifters. Crankcase capacity: 4 qt. (Add 1 qt. with new filter). Cooling system capacity with heater: 22.4 qt. Carburetor: Rochester two-barrel Model 2GC. Engine code: F.

Ventura optional V-8 (made by Buick)
Overhead valve. Cast-iron block. Bore & stroke: 3.80 x 3.85 in. Displacement: 350 cid. Compression ratio: 8.0:1. Net hp: 145 at 3200 rpm. Taxable hp: 48.00. Net torque: 270 lbs.-ft. at 2000 rpm. Five main bearings. Hydraulic valve lifters. Crankcase capacity: 4 qt. Cooling system capacity with heater: 20 qt. Carburetor: Two-barrel. Engine code: E.

LeMans (federal), Firebird base V-8
Overhead valve. Cast-iron block. Bore & stroke: 3.88 x 3.53 in. Displacement: 350 cid. Compression ratio: 7.6:1. Net hp: 155 at 3600 rpm. Taxable hp: 48.00. Net torque: 280 lbs.-ft. at 2000 rpm. Five main bearings. Hydraulic valve lifters. Crankcase capacity: 5 qt. Cooling system capacity with heater: (LeMans) 21.3 qt.; (Firebird) 22 qt. Carburetor: Two-barrel. Engine code: M.

Ventura optional V-8 (made by Buick)
Overhead valve. Cast-iron block. Bore & stroke: 3.80 x 3.85 in. Displacement: 350 cid. Compression ratio: 8.0:1. Net hp: 165 at 3800 rpm. Taxable hp: 48.00. Net torque: 260 lbs.-ft. at 2200 rpm. Five main bearings. Hydraulic valve lifters. Crankcase capacity: 4 qt. Cooling system capacity with heater: 20 qt. Carburetor: Four-barrel. Engine code: H.

Catalina, Bonneville, Grand Am base V-8; Grand Ville optional V-8
Overhead valve. Cast-iron block. Bore & stroke: 4.12 x 3.75 in. Displacement: 400 cid. Compression ratio: 7.6:1. Net hp: 170 at 4000 rpm. Taxable hp: 54.30. Net torque: 305 lbs.-ft. at 2000. Five main bearings. Hydraulic valve lifters. Crankcase capacity: 5 qt. Cooling system capacity with heater: (Grand Am) 21.3 qt.; (Others) 21.6 qt. Carburetor: Two-barrel. Engine code: R.

LeMans (California), Firebird, Formula base V-8
Overhead valve. Cast-iron block. Bore & stroke: 3.88 x 3.53 in. Displacement: 350 cid. Compression ratio: 7.6:1. Net hp: 175 at 4000 rpm. Taxable hp: 48.00. Net torque: 280 lbs.-ft. at 2000 rpm. Five main bearings. Hydraulic valve lifters. Crankcase capacity: 5 qt. Cooling system capacity with heater: (LeMans) 21.3 qt.; (Firebird) 22 qt. Dual exhaust. Carburetor: Four-barrel. Engine code: J.

Grand Ville, Grand Prix, Trans Am base V-8; Formula, LeMans, Catalina, Bonneville optional V-8
Overhead valve. Cast-iron block. Bore & stroke: 4.12 x 3.75 in. Displacement: 400 cid. Compression ratio: 7.6:1. Net hp: 185 at 3600 rpm. Taxable hp: 54.30. Net torque: 310 lbs.-ft. at 1600 rpm. Five main bearings. Hydraulic valve lifters. Crankcase capacity: 5 qt. Cooling system capacity with heater: (Firebird) 21.3 qt.; (Grand Prix) 21.5 qt.; (Others) 21.6 qt. Carburetor: Two-barrel. Engine code: S.

Grand Am, Grand Prix, Catalina, Bonneville, Grand Ville, Trans Am optional V-8
Overhead valve. Cast-iron block. Bore & stroke: 4.15 x 4.21 in. Displacement: 455 cid. Compression ratio: 8.00:1. Net hp: 200 at 3500 rpm. Taxable hp: 55.20. Net torque: 330 lbs.-ft. at 2000 rpm. Five main bearings. Hydraulic valve lifters. Crankcase capacity: 5 qt. Cooling system capacity with heater: (LeMans) 19.9 qt.; (Grand Prix) 20.2 qt.; (Firebird) 23.3 qt.; (Pontiac) 19.8 qt. Dual exhaust. Carburetor: Four-barrel. Engine code: Y.

1976 PONTIAC

Ventura, LeMans base V-8
Overhead valve. Cast-iron block and head. Bore & stroke: 3.50 x 3.39 in. Displacement: 260 cid (4.3 liters). Compression ratio: 7.5:1. Brake horsepower: 110 at 3400 rpm. Taxable hp: 39.20. Torque: 205 lbs.-ft. at 1600 rpm. Five main bearings. Hydraulic valve lifters. Crankcase capacity: 4 qt. Cooling system capacity: (Ventura) 20.6 qt.; (LeMans) 23.5 qt. Carburetor: Rochester two-barrel. VIN code: F.

Ventura optional V-8
Overhead valve. Cast-iron block and head. Bore & stroke: 3.80 x 3.85 in. Displacement: 350 cid (5.7 liters). Compression ratio: 8.0:1. Brake horsepower: 140 at 3200 rpm. Taxable hp: 48.00. Torque: 280 lbs.-ft. at 1600 rpm. Five main bearings. Hydraulic valve lifters. Crankcase capacity: 4 qt. Cooling system capacity: 17.3 qt. Carburetor: Rochester two-barrel.

Catalina, Bonneville, LeMans Safari base V-8; Ventura optional V-8
Overhead valve. Cast-iron block and head. Bore & stroke: 3.80 x 3.85 in. Displacement: 350 cid (5.7 liters). Compression ratio: 8.0:1. Brake horsepower: 155 at 3400 rpm. Taxable hp: 48.00. Torque: 280 lbs.-ft. at 1800 rpm. Five main bearings. Hydraulic valve lifters. Crankcase capacity: 5 qt. Cooling system capacity: (Pontiac) 21.6 qt.; (LeMans) 21.4 qt.; (Ventura) 17.3 qt. Carburetor: Rochester four-barrel.

Grand Prix, Firebird, Formula base V-8; Firebird, LeMans optional V-8
Overhead valve. Cast-iron block and head. Bore & stroke: 3.88 x 3.75 in. Displacement: 350 cid (5.7 liters). Compression ratio: 7.6:1. Brake horsepower: 160 at 4000 rpm. Taxable hp: 48.00. Torque: 280 lbs.-ft. at 2000 rpm. Five main bearings. Hydraulic valve lifters. Crankcase capacity: 5 qt. Cooling system capacity: (Grand Prix) 21.6 qt.; (Firebird) 21.2 qt.; (LeMans) 21.4 qt. Carburetor: Rochester two-barrel. VIN code: H or M.

LeMans, Grand Prix optional V-8
Overhead valve. Cast-iron block and head. Bore & stroke: 4.12 x 3.75 in. Displacement: 400 cid (6.6 liters). Compression ratio: 7.6:1. Brake horsepower: 170 at 4000 rpm. Taxable hp: 54.30. Torque: 310 lbs.-ft. at 1600 rpm. Five main bearings. Hydraulic valve lifters. Crankcase capacity: 5 qt. Cooling system capacity with heater: (Grand Prix) 22.2 qt.; (LeMans) 21.4 qt. Carburetor: Rochester two-barrel. VIN code: R.

LeMans (California), Firebird, Formula base V-8
Overhead valve. Cast-iron block. Bore & stroke: 3.88 x 3.53 in. Displacement: 350 cid. Compression ratio: 7.6:1. Net hp: 165 at 4000 rpm. Taxable hp: 48.00. Net torque: 260 lbs.-ft. at 2400 rpm. Five main bearings. Hydraulic valve lifters. Crankcase capacity: 5 qt. Cooling system capacity with heater: (LeMans) 21.3 qt.; (Firebird) 22 qt. Dual exhaust. Carburetor: Four-barrel. Engine code: J.

Trans Am, Bonneville Brougham, Grand Prix SJ base V-8; LeMans, Firebird, Grand Prix, Catalina, Bonneville optional V-8
Overhead valve. Cast-iron block and head. Bore & stroke: 4.12 x 3.75 in. Displacement: 400 cid (6.6 liters). Compression ratio: 7.6:1. Brake horsepower: 185 at 3600 rpm. Taxable hp: 54.30. Torque: 310 lbs. ft. at 1600 rpm. Five main bearings. Hydraulic valve lifters. Crankcase capacity: 5 qt. Cooling system capacity with heater: (Grand Prix) 22.2 qt.; (LeMans) 21.4 qt; (Firebird) 21.6 qt.; (Pontiac) 21.6 qt. Carburetor: Rochester four-barrel. VIN code: S.

Catalina, Grand Safari base V-8; LeMans, Firebird, Grand Prix, Catalina, Bonneville optional V-8
Overhead valve. Cast-iron block and head. Bore & stroke: 4.15 x 4.21 in. Displacement: 455 cid (7.5 liters). Compression ratio: 7.6:1. Brake horsepower: 200 at 3500 rpm. Taxable hp: 55.20. Torque: 330 lbs.-ft. at 2000 rpm. Five main bearings. Hydraulic valve lifters. Crankcase capacity: 5 qt. Cooling system capacity with heater: (Grand Prix) 22.2 qt.; (LeMans) 21.6 qt; (Firebird) 23.3 qt.; (Pontiac) 22.1 qt. Carburetor: Rochester four-barrel. VIN code: W.

1977 PONTIAC

Formula, Bonneville, Grand Prix base V-8; Ventura, Firebird, LeMans, Catalina optional V-8
Overhead valve. Cast-iron block and head. Bore & stroke: 4.00 x 3.00 in. Displacement: 301 cid (4.9 liters). Compression ratio: 8.2:1. Brake horsepower: 135 at 4000 rpm. Taxable hp: 51.2. Torque: 235-245 lbs.-ft. at 2000 rpm. (full-size: 250 at 1600). Five main bearings. Hydraulic valve lifters. Carburetor: two-barrel. VIN code: Y.

Ventura, Firebird optional V-8
Overhead valve. Cast-iron block and head. Bore & stroke: 3.74 x 3.48 in. Displacement: 305 cid (5.0 liters). Taxable hp: 44.7. Compression ratio: 8.5:1. Brake horsepower: 140 at 3800 rpm. Taxable hp: 44.7. Torque: 245 lbs.-ft. at 2400 rpm. Five main bearings. Hydraulic valve lifters. Carburetor: two-barrel Rochester 2GC. VIN code: U.

Firebird, LeMans, Catalina, Bonneville, Grand Prix optional V-8
Overhead valve. Cast-iron block and head. Bore & stroke: 3.88 x 3.75 in. Displacement: 350 cid (5.7 liters). Compression ratio: 7.6:1. Brake horsepower: 170 at 4000 rpm. Taxable hp: 51.2. Torque: 280 lbs.-ft. at 1800 rpm. Five main bearings. Hydraulic valve lifters. Carburetor: Four-barrel Rochester M4MC. VIN code: P.

Firebird, LeMans, Catalina, Bonneville, Grand Prix alternate V-8
Overhead valve. Cast-iron block and head. Bore & stroke: 4.00 x 3.48 in. Displacement: 350 cid (5.7 liters). Compression ratio: 8.0:1. Brake horsepower: 170 at 3800 rpm. Taxable hp: 51.2. Torque: 275 lbs.-ft. at 2000 rpm. Five main bearings. Hydraulic valve lifters. Carburetor: Four-barrel Rochester M4MC. VIN code: L. Chevrolet-built.

Alternate Firebird, LeMans, Catalina, Bonneville, Grand Prix V-8
Overhead valve. Cast-iron block and head. Bore & stroke: 4.06 x 3.38 in. Displacement: 350 cid (5.7 liters). Compression ratio: 8.0:1. Brake horsepower: 170 at 3800 rpm. Taxable hp: 51.2. Torque: 275 lbs.-ft. at 2000 rpm. Five main bearings. Hydraulic valve lifters. Carburetor: Four-barrel Rochester M4MC. VIN code: R. Oldsmobile-built.

Grand Prix SJ base V-8; Firebird, LeMans, Catalina, Bonneville, Grand Prix optional V-8
Overhead valve. Cast-iron block and head. Bore & stroke: 4.12 x 3.75 in. Displacement: 400 cid (6.6 liters). Compression ratio: 7.6:1. Brake horsepower: 180 at 3600 rpm. Taxable hp: 54.30. Torque: 325 lbs.-ft. at 1600 rpm. Five main bearings. Hydraulic valve lifters. Carburetor: Four-barrel Rochester M4MC. VIN code: Z.

Firebird, LeMans, Catalina, Bonneville, Grand Prix alternate V-8
Overhead valve. Cast-iron block and head. Bore & stroke: 4.35 x 3.38 in. Displacement: 403 cid (6.6 liters). Compression ratio: 8.0:1. Brake horsepower: 185 at 3600 rpm. Taxable hp: 60.6. Torque: 320 lbs.-ft. at 2200 rpm. Five main bearings. Hydraulic valve lifters. Carburetor: Four-barrel Rochester M4MC. VIN code: K.

Trans Am base V-8
Overhead valve. Cast-iron block and head. Bore & stroke: 4.12 x 3.75 in. Displacement: 400 cid (6.6 liters). Compression ratio: 8.0:1. Brake horsepower: 200 at 3600 rpm. Taxable hp: 54.30. Torque: 325 at 2400 rpm. Five main bearings. Hydraulic valve lifters. Carburetor: Four-barrel Rochester M4MC. VIN code: Z.

1978 PONTIAC

Grand Am, Bonneville, Grand Prix LJ base V-8; Grand Prix, Catalina optional V-8
Overhead valve. Cast-iron block and head. Bore & stroke: 4.00 x 3.00 in. Displacement: 301 cid (4.9 liters). Compression ratio: 8.2:1. Brake horsepower: 140 at 3600 rpm. Taxable hp: 51.2. Torque: 235 lbs.-ft. at 2000 rpm. Five main bearings. Hydraulic valve lifters. Carburetor: Two-barrel Rochester M2MC. VIN code: Y.

Formula base V-8; Phoenix, Firebird, LeMans, Grand Prix SJ, Grand Prix LJ optional V-8
Overhead valve. Cast-iron block and head. Bore & stroke: 3.74 x 3.48 in. Displacement: 305 cid (5.0 liters). Compression ratio: 8.4:1. Brake horsepower: 145 at 3800 rpm. Taxable hp: 44.7. Torque: 245 lbs.-ft. at 2400 rpm. Five main bearings. Hydraulic valve lifters. Carburetor: Two-barrel Rochester 2GC. VIN code: U.

Grand Prix SJ base V-8; Grand Am, Grand Prix optional V-8
Overhead valve. Cast-iron block and head. Bore & stroke: 4.00 x 3.00 in. Displacement: 301 cid (4.9 liters). Compression ratio: 8.2:1. Brake horsepower: 150 at 4000 rpm. Taxable hp: 51.2. Torque: 239 lbs.-ft. at 2000 rpm. Five main bearings. Hydraulic valve lifters. Carburetor: Two-barrel Rochester M2MC. VIN code: W.

Firebird, Catalina, Bonneville optional V-8
Overhead valve. Cast-iron block and head. Bore & stroke: 3.80 x 3.85 in. Displacement: 350 cid (5.7 liters). Compression ratio: 8.0:1. Brake

horsepower: 155 at 3400 rpm. Taxable hp: 51.2. Torque: 280 at 1800 rpm. Five main bearings. Hydraulic valve lifters. Carburetor: Four-barrel Rochester M4MC. VIN code: X. Buick-built.

Firebird, Catalina, Bonneville alternate V-8
Overhead valve. Cast-iron block and head. Bore & stroke: 4.00 x 3.48 in. Displacement: 350 cid (5.7 liters). Compression ratio: 8.2:1. Brake horsepower: 170 at 3800 rpm. Taxable hp: 51.2.Torque: 270 lbs.-ft. at 2400 rpm. Five main bearings. Hydraulic valve lifters. Carburetor: Four-barrel Rochester M4MC. VIN code: L. Chevrolet-built.

Firebird, Catalina, Bonneville alternate V-8
Overhead valve. Cast-iron block and head. Bore & stroke: 4.06 x 3.38 in. Displacement: 350 cid (5.7 liters). Compression ratio: 7.9:1. Brake horsepower: 170 at 3800 rpm. Taxable hp: 51.2. Torque: 275 lbs.-ft. at 2000 rpm. Five main bearings. Hydraulic valve lifters. Carburetor: Four-barrel Rochester M4MC. VIN code: R. Oldsmobile-built.

Trans Am base V-8; Formula, Catalina, Bonneville optional V-8
Overhead valve. Cast-iron block and head. Bore & stroke: 4.12 x 3.75 in. Displacement: 400 cid (6.6 liters). Compression ratio: 7.7:1. Brake horsepower: 180 at 3600 rpm. Taxable hp: 54.3. Torque: 325 lbs.-ft. at 1600 rpm. Five main bearings. Hydraulic valve lifters. Carburetor: Four-barrel Rochester M4MC. VIN code: Z.

Trans Am, Formula optional V-8
Overhead valve. Cast-iron block. Bore & stroke: 4.35 x 3.385 in. Displacement: 403 cid. Compression ratio: 8.0:1. Brake horsepower: 185 at 3600 rpm. Taxable hp: 60.6. Torque: 320 lbs.-ft. at 2000 rpm. Five main bearings. Hydraulic valve lifters. Carburetor: Four-barrel. Mandatory standard equipment in Trans Ams sold in California and high-altitude counties optional equipment in Formulas sold in California and high-altitude counties. VIN code: K.

Trans Am, Formula optional V-8
Overhead valve. Cast-iron block and head. Bore & stroke: 4.12 x 3.75 in. Displacement: 400 cid (6.6 liters). Compression ratio: 8.1:1. Brake horsepower: 220 at 4000 rpm. Taxable hp: 54.30. Torque: 320 lbs.-ft. at 2800 rpm. Five main bearings. Hydraulic valve lifters. Carburetor: Four-barrel Rochester M4MC. VIN code: Z.

1979 PONTIAC

Bonneville, Grand Prix LJ, Safari base V-8; Firebird, LeMans, Grand Prix, Catalina optional V-8
Overhead valve. Cast-iron block and head. Bore & stroke: 4.06 x 3.04 in. Displacement: 301 cid (4.9 liters). Compression ratio: 8.1:1. Brake horsepower: 140 at 3600 rpm. Taxable hp: 51.2. Torque: 235 lbs.-ft. at 2000 rpm. Five main bearings. Hydraulic valve lifters. Carburetor: Two-barrel Rochester M2MC. VIN code: Y.

Sunbird, Phoenix optional V-8
Overhead valve. Cast-iron block and head. Bore & stroke: 3.74 x 3.48 in. Displacement: 305 cid (5.0 liters). Compression ratio: 8.4:1. Brake horsepower: 145 at 3800 rpm. Taxable hp: 44.7. Torque: 245 lbs.-ft. at 2400 rpm. Five main bearings. Hydraulic valve lifters. Carburetor: Two-barrel Rochester 2GC. VIN code: U.

Firebird, Formula, Grand Prix SJ base V-8; Firebird, Trans Am, LeMans, Grand Am, Bonneville, Catalina, Grand Prix optional V-8
Overhead valve. Cast-iron block and head. Bore & stroke: 4.06 x 3.04 in. Displacement: 301 cid (4.9 liters). Compression ratio: 8.1:1. Brake horsepower: 150 at 4000 rpm. Taxable hp: 51.2. Torque: 240 lbs.-ft. at 2000 rpm. Five main bearings. Hydraulic valve lifters. Carburetor: Four-barrel Rochester M4MC. VIN code: W.

Sunbird, Phoenix, Firebird alternate V-8
Overhead valve. Cast-iron block and head. Bore & stroke: 3.74 x 3.48 in. Displacement: 305 cid (5.0 liters). Compression ratio: 8.4:1. Brake horsepower: 150 at 4000 rpm. Taxable hp: 44.7. Torque: 225 lbs.-ft. at 2400 rpm. Five main bearings. Hydraulic valve lifters. Carburetor: Two-barrel Rochester 2GC. VIN code: H.

Phoenix, Firebird, Catalina, Bonneville, LeMans Safari alternate V-8
Overhead valve. Cast-iron block and head. Bore & stroke: 3.80 x 3.85 in. Displacement: 350 cid (5.7 liters). Compression ratio: 8.0:1. Brake horsepower: 155 at 3400 rpm. Taxable hp: 51.2. Torque: 280 lbs.-ft. at 1800 rpm. Five main bearings. Hydraulic valve lifters. Carburetor: Four-barrel Rochester M4MC. VIN code: X. Built by Buick.

Phoenix, Firebird, Catalina, Bonneville, LeMans Safari optional V-8
Overhead valve. Cast-iron block and head. Bore & stroke: 4.00 x 3.48 in. Displacement: 350 cid (5.7 liters). Compression ratio: 8.2:1. Brake horsepower: 160 at 3800 rpm. Taxable hp: 51.2. Torque: 260 lbs.-ft. at 2400 rpm. Five main bearings. Hydraulic valve lifters. Carburetor: Four-barrel Rochester M4MC. VIN code: L. Chevrolet-built.

Here's a view of the right-hand side of the 1979 TA 6.6-litre V-8 in a white car. The "shaker" type air scoop that rose up through the hood was painted to match each car's body color and the engine call-out decals were color-coordinated. The engine block, cylinder heads, intake manifold, and valve covers were finished in a medium blue color.

Displacing 400-cid, the "TA 6.6-litre" V-8 of 1979 was available only as an option in Trans Am and Formula models. This Pontiac-built V-8 had a 4.12 x 3.75-inch bore and stroke, a four-barrel carb, 8.1:1 compression ratio, and 220 hp at 4000 rpm. It should not be confused with the Oldsmobile-built 6.6-liter V-8 offered the same year. The latter engine had a 4.35 x 3.38-inch bore and stroke and developed only 185 hp.

Here's a view of the left-hand side of the 1979 TA 6.6-litre V-8 in the white car. On this side, the numerical portion of the engine call-out decal is towards the front of the car. This in-the-car view depicts a nicely-detailed engine compartment.

Phoenix, Firebird, Catalina, Bonneville, LeMans Safari alternate V-8
Overhead valve. Cast-iron block and head. Bore & stroke: 4.06 x 3.38 in. Displacement: 350 cid (5.7 liters). Compression ratio: 7.9:1. Brake horsepower: 160 at 3600 rpm. Taxable hp: 51.2. Torque: 270 lbs.-ft. at 2000 rpm. Five main bearings. Hydraulic valve lifters. Carburetor: Four-barrel Rochester M4MC. VIN code: R. Built by Oldsmobile.

Trans Am base V-8; Formula optional V-8
Overhead valve. Cast-iron block and head. Bore & stroke: 4.35 x 3.38 in. Displacement: 403 cid (6.6 liters). Compression ratio: 7.9:1. Brake horsepower: 185 at 3600 rpm. Taxable hp: 60.6. Torque: 320 lbs.-ft. at 2000 rpm. Five main bearings. Hydraulic valve lifters. Carburetor: Four-barrel Rochester M4MC. VIN code: K. Oldsmobile-built.

Formula, Trans Am optional V-8
Overhead valve. Cast-iron block and head. Bore & stroke: 4.12 x 3.75 in. Displacement: 400 cid (6.6 liters). Compression ratio: 8.1:1. Brake horsepower: 220 at 4000 rpm. Taxable hp: 54.30. Torque: 320 lbs.-ft. at 2800 rpm. Five main bearings. Hydraulic valve lifters. Carburetor: Four-barrel Rochester M4MC. VIN code: Z.

1980 PONTIAC

Catalina, Bonneville, full-size Safari optional diesel V-8
Overhead valve. Cast-iron block and head. Bore & stroke: 4.06 x 3.39 in. Displacement: 350 cid (5.7 liters). Compression ratio: 22.5:1. Brake horsepower: 105 at 3200 rpm. Taxable hp: 52.7. Torque: 205 lbs.-ft. at 1600 rpm. Five main bearings. Hydraulic valve lifters. Induction: Fuel injection. VIN code: N. Oldsmobile-built.

One of the most desirable modern Pontiac V-8s is the 1980 Trans Am Turbocharged V-8. This engine was a $350 option in Trans Ams and a $530 option in Formula Firebirds. The motor is a 301-cid V-8 fitted with a AirResearch Corporation IBO305 turbocharger. Only 23,422 cars were built with this engine and this one is seen here on display at the Indianapolis Speedway "Hall of Fame" Museum.

Firebird, LeMans, Grand Prix, Catalina, Bonneville optional V-8
Overhead valve. Cast-iron block and head. Bore & stroke: 3.75 x 3.00 in. Displacement: 265 cid (4.3 liters). Compression ratio: 8.3:1. Brake horsepower: 120 at 3600 rpm. Taxable hp: 45.00. Torque: 210 lbs.-ft. at 1600 rpm. Five main bearings. Hydraulic valve lifters. Carburetor: Two-barrel. VIN code: S.

Introduced on certain Pontiac models in 1980, the General Motors Computer Command Control (CCC) engine management system became standard on all Pontiacs in 1981, except those with the diesel V-8 engine. This drawing illustrates the main components of the system.

Formula, Trans Am base V-8; Firebird, LeMans, Grand Prix, Catalina, Bonneville optional V-8
Overhead valve. Cast-iron block and head. Bore & stroke: 4.06 x 3.04 in. Displacement: 301 cid (4.9 liters). Compression ratio: 8.1:1. Brake horsepower: 140 at 4000 rpm. Taxable hp: 51.2. Torque: 240 lbs.-ft. at 1800 rpm. Five main bearings. Hydraulic valve lifters. Carburetor: Four-barrel Rochester. VIN code: W.

Firebird, LeMans, Grand Am, California Grand Prix base V-8
Overhead valve. Cast-iron block and head. Bore & stroke: 3.74 x 3.48 in. Displacement: 305 cid (5.0 liters). Compression ratio: 8.4:1. Brake horsepower: 150 at 3800 rpm. Taxable hp: 44.7. Torque: 230 lbs.-ft. at 2400 rpm. Five main bearings. Hydraulic valve lifters. VIN code: H.

Formula, Trans Am, Grand Am, Grand Prix SJ Electronic Engine Control optional V-8
Overhead valve. Cast-iron block and head. Bore & stroke: 4.06 x 3.04 in. Displacement: 301 cid (4.9 liters). Compression ratio: 8.1:1. Brake horsepower: 155 at 4400 rpm. Taxable hp: 51.2. Torque: 240 lbs.-ft. at 2200 rpm. Five main bearings. Hydraulic valve lifters. Carburetor: Four-barrel Carburetor: Four-barrel Rochester. VIN code: T.

Safari and California full-size Pontiac optional V-8
Overhead valve. Cast-iron block and head. Bore & stroke: 4.06 x 3.39 in. Displacement: 350 cid (5.7 liters). Compression ratio: 8.0:1. Brake horsepower: 160 at 3600 rpm. Taxable hp: 52.7. Torque: 270 lbs.-ft. at 2000 rpm. Five main bearings. Hydraulic valve lifters. Carburetor: Four-barrel Rochester E4MC. VIN code: R. Oldsmobile-built.

Formula, Trans Am optional Turbocharged V-8
Overhead valve. Cast-iron block and head. Bore & stroke: 4.06 x 3.04 in. Displacement: 301 cid (4.9 liters). Compression ratio: 7.6:1. Brake horsepower: 210 at 4000 rpm. Taxable hp: 51.2. Torque: 345 lbs.-ft. at 1800 rpm. Five main bearings. Hydraulic valve lifters. Carburetor: Rochester. VIN code: W.

1981 PONTIAC

Grand Prix, Catalina, Bonneville, Full-size Safari optional diesel V-8
Overhead valve. Cast-iron block and head. Bore & stroke: 4.06 x 3.39 in. Displacement: 350 cid (5.7 liters). Compression ratio: 22.5:1. Brake horsepower: 105 at 3200 rpm. Taxable hp: 52.7. Torque: 205 lbs.-ft. at 1600 rpm. Five main bearings. Hydraulic valve lifters. Induction: Fuel injection. VIN code: N. Oldsmobile-built.

Formula base V-8; Firebird, LeMans, Grand Prix, Catalina, Bonneville optional V-8
Overhead valve. Cast-iron block and head. Bore & stroke: 3.75 x 3.00 in. Displacement: 265 cid (4.3 liters). Compression ratio: 8.3:1. Brake horsepower: 120 at 4000 rpm. Taxable hp: 45.00. Torque: 205 lbs.-ft. at 2000 rpm. Five main bearings. Hydraulic valve lifters. Carburetor: Two-barrel Rochester E2ME. VIN code: S.

Safari base V-8; Firebird, Catalina, Bonneville, LeMans Safari optional V-8
Overhead valve. Cast-iron block and head. Bore & stroke: 4.00 x 3.00 in. Displacement: 301 cid (4.9 liters). Compression ratio: 8.1:1. Brake horsepower: 135 at 3600 rpm. Taxable hp: 51.2. Torque: 235 lbs.-ft. at 1600 rpm. Five main bearings. Hydraulic valve lifters. Carburetor: Four-barrel Rochester E4ME. VIN code: W.

Safari base V-8; Catalina, Bonneville optional V-8
Overhead valve. Cast-iron block and head. Bore & stroke: 3.80 x 3.38 in. Displacement: 307 cid (5.0 liters). Compression ratio: 8.0:1. Brake horsepower: 145 at 3800 rpm. Taxable hp: 46.2. Torque: 240 lbs.-ft. at 2400 rpm. Five main bearings. Hydraulic valve lifters. Carburetor: Four-barrel Rochester E4ME. VIN code: Y. Oldsmobile-built.

Formula, Trans Am optional V-8
Overhead valve. Cast-iron block and head. Bore & stroke: 3.74 x 3.48 in. Displacement: 305 cid (5.0 liters). Compression ratio: 8.4:1. Brake horsepower: 145 at 3800 rpm. Taxable hp: 44.7. Torque: 240 lbs.-ft. at 2400 rpm. Five main bearings. Hydraulic valve lifters. Carburetor: Four-barrel Rochester E4ME. VIN code: H. Chevrolet-built.

Trans Am base V-8; Firebird optional V-8
Overhead valve. Cast-iron block and head. Bore & stroke: 4.00 x 3.00 in. Displacement: 301 cid (4.9 liters). Compression ratio: 8.1:1. Brake horsepower: 150 at 4000 rpm. Taxable hp: 51.2. Torque: 245 lbs.-ft. at 2000 rpm. Five main bearings. Hydraulic valve lifters. Carburetor: Four-barrel Rochester E4ME. VIN code: W.

Formula, Trans Am optional Turbocharged V-8
Overhead valve. Cast-iron block and head. Bore & stroke: 4.00 x 3.00 in. Displacement: 301 cid (4.9 liters). Compression ratio: 7.5:1. Brake horsepower: 200 at 4000 rpm. Taxable hp: 51.2. Torque: 340 lbs.-ft. at 2000 rpm. Five main bearings. Hydraulic valve lifters. Carburetor: Four-barrel Rochester E4ME. VIN code: T.

1982 PONTIAC

Trans Am base V-8; Firebird optional V-8
Overhead valve. Cast-iron block and head. Bore & stroke: 3.74 x 3.48 in. Displacement: 305 cid (5.0 liters). Compression ratio: 8.6:1. Brake horsepower: 145 at 4000 rpm. Taxable hp: 44.7. Torque: 240 lbs.-ft. at 2000 rpm. Five main bearings. Hydraulic valve lifters. Carburetor: Four-barrel Rochester E4ME. VIN code: H. Chevrolet-built.

Trans Am optional V-8
Overhead valve. Cast-iron block and head. Bore & stroke: 3.74 x 3.48 in. Displacement: 305 cid (5.0 liters). Compression ratio: 9.5:1. Brake horsepower: 165 at 4200 rpm. Taxable hp: 44.7. Torque: 240 lbs.-ft. at 2400 rpm. Five main bearings. Hydraulic valve lifters. Induction: Throttle Body Injected. VIN code: 7. Chevrolet-built.

Grand Prix, Bonneville optional diesel V-8
Overhead valve. Cast-iron block and head. Bore & stroke: 4.06 x 3.39 in. Displacement: 350 cid (5.7 liters). Compression ratio: 21.6:1. Brake horsepower: 105 at 3200 rpm. Taxable hp: 52.7. Torque: 205 lbs.-ft. at 1600 rpm. Five main bearings. Hydraulic valve lifters. Induction: Fuel injection. VIN code: N. Oldsmobile-built.

1983 PONTIAC

Trans Am base V-8; Firebird, Grand Prix, Bonneville optional V-8
Overhead valve. Cast-iron block and head. Bore & stroke: 3.74 x 3.48 in. Displacement: 305 cid (5.0 liters). Compression ratio: 8.6:1. Brake horsepower: 150 at 4000 rpm. Taxable hp: 44.7. Torque: 240 lbs.-ft. at 2400 rpm. Five main bearings. Hydraulic valve lifters. Carburetor: Four-barrel Rochester E4ME. VIN code: H. Chevrolet-built.

Trans Am optional V-8
Overhead valve. Cast-iron block and head. Bore & stroke: 3.74 x 3.48 in. Displacement: 305 cid (5.0 liters). Compression ratio: 9.5:1. Brake horsepower: 175 at 4200 rpm. Taxable hp: 44.7. Torque: 250 lbs.-ft. at 2800 rpm. Five main bearings. Hydraulic valve lifters. Induction: Crossfire Fuel Injection. VIN code: 7. Chevrolet-built.

Grand Prix, Bonneville, Parisienne optional diesel V-8
Overhead valve. Cast-iron block and head. Bore & stroke: 4.06 x 3.39 in. Displacement: 350 cid (5.7 liters). Compression ratio: 21.6:1. Brake horsepower: 105 at 3200 rpm. Taxable hp: 52.7. Torque: 200 lbs.-ft. at 1600 rpm. Five main bearings. Hydraulic valve lifters. Induction: Fuel injection. VIN code: N. Oldsmobile-built.

1984 PONTIAC

Trans Am base V-8; Firebird, Grand Prix, Bonneville optional V-8
Overhead valve. Cast-iron block and head. Bore & stroke: 3.74 x 3.48 in. Displacement: 305 cid (5.0 liters). Compression ratio: 8.6:1. Brake horsepower: 150 at 4000 rpm. Taxable hp: 44.7. Torque: 240 lbs.-ft. at 2400 rpm. Five main bearings. Hydraulic valve lifters. Carburetor: Four-barrel Rochester E4ME. VIN code: H. Chevrolet-built.

Trans Am optional H.O. V-8
Overhead valve. High Output. Cast-iron block and head. Bore & stroke: 3.74 x 3.48 in. Displacement: 305 cid (5.0 liters). Compression ratio: 9.5:1. Brake horsepower: 190 at 4800 rpm. Taxable hp: 44.7. Torque: 240 lbs.-ft. at 3200 rpm. Five main bearings. Hydraulic valve lifters. Carburetor: Four-barrel Rochester E4ME. VIN code: G. Chevrolet-built.

Grand Prix, Bonneville, Parisienne optional diesel V-8
Overhead valve. Cast-iron block and head. Bore & stroke: 4.06 x 3.39 in.

Displacement: 350 cid (5.7 liters). Compression ratio: 22.5:1. Brake horsepower: 105 at 3200 rpm. Taxable hp: 52.7. Torque: 200 lbs.-ft. at 1600 rpm. Five main bearings. Hydraulic valve lifters. Induction: Fuel injection. VIN code: N. Oldsmobile-built.

1985 PONTIAC

Trans Am base V-8; Firebird, Grand Prix, Bonneville, Parisienne optional V-8
Overhead valve. Cast-iron block and head. Bore & stroke: 3.74 x 3.48 in. Displacement: 305 cid (5.0 liters). Compression ratio: 9.5:1. Brake horsepower: 165 at 4200 rpm. Taxable hp: 44.7. Torque: 250 lbs.-ft. at 2400 rpm. Five main bearings. Hydraulic valve lifters. Carburetor: Four-barrel. VIN code: H. Chevrolet-built.

Trans Am optional V-8
Overhead valve. High Output. Cast-iron block and head. Bore & stroke: 3.74 x 3.48 in. Displacement: 305 cid (5.0 liters). Compression ratio: 9.5:1. Horsepower: 190 at 4800 rpm. Taxable hp: 44.7. Torque: 240 lbs.-ft. at 3200 rpm. VIN code: G. Optional V-8 Trans Am Fuel-injected version of 305-cid Horsepower: 205 at 4400 rpm. Torque: 275 lbs.-ft. at 3200 rpm. VIN code: F. Chevrolet-built.

Parisienne optional diesel V-8
Overhead valve. Cast-iron block and head. Bore & stroke: 4.06 x 3.39 in. Displacement: 350 cid (5.7 liters). Compression ratio: 22.5:1. Brake horsepower: 105 at 3200 rpm. Taxable hp: 52.7. Torque: 200 lbs.-ft. at 1600 rpm. Five main bearings. Hydraulic valve lifters. Induction: Fuel injection. VIN code: N. Oldsmobile-built.

1986 PONTIAC

Firebird, Grand Prix, Bonneville, Parisienne optional V-8
Overhead valve. Cast-iron block and head. Bore & stroke: 3.74 x 3.48 in. Displacement: 305 cid (5.0 liters). Compression ratio: 9.5:1. Brake horsepower: 150 at 4200 rpm. Taxable hp: 44.7. Torque: 235-245 lbs.-ft. at 2000-2400 rpm. Five main bearings. Hydraulic valve lifters. Carburetor: Four-barrel. VIN code: H. Chevrolet-built.

Trans Am base V-8
Overhead valve. Cast-iron block and head. Bore & stroke: 3.74 x 3.48 in. Displacement: 305 cid (5.0 liters). Compression ratio: 9.5:1. Brake horsepower: 155 at 4000. Taxable hp: 44.7. Torque: 235-245 lbs.-ft. at 2000-2400 rpm. Five main bearings. Hydraulic valve lifters. Carburetor: Four-barrel. VIN code: H. Chevrolet-built.

Parisienne wagon base V-8
Overhead valve. Cast-iron block and head. Bore & stroke: 3.74 x 3.48 in. Displacement: 305 cid (5.0 liters). Compression ratio: 9.5:1. Brake horsepower: 165 at 4200. Taxable hp: 44.7. Torque: 235-245 lbs.-ft. at 2000-2400 rpm. Five main bearings. Hydraulic valve lifters. Carburetor: Four-barrel. VIN code: H. Chevrolet-built.

Trans Am optional V-8
Overhead valve. Cast-iron block and head. Bore & stroke: 3.74 x 3.48 in. Displacement: 305 cid (5.0 liters). Compression ratio: 9.5:1. Horsepower: 190 at 4800 rpm. Taxable hp: 44.7. Torque: 240 lbs.-ft. at 3200 rpm. Five main bearings. Hydraulic valve lifters. Carburetor: Four-barrel. VIN code: G. Chevrolet-built.

Trans Am optional V-8
Overhead valve. Cast-iron block and head. Bore & stroke: 3.74 x 3.48 in. Displacement: 305 cid (5.0 liters). Compression ratio: 9.5:1. Horsepower: 205 at 4000 rpm. Taxable hp: 44.7. Torque: 270 lbs.-ft. at 3200 rpm. Five main bearings. Hydraulic valve lifters. Induction: Fuel-injected. VIN code: F. Chevrolet-built.

1987 PONTIAC

Pontiac Safari base V-8
Overhead valve. Cast-iron block. Bore & stroke: 3.80 x 3.85 in. Displacement: 307 cid (5.0L). Brake horsepower: 140 at 3200 rpm. Taxable hp: 46.20. Carburetor: Four-barrel. RPO code: LV2. VIN code: Y. Produced in U.S. and Canada.

Grand Prix base V-8
Overhead valve. Cast-iron block. Bore & stroke: 3.74 x 3.48 in. Displacement: 305 cid (5.0L). Brake horsepower: 150 at 4000 rpm. Taxable hp: 44.8. Carburetor: Four-barrel. RPO code: LG4. VIN code: H. Standard V-8 in all Grand Prix including 2+2. Produced in U.S. and Canada.

Firebird, Formula, Trans Am base V-8
Overhead valve. Cast-iron block and head. Displacement: 305 cid (5.0L). Bore & stroke: 3.74 x 3.48 in. Brake horsepower: 155 at 4200 rpm. Taxable hp: 44.8. Torque: 235-245 lbs.-ft. at 2000-2400 rpm. Carburetor: Four-barrel. RPO code: LG4. VIN code: E. Standard with five-speed manual in base Firebird V-8, Formula and Trans Am. Available with four-speed automatic in same models. Produced in U.S. or Canada.

Formula, Trans Am optional V-8; GTA delete-option V-8
Overhead valve. Cast-iron block and head. Bore & stroke: 3.74 x 3.48 in. Displacement: 305 cid (5.0L). Brake horsepower: 165 at 4400 rpm. Taxable hp: 44.8. Torque: 235 lbs.-ft. at 2000 rpm. Carburetor: Four-barrel. RPO code: LB9. VIN code: F. Produced in U.S. or Canada.

GTA base V-8
Overhead valve. Cast-iron block and head. Bore & stroke: 3.74 x 3.48 in. Displacement: 350 cid (5.7L). Brake horsepower: 210 at 4000 rpm. Taxable hp: 51.2. Torque: 315 lbs.-ft. at 3200 rpm. Carburetor: Four-barrel. RPO code: B2L. VIN code: 8. Standard with four-speed automatic in GTA (optional in Formula and Trans Am). Includes low-profile air induction system with aluminum plenum and individual aluminum tuned runners, an extruded dual fuel rail assembly with computer controlled fuel injectors, and a special low-restriction exhaust system.

1988 PONTIAC

Pontiac Safari base V-8
Overhead valve. Cast-iron block. Bore & stroke: 3.80 x 3.39 in. Displacement: 307 cid (5.0L). Compression ratio: 8.0.1. Brake horsepower: 140 at 3200 rpm. Taxable hp: 46.20. Torque: 255 lbs.-ft. at 2000 rpm. Carburetor: Four-barrel. RPO code: LV2. VIN code: Y.

Formula, Trans Am base V-8; Firebird optional V-8
Overhead valve. Cast-iron block and head. Aluminum intake manifold. Bore & stroke: 3.74 x 3.48 in. Displacement: 305 cid (5.0L). Brake horsepower: 170 at 4000 rpm. Taxable hp: 44.8. Torque: 225 lbs.-ft. at 2400 rpm. Compression ratio: 9.3:1. Induction: EFI/TBI. UPC code: L03. VIN code: E. Produced in U.S. or Canada.

Formula, Trans Am optional V-8; GTA delete option V-8
Overhead valve. Cast-iron block and head. Aluminum intake manifold. Bore & stroke: 3.74 x 3.48 in. Displacement: 305 cid (5.0L). Brake horsepower: 190 at 4000 rpm (automatic). Taxable hp: 44.8. Torque: 295 lbs.-ft. at 2800 rpm (automatic). Compression ratio: 9.3:1. Induction: TPI. UPC code: LB9. VIN code: F. Available with four-speed automatic in Formula and Trans Am (delete option in GTA).

Formula, Trans Am optional H.O. V-8; GTA delete option H.O. V-8
Overhead valve. Cast-iron block and head. Aluminum intake manifold. Bore & stroke: 3.74 x 3.48 in. Displacement: 305 cid (5.0L). Brake horsepower: 215 at 4400 rpm (manual). Taxable hp: 44.8. Torque: 285 lbs.-ft. at 3200 rpm (manual). Compression ratio: 9.3:1. Induction: TPI. UPC code: LB9. VIN code: F. Available with five-speed manual in Formula and Trans Am (delete option in GTA).

GTA base V-8
Overhead valve. Cast-iron block and head. Aluminum intake manifold. Bore & stroke: 4.00 x 3.48 in. Displacement: 350 cid (5.7L). Brake horsepower: 225 at 4400 rpm. Taxable hp: 51.2. Torque: 330 lbs.-ft. at 3200 rpm. Compression ratio: 9.3:1. Induction: EFI/TPI. UPC code: B2L. VIN code: 8. Standard with four-speed automatic in GTA (optional in Formula and Trans Am). Includes low-profile air induction system with aluminum plenum and individual aluminum tuned runners, an extruded dual fuel rail assembly with computer controlled fuel injectors and a special low-restriction single exhaust system.

1989 PONTIAC

Pontiac Safari base V-8
Overhead valve. Cast-iron block. Bore & stroke: 3.80 x 3.39 in. Displacement: 307 cid. (5.0L). Compression ratio: 8.0:1. Brake horsepower: 140 at 3200 rpm. Taxable hp: 46.20. Torque: 255 lbs.-ft. at 2000 rpm. Induction: Four-barrel carburetor. RPO code: LV2. Standard V-8 in all Safaris. VIN code: Y.

Firebird base V-8
Overhead valve. Cast-iron block and head. Aluminum intake manifold. Bore & stroke: 3.74 x 3.48 in. Displacement: 305 cid. (5.0L). Brake horsepower: 170 at 4000 rpm. Taxable hp: 44.8. Torque: 255 lbs.-ft. at 2400 rpm. Compression ratio: 9.3:1. Induction: EFI/TBI. RPO code: L03. VIN code: E or F. Produced in U.S. or Canada. Standard Formula and Trans Am. Available in base Firebird.

Formula, Trans Am optional V-8; GTA delete option V-8
Overhead valve. Cast-iron block and head. Aluminum intake manifold. Bore & stroke: 3.74 x 3.48 in. Displacement: 305 cid. (5.0L). Brake horsepower: 190 at 4000 rpm (automatic) or 225 at 4400 rpm (manual/dual exhausts). Taxable hp: 44.8. Torque: 295 lbs.-ft. at 2800 rpm (automatic) or 285 lbs.-ft. at 3200 rpm (manual). Compression ratio: 9.3:1. Induction: TPI. RPO code: LB9. VIN code: E or F. Available with five-speed manual in Formula and Trans Am (delete option in GTA). Available with four-speed automatic in same applications.

GTA base V-8
Overhead valve. Cast-iron block and head. Aluminum intake manifold. Bore & stroke: 4.00 x 3.48 in. Displacement: 350 cid. (5.7L). Brake horsepower: 235 at 4400 rpm. Taxable hp: 51.2. Torque: 330 lbs.-ft. at 3200 rpm. Compression ratio: 9.3:1. Induction: EFI/TPI. RPO code: B2L. VIN code: 8. Includes low-profile air induction system with aluminum plenum and individual aluminum tuned runners, an extruded dual fuel rail assembly with computer controlled fuel injectors and a special low-restriction single exhaust system.

1990 PONTIAC

Formula base V-8; Firebird optional V-8
Overhead valve. Cast-iron block and head. Aluminum intake manifold. Bore & stroke: 3.74 x 3.48 in. Displacement: 305 cid (5.0L). Brake horsepower: 170 at 4400 rpm. Taxable hp: 44.8. Torque: 255 lbs.-ft. at 2400 rpm. Compression ratio: 9.3:1. Induction: EFI/TBI. RPO code: L03. VIN code: E or F. Produced in U.S. or Canada. Standard Formula. Available in base Firebird.

Trans Am base V-8; Formula optional V-8; GTA delete option V-8
Overhead valve. Cast-iron block and head. Aluminum intake manifold. Bore & stroke: 3.74 x 3.48 in. Displacement: 305 cid (5.0L). Brake horsepower: 200 at 4400 rpm (Formula with automatic and Trans Am with five-speed manual) or 225 at 4600 rpm (GTA and Formula with five-speed manual). Taxable hp: 44.8. Torque: 300 at 3200 rpm (Formula with automatic and Trans Am with five-speed manual) or 290 at 3200 rpm (GTA and Formula with five-speed manual). Compression ratio: 9.3:1. Induction: EFI/TPI. RPO code: LB9. VIN code: E or F. Available with five-speed manual in Formula and Trans Am (delete option in GTA).

GTA base V-8; Formula, Trans Am optional V-8
Overhead valve. Cast-iron block and head. Aluminum intake manifold. Bore & stroke: 4.00 x 3.48 in. Displacement: 350 cid (5.7L). Brake horsepower: 235 at 4400 rpm. Taxable hp: 51.2. Torque: 340 lbs.-ft. at 3200 rpm. Compression ratio: 9.3:1. Induction: EFI/TPI. RPO code: B2L. VIN code: 8. Standard with four-speed automatic in GTA (optional in Formula and Trans Am). Includes low-profile air induction system with aluminum plenum and individual aluminum tuned runners, an extruded dual fuel rail assembly with computer controlled fuel injectors and a special low-restriction single exhaust system.

1991 PONTIAC

Formula base V-8; Firebird optional V-8
Overhead valve. Cast-iron block and head. Aluminum intake manifold. Bore & stroke: 3.74 x 3.48 in. Displacement: 305 cid (5.0L). Brake horsepower: 170 at 4400 rpm. Taxable hp: 44.8. Torque: 255 lbs.-ft. at 2400 rpm. Compression ratio: 9.3:1. Induction: EFI/TBI. RPO code: L03. VIN code: E or F. Standard Formula. Available in base Firebird.

Trans Am base V-8; Formula optional V-8; GTA delete option V-8
Overhead valve. Cast-iron block and head. Aluminum intake manifold. Bore & stroke: 3.74 x 3.48 in. Displacement: 305 cid (5.0L). Brake horsepower: 205 at 4200 rpm (Formula, Trans Am and GTA with automatic); 230 hp at 4400 rpm (all others). Taxable hp: 44.8. Torque: 300 lbs.-ft. at 3200 rpm (GTA/Formula with manual; optional Trans Am); Torque: 285 lbs.-ft. at 3200 rpm (GTA, Trans Am and Formula with five-speed manual). Compression ratio: 9.3:1. Induction: EFI/TPI. RPO code: LB9. VIN code: E or F. Available with five-speed manual in Formula and Trans Am (delete option in GTA).

GTA base V-8
Overhead valve. Cast-iron block and head. Aluminum intake manifold. Bore & stroke: 4.00 x 3.48 in. Displacement: 350 cid (5.7L). Brake horsepower: 240 at 4400 rpm. Taxable hp: 51.2. Torque: 340 lbs.-ft. at 3200 rpm. Compression ratio: 9.3:1. Induction: EFI/TPI. RPO code: B2L. VIN code: 8. Standard with four-speed automatic in GTA (optional in Formula and Trans Am). Includes Performance Enhancement group with dual catalytic converter exhausts, engine oil cooler, and performance axle ratio.

1992 PONTIAC

Formula base V-8; Firebird optional V-8
Overhead valve. Cast-iron block and head. Aluminum intake manifold. Bore & stroke: 3.74 x 3.48 in. Displacement: 305 cid (5.0L). Brake horsepower: 170 at 4000 rpm. Taxable hp: 44.8. Torque: 255 lbs.-ft. at 2400 rpm. Compression ratio: 9.3:1. Induction: EFI/TBI. RPO code: L03. VIN code: E or F. Standard Formula. Available in base Firebird.

Trans Am base V-8; Formula optional V-8; GTA delete option V-8
Overhead valve. Cast-iron block and head. Aluminum intake manifold. Bore & stroke: 3.74 x 3.48 in. Displacement: 305 cid (5.0L). Brake horsepower: 205 at 4200 rpm (Formula, Trans Am and GTA with automatic); 230 hp at 4400 rpm (all others). Taxable hp: 44.8. Torque: 300 lbs.-ft. at 3200 rpm (GTA/Formula with manual; optional Trans Am); Torque: 285 lbs.-ft. at 3200 rpm (GTA, Trans Am and Formula with five-speed manual). Compression ratio: 9.3:1. Induction: EFI/TPI. RPO code: LB9. VIN code: E or F. Available with five-speed manual in Formula and Trans Am (delete option in GTA).

GTA base V-8
Overhead valve. Cast-iron block and head. Aluminum intake manifold. Bore & stroke: 4.00 x 3.48 in. Displacement: 350 cid (5.7L). Brake horsepower: 240 at 4400 rpm. Taxable hp: 51.2. Torque: 340 lbs.-ft. at 3200 rpm. Compression ratio: 9.3:1. Induction: EFI/TPI. RPO code: B2L. VIN code: 8. Standard with four-speed automatic in GTA (optional in Formula and Trans Am). Includes Performance Enhancement group with dual catalytic converter exhausts, engine oil cooler and performance axle ratio.

1993 PONTIAC

Formula, Trans Am base V-8 (not available in base Firebird)
Overhead valve. Cast-iron block and head. Aluminum intake manifold. Bore & stroke: 4.00 x 3.48 in. Displacement: 350 cid (5.7 liter). Compression ratio: 10.5:1. Brake horsepower: 270 at 4800 rpm. Taxable hp: 51.2. Torque: 325 lbs.-ft. at 2400 rpm. Induction: PFI. UPC code: LT1. VIN code: P.

1994 PONTIAC

Formula, Trans Am base V-8 (not available in base Firebird)
Overhead valve. Cast-iron block and head. Aluminum intake manifold.

Bore & stroke: 4.00 x 3.48 in. Displacement: 350 cid (5.7 liter). Compression ratio: 10.5:1. Brake horsepower: 275 at 5000 rpm. Taxable hp: 51.2. Torque: 325 lbs.-ft. at 2400 rpm. Induction: SPFI. UPC code: LT1. VIN code: P.

1995 PONTIAC

Formula, Trans Am base V-8 (not available in base Firebird)
Overhead valve. Cast-iron block and head. Aluminum intake manifold. Bore & stroke: 4.00 x 3.48 in. Displacement: 350 cid (5.7 liter). Compression ratio: 10.5:1. Brake horsepower: 275 at 5000 rpm. Taxable hp: 51.2. Torque: 325 lbs.-ft. at 2400 rpm. Induction: SPFI. UPC code: LT1. VIN code: P.

1996 PONTIAC

Formula, Trans Am base V-8 (not available in base Firebird)
Overhead valve. Cast-iron block and head. Aluminum intake manifold. Bore & stroke: 4.00 x 3.48 in. Displacement: 350 cid (5.7L). Brake horsepower: 285 at 5000 rpm. Taxable hp: 51.2. Torque: 325 lbs.-ft. at 2400 rpm. Compression ratio: 10.5:1. Induction: SFI. UPC code: LT1. VIN code: P.

Formula, Trans Am optional Ram Air V-8
Overhead valve. Cast-iron block and head. Aluminum intake manifold. Bore & stroke: 4.00 x 3.48 in. Displacement: 350 cid (5.7L). Brake horsepower: 305 at 5000 rpm. Taxable hp: 51.2. Torque: 325 lbs.-ft. at 2400 rpm. Compression ratio: 10.5:1. Induction: SFI. UPC code: LT1. VIN code: 5. (With WS6 Ram Air Performance package.)

1997 PONTIAC

Formula, Trans Am base V-8 (not available in base Firebird)
Overhead valve. Cast-iron block and head. Aluminum intake manifold. Bore & stroke: 4.00 x 3.48 in. Displacement: 350 cid (5.7L). Brake horsepower: 285 at 5000 rpm. Taxable hp: 51.2. Torque: 325 lbs.-ft. at 2400 rpm. Compression ratio: 10.5:1. Induction: SFI. VIN code: 5. UPC code: LT1.

Formula, Trans Am optional Ram Air V-8
Overhead valve. Cast-iron block and head. Aluminum intake manifold. Bore & stroke: 4.00 x 3.48 in. Displacement: 350 cid (5.7L). Brake horsepower: 305 at 5000 rpm. Taxable hp: 51.2. Torque: 325 lbs.-ft. at 2400 rpm. Compression ratio: 10.5:1. Induction: SFI. VIN code: 5. UPC code: LT1. (With WS6 Ram Air Performance package.)

1998 PONTIAC

Formula, Trans Am base V-8 (not available in base Firebird)
Overhead valve. Aluminum block and head. Aluminum intake manifold. Bore & stroke: 3.90 x 3.62 in. Displacement: 346 cid (5.7L). Brake horsepower: 305 at 5200 rpm. Taxable hp: 49.00. Torque: 335 lbs.-ft. at 4000 rpm. Compression ratio: 10.5:1. Induction: SFI. VIN code: G. UPC code: LS1.

Formula, Trans Am optional Ram Air V-8
Overhead valve. Ram Air. Aluminum block and head. Aluminum intake manifold. Bore & stroke: 3.90 x 3.62 in. Displacement: 346 cid (5.7L). Brake horsepower: 320 at 5200 rpm. Taxable hp: 49.00. Torque: 345 lbs.-

By 1996, the only Pontiacs available with a V-8 engine were the Formula Firebird and the Trans Am. The base V-8 was a 5.7-liter (350-cid) 285-hp Chevy-built LT1 engine. Both models could also be had with the optional WS6 Ram Air Performance package seen here. The WS6 package increased the cars' output to 305 hp at 5000 rpm.

ft. at 4400 rpm. Compression ratio: 10.5:1. Induction: SFI. VIN code: G. UPC code: LS1.

1999 PONTIAC

Formula, Trans Am base V-8 (not available in base Firebird)
Overhead valve. Aluminum block and head. Aluminum intake manifold. Bore & stroke: 3.90 x 3.62 in. Displacement: 346 cid (5.7L). Brake horsepower: 305 at 5200 rpm. Taxable hp: 49.00. Torque: 335 lbs.-ft. at 4000 rpm. Compression ratio: 10.5:1. Induction: SFI. VIN code: G. UPC code: LS1.

Formula, Trans Am optional Ram Air V-8
Overhead valve. Ram Air. Aluminum block and head. Aluminum intake manifold. Bore & stroke: 3.90 x 3.62 in. Displacement: 346 cid (5.7L). Brake horsepower: 320 at 5200 rpm. Taxable hp: 49.00. Torque: 345 lbs.-ft. at 4400 rpm. Compression ratio: 10.5:1. Induction: SFI. VIN code: G. UPC code: LS1.

2000 PONTIAC

Formula, Trans Am base V-8 (not available in base Firebird)
Overhead valve. Aluminum block and head. Aluminum intake manifold. Bore & stroke: 3.90 x 3.62 in. Displacement: 346 cid (5.7L). Brake horsepower: 305 at 5200 rpm. Taxable hp: 49.00. Torque: 335 at 4000 rpm. Compression ratio: 10.5:1. Induction: SFI. VIN code: G. UPC code: LS1.

Formula, Trans Am optional Ram Air V-8
Overhead valve. Ram Air. Aluminum block and head. Aluminum intake manifold. Bore & stroke: 3.90 x 3.62 in. Displacement: 346 cid (5.7L). Brake horsepower: 320 at 5200 rpm. Taxable hp: 49.00. Torque: 345 at 4400 rpm. Compression ratio: 10.5:1. Induction: SFI. VIN code: G. UPC code: LS1.

2001 PONTIAC

Formula, Trans Am base V-8 (not available in base Firebird)
Overhead valve. Aluminum block and cylinder head. Aluminum intake manifold. Bore & stroke: 3.90 x 3.62 in. Displacement: 350 cid (5.7L). Brake horsepower: 310 at 5200 rpm. Taxable hp: 49.00. Torque: 340 lbs.-ft. at 4000 rpm. Compression ratio: 10.5:1. Induction: SFI. VIN code: G. UPC code: LS1.

Formula, Trans Am optional Ram Air V-8
Overhead valve. Aluminum block and cylinder head. Aluminum intake manifold. Ram-air induction. Bore & stroke: 3.90 x 3.62 in. Displacement: 350 cid (5.7L). Brake horsepower: 325 at 5200 rpm. Taxable hp: 49.00. Torque: 350 lbs.-ft. at 4000 rpm. Compression ratio: 10.5:1. Induction: SFI. VIN code: G. UPC code: LS1.

2002 PONTIAC

Formula, Trans Am base V-8 (not available in base Firebird)
Overhead valve. Aluminum block and cylinder head. Aluminum intake manifold. Bore & stroke: 3.90 x 3.62 in. Displacement: 350 cid (5.7L). Brake horsepower: 310 at 5200 rpm. Taxable hp: 49.00. Torque: 340 lbs.-ft. at 4000 rpm. Compression ratio: 10.5:1. Induction: SFI. VIN code: G. UPC code: LS1.

Formula, Trans Am optional Ram Air V-8
Overhead valve. Aluminum block and cylinder head. Aluminum intake manifold. Ram-air induction. Bore & stroke: 3.90 x 3.62 in. Displacement: 350 cid (5.7L). Brake horsepower: 325 at 5200 rpm. Taxable hp: 49.00. Torque: 350 lbs.-ft. at 4000 rpm. Compression ratio: 10.5:1. Induction: SFI. VIN code: G. UPC code: LS1.

Selected Pontiac V-8 Tune-Up Specs 1955-1965

Year	Model	Spark Plugs Type	Spark Plugs Gap	Point Gap	Deg Cam Angle	Timing	Idle Speed	Compression Pressure	Battery (Delco)
All V-8		AC44-5	.035	.016	30	5B	475 rpm	140	11MS/2SM

Note 1: Add 25 rpm to idle speed for cars with Hydra-Matic.

Year	Model	Spark Plugs Type	Spark Plugs Gap	Point Gap	Deg Cam Angle	Timing	Idle Speed	Compression Pressure	Battery (Delco)
1956	All V-8	AC44	.035	.016	30	5B	475 rpm	150	11MS/2SM

Note 2: Add 25 rpm to idle speed for cars with Hydra-Matic.

Year	Model	Spark Plugs Type	Spark Plugs Gap	Point Gap	Deg Cam Angle	Timing	Idle Speed	Compression Pressure	Battery (Delco)
1957	Normal V-8	AC44-5	.035	.016	30	6B	Note 3	160	11MS/2SM
	Extra HP V-8	AC44	.035	.016	30	6B	Note 3	160	11MS/2SM

Note 3: Idle speed in neutral is 460 rpm for cars with manual (synchromesh) transmission and 500 rpm for cars with Hydra-Matic transmission.

Year	Model	Spark Plugs Type	Spark Plugs Gap	Point Gap	Deg Cam Angle	Timing	Idle Speed	Compression Pressure	Battery (Delco)
1958	All V-8s	AC45	.033-.038	Note 4	28-32	6B	Note 5	NA	558

Note 4: Turn point adjusting screw in clockwise direction until engine misfires, then back off screw one-half turn.
Note 5: Idle speed in neutral is 460 rpm for cars with manual (synchromesh) transmission and 500 rpm for cars with Hydra-Matic transmission.

Year	Model	Spark Plugs Type	Spark Plugs Gap	Point Gap	Deg Cam Angle	Timing	Idle Speed	Compression Pressure	Battery (Delco)
1959	All V-8s	AC45	.033-.038	Note 6	28-32	6B	Note 7	NA	358

Note 6: Turn point adjusting screw in clockwise direction until engine misfires, then back off screw one-half turn..
Note 7: Set idle speed to 480-500 rpm for cars with manual transmission and Rochester 2GC carburetor; 480-500 rpm in drive for cars with Hydra-Matic and Rochester 2GC carburetor; 475 rpm in neutral for cars with Rochester 4GC carburetor; 550 rpm for cars with manual transmission and Carter AFB carburetor; 480-500 rpm for cars with Hydra-Matic and Carter AFB carburetor; and 480-500 rpm for cars with Tri-Power (in neutral for cars with manual transmission and in drive for cars with Hydra-Matic transmission).

Year	Model	Spark Plugs Type	Spark Plugs Gap	Point Gap	Deg Cam Angle	Timing	Idle Speed	Compression Pressure	Battery (Delco)
1960	All V-8s	AC45S	.033-.038	.016	28-32	6B	Note 8	NA	Note 9

Note 8: Set idle speed to 480-500 rpm for cars with manual transmission and Rochester 2GC carburetor; 480-500 rpm in drive for cars with Hydra-Matic and Rochester 2GC carburetor; 480-500 rpm for cars with manual transmission and Carter AFB carburetor; 480 rpm for cars with Hydra-Matic and Carter AFB carburetor and 480-500 rpm for cars with Tri-Power (in neutral for cars with manual transmission and in drive for cars with Hydra-Matic transmission).
Note 9: Delco Type 458 12-volt battery for cars with manual transmission; Delco Type 558 12-volt battery for cars with Hydra-Matic transmission and Delco Type 670 for cars requiring extra-duty starting power.

1961	All Tempest V-8s	AC45FFS	.033-.038	.013-.019	28-32	5B	160	580-600	1980454
	All Pontiac V-8s	AC45S	.035	.016	28-32	6B	Note 10	Note 11	Note 12

Note 10: Set idle speed to 480-500 rpm for cars with manual transmission and Rochester 2GC carburetor; 480-500 rpm in drive for cars with Hydra-Matic and Rochester 2GC carburetor; 480-500 rpm for cars with manual transmission and Carter AFB carburetor; 500-520 rpm for cars with Hydra-Matic and Carter AFB carburetor; and 480-500 rpm for cars with Tri-Power (in neutral for cars with manual transmission and in drive for cars with Hydra-Matic transmission).
Note 11: Compression (cranking) pressure is 140-150 lbs. for V-8s with 8.6:1 compression ratio, and 155-165 lbs. for V-8s with 10.25:1 or 10.75:1 compression ratios.
Note 12: Delco Type 458 (wet) or 459 (dry) 12-volt battery for cars with manual transmission, and Delco Type 558 (wet) or 559 (dry) 12-volt battery for cars with Hydra-Matic transmission.

1962	All Tempest V-8s	AC45FFS	.035	.013-.019	31-34	5B	155-165	525	1980454
	All Pontiac V-8s	AC45S	.035	.016	28-32	6B	Note 13	Note 14	Note 15

Note 13: Set idle speed to 480-500 rpm for cars with manual transmission and Rochester 2GC carburetor; 480-500 rpm in drive for cars with Hydra-Matic and Rochester 2GC carburetor; 480-500 rpm for cars with Hydra-Maticl transmission and Carter AFB carburetor; 1000 rpm for cars with Super-Duty engines; and 480-500 rpm for cars with Tri-Power (in neutral for cars with manual transmission and in drive for cars with Hydra-Matic transmission).

Note 14: Compression (cranking) pressure is 140-150 lbs. for V-8s with 8.6:1 compression ratio, and 155-165 lbs. for V-8s with 10.25:1 or 10.75:1 compression ratios.
Note 15: Delco Type 458 (wet) or 459 (dry) 12-volt battery for cars with manual transmission, and Delco Type 558 (wet) or 559 (dry) 12-volt battery for cars with Hydra-Matic transmission.

1963	326 V-8	AC45S	.035	.016	30	6B	160	500	2SM	
	389 V-8	AC45S	.035	.016	30	6B	160	500	2SM	
	389 V-8 (3 x 2)	AC45S	.035	.016	30	6B	160	500	2SM	
	421 V-6 (3 x 2)	AC45S	.035	.016	30	6B	160	600	2SM	
1964	326 V-8	AC45S	.035	.016	30	6B	160	500	2SM	
	389 V-8		AC45S	.035	.016	30	6B	160	500	2SM
	389 V-8 (3 x 2)	AC45S	.035	.016	30	6B	160	500	2SM	
	421 V-6 (3 x 2)	AC45S	.035	.016	30	6B	160	600	2SM	
1965	389 V-8	AC45S	.035	.016	30	6B	160	500	2SM	
	389 V-8 (3 x 2)	AC45S	.035	.016	30	6B	160	500	2SM	
	421 V-6 (3 x 2)	AC45S	.035	.016	30	6B	160	600	2SM	

Pontiac V-8 Firing Order (1955-1976)

1955-1976
(All V-8) 1-8-4-3-6-5-7-2

Note: Cylinders are numbered front to back (left) 1-3-5-7; (right) 2-4-6-8.

Base V-8 in the '62 Grand Prix was the 389-cid V-8 with a Carter AFB four-barrel carburetor. With stick shift you got a 235-hp version with 8.6:1 compression ratio. The automatic version had 10.25:1 compression heads and 303 hp at 4600 rpm. (John Lee photo)

STUDEBAKER
1951-1966

1951 STUDEBAKER

Studebaker's first V-8 engine arrived in 1951 and was available only in Commanders. This advertisement from *The Saturday Evening Post* heralded the new engine as a "Great New V-8." Studebaker promised V-8 buyers more power, but better gas mileage and pointed out that premium fuel was not required.

Commander, Land Cruiser Series V-8
V-8. Overhead valves. Cast-iron block. Bore & stroke: 3.375 x 3.25 in. Displacement: 232.6 cid. Compression ratio: 7.00:1. Brake hp: 120 at 4000 rpm. Torque: 190 lbs.-ft. at 2000 rpm. Five main bearings. Solid valve lifters. Carburetor: Stromberg AAUVB two-barrel. Cooling system capacity: 17-1/2 qt. (add 1-12 qt. for heater). Crankcase capacity: 6 qt. (add 1 qt. if new filter installed).

1952 STUDEBAKER

It's hard to say if this 1952 Commander V-8 chassis was on display at a real auto show, or whether this photo was staged only for the year's sales catalog. The 232.6-cid Commander V-8 had a 7.0:1 compression ratio and developed 120 hp. A 7.5:1 "high-compression" head was available for cars used in high-altitudes. The brochure showed a light green block and intake, silver valve covers, and a yellow carburetor, but this was done for show.

Commander, Land Cruiser Series V-8
V-8. Overhead valves. Cast-iron block. Bore & stroke: 3.375 x 3.25 in. Displacement: 232.6 cid. Compression ratio: 7.00:1. Brake hp: 120 at 4000 rpm. Torque: 190 lbs.-ft. at 2000 rpm. Five main bearings. Solid valve lifters. Carburetor: Stromberg AAUVB two-barrel. Cooling system capacity: 17-1/2 qt. (add 1-1/2 qt. for heater). Crankcase capacity: 6 qt. (add 1 qt. if new filter installed).

1953 STUDEBAKER

Commander, Land Cruiser Series V-8
V-8. Overhead valves. Cast-iron block. Bore & stroke: 3.375 x 3.25 in. Displacement: 232.6 cid. Compression ratio: 7.50:1. Brake hp: 120 at 4000 rpm. Torque: 190 lbs.-ft. at 2000 rpm. Five main bearings. Solid valve lifters. Carburetor: Stromberg WWUVL two-barrel. Cooling system capacity: 17-1/2 qt. (add 1-1/2 qt. for heater). Crankcase capacity: 6 qt. (add 1 qt. if new filter installed).

1954 STUDEBAKER

Commander, Land Cruiser Series V-8
V-8. Overhead valves. Cast-iron block. Bore & stroke: 3.375 x 3.25 in. Displacement: 232.6 cid. Compression ratio: 7.50:1. Brake hp: 127 at 4000 rpm. Torque: 202 lbs.-ft. at 2000 rpm. Five main bearings. Solid valve lifters. Carburetor: Stromberg WWUVL two-barrel. Cooling system capacity: 17-1/2 qt. (add 1-1/2 qt. for heater). Crankcase capacity: 6 qt. (add 1 qt. if new filter installed).

1955 STUDEBAKER

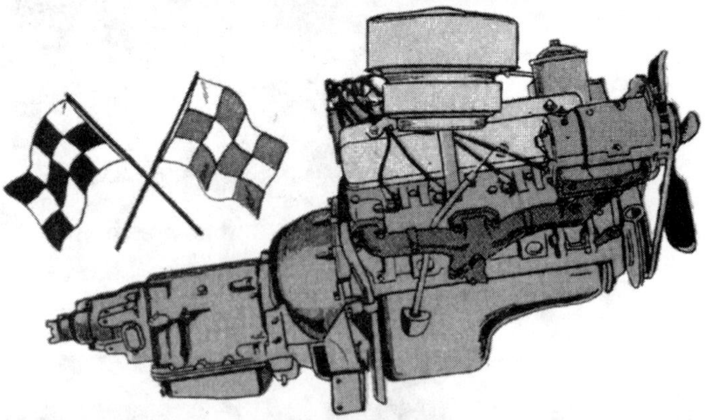

Studebaker introduced a 259-cid V-8 in 1955 and it returned as the smaller V-8 for 1956. But this engine was also bored out to create an optional 289-cid V-8 for 1956. The largest engine offered that year was a Packard-built 352-cid 275-hp V-8 fitted to the sporty Golden Hawk. This advertisement illustrated only the Studebaker-built motor and noted options up to 210 hp.

Commander Series base V-8 (early production; prior to Jan. 3, 1955)
V-8. Overhead valves. Cast-iron block. Bore & stroke: 3.562 x 2.81 in. Displacement: 224.3 cid. Compression ratio: 7.50:1. Brake hp: 140 at 4500 rpm. Torque: 202 lbs.-ft. at 2800 rpm. Five main bearings. Solid valve lifters. Carburetor: Stromberg WW6 two-barrel. Cooling system

capacity: 17-1/4 qt. (add 1-1/2 qt. for heater). Crankcase capacity: 6 qt. (add 1 qt. if new filter installed).

Commander Series optional V-8 (early production; prior to Jan. 3, 1955)
V-8. Overhead valves. Cast-iron block. Bore & stroke: 3.562 x 2.81 in. Displacement: 224.3 cid. Compression ratio: 7.50:1. Brake hp: 160 at 4500 rpm. Torque: 260 lbs.-ft. at 2800 rpm. Five main bearings. Solid valve lifters. Carburetor: Carter WCFB four-barrel. Cooling system capacity: 17-1/4 qt. (add 1-1/2 qt. for heater). Crankcase capacity: 6 qt. (add 1 qt. if new filter installed).

Commander Series base V-8 (late Production; beginning Jan. 3, 1955)
V-8. Overhead valves. Cast-iron block. Bore & stroke: 3.562 x 3.25 in. Displacement: 259.2 cid. Compression ratio: 7.50:1. Brake hp: 162 at 4500 rpm. Torque: 250 lbs.-ft. at 2800 rpm. Five main bearings. Solid valve lifters. Carburetor: Stroimberg WW6 two-barrel. Cooling system capacity: 17-1/4 qt. (add 1-1/2 qt. for heater). Crankcase capacity: 6 qt. (add 1 qt. if new filter installed). Dual exhausts.

Commander Series optional V-8 (late Production; beginning Jan. 3, 1955)
V-8. Overhead valves. Cast-iron block. Bore & stroke: 3.562 x 3.25 in. Displacement: 259.2 cid. Compression ratio: 7.50:1. Brake hp: 175 at 4500 rpm. Torque: 250 lbs.-ft. at 3000 rpm. Five main bearings. Solid valve lifters. Carburetor: Carter WCFB four-barrel. Cooling system capacity: 17-1/4 qt. (add 1-1/2 qt. for heater). Crankcase capacity: 6 qt. (add 1 qt. if new filter installed). Dual exhausts.

President Series V-8
V-8. Overhead valves. Cast-iron block. Bore & stroke: 3.562 x 3.25 in. Displacement: 259.2 cid. Compression ratio: 7.50:1. Brake hp: 185 at 4500 rpm. Torque: 258 lbs.-ft. at 2800 rpm. Five main bearings. Solid valve lifters. Carburetor: Carter WCFB four-barrel. Cooling system capacity: 17-1/4 qt. (add 1-1/2 qt. for heater). Crankcase capacity: 6 qt. (add 1 qt. if new filter installed).
Note: In early 1955, the 224-cid V-8 was used in the Studebaker Commanders. Effective with car serial numbers 8,397,201; 8,843,001; and 8,958,101 the 162-hp version of the 259-cid V-8 was used in Studebaker Commanders as the base engine with the 175-hp four-barrel version optional.

1956 STUDEBAKER

Commander, Power Hawk Series base V-8
V-8. Overhead valves. Cast-iron block. Bore & stroke: 3.562 x 3.25 in. Displacement: 259.2 cid. Compression ratio: 7.80:1. Brake hp: 170 at 4500 rpm. Torque: 260 lbs.-ft. at 2800 rpm. Five main bearings. Solid valve lifters. Carburetor: Two-barrel. Cooling system capacity: 17-1/4 qt. (add 1-1/2 qt. for heater). Crankcase capacity: 5 qt. (add 1 qt. if new filter installed).

Commander, Power Hawk Series optional V-8
V-8. Overhead valves. Cast-iron block. Bore & stroke: 3.562 x 3.25 in. Displacement: 259.2 cid. Compression ratio: 7.80:1. Brake hp: 185 at 4500 rpm. Torque: 265 lbs.-ft. at 2800 rpm. Five main bearings. Solid valve lifters. Carburetor: Four-barrel. Cooling system capacity: 17 qt. (add 1-1/2 qt. for heater). Crankcase capacity: 5 qt. (add 1 qt. if new filter installed). Dual exhausts.

President Series base V-8 (except President Classic and Skyhawk)
V-8. Overhead valves. Cast-iron block. Bore & stroke: 3.562 x 3.625 in. Displacement: 289.1 cid. Compression ratio: 7.80:1. Brake hp: 195 at 4500 rpm. Torque: 286 lbs.-ft. at 2800 rpm. Five main bearings. Solid valve lifters. Carburetor: Two-barrel. Cooling system capacity: 17 qt. (add 1-1/2 qt. for heater). Crankcase capacity: 5 qt. (add 1 qt. if new filter installed). Dual exhausts.

President Classic and Skyhawk base V-8; other President series models optional V-8
V-8. Overhead valves. Cast-iron block. Bore & stroke: 3.562 x 3.625 in. Displacement: 289.1 cid. Compression ratio: 7.80:1. Brake hp: 210 at 4500 rpm. Torque: 292 lbs.-ft. at 2800 rpm. Five main bearings. Solid valve lifters. Carburetor: Four-barrel. Cooling system capacity: 17 qt. (add 1-1/2 qt. for heater). Crankcase capacity: 5 qt. (add 1 qt. if new filter installed). Dual exhausts.

Golden Hawk V-8 (Manufactured by Packard Div.)
V-8. Overhead valves. Cast-iron block. Bore & stroke: 4.00 x 3.50 in. Displacement: 351.8 cid. Compression ratio: 9.50:1. Brake hp: 275 at 4600 rpm. Torque: 380 lbs.-ft. at 2800 rpm. Five main bearings. Non-adjustable hydraulic valve lifters. Carburetor: Four-barrel. Cooling system capacity: 25 qt. (add 1-1/2 qt. for heater). Crankcase capacity: 5 qt. (add 1 qt. if new filter installed). Dual exhausts.

1957 STUDEBAKER

In 1957, the Packard V-8 was dropped from the Hawk series. The top-of-the-line Golden Hawk featured a supercharged version of the 289-cid Studebaker V-8 that generated the same 275 hp as the previous engine. It carried a McCulloch VS-57 supercharger and had a dual exhaust system. (Photo by Tom Collins)

Commander Series base V-8
V-8. Overhead valves. Cast-iron block. Bore & stroke: 3.562 x 3.25 in. Displacement: 259.2 cid. Compression ratio: 8.00:1. Brake hp: 180 at 4500 rpm. Torque: 260 lbs.-ft. at 2800 rpm. Five main bearings. Solid valve lifters. Carburetor: Two-barrel. Cooling system capacity: 17 qt. (add 1-1/2 qt. for heater). Crankcase capacity: 5 qt. (add 1 qt. if new filter installed).

Commander Series optional V-8
V-8. Overhead valves. Cast-iron block. Bore & stroke: 3.562 x 3.25 in. Displacement: 259.2 cid. Compression ratio: 8.00:1. Brake hp: 195 at 4500 rpm. Torque: 265 lbs.-ft. at 3000 rpm. Five main bearings. Solid valve lifters. Carburetor: Four-barrel. Cooling system capacity: 17 qt. (add 1-1/2 qt. for heater). Crankcase capacity: 5 qt. (add 1 qt. if new filter installed). Dual exhausts.

President Series base V-8 (except President Classic)
V-8. Overhead valves. Cast-iron block. Bore & stroke: 3.562 x 3.625 in. Displacement: 289.1 cid. Compression ratio: 8.00:1. Brake hp: 210 at 4500 rpm. Torque: 300 lbs.-ft. at 2800 rpm. Five main bearings. Solid valve lifters. Carburetor: Two-barrel. Cooling system capacity: 17 qt. (add 1-1/2 qt. for heater). Crankcase capacity: 5 qt. (add 1 qt. if new filter installed). Dual exhausts.

President Classic standard V-8; President, Silver Hawk Series optional V-8
V-8. Overhead valves. Cast-iron block. Bore & stroke: 3.562 x 3.625 in. Displacement: 289.1 cid. Compression ratio: 8.00:1. Brake hp: 225 at 4500 rpm. Torque: 305 lbs.-ft. at 3000 rpm. Five main bearings. Solid valve lifters. Carburetor: Four-barrel. Cooling system capacity: 17 qt. (add 1-1/2 qt. for heater). Crankcase capacity: 5 qt. (add 1 qt. if new filter installed). Dual exhausts.

Golden Hawk Supercharged V-8
V-8. Overhead valves. Cast-iron block. Bore & stroke: 3.562 x 3.625 in. Displacement: 289.1 cid. Compression ratio: 7.50:1. Brake hp: 275 at 4800 rpm. Torque: 333 lbs.-ft. at 3200 rpm. Five main bearings. Solid valve lifters. Carburetor: Two-barrel supercharged (McCulloch VS-57 supercharger). Cooling system capacity: 17 qt. (add 1-1/2 qt. for heater). Crankcase capacity: 5 qt. (add 1 qt. if new filter installed).

1958 STUDEBAKER

Commander Series base V-8
V-8. Overhead valves. Cast-iron block. Bore & stroke: 3.562 x 3.25 in. Displacement: 259.2 cid. Compression ratio: 8.30:1. Brake hp: 180 at 4500 rpm. Torque: 260 lbs.-ft. at 2800 rpm. Five main bearings. Solid valve lifters. Carburetor: Stromberg WW6 two-barrel. Cooling system capacity: 17 qt. (add 1-1/2 qt. for heater). Crankcase capacity: 5 qt. (add 1 qt. if new filter installed).

Commander Series optional V-8
V-8. Overhead valves. Cast-iron block. Bore & stroke: 3.562 x 3.25 in. Displacement: 259.2 cid. Compression ratio: 8.30:1. Brake hp: 195 at 4500 rpm. Torque: 265 lbs.-ft. at 3000 rpm. Five main bearings. Solid valve lifters. Carburetor: Carter WCFB four-barrel. Cooling system capacity: 17 qt. (add 1-1/2 qt. for heater). Crankcase capacity: 5 qt. (add 1 qt. if new filter installed). Dual exhausts.

Silver Hawk Series base V-8
V-8. Overhead valves. Cast-iron block. Bore & stroke: 3.562 x 3.625 in. Displacement: 289.1 cid. Compression ratio: 8.30:1. Brake hp: 210 at 4500 rpm. Torque: 300 lbs.-ft. at 2800 rpm. Five main bearings. Solid valve lifters. Carburetor: Stromberg WW6 two-barrel. Cooling system capacity: 17 qt. (add 1-1/2 qt. for heater). Crankcase capacity: 5 qt. (add 1 qt. if new filter installed). Dual exhausts.

President Series base V-8; Silver Hawk Series optional V-8
V-8. Overhead valves. Cast-iron block. Bore & stroke: 3.562 x 3.625 in. Displacement: 289.1 cid. Compression ratio: 8.30:1. Brake hp: 225 at 4500 rpm. Torque: 305 lbs.-ft. at 3000 rpm. Five main bearings. Solid valve lifters. Carburetor: Carter WCFB four-barrel. Cooling system capacity: 17 qt. (add 1-1/2 qt. for heater). Crankcase capacity: 5 qt. (add 1 qt. if new filter installed). Dual exhausts.

Golden Hawk Series Supercharged V-8
V-8. Overhead valves. Cast-iron block. Bore & stroke: 3.562 x 3.625 in. Displacement: 289.1 cid. Compression ratio: (Standard) 7.80:1; (Optional) 7.00:1. Brake hp: 275 at 4800 rpm. Torque: 333 lbs.-ft. at 3200 rpm. Five main bearings. Solid valve lifters. Carburetor: Stromberg WW6 two-barrel. Cooling system capacity: 17 qt. (add 1-1/2 qt. for heater). Crankcase capacity: 5 qt. (add 1 qt. if new filter installed). Dual exhausts.

1959 STUDEBAKER

Lark Regal, Silver Hawk Series base V-8
V-8. Overhead valves. Cast-iron block. Bore & stroke: 3.562 x 3.25 in. Displacement: 259.2 cid. Compression ratio: 8.80:1. Brake hp: 180 at 4500 rpm. Torque: 260 lbs.-ft. at 2800 rpm. Five main bearings. Solid valve lifters. Carburetor: Stromberg WW6 two-barrel. Cooling system capacity: 17 qt. (add 1 qt. for heater in Lark; 1-1/2 qt. for heater in Hawk). Crankcase capacity: 5 qt. (add 1 qt. if new filter installed).

Lark Regal, Silver Hawk Series optional "power kit" V-8
V-8. Overhead valves. Cast-iron block. Bore & stroke: 3.562 x 3.25 in. Displacement: 259.2 cid. Compression ratio: 8.80:1. Brake hp: 195 at 4500 rpm. Torque: 265 lbs.-ft. at 2800 rpm. Five main bearings. Solid valve lifters. Carburetor: Carter WCFB four-barrel. Cooling system capacity: 17 qt. (add 1 qt. for heater in Lark Regal; 1-1/2 qt. for heater in Hawk). Crankcase capacity: 5 qt. (add 1 qt. if new filter installed). Dual exhausts.

1960 STUDEBAKER

Lark Deluxe, Regal Series base V-8; also base V-8 in Hawks sold outside North America
V-8. Overhead valves. Cast-iron block. Bore & stroke: 3.562 x 3.25 in. Displacement: 259.2 cid. Compression ratio: 8.50:1. Brake hp: 180 at 4500 rpm. Torque: 260 lbs.-ft. at 2800 rpm. Five main bearings. Solid valve lifters. Carburetor: Stromberg WW6 two-barrel. Cooling system capacity: 17 qt. (add 1 qt. for heater in Lark; 1-1/2 qt. for heater in Hawk). Crankcase capacity: 5 qt. (add 1 qt. if new filter installed).

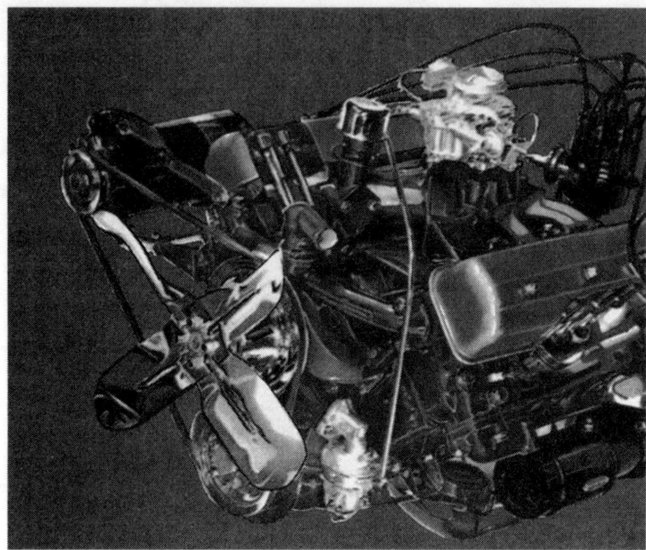

Illustrated in the 1960 Studebaker Lark sales catalog was the optional 180-hp Studebaker two-barrel V-8. This engine was also available in a four-barrel version that developed 195 hp. The latter engine could speed the Lark from 0 to 60 mph in under 10 seconds. The engine had mechanical valve lifters and five main bearings. The 180-hp version could be safely operated on regular-grade fuel.

Two versions of the 289-cid V-8 were offered in 1960 Studebaker Hawk models. The standard engine had a two-barrel carburetor and cranked up 210 hp at 4500 rpm. A 225-hp version with a "power kit" option was also available. It utilized a Carter WCFB four-barrel carburetor. Both engines had solid valve lifters and an 8.5:1 compression ratio. Dual exhausts were part of the power kit.

Lark Deluxe, Lark Regal, Silver Hawk Series optional "power kit" V-8
V-8. Overhead valves. Cast-iron block. Bore & stroke: 3.562 x 3.25 in. Displacement: 259.2 cid. Compression ratio: 8.50:1. Brake hp: 195 at 4500 rpm. Torque: 265 lbs.-ft. at 3000 rpm. Five main bearings. Solid valve lifters. Carburetor: Carter WCFB four-barrel. Cooling system capacity: 17 qt. (add 1 qt. for heater in Lark; 1-1/2 qt. for heater in Hawk). Crankcase capacity: 5 qt. (add 1 qt. if new filter installed). Dual exhausts.

Hawk Series base V-8 (U.S. and Canada)
V-8. Overhead valves. Cast-iron block. Bore & stroke: 3.562 x 3.625 in. Displacement: 289.1 cid. Compression ratio: 8.50:1. Brake hp: 210 at 4500 rpm. Torque: 300 lbs.-ft. at 2800 rpm. Five main bearings. Solid valve lifters. Carburetor: Two-barrel. Cooling system capacity: 17 qt. (add 1-1/2 qt. for heater). Crankcase capacity: 5 qt. (add 1 qt. if new filter installed). Dual exhausts.

Hawk Series optional "power kit" V-8
V-8. Overhead valves. Cast-iron block. Bore & stroke: 3.562 x 3.625 in. Displacement: 289.1 cid. Compression ratio: 8.50:1. Brake hp: 225 at 4500 rpm. Torque: 305 lbs.-ft. at 3000 rpm. Five main bearings. Solid valve lifters. Carburetor: Carter WCFB four-barrel. Cooling system capacity: 17 qt. (add 1-1/2 qt. for heater). Crankcase capacity: 5 qt. (add 1 qt. if new filter installed). Dual exhausts.

1961 STUDEBAKER

Lark Deluxe, Regal, Cruiser Series base V-8; base V-8 in Hawk sold outside North America
V-8. Overhead valves. Cast-iron block. Bore & stroke: 3.562 x 3.25 in. Displacement: 259.2 cid. Compression ratio: 8.80:1. Brake hp: 180 at 4500 rpm. Torque: 260 lbs.-ft. at 2800 rpm. Five main bearings. Solid valve lifters. Carburetor: Stromberg WW6 two-barrel. Cooling system capacity: 17 qt. (add 1 qt. for heater). Crankcase capacity: 5 qt. (add 1 qt. if new filter installed).

Lark Deluxe, Regal, Cruiser Series optional "power kit" V-8
V-8. Overhead valves. Cast-iron block. Bore & stroke: 3.562 x 3.25 in. Displacement: 259.2 cid. Compression ratio: 8.80:1. Brake hp: 195 at 4500 rpm. Torque: 265 lbs.-ft. at 3000 rpm. Five main bearings. Solid valve lifters. Carburetor: Carter WCFB four-barrel. Cooling system capacity: 17 qt. (add 1 qt. for heater). Crankcase capacity: 5 qt. (add 1 qt. if new filter installed). Dual exhausts.

Hawk Series base V-8 (U.S. and Canada); Cruiser Series optional V-8
V-8. Overhead valves. Cast-iron block. Bore & stroke: 3.562 x 3.625 in. Displacement: 289.1 cid. Compression ratio: 8.80:1. Brake hp: 210 at 4500 rpm. Torque: 300 lbs.-ft. at 2800 rpm. Five main bearings. Solid valve lifters. Carburetor: Two-barrel. Cooling system capacity: 17 qt. (add 1-1/2 qt. for heater in Hawk; 1 qt. for heater in Cruiser). Crankcase capacity: 5 qt. (add 1 qt. if new filter installed). Dual exhausts.

Hawk, Cruiser Series optional "power kit" V-8
V-8. Overhead valves. Cast-iron block. Bore & stroke: 3.562 x 3.625 in. Displacement: 289.1 cid. Compression ratio: 8.80:1. Brake hp: 225 at 4500 rpm. Torque: 305 lbs.-ft. at 3000 rpm. Five main bearings. Solid valve lifters. Carburetor: Carter WCFB four-barrel. Cooling system capacity: 17 qt. (add 1-1/2 qt. for heater in Hawk; 1 qt. for heater in Cruiser). Crankcase capacity: 5 qt. (add 1 qt. if new filter installed). Dual exhausts.

1962 STUDEBAKER

The 1962 Studebaker Grand Turismo Hawk was available with two versions of the 289-cid V-8, one having a two-barrel carburetor and one having the optional power kit, including a four-barrel carburetor. This year, both of these engines featured a dual exhaust system when used in the GT Hawk model. The horsepower ratings were 210 and 225, respectively. (Tom Collins photo)

Lark Deluxe, Regal, Daytona, Cruiser Series base V-8
V-8. Overhead valves. Cast-iron block. Bore & stroke: 3.562 x 3.25 in. Displacement: 259.2 cid. Compression ratio: 8.25:1. Brake hp: 180 at 4500 rpm. Torque: 260 lbs.-ft. at 2800 rpm. Five main bearings. Solid valve lifters. Carburetor: Stromberg WW6 two-barrel. Cooling system capacity: 17 qt. (add 1 qt. for heater). Crankcase capacity: 5 qt. (add 1 qt. if new filter installed).

Lark Deluxe, Regal, Daytona, Cruiser Series optional "power kit" V-8
V-8. Overhead valves. Cast-iron block. Bore & stroke: 3.562 x 3.25 in. Displacement: 259.2 cid. Compression ratio: 8.25:1. Brake hp: 195 at 4500 rpm. Torque: 265 lbs.-ft. at 3000 rpm. Five main bearings. Solid valve lifters. Carburetor: Carter WCFB four-barrel. Cooling system capacity: 17 qt. (add 1 qt. for heater). Crankcase capacity: 5 qt. (add 1 qt. if new filter installed). Dual exhausts.

Gran Turismo Hawk, Lark Pursuit Marshal Series base V-8; Optional Lark (all) Series V-8
V-8. Overhead valves. Cast-iron block. Bore & stroke: 3.562 x 3.625 in. Displacement: 289.1 cid. Compression ratio: 8.25:1. Brake hp: 210 at 4500 rpm. Torque: 300 lbs.-ft. at 2800 rpm. Five main bearings. Solid valve lifters. Carburetor: Two-barrel. Cooling system capacity: 17 qt. (add 1-1/2 qt. for heater in Hawk; 1 qt. for heater in Lark). Crankcase capacity: 5 qt. (add 1 qt. if new filter installed). Dual exhausts included on Hawk only.

Gran Turismo Hawk Series optional "power kit" V-8; Optional Lark (all) Series V-8
V-8. Overhead valves. Cast-iron block. Bore & stroke: 3.562 x 3.625 in. Displacement: 289.1 cid. Compression ratio: 8.25:1. Brake hp: 225 at 4500 rpm. Torque: 305 lbs.-ft. at 3000 rpm. Five main bearings. Solid valve lifters. Carburetor: Carter WCFB four-barrel. Cooling system capacity: 17 qt. (add 1-1/2 qt. for heater in Hawk; 1 qt. for heater in Lark). Crankcase capacity: 5 qt. (add 1 qt. if new filter installed).

The 1962 Studebaker Lark offered three engine choices in five horsepower ranges. The first engine was the 112-hp "Skybolt Six." Two different V-8s could also be ordered: 259- and 289-cid. Horsepower options for the two V-8s were 180, 195, 210 and 225. All of the engines were said to give remarkable economy on regular gas.

1963 STUDEBAKER

Lark Standard, Regal, Custom Daytona Series base V-8
V-8. Overhead valves. Cast-iron block. Bore & stroke: 3.562 x 3.25 in. Displacement: 259.2 cid. Compression ratio: 8.25:1. Brake hp: 180 at 4500 rpm. Torque: 260 lbs.-ft. at 2800 rpm. Five main bearings. Solid valve lifters. Carburetor: Stromberg WW6 two-barrel. Cooling system capacity: 17 qt. (add 1 qt. for heater). Crankcase capacity: 5 qt. (add 1 qt. if new filter installed).

Lark Standard, Regal, Custom Daytona Series optional "power kit" V-8
V-8. Overhead valves. Cast-iron block. Bore & stroke: 3.562 x 3.25 in. Displacement: 259.2 cid. Compression ratio: 8.25:1. Brake hp: 195 at 4500 rpm. Torque: 265 lbs.-ft. at 3000 rpm. Five main bearings. Solid valve lifters. Carburetor: Carter AFB four-barrel. Cooling system capacity: 17 qt. (add 1 qt. for heater). Crankcase capacity: 5 qt. (add 1 qt. if new filter installed). Dual exhausts.

Gran Turismo Hawk, Cruiser, Lark Pursuit Marshal Series base V-8; Optional Lark V-8
V-8. Overhead valves. Cast-iron block. Bore & stroke: 3.562 x 3.625 in. Displacement: 289.1 cid. Compression ratio: 8.25:1. Brake hp: 210 at 4500 rpm. Torque: 300 lbs.-ft. at 2800 rpm. Five main bearings. Solid valve lifters. Carburetor: Stromberg WW6 two-barrel. Cooling system capacity: 17 qt. (add 1 qt. for heater). Crankcase capacity: 5 qt. (add 1 qt. if new filter installed). Includes dual exhausts on Hawk only.
Note: Beginning in 1963, the Hawk got a smaller, more compact heater and only one additional quart of coolant was needed when the Hawk was ordered with a heater.

GT Hawk, Cruiser Series optional "power kit" V-8; Optional Lark (all other) Series V-8
V-8. Overhead valves. Cast-iron block. Bore & stroke: 3.562 x 3.625 in. Displacement: 289.1 cid. Compression ratio: 8.25:1. Brake hp: 225 at 4500 rpm. Torque: 305 lbs.-ft. at 3000 rpm. Five main bearings. Solid valve lifters. Carburetor: Carter AFB four-barrel. Cooling system capacity: 17 qt. (add 1 qt. for heater). Crankcase capacity: 5 qt. (add 1 qt. if new filter installed). Dual exhausts.

Avanti Series "R1" base V-8; optional V-8 in Lark and Hawk models
V-8. Overhead valves. Cast-iron block. Bore & stroke: 3.562 x 3.625 in. Displacement: 289.1 cid. Compression ratio: 10.25:1. Brake hp: 240 at 5000 rpm. Torque: 280 lbs.-ft. at 4200 rpm. Five main bearings. Solid valve lifters. Carburetor: Carter AFB four-barrel. Cooling system capacity: 17 qt. (add 1 qt. for heater). Crankcase capacity: 5 qt. (add 1 qt. if new filter installed). Dual exhausts.

Avanti Series "R2" optional V-8; optional V-8 in Lark and Hawk models
V-8. Overhead valves. Cast-iron block. Bore & stroke: 3.562 x 3.625 in. Displacement: 289.1 cid. Compression ratio: 9.00:1. Brake hp: 289 at 4800 rpm. Torque: 303 lbs.-ft. at 5000 rpm. Five main bearings. Solid valve lifters. Carburetor: Sealed Carter AFB four-barrel carburetor used in conjunction with a Paxton SN-60 centrifugal supercharger. Cooling system capacity: 17 qt. (add 1 qt. for heater). Crankcase capacity: 5 qt. (add 1 qt. if new filter installed). Dual exhausts.
Note: Torque rating, which was not specified by Studebaker, is estimated through data analysis.

The top engine option in the Studebaker Avanti, introduced in 1963, was the R2 version. This 289-cid Studebaker-built V-8 carried a Paxton SN-60 centrifugal supercharger and a sealed Carter AFB carb. This helped it develop one horsepower per cubic inch. The R2-powered Avanti did the standing-start quarter mile in 15.8 seconds at 91 mph. It could move from 0-to-60 mph in a flat eight seconds.

1964 STUDEBAKER

Lark Challenger, Lark Commander Series base V-8
V-8. Overhead valves. Cast-iron block. Bore & stroke: 3.562 x 3.25 in. Displacement: 259.2 cid. Compression ratio: 8.25:1. Brake hp: 180 at 4500 rpm. Torque: 260 lbs.-ft. at 2800 rpm. Five main bearings. Solid valve lifters. Carburetor: Stromberg WW6 two-barrel. Cooling system capacity: 17 qt. (add 1 qt. for heater). Crankcase capacity: 5 qt. (add 1 qt. if new filter installed).

Lark Challenger, Lark Commander Series optional "power kit" V-8
V-8. Overhead valves. Cast-iron block. Bore & stroke: 3.562 x 3.25 in. Displacement: 259.2 cid. Compression ratio: 8.25:1. Brake hp: 195 at 4500 rpm. Torque: 265 lbs.-ft. at 3000 rpm. Five main bearings. Solid valve lifters. Carburetor: Carter AFB four-barrel. Cooling system capacity: 17 qt. (add 1 qt. for heater). Crankcase capacity: 5 qt. (add 1 qt. if new filter installed). Dual exhausts.

Gran Turismo Hawk, Cruiser, Lark Pursuit Marshal Series base V-8; Optional Lark (all) and Daytona Series V-8
V-8. Overhead valves. Cast-iron block. Bore & stroke: 3.562 x 3.625 in. Displacement: 289.1 cid. Compression ratio: 8.25:1. Brake hp: 210 at 4500 rpm. Torque: 300 lbs.-ft. at 2800 rpm. Five main bearings. Solid valve lifters. Carburetor: Stromberg WW6 two-barrel. Cooling system capacity: 17 qt. (add 1 qt. for heater). Crankcase capacity: 5 qt. (add 1 qt. if new filter installed). Dual exhausts included on Hawk.

GT Hawk, Cruiser, Pursuit Marshal Series optional "power kit" V-8; Optional Lark (all) and Daytona Series V-8
V-8. Overhead valves. Cast-iron block. Bore & stroke: 3.562 x 3.625 in. Displacement: 289.1 cid. Compression ratio: 8.25:1. Brake hp: 225 at 4500 rpm. Torque: 305 lbs.-ft. at 3000 rpm. Five main bearings. Solid valve lifters. Carburetor: Carter AFB four-barrel. Cooling system capacity: 17 qt. (add 1 qt. for heater). Crankcase capacity: 5 qt. (add 1 qt. if new filter installed). Dual exhausts.

Avanti Series "R1" base V-8; GT Hawk, Lark Challenger, Lark Commander, Daytona and Cruiser optional V-8
V-8. Overhead valves. Cast-iron block. Bore & stroke: 3.562 x 3.625 in. Displacement: 289.1 cid. Compression ratio: 10.25:1. Brake hp: 240 at 5000 rpm. Torque: 280 lbs.-ft. at 4200 rpm. Five main bearings. Solid valve lifters. Carburetor: Carter AFB four-barrel. Cooling system capacity: 17 qt. (add 1 qt. for heater). Crankcase capacity: 5 qt. (add 1 qt. if new filter installed). Dual exhausts.

All series "R2" optional V-8
V-8. Overhead valves. Cast-iron block. Bore & stroke: 3.562 x 3.625 in. Displacement: 289.1 cid. Compression ratio: 9.00:1. Brake hp: 289 at 4800 rpm. Torque: 303 lbs.-ft. at 5000 rpm. Five main bearings. Solid valve lifters. Carburetor: Sealed Carter AFB four-barrel carburetor used in conjunction with a Paxton SN-60 centrifugal supercharger. Cooling system capacity: 17 qt. (add 1 qt. for heater). Crankcase capacity: 5 qt. (add 1 qt. if new filter installed). Dual exhausts.
Note: Torque rating, which was not specified by Studebaker, is estimated through data analysis.

All series "R3" optional V-8 (available late 1963 model-year production)
V-8. Overhead valves. Cast-iron block. Bore & stroke: 3.656 x 3.625 in. Displacement: 304.5 cid. Compression ratio: 9.75:1. Brake hp: 335 at 5350 rpm. Torque: 320 lbs.-ft. at 4000 rpm. Five main bearings. Solid valve lifters. Carburetor: Carter AFB four-barrel carburetor fully-enclosed in a pressure box used in conjunction with a Paxton SN-60 centrifugal supercharger. Cooling system capacity: 17 qt. (add 1 qt. for heater). Crankcase capacity: 6 qt. (add 1 qt. if new filter installed). Dual exhausts.
Note: Some pre-production data states that the R3 engine's compression ratio was 9.60:1.

All series "R4" optional V-8
V-8. Overhead valves. Cast-iron block. Bore & stroke: 3.656 x 3.625 in. Displacement: 304.5 cid. Compression ratio: 12.0:1. Brake hp: 280 at 5000 rpm. Torque: Not specified by Studebaker. Five main bearings. Solid valve lifters. Carburetor: Two Carter AFB four-barrels. Cooling system capacity: 17 qt. (add 1 qt. for heater). Crankcase capacity: 6 qt. (add 1 qt. if new filter installed). Dual exhausts.
Note: Some pre-production data states that the R4 engine's compression ratio was 11.00:1.

1965 STUDEBAKER

Commander, Daytona, Cruiser V-8
V-8. Overhead valves. Cast-iron block. Bore & stroke: 3.875 x 3.00 in. Displacement: 283 cid. Compression ratio: 9.25:1. Brake hp: 195 at 4800 rpm. Torque: 285 lbs.-ft. at 2400 rpm. Five main bearings. Adjustable hydraulic valve lifters. Carburetor: Rochester 2GV two-barrel. Cooling system capacity: 15 qt. (add 1 qt. for heater). Crankcase capacity: 4 qt. (add 1 qt. if new filter installed).

1966 STUDEBAKER

Commander, Daytona, Cruiser V-8
V-8. Overhead valves. Cast-iron block. Bore & stroke: 3.875 x 3.00 in. Displacement: 283 cid. Compression ratio: 9.25:1. Brake hp: 195 at 4800 rpm. Torque: 285 lbs.-ft. at 2400 rpm. Five main bearings. Adjustable hydraulic valve lifters. Carburetor: Rochester 2GV two-barrel. Cooling system capacity: 15 qt. (add 1 qt. for heater). Crankcase capacity: 4 qt. (add 1 qt. if new filter installed).

Studebaker V-8 Firing Order

1951-1964
1-8-4-3-6-5-7-2

All Studebaker V-8s are numbered from front to back: left bank 1-3-5-7; right bank: 2-4-6-8. The spark plug wires go into the distributor cap in the firing order and in a counterclockwise direction.

Studebaker section checked by Bob Palma, Technical Editor, Studebaker Drivers Club.

Selected Pontiac V-8 Tune-Up Specs 1955-1965

Year	Model	Spark Plugs Type	Gap	Point Gap	Deg Cam Angle	Timing	Idle Speed	Compression Pressure	Battery (Delco)
1951	Commander	CH-H8	.035	.013	28-34	IGN	500	130	Group 1
1952	Commander	CH-H8	.035	.013	28-34	IGN	500	130	Group 1
1953	Commander	CH-H10	.035	.013	28-34	4B	550	130	Group 1
1954	Commander	CH-H11	.035	.013	28-34	4B	550	130	Group 1
1955	Commander	CH-H11	.035	.013	28-34	4B	550	130	Group 1
	President	CH-H11	.035	.013	28-34	4B	550	130	Group 1
1956	Commander	CH-H11	.035	.015	28-34	4B	550	130-150	11HS/2SM
	President	CH-H11	.035	.015	28-34	4B	550	130-150	11HS/2SM
	Golden Hawk	CH-N8	.035	.016	28-34	5B	450	120	11HS/2SM
1957	Commander	CH-H11	.035	.015	28-34	4B	550	130-150	11HS/2SM
	President	CH-H11	.035	.015	28-34	4B	550	130-150	11HS/2SM
	Golden Hawk	CH-H10	.035	.015	28-34	4B	550	130	11HS/2SM
1958	259-cid	CH-H18Y	.035	.016	28-34	4B	550-575	130-150	HO-11-50
	289-cid	CH-H18Y	.035	.016	28-34	4B	550-575	130-150	HO-11-50
	Golden Hawk	CH-H10	.035	.016	28-34	4B	550-575	130-150	HO-11-50
1959	V-8	CH-H18Y	.035	.016	28-34	4B	550	140-160	HO-11-50
1960	V-8	CH-H18Y	.035	.016	28-34	4B	550	140-160	HO-11-50
1961	V-8	CH-H14Y	.035	.016	28-32	4B	550	140-160	HO-11-50
1962	V-8	CH-H14Y	.035	.016	27-31	4B	550	140-160	HO-11-50
1963	V-8	CH-H14Y	.035	.017	27-31	4B	550	140-160	HO-11-50
	Avanti R1 V-8	CH-J12Y	.035	.017	39 (*)	4B	650	185-195	HDD-3EE
	Avanti R2 V-8	CH-J12Y	.035	.017	39 (*)	24B	650	160-170	HDD-3EE

Note: R1 ignition timing at 650 rpm; R2 ignition timing at 1600 rpm. All Avanti cam dwell settings are overall, for both sets of points; HDD-3EE battery used only in Avantis.

Year	Model	Spark Plugs Type	Gap	Point Gap	Deg Cam Angle	Timing	Idle Speed	Compression Pressure	Battery (Delco)
1964	V-8	CH-H14Y	.035	.017	27-31	4B	550	140-160	HO-11-50
	Avanti R1 V-8	CH-J12Y	.035	.019	32-36	4B	650	185-195	HO-11-50
	Avanti R2 V-8	CH-J12Y	.035	.019	32-36	24B	650	160-170	HO-11-50

Note: R1 ignition timing at 650 rpm; R2 ignition timing at 1600 rpm. All Avanti cam dwell settings are overall, for both sets of points; HO-11-50 battery used in all models.
Note: Set timing for all models with vacuum advance disconnected and hose plugged

MINOR MAKES

ABBOTT

The Abbott-Detroit was an assembled car built in Michigan from 1909-1916. Later on, it was built in Cleveland, Ohio, as the Abbott. The company lasted until 1918. The V-8-powered Model 880 was offered only in 1916. Like other Abbott-Detroit and Abbott models, it used a proprietary engine made by another firm, in this case Herschell-Spillman of North Tonawanda, N.Y.

1916 Abbott 880 V-8
V-8. Cast-iron blocks of four. Valves located in V. Bore & stroke: 3-1/4 x 5 in. Displacement: 331.8 cid. Taxable hp: 33.80. Three crankshaft bearings. Pump cooling. Force feed oiling. Zenith carburetor. Zenith system. Remy ignition. Willard 120 battery.

APPERSON

Elmer Apperson started in the automobile business in 1894 and had his first car on the market by 1902. He introduced his first V-8 in 1916. The engine used in this car was made by Apperson in Kokomo, Ind.

1916 Apperson 8-16 V-8
V-8. Cast-iron blocks of four. Side valves. Bore & stroke: 3-1/8 x 5 in. Displacement: 306.8 cid. Taxable hp: 31.25. Three crankshaft bearings. Thermo siphon cooling. Force feed oiling. Carburetor fed by vacuum. Dual ignition.

1917 Apperson 8-17 V-8
V-8. Cast-iron blocks of four. Valves located in V. Bore & stroke: 3-1/8 x 5 in. Displacement: 306.8 cid. Taxable hp: 31.25. Three crankshaft bearings. Thermo siphon cooling. Force feed oiling. Rayfield V carburetor fed by vacuum. Delco ignition.

1918 Apperson 8-18 V-8
V-8. Cast-iron blocks of four. Valves located in V. Bore & stroke: 3-1/8 x 5 in. Displacement: 306.8 cid. Taxable hp: 31.25. Three crankshaft bearings. Thermo siphon cooling. Force feed oiling. Johnson V carburetor fed by Stewart-Warner vacuum system. Remy ignition.

1919 Apperson 8-19 V-8
V-8. Cast-iron blocks of four. Side valves. Bore & stroke: 3-1/4 x 5 in. Displacement: 331.8 cid. Brake hp: 60. Taxable hp: 33.8. Three crankshaft bearings. Thermo siphon cooling. Force feed oiling. Crankcase capacity: 8 qt. Johnson V carburetor fed by Stewart-Warner vacuum system. Remy ignition. Willard battery.

1920 Apperson 8-20 V-8
V-8. Cast-iron blocks of four. Side valves. Bore & stroke: 3-1/4 x 5 in. Displacement: 331.8 cid. Brake hp: 60. Taxable hp: 33.8. Three crankshaft bearings. Thermo siphon cooling. Force feed and splash oiling. Crankcase capacity: 8 qt. Johnson 1-1/4-inch carburetor fed by Stewart-Warner vacuum system. Remy ignition.

1920 Apperson Anniversary V-8
V-8. Cast-iron blocks of four. Side valves. Bore & stroke: 3-1/4 x 5 in. Displacement: 331.8 cid. Brake hp: 60 at 2100 rpm. Taxable hp: 33.8. Three crankshaft bearings. Thermo siphon cooling. Force feed and splash oiling. Crankcase capacity: 8 qt. Johnson 1-1/4-inch carburetor fed by Stewart-Warner vacuum system. Remy ignition.

1921 Apperson 8-21 V-8
V-8. Cast-iron blocks of four. Side valves. Bore & stroke: 3-1/4 x 5 in. Displacement: 331.8 cid. Brake hp: 70 at 2200 rpm. Taxable hp: 33.8. Three crankshaft bearings. Thermo siphon cooling. Crankshaft drilled for oil. Force feed and splash oiling. Crankcase capacity: 8 qt. Johnson 1-1/4-inch carburetor fed by Stewart-Warner vacuum system. Remy ignition.

1922 Apperson 8-22 V-8
V-8. Cast-iron blocks of four. Side valves. Bore & stroke: 3-1/4 x 5 in. Displacement: 331.8 cid. Brake hp: 70 at 2200 rpm. Taxable hp: 33.8. Three crankshaft bearings. Thermo siphon cooling. Crankshaft drilled for oil. Force feed and splash oiling. Crankcase capacity: 8 qt. Johnson 1-1/4-inch carburetor fed by Stewart-Warner vacuum system. Remy ignition.

1923 Apperson 8-23 V-8
V-8. Cast-iron blocks of four. Side valves. Bore & stroke: 3-1/4 x 5 in. Displacement: 331.8 cid. Brake hp: 70 at 2200 rpm. Taxable hp: 33.8. Three crankshaft bearings. Thermo siphon cooling. Crankshaft drilled for oil. Force feed and splash oiling. Crankcase capacity: 8 qt. Johnson 1-1/4-inch carburetor fed by Stewart-Warner vacuum system. Remy ignition.

1924 Apperson 8-24 V-8
V-8. Cast-iron blocks of four. Side valves. Bore & stroke: 3-1/4 x 5 in. Displacement: 331.8 cid. Brake hp: 70 at 2200 rpm. Taxable hp: 33.8. Three crankshaft bearings. Thermo siphon cooling. Crankshaft drilled for oil. Force feed and splash oiling. Crankcase capacity: 8 qt. Johnson 1-1/4-inch carburetor fed by Stewart-Warner vacuum system. Remy ignition.

1925 Apperson 8-25 V-8
V-8. Cast-iron blocks of four. Side valves. Bore & stroke: 3-1/4 x 5 in. Displacement: 331.8 cid. Brake hp: 60 at 2800 rpm. Taxable hp: 33.8. Three crankshaft bearings. Thermo siphon cooling. Crankshaft drilled for oil. Force feed and splash oiling. Crankcase capacity: 8 qt. Shebler S11 1-3/4-inch carburetor fed by Stewart-Warner vacuum system. Remy ignition. Presto 112-amp 6-volt battery.

COLE

Manufacturing of Coles began in Indianapolis, Ind., in 1909. When Cadillac revolutionized the industry with its V-8 in 1915, other automakers were forced to use V-8s. Cole purchased its version from the Northway Division of General Motors, which also built the Cadillac V-8. J.J. Cole believed that companies specializing in engine manufacture provided better power plants than independent automakers could produce themselves. He described his vehicles as "standardized cars."

1916 Cole 8-60 V-8
Northway V-8. Cast-iron blocks of four. Side valves. Bore & stroke: 3-1/2 x 4-1/2 in. Displacement: 346.3 cid. Taxable hp: 39.20. Three crankshaft bearings. Pump cooling. Force feed oiling. Stromberg carburetor. Carburetor fed by Stewart vacuum system. Delco dual ignition. Willard 6-volt battery.

1917 Cole 8-60 V-8
Northway V-8. Cast-iron blocks of four. Side valves. Bore & stroke: 3-1/2 x 4-1/2 in. Displacement: 346.3 cid. Taxable hp: 39.20. Three crankshaft bearings. Pump cooling. Force feed oiling. Stromberg V carburetor. Carburetor fed by Stewart vacuum system. Delco dual ignition. Willard 6-volt battery.

1918 Cole 8-70 V-8
Northway V-8. Cast-iron blocks of four. Side valves. Bore & stroke: 3-1/2 x 4-1/2 in. Displacement: 346.3 cid. Taxable hp: 39.20. Three crankshaft bearings. Pump cooling. Force feed oiling. Stromberg V carburetor. Carburetor fed by Stewart vacuum system. Delco dual ignition. Willard 6-volt battery.

1919 Cole 8-70 V-8
Northway V-8. Cast-iron blocks of four. Side valves. Bore & stroke: 3-1/2 x 4-1/2 in. Displacement: 346.3 cid. Taxable hp: 39.20. Three crankshaft bearings. Pump cooling. Force feed oiling. Stromberg V carburetor. Carburetor fed by Stewart vacuum system. Delco dual ignition. Willard 6-volt battery.

1920 Cole Aero-Eight 8-70 V-8
Northway V-8. Cast-iron blocks of four. Side valves. Bore & stroke: 3-1/2 x 4-1/2 in. Displacement: 346.3 cid. Brake hp: 80 at 2800 rpm. Taxable hp: 39.20. Three crankshaft bearings. Pump cooling. Force feed oiling. Stromberg V carburetor. Carburetor fed by Stewart vacuum system. Delco dual ignition. Willard 6-volt battery.

1921 Cole Aero-Eight V-8
Northway V-8. Cast-iron blocks of four. Side valves. Bore & stroke: 3-1/2 x 4-1/2 in. Displacement: 346.3 cid. Brake hp: 80 at 2800 rpm. Taxable hp: 39.20. Three crankshaft bearings. Pump cooling. Force feed oiling. Stromberg V carburetor. Carburetor fed by Stewart vacuum system. Delco dual ignition. Willard 6-volt battery.

1922 Cole 890 V-8
Northway V-8. Cast-iron blocks of four. Side valves. Bore & stroke: 3-1/2 x 4-1/2 in. Displacement: 346.3 cid. Brake hp: 80 at 2800 rpm. Taxable hp: 39.20. Three crankshaft bearings. Pump cooling. Force feed oiling. Johnson 1-1/2-inch carburetor. Carburetor fed by Stewart vacuum system. Delco dual ignition. Willard 6-volt battery.

1923 Cole 890 V-8
Northway V-8. Cast-iron blocks of four. Side valves. Bore & stroke: 3-1/2 x 4-1/2 in. Displacement: 346.3 cid. Brake hp: 80 at 2800 rpm. Taxable hp: 39.20. Three crankshaft bearings. Pump cooling. Force feed oiling. Johnson 1-1/2-inch carburetor. Carburetor fed by Stewart vacuum system. Delco dual ignition. Willard 6-volt battery.

1924 Cole 890 V-8
Northway V-8. Cast-iron blocks of four. Side valves. Bore & stroke: 3-1/2 x 4-1/2 in. Displacement: 346.3 cid. Brake hp: 80 at 2800 rpm. Taxable hp: 33.80. Three crankshaft bearings. Pump cooling. Force feed oiling. Johnson 1-1/2-inch carburetor. Carburetor fed by Stewart vacuum system. Delco dual ignition. Willard 6-volt battery.

1925 Cole 890 V-8
Northway V-8. Cast-iron blocks of four. Side valves. Bore & stroke: 3-1/2 x 4-1/2 in. Displacement: 296.0 cid. Brake hp: 90 at 2600 rpm. Taxable hp: 33.80. Compression ratio: 3.6:1. Three crankshaft bearings. Pump cooling. Force feed oiling. Johnson 1-1/2-inch carburetor. Carburetor fed by Stewart vacuum system. Delco dual ignition. Willard 6-volt battery.

CORD

1936 Model 810 and 1937 Model 812
Manufacturer: Lycoming Motors (a subsidiary of the Auburn Automobile Company)

Unsupercharged, 1936 and 1937Model: FB
Type: 90 degree V-8 Displacement: 288.6 cid Bore and stroke: 3-1/2 x 3-3/4 in. Compression ratio: 6.5:1 Valves: L head Crankshaft: Counterbalanced, 3 main bearings, 2.50 diameter Firing order: 1L, 3L, 4L, 2L, 2R, 1R, 3R, 4R Crankcase capacity: 8 quarts Coolant capacity: 28 quarts Cylinder heads: aluminum Pistons: aluminum, four rings Ignition: Autolite, single coil, single points, centrifugal advance only Oil filtration: none Brake horsepower: 117 at 3600 rpm Torque: 223 lbs.-ft. at 1800 rpm Carburetor: Stromberg EE-15, dual 1" throats Choke: Automatic

Supercharged, 1937 only
Same, except: Model: FC Compression ratio: 6.32:1 Firing order: 1L, 3L, 3R, 2L, 2R, 1R, 4L, 4R Ignition: Autolite, single coil, dual points, centrifugal advance only Brake horsepower: 186 -195 at 4200 rpm (A) Torque: 258 lbs.-ft. 2800 rpm – 273 lbs.-ft. at 3000 rpm Carburetor: Stromberg AA-25, dual 1-1/4" throats

NOTES: The range of horsepower and torque come from Lycoming charts. The lower figures were for test and very early engines. Lycoming tweaked the cam profiles and timing after these earliest engines, so nearly all production engines boasted the higher outputs.

The Cord V-8 was the only V-8 engine ever built by Lycoming, which manufactured engines for all the cars of the Cord Corporation (Auburn, Cord, and Duesenberg). It is believed that this engine was created in response to a request by Harold Ames, then president of Duesenberg, Incorporated, for a V-8 engine to power a new, smaller Duesenberg.

When Auburn decided that the new car would have front-wheel drive and be called a Cord, the engine was reversed for use with front drive.

This engine was designed from the beginning for the later addition of a centrifugal supercharger. The center cam journal was wide enough to accommodate the bevel gear that drove the supercharger, and bosses were provided in the block for future cooling passages. Like Lycoming's earlier Auburn V-12, the Cord engine's valves were nearly horizontal, and operated by roller-equipped rocker arms and a high-mounted camshaft. The FB engine fired one complete bank then the other. The supercharged FC engine used a more conventional back-and-forth order.

CUNNINGHAM

The Cunningham company, of Rochester, N.Y., competed with Cadillac, Cole, and others in the luxury-car market. After Cadillac brought out its V-8, Cunningham tried to match it with its own massive 442-cid, 45-hp V-8. Race driver Ralph DePalma drove a V-8-powered Cunningham Special Roadster 98 mph in a six-mile contest at Sheepshead Bay, N.Y.

1917 Cunningham V (V-8)
V-8. Cast-iron blocks of four. Side valves. Bore & stroke: 3-3/4 x 5 in. Displacement: 441.7 cid. Taxable hp: 45.00. Three crankshaft bearings. Pump cooling. Force feed oiling. Drilled crankshaft. Crankcase capacity: 12 qt. Cooling system capacity: 10 gal. Stromberg V carburetor. Carburetor fed by vacuum system. Westinghouse ignition. Willard battery.

1918 Cunningham V-2 (V-8)
V-8. Cast-iron blocks of four. Side valves. Bore & stroke: 3-3/4 x 5 in. Displacement: 441.7 cid. Taxable hp: 45.00. Three crankshaft bearings. Pump cooling. Force feed oiling. Drilled crankshaft. Crankcase capacity: 12 qt. Cooling system capacity: 10 gal. Stromberg V carburetor. Carburetor fed by vacuum system. Delco ignition. Westinghouse ignition. Willard battery.

1919 Cunningham V-3 (V-8)
V-8. Cast-iron blocks of four. Side valves. Bore & stroke: 3-3/4 x 5 in. Displacement: 441.7 cid. Taxable hp: 45.00. Three crankshaft bearings. Pump cooling. Force feed oiling. Drilled crankshaft. Crankcase capacity: 12 qt. Cooling system capacity: 10 gal. Stromberg V carburetor. Carburetor fed by vacuum system. Delco ignition. Willard battery.

1920 Cunningham (V-8)
V-8. Cast-iron blocks of four. Side valves. Bore & stroke: 3-3/4 x 5 in. Displacement: 441.7 cid. Brake hp: 95 at 2200 rpm. Taxable hp: 45.00. Three crankshaft bearings. Pump cooling. Force feed oiling. Drilled crankshaft. Crankcase capacity: 12 qt. Cooling system capacity: 10 gal. Zenith carburetor. Carburetor fed by vacuum system. Delco ignition. Willard 100-amp 6-volt battery.

1921 Cunningham (V-8)
V-8. Cast-iron blocks of four. Side valves. Bore & stroke: 3-3/4 x 5 in. Displacement: 441.7 cid. Brake hp: 95 at 2200 rpm. Taxable hp: 45.00. Three crankshaft bearings. Pump cooling. Force feed oiling. Drilled crankshaft. Crankcase capacity: 12 qt. Cooling system capacity: 10 gal. Zenith carburetor. Carburetor fed by vacuum system. Delco ignition. Willard 100-amp 6-volt battery.

1922 Cunningham V-5 (V-8)
V-8. Cast-iron blocks of four. Side valves. Bore & stroke: 3-3/4 x 5 in. Displacement: 441.7 cid. Brake hp: 95 at 2200 rpm. Taxable hp: 45.00. Three crankshaft bearings. Pump cooling. Force feed oiling. Drilled crankshaft. Crankcase capacity: 12 qt. Cooling system capacity: 10 gal. Stromberg 1-1/2-inch carburetor. Carburetor fed by vacuum system. Delco ignition. Willard 100-amp 6-volt battery.

1923 Cunningham V-5 (V-8)
V-8. Cast-iron blocks of four. Side valves. Bore & stroke: 3-3/4 x 5 in. Displacement: 441.7 cid. Brake hp: 95 at 2200 rpm. Taxable hp: 45.00. Three crankshaft bearings. Pump cooling. Force feed oiling. Drilled crankshaft. Crankcase capacity: 12 qt. Cooling system capacity: 10 gal. Stromberg 1-1/4-inch carburetor. Carburetor fed by vacuum system. Delco ignition. Willard 120-amp 6-volt battery.

1924 Cunningham V-5 (V-8)
V-8. Cast-iron blocks of four. Side valves. Bore & stroke: 3-3/4 x 5 in. Displacement: 441.7 cid. Brake hp: 95 at 2200 rpm. Taxable hp: 45.00. Three crankshaft bearings. Pump cooling. Force feed oiling. Drilled crankshaft. Crankcase capacity: 12 qt. Cooling system capacity: 10 gal. Stromberg 03 1-1/2-inch carburetor. Carburetor fed by vacuum system. Delco ignition. Willard 120-amp 6-volt battery.

1925 Cunningham V-6 (V-8)
V-8. Cast-iron blocks of four. Side valves. Bore & stroke: 3-3/4 x 5 in. Displacement: 441.7 cid. Compression ratio: 4.3:1. Brake hp: 90 at 2400 rpm. Taxable hp: 45.00. Three crankshaft bearings. Pump cooling. Force feed oiling. Drilled crankshaft. Crankcase capacity: 8 qt. Cooling system capacity: 10 gal. Stromberg 1-1/4-inch carburetor. Carburetor fed by vacuum system. Delco ignition. Willard 120-amp 6-volt battery.

1926 Cunningham V-6 (V-8)
V-8. Cast-iron blocks of four. Side valves. Bore & stroke: 3-3/4 x 5 in. Displacement: 441.7 cid. Compression ratio: 4.3:1. Brake hp: 90 at 2400 rpm. Taxable hp: 45.00. Three crankshaft bearings. Sylphon cooling with pump. Force feed oiling. Drilled crankshaft. Crankcase capacity: 8 qt. Cooling system capacity: 7 gal. Stromberg 1-1/4-inch carburetor. Carburetor fed by Stewart vacuum system. Delco ignition. Willard 120-amp 6-volt battery.

1927 Cunningham V-6 (V-8)
V-8. Cast-iron blocks of four. Side valves. Bore & stroke: 3-3/4 x 5 in. Displacement: 441.7 cid. Compression ratio: 4.3:1. Brake hp: 90 at 2400 rpm. Taxable hp: 45.00. Three crankshaft bearings. Sylphon cooling with pump. Force feed oiling. Drilled crankshaft. Crankcase capacity: 8 qt. Cooling system capacity: 7 gal. Stromberg 1-1/4-inch carburetor. Carburetor fed by Stewart vacuum system. Delco ignition. Willard 132-amp 6-volt battery.

1928 Cunningham V-7 (V-8)
V-8. Cast-iron blocks of four. Side valves. Bore & stroke: 3-3/4 x 5 in. Displacement: 441.7 cid. Compression ratio: 4.3:1. Brake hp: 95 at 2400 rpm. Taxable hp: 45.00. Three crankshaft bearings. Sylphon cooling with pump. Force feed oiling. Drilled crankshaft. Crankcase capacity: 8 qt. Cooling system capacity: 7-1/4 gal. Stromberg 1-1/2-inch carburetor. Carburetor fed by Stewart vacuum system. Delco-Remy ignition. Willard 135-amp 6-volt battery.

1929 Cunningham V-8
V-8. Cast-iron blocks of four. Side valves. Bore & stroke: 3-3/4 x 5 in. Displacement: 441.7 cid. Compression ratio: 5.0:1. Brake hp: 106 at 2400 rpm. Taxable hp: 45.00. Three crankshaft bearings. Sylphon cooling with pump. Force feed oiling. Drilled crankshaft. Crankcase capacity: 8 qt. Cooling system capacity: 7-1/4 gal. Stromberg 002 1-1/4-inch two-barrel carburetor. Carburetor fed by Stewart vacuum system. Delco-Remy ignition. Willard 135-amp 6-volt battery.

1930 Cunningham V-8
V-8. Cast-iron blocks of four. Side valves. Bore & stroke: 3-3/4 x 5 in. Displacement: 441.7 cid. Compression ratio: 5.0:1 (5.8:1 optional). Brake hp: 110 at 2500 rpm. Taxable hp: 45.00. Three crankshaft bearings. Sylphon cooling with pump. Force feed oiling. Drilled crankshaft. Crankcase capacity: 8 qt. Cooling system capacity: 7-1/4 gal. Stromberg 002 1-1/4-inch two-barrel carburetor. Carburetor fed by Stewart vacuum system. Delco-Remy ignition. Willard 135-amp 6-volt battery.

DANIELS

Daniels Motor Car Co. of Reading, Pa., was in operation from 1916-1924. The firm produced only V-8s. They initially used the same 331.8-cid Herschell-Spillman motor employed in the Abbott. A larger, more powerful V-8, built by Daniels, arrived in 1920. It was quite a "mill" and the Daniels was quite a car.

1916 Daniels Model A (V-8)
Hershell-Spillman V-8. Cast-iron blocks of four. Side valves. Bore & stroke: 3-1/4 x 5 in. Displacement: 331.8 cid. Taxable hp: 33.80. Three crankshaft bearings. Pump cooling. Force feed oiling. Zenith carburetor. Carburetor fed by Stewart vacuum system. Westinghouse ignition. Willard 100-amp 6-volt battery.

1917 Daniels Model A (V-8)
Hershell-Spillman V-8. Cast-iron blocks of four. Side valves. Bore & stroke: 3-1/4 x 5 in. Displacement: 331.8 cid. Taxable hp: 33.80. Three crankshaft bearings. Pump cooling. Force feed oiling. Zenith Model V carburetor. Carburetor fed by Stewart vacuum system. Westinghouse ignition. Willard 100-amp 6-volt battery.

1918 Daniels Model B (V-8)
Hershell-Spillman V-8. Cast-iron blocks of four. Side valves. Bore & stroke: 3-1/4 x 5 in. Displacement: 331.8 cid. Taxable hp: 33.80. Three crankshaft bearings. Pump cooling. Force feed oiling. Zenith Model V carburetor. Carburetor fed by Stewart vacuum system. Westinghouse ignition. Willard 100-amp 6-volt battery.

1919 Daniels Model B (V-8)
Hershell-Spillman V-8. Cast-iron blocks of four. Side valves. Bore & stroke: 3-1/4 x 5 in. Displacement: 331.8 cid. Taxable hp: 33.80. Three crankshaft bearings. Pump cooling. Force feed oiling. Zenith Model V carburetor. Carburetor fed by Stewart vacuum system. Westinghouse ignition. Willard 100-amp 6-volt battery.

1920 Daniels Model D-8/D-19 (V-8)
V-8. Cast-iron blocks of four. Side valves. Bore & stroke: 3-1/2 x 5-1/4 in. Displacement: 404.1 cid. Brake hp: 90 at 2000 rpm. Taxable hp: 39.2. Three crankshaft bearings. Pump cooling. Force feed oiling. Crankcase capacity: 8 qt. Cooling system capacity: 10-1/2 gal. Zenith Model V carburetor. Delco ignition. Willard 145-amp 6-volt battery.

1921 Daniels Model D-19 (V-8)
V-8. Cast-iron blocks of four. Side valves. Bore & stroke: 3-1/2 x 5-1/4 in. Displacement: 404.1 cid. Brake hp: 90 at 2150 rpm. Taxable hp: 39.2. Three crankshaft bearings. Pump cooling. Force feed oiling. Zenith 1-1/4-inch carburetor. Crankcase capacity: 8 qt. Cooling system capacity: 10-1/2 gal. Delco ignition. Willard 145-amp 6-volt battery.

1922 Daniels Model D-19/D-32 (V-8)
V-8. Cast-iron blocks of four. Side valves. Bore & stroke: 3-1/2 x 5-1/4 in. Displacement: 404.1 cid. Brake hp: 90 at 2150 rpm. Taxable hp: 39.2. Three crankshaft bearings. Pump cooling. Force feed oiling. Zenith 1-1/4-inch carburetor. Crankcase capacity: 8 qt. Cooling system capacity: 10-1/4 gal. Delco ignition. Willard 145-amp 6-volt battery.

1923 Daniels Model D-19 (V-8)
V-8. Cast-iron blocks of four. Side valves. Bore & stroke: 3-1/2 x 5-1/4 in. Displacement: 404.1 cid. Compression ratio: 4.44:1. Brake hp: 90 at 2150 rpm. Taxable hp: 39.2. Three crankshaft bearings. Pump cooling. Force feed oiling. Stromberg 1-11/16-inch carburetor. Crankcase capacity: 6 qt. Cooling system capacity: 10 gal. Delco ignition. Willard 140-amp 6-volt battery.

DOUGLAS

Omaha, Nebraska, was the home of the Douglas automobile. The Douglas Motor Corporation succeeded the Drummond Motor Co. and used the same engine. It was the 331.8-cid Herschell-Spillman power plant. The company advertised that it was "supplying the Middlewest with a high-powered car at a reasonable price."

1919 Douglas 8 (V-8)
Hershell-Spillman V-8. Cast-iron block. Cylinders cast in blocks of four. Valves in V. Bore & stroke: 3-1/4 x 5 in. Displacement: 331.8 cid. Taxable hp: 33.80. Three crankshaft bearings. Pump cooling. Force feed oiling. Zenith V carburetor. Carburetor fed by Stewart vacuum system. Westinghouse ignition. Willard 100-amp 6-volt battery.

DRUMMOND

The Drummond Motor Co. proceeded Douglas in Omaha. Both the 1917 Drummond Model 17 and the 1918-1919 Douglas Eight used the same engine. It was the 331.8-cid Herschell-Spillman power plant.

1919 Drummond Model 17 (V-8)
Hershell-Spillman V-8. Cast-iron block. Cylinders cast in blocks of four. Valves in V. Bore & stroke: 3-1/4 x 5 in. Displacement: 331.8 cid. Taxable hp: 33.80. Three crankshaft bearings. Pump cooling. Force feed oiling. Zenith V carburetor. Carburetor fed by Stewart vacuum system. Westinghouse ignition. Willard 100-amp 6-volt battery.

ECONOMY

Based in Tiffin, Ohio, the Economy Motor Co. produced mid-priced assembled automobiles from 1916-1922. The Economy Eight was marketed in 1918. The engine came from the Ferro engine works.

1918 Economy C8-48 (V-8)
Ferro V-8. Cast-iron block. Cylinders cast in blocks of four. Valves in V. Bore & stroke: 3 x 3-1/2 in. Taxable hp: 26.50. Two crankshaft bearings. Pump cooling. Force and splash. Schebler V carburetor. Carburetor fed by Stewart vacuum pump. Connecticut two-unit ignition. Willard battery.

HOLLIER

The man behind this car was Charles Lewis, the ex-boss at Jackson Automobile Co. The firm started naturally enough in Jackson, Mich., and later moved to Chelsea. The Hollier was fitted with a 40-hp V-8 of the maker's own design. In 1917 a six-cylinder Hollier was added. It outlasted the V-8, surviving until 1921.

1916 Hollier 8 (V-8)
V-8. Cast iron. Cylinders cast in blocks of four. Valves in V. Bore & stroke: 3 x 4-1/4 in. Displacement: 240.3 cid. Taxable hp: 28.80. Three crankshaft bearings. Thermo Syphon cooling. Force feed oiling. Stewart carburetor. Carburetor fed by gravity. Atwater Kent ignition. Willard 100-amp 6-volt battery.

1917 Hollier 178 (V-8)
V-8. Cast iron. Cylinders cast in blocks of four. Valves in V. Bore & stroke: 3 x 4-1/4 in. Displacement: 240.3 cid. Taxable hp: 28.80. Three crankshaft bearings. Thermo Syphon cooling. Force feed oiling. Stewart V carburetor. Carburetor fed by gravity. Atwater Kent ignition. Willard 100-amp 6-volt battery.

1918 Hollier 188 (V-8)
V-8. Cast iron. Cylinders cast in blocks of four. Valves in V. Bore & stroke: 3 x 4-1/4 in. Displacement: 240.3 cid. Taxable hp: 28.80. Three crankshaft bearings. Thermo Syphon cooling. Force feed oiling. Stewart V carburetor. Carburetor fed by gravity. Atwater Kent ignition. Willard 100-amp 6-volt battery.

1919 Hollier 198 (V-8)
V-8. Cast iron. Cylinders cast in blocks of four. Valves in V. Bore & stroke: 3 x 4-1/4 in. Displacement: 240.3 cid. Taxable hp: 28.80. Three crankshaft bearings. Thermo Syphon cooling. Force feed oiling. Stewart V carburetor. Carburetor fed by gravity. Atwater Kent ignition. Willard 100-amp 6-volt battery.

HOMER/LAUGHLIN

One of few makes indigenous to the West Coast, this car was powered by a 16-hp V-8 of the company's own design. The car and company lasted just two years (See 1918 Laughlin). Homer Laughlin Engineering went on to make auxiliary transmissions for Model T Fords.

1917 Homer Laughlin Model D (V-8)
V-8. Cast iron. Cylinders cast in blocks of four. Valves in V. Bore & stroke: 2-1/4 x 3-3/4 in. Displacement: 119.3 cid. Taxable hp: 16.20. Two crankshaft bearings. Thermo Syphon cooling. Splash oiling. Ensign V carburetor. Carburetor fed by pressure. Atwater Kent dual ignition.

JACKSON

"No Hill Too Steep, No Sand Too Deep," was the slogan of this Jackson, Mich., automaker. It started in 1903 and got into the V-8 market niche only a year after Cadillac did. The engines used in these cars were sourced from Ferro, a maker of proprietary engines also used in the Economy Model 48 and the Scripps-Booth model H.

1916 Jackson 348 (V-8)
Ferro V-8. Cast iron. Cylinders cast in blocks of four. Valves in V. Bore & stroke: 2-7/8 x 4-3/4 in. Displacement: 246.7 cid. Taxable hp: 26.50. Two crankshaft bearings. Pump cooling. Force oiling. Schebler carburetor. Carburetor fed by Stewart vacuum pump. Connecticut two-unit ignition. Willard battery.

1916 Jackson 68 (V-8)
Ferro V-8. Cast iron. Cylinders cast in blocks of four. Valves in V. Bore & stroke: 3-1/2 x 4-1/2 in. Displacement: 327.1 cid. Taxable hp: 39.20. Three crankshaft bearings. Pump cooling. Force oiling. Schebler carburetor. Carburetor fed by Stewart vacuum pump. Connecticut two-unit ignition. Willard battery.

1917 Jackson Wolverine Model 349 (V-8)
Ferro V-8. Cast iron. Cylinders cast in blocks of four. Valves in V. Bore & stroke: 3 x 3-1/2 in. Displacement: 197.9 cid. Taxable hp: 28.80. Three crankshaft bearings. Thermo-Syphon cooling. Force oiling. Zenith V carburetor. Carburetor fed by Stewart vacuum pump. Remy single-unit ignition. Willard battery.

1918 Jackson Wolverine VIII (V-8)
Ferro V-8. Cast iron. Cylinders cast in blocks of four. Valves in V. Bore & stroke: 3 x 3-1/2 in. Displacement: 197.9 cid. Taxable hp: 28.80. Three crankshaft bearings. Thermo-Syphon cooling. Force oiling. Zenith V carburetor. Carburetor fed by Stewart vacuum pump. Autolite single-unit ignition. Willard six-volt battery.

KING

The King was built in Detroit for most of its life and in Buffalo, N.Y. its last two years (1923-1924). It was in December 1914, only two months after the launch of the Cadillac V-8, that the King Model D hit the market as "the world's first popular-priced eight." It was $1,350. The V-8 was its own design. Charles Brady King was also responsible for the design of the King-Bugatti aero engine. Unfortunately, his car company's fortunes went downhill after 1916 when only V-8s were built.

1915 King Model D (V-8)
V-8. Cast iron. Cylinders cast in blocks of four. Valves in V. Bore & stroke: 2-3/4 x 5 in. Displacement: 236 cid. Brake hp: 40/50. Taxable hp: 26.50. Three crankshaft bearings. Thermo-Syphon cooling. Force feed and splash oiling. Zenith carburetor. Carburetor fed by Stewart vacuum pump. Atwater-Kent single two-unit ignition. Willard 80 battery.

1916 King Model D (V-8)
V-8. Cast iron. Cylinders cast in blocks of four. Valves in V. Bore & stroke: 2-7/8 x 5 in. Displacement: 259.7 cid. Taxable hp: 26.50. Three crankshaft bearings. Thermo-Syphon cooling. Force feed and splash oiling. Zenith carburetor. Carburetor fed by Stewart vacuum pump. Atwater-Kent single two-unit ignition. Willard 80 battery.

1916 King Model E (V-8)
V-8. Cast iron. Cylinders cast in blocks of four. Valves in V. Bore & stroke: 3 x 5 in. Displacement: 282.7 cid. Brake hp: 60 at 2400 rpm. Taxable hp: 8.80. Three crankshaft bearings. Thermo-Syphon cooling. Force feed and splash oiling. Ball & Ball carburetor. Carburetor fed by Stewart vacuum pump. Atwater-Kent single two-unit ignition. Willard 80 battery.

1917 King Model E (V-8)
V-8. Cast iron. Cylinders cast in blocks of four. Valves in V. Bore & stroke: 3 x 5 in. Displacement: 282.7 cid. Brake hp: 60 at 2400 rpm. Taxable hp: 28.80. Three crankshaft bearings. Thermo-Syphon cooling. Force feed oiling. Ball & Ball V carburetor. Carburetor fed by vacuum. Atwater-Kent single H ignition. Willard 80 battery.

1918 King Model EE (V-8)
V-8. Cast iron. Cylinders cast in blocks of four. Valves in V. Bore & stroke: 3 x 5 in. Displacement: 282.7 cid. Brake hp: 60 at 2400 rpm. Taxable hp: 28.80. Three crankshaft bearings. Thermo-Syphon cooling. Force feed oiling. Ball & Ball V carburetor. Carburetor fed by vacuum. Atwater-Kent single H ignition. Willard 80 battery.

1919 King Model F (V-8)
V-8. Cast iron. Cylinders cast in blocks of four. Valves in head. Bore & stroke: 3 x 5 in. Displacement: 282.7 cid. Brake hp: 60 at 2400 rpm. Taxable hp: 28.80. Three crankshaft bearings. Thermo-Syphon cooling. Force feed oiling. Ball & Ball V carburetor. Carburetor fed by vacuum. Atwater-Kent single H ignition. Willard 80 battery.

1920 King Model H (V-8)
V-8. Cast iron. Cylinders cast in blocks of four. Side valve. Bore & stroke: 3 x 5 in. Displacement: 282.7 cid. Brake hp: 60 at 2400 rpm. Taxable hp: 28.80. Three crankshaft bearings. Thermo-Syphon cooling. Force feed oiling. Ball & Ball 1-1/4-in. carburetor. Carburetor fed by Stewart vacuum tank. Atwater-Kent single H ignition. Prestolite battery.

1921 King Model H (V-8)
V-8. Cast iron. Cylinders cast in blocks of four. Side valve. Bore & stroke: 3 x 5 in. Displacement: 282.7 cid. Brake hp: 60 at 2400 rpm. Taxable hp:

28.80. Three crankshaft bearings. Thermo-Syphon cooling. Force feed oiling. Ball & Ball 1-1/4-in. carburetor. Carburetor fed by Stewart vacuum tank. Atwater-Kent single H ignition. Prestolite battery.

1922 King Model J (V-8)
V-8. Cast iron. Cylinders cast in blocks of four. Side valve. Bore & stroke: 3 x 5 in. Displacement: 282.7 cid. Brake hp: 60 at 2400 rpm. Taxable hp: 28.80. Three crankshaft bearings. Thermo-Syphon cooling. Force feed oiling. Ball & Ball 1-1/4-in. carburetor. Carburetor fed by Stewart vacuum tank. Westinghouse ignition. Prestolite battery.

1923 King Model L (V-8)
V-8. Cast iron. Cylinders cast in blocks of four. Side valve. Bore & stroke: 3 x 5 in. Displacement: 282.7 cid. Brake hp: 60 at 2400 rpm. Taxable hp: 28.80. Three crankshaft bearings. Thermo-Syphon cooling. Force feed oiling. Ball & Ball 1-1/4-in. carburetor. Carburetor fed by Stewart vacuum tank. Atwater-Kent ignition. Prestolite battery.

1924 King Model L (V-8)
V-8. Cast iron. Cylinders cast in blocks of four. Side valve. Bore & stroke: 3 x 5 in. Displacement: 282.7 cid. Brake hp: 60 at 2400 rpm. Taxable hp: 28.80. Three crankshaft bearings. Thermo-Syphon cooling. Force feed oiling. Ball & Ball 1-1/4-in. carburetor. Carburetor fed by Stewart vacuum tank. Atwater-Kent ignition. Prestolite battery. Oil capacity: 8 qt. Cooling system capacity: 9 gal.

LA FAYETTE

With many ex-Cadillac employees working for it, there's little surprise this automaker preferred V-8s and powerful 90/100-hp versions at that. Charlie Nash was working, behind the scenes at first, to produce La Fayettes with facilities in Indianapolis, Ind., and Milwaukee, Wis. The engine was of the company's own design and got just 10 miles per gallon of gas.

1921 La Fayette Model 134 (V-8)
V-8. L-head. Cast iron. Cylinders cast in blocks of four. Valves in block. Bore & stroke: 3-1/4 x 5-1/4 in. Displacement: 348.4 cid. Brake hp: 90 at 2700 rpm. Taxable hp: 33.8. Five crankshaft bearings. Pump cooling. Cooling system capacity: 6 gal. Force feed and splash oiling. Crankcase capacity: 7 qt. Johnson 2-in. carburetor. Delco ignition. Exide 6-130 battery.

1922 La Fayette Model 134 (V-8)
V-8. L-head. Cast iron. Cylinders cast in blocks of four. Valves in block. Bore & stroke: 3-1/4 x 5-1/4 in. Displacement: 348.4. Brake hp: 100 at 2750 rpm. Five crankshaft bearings. Pump cooling. Cooling system capacity: 6 gal. Force feed and splash oiling. Crankcase capacity: 7 qt. Johnson 2-in. carburetor. Delco ignition. Exide 6-150 battery.

1923 La Fayette Model 134 (V-8)
V-8. L-head. Cast iron. Cylinders cast in blocks of four. Valves in block. Bore & stroke: 3-1/4 x 5-1/4 in. Displacement: 348.4. Compression ratio: 4.8:1. Brake hp: 100 at 2750 rpm. Five crankshaft bearings. Pump cooling. Cooling system capacity: 6 qt. Force feed and splash oiling. Crankcase capacity: 7 qt. Johnson 2-in. carburetor. Delco ignition. Exide 6-150 battery.

1924 La Fayette Model 134 (V-8)
V-8. L-head. Cast iron. Cylinders cast in blocks of four. Valves in block. Bore & stroke: 3-1/4 x 5-1/4 in. Displacement: 348.4. Compression ratio: 4.8:1. Brake hp: 100 at 2750 rpm. Five crankshaft bearings. Pump cooling. Cooling system capacity: 6 qt. Force feed and splash oiling. Crankcase capacity: 7 qt. Johnson 2-in. carburetor. Delco ignition. Exide 6-150 battery.

LAUGHLIN

One of few makes indigenous to the West Coast, this car was powered by a 16-hp V-8 of the company's own design. It was known as the Homer Laughlin in 1917 and the Laughlin in 1918. Homer Laughlin Engineering went on to make auxiliary transmissions for Model T Fords.

1918 Laughlin (V-8)
V-8. Cast iron. Cylinders cast in blocks of four. Valves in V. Bore & stroke: 2-1/4 x 3-3/4 in. Displacement: 119.3 cid. Taxable hp: 16.20. Two crankshaft bearings. Thermo Syphon cooling. Splash oiling. Zenith V carburetor. Carburetor fed by Stewart vacuum tank. Atwater Kent dual ignition.

MONARCH

There were numerous Monarch automobiles but only one Monarch Eight. And it was made for just one year, in Detroit, Mich. In fact, it was in production for just a short time during that year. The engine was designed by Robert Craig Hupp, of Hupmobile fame. Robert was also the brother-in-law of Joseph Bloom, who organized Monarch Motor Car Co.

1916 Monarch (V-8)
V-8. Cast iron. Cylinders cast in blocks of four. Valves in V. Bore & stroke: 3 x 5 in. Displacement: 282.7 cid. Taxable hp: 28.40. Three crankshaft bearings. Pump cooling. Force feed oiling. Zenith V carburetor. Carburetor fed by Stewart vacuum tank. Atwater Kent single ignition.

MURRAY

A Rolls-like radiator and a Herschell-Spillman-supplied V-8 engine characterized the early products of William M. Murray's company, which operated in Pittsburgh, Pa., during this time span (and later in Newark, N.J. and Boston, Mass.). The V-8 model was introduced at the New York Automobile Salon in December 1916 and survived only two model years.

1917 Murray Model 70-T (V-8)
Herschell-Spillman V-8. Cast iron. Cylinders cast in blocks of four. Valves in V. Bore & stroke: 3-1/4 x 5 in. Displacement: 331.8 cid. Taxable hp: 33.80. Three crankshaft bearings. Thermo Syphon cooling. Force and splash oiling. Zenith carburetor. Carburetor fed by Stewart vacuum tank. Splitdorf Dual-H ignition.

1918 Murray Model 70-T (V-8)
Herschell-Spillman V-8. Cast iron. Cylinders cast in blocks of four. Valves in V. Bore & stroke: 3-1/4 x 5 in. Displacement: 331.8 cid. Taxable hp: 33.80. Three crankshaft bearings. Thermo Syphon cooling. Force and splash oiling. Zenith V carburetor. Carburetor fed by Stewart vacuum tank. Splitdorf Dual-H ignition.

OAKLAND

Carriage maker Edward M. Murphy died soon after forming Oakland and General Motors took it over. The new-for-1916 Model 50 featured the company's first V-8. This 346.3-cid monster produced 71 hp and featured aluminum pistons and a counterbalanced crankshaft. V-8 serial numbers ran from 500000 to 502000, indicating that some 2,000 were built in 1916 and 1917. In 1930, four years after launching the Pontiac "Chief of the Sixes," Oakland released a new V-8 model with an engine from GM's Northway Division. This 100-hp car failed to sell well and became the senior Pontiac model in 1932.

1916 Oakland 50 (V-8)
GM Northway V-8. Cast iron. Cylinders cast in blocks of four. Valves in V. Bore & stroke: 3-1/2 x 4-1/2 in. Displacement: 346.3 cid. Brake hp: 71 at 2600 rpm. Taxable hp: 39.20. Three crankshaft bearings. Pump cooling. Force and splash oiling. Stromberg carburetor. Carburetor fed by Stewart vacuum tank. Delco single ignition. This V-8 featured aluminum pistons and a counter-balanced crankshaft.

1917 Oakland 50 (V-8)
GM Northway V-8. Cast iron. Cylinders cast in blocks of four. Valves in V. Bore & stroke: 3-1/2 x 4-1/2 in. Displacement: 346.3 cid. Brake hp: 71 at 2600 rpm. Taxable hp: 39.20. Three crankshaft bearings. Pump cooling. Force and splash oiling. Stromberg carburetor. Carburetor fed by Stewart vacuum tank. Delco single ignition. This V-8 featured aluminum pistons and a counter-balanced crankshaft.

1930 Oakland 101 V-8
L-head. V-block. Eight. Cast-iron block. Bore & stroke: 3-7/16 x 3-3/8 in. Displacement: 251 cid. Compression ratio: 5.1:1. Brake hp: 85 at 3200

rpm. NACC hp: 37.80. Main bearings: three. Valve lifters: Solid. Pump cooling. Carburetor: Marvel 1-3/4-in. one-barrel.

1931 Oakland V-8
L-head. V-block. Eight. Cast-iron block. Bore & stroke: 3-7/16 x 3-3/8 in. Displacement: 251 cid. Compression ratio: 5.0:1. Brake hp: 85 at 3200 rpm. NACC hp: 37.80. Main bearings: three. Valve lifters: Solid. Pump cooling. Carburetor: Marvel 1-3/4-in. one-barrel.
Note: In 1932 the Oakland V-8 became the Pontiac V-8.

PEERLESS

With this Cleveland company, the name said it all. It was the third "P" in the Packard-Pierce-Arrow-Peerless triangle of luxury marques. Its history dates from 1900, so the V-8 was a later, keep-up-with-Caddy, development. Trade journals listed this power plant as being of Peerless manufacture, though it was dimensionally identical to the popular Herschell-Spillman V-8. It was known as the "Equipoised Eight" after 1925, when it gained detachable cylinder heads.

1916 Peerless Model 56 (V-8)
V-8. Cast iron. Cylinders cast in blocks of four. Valves in V. Bore & stroke: 3-1/4 x 5 in. Displacement: 331.8 cid. Brake hp: 80 at 2700 rpm. Taxable hp: 33.84. Three crankshaft bearings. Pump cooling. Force oiling. Peerless carburetor. Carburetor fed by Stewart vacuum tank. Atwater-Kent single ignition.

1917 Peerless Model 56-2 (V-8)
V-8. Cast iron. Cylinders cast in blocks of four. Valves in V. Bore & stroke: 3-1/4 x 5 in. Displacement: 331.8 cid. Brake hp: 80 at 2700 rpm. Taxable hp: 33.84. Three crankshaft bearings. Pump cooling. Force oiling. Ball & Ball 1-1/4-in. carburetor. Carburetor fed by Stewart vacuum tank. Atwater-Kent single ignition.

1918 Peerless Model 56-3 (V-8)
V-8. Cast iron. Cylinders cast in blocks of four. Valves in V. Bore & stroke: 3-1/4 x 5 in. Displacement: 331.8 cid. Brake hp: 80 at 2700 rpm. Taxable hp: 33.84. Three crankshaft bearings. Pump cooling. Force oiling. Ball & Ball 1-1/4-in. carburetor. Carburetor fed by Stewart vacuum tank. Atwater-Kent single ignition.

1919 Peerless Model 56-4 (V-8)
V-8. Cast iron. Cylinders cast in blocks of four. Valves in V. Bore & stroke: 3-1/4 x 5 in. Displacement: 331.8 cid. Brake hp: 80 at 2700 rpm. Taxable hp: 33.84. Three crankshaft bearings. Pump cooling. Force oiling. Ball & Ball 1-1/4-in. carburetor. Carburetor fed by Stewart vacuum tank. Atwater-Kent single ignition.

1920 Peerless Model 56-5 (V-8)
V-8. Cast iron. Cylinders cast in blocks of four. Valves in V. Bore & stroke: 3-1/4 x 5 in. Displacement: 331.8 cid. Brake hp: 80 at 2700 rpm. Taxable hp: 33.84. Three crankshaft bearings. Pump cooling. Force oiling. Ball & Ball 1-1/4-in. carburetor. Carburetor fed by Stewart vacuum tank. Atwater-Kent single ignition.

1921 Peerless Model 56-6 (V-8)
V-8. Cast iron. Cylinders cast in blocks of four. Valves in V. Bore & stroke: 3-1/4 x 5 in. Displacement: 331.8 cid. Brake hp: 80 at 2800 rpm. Taxable hp: 33.84. Three crankshaft bearings. Pump cooling. Force oiling. Ball & Ball 1-1/8-in. carburetor. Carburetor fed by Stewart vacuum tank. Atwater-Kent single ignition.

1922 Peerless Model 56-7 (V-8)
V-8. Cast iron. Cylinders cast in blocks of four. Valves in V. Bore & stroke: 3-1/4 x 5 in. Displacement: 331.8 cid. Brake hp: 80 at 2800 rpm. Taxable hp: 33.84. Three crankshaft bearings. Pump cooling. Force oiling. Ball & Ball 1-1/8-in. carburetor. Carburetor fed by Stewart vacuum tank. Atwater-Kent single ignition.

1923 Peerless Model 66 (V-8)
V-8. Cast iron. Cylinders cast in blocks of four. Valves in V. Bore & stroke: 3-1/4 x 5 in. Displacement: 331.8 cid. Brake hp: 70 at 2700 rpm. Taxable hp: 33.84. Three crankshaft bearings. Pump cooling. Force oiling. Ball & Ball 1-1/8-in. carburetor. Carburetor fed by Stewart vacuum tank. Delco ignition.

1924 Peerless Model 66 (V-8)
V-8. Cast iron. Cylinders cast in blocks of four. Valves in V. Bore & stroke: 3-1/4 x 5 in. Displacement: 331.8 cid. Brake hp: 75 at 2000 rpm. Taxable hp: 33.84. Three crankshaft bearings. Pump cooling. Force oiling. Ball & Ball 1-1/8-in. carburetor. Carburetor fed by Stewart vacuum tank. Delco ignition.

1925 Peerless Model 8-67 (V-8)
V-8. Cast iron. Cylinders cast in blocks of four. Valves in V. Bore & stroke: 3-1/4 x 5 in. Displacement: 331.8 cid. Brake hp: 70 at 2700 rpm. Taxable hp: 33.84. Three crankshaft bearings. Pump cooling. Force oiling. Stromberg OE3 carburetor. Carburetor fed by Stewart vacuum tank. Delco ignition.

1926 Peerless Model 8-67 (V-8)
V-8. Cast iron. Cylinders cast in blocks of four. Valves in V. Bore & stroke: 3-1/4 x 5 in. Displacement: 331.8 cid. Compression ratio: 4.9:1. Brake hp: 70 at 2700 rpm. Taxable hp: 33.84. Three crankshaft bearings. Pump cooling. Force oiling. Stromberg 1-1/4-in. carburetor. Carburetor fed by Stewart vacuum tank. Delco ignition.

1927 Peerless Model 8-69 (V-8)
V-8. Cast iron. Cylinders cast in blocks of four. Valves in V. Bore & stroke: 3-1/4 x 5 in. Displacement: 331.8 cid. Compression ratio: 4.55:1. Brake hp: 80 at 2800 rpm. Taxable hp: 33.84. Three crankshaft bearings. Pump cooling. Force oiling. Stromberg OF3 1-1/2-in. carburetor. Carburetor fed by Stewart vacuum tank. Delco ignition.

1928 Peerless Model 8-69 (V-8)
V-8. Cast iron. Cylinders cast in blocks of four. Valves in V. Bore & stroke: 3-1/4 x 5 in. Displacement: 331.8 cid. Compression ratio: 4.55:1. Brake hp: 80 at 2800 rpm. Taxable hp: 33.84. Three crankshaft bearings. Pump cooling. Force oiling. Stromberg OE3 1-1/2-in. carburetor. Carburetor fed by Stewart vacuum tank. Delco ignition.

PILOT

"The Car Ahead" was made in Richmond, Ind. The 8-55 model had a one-year run in 1916. Like the Pilot automobile itself, there was nothing really special about the new engine and only sixes were offered afterwards, until the firm folded in 1924.

1916 Pilot 8-55 V-8
V-8. Cast-iron blocks of four. Valves located in V. Bore & stroke: 3 x 5-1/8 in. Displacement: 289.8 cid. Taxable hp: 28.80. Three crankshaft bearings. Pump cooling. Force feed oiling. Zenith carburator. Carburetor fed by Stewart Warner vacuum tank. Westinghouse single ignition.Westinghouse battery.

REGAL

This Detroit-built beauty went to a V-8 in 1916. The Model F-8, along with the company's four-cylinder models (Regal never made a six), was designed by S.G. Jenks and supplied by the Port Huron Construction Co.

1916 Regal F-8 (V-8)
V-8. Cast-iron blocks of four. Valves located in V. Bore & stroke: 3 x 4-1/2 in. Displacement: 254.4 cid. Taxable hp. 28.80. Two crankshaft bearings. Pump cooling. Force and splash feed oiling. Stewart carburetor. Carburetor fed by Stewart vacuum tank. Connecticut single ignition. Gould 50 battery.

1917 Regal F-8 (V-8)
V-8. Cast-iron blocks of four. Valves located in V. Bore & stroke: 3 x 4-1/2 in. Displacement: 254.4 cid. Taxable hp: 28.80. Two crankshaft bearings. Pump cooling. Force and splash feed oiling. Johnson V carburetor. Carburetor fed by Stewart vacuum tank. Heinz single ignition.

ROCK FALLS

The Rock Falls Manufacturing Co., of Sterling, Ill., started out as a custom body maker and hearse builder in 1878. It switched to motorized hearses in 1909. The only V-8 the company tried was the Model 10, which employed Herschell-Spillman's 332-cid V-8. The firm made passenger cars from 1919-1925, but none of these were V-8 powered.

1916 Rock Falls Model 10 (V-8)
Hershell-Spillman V-8. Cast-iron blocks of four. Valves located in V. Bore

& stroke: 3-1/4 x 5 in. Displacement: 331.8 cid. Taxable hp: 28.80. Three crankshaft bearings. Pump cooling. Force and splash feed oiling. Zenith carburater. Carburetor fed by Stewart vacuum tank. Westinghouse ignition.

ROSS

The Ross was built in Detroit between 1915 and 1918. The firm's first cars used the Herschell-Spillman V-8. In October 1916, a Ross Eight climbed Fillmore St. Hill in high gear, making it the first Detroit-built car to do so.

1916 Ross Model A (V-8)
V-8. Cast-iron blocks of four. Valves located in V. Bore & stroke: 3 x 4-1/2 in. Displacement: 254.4 cid. Taxable hp: 28.80. Three crankshaft bearings. Thermo Syphon cooling. Force feed oiling. Zenith carburetor. Carburetor fed by Stewart vacuum tank. Atwater Kent single ignition. Detroit 80 battery.

1916 Ross Model C (V-8)
Hershell-Spillman V-8. V-8. Cast-iron blocks of four. Valves located in V. Bore & stroke: 3-1/4 x 5 in. Displacement: 331.8 cid. Taxable hp: 33.84. Three crankshaft bearings. Pump cooling. Force feed oiling. Zenith carburetor. Carburetor fed by Stewart vacuum tank. Atwater Kent single ignition.

1917 Ross Model C (V-8)
Hershell-Spillman V-8. V-8. Cast-iron blocks of four. Valves located in V. Bore & stroke: 3-1/4 x 5 in. Displacement: 331.8 cid. Taxable hp: 33.84. Three crankshaft bearings. Pump cooling. Force feed oiling. Zenith V carburetor. Carburetor fed by Stewart vacuum tank. Atwater Kent single ignition.

1918 Ross 8 (V-8)
Hershell-Spillman V-8. V-8. Cast-iron blocks of four. Valves located in V. Bore & stroke: 3-1/4 x 5 in. Displacement: 331.8 cid. Taxable hp: 33.84. Three crankshaft bearings. Pump cooling. Force and splash oiling. Zenith V carburetor. Carburetor fed by Stewart vacuum tank. Connecticut single ignition.

SCRIPPS-BOOTH

Newspaper scion James Scripps-Booth launched his Detroit-built car in 1912 and added a V-8 just after Cadillac made the engine type popular in fancy production cars. (Scripps-Booth was the second maker to have one). The 162.3-cid engine, designed by auto pioneer Alanson P. Brush, was manufactured by Ferro and could power the Model D to 75 mph. In 1917, Scripps-Booth joined the GM family and soon a Northway six was the only available engine. The final year for the V-8 was 1918.

1916 Scripps-Booth Model D (V-8)
Ferro V-8. Cast-iron blocks of four. Valves located in V. Bore & stroke: 2-5/8 x 3-3/4 in. Displacement: 162.3 cid. Horsepower: 35. Taxable hp: 22.00. Two crankshaft bearings. Thermo Syphon cooling. Force feed oiling. Zenith carburetor. Carburetor fed by vacuum. Atwater Kent single ignition. Willard 35 battery.

1917 Scripps-Booth Model D (V-8)
Ferro V-8. Cast-iron blocks of four. Valves located in head. Bore & stroke: 2-5/8 x 3-3/4 in. Displacement: 162.3 cid. Horsepower: 35. Taxable hp: 22.00. Two crankshaft bearings. Thermo Syphon cooling. Force feed oiling. Zenith carburetor. Carburetor fed by vacuum. Remy single A ignition. Willard 35 battery.

1918 Scripps-Booth Model H (V-8)
Ferro V-8. Cast-iron blocks of four. Valves located in head. Bore & stroke: 2-5/8 x 3-3/4 in. Displacement: 162.3 cid. Horsepower: 35. Taxable hp: 22.00. Two crankshaft bearings. Thermo Syphon cooling. Force feed and splash oiling. Zenith carburetor. Carburetor fed by vacuum. Remy single A ignition. Willard 35 battery.

SHERIDAN

Sheridan Motor Car Co. of Muncie, Ind., was a wholly-owned subsidiary of General Motors. The Model D, which was offered only one year, utilized an "80-hp" Northway V-8. In reality, it is believed that only a few pilot versions of this car were built.

1921 Sheridan Model D (V-8)
GM Northway V-8. Cast iron. Cylinders cast in blocks of four. Valves in V. Bore & stroke: 3-1/2 x 4-1/2 in. Displacement: 346.3 cid. Brake hp: 71 at 2600 rpm. Taxable hp: 39.20. Three crankshaft bearings. Pump cooling. Force oiling. Stromberg carburetor. Carburetor fed by Stewart vacuum tank. Delco single ignition. This V-8 featured aluminum pistons and a counter-balanced crankshaft.

Many American cars used the name Standard, but just one offered a V-8 for awhile. This Standard was made by the Standard Steel Car Co. of Butler, Pa. between 1914-1923. The V-8 versions were offered from 1916 onwards. Herschell-Spillman V-8 engines were used. The Standard Eight was known as "the Monarch of the Mountains."

STANDARD

1916 Standard 8 (V-8)
V-8. Cast-iron block. Cast in blocks of four. Valves located in V. Bore & stroke: 3 x 5 in. Displacement: 282.7 cid. Taxable hp: 28.80. Three crankshaft bearings. Pump cooling. Force feed oiling. Zenith carburetor. Carburetor fed by Stewart vacuum tank. Westinghouse single ignition. Gould 80 battery.

1917 Standard Model F (V-8)
Hershell-Spillman V-8. Cast-iron block. Cast in blocks of four. Valves located in V. Bore & stroke: 3-1/4 x 5 in. Displacement: 331.8 cid. Brake hp: 70 at 2200 rpm. Taxable hp: 33.80. Three crankshaft bearings. Pump cooling. Force feed oiling. Zenith V carburetor. Carburetor fed by Stewart vacuum tank. Splitdorf ignition.

1917 Standard Model E (V-8)
V-8. Cast-iron block. Cast in blocks of four. Valves located in V. Bore & stroke: 3 x 5 in. Displacement: 282.7 cid. Taxable hp: 28.80. Three crankshaft bearings. Pump cooling. Force feed oiling. Zenith carburetor. Carburetor fed by Stewart vacuum tank. Westinghouse single ignition.

1918 Standard Model G (V-8)
Hershell-Spillman V-8. Cast-iron block. Cast in blocks of four. Valves located in V. Bore & stroke: 3-1/4 x 5 in. Displacement: 331.8 cid. Brake hp: 70 at 2200 rpm. Taxable hp: 33.80. Three crankshaft bearings. Pump cooling. Force feed oiling. Zenith V carburetor. Carburetor fed by Stewart vacuum tank. Splitdorf ignition. Willard battery.

1919 Standard Model 8-G (V-8)
Hershell-Spillman V-8. Cast-iron block. Cast in blocks of four. Valves located in V. Bore & stroke: 3-1/4 x 5 in. Displacement: 331.8 cid. Brake hp: 70 at 2200 rpm. Taxable hp: 33.80. Three crankshaft bearings. Pump cooling. Force feed oiling. Zenith S carburetor. Carburetor fed by Stewart vacuum tank. Splitdorf ignition. Willard battery.

1920 Standard Model 8 (V-8)
Hershell-Spillman V-8. Cast-iron block. Cast in blocks of four. Valves located in V. Bore & stroke: 3-1/4 x 5 in. Displacement: 331.8 cid. Brake hp: 70 at 2200 rpm. Taxable hp: 33.80. Three crankshaft bearings. Pump cooling. Force feed oiling. Zenith 1-1/4 carburetor. Carburetor fed by Stewart vacuum tank. Atwater-Kent ignition.

1921 Standard Model 8-1 (V-8)
Hershell-Spillman V-8. Cast-iron block. Cast in blocks of four. Valves located in V. Bore & stroke: 3-1/4 x 5 in. Displacement: 331.8 cid. Brake hp: 65 at 2400 rpm. Taxable hp: 33.80. Three crankshaft bearings. Pump cooling. Force feed oiling. Zenith 1-1/4 carburetor. Carburetor fed by Stewart vacuum tank. Atwater-Kent ignition. Willard battery.

1922 Standard Model 11 (V-8)
Hershell-Spillman V-8. Cast-iron block. Cast in blocks of four. Valves located in V. Bore & stroke: 3-1/4 x 5 in. Displacement: 331.8 cid. Brake hp: 65 at 2200 rpm. Taxable hp: 33.80. Three crankshaft bearings. Pump cooling. Force feed oiling. Zenith 1-1/4 carburetor. Carburetor fed by Stewart vacuum tank. Splitdorf ignition. Willard 6-100 battery.

1923 Standard Model 11 (V-8)
Hershell-Spillman V-8. Cast-iron block. Cast in blocks of four. Valves located in V. Bore & stroke: 3-1/4 x 5 in. Displacement: 331.8 cid. Brake hp: 65 at 2200 rpm. Taxable hp: 33.80. Three crankshaft bearings. Pump cooling. Force feed oiling. Zenith 1-1/4 carburetor. Carburetor fed by Stewart vacuum tank. Splitdorf ignition. Willard 6-100 battery.

VIKING

This was a small "companion car" to the contemporary Oldsmobile. Its stylized "V" emblem was meant to denote both Viking and V-8.The power plant was far different from that used in the Olds Six. The 90-degree V-8 had a chain-driven camshaft operated by horizontal valves between the cylinder banks. This made the valve assemblies easy to get at by removing a simple cover plate. The car came out in March of 1929, but the Great Depression killed any chance for it to be a sales success.

1929-1930 Viking 8 (V-8)
V-8. Cast-iron block. Cast in blocks of four. Valves located in V. Bore & stroke: 3-3/8 x 3-5/8 in. Displacement: 259.5 cid. Compression ratio: 5.10:1. Brake hp: 81 at 3200 rpm. Taxable hp: 36.50. Three crankshaft bearings. Pump cooling. Force feed oiling. Johnson Model R 1-1/2-in. carburetor. Fuel feed: AC. Delco-Remy ignition. Willard 100 battery.

WILLIS SAINTE CLAIRE

Built in Marysville, Mich. from 1921-1927, this car was made by Childe Harold Wills, a renowned engineer who left Ford to start his own firm. The Wills V-8 was inspired by a famous World War I Hispano-Suiza airplane engine and had similar one-piece head-and-block construction and twin overhead camshafts. C. Harold Wills (who hated his "poetic" first name) changed the Hisso engine's straight-cut gears for quieter spiral-bevel gears.

1922 Wills Sainte Claire (V-8)
V-8. Cast-iron block. Cast in blocks of four. Valves located in V. Bore & stroke: 3-1/4 x 4 in. Displacement: 265.4. Brake hp: 67 at 2700 rpm. Taxable hp: 33.80. Three crankshaft bearings. Thermo Syphon cooling. Force feed and splash oiling. Zenith 1-1/8-in. carburetor. Fuel feed: vacuum tank. Delco ignition. USL battery.

1923 Wills Sainte Claire (V-8)
V-8. Cast-iron block. Cast in blocks of four. Valves located in V. Bore & stroke: 3-1/4 x 4 in. Displacement: 265.4. Brake hp: 67 at 2700 rpm. Taxable hp: 33.80. Three crankshaft bearings. Thermo Syphon cooling. Force feed oiling. Zenith 1-1/8-in. carburetor. Fuel feed: Stewart vacuum tank. Delco ignition. Willard battery.

1924 Wills Sainte Claire (V-8)
V-8. Cast-iron block. Cast in blocks of four. Valves located in V. Bore & stroke: 3-1/4 x 4 in. Displacement: 265.4. Brake hp: 67 at 2700 rpm. Taxable hp: 33.80. Three crankshaft bearings. Thermo Syphon cooling. Force feed oiling. Holley 1-1/4-in. carburetor. Fuel feed: Stewart vacuum tank. Delco ignition. Willard battery.

1925 Wills Sainte Claire (V-8)
V-8. Cast-iron block. Cast in blocks of four. Valves located in V. Bore & stroke: 3-1/4 x 4 in. Displacement: 265.4. Brake hp: 67 at 2700 rpm. Taxable hp: 33.80. Three crankshaft bearings. Thermo Syphon cooling. Force feed oiling. Schebler 1-1/4-in. carburetor. Fuel feed: Stewart vacuum tank. Delco ignition. Willard battery.

1926 Wills Sainte Claire (V-8)
V-8. Cast-iron block. Cast in blocks of four. Valves located in V. Bore & stroke: 3-1/4 x 4 in. Displacement: 265.4. Brake hp: 67 at 2700 rpm. Taxable hp: 33.80. Three crankshaft bearings. Thermo Syphon cooling. Pressure oiling. Schebler Model S 1-1/4-in. carburetor. Fuel feed: Stewart vacuum tank. Delco ignition. Willard Model D 140 battery.

1927 Wills Sainte Claire (V-8)
V-8. Cast-iron block. Cast in blocks of four. Valves located in V. Bore & stroke: 3-1/4 x 4 in. Displacement: 265.4. Brake hp: 67 at 2700 rpm. Taxable hp: 33.80. Three crankshaft bearings. Thermo Syphon cooling. Pressure oiling. Schebler Model EJ 1-1/4-in. carburetor. Fuel feed: Stewart vacuum tank. Delco ignition. Willard 132 battery.

YALE

This Yale was built in Saginaw, Mich. The company dates are 1916-1918, but Motor listed the brand for 1917-1918 only. A V-8 with 31 NACC horsepower was the only available engine in the first-year Model K. The Model M of 1918 had a totally different V-8. Yale manufactured its own V-8s, but not for very long.

1917 Yale Model K (V-8)
V-8. Cast-iron block. Cast in blocks of four. Valves located in V. Bore & stroke: 3-1/8 x 4-1/2 in. Displacement: 276.1 cid. Taxable hp: 31.25. Two crankshaft bearings. Thermo Syphon cooling. Force feed and splash oiling. Rayfield V carburetor. Fuel feed: Stewart vacuum tank. Atwater-Kent ignition.

1918 Yale Model M (V-8)
V-8. Cast-iron block. Cast in blocks of four. Valves located in V. Bore & stroke: 3 x 5 in. Displacement: 283 cid. Taxable hp: 28.80. Three crankshaft bearings. Thermo Syphon cooling. Splash oiling. Zenith V carburetor. Fuel feed: Stewart vacuum tank. Atwater-Kent ignition.

This 1953 Ethyl Corporation advertisement compared the performance of early cars to that of the V-8-powered '53 Mercury. While even the 63-hp Kissel straight 8 could operate on additive-free gasoline, the 125-hp Mercury V-8 with its 7.20:1 compression ratio needed the anti-knock fuel compound developed in 1921.

$5.00 "V8 OWNERS" DISCOUNT

*if you subscribe now to **Old Cars Weekly!***

Old Cars Weekly is the perfect marketplace for information, parts, vehicles and national auctions. $39.98 is the regular cost for 52 issues (1 yr.) – but if you act now on this hot offer you can **get $5.00** off and get **52 issues for only $34.98**!

SUPER CHARGED ***books to fuel your V-8 fire!***

Created by Krause Publications 25 years ago, *Standard Catalogs* have become the ultimate automotive reference books. They combine history, photos, buyer's guide information, technical specifications and pricing into one handy, easy-to-use publication that tells you all there is to know about cars and trucks. You get original equipment, serial numbers, factory prices, production totals, engine data, technical specifications and current collector prices. Standard Catalogs are available by era (prewar, postwar, etc.); by type of vehicle (domestic, import, truck, etc.); by brand (Chevrolet, Ford, Chrysler, etc.) or by model (Camaro, Chevelle, Corvette, Firebird, GTO and Mustang).

Standard Catalog of® 4x4s 1945-2000
2nd Edition
Softcover • 8-1/2x11 • 736 pages
300+ b&w photos • 8-page color section
Item# FX02 • $34.95

Standard Catalog of® American Cars 1805-1942
3rd Edition
Softcover • 8-1/2x11 • 1,598 pages
5,000+ b&w photos
Item# AB03 • $55.00

Standard Catalog of® American Cars 1946-1975
Revised 4th Edition
Softcover • 8-1/2x11 • 976 pages
2,800 b&w photos • 8-page color section
Item# AC04R • $34.95

Standard Catalog of® American Cars 1976-1999
3rd Edition
Softcover • 8-1/2x11 • 976 pages
2,000 b&w photos • 16-page color section
Item# AD03 • $34.95

Standard Catalog of® American Light-Duty Trucks
3rd Edition
Softcover • 8-1/2x11 • 944 pages
3,600+ b&w photos
Item# PT03 • $34.95

Standard Catalog of® Buick 1903-2000
2nd Edition
Softcover • 8-1/2x11 • 304 pages
500 b&w photos • 8-page color section
Item# AK02 • $21.95

Standard Catalog of® Cadillac 1903-2000
2nd Edition
Softcover • 8-1/2x11 • 304 pages
600+ b&w photos • 16-page color section
Item# AL02 • $21.95

Standard Catalog of® Camaro 1967-2002
Softcover • 8-1/4x10-7/8 • 224 pages
200+ color photos
Item# CMRO1 • $24.95

Standard Catalog of® Chevelle 1964-1987
Softcover • 8-1/4x10-7/8 • 256 pages
150+ color photos
Item# CHVL1 • $24.99

Standard Catalog of® Chevrolet 1912-2003
3rd Edition
Softcover • 8-1/2x11 • 520 pages
1,000+ b&w photos • 8-page color section
Item# AV03 • $24.99

Standard Catalog of® Chrysler 1914-2000
2nd Edition
Softcover • 8-1/2x11 • 552 pages
500 b&w photos • 8-page color section
Item# AY02 • $22.95

Standard Catalog of® Corvette 1953-2001
Softcover • 8-1/4x10-7/8 • 224 pages
225 color photos
Item# VET01 • $24.95

Standard Catalog of® Ferrari 1947-2003
Softcover • 8-1/4x10-7/8 • 224 pages
250+ color photos
Item# FERI1 • $24.99

Standard Catalog of® Firebird 1967-2002
Softcover • 8-1/4x10-7/8 • 224 pages
200+ color photos
Item# BIRD1 • $24.95

Standard Catalog of® Ford 1903-2003
3rd Edition
Softcover • 8-1/2x11 • 640 pages
1,800 b&w photos • 16-page color section
Item# AF03 • $24.95

Standard Catalog of® Imported Cars 1946-2002
2nd Edition
Softcover • 8-1/2x11 • 912 pages
1,500+ b&w photos
Item# AI02 • $32.95

Standard Catalog of® Light-Duty Dodge Trucks 1917-2002
Softcover • 8-1/2x11 • 264 pages
500+ b&w photos • 16-page color section
Item# LDDG1 • $24.95

Standard Catalog of® Light-Duty Ford Trucks 1905-2002
Softcover • 8-1/2x11 • 360 pages
500+ b&w photos
Item# LDFT1 • $24.95

Standard Catalog of® Mustang 1964-2001
Softcover • 8-1/4x10-7/8 • 256 pages
300 color photos
Item# MUS01 • $24.95

Standard Catalog of® Oldsmobile, 1897-1997
Softcover • 8-1/2x11 • 304 pages
800 b&w photos
Item# OLDS • $21.95

Standard Catalog of® Pontiac 1926-2002
2nd Edition
Softcover • 8-1/2x11 • 368 pages
700+ b&w photos • 16-page color section
Item# APO02 • $21.95